Methods in Enzymology

Volume 192
BIOMEMBRANES
Part W
Cellular and Subcellular Transport:
Epithelial Cells

METHODS IN ENZYMOLOGY

EDITORS-IN-CHIEF

John N. Abelson   Melvin I. Simon

DIVISION OF BIOLOGY
CALIFORNIA INSTITUTE OF TECHNOLOGY
PASADENA, CALIFORNIA

FOUNDING EDITORS

Sidney P. Colowick and Nathan O. Kaplan

Methods in Enzymology

Volume 192

Biomembranes

Part W

Cellular and Subcellular Transport:
Epithelial Cells

EDITED BY

Sidney Fleischer
Becca Fleischer

DEPARTMENT OF MOLECULAR BIOLOGY
VANDERBILT UNIVERSITY
NASHVILLE, TENNESSEE

Editorial Advisory Board

ACADEMIC PRESS, INC.
Harcourt Brace Jovanovich, Publishers

San Diego   New York   Boston
London   Sydney   Tokyo   Toronto

This book is printed on acid-free paper. $\textcircled{\infty}$

Academic Press, Inc.
San Diego, California 92101

United Kingdom Edition published by
Academic Press Limited
24-28 Oval Road, London NW1 7DX

Library of Congress Catalog Card Number: 54-9110

ISBN   0-12-182093-9   (alk. paper)

Printed in the United States of America
90 91 92 93   9 8 7 6 5 4 3 2 1

Table of Contents

Section I. Gastrointestinal System

A. Salivary Gland

B. Stomach

E. Liver

Section II. Other Epithelia

A. Trachea

B. Cornea and Eye

C. Sweat Glands

D. Insect Epithelia

E. Urinary Bladder

F. Gallbladder

G. Amphibian and Reptilian Epithelia

F. Fish Epithelia

Contributors to Volume 192

Article numbers are in parentheses following the names of contributors.
Affiliations listed are current.

ZAHUR AHMED (21), *Department of Physiology, School of Medicine, State University of New York at Buffalo, Buffalo, New York 14214*

G. ALTENBERG (43), *Department of Physiology and Biophysics, The University of Texas Medical Branch, Galveston, Texas 77550*

INDU S. AMBUDKAR (3), *Clinical Investigations and Patient Care Branch, National Institute of Dental Research, National Institutes of Health, Bethesda, Maryland 20892*

B. E. ARGENT (17), *Department of Physiological Sciences, University of Newcastle upon Tyne, Newcastle upon Tyne NE2 4HH, England*

BRUCE J. BAUM (3), *Clinical Investigations and Patient Care Branch, National Institute of Dental Research, National Institutes of Health, Bethesda, Maryland 20892*

THOMAS BERGLINDH (6), *Department of Biomedical Research, Pharmaia AB, S-751 82 Uppsala, Sweden*

A. BERTELOOT (26), *Department of Physiology, Faculty of Medicine, University of Montreal, Montreal, Quebec H3C 3T8, Canada*

RICHARD C. BOUCHER (36), *Department of Medicine, The University of North Carolina at Chapel Hill, Chapel Hill, North Carolina 27599*

J. L. BOYER (32, 33, 34), *Department of Medicine and Liver Center, Yale University School of Medicine, New Haven, Connecticut 06510*

R. M. CASE (17), *Department of Physiological Sciences, University of Manchester, Manchester M13 9PT, England*

DEAN CHANG (47), *Department of Physiology and the Cellular and Molecular Biology Program, University of Michigan Medical School, Ann Arbor, Michigan 48109*

MOIRA CIOFFI (40), *Department of Biology, Temple University, Philadelphia, Pennsylvania 19122*

MORTIMER M. CIVAN (44), *Departments of Physiology and Medicine, University of Pennsylvania School of Medicine, Philadelphia, Pennsylvania 19104*

MICHEL CLARET (31), *INSERM U 274, Université Paris-Sud, Orsay, 91405 Cedex, France*

LAURENT COMBETTES (31), *INSERM U 274, Université Paris-Sud, Orsay, 91405 Cedex, France*

J. COPELLO (43), *Department of Physiology and Biophysics, The University of Texas Medical Branch, Galveston, Texas 77550*

C. COTTON (43), *Department of Pediatrics, Case Western Reserve University, Cleveland, Ohio 44106*

DWIGHT N. CRAWFORD (39), *Department of Biology, Temple University, Philadelphia, Pennsylvania 19122*

S. CURCI (5), *Istituto di Fisiologia Generale, Università di Bari, 70126 Bari, Italy*

CATHERINE DARGEMONT (31), *Ecole Normale Supérieure, Paris 75006, France*

DAVID C. DAWSON (47), *Department of Physiology and the Cellular and Molecular Biology Program, University of Michigan Medical School, Ann Arbor, Michigan 48109*

K. DAWSON (43), *Department of Physiology and the Cellular and Molecular Biology Program, University of Michigan Medical School, Ann Arbor, Michigan 48109.*

DANIEL C. DEVOR (21), *Department of Physiology, School of Medicine, State University of New York at Buffalo, Buffalo, New York 14214*

ix

KIERTISIN DHARMSATHAPHORN (24), Department of Medicine, University of California, San Diego, San Diego, California 92103

P. DIETL (46), Department of Physiology, Dartmouth Medical School, Hanover, New Hampshire 03756

MICHAEL E. DUFFEY (21), Department of Physiology, School of Medicine, State University of New York at Buffalo, Buffalo, New York 14214

FRANKLIN H. EPSTEIN (49), The Mount Desert Island Biological Laboratory, Salsbury Cove, Maine 04672

MICHAEL FIELD (48), Department of Medicine and Physiology, College of Physicians and Surgeons, Columbia University, New York, New York 10032

GUNNAR FLEMSTRÖM (9), Department of Physiology and Medical Biophysics, Uppsala University Biomedical Center, S-751 23 Uppsala, Sweden

HENRIK FORSSELL (9), Department of Surgery, Sahlgren's Hospital, University of Gothenburg, S-143 45 Gothenburg, Sweden

J. G. FORTE (10), Department of Molecular and Cell Biology, University of California, Berkeley, Berkeley, California 94720

MICHAEL FROMM (29), Institut für Klinische Physiologie, Universitätsklinikum Steglitz, Freie Universitat Berlin, D-1000 Berlin 45, Federal Republic of Germany

E. FRÖMTER (5, 45), Zentrum der Physiologie, Johann Wolfgang Goethe-Universität, D-6000 Frankfurt am Main, Federal Republic of Germany

D. V. GALLACHER (20), Department of Physiology, University of Liverpool, Liverpool L69 3BX, England

JERRY D. GARDNER (16, 18), The Digestive Diseases Branch, National Institute of Diabetes, Digestive and Kidney Diseases, National Institutes of Health, Bethesda, Maryland 20892

HAIM GARTY (44), Department of Membrane Research, The Weizmann Institute of Science, Rehovot, Israel

B. GASSNER (46), Department of Physiology, University of Würzburg, D-8700 Würzburg, Federal Republic of Germany

L. G. M. GORDON (45), Department of Physiology, University of Otago Medical School, Dunedin, New Zealand

J. GRAF (32), Department of General and Experimental Pathology, University of Vienna, Vienna, Austria

K. HAGG (2), Medizinische Universitätsklinik, D-7800 Freiburg, Federal Republic of Germany

PHILIPPE A. HALBAN (13, 14), Jeantet Research Laboratories, University of Geneva School of Medicine, CH-1211 Geneva 4, Switzerland

JOHN W. HANRAHAN (42), Department of Physiology, McGill University, Montreal, Quebec H3G 1Y6, Canada

WILLIAM R. HARVEY (39, 40), Department of Biology, Temple University, Philadelphia, Pennsylvania 19122

ULRICH HEGEL (29), Institut für Klinische Physiologie, Universitätsklinikum Steglitz, Freie Universitat Berlin, D-1000 Berlin 45, Federal Republic of Germany

JOSEPH HELMAN (3), Department of Oral and Maxillofacial Surgery, Rambam Medical Center, Haifa, Israel 31096

STEPHEN HERSEY (8), Department of Physiology, Emory University, Atlanta, Georgia 30322

ULRICH HOPFER (25), Department of Physiology and Biophysics, Case Western Reserve University, Cleveland, Ohio 44106

VALERIE J. HORN (3), Clinical Investigations and Patient Care Branch, National Institute of Dental Research, National Institutes of Health, Bethesda, Maryland 20892

ROBERT T. JENSEN (16, 18), The Digestive Diseases Branch, National Institute of Diabetes, Digestive and Kidney Diseases, National Institutes of Health, Bethesda, Maryland 20892

THOMAS J. JENTSCH (37), Zentrum für Molekulare Neurobiologie, Universität Hamburg, D-2000 Hamburg 20, Federal Republic of Germany

NEIL KAPLOWITZ (30), *Department of Medicine, USC School of Medicine, Los Angeles, California 90033*

GEORGE A. KIMMICH (22), *Department of Biochemistry, School of Medicine and Dentistry, University of Rochester, Rochester, New York 14642*

ULLA KLEIN (40), *Zoologisches Institut der Universität, D-8000 Munchen 2, Federal Republic of Germany*

H. KNAUF (2), *Medizinische Universitätsklinik, D-7800 Freiburg, Federal Republic of Germany*

G. KOTTRA (45), *Zentrum der Physiologie, Klinikum der J. W. Goethe-Universität, 6000 Frankfurt 70, Federal Republic of Germany*

SIMON A. LEWIS (42), *Department of Physiology and Biophysics, University of Texas Medical Branch, Galveston, Texas 77550*

CAROLE M. LIEDTKE (35), *Department of Pediatrics, Case Western Reserve University, Cleveland, Ohio 44106*

TERRY E. MACHEN (4), *Department of Molecular and Cell Biology, University of California, Berkeley, Berkeley, California 94720*

JAMES L. MADARA (24), *Department of Pathology, Harvard Medical School, Boston, Massachusetts 02115*

S. H. P. MADDRELL (41), *Department of Zoology, University of Cambridge, Cambridge CB2 3EJ, England*

PARAMESWARA MALATHI (27), *Department of Physiology and Biophysics, University of Medicine and Dentistry of New Jersey–Robert Wood Johnson Medical School, Piscataway, New Jersey 08854*

DANUTA H. MALINOWSKA (7), *Department of Physiology and Biophysics, University of Cincinnati College of Medicine, Cincinnati, Ohio 45267*

JEAN-PIERRE MAUGER (31), *INSERM U 274, Université Paris-Sud, Orsay, 91405 Cedex, France*

PAOLO MEDA (13, 14), *Department of Morphology, University of Geneva School of Medicine, CH-1211 Geneva 4, Switzerland*

PETER J. MEIER (33, 34), *Division of Clinical Pharmacology, Department of Medicine, University Hospital, CH-8091 Zurich, Switzerland*

HANS PETER MEISSNER (15), *Free University of Berlin, D-1000 Berlin 21, Federal Republic of Germany*

JAMES E. MELVIN (3), *Department of Dental Research, University of Rochester Medical Center, Rochester, New York 14642*

DELIA MENOZZI (18), *The Digestive Diseases Branch, National Institute of Diabetes, Digestive and Kidney Diseases, National Institutes of Health, Bethesda, Maryland 20892*

LAWRENCE M. MERTZ (3), *Endocrinology and Reproduction Research Branch, National Institute of Child Health and Human Development, National Institutes of Health, Bethesda, Maryland 20892*

MELISSA MILLER (8), *Department of Physiology, Emory University, Atlanta, Georgia 30322*

AUSTIN K. MIRCHEFF (23), *Department of Physiology and Biophysics, University of Southern California, School of Medicine, Los Angeles, California 90033*

MARK W. MUSCH (48), *Department of Medicine, The University of Chicago, Chicago, Illinois 60637*

PAUL A. NEGULESCU (4), *Department of Molecular and Cell Biology, University of California, Berkeley, Berkeley, California 94720*

I. NOVAK (1), *Physiologisches Institut, Albert-Ludwigs-Universität, D-7800 Freiburg, Federal Republic of Germany*

SCOTT M. O'GRADY (48), *Department of Veterinary Biology, University of Minnesota, St. Paul, Minnesota 55108*

H. OBERLEITHNER (46), *Department of Physiology, University of Würzburg, D-8700 Würzburg, Federal Republic of Germany*

MURAD OOKHTENS (30), *Department of Medicine, USC School of Medicine, Los Angeles, California 90033*

Y. OSIPCHUK (20), *Bogomoletz Institute of Physiology, Ukranian Academy of Sciences, 252601 Kiev 24, U.S.S.R.*

J. A. OVERTON (41), *Department of Zoology, University of Cambridge, Cambridge CB2 3EJ, England*

BRIAN E. PEERCE (11), *Department of Physiology and Biophysics, University of Texas Medical Branch, Galveston, Texas 77550*

O. H. PETERSEN (20), *Department of Physiology, University of Liverpool, Liverpool L69 3BX, England*

J. M. PHILLIPS (32), *Research Institute and Department of Pathology, Hospital for Sick Children and University of Toronto, Toronto, Ontario, Canada*

JAN POHL (8), *Microchemical Facility, Emory University, Atlanta, Georgia 30322*

W. W. REENSTRA (10), *Research Institute, Children's Hospital of Oakland, Oakland, California 94609*

L. REUSS (43), *Department of Physiology and Biophysics, The University of Texas Medical Branch, Galveston, Texas 77550*

FUSAKO SATO (38), *Department of Dermatology, University of Iowa College of Medicine, Iowa City, Iowa 52242*

KENZO SATO (38), *Department of Dermatology, University of Iowa College of Medicine, Iowa City, Iowa 52242*

IRENE SCHULZ (19), *Max-Planck-Institut für Biophysik, D-6000 Frankfurt am Main 70, Federal Republic of Germany*

HELMUT SCHWEIKL (40), *Zoologisches Institut der Universität, D-8000 Munchen 2, Federal Republic of Germany*

Y. SEGAL (43), *Department of Physiology and Biophysics, The University of Texas Medical Branch, Galveston, Texas 77550*

G. SEMENZA (26), *Department of Biochemistry, Swiss Institute of Technology, ETH-Zentrum, CH-8092 Zurich, Switzerland*

PATRICIO SILVA (49), *Department of Medicine, Beth Israel Hospital, Boston, Massachusetts 02215*

STEVEN M. SIMASKO (21), *Department of Physiology, School of Medicine, State University of New York at Buffalo, Buffalo, New York 14214*

RICHARD J. SOLOMON (49), *Renal Unit, Westchester County Medical Center and New York Medical College, Valhalla, New York 10595*

DANIEL D. SPAETH (39), *Department of Biology, Temple University, Philadelphia, Pennsylvaia 19122*

MARK TAKAHASHI (27), *Department of Physiology and Biophysics, University of Medicine and Dentistry of New Jersey–Robert Wood Johnson Medical School, Piscataway, New Jersey 08854*

LAURA TANG (8), *Department of Physiology, Emory University, Atlanta, Georgia 30322*

R. JAMES TURNER (3), *Clinical Investigations and Patient Care Branch, National Institute of Dental Research, Bethesda, Maryland 20892*

EMILE J. J. M. VAN CORVEN (23), *Department of Physiology and Biophysics, University of Southern California, School of Medicine, Los Angeles, California 90033*

M. WAKUI (20), *Department of Physiology, Tohoku University, School of Medicine, Sendai 980, Japan*

JULIAN R. F. WALTERS (28), *Department of Medicine, Royal Postgraduate Medical School, Hammersmith Hospital, London W12 OHS, England*

W. WANG (46), *Department of Cellular and Molecular Physiology, Yale Medical School, New Haven, Connecticut 06510*

STEPHEN A. WANK (16), *The Digestive Diseases Branch, National Institute of Diabetes, Digestive and Kidney Diseases, National Institutes of Health, Bethesda, Maryland 20892*

F. WEHNER (43), *Max-Plank-Institut für Sytemphysiologie, D-4600 Dortmund 1, Federal Republic of Germany*

MILTON M. WEISER (28), *Department of Medicine, State University of New York at Buffalo, Buffalo General Hospital, Buffalo, New York 14203*

HELMUT WIECZOREK (40), *Zoologisches Institut der Universität, D-8000 Munchen 2, Federal Republic of Germany*

MICHAEL WIEDERHOLT (37), *Institut für Klinische Physiologie, Klinikum Steglitz, Freie Universität Berlin, D-1000 Berlin 45, Federal Republic of Germany*

MICHAEL G. WOLFERSBERGER (40), *Department of Biology, Temple University, Philadelphia, Pennsylvania 19122*

CLAES B. WOLLHEIM (13, 14), *Department of Medicine, Division of Clinical Biochem-istry, University of Geneva School of Medicine, CH-1211 Geneva 4, Switzerland*

TADATAKA YAMADA (12), *Department of Internal Medicine, University of Michigan Hospitals, Ann Arbor, Michigan 48109*

JAMES R. YANKASKAS (36), *Department of Medicine, The University of North Carolina at Chapel Hill, Chapel Hill, North Carolina 27599*

D. YULE (20), *Department of Physiology, University of Liverpool, Liverpool L69 3BX, England*

Preface

Biological transport is part of the Biomembranes series of *Methods in Enzymology*. It is a continuation of methodology concerned with membrane function. This is a particularly good time to cover the topic of biological membrane transport because there is now a strong conceptual basis for its understanding. The field of transport has been subdivided into five topics.

1. Transport in Bacteria, Mitochondria, and Chloroplasts
2. ATP-Driven Pumps and Related Transport
3. General Methodology of Cellular and Subcellular Transport
4. Cellular and Subcellular Transport: Eukaryotic (Nonepithelial) Cells
5. Cellular and Subcellular Transport: Epithelial Cells

Topic 1 covered in Volumes 125 and 126 initiated the series. Topic 2 is covered in Volumes 156 and 157, Topic 3 in Volumes 171 and 172, and Topic 4 in Volumes 173 and 174. The remaining Topic 5 is now covered in 191 and 192.

Topic 5 is divided into two parts: this volume (Part W) which deals with gastrointestinal and a diversity of other epithelial cells, and Volume 191 (Part V) which covers transport in kidney, hormonal modulation, stimulus secretion coupling, pharmacological agents and targeting, and intracellular trafficking in epithelial cells.

We are fortunate to have the good counsel of our Advisory Board. Their input insures the quality of these volumes. The same Advisory Board has served for the complete transport series. Valuable input on the outlines of the five topics was also provided by Qais Al-Awqati, Ernesto Carafoli, Halvor Christensen, Isadore Edelman, Joseph Hoffman, Phil Knauf, and Hermann Passow. Additional valuable input for Volumes 191 and 192 was obtained from Michael Berridge, Eberhard Fromter, Ari Helenius, and Heine Murer.

The names of our advisory board members were inadvertently omitted in Volumes 125 and 126. When we noted the omission, it was too late to rectify the problem. For volumes 125 and 126, we are also pleased to acknowledge the advice of Angelo Azzi, Youssef Hatefi, Dieter Oesterhelt, and Peter Pedersen.

The enthusiasm and cooperation of the participants have enriched and made these volumes possible. The friendly cooperation of the staff of Academic Press is gratefully acknowledged. We are pleased to acknowledge Ms. Laura Taylor for her tireless efforts and secretarial skills.

These volumes are dedicated to Professor Sidney Colowick, a dear friend and colleague, who died in 1985. We shall miss his wise counsel, encouragement, and friendship.

SIDNEY FLEISCHER
BECCA FLEISCHER

METHODS IN ENZYMOLOGY

VOLUME 68. Recombinant DNA
Edited by RAY WU

VOLUME 69. Photosynthesis and Nitrogen Fixation (Part C)
Edited by ANTHONY SAN PIETRO

VOLUME 70. Immunochemical Techniques (Part A)
Edited by HELEN VAN VUNAKIS AND JOHN J. LANGONE

VOLUME 71. Lipids (Part C)
Edited by JOHN M. LOWENSTEIN

VOLUME 72. Lipids (Part D)
Edited by JOHN M. LOWENSTEIN

VOLUME 73. Immunochemical Techniques (Part B)
Edited by JOHN J. LANGONE AND HELEN VAN VUNAKIS

VOLUME 74. Immunochemical Techniques (Part C)
Edited by JOHN J. LANGONE AND HELEN VAN VUNAKIS

VOLUME 75. Cumulative Subject Index Volumes XXXI, XXXII, and XXXIV–LX
Edited by EDWARD A. DENNIS AND MARTHA G. DENNIS

VOLUME 76. Hemoglobins
Edited by ERALDO ANTONINI, LUIGI ROSSI-BERNARDI, AND EMILIA CHIANCONE

VOLUME 77. Detoxication and Drug Metabolism
Edited by WILLIAM B. JAKOBY

VOLUME 78. Interferons (Part A)
Edited by SIDNEY PESTKA

VOLUME 79. Interferons (Part B)
Edited by SIDNEY PESTKA

VOLUME 80. Proteolytic Enzymes (Part C)
Edited by LASZLO LORAND

VOLUME 183. Molecular Evolution: Computer Analysis of Protein and Nucleic Acid Sequences
Edited by RUSSELL F. DOOLITTLE

VOLUME 184. Avidin-Biotin Technology
Edited by MEIR WILCHEK AND EDWARD A. BAYER

VOLUME 185. Gene Expression Technology
Edited by DAVID V. GOEDDEL

VOLUME 186. Oxygen Radicals in Biological Systems (Part B: Oxygen Radicals and Antioxidents)
Edited by LESTER PACKER AND ALEXANDER N. GLAZER

VOLUME 187. Arachidonate Related Lipid Mediators
Edited by ROBERT C. MURPHY AND FRANK A. FITZPATRICK

VOLUME 188. Hydrocarbons and Methylotrophy
Edited by MARY E. LIDSTROM

VOLUME 189. Retinoids (Part A: Molecular and Metabolic Aspects)
Edited by LESTER PACKER

VOLUME 190. Retinoids (Part B: Cell Differentiation and Clinical Applications)
Edited by LESTER PACKER

VOLUME 191. Biomembranes (Part V: Cellular and Subcellular Transport: Epithelial Cells)
Edited by SIDNEY FLEISCHER AND BECCA FLEISCHER

VOLUME 192. Biomembranes (Part W: Cellular and Subcellular Transport: Epithelial Cells)
Edited by SIDNEY FLEISCHER AND BECCA FLEISCHER

VOLUME 193. Mass Spectrometry
Edited by JAMES A. MCCLOSKEY

VOLUME 194. Guide to Yeast Genetics and Molecular Biology
Edited by CHRISTINE GUTHRIE AND GERALD R. FINK

VOLUME 195. Adenylyl Cyclase, G Proteins, and Guanylyl Cyclase (in preparation)
Edited by ROGER A. JOHNSON AND JACKIE D. CORBIN

VOLUME 196. Molecular Motors and the Cytoskeleton (in preparation)
Edited by RICHARD B. VALLEE

VOLUME 197. Phospholipases (in preparation)
Edited by EDWARD A. DENNIS

VOLUME 198. Peptide Growth Factors (Part C) (in preparation)
Edited by DAVID BARNES, J. P. MATHER, AND GORDON H. SATO

Section I

Gastrointestinal System

A.   Salivary Gland
Chapters 1 through 3

B.   Stomach
Chapters 4 through 12

C.   Pancreas
Chapters 13 through 20

D   Intestine
Chapters 21 through 29

E   Liver
Chapters 30 through 34

[1] Salivary Secretion: Studies on Intact Glands *in Vivo* and *in Vitro*

By I. NOVAK

Introduction

Unlike most other glands, the major salivary glands, parotid, mandibular, and sublingual, show a bewildering variety of structural and physiological patterns.[1,2] As a result they have become a useful investigative tool for the study of some basic problems in physiology, such as transepithelial salt and water transport, protein synthesis and exocytosis, the neuropharmacology of autonomic nerves and receptors, and stimulus–secretion coupling.

Most of our knowledge about salivary gland function and saliva formation is based on studies of the glands *in vivo*. One of its most attractive features, which has rendered salivary glands ideal for whole *in vivo* organ studies, is the fact that saliva drains into the oral cavity, and thus is accessible for easy collection. Utilizing this property, Thaysen and co-workers[3,4] studied the relationship between salivary secretory rate and electrolyte composition in human parotid saliva. They described that saliva is formed in two stages by two functionally and morphologically distinct epithelia comprising the glands. As we understand it today, secretory endpieces secrete fluid and electrolytes in plasma-like proportions, and they also secrete organic components. Series of convergent ducts convey the formed saliva and may modify it by reabsorbing and secreting various electrolytes, leaving the fluid volume essentially unaltered.

Other *in vivo* studies, conducted on various animals, which have contributed to our knowledge about salivary gland function include (1) investigation of fluid and protein secretory responses to various drugs, or to direct nerve stimulation, (2) micropuncture sampling of the fluid from the endpiece-intercalated duct region and from other duct segments, and (3)

[1] J. A. Young and E. W. van Lennep, "The Morphology of Salivary Glands." Academic Press, London, 1978.
[2] J. A. Young and E. W. van Lennep, *in* "Membrane Transport in Biology" (G. Giebisch, D. C. Tosteson, and H. H. Ussing, eds.), Vol. 4B, p. 563. Springer-Verlag, Berlin, 1979.
[3] J. H. Thaysen, N. A. Thorn, and I. L. Schwartz, *Am. J. Physiol.* **178**, 155 (1954).
[4] J. H. Thaysen, *in* "The Alkali Metal Ions in Biology. Handbuch der Experimentallen Pharmacologie" (H. H. Ussing, P. Kruhöffer, and N. A. Thorn, eds.), Vol. 13, p. 424. Springer, Berlin, 1960.

microperfusion of various ductal segments.[2,5,6] *In vivo* experiments, however, have some important limitations. Gland function is influenced by changes in blood flow, blood pressure, and endogenous blood-borne agonists. Furthermore, concentrations of autonomic agonists in plasma cannot be determined precisely. *In vitro* perfused gland preparations eliminate many of these variables, and they have the advantage over other *in vitro* preparations, such as isolated cells and membrane vesicles, in that the polarity of the epithelium is preserved and the actual secretion can be monitored. Although isolated gland preparations have been used quite successfully for electrophysiological studies for many years,[7-9] it was not until the 1980s that successful preparations of the rat and rabbit mandibular glands were developed for the study of electrolyte and fluid secretion.[10,11] The remainder of this chapter focuses primarily on the techniques used for the *in vivo* and *in vitro* studies of intact rat and rabbit mandibular glands. *In vivo* studies dealing with other salivary glands, including those of humans, are described elsewhere.[2,3,5,12-14]

Animals and Anesthesia

Male New Zealand white rabbits (body weight 2.5–3.0 kg) are anesthetized with an intraperitoneal injection of urethane (ethyl carbamate, 1.5 g/kg body wt; Sigma, St. Louis, MO). Wistar rats (200–300 g) are anesthetized with an intraperitoneal injection of inactin (sodium-5-ethyl-1-methylpropyl-2-thiobarbiturate, 120 mg/kg body wt; Byk Gulden, Konstanz, FRG). Supplementary doses of anesthetic are given intravenously via a cannula placed in the femoral vein in rats and in the marginal ear vein in rabbits. To maintain the normal body temperature during surgery and during *in vivo* experiments, animals should be placed on a thermostatically controlled heating table or blanket. For longer lasting

[5] L. H. Schneyer, J. A. Young, and C. A. Schneyer, *Physiol. Rev.* **52**, 720 (1972).
[6] J. A. Young, D. I. Cook, E. W. van Lennep, and M. Roberts, *in* "Physiology of the Gastrointestinal Tract" (L. R. Johnson, ed.), p. 773. Raven, New York, 1987.
[7] A. Lundberg, *Physiol. Rev.* **38**, 21 (1958).
[8] H. Yoshimura and Y. Imai, *Jpn. J. Physiol.* **17**, 280 (1967).
[9] O. H. Petersen and J. H. Poulsen, *Acta Physiol. Scand.* **70**, 293 (1967).
[10] R. M. Case, A. D. Conigrave, I. Novak, and J. A. Young, *J. Physiol.* **300**, 467 (1980).
[11] J. Compton, J. R. Martinez, A. M. Martinez, and J. A. Young, *Arch. Oral. Biol.* **26**, 555 (1981).
[12] L. M. Sreebny, "The Salivary System." CRC Press, Boca Raton, Florida, 1987.
[13] A. C. Kerr, "The Physiological Regulation of Salivary Secretion in Man." Pergamon, Oxford, 1961.
[14] G. N. Jenkins, "The Physiology and Biochemistry of the Mouth." Blackwell, Oxford, 1978.

experiments it is advisable to monitor blood pressure and to perform a tracheostomy, for easy removal of mucous secretion. After a ventral midline incision in the neck, the trachea is cleared and cannulated with polyethylene or vinyl tubing (rabbit: 4 mm i.d., 5 mm o.d.; rat: 1.5 mm i.d., 2.5 mm o.d.). Femoral artery can be cannulated with a cannula fitted with a three-way tap enabling monitoring of blood pressure and collection of blood samples.

In Vivo Glands

Collection of Saliva

The main duct can be cannulated either at the oral opening or through a neck incision.[15,16] The operative procedures are best carried out under a dissection microscope. Special instruments required are No. 4 and 5 jewelers forceps, iris scissors, and braided silk for ties. The paired mandibular ducts terminate on the ventral surface of two fleshy papillae lying in the midline behind the incisors in the floor of the mouth. In the rat, which also has a major sublingual gland, the sublingual duct openings are located posterolaterally to the mandibular ducts. The sublingual duct openings, however, are smaller and saliva is more viscous. A fine heat-drawn polyethylene catheter is inserted into the opening of the mandibular duct and secured in situ with a silk tie.

The extralobular main duct of the rat and rabbit mandibular glands can be exposed and easily cannulated through a neck midline incision. In the rat, mandibular glands are located superficially close to the midline. Each gland is surrounded by a capsule, which also encloses the major sublingual gland. In the rabbit, mandibular glands are positioned underneath the mandible. The main duct is exposed and a small incision is made with iris scissors. The duct is then cannulated in the rabbit with vinyl tubing (0.4 mm i.d., 0.8 mm o.d.), or in the rat with heat-drawn polyethylene cannula (0.2—0.4 mm o.d.). The cannula is placed so that the beveled tip lies just outside the gland parenchyma. The duct cannula should be kept as short as possible to reduce the dead space of the collection system.

Saliva can be collected in fixed time intervals into tared polypropylene vials, and salivary volumes are determined by weighing. To prevent evaporation of small samples it is advisable to collect saliva under oil. Samples can be stored below 4° to await analysis. Saliva can be routinely analyzed

[15] J. A. Young and E. Schögel, *Pfluegers Arch.* **291**, 85 (1966).
[16] I. Novak, "Electrolyte and Fluid Transport in Salivary Glands" Ph. D. thesis, University of Sydney, Sydney, 1984.

for Na^+ and K^+ using a flame photometer (Eppendorf, Hamburg, FRG) and for Cl^- using a potentiometric titrator (Radiometer, CMT 10 chloride titrator, Copenhagen, Denmark). Bicarbonate can be measured (as total CO_2) using a Natelson microgasometer. Protein content of saliva can be estimated by the Lowry method.[17]

Stimulation of Secretion

Both rabbit and rat mandibular glands are well supplied with sympathetic and parasympathetic nerve fibers.[1,5,18] Preganglionic sympathetic fibers arise in the first two thoracic segments of the spinal chord. Fibers ascend in the sympathetic trunk running along the common carotid artery and the vagus nerve, and synapse with the postganglionic fibers in the superior cervical ganglion close to the carotid bifurcation. Postganglionic fibers run into the gland along the major blood vessels. Preganglionic parasympathetic fibers arise in the superior and inferior salivary nuclei in the lower brain stem, continue in the facial nerve, diverge from it in the chorda tympani, and then join the lingual nerve. The parasympathetic ganglia are rather diffuse and lie along the main excretory ducts and within the gland substance.

Electrical Stimulation. Sympathetic stimulation can be most easily achieved by placing electrodes under the superior cervical ganglion.[19] In this case both pre- and postganglionic fibers would be stimulated. It is also possible to stimulate pre- and postganglionic fibers separately by placing electrodes under the sympathetic trunk[19-21] or, in larger animals, under the nerve bundle running along the glandular artery.[22] Direct stimulation of parasympathetic postganglionic nerve fibers in rat and rabbit mandibular glands is difficult due to the diffuse location of the ganglia. In the rat, preganglionic fibers can be stimulated with electrodes placed under the main duct.[19] In the rabbit, the chorda tympani nerve is a discernible structure running along the main duct, and it can be dissected and stimulated directly.[21]

Electrical stimulation is evoked by means of square pulses from a pulse generator, e.g., Grass stimulator (Grass Instruments, Quincy, MA). Pulses

[17] O. H. Lowry, N. J. Rosenbrough, A. L. Farr, and R. J. Randall, *J. Biol. Chem.* **193**, 265 (1951).
[18] N. Emmelin, *Philos. Trans. R. Soc. London, Ser. B* **296**, 27 (1981).
[19] Y. Yoshida, R. L. Sprecher, C. A. Schneyer, and L. H. Schneyer, *Proc. Soc. Exp. Biol. Med.* **126**, 912 (1967).
[20] J. R. Garrett and A. Thulin, *Cell Tissue Res.* **159**, 179 (1975).
[21] L. H. Smaje, *J. Physiol.* **231**, 179 (1973).
[22] K. E. Creed and J. A. F. Wilson, *Aust. J. Exp. Biol. Med. Sci.* **47**, 135 (1969).

of 2–10 V at a frequency of 0.5–20 Hz and 1- to 5-msec duration have been used by various workers to produce a range of secretory responses.[9,18,19,21]

Secretagogues. One of the ways to stimulate salivary glands to secrete in response to a cholinergic agonist is to infuse acetylcholine into the bloodstream close to the gland, e.g., into the external maxillary artery. Assuming that the normal stimulated blood flow to the rat mandibular gland is 1 ml/min,[23] acetylcholine infusion of 10^{-3}–10^{-2} μmol/min would result in a 10^{-6}–10^{-5} M concentration of the agonist at the gland.[16] A more widely used method of cholinergic stimulation is to use drugs which are not degraded by cholinesterase and introduce them intraperitoneally (ip), or, in lower doses, intravenously (iv). Carbachol used in doses of 0.01 to 0.60 mg/kg body wt evokes a wide range of secretory responses.[15,24–27] Pilocarpine is used in doses of 0.2 to 10 mg/kg.[15,25,28,29]

Apart from adrenaline,[30] other secretagogues employed to stimulate the adrenergic receptors in salivary glands are phenylephrine (α), infused close-arterially (5 mg/kg), and isoproterenol (β), given ip (10–40 mg/kg).[24,27,28]

In an attempt to selectively stimulate the parasympathetic system, or one part of the sympathetic system, secretagogues and electrical stimulation are sometimes used in combination with antagonists, e.g., propranolol (β), 1–5 mg/kg, phenoxybenzamine (α), 5–10 mg/kg, or atropine, 1 mg/kg.[19,21]

The techniques for the *in vivo* studies of intact glands are relatively simple and do not require special equipment. However, as mentioned above, there are some limitations. For example, sympathetic nerve stimulation or α-agonist infusion produces vasoconstriction. Parasympathomimetic drugs (carbachol, pilocarpine) could, due to their nicotinic effect, also stimulate postganglionic sympathetic fibers.[31] Thus, should one wish to study the mechanism of salivary secretion in more detail, isolation of glands is desirable.

[23] M. T. Coroneo, A. R. Denniss, and J. A. Young, *Pfluegers Arch.* **381**, 223, (1979).

[24] J. A. Young, C. J. Martin, and F. D. Weber, *Pfluegers Arch.* **327**, 285 (1971).

[25] J. A. Mangos, N. R. McSherry, K. Irwin, and R. Hong, *Am. J. Physiol.* **225**, 450 (1973).

[26] C. J. Martin and J. A. Young, *Pfluegers Arch.* **327**, 303 (1971).

[27] J. R. Martinez, D. O. Quissell, D. L. Wood, and M. Giles, *J. Pharmacol. Exp. Ther.* **194**, 384 (1975).

[28] J. R. Martinez and J. Camden, *J. Dent. Res.* **62**, 543 (1983).

[29] J. A. Mangos, G. Braun, and K. F. Hamann, *Pfluegers Arch.* **291**, 99 (1966).

[30] A. S. V. Burgen and N. G. Emmelin, "Physiology of the Salivary Glands." Arnold, London, 1961.

[31] C. A. Schneyer and H. D. Hall, *Proc. Soc. Exp. Biol. Med.* **121**, 96 (1966).

In Vitro Glands

Isolation of Salivary Glands

The technique for isolation of mandibular glands was adapted from the original work of Conigrave and Case,[10,32] which was itself based on Case's original work with the isolated cat pancreas.[33] On adapting this technique to the rat mandibular gland, a reference was also made to the preparation described by Compton et al.[11] and Novak[16] for this gland.

Rabbit Mandibular Gland. The first step in the isolation procedures for the rabbit mandibular gland is the cannulation of the main duct, as described above, followed by cannulation of the external maxillary artery. The glandular artery is supplied by the external maxillary artery, which itself is derived from the external carotid artery. The glandular artery and the external maxillary artery are cleared of lymphatic, adipose, and connective tissue. Extraneous branches are cut after they have been ligated. The external maxillary artery is tied off at its most cranial location. Then the artery is temporarily occluded cranial to the glandular branch. This stops the blood flow in the area of the artery which is to be cannulated, but the blood flow to the gland is maintained. The external maxillary artery is then cut and cannulated retrogradely with vinyl tubing (0.58 mm i.d., 0.96 mm o.d.) filled with isotonic saline. A provisional tie is placed around the cannula, the occlusion is released, and the tip of the cannula is advanced toward the junction of the glandular artery. The tie is then secured. Two loose ties are placed around the maxillary artery caudal to the junction of the glandular artery.

The glandular vein drains into the anterior facial vein. Both veins are cleared and the anterior facial vein is prepared for cannulation cranially to the junction of the glandular vein. After the duct and the blood vessels have been prepared, the mandibular gland itself is cleared of the surrounding tissue and exposed. The rabbit is then moved close to the perfusion apparatus. The arterial cannula is connected to the perfusion apparatus with a 23-gauge hypodermic needle. The perfusion rate is gradually increased to 4 ml/min. The anterior facial vein is cannulated immediately with vinyl tubing (0.96 mm i.d., 1.27 mm o.d.), or the vein is cut. This is done to prevent *in situ* perfusion of the whole animal when one wishes to use the second gland of the animal. As soon as the gland is perfused at 4 ml/min, the two loose ties around the external maxillary artery are tightened. Both the artery and the vein are cut beyond the cannulated segments. The gland is then removed from the animal and placed in an adjacent

[32] A. D. Conigrave, B.Sc. Med. thesis, University of Sydney, Sydney (1978).
[33] R. M. Case, A. A. Harper, and T. Scratcherd, *J. Physiol.* **196,** 133 (1968).

humidified organ bath, which is maintained at 37°. If sampling of the venous effluent with syringe is required (e.g., for O_2 measurements), it is advisable to cannulate the facial vein both proximal and distal to the junction with the glandular artery. This avoids obstructing the venous outflow in case the rate of collection is not matched exactly by the rate of outflow. Alternatively, it is possible to cannulate the glandular vein, rather than the anterior facial vein (vinyl tubing 0.58 mm i.d., 0.96 mm i.d.). This is best done after the gland is perfused *in vitro*.

The technique of *in situ* followed by *in vitro* perfusion ensures constant perfusion of the gland with blood or perfusate and it therefore avoids any period of anoxia. It also enables one to secure any leaks in the perfusion arrangement at a time when the gland can still be returned to its normal blood supply. The contralateral gland of the animal can also be isolated and used as a paired control or a test gland. In that case, it is desirable to prepare the second gland within 15–30 min.

Rat Mandibular Gland. The operative procedure for preparing the mandibular gland of the rat is slightly different from the one used for the rabbit. The first step of the operation is to clear the arterial blood supply of the gland. This should be begun at the cranial end, where the external maxillary artery runs almost parallel to the anterior facial vein between the digastric and the masseter muscles. Dissection and clearing of the artery is continued toward the gland. The submental artery, the tonsillar artery and the vein, and the arterial branches to lymph nodes in the region are ligated and sectioned. The digastric muscle is cut at its tendinous part to expose the region of the glandular artery and the salivary duct. In contrast to the rabbit, the rat has a major sublingual gland, which lies in the same capsule as the mandibular gland, but lateral and caudal to it. For convenience it does not need to be dissected away from the mandibular gland until the experiment is finished. The main duct of the mandibular gland is cannulated, as described above, and then reflected away from the underlying glandular artery. Preparation of the external maxillary artery is the same as in the rabbit. A small incision is made just below the lymph node branch. A cannula is placed in the lumen and advanced slowly, while the artery is held with forceps to prevent it from peeling and folding away from the adventitia, a problem which is not encountered with the rabbit maxillary artery. The cannula is saline-filled vinyl tubing (0.4 mm i.d., 0.8 mm o.d.) fitted with a metal tip which is about 5 mm long. The metal tip can be made from a shortened 25-gauge needle which has a shallow, reground bevel.

The rest of the procedures for isolation and perfusion of the gland complex is the same as for the rabbit, except that for the rat a slower perfusion rate of 2 ml/min is sufficient.[16]

Perfusion System

The "control" perfusion fluid which we routinely use contains the following ions (in mmol/liter): $Na^+ = 146$, $K^+ = 4.3$, $Ca^{2+} = 1.0$, $Mg^{2+} = 1.0$, $Cl^- = 121.3$, $HCO_3^- = 25.0$, $HPO_4^{2-}/H_2PO_4^- = 1.0$, $SO_4^{2-} = 1.0$, and a buffer HEPES = 10. Solutions are equilibrated with 5% CO_2 in O_2. The only substrate provided is glucose (5 mmol/liter). Without exogenous substrate, secretion is only about 50% of the control. Increase in glucose concentration above 5 mmol/liter has no further effect on secretion.

The perfusion solution is placed in a reservoir and equilibrated with 5% CO_2 in O_2. From there it is led through heat-exchange coils and pumped into the glandular artery using a peristaltic pump (e.g., Cole-Palmer). The solution is at 37° when it reaches the glandular artery. Isolated glands are placed into an organ bath which is humidified and maintained at 37°. The temperature of the glands can be monitored continuously with the aid of a thermistor. A bubble trap is placed in the line just before the arterial catheter. This can be equipped with a three-way tap so that monitoring of the perfusion pressure is possible. The normal perfusion pressure (corrected for pressure due to cannula) encountered in cholinergically stimulated rabbit glands is 50–70 mmHg, and in the rat glands it is about 70–90 mmHg higher. Increase in the perfusion pressure up to about 350 mmHg has no effect on fluid secretion. However, at about 500 mmHg, fluid secretion increases, and the effect is irreversible, indicating that pathological damage has occurred.[34]

Agonists, other stimulating agents, and antagonists are dissolved either in isotonic saline or in perfusate and then administered at 0.1 ml/min with the aid of a constant infusion pump (e.g., Braun-Melsungen, Eschborn, FRG) via a three-way tap placed in front of the bubble trap.

Edema and Gland Weight

Perfusion solutions do not normally contain any osmotically active macromolecular substances. As a consequence, both rabbit and rat mandibular glands become edematous during perfusion, but edema is restricted to the interstitial spaces. Inclusion of a macromolecular substance such as dextran (7%) in the perfusate reduces the edema markedly.[16] Dextran improves vascular as opposed to transcapsular flow. This can be measured by collecting venous effluent and comparing it to the arterial input (Fig. 1).

[34] I. Novak, J. M. Lingard, and J. A. Young, *in* "Fluid and Electrolyte Abnormalities in Exocrine Glands in Cystic Fibrosis" (P. M. Quinton, J. R. Martinez, and U. Hopfer, eds.), p. 103. San Francisco Press, San Francisco, 1982.

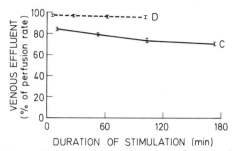

FIG. 1. Venous effluent of rabbit mandibular glands perfused *in vitro* with control solutions (C) or solutions containing 7% dextran (D). Glands were stimulated with continuous infusion of acetylcholine (8×10^{-7} M). Each point represents the mean $\pm$SEM of 5–55 samples from 4 to 20 glands.

Dextran, however, has no effect on the rate of salivary secretion or salivary electrolyte composition. Hence, its use in reducing the edema offers no advantage.

In order to compare secretory responses of various glands, it is customary to express salivary secretory rates corrected for the gland weight. Development of edema would, therefore, seem to pose a problem. However, the weight gain due to edema can be eliminated at the end of the experiment simply by pressing the gland firmly between sheets of blotting paper. The feasibility of this method was established from an observation of a highly significant correlation between the weights of perfused glands (after blotting to eliminate edema) and the nonperfused contralateral glands from the same animal.[16] At this stage the rat sublingual gland is dissected free from the mandibular gland.

Fluid Secretion and Saliva Composition

Secretory responses of *in vitro* and *in vivo* rat and rabbit mandibular glands are quite comparable. Rat glands do not secrete unless stimulated, but rabbit glands can sometimes produce so-called "spontaneous secretion."[6,10,16,30] Nevertheless, this secretion is low (~ 4 μl/g-min)[16] compared to the stimulated response. Both rabbit and rat mandibular glands secrete vigorously in response to parasympathetic stimulation (Fig. 2). In the rat mandibular gland α- and β-adrenergic stimulation produces a more moderate response in the volume of saliva (Fig. 3). Interestingly, protein-rich β-adrenergic saliva is rather high in K^+ and HCO_3^-, an effect which has been ascribed to stimulation of the ductal transport.[26] In the rabbit mandibular, β-adrenergic response is rather small, and stimulation of the α-adrenergic pathway evokes no fluid secretion of its own.[10,21]

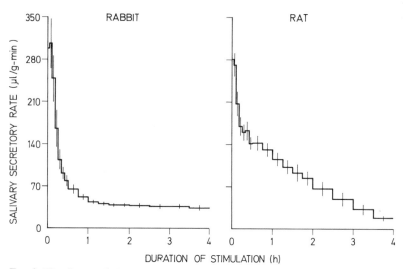

FIG. 2. The characteristic secretory response of the rabbit and the rat mandibular glands perfused with control solution and stimulated continuously with a maximal dose of acetylcholine (8×10^{-7} M in rabbit, 10^{-6} M in rat). Each horizontal bar is the mean ± SEM of data obtained from 10 to 14 glands. Acetylcholine stimulation was begun at zero time and continued throughout the experiment.

For *in vitro* studies of salivary glands (Figs. 1–4) the following protocols are used. Each gland is perfused with one solution and then, after 15 min of continuous perfusion, stimulation is begun by continuous infusion of acetylcholine or other agonists. The maximal dose of acethylcholine for the rabbit mandibular gland is 0.8 μmol/liter and for the rat mandibular 1.0 μmol/liter.[10,35] Characteristic secretory responses for each gland are shown in Fig. 2. At maximal doses, acetylcholine induces a characteristic tachyphylactic response: the initial phase secretion peaks at about 300–400 μl/g-min, and thereafter the secretory rate declines over a period of about 1 hr to reach a plateau of 60–100 μl/g-min. The maximal dose of the α-adrenergic agonist, phenylephrine, is 1 μmol/liter for the rat.[16,36] In the rabbit mandibular gland α-adrenergic stimulation does not evoke fluid secretion, although it may modify the cholinergic response.[10] The maximal dose of the β-adrenergic agonist, isoproterenol, is 5 μmol/liter for the rabbit[10] and 1–10 μmol/liter for the rat gland.[36] Figure 3 shows characteristic secretory response of the rat mandibular gland stimulated with acetylcholine, phenylephrine or isoproterenol.

[35] M. Murakami, I. Novak, and J. A. Young, *Am. J. Physiol.* **252**, G84 (1986).
[36] J. R. Martinez and N. Cassity, *Arch. Oral Biol.* **28**, 1101 (1983).

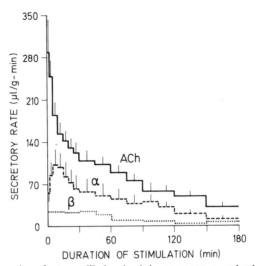

FIG. 3. Fluid secretion of rat mandibular glands in response to sustained stimulation with cholinergic or adrenergic agonists. The α-adrenergic agonist, phenylephrine (10^{-6} M), was infused with atropine (10^{-6} M) and propranolol (10^{-6} M) (α). The β-adrenergic agonist, isoproterenol (10^{-6} M), was infused with atropine (10^{-6} M) and phentolamine (10^{-6} M) (β). For comparison, the secretory response of the glands perfused with acetylcholine (10^{-6} M) is included in the figure (ACh). Horizontal bars for acetylcholine- and phenylephrine-stimulated glands represent the mean ±SEM of six and four experiments; for isoproterenol-stimulated glands the mean of two experiments is shown.

Saliva is collected into tared vials at fixed time intervals after stimulation (at 2, 3, and 5 min, then at 5-min intervals for the first 30 min, then at 15- and 30-min intervals for the rest of the experiment), volumes are determined by weighing, and electrolyte and protein composition can be determined as described above. Figure 4 shows the composition of saliva, collected from the rat mandibular gland stimulated with acetylcholine, as a function of salivary secretory rate. The secretory rate–electrolyte concentration relationships reflect the ductal modification of primary plasma-like secretion, i.e., ducts reabsorb Na^+ and Cl^- and secrete K^+ and HCO_3^-.[2,5,6,26]

Conclusions

Whole gland preparation is a relatively simple and inexpensive technique which has one important advantage in that one can study a real physiological response — secretion. Studies of *in vivo* salivary glands have provided us with a basic understanding of saliva formation and its control.[2,6] Studies of *in vitro* perfused glands in the past 10 years give us an

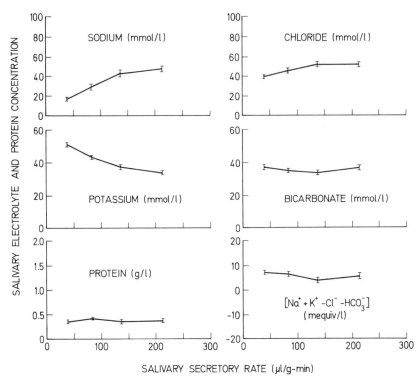

FIG. 4. The electrolyte and protein concentration, in saliva collected from isolated rat mandibular glands stimulated with acetylcholine (10^{-6} M), is plotted as a function of salivary secretory rate. Values are means ±SEM of measurements made in 26–33 samples obtained from 10 glands.

insight into the cellular mechanism of water and electrolyte secretion. Using this preparation one can replace various ions in perfusing solutions and add ion transport inhibitors, and by measuring salivary secretory rate, ionic composition of saliva, and intracellular pH, it is possible to determine the ion transport systems involved in secretion, e.g., secretory endpieces possess the $Na^+/K^+/2Cl^-$ cotransporter and the Na^+/H^+ and Cl^-/HCO_3^- exchangers.[37,38] *In vitro* perfused glands have also been used to study the mechanism of nonelectrolyte secretion, which can be resolved into two components: passive diffusion and solvent drag.[34,39,40] Energy metabolism

[37] I. Novak and J. A. Young, *Pfluegers Arch.* **407**, 649 (1986).

[38] D. Pirani, L. A. R. Evans, D. I. Cook, and J. A. Young, *Pfluegers Arch.* **408**, 178 (1987).

[39] R. M. Case, D. I. Cook, M. Hunter, M. C. Steward, and J. A. Young, *J. Membr. Biol.* **84**, 239 (1985).

[40] A. J. Howorth, R. M. Case, and M. C. Steward, *Pfluegers Arch.* **408**, 209 (1987).

and tissue content of certain electrolytes (e.g., Na^+ and Cl^-) can be studied noninvasively on intact glands by means of nuclear magnetic resonance (NMR).[41-43] Even though studies of intact salivary glands have wide applications, there are some limitations. Salivary glands comprise several morphologically and functionally distinct epithelia — secretory endpieces and a series of ducts.[1,6] Thus, if one wishes to study, for example, ion transport processes and their regulation at the cellular level, it becomes desirable to separate the epithelia and to gain access to the basolateral and lumenal membranes and to the cellular compartment.

Acknowledgments

Studies of salivary glands described in this chapter were carried out in the Department of Physiology, University of Sydney, Australia. I would like to thank C. P. Hansen for careful criticism of the manuscript.

[41] M. Murakami, Y. Seo, H. Watari, H. Ueda, T. Hashimoto, and K. Tagawa, *Jpn. J. Physiol.* **37**, 411 (1987).
[42] M. Murakami, Y. Seo, T. Matsumoto, O. Ichikawa, A. Ikeda, and H. Watari, *Jpn. J. Physiol.* **36**, 1267 (1986).
[43] Y. Seo, M. Murakami, T. Matsumoto, H. Nishikawa, and H. Watari, *Pfluegers Arch.* **409**, 343 (1987).

[2] Determination of Short-Circuit Current in Small Tubular Structures via Cable Analysis

By K. Haag and H. Knauf

Introduction

In the classical sense of Ussing and Zerahn[1] *in vitro* determination of short-circuit current (I_{sc}) is performed on planar tissue preparations separating two compartments of fluids. In small tubular structures short circuiting of the tissue may give rise to both technical and theoretical problems. In these cases cable analysis has proved to be the method of choice for the direct determination of specific transepithelial resistance, R_m. The I_{sc} is then derived from the open circuit potential difference (PD) and R_m.

Application of cable analysis on biological structures was originally performed on nerve fibers by Hodgkin and Rushton.[2] Later the method

[1] H. H. Ussing and K. Zerahn, *Acta Physiol. Scand.* **23**, 110 (1951).
[2] A. L. Hodgkin and W. A. H. Rushton, *Proc. R. Soc. London, Ser. B* **133**, 444 (1946).

was modified by Hegel *et al.*[3] for the proximal tubule of the kidney and by Knauf *et al.*[4-6] for the salivary duct and the rat colon. The method can be applied *in vitro* as well as *in vivo*,[5,7] and even to humans.[4] With these techniques, it was possible to develop a new way of directly determining I_{sc} via clamping of larger tubular structures such as rat colon *in vivo*.

Using cable analysis, the specific electrical resistance of the epithelium (R_m) must first be determined, and then the I_{sc} calculated from the open-circuit potential difference (PD_{oc}) and R_m according to Ohm's law, i.e., $I_{sc} = PD_{oc}/R_m$. This extrapolation from PD_{oc} to zero PD is feasible if R_m is proved to be an ohmic resistor in the range of experimental relevance (see Appendix).

In the physical model the salivary duct is considered a tubular structure comparable to a poorly insulated conducting cable. The core is represented by the lumenal electrolyte solution with an electrical conductivity of R_i^{-1} (Ω^{-1} cm^{-1}), whereas the insulating resistance is represented by the specific transepithelial resistance of R_m (Ω cm^2). When DC current of constant magnitude (I) is injected intralumenally, the transepithelial potential difference (PD) depolarizes (or hyperpolarizes), thereby decreasing the corresponding changes of PD (ΔPD) along the longitudinal axis (the x axis) of the tubular structure exponentially. The circuit is closed via an Ag-AgCl plate adjacent to the tubular segment or by an electrode far away from it in a well-conducting medium on the serosal side to keep the potential on the outside diameter constant and to avoid disturbances by the geometrical arrangement of the current-supplying electrodes.

The length constant, λ, must be experimentally determined from the exponential decrease of current-induced voltage deflections, ΔPD, along the x axis of the tubular structure. Together with ΔPD_0, i.e., the voltage deflection at the origin of current injection ($x = 0$), the calculation of the transepithelial resistance R_m is feasible. Conditions for the applicability of cable analysis, as well as computation formulas, are given in the Appendix.

Methods

In Vitro Determination of I_{sc} in the Salivary Duct

The technique is best illustrated in the study of the main duct of rabbit submaxillary gland. Rabbits are anesthetized with urethane injected subcu-

[3] U. Hegel, E. Frömter, and T. Wick, *Pfluegers Arch.* **294**, 274 (1967).
[4] H. Knauf and E. Frömter, *Pfluegers Arch.* **316**, 259 (1970).
[5] H. Knauf, *Pfluegers Arch.* **333**, 82 (1972).
[6] H. Knauf, K. Haag, R. Lübcke, E. Berger, and W. Gerok, *Am. J. Physiol.* **246**, G151 (1984).
[7] W. D. Gruber, H. Knauf, and E. Frömter, *Pfluegers Arch.* **312**, 91 (1969).

taneously (2.5 g/kg body wt, 25 g/100 ml solution). Two hours later they are tracheostomized. The submaxillary glands are exposed through a midline neck incision. The main excretory duct crosses behind the tendon of the digastric muscle, which must be cut so that free access may be gained to the oral part of the duct. The duct is then dissected with watchmaker forceps and iris scissors under a stereomicroscope. The connective tissue directly surrounding the epithelium cannot be removed without rupturing the duct. The duct is severed at the point where it penetrates the floor of the mouth and at the hilus of the gland. In this way 8- to 12-mm-long segments of the main excretory duct may be separated, as compared to the total length of the duct of about 20 mm. The diameter of the lumen is about 0.5 mm.

The ends of the isolated duct are mounted over a pair of glass capillaries (B and G in Fig. 1) with an outer diameter of about 0.5 mm and tied in place with silk thread. The capillaries are held in the wall of a Perspex chamber by means of Teflon screws and rubber gaskets. The chamber is filled with 7.5 ml rabbit serum and equilibrated with 95% O_2 – 5% CO_2. A multibarreled catheter is built by heating and then pulling polyetyhlene tubing. This catheter is advanced into the holding capillary G and fixed by a polymerizing adhesive (UHU plus). One barrel of catheter F is connected to a current source for the injection of current pulses with a constant magnitude of 10 μA and duration of 5 min. The circuit is closed via an agar bridge leading to the bath fluid. For different perfusion solutions, e.g., drugs added to Ringer's solution, the remaining barrels of catheter F could be connected to its own syringes, which are driven by Braun (Melsungen,

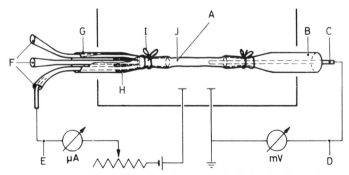

FIG. 1. Measurement of specific wall resistance in the isolated salivary duct. A, Tissue compartment; B, outflow holding capillary; C, movable glass recording capillary; D, voltage recording circuit; E, current injection circuit; F, individual polyethylene barrels; G, glass casing of multibarreled perfusion capillary; H, sealing cement. [From H. Knauf, *Pfluegers Arch.* **333**, 82 (1972). Reproduced with permission.]

FRG) perfusion pumps (rate 12.5 μl/min). The lumenal perfusate flows out via capillary B.

For the PD measurement a thin glass capillary (C) filled with perfusion fluid and connected to a calomel half-cell is advanced into the right holding capillary (B) using a micromanipulator (RP4; Brinkmann Instruments, Westbury, NY). A symmetrical reference electrode is connected to earth and to an agar bridge leading into the bathing solution. Electrical PD values, measured with a high impedance electrometer amplifier (Knick, type 27), are recorded with a pen recorder (Servogor, Metrawatt). The open-circuit PD during continuous perfusion was as high in the holding capillary as in the duct lumen, thus indicating that there was no leak between the holding capillary and the duct epithelium at the point of fixation. By the injection of current pulses via capillary G voltage deflections (ΔPD) are superimposed on the spontaneous transepithelial PD, which decreases along the tubular axis. In order to measure this longitudinal voltage attenuation, the PD-sensitive capillary C is advanced along the duct axis in steps of 0.1 mm using a micromanipulator, as mentioned above. After logarithmic transformation of the ΔPD data, the length constant (λ) was calculated as 0.055 ± 0.005 cm using linear regression analysis. The specific transepithelial resistance R_m was calculated according to Eq. (1) (see Appendix) yielding 11.2 ± 1.6 Ω cm^2. With an open-circuit PD$_{oc}$ of 17 ± 3 mV (lumen negative), the I_{sc} was calculated according to Ohm's law, i.e., PD$_{oc}$/R_m, as 1.5 ± 0.3 mA cm^{-2}. There seem to be great species differences. For instance, in the main excretory duct of rat submandibular gland R_m was determined to be 400 ± 32 Ω cm^2 by Gruber *et al.*[7] With a PD$_{oc}$ of 70 ± 2 mV the I_{sc} was calculated to be equal to 187 ± 19 μA cm^{-2}. It should be noted that in both tissue preparations the conditions for the applicability of cable analysis mentioned in the Appendix were fairly satisfied.

In Vivo Determination I_{sc} in the Human Salivary Duct

The following technique also allows the determination of the I_{sc} in humans. A multibarreled polyethylene catheter was prepared as described above with six barrels; the outer diameter of the catheter was of the order of 0.5 mm. Under a stereomicroscope the barrels (i.d. 40 μm) were opened at 2, 6, 10, and 12 mm from the tip using splinters of a razor blade (Fig. 2). The catheter is then introduced in the salivary duct and each barrel is filled with Ringer's solution. the duct is perfused via barrel 1 (rate 0.15 ml/min) and all other barrels are connected to calomel half-cells for the PD measurement (V_1, V_2). In this way, the additional openings at the tips of barrels 2, 4, 5, and 6 do not impair measurement. An intravenous or subcutaneous catheter perfused with Ringer's solution serves as the reference point.

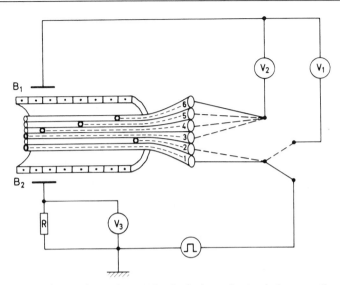

FIG. 2. Schematic experimental set-up for the I_{sc} determination in human salivary duct *in vivo*. For details see text. [From H. Knauf and E. Frömter, *Pfluegers Arch.* **316**, 259 (1970). Reproduced with permission.]

The constancy of the open-circuit PD along the duct axis can easily be checked by the individual PD-sensitive barrels of the polyethylene catheter (see Appendix). Current pulses are introduced via barrel 1. Direct current ($25-200$ μA) is obtained from a battery and measured as the voltage drop across a precision resistor (V_3). The circuit is closed via an Ag-AgCl plate adjacent to the buccal mucosa. As a check the electrical functions of barrels 1 and 2 can be reversed to demonstrate a symmetrical voltage attenuation along the duct axis independent of whether the current is injected near, or away from, the hilus of the gland. Keeping in mind that the experimental set-up is equivalent to a both-sides-open tubular structure, Eq.(6) (see Appendix) is used for the calculation of the specific transepithelial resistance R_m. In the main excretory duct of the submandibular gland and of the parotid, R_m was determined to be 365 ± 33 Ω cm^2 and 3331 ± 26 Ω cm^2, respectively. The I_{sc} was calculated as 178 ± 24 μA cm^{-2} and 160 ± 22 μA cm^{-2}, respectively. The length constant λ was about 0.6 cm in both ducts. The conditions for the applicability of cable analysis mentioned in the Appendix were satisfied; and, especially, a linear "U/I ratio" was demonstrated in both ducts, indicating ohmic behavior of R_m.

The R_m data corresponded well to the data from the rat duct. This is in accordance with histological observations demonstrating more similarities between human and rat than human and rabbit salivary duct epithelium.

In Vivo Determination of I_{sc} in the Rat Colon

As an example for the *in vivo* application of cable analysis in the rat intestinal tract the determination of R_m is demonstrated in the rat proximal colon (i.d. 0.3 cm). The technical set-up for the injection of current pulses and the PD measurement is depicted schematically in Fig. 3. After anesthesia with inactin [sodium salt of 5-ethyl-5-(1'-methylpropyl)-2-thiobarbituric acid, 20 mg/100 g body wt ip] a 2.5-cm segment of the rat proximal colon is opened at both the oral and aboral end *in vivo*, taking care that blood supply of the tissue is maintained. Two glass capillaries, A and B, are introduced into the colonic lumen, moved forward approximately 0.3 cm, and fixed with thread. The glass capillary A is connected to both a constant-perfusion pump (Unita; Braun) and a current source via an Ag-AgCl electrode. The circuit is closed via an Ag-AgCl plate adjacent to the colonic segment, and the peritoneal cavity is filled with Ringer's solution. Direct current pulses have a magnitude of 20 μA and a duration of 5 sec. For the measurement of the transepithelial PD, a thin glass pipet (o.d. 0.1 cm) filled with 2% agar in 3 M KCl is inserted from the aboral capillary B into the colonic lumen using a micromanipulator as described above. A small lateral opening about 2 cm from the tip represents the PD-sensing port of the electrode. The tip itself is closed by melting and is therefore insensitive to the PD. The stripped rat tail serves as the reference point. The registration of the PD is as described above. During the measurement of ΔPD along the colon axis the PD-insensitive tip of the electrode never leaves capillary A, thereby guaranteeing that the space occupied by the electrode in the colonic lumen, i.e., the electrical properties, remain exactly constant. In the lower panel of Fig. 3 the current-induced voltage deflections along the colon axis and the determination of length λ and ΔPD_0 are shown. The experimental data are analyzed according to Eq. (Appendix) (4), yielding 128 ± 16 Ω cm^2 for R_m. The I_{sc} was determined to 70 ± 16 μAcm^{-2}. Therefore the regression coefficient (for the determination of λ) from the exponential voltage decay within the colon is always greater than 0.995 ($n \geq 8$).

Alternative Procedures

In tubular structures with an inside radius greater than 0.1 cm, such as that of the rat lower digestive tract, the I_{sc} can also be determined by direct short circuiting. For the *in vitro* determination, the tubular structures are cut open along the longitudinal axis and treated as a planar tissue preparation separating two compartments of fluids, which are short circuited by DC current. This technique was originally described by Ussing and Zer-

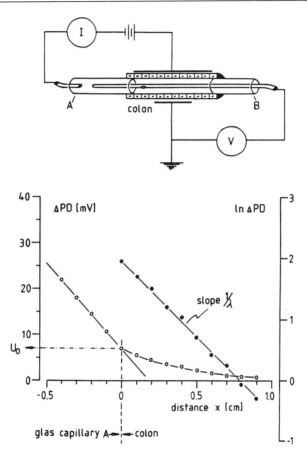

Fig. 3. *Top:* Schematic circuit diagram illustrating experimental set-up for monitoring transcolonic potential difference *(V)* and for introduction of current pulses *(I)*. Sensing port of potential difference-measuring pipet is represented by a small lateral opening about 2 cm from tip. Its axial position can be changed by moving the pipet forward or backward by means of a micromanipulator. A and B, Glass capillaries A and B. *Bottom:* Current-induced voltage deflection, ΔPD (O), in holding capillary A ($x < 0$) and in rat colon ($x > 0$). In right section exponential voltage decay along the colon axis is demonstrated. After logarithmic transformation of ΔPD (right scale), data (●) form a straight line, with slope equal to λ^{-1}. U_0 means ΔPD at the current entry into the colon segment (ΔPD$_0$). [From H. Knauf, K. Haag, R. Lübcke, E. Berger, and W. Gerok, *Am. J. Physiol.* **246,** G151 (1984) and from K. Haag, R. Lübcke, E. Berger, H. Knauf, and W. Gerok, *in* "Intestinal Absorption and Secretion" (E. Skadhauge and K. Heintze, eds.), p. 285. MTP Press, Lancaster, 1984. Reproduced with permission.]

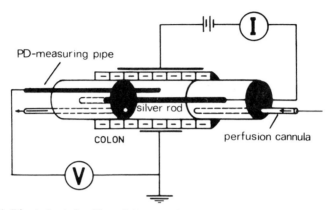

FIG. 4. Direct short circuiting of the rat colon *in vivo* using an axial silver rod for the current injection. Two holding plastic cylinders with several drill holes were introduced and the colon segment was perfused at a rate of 0.5 ml/min with Ringer's solution. [From K. Haag, R. Lübcke, H. Knauf, E. Berger, and W. Gerok, *Pfluegers Arch.* **405** (Suppl. 1), 67 (1985). Reproduced with permission.]

ahn.[1] *In vivo* a segment of a tubular structure can be short circuited as shown in Fig. 4 for the rat proximal colon using an axial silver rod as the intralumenal current-introducing electrode. As reported recently the I_{sc} and R_m values determined via cable analysis agree fairly well with that gained by direct short circuiting in the rat colon *in vivo*.[8] Direct I_{sc} determination *in vivo* in the main excretory duct of human parotid ($r \leq 0.1$ cm) was first tried by Knauf and Frömter.[4] The authors used an Ag-AgCl wire around which the perfusion catheter was wound (see Fig. 5). By this direct clamping only relative changes of I_{sc} could be monitored. An exact reference to a defined area of the tissue being short circuited was not possible due to inconstant electrical field. In a similar manner short circuiting of the salivary duct was tried *in vitro*; however, we found in our experiments that the surface area of the tubular segment could not be exactly defined by such an experimental set-up, and therefore only qualitative results could be obtained.

Summary

Use of cable analysis is a time-consuming maneuver. On the other hand, the advantage of the cable method consists in obtaining the I_{sc} and R_m related to unit area without the explicit measurement of inside radius r of the tubular structure. Obviously, application of the clamping technique

[8] K. Haag, R. Lübcke, H. Knauf, E. Berger, and W. Gerok, *Pfluegers Arch.* **405** (Suppl. 1), 67 (1985).

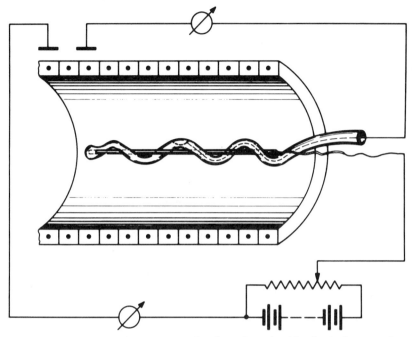

Fig. 5. Schematic experimental set-up for direct short circuiting in the human salivary duct *in vivo*. The current was obtained from a battery using a variable resistor and was introduced by an Ag-AgCl wire. The PD was measured via a lateral opening of the catheter wound around the axial wire. Using this technique in the salivary duct only relative changes of the I_{sc} can be monitored.

requires, in addition, the determination of the surface area. In summary, for small tubular structures, such as the salivary ducts of rats and rabbits, and for human experiments, cable analysis is the method of choice for the I_{sc} determination *in vivo* as well as *in vitro*. For larger tubular structures such as the rat colon the I_{sc} should be determined *in vivo* by the clamping technique described above, whereas the *in vitro* measurements should be done in an Ussing-type chamber. In the intermediate range of size both *in vivo* techniques should be applied, in which case one method may serve as a check of the other.

Appendix

Calculation of the Transepithelial Resistance R_m Based on the Experimental Data

One-Side-Open Tubular Structure. If the current, I, is introduced from one end of a tubular segment (see Fig. 1), R_m must be calculated according to Eq. (1):

$$R_m^2 = 4\pi\lambda^3\Delta PD_0 R_i/\alpha I \tag{1}$$

where R_i is the resistivity of the lumenal fluid, which can be assumed to be 50 Ω cm for Ringer solution.[9]

The quantity α accounts for the finite length (l) of the tubular segment and is given by

$$\alpha = (\coth)(l/\lambda) \tag{2}$$

The quantity α can b set to 1 if $l > 2\lambda$ and a systematic error of 2% is tolerated.

As a check of the measurement and applicability of cable analysis, the inside radius r of the tubular structure can be calculated from the electrical data and can be compared with the anatomical radius:

$$r^2 = \alpha\lambda R_i I/\pi\Delta PD_0 \tag{3}$$

Equation (1) is valid only if the outside radius (r_c) of the axial PD-measuring electrode can be neglected compared with r. If the ratio r_c/r is smaller than 0.5 an underestimation of R_m of the order of 15% must be tolerated. The effect of the PD-measuring electrode can be calculated exactly if the space occupied by the electrode remains constant during the measurement, as shown in the experimental set-up of Fig. 3. Equations (1) and (3) then have to be modified.[6]

$$R_m^2 = (4\pi\lambda^3\Delta PD_0 R_i/\alpha I)\,[r^2/(r^2 - r_c^2)] \tag{4}$$

$$r^2 = (\alpha\lambda R_i I/\pi\Delta PD_0) + r_c^2 \tag{5}$$

Both-Sides-Open Tubular Structure. When introducing current I intralumenally in the middle of a tubular segment (see Fig. 2), only 50% of the total amount of I spreads out to each end of the tubular structure. Therefore, Eq. (1) must be modified:

$$R_m^2 = 8\pi\lambda^3\Delta PD_0 R_i/I \tag{6}$$

In Eq. (6) it is assumed that the distance between the current injection point and the end of the tubular segment is large compared with λ. Analogous to Eq. (3), inside radius r can be calculated according to Eq. (7):

$$r^2 = \lambda R_i I/2\pi\Delta PD_0 \tag{7}$$

---

[9] H. Falkenhager, "Theorie der Elektrolyte." Hirzel, Federal Republic of Germany, 1971.

Conditions for the Applicability of Cable Analysis

First, the transepithelial resistance R_m must be independent of the axial position of the tubular segment. Second, R_m must be an ohmic resistor in the voltage range of open-circuit PD (PD_{oc}) and $PD_{oc} + \Delta PD_0$. In any case, zero PD must be included in this range. Last, the inside radius of the tubular segment must be fairly constant, with length l being at least of the order of the electrical length constant λ.

Check of the Applicability of Cable Analysis and of the Measurement

The conditions mentioned above can be checked by the following experimental tests:

1. The open-circuit PD spontaneously generated by the epithelium should be fairly constant along the longitudinal axis of the tubular segment, indicating a uniform epithelium.
2. The voltage deflection ΔPD at any fixed position x should be proportional to total current I injected at varying magnitude. This linear U/I ratio is equivalent to ohmic behavior of R_m,[6] which is a basic condition for the determination of both R_m and I_{sc} via cable analysis (see above).
3. Voltage deflections (ΔPD) should exhibit a simple exponential decrease along the x axis of the tubular segment at least for one λ interval.
4. The radius r calculated from the electrical data [Eqs. (3), (5), or (7)], should be in accordance with the anatomical inside radius of the tubular segment.

Origin Distortion

If the depolarizing current is injected intralumenally through a small opening of a catheter in the middle of a tubular segment, the voltage deflection ΔPD no longer exhibits a simple exponential decrease in the vicinity of the origin of current injection. This so-called "origin distortion" is caused by the high current density at the origin producing an artificially high ΔPD_0. Therefore, ΔPD_0 must be extrapolated from the more distant measurement points to the origin. Using a multibarreled catheter for the current injection and for the PD measurement (see Fig. 2), ΔPD_0 can be immediately extrapolated based on the known distances between current injection and PD measurement positions.

If the current is injected from one side via a holding capillary origin distortion can be minimized by choosing the diameter of the holding capillary similar to that of the tubular structure. In this case the position $x = 0$, i.e., at the current entry in the tubular structure, and ΔPD_0 are defined by the change from linear (in the holding capillary) to exponential voltage decay (see Fig. 3). To avoid the problem of origin distortion, R_m can be calculated without the determination of ΔPD_0, e.g., using Eqs. (1) and (3), but the inside radius r of the tubular structure must then be exactly measured. This may be difficult in small tubular structures such as the salivary duct.

[3] Dispersed Salivary Gland Acinar Cell Preparations for Use in Studies of Neuroreceptor-Coupled Secretory Events

By BRUCE J. BAUM, INDU S. AMBUDKAR, JOSEPH HELMAN, VALERIE J. HORN, JAMES E. MELVIN, LAWRENCE M. MERTZ, and R. JAMES TURNER

Introduction

Mammalian salivary glands offer excellent models for studying hormone receptor-coupled mechanisms. The most frequently chosen animal source for study is the rat. Acinar cells represent approximately 60% of rat submandibular gland and 85–90% of rat parotid gland preparations. Acinar cells are considered the only site of water transport in salivary glands and contribute about 85% of the protein constituents of saliva. These cells are also heavily endowed with hormone receptors, including α and β-adrenergic, muscarinic–cholinergic, vasoactive intestinal peptide (VIP) and substance P peptidergic, purinergic, insulin, parathormone, vitamin D, etc., i.e., they are a very responsive cell type. Previously, in this series, Schramm and Selinger described a slice preparation from the rat parotid gland for use in secretory studies.[1] It is the intent of this chapter to describe the dispersed acinar cell preparations currently (mid-1989) in usage as well as to provide information on how these preparations may be used in evaluating neurotransmitter coupling to exocrine secretory events. We will focus only on the utilization of the rat parotid and submandibular glands because relatively less study of other salivary glands has occurred.

Salivary Gland Cell Preparations

In general a pair of glands from one or two healthy male rats (250–350 g) is ample to perform single experiments. Infections of animals with

---

[1] M. Schramm and Z. Selinger, this series, Vol. 34, p. 461.

METHODS IN ENZYMOLOGY, VOL. 192

sialodacroadenitis virus are common and are an absolute contraindication for experimental use. Animals used for acinar preparations should be certified free of this virus. We routinely allow animals access to food and water *ad libitum* prior to sacrifice. Animals are sacrificed by ether anesthesia and subsequent cardiac puncture with a scalpel blade. Glands are dissected free of fat and connective tissues. The parotid glands lie on the sides of the neck. They are almost white in color, diffuse, and nonencapsulated. The submandibular glands are found beneath the lower jaw. They are reddish brown in color, compact, encapsulated, and much larger than parotid glands. Care must be taken to remove lymph nodes (tan color) and to separate the small sublingual gland from each submandibular gland.

Media

Incubation medium: We utilize a modified Hanks' balanced salt solution containing 137 mM NaCl, 5.4 mM KCl, 0.81 mM MgSO$_4$, 1.28 mM CaCl$_2$, 0.33 mM Na$_2$HPO$_4$, 0.44 mM KH$_2$PO$_4$, 5.6 mM glucose, 33 mM HEPES (pH 7.4), 0.1% bovine serum albumin

Dispersion medium 1: Incubation medium containing 100 U/ml collagenase (type CLSPA; Worthington Biochemical Corp., Freehold, NJ) and 0.1–0.2 mg/ml hyaluronidase (type 1-S; Sigma Chemical Co., St. Louis, MO). The CLSPA collagenase is quite expensive. We have tried many other types of collagenase from different companies. However, all other types result in acini with at least two-fold higher basal secretion and one-third lower stimulated responses than acini prepared with CLSPA collagenase. Differences may exist between batches of CLSPA collagenase and consequently each lot should be thoroughly tested. On occasion a batch of CLSPA collagenase has not yielded a highly responsive acinar preparation.

Dispersion medium 2: Eagle's minimal essential medium (Biofluids, Rockville, MD) containing 100 U/ml collagenase, 0.15 mg/ml hyaluronidase, and 1% bovine serum albumin. This dispersion medium is utilized when very finely dispersed acini (particularly from parotid glands) are desired for use in studies employing a filtration step (below).

Procedure 1

Glands from one rat are placed in 1–2 ml incubation medium on Parafilm or in a plastic weighing boat. Any remaining extraneous tissue is quickly removed and the glandular tissue is minced (300 times) with iris scissors. Minced tissue is transferred to a 50-ml polypropylene test tube (No. 2070; Falcon Plasticware, Oxnard, CA) containing 10 ml dispersion medium 1, gassed with 95% O_2–5% CO_2 for 10 sec, capped, and placed in a Dubnoff-type metabolic shaker at 37° and 110 rpm, in a custom-made Plexiglas holder which allows tubes to rest in water at an angle of ~25° to the bottom of the shaker platform. At 20-min intervals the tissue is regassed and triturated with a 10-ml plastic pipet (up and down 10 times) to aid in dispersion. After 80 min the resulting acinar cell suspension is washed three times in incubation medium by centrifugation at ~400 g for 10 sec, and then maintained at 37° in incubation medium containing 0.1 mg/ml lima bean trypsin inhibitor with gassing every 20 min until experimental use. Any large, undispersed, tissue clumps are allowed to settle out for ~10 sec and removed. In general this procedure yields submandibular acinar preparations which are more finely dispersed than parotid acini. We utilize this procedure for all parotid experiments except those involving a filtration step for rapid ion flux analysis (below). For these latter experiments procedure 2 is employed.

Procedure 2

Parotid glands from two rats are placed in a weighing boat in 1.5 ml of of dispersion medium 2 and are injected, using a 30-gauge needle, with 8.5 ml of the same medium. After removal of nonparenchymal elements, the tissue is finely minced and incubated in a shaker bath as above. After 80 min the dispersed tissue is centrifuged (~400 g for 30 sec) and resuspended in 10 ml fresh dispersion medium 2. After an additional 40 min of incubation (again pipetting and gassing at 20-min intervals), acini are washed three times as above in incubation medium, resuspended in 13.5 ml incubation medium, and preincubated for an additional 30 min.

Salivary Gland Neurotransmitter Receptors

There are, at present, no problems associated with measurement of neurotransmitter receptors that are unique to salivary glands. Many laboratories have examined autonomic receptor characteristics in membrane preparations from parotid and submandibular glands. As with other tissues there are very few studies which have examined receptors in intact salivary acinar cells. Studies with other cell types have pointed out that characteristics of receptors measured in intact cells may yield impressions quite different from those derived from studies with isolated membranes, particularly when radiolabeled agonists are employed as ligands.[2,3] Our own work with α-adrenoreceptors[4] has shown this to be true for parotid acini.

[2] R. N. Pittman and P. B. Molinoff, *J. Cyclic Nucleotide Res.* **6,** 421 (1980).
[3] F. Sladeczek, J. Bockaert, and J.-P. Mauger, *Mol. Pharmacol.* **24,** 392 (1983).

In this chapter only measurement of receptors in membranes will be addressed, because it is most commonly employed. However, caution is urged regarding physiological interpretation.

Preparation of Membranes

Minced gland tissue or acinar preparations are washed once, by centrifugation (400 g, 30 sec), with incubation medium and then twice with ice-cold 50 mM Tris buffer, pH 7.4 (50 mM sodium phosphate buffer, pH 7.4, also has been used for measuring muscarinic receptors). Tissue fragments or acini are then resuspended in buffer (20 vol; w/v) and homogenized with a Polytron PT-10 homogenizer (Brinkmann Instruments, Westbury, NY) for three 10-sec bursts at setting 5. The homogenate is then filtered through four layers of medical cotton gauze and the resulting filtrates centrifuged for 10 min at 40,000 g. The supernatant is discarded and the pellet is resuspended in 20 vol buffer, rehomogenized once for 10 sec, and recentrifuged as above. After discarding the supernatant, the pellet is rehomogenized once more for 10 sec in sufficient buffer to yield a protein concentration of 1 μg/μl and used as the membrane source for radioligand binding studies. We typically prepare membranes from the glands of two rats, use a portion that day, and save the remaining membranes in 1- or 2-ml aliquots, frozen in liquid N_2. The yield of membranes is about 15 mg protein from two rats. Frozen membranes are stable for at least a month and are thawed slowly in cold water and resuspended thoroughly prior to use.

Radioligand Binding Assay

Measurement of the three major neurotransmitter receptors of salivary glands is usually accomplished with antagonists as radioligands. Incubations are performed in a Dubnoff metabolic shaker in a total volume of 1 ml in 12 × 75 mm glass test tubes. Reaction mixtures contain buffer, 150–300 μg membrane protein, appropriate radioligand (see Table I) in a volume of 50–100 μl and, for determination of nonspecific binding, a competing, autonomic antagonist at a final concentration of 10 μM. In addition, β-adrenoreceptor reaction mixtures contain 10 mM MgCl$_2$. The assay is initiated by addition of protein and stopped by addition of 1–2 ml cold buffer, followed by immediate filtration through GF/C filters (Whatman, Clifton, NJ) on a multiport vacuum apparatus (Amicon Corp., Danvers, MA). Filters are air dried (4–16 hr) and assayed for radioactivity in 10 ml of Ready-Solv (Beckman Instruments, Fullerton, CA). Nonspecific binding, i.e., that observed in the presence of 10 μM competing antagonist, ranges from 20 to 40% of total radioligand binding. The radioligand binding assays described here can be performed with intact acini, though use of the incubation medium described earlier, not hypotonic buffer, is recommended. In our experience studying α-adrenoreceptors in acini, [^{3}H]prazosin binding conditions (incubation time, ligand concentration) are similar to those used for membranes (Table I), except measurement is at 37°. When agonist binding (epinephrine) is studied, very rapid time points are used (equilibrium is reached within 1 min).[4] There is much less information and experience with intact cell receptor measurements (especially agonist) and most incubation conditions will have to be individually worked out.

Neuroreceptor Coupling to Secretory Events

The use of the term coupling here refers to the signal transduction mechanism linking agonist binding at the receptor site to the physiological response (secretion). Salivary gland acinar cells have two functional signal transduction mechanisms common to most cell types. The β-adrenergic receptor (and the VIP receptor) are linked primarily to cyclic AMP formation and activation of cyclic AMP-dependent protein kinases. The α_1-adrenergic and muscarinic receptors (and the substance P receptor) are linked to the hydrolysis of phosphatidylinositol 4,5-bisphosphate and the generation of inositol trisphosphate (which mobilizes intracellular Ca^{2+}) and diacylglycerol (which activates protein kinase C). These two signal pathways in salivary cells can interact.[5] Previous chapters in this series have described in detail the measurements of these signal transduction events in other cell types (e.g., Steiner,[6] Corbin,[7] Wallace and Fain,[8] and Martin[9]). Since the methods are readily transferred to salivary cells (e.g., Baum *et al.*[10] and Aub and Putney[11]), they will not be described here. Rather we will present ways to measure the appropriate coupled secretory events in these cells. There are two major types of responses associated with secretion: (1) the release of prepackaged protein and (2) the activation of ion fluxes required for transepithelial water movement to occur.

[4] Y. Ishikawa, M. V. Gee, B. J. Baum, and G. S. Roth, *Exp. Gerontol.* **24**, 25 (1989).

[5] V. J. Horn, B. J. Baum, and I. S. Ambudkar, *J. Biol. Chem.* **263**, 12454 (1988).

[6] A. L. Steiner, this series, Vol. 10, p. 96.

[7] J. D. Corbin, this series, Vol. 99, p. 227.

[8] M. Wallace and J. N. Fain, this series, Vol. 109, p. 469.

[9] T. F. J. Martin, this series, Vol. 124, p. 424.

[10] B. J. Baum, J. M. Freiberg, H. Ito, G. S. Roth, and C. R. Filburn, *J. Biol. Chem.* **256**, 9731 (1981).

[11] D. L. Aub and J. W. Putney, Jr., *Biochem. J.* **225**, 263 (1985).

	α_1-Adrenergic	β-Adrenergic	Muscarinic
Ligand:	[³H]Prazosin	[³H]Dihydroalprenolol	[³H]Quinuclidinyl benzilate
Concentrations:	0.1–4 nM	0.25–20 nM	0.05–1 nM
Competing antagonist:	Phentolamine	Propranolol	Atropine
Temperature:	20°	37°	37°
Time:	30 min	10 min	90 min
K_D (nM):	0.8	9.0	0.3
B_{max} (fmol/mg protein):	13.2	207	364

[a] These conditions are taken from previous studies in our laboratory.[b–d] [³H]Prazosin (28 Ci/mmol) was from Amersham (Arlington Heights, IL); [³H]dihydroalprenolol (41.3 Ci/mmol) and [³H]quinuclidinyl benzilate (33.1 Ci/mmol) were from New England Nuclear (Boston, MA). Other useful ligands are available. Phentolamine was from Ciba Pharmaceuticals (Summit, NJ) and propranolol and atropine were from Sigma Chemical Company (St. Louis, MO). The ability of the receptor site to bind the radioligand (K_D) and the maximum number of receptor binding sites (B_{max}) were determined by the methods of Cheng and Prusoff[e] and Scatchard,[f] respectively.

[b] H. Ito, B. J. Baum, and G. S. Roth, *Mech. Ageing Dev.* **15**, 177 (1981).

[c] H. Ito, M. T. Hoopes, B. J. Baum, and G. S. Roth, *Biochem. Pharmacol.* **31**, 567 (1982).

[d] P. F. van der Ven, T. Takuma, and B. J. Baum, *J. Dent. Res.* **65**, 382 (1986).

[e] Y. Cheng and W. H. Prusoff, *Biochem. Pharmacol.* **22**, 3099 (1973).

[f] G. Scatchard, *Ann. N.Y. Acad. Sci.* **51**, 660 (1949).

Measurement of Protein Secretion

Protein secretion from parotid acini is typically followed as release of amylase after β-adrenergic stimulation. Note that muscarinic and α_1-adrenergic receptor activation also results in amylase release, which is ~25–30% of that seen after β-adrenoreceptor activation. Schramm and Selinger[1] in this series previously described measurement of this secretory response with minced parotid tissue (slices). We perform experiments with parotid acinar cells in 17 × 100 mm polypropylene tubes (No. 2059; Falcon Plasticware). Acinar cells from 2 rats are dispensed in 8–12 tubes in 1 ml incubation media and gassed for 10 sec. Tubes are returned to the metabolic shaker at 37°, being placed in an appropriately sized, custom-made Plexiglas holder. At 1-min intervals tubes are centrifuged and fresh incubation medium (± isoproterenol bitartrate; Sigma Chemical Co., St. Louis, MO) is added. The tubes are then gassed and returned to the metabolic shaker. At an appropriate time, the tubes are removed, centrifuged (~400 g, 10 sec) and 100 μl of the supernatant is withdrawn and placed in a 12 × 75 glass test tube. The medium aliquot and the remaining acinar incubation mixture are placed on ice until all tubes are removed. Incubation mixtures (i.e., 900 µl medium plus acini) are homogenized with a Polytron PT-10 (speed 5, 10 sec). Thereafter the medium aliquots and homogenates are diluted ~1 : 1000 with amylase assay buffer:

Amylase assay buffer: 20 mM sodium phosphate, pH 6.9; 6 mM NaCl; 1 mg/ml bovine serum albumin

The amylase assay is performed according to a modification of the Bernfeld technique.[12] Reaction mixtures in 12 × 75 mm glass tubes contain 10–50 µl diluted sample, 50 µl starch (10 mg/ml in amylase assay buffer, dissolved with gentle heating), and assay buffer as needed to give a final volume of 100 µl. The assay is linear with time for at least 8 min at 30° in a metabolic shaker and is proportional to protein with dilutions from ~1 : 300 – 1 : 1500 with a typical preparation in our laboratory. However, due to variations in rat size, acinar yield, etc., the linear range of enzyme activity may vary and therefore must be carefully determined. The reaction is quenched with 100 µl stop solution. This is prepared from two stock solutions, mixed equally, just prior to assay:

Amylase stop solution stocks: 2% dinitrosalicylate in 0.4 N NaOH (store at 4°, bring to room temperature before use) and 60% Na$^+$,K$^+$-tartrate in 0.4 N NaOH (store at room temperature)

Next the samples are placed in boiling H$_2$O for 5 min. A linear, standard curve for color development consists of 0.024–0.3 µmol maltose. After cooling at room temperature for 10 min, 1 ml H$_2$O is added per tube, tubes are vortexed, and absorbance at 540 nm is measured. Any enzyme samples yielding a maltose equivalent color absorbance of greater than 0.35 A_{540} units should be repeated at a greater dilution. Data are expressed as percentage of total amylase (medium plus acini) as described by Schramm and Selinger.[1] Amylase release is linear with time for 30–40 min. Typically, 10 µM isoproterenol will cause release of 50–70% amylase from acini in 30 min while control incubations will show 5–8% release over this same time.

To evaluate protein secretion from submandibular acinar cells we utilize an assay measuring release of protein radiolabeled by a pulse-chase protocol similar to that described by Quissell and Barzen.[13] Acinar preparations are placed in 17 × 100 mm polypropylene tubes in incubation medium containing [^{14}C]glucosamine (60 mCi/mmol; Amersham, Arling-

[12] P. Bernfeld, this series, Vol. 1, p. 149.
[13] D. O. Quissell and K. A. Barzen, *Am. J. Physiol.* **238,** C99 (1980).

ton Heights, IL; 1 μCi/ml) for 1 hr or [^{14}C]leucine (342 mCi/mmol; New England Nuclear, Boston, MA; 4 μCi/ml) for 10 min, washed by centrifugation three times (~400 g, 10 sec), gassed, and reincubated in incubation medium containing 1 mM unlabeled glucosamine or leucine. After 40–60 min (gassing at 20-min intervals) this medium is replaced with fresh medium $\pm$ agonist, acini are gassed and returned to the metabolic shaker for up to 1 hr. At appropriate intervals tubes are removed, centrifuged (~400 g, 10 sec) and 200 μl medium is withdrawn. Media and incubation mixtures are then placed on ice until all tubes are removed. The reaction mixtures are homogenized as above for parotid amylase release experiments and triplicate aliquots of media and homogenates (50 μl) are made 10% (w/v) in CCl$_3$COOH containing 25 μM unlabeled glucosamine or leucine. To each medium aliquot subjected to acid precipitation, 10 μg of bovine serum albumin is added. Samples are chilled for at least 30 min and precipitated material collected on cellulose filters (0.45-μm pore; Millipore, Bedford, MA). Filters are placed in 10 ml Ready-Solv and assayed for radioactivity. Percentage of total precipitated radioactivity secreted is calculated in a manner similar to the amylase release results. Protein release is also linear for 30–40 min. It is important to include a 0-min incubation for assessment of nonspecific radioactivity precipitated, and to subtract this when calculating secretion data. Typically 10 μM isoproterenol or epinephrine (Sigmal Chemical Co., St. Louis, MO) will cause release of 20–40% of acid-precipitable radiolabel in 40 min while control release will be ~5–15%. Alternatively, several laboratories have measured protein release from submandibular acinar cells by following secretion of lactoperoxidase.[14] We, however, have not found this as useful as the assay described.

Measurement of Ion Fluxes Related to Water Transport

The movements of two ions, Ca^{2+} and Cl$^-$, are closely tied to water movement across salivary acini following α-adrenergic, muscarinic, or substance P stimuli. Calcium ions are mobilized from an intracellular store[15] and function as a second messenger involved in protein secretion events and especially in K$^+$ and Cl$^-$ fluxes. The availability of Ca^{2+}-sensitive fluorescent dyes has greatly facilitated measurements of intracellular Ca^{2+} mobilization in many cells. Previously, in this series, Capponi *et al.*[16] described in detail the use of one such probe, quin2. We have modified their methods to measure Ca^{2+} mobilization in salivary acinar preparations

[14] B. I. Bogart and J. Picarelli, *Am. J. Physiol.* **235**, C256 (1978).
[15] J. W. Putney, Jr., *Annu. Rev. Physiol.* **48**, 75 (1986).
[16] A. M. Capponi, P. D. Lew, W. Schlegel, and T. Pozzan, this series, Vol. 124, p. 116.

using either quin2 or Fura-2 and important considerations are given here. The reader is referred to the Capponi et al.[16] chapter for additional details. Submandibular and parotid acinar cell preparations are placed in incubation medium containing either $10-20 \mu M$ quin2-AM or $1-2 \mu M$ Fura-2-AM (esterified forms obtained from Calbiochem, La Jolla, CA; made up as a 50 mM stock is dimethyl sulfoxide and stored at $-70°$).[17,18] After incubation for 45 min at $30°$ in a metabolic shaker (gassing at 20-min intervals), cells are washed by centrifugation ($\sim$400 g, 10 sec) three times in incubation medium to remove extracellular dye. Cells are then resuspended in the same medium containing 0.5 mM probenacid, which retards dye leakage (see below) but has no effect on cell response,[19] and kept at $30°$ until used for fluorescence experiments. Dye leakage from salivary cells is estimated by adding either 10 mM EGTA or $50 \mu M$ MnCl$_2$ followed by $100 \mu M$ diethylenetriaminepentacetic acid to the extracellular medium. Fluorescence measurements are performed in an SLM-8000 spectrofluorimeter (SLM Instruments, Urbana, IL). The excitation and emission wavelengths are set to 339 and 495 nm for quin2 and 340 and 510 nm for Fura-2, and the slit width adjusted to 4 nm. Cells are gently stirred during the fluorescence assay and the temperature maintained at $37°$. At the end of each experiment, F_{max} is determined in quin2-loaded cells by permeabilization of cells with 0.007% Triton X-100 in the presence of 1.28 mM Ca^{2+} in the medium. With Fura-2-loaded cells, $2 \mu M$ ionomycin is added to obtain F_{max}. F_{min} is determined by the addition of 10 mM EGTA and 20 mM Tris base. Cytosolic free Ca^{2+} ($[Ca^{2+}]_i$), equivalent to fluorescence (F) for any measurement point, can be calculated according to Tsien et al.[20]:

$$[Ca^{2+}]_i = K_d[(F - F_{min})/(F_{max} - F)]$$

where K_d is the binding constant of the dye for Ca^{2+}. The fluorescence values are corrected for the fluorescence due to extracellular dye.[21] A representative experiment, showing the rapid elevation of intracellular Ca^{2+} in quin2- and Fura-2-loaded acinar cells in response to the muscarinc agonist carbachol ($10 \mu M$, Sigma Chemical Co., St. Louis, MO), is presented in Fig. 1. α_1-Adrenergic stimulation ($10 \mu M$ epinephrine) of these cells results in comparable changes in $[Ca^{2+}]_i$. Using these cell preparations, we find that neuroreceptor-elicited, short-term $[Ca^{2+}]_i$ mobilization events

[17] J. Helman, I. S. Ambudkar, and B. J. Baum, Eur. J. Pharmacol. 143, 65 (1987).
[18] I. S. Ambudkar, J. E. Melvin, and B. J. Baum, Pfluegers Arch. 412, 75 (1988).
[19] T. H. Steinberg, A. S. Newman, J. A. Swanson, and S. C. Silverstein, J. Cell Biol. 105, 2695 (1987).
[20] R. Y. Tsien, T. Pozzan, and J. J. Rink, J. Cell Biol. 94, 325 (1982).
[21] P. M. McDonough and D. C. Button, Cell Calcium 10, 171 (1989).

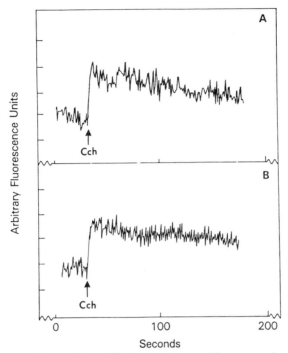

FIG. 1. Elevation of intracellular Ca^{2+} (shown as increased fluorescence in arbitrary units) after muscarinic–cholinergic [10 μM carbachol (Cch)] stimulation of rat parotid acinar cells preloaded using either 13 μM quin2-AM (A) or 2 μM Fura-2-AM (B). Peak $[Ca^{2+}]_i$ values, estimated from three to eight similar experiments, were 745 ± 73 nM with quin2 and 878 ± 98 nM with Fura-2. The apparent slight increase in the sustained response with Fura-2 loaded cells can be accounted for by dye leakage.

assessed using Fura-2 are generally similar in pattern and magnitude to those observed with quin2.

Before deciding which dye should be used for studies of $[Ca^{2+}]_i$ with salivary epithelial cells, several points should be considered. Quin2 is used at higher concentrations than Fura-2 and it can, under certain conditions, buffer $[Ca^{2+}]_i$. With our cell preparations as shown above, we find that using ≤ 25 μM quin2-AM to load parotid acinar cells results in no significant alterations in agonist-stimulated $[Ca^{2+}]_i$ changes from those observed with Fura-2. If, however, higher concentrations of quin2-AM are used for loading, an attenuation of the $[Ca^{2+}]_i$ changes is seen. Importantly, loading of parotid acinar cells with ≤ 25 μM quin2-AM also has no effect on the water-transporting capacity (Cl⁻ efflux; see below) of these cells (Fig. 2). Parotid acinar cells exhibit the same qualitative and quantitative pattern of carbachol-stimulated Cl⁻ efflux without or with 25 μM quin2-AM loading.

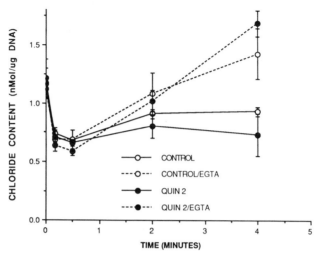

FIG. 2. Effect of muscarinic–cholinergic stimulation on Cl^- content of rat parotid acinar cells. Cells exposed to 10 μM carbachol (at time zero) rapidly lose ~50% of their Cl^- in the presence or absence (+EGTA, 5 mM) of extracellullar Ca^{2+}. Preloading cells with 25 μM quin2 AM has no effect on these responses.

Generally, Fura-2 is now more widely used (for $[Ca^{2+}]_i$ studies) than quin2 with most cell types. However, we find use of Fura-2 with parotid acinar cells (versus that of quin2) subject to an important limitation. In the presence of 0.5 mM probenacid parotid cells retain quin2 over an extended time (at least 30 min at 37°) with negligible dye leak. Conversely, even with 2 mM probenacid (higher concentrations are toxic), parotid acinar cells do not adequately retain Fura-2. Thus, for long-term studies of $[Ca^{2+}]_i$ (e.g., as related to sustained fluid secretory capacity), quin2 is a more suitable dye choice with salivary cell preparations. Freshly washed Fura-2-loaded cells can be successfully used, however, for short-term (<500 sec) measurements of $[Ca^{2+}]_i$ changes.

At present, it is felt that Cl^- provides the driving force for transepithelial water movement in salivary acini.[22,23] Uptake of Cl^- in rat parotid acinar cells appears to be primarily mediated by the $Na^+/K^+/Cl^-$ cotransport system in the basolateral membrane.[23,24] It is not clear whether Cl^- uptake via this mechanism is regulated directly by neuroreceptor stimuli. Efflux of

[22] J. A. Young, in "Electrolyte and Water Transport Across Gastrointestinal Epithelia" (R. M. Case, A. Garner, L. A. Turnberg, and J. A. Young, eds.), p. 181, Raven, New York, 1982.
[23] R. J. Turner, J. N. George, and B. J. Baum, J. Membr. Biol. 94, 143 (1986).
[24] M. Kawaguchi, R. J. Turner, and B. J. Baum, Arch. Oral Biol. 31, 679 (1986).

Cl^- is thought to occur via an apical membrane Cl^- conductance which is activated following stimulation of a Ca^{2+}-mobilizing receptor, i.e., muscarinic, α_1-adrenergic, and substance P. To measure such stimulated Cl^- efflux, parotid acini from two rats are preincubated for at least 30 min at $37°$ in ~10 ml incubation medium containing 3 μCi $^{36}Cl^-$/ml ($Na^{36}Cl$, 16.1 mCi/g; ICN, Irvine, CA). The $^{36}Cl^-$ content of the acinar preparation (using procedure 2) reaches equilibrium within 5 min and remains constant for at least 5 hr at $37°$ when gassed at 10- to 20-min intervals. Aliquots of acini (190 μl) are removed and placed in 17 × 100 mm polypropylene tubes, 10 μl of carbachol (at 20 × desired final concentration) is added and the tubes put in a metabolic shaker. At appropriate times, Cl^- flux is quenched with the addition of 3 ml ice-cold stop solution [150 mM sodium gluconate, 10 mM HEPES, pH 7.4, 1 mM mannitol with 2 μCi [^{3}H]mannitol (19.1 mCi/mmol; New England Nuclear, Boston, MA)]. The reaction mixture is then immediately filtered over a 0.45-μm cellulose filter and washed twice with ice-cold stop solution without [^{3}H]mannitol. Under these conditions [^{3}H]mannitol serves as marker for trapped extracellular fluid. This nonspecific trapping typically accounts for less than 15% of radioactivity associated with the acini.

Filters are next transferred to 2.2 ml 1.5 M perchloric acid and heated at $70-75°$ for 15 min. Aliquots (2 × 600 μl) of the cooled (10 min, on ice) perchlorate extract are assayed for DNA by the diphenylamine procedure,[25] while other aliquots are taken for measurement of $^{36}Cl^-$ radioactivity by liquid scintillation spectrometry. Data are expressed as nanomoles Cl^- present in acini per microgram DNA.[26] As shown in Fig. 2, 10 μM carbachol results in a rapid (<10 sec), 50% loss of acinar Cl^-. This initial release phase is followed by a slow reuptake of Cl^- (mediated by the $Na^+/K^+/Cl^-$ cotransporter[26]). In the absence of extracellular Ca^{2+}, the Cl^- contents of cells rapidly returns to prestimulus levels.

---

[25] G. M. Richards, *Anal. Biochem.* **57**, 369 (1974).
[26] J. E. Melvin, M. Kawaguchi, B. J. Baum, and R. J. Turner, *Biochem. Biophys. Res. Commun.* **145**, 754 (1987).

[4] Intracellular Ion Activities and Membrane Transport in Parietal Cells Measured with Fluorescent Dyes

By PAUL A. NEGULESCU and TERRY E. MACHEN

Introduction

There has been rapid progress in the development of fluorescent dyes for measuring intracellular ion activities. Dyes are now available for measuring pH,[1] $[Ca^{2+}]$,[2] $[Na^+]$,[3] $[K^+]$,[4] $[Mg^{2+}]$,[5] and $[Cl^-]$.[6] Most work has been with the Ca^{2+} and pH dyes because these were the first to be developed, but it is expected that the other ions will be studied just as intensely now that the other indicators are also being introduced. With the exception of the Cl^- probe, these dyes are presently available as membrane-permeable acetoxymethyl ester (AM ester) derivatives that enter cells readily. Intracellular esterases then cleave the dyes into charged, membrane-impermeant forms which remain trapped in the cytoplasm. These dyes have allowed rapid progress in cellular physiology because it has now become possible to measure intracellular ion activities in small cells with good spatial and temporal resolution. Many of these dyes exhibit a shift in their excitation or emission spectra when they bind their specific ions (e.g., see Figs. 6 and 11), allowing them to be used in the so-called ratio mode, in which the ratio of the fluorescence intensity at two different wavelengths is measured. The advantage of fluorescence ratioing is that changes in brightness due to variation in dye concentration or path length affect both wavelengths equally. By ratioing the fluorescence intensities one obtains an estimate of ionic concentration that is independent of the dye concentration or amount. It is now possible to make intensity ratio measurements on the single cell level by using a microscope, some mechanism for alternating the light between two fixed excitation or (more rarely) emission wavelengths, and a measuring device such as a photomultiplier or a television camera and a digital image processing hardware.

The purpose of the chapter is to summarize the methods we have used for measuring intracellular pH (pH$_i$, with BCECF), $[Ca_i^{2+}]$ (Ca_i^{2+}, with

[1] T. J. Rink, Y. Tsien, and T. Pozzan, *J. Cell Biol.* **95**, 189 (1982).

[2] G. Grynkiewicz, M. Poenie, and R. Y. Tsien, *J. Biol. Chem.* **260**, 3440 (1985).

[3] A. Minta, and R. Y. Tsien, *J. Biol. Chem.* **264**, 19449 (1989).

[4] R. P. Haughland, "Molecular Probes Handbook" p. 95. 1989.

[5] B. Raju, E. Murphy, L. A. Levy, R. D. Hall, and R. E. London, *Am. J. Physiol.* **256**, C540 (1989).

[6] N. I. Illsley and A. S. Verkman, *Biochem. J.* **26**, 1215, (1987).

Fura-2) and [Na$^+$] (Na$_i^+$, with SBFI) in single, identified cells of gastric glands isolated from rabbits. (The chemical names for these dyes are given in Table I). General methods for loading the dyes into cells, attaching the glands to coverslips and perfusing them with different solutions using flow-through chambers will be described. Techniques for assessing intracellular dye behavior and calibrating the dye fluorescence in cells will be discussed. In addition, approaches for identifying specific membrane transport systems and for measuring rates of transport by these mechanisms will be presented.

General Methods for Loading Cells with Dyes, Perfusing with Solutions, and Measuring Fluorescence Excitation Ratios

Loading the Dyes into Gland Cells

With the development of the acetoxymethyl ester derivatives of BCECF, Fura-2, and SBFI, loading the dyes into cells has become routine. Each of the dyes is first dissolved in dimethylsulfoxide (DMSO) to make a 10 mM stock solution, which can be stored at $-20°$ for 4 weeks (sometimes longer) with no apparent loss of activity.

For experiments on isolated glands, the glands are prepared using the high-pressure perfusion and collagenase digestion techniques that were originally developed by Berglindh and Obrink.[7] The dyes are added (2 μM for BCECF, 5 μM for Fura-2, and 7 μM for SBFI) to a 5% "cytocrit," and the glands are incubated for 10–45 min. For SBFI (but not the others) it is necessary to mix the 10 mM stock with a 25% (w/v) solution of Pluronic F-127 (Molecular Probes, Eugene, OR) before addition to the gland loading solution. For experiments on intact epithelial sheets higher concentrations of dye (e.g., 10–50 μM) and longer incubation times (up to 3 hr for the intact salamander gall bladder or stomach) are required. Most cell types, including rabbit gastric glands, exhibit a more uniform loading if the incubation is done at room temperature rather than at 37°. The reason for this is unclear, but one possibility is that permeation of the dye esters across the plasma membranes is so fast at 37° that cytoplasmic esterases are overwhelmed, allowing accumulation of dye in intracellular, membrane-bound compartments. At room temperature the dye esters get into the cytoplasm more slowly, are cleaved to their impermeable forms, and remain trapped there. Another problem is that once the esters have been cleaved from the parent molecule, the dye leaks (or is transported) out of the cells more quickly at 37° as compared with room temperature.

---

[7] T. Berglindh T. and K. J. Obrink, *Acta Physiol. Scand.* **96,** 150 (1976).

Determining How Much Dye Is Trapped in the Cells

The concentration of intracellular, trapped dye has been determined by taking a known amount of glands in a cuvette, lysing them with digitonin into a defined volume, and comparing the fluorescence with a standard curve. The intracellular concentration of the dyes can then be calculated by knowing the cell water/dry weight ratio for glands, which has been found to be 2 μl/mg by many investigators.[7] We have found that the glands contain 10-50 μM BCECF[8] and 50-150 μM Fura-2.[9] However, these dyes are not uniformly distributed in the two cell types of the glands. Both parietal cells and chief cells load very well with all the dyes, but after about an hour parietal cells retain much higher concentrations of Fura-2 and SBFI as compared with the parietal cells, while chief cells contain more BCECF. The reason for the differences in dye handling by the cells is unknown.

Determining How Large the Fluorescence Signals Should Be

Using the loading protocols described above, most cells yield fluorescence intensities that are somewhere between 10 and 30 times greater than background. "Background" refers here to the emission obtained at a given excitation wavelength from the same experimental set-up with unloaded cells. This includes the dark current of the photomultiplier or camera, stray light from the experimental set-up, the signal from extracellular dye, and light scatter and autofluorescence from the cells. In order for the dye ratio to be independent of the background ratio (which is usually close to 1.0), the fluorescence intensity of the probe at each wavelength must remain significantly above background levels under all experimental conditions. By taking steps to minimize and correct for any factors not related to dye fluorescence, we have found that signals at least 3× greater than background are sufficient. In practical terms this means that the minimum acceptable dye signal must be 3× greater than the fluorescence from the dimmer of the two wavelengths. For example, BCECF emits less light when excited at 440 nm than when excited at 490 nm (see Fig. 3). Thus, the emission at 440 nm should remain at least 3× above background. The situation for Fura-2 is more complicated because an increase in Ca^{2+} causes increased emission at 350 nm and a decrease at 385 nm (see Fig. 6A-C). Since the intensity at 385 nm decreases approximately threefold going from 0 Ca^{2+} to saturating Ca^{2+} *in situ* (Fig. 6C), the intensity at 385 nm should be at least ninefold greater than background at low Ca_i^{2+} (i.e., in the resting cell) in order for the 385-nm signal to remain threefold above background at high Ca_i^{2+}. *In situ*, SBFI decreases at all wavelengths as Na^+ increases (see Fig. 11C) but the largest decreases occurs at 380 nm and is a drop of ~ twofold between 0 and 150 mM Na^+. Thus, SBFI emission at

380 nm should be $6\times$ above background at low Na_i^+. Allowances should be made for the tendency of the cells to lose dye during the experiment.

Glands and Cells Suspended in a Cuvette

When cells are suspended in solution in a cuvette, measurements are often made by exciting the dyes only at the ion-sensitive regions of the spectra (e.g., at 500 nm for BCECF or at 385 nm for Fura-2). This cuvette approach has the major advantage of being able to perform replicates on many cells very quickly. In cases where the fluorescence of the probe is too dim for measurements on single cells (e.g., quin 2) the cuvette may be the only option.

The cuvette technique has at least four drawbacks, however. First, these measurements are often performed at a single excitation wavelength. This sacrifices about 50% of the fluorescence response of Fura-2 and SBFI because of their shifts in excitation spectra upon binding the ions. Second, a finite amount of dye leaks from the cells, and therefore the signals measured are a function of both intracellular and extracellular concentrations of the ions. This dye leakage is often rather small, but when it occurs it can be troublesome. The standard method for assessing dye leakage is to spin the cells out of suspension using a centrifuge and measure the dye remaining in the solution. However, this correction is not perfect because centrifuging cells can damage them, causing release of dye and a consequent overestimate of leakage. It is fortunate that in situations where this problem has been addressed, leaked dye contributes less than 5% to the overall signal (i.e., cells plus extracellular dye).[8] A third problem with cuvette experiments is that the number of possible solutions that can be added to the cells is usually limited to two or three, e.g., the first solution the cells are suspended in plus one or two additions. Finally, cuvette experiments necessarily require the use of millions of cells, so there is no information about the heterogeneity of cell responses. In addition, the cells should be purified to homogeneity in order to avoid the ambiguity of responses from more than one cell type. Although the cuvette approach has been profitably applied to the study of purified cells,[10,11] the limitations listed above have encouraged the development of techniques for measuring fluorescence from single cells using microfluorimetry.

[8] A. M. Paradiso, P. A. Negulescu, and T. E. Machen, *Am. J. Physiol.* **250,** G524 (1986).

[9] P. A. Negulescu, W. W. Reenstra, and T. E. Machen, *Am. J. Physiol.* **256,** C241 (1989).

[10] C. S. Chew, and M. Brown, *Biochim. Biophys. Acta* **888,** 116 (1986).

[11] S. J. Pandol, M. S. Schoeffield, G. Sachs, and S. Muallem, *J. Biol. Chem.* **260,** 10081 (1985).

Perfusing the Basolateral Surface of Glands on the Stage of a Microscope

The arrangement we have adopted for investigating the transport properties of single cells in glands is to mount the glands in a special chamber and then to perfuse different solutions over the cells' surfaces so that any leaked dye is quickly washed away. This chamber is mounted on the stage of a microscope so that we can collect emitted light from single, identified parietal or chief cells within the intact gastric gland. Because the gland forms a tube that is sealed at both ends, the perfusion in the chamber occurs only over the basolateral surface of the cells, and the lumenal surface remains unstirred.

The chamber is a stainless steel wafer with a slot drilled out of the center. Holes are drilled from the outside edge to the long ends of the slot to accommodate 22-guage needles that serve as input and output ports. A round coverslip is epoxied into a recess on one surface of the water. A coverslip with attached glands is placed in a vacuum-greased recess on the other side of the wafer and sealed by means of an "O" ring and screw-down holder. The perfusion chamber has a volume of about 50 μl. The chamber then fits into a water-jacketed, heated (37°) aluminum block on the microscope stage. For measurements on cultured cells, the cells are grown on the appropriate coverglasses. For isolated cells, the coverglasses are usually coated with poly-D-lysine (1%) poly-D-lysine in water, a small drop in the middle of each coverglass for 2 min followed by washes with 1 M NaCl (or KCl) and then with Ringer's. Different perfusion solutions are maintained at 37° by placing them in hanging plastic syringes (60 ml) which are connected to an eight-point manifold solution selector (Hamilton Co., Reno, NV) via one-way stopcocks and polyethylene tubing. All of this is contained in a Plexiglas "warming box" that is heated to 40°. The manifold output is connected to the perfusion chamber via another 15-cm length of tubing. The height of the warming box above the stage controls the rate of perfusion through the chamber.

Controlling pH in HCO_3^-/CO_2 Solutions

Measurements of cellular activities should ideally be performed under physiological conditions, and this usually means in solutions containing HCO_3^- and saturated with a gas mixture containing CO_2. The main problems that arise result from the loss of CO_2 through gas-permeable polyethylene tubing. These losses can be quite substantial, especially if the distances are long and the tubing is small in diameter. It is best to use gas-impermeable tubing or stainless steel tubes that are connected to the chamber and solution storage syringes by very short polyethylene or rubber connectors.

*Measuring Fluorescence with a Photomultiplier or TV Camera and
Digital Image Processor*

Once the cells have been attached to a coverslip in a chamber on the stage of a microscope, the next stage is to excite the dye at two different wavelengths with a xenon lamp. The excitation light in one of our set-ups is provided by the microscope 75-W xenon bulb, and the appropriate wavelength is selected by alternating two filters (440 ± 5 nm and 490 ± 5 nm for BCECF and 345 ± 10 nm and 385 ± 10 nm for Fura-2 and SBFI; Omega Opital, Brattleboro, VT) mounted in a paddle wheel driven by a stepping motor (SLO-SYN; The Superior Electric Company, Bristol, CT) at 2 Hz. The excitation intensity is regulated by neutral density filters to reduce bleaching of the dye and exposure of the cells to unnecessary light intensities. The excitation beam passes through an iris aperture (to confine the illuminated field to the desired size), a dichroic reflector (510 nm for BCECF and 395 nm for Fura-2 and SBFI), and either a Zeiss Neofluor objective (×63, 1.25 NA) or Nikon Fluorite objective (×40, 1.30 NA) before hitting the dye-loaded cells. The Nikon lens is best for Fura-2 and SBFI because of its excellent UV transmission. However, the Zeiss ×63 is also very good, and its larger working distance (0.6 mm) offers some flexibility in chamber design. The emitted fluorescence passes through a filter (520 to 550 nm bandpass for BCECF and 450-nm longpass for Fura-2 and SBFI) and an image plan pinhole (which limits the collected light to a single parietal cell) before hitting a photomultiplier tube and photon counter (QL 30-F, APED-2; Thorn Emi Gencom, Inc., Fairfield, NJ). Parietal cells are identified according to their localization and morphological appearance.[12,13] To assure that the signal is coming only from a specific cell it is often necessary to focus through the gland to make sure that there are no cells underneath the one of interest.

An IBM/AT computer and UMANS software (User-Friendly Microspectrofluorimetry Data Acquisition and Analysis System, by Chester M. Regen, Urbana, IL) controls the stepping motor and collects, stores, and processes the data. Single-wavelength intensities and calculated ratios 490 nm/440 nm (once per second) are displayed on-line on the monitor. Further data processing is performed using either Lotus 1-2-3 2.0 or Plotit 1.4 (Scientific Programming Enterprises, Haslett, MI) and a customized data analysis program (by Joseph A. Bonanno, Optometry School, and William Weintraub, Department of Molecular and Cell Biology, University of California, Berkeley) on an IBM/AT or Excel 1.04 (Microsoft Corp.,

[12] A. M. Paradiso, R. Y. Tsien, and T. E. Machen, *Nature (London)* **325**, 447 (1987).
[13] A. M. Paradiso, R. Y. Tsien, J. R. Demarest, and T. E. Machen, *Am. J. Physiol.* **253**, C30 (1987).

Redmont, WA), Cricketgraph 1.2 (Cricket Software, Malvern, PA), and SuperPaint 1.0 (Silicon Beach Software, Inc., San Diego, CA) on a Macintosh II (Apple Computer, Inc., Cupertino, CA).

In a second set-up alternating excitation light is provided by the xenon lamp, beam splitter, two monochromators (440 and 490 nm), and rotating chopper mirror (20 Hz) of a Spex Fluorolog spectrofluorimeter. Excitation slit widths are normally set at ~0.25 mm, corresponding to 0.9-nm bandpass, to prevent photodamage and bleaching of the dye. The other optical components are the same as the first set-up. The output from the photomultiplier is passed to the Spex Datamate microcomputer, which averages the emission intensities collected over a 0.5- or 1.0-sec time period at each excitation wavelength.

Fluorescence from all the cells in single glands can also be measured using digital image processing of video images of the fluorescence of each excitation wavelength.[12,14] Black and white fluorescence images of whole glands are acquired using a silicon-intensified target camera (Dage 66) and relayed to a Gould FD5000 image processor which is controlled by a DEC computer (PDP 11/73) using software written by Roger Y. Tsien (Department of Pharmacology, University of California at San Diego, La Jolla, CA). After correcting for background and dark current of the camera, the fluorescence intensity ratio is calculated for each pixel and displayed as 1 of 64 pseudocolors. These ratios are then calibrated as discussed below for each dye. Analysis and plotting of ratio vs time for the individual cells is accomplished using a graphics emulation terminal (Smarterm 240). This imaging system allows the collection of data from all the parietal and chief cells in a gland simultaneously, though the same problems of cellular identification pertain.

Calibrating the Dye Signal

At the end of an experimental protocol it is necessary to calibrate the fluorescence signal in terms of free ionic concentration or activity. Ideally one performs a so-called *in-situ* calibration in which the cell membranes treated with an ionophore that equilibrates intracellular and extracellular concentrations of the specific ion of choice, while retaining the dye in the cell. Then, the extracellular solution is titrated to different ionic concentrations, fluorescence intensity (or intensity ratio) is measured, and a calibration curve can then be constructed. With this approach one can be assured that the calibration is being performed on dye experiencing the same cytoplasmic environment that it experiences during an experiment. This is

---

[14] R. Y. Tsien, and M. Poenie, *Trends Biochem. Sci.* **11**, 450 (1986).

important because all of the dyes exhibit different excitation spectra and [ion]–fluorescence relationships when they are in the cytoplasm *vs* in free solution (see Table I and Figs. 3, 6, and 11). For example, the excitation spectra of BCECF[15,16] and Fura-2[17] and SBFI (see Fig. 11) show increased emission at longer wavelengths when compared to their behavior in free solution. As a result of the shift, a calibration based on the 490/440 ratio of BCECF in free solution usually causes an underestimate of pH_i by 0.1–0.2 pH unit. Even worse problems can arise with the calibrations of Fura-2 and SBFI. The specific protocols for calibrating each of the eye signals will be discussed below.

Potential Problems with Membrane-Permeant Dyes

Toxic Effects of Dye Loading

The fluorescent, impermeable forms of the dyes accumulate intracellularly because cells have esterases that cleave the acetoxymethylester bonds. In the process, several molecules each of formaldehyde and acetic acid are generated for each molecule of dye that is trapped in the cytoplasm (five each for Fura-2, four each for SBFI, and three formaldehyde and two acetic acid for BCECF). Because cellular concentrations of the dyes reach 10–200 μM, the [formaldehyde] could be as high as 1.0 mM. These products appear to be toxic to the parietal cells, for it has been found that their O_2 consumption and ability to secrete H^+ (as judged by the accumulation of the weak base aminopyrine) decrease quite dramatically during BCECF and Fura-2 loading.[9] This can be seen from Fig. 1, which shows that parietal cell function, as measured by O_2 consumption and accumulation of the weak base aminopyrine, is rapidly inhibited by dye loading. Similar inhibitory effects on O_2 consumption are seen with 0.5 mM formaldehyde.[9] These inhibitory effects on aminopyrine accumulation are reversible because the cells recover normal secretory function within about 80 min of being washed with fresh solution (thereby removing the extracellular BCECF/AM or Fura-2/AM as well as accumulated formaldehyde).[9] These dye-loaded cells then secrete HCl at normal rates (Fig. 1B). Dye-loaded glands can often (but not always) be stored at room temperature or at 4° for up to 10 hr with no apparent loss in function.

[15] S. Muallem, C. Burnham, D. Blissard, T. Berglindh, and G. Sachs, *J. Biol. Chem.* **260**, 6641 (1985).

[16] S. Grinstein, C. A. Clarke, and A. Rothstein, *J. Gen. Physiol.* **82**, 619, (1983).

[17] W. Almers, and E. Neher, *FEBS Lett.* **192**, 13 (1985).

TABLE I
COMPARISON OF AFFINITY AND FLUORESCENT RESPONSES OF
BCECF, FURA-2, AND SBFI *in Vitro* AND *in Situ*

	In vitro[a]		In situ[b]	
	K_d or pK_a[c]	R_{max}/R_{min}[d]	K_d or pK_a	$R_{max}R_{min}$
BCECF[a]	7.1	6.5	7.0	4.0
Fura-2[f]	200 nM	30.0	350 nM	5.0
SBFI[g]	17 mM	3.0	>60 mM	2.2

[a] Calcium ion values determined in solutions containing 115 mM KCl, 15 mM NaCl, 10 mM HEPES, and 10 mM EGTA with various [Ca^{2+}] (as described in Ref. 2.), pH 7.1, at 24° in a cuvette as described in the text. pH values obtained in solutions containing 120 mM potassium gluconate, 30 mM KCl, 10 mM HEPES, pH 7.1, at 24°. Sodium ion values obtained in solutions containing $Na^+ + K^+ = 150$ mM. Anionic balance was 120 mM gluconate and 30 mM Cl^- (Na^+ or K^+ salt) as described in text.

[b] Values determined for dye in cells on the microscope stage. Calibration solutions identical to those for *in vitro* measurements except for the presence of ionophores and permeabilizing agents (see text). These values can vary depending on both the excitation system (filters vs monochromators) and the other optical components.

[c] K_d or pK_a determined at 24° for both *in vitro* and *in situ* measurements. Affinities may be slightly higher at 37°, but by a small amount.

[d] Value obtained by dividing the intensity ratio at saturating concentrations of H^+, Ca^{2+}, and Na^+ (R_{max} for SBFI determined at 140 mM Na^+) by the intensity ratio at minimum ionic concentration. This is a normalized value for the total fluorescence response.

[e] BCECF, 2′,7′-bis[2-carboxyethyl-5 (and 6)-carboxyfluorescein.

[f] Fura-2, 1-[2-(5 carboxyoxazol-2-yl)-6-aminobenzofuran-5-oxy]-2-(2′-amino-5′-methylphenoxy)-ethane-N,N,N',N'-tetraacetic acid.

[g] SBFI, Sodium-binding benzofuran isophthalate.

Incomplete Hydrolysis of the Dye

This issue must be considered whenever AM ester derivatives are loaded into cells. The problem may be more severe with Fura-2 than with BCECF because five AM ester groups must be released from Fura-2, compared with three from BCECF. An excellent method to determine whether there is a significant amount of partially cleaved Fura-2 in the cells is to take advantage of the fact that Mn^{2+} quenches the Ca^{2+}-sensitive Fura-2. As shown in Fig. 7A, addition of excess $MnCl_2$ (e.g., 1 mM) to

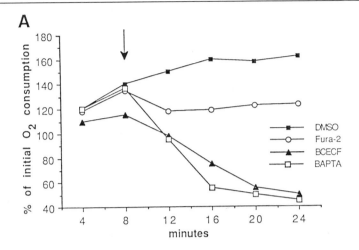

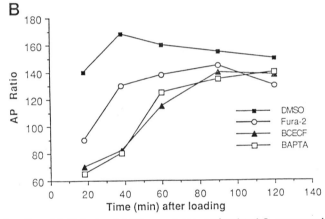

FIG. 1. Loading with AM esters inhibits secretagogue-stimulated O_2 consumption and H^+ secretion. Recovery occurs in 80 min. (A) A suspension of glands was stimulated with 10^{-4} M dibutyryl-cAMP (dbcAMP) plus 10^{-4} M isobutylmethylxanthine (IBMX) at time 0, and 8 min later AM ester derivatives of Fura-2 (O) BCEF (▲), and BAPTA (□) were added (10 μM final concentration). Controls contained 10 μM DMSO (■). Note that all of the esters inhibited O_2 consumption induced by the secretagogues. Formaldehyde (0.5 mM) had very similar inhibitory effects to those exhibited by BCECF and BAPTA (not shown[9]). (B) At various times after loading glands with the AM esters (10 min at 10 μM), glands were washed for the times shown on the x axis and then stimulated with dbcAMP + IBMX; uptake of the radioactive weak base aminopyrine (AP) was then measured. Full recovery of ability to secrete H^+ occurred after 80 min. (Taken from Ref. 9 with permission of the American Physiological Society.)

ionomycin-treated glands caused fluorescence to decrease to nearly auto-fluorescence levels, indicating that most dye was Ca^{2+} sensitive. A more quantitative indication of the quality of loading is obtained by lysing cells with Triton X-100 and titrating the supernatant to check the dye spectrum in free solution.[18,19] Any differences in affinity or spectra would be indicative of partially cleaved dye. This problem may be alleviated by warming cells after loading, which increases esterase activity while minimizing compartmentation due to rapid entry of extracellular dye. If the problem is really severe, the dye may have to be injected or loaded in another "disruptive" manner (e.g., hypotonic or electric shock or low external pH).

Compartmentation of Dyes

Compartmentation of dye can often be determined visually as the presence of distinct, punctate, bright fluorescent spots. The uniformity of dye loading should be checked by looking at the cells under a fluorescence microscope. Although none of the dyes has been found to be trapped in the intracellular vesicles of parietal cells or the zymogen granules of chief cells, it is difficult to tell unambiguously that intracellular compartments are completely dye free. The dyes often accumulate in the extracellular space of the mucus-secreting surface cells, presumably trapped by some mucus. Care must be taken to prevent such extracellular trapping, for this drastically affects results.

A quantitative estimate of the amount of signal due to compartmentalized probe can be made by selective permeabilization of different intracellular compartments. For example, digitonin at concentrations up to $20 \ \mu M$ should release only cytoplasmic contents.[18,20] We show below (Fig. 7B) that this treatment released about 85% of Fura-2 from parietal cells, so compartmentation was small.

More rigorous determination of compartmentation can be achieved by monitoring the release of acridine orange (from acidic compartments) and rhodamine 123 (from mitochondria) with higher doses of digitonin and correlating the fluorescence decrease at appropriate wavelengths with release of the dyes as was done by Kawanishi *et al.*[18] in their study on Fura-2-loaded hepatocytes. Although gastric cells and hepatocytes do not demonstrate significant compartmentation of Fura-2, colocalization of rhodamine 123 and Fura-2 fluorescence has been observed in endothelial

[18] T. Kawanishi, L. M. Blank, A. T. Harootunian, M. T. Smith, and R. Y. Tsien, *J. Biol. Chem.* **264**, 12859 (1989).
[19] M. Scanlon, D. A. Williams, and F. S. Fay, *J. Biol. Chem.* **262**, 6308 (1987).
[20] T. J. Rink, C. Montecucco, T. R. Hesketh, and R. Y. Tsien, *Biochim. Biophys. Acta* **595**, 15 (1980).

cells.[21] This pattern was specific to Fura-2; BCECF and indo-1 showed diffuse cytoplasmic staining. Thus, it may be possible to avoid the problem by screening among the various probes now available. In those cases where significant compartmentation is unavoidable, several approaches have been suggested to minimize its severity, including mixing of the dye with dispersing agents, loading at low temperature, and treatment with a variety of intracellular transport inhibitors.[17,22,23,24]

Variability of Commercially Available Dyes

Because the synthesis of many probes is complex, most investigators purchase dyes from commercial vendors. It has been our experience that the performance among different lots or batches of dyes can vary, both in loading efficacy and sensitivity. This was particularly true of Fura-2 and SBFI. As a result, one should keep track of when different lots are being used and new calibrations should be performed to test new lots.

Measuring pH$_i$ with BCECF

BCECF and Other pH-Sensitive Dyes

A variety of probes are available for measuring pH$_i$ (write Molecular Probs, P.O. Box 22010, Eugene, OR 97402-0414 for catalog), but most recent work has been performed using the fluorscein derivative BCECF {2′,7′-bis[carboxyethyl-5 (and 6)-carboxy] fluorescein} and this is the dye that we have used almost exclusively. The excitation spectrum of BCECF has a pH-sensitive region (490–500 nm) and a pH-insensitive isoexcitation point at 440 nm. BCECF has three main advantages: (1) It has a pK_a of approximately 7.0, which is close to the pH$_i$ normally observed in most cells; (2) it has four carboxyl groups, so it is rather impermeant through cell membranes, and it therefore remains trapped in the cytoplasm; and (3) it is available as an AM ester, so loading the cells with dye is easy.

In addition to BCECF, other fluorescein derivatives can also be used [e.g., carboxyfluorescein diacetate (pK_a 6.4) and dimethylcarboxyfluorescein (pK_a 7.0)], and new pH indicators SNARF and SNAFL (Molecular Probes), which are more sensitive (i.e., they exhibit larger changes in

---

[21] S. F. Steinberg, J. P. Bilezikan, and Q. Al-Awqati, *Am. J. Physiol.* **253**, C744 (1987).

[22] R. Y. Tsien, *in* "Methods in Cell Biology," Vol. 30. p. 127. Academic Press, New York, 1989.

[23] R. Y. Tsien, *Annu. Rev. Neurosci.* **12**, 227 (1989).

[24] F. Di Virgilio, T. H. Steinberg, J. A. Swanson, and S. C. Silverstein, *J. Immunol.* **140**, 915 (1988).

fluorescence for a given change of pH_i) and have a higher pK_a (ranging from 7.65 to 7.90), and now available.

Loading BCECF/AM and Calibrating pH_i

BCECF/AM is nonfluorescent, and penetration into the cells and cleavage by intracellular estrases into the fluorescent, impermeant form has been shown to be very rapid (Fig. 2).[25] Most workers load with $2-5$ μM BCECF/AM for 30–45 min, followed by a 90-min wash.[9]

The best method for calibrating the dye is to incubate cells with a high $[K^+]$ Ringer's solution plus 10–50 μM nigericin, the artificial K^+/H^+ exchange ionophore at the end of an experimental run. This treatment allows H^+ to equilibrate freely across the plasma membrane. By using K^+–Ringer's with several different pH values, it is possible to construct a titration curve of fluorescence intensity ratio vs pH. The K^+–Ringer's should ideally approximate the cellular ionic contents. Usually these concentrations are not known precisely, but a good approximation is to use 120 mM K^+, 10 mM Na^+, 1 mM Mg^{2+}, 30 mM Cl^-, 100 nM Ca^{2+} (or Ca^{2+} free), 100 mM gluconate (or cyclamate or some other impermeant anion), 10 mM glucose, and 10 mM HEPES. This solution is then titrated to several (at least three) different pH values.

The K^+ concentration is the most crucial for pH_i calibration. In principle, if $[K^+]_o$ is too high (i.e., $[K^+]_o > [K^+]_i$) during the calibration, pH_i will be underestimated, while if $[K^+]_o$ is too low ($[K^+]_o < [K^+]_i$), pH_i will be overestimated. The magnitude of this effect depends on $[K^+]_o$, $[K^+]_i$, pH_o, and pH_i. As an example, if $[K^+]_o = 150$ mM while cellular $[K^+] = 75$ mM, then it is possible that calibrations will be underestimated by 0.3 pH units. In most cells there are multiple pathways besides the added nigericin for shuttling K^+ and H^+ across the membranes, and different $[K^+]_o$ have little effect on the calibration curve of fluorescence intensity ratio vs pH_o.[26] However, if different $[K^+]_o$ affects calibration curves markedly, it is necessary to determine $[K^+]_i$ for accurate dye calibration. Also, it should be noted that it occasionally takes up to 30 min to achieve equilibrium of pH_i with pH_o. By performing careful calibrations at the end of each experimental run, it is possible to use BCECF to obtain measurements of pH_i from about pH_i 6.2–7.8.

BCECF calibration curves performed *in situ* have shown that BCECF retains a pK_a of approximately 7 inside many cells,[27,28] and the 490/440

[25] J. D. Kaunitz, *Am. J. Physiol.* **254**, G502 (1988).

[26] J. R. Chaillet, and W. F. Boron, *J. Gen. Physiol.* **86**, 765 (1985).

[27] G. Boyarsky, M. B. Ganz, B. R. Sterzel, and W. F. Boron, *Am. J. Physiol.* **255**, C844 (1988).

[28] J. A. Bonanno, and T. E. Machen, *Exp. Eye Res.* **49**, 129 (1989).

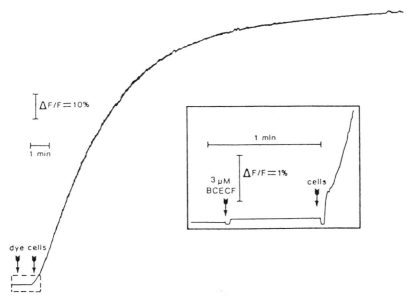

Fig. 2. Loading of the AM ester of BCECF into colonic glands occurs rapidly. At the first arrow, 3 μM BCECF was added to the cuvette. Note that the fluorescence is low at this time because BCECF is relatively nonfluorescent. At the second arrow colonic glands (3 million cells in 2 ml) were added to the cuvette, and fluorescence increased rapidly due to the rapid entry and cleavage of the ester to the fluorescent free acid by cellular esterases. *Inset:* Fluorescence increases very quickly by 4000-fold by the cleavage of the esterified dye. (Taken from Ref. 20 with permission of the American Physiological Society.)

ratio is a linear function of pH over a fairly wide range (Fig. 3C) although the fluorescence response to the dye is slightly reduced *in situ* (see Fig. 3C and Table 1). In parietal cells, the pK_a of the dye was calculated to be 7.2 compared with 7.0 *in vitro* (Table 1). This can lead to underestimation of pH_i of up to 0.2 pH units if calibration is based on the *in vitro* calibration curve and this effect is most marked in the physiological range.

The same protocol is also useful for cuvette experiments. It is, though, occasionally necessary to titrate the extracellular solution with concentrated acid or base (to attain a new pH) rather than removing the entire bathing solution and replacing it with a fresh one. This procedure is sometimes required because the cells become fragile in the presence of the ionophore. Another alternative is to treat cells with digitonin, liberating the dye into the extracellular space, and then titrate this dye to different pH values. This technique requires correction for the above-mentioned red shift that the dye experiences when it is trapped in the cytoplasm of most cells.[1,15,16]

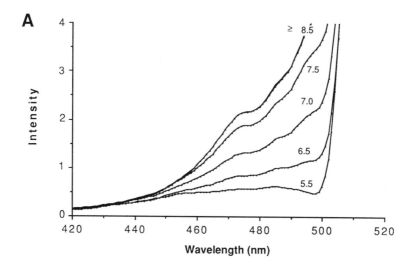

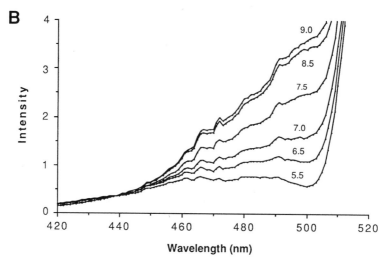

FIG. 3. Excitation spectra and pH sensitivity of BCECF *in situ* and *in vitro*. (A) *In vitro* excitation spectra of BCECF-free acid on the stage of the microscope. Increased emission above 500 nm is due to excitation light passing through the 520-nm long-pass emission filter. (B) *In situ* excitation spectra of BCECF in a single parietal cell on the microscope. Cells were treated with 20 μM nigericin in a high K^+ calibration buffer (see Table I legend), and pH_o was titrated in steps to different values. (C) Comparison of 490/440 nm excitation ratio *in vitro* and *in situ*. Note linearity of response between pH 6.5 and 7.5.

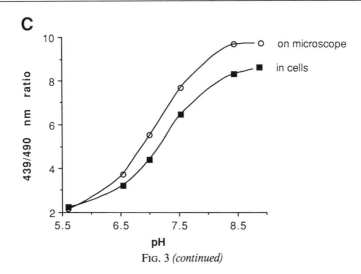

FIG. 3 *(continued)*

Dye Loss Caused by Bleaching and Leakage

During experiments in a perfusion chamber, it is common for the signals at both 490 and 440 nm to decrease.[13] A typical rate of decrease is 30–50%/hr. This loss of signal is due both to leakage of dye from the cells as well as to bleaching by the xenon lamp. By setting the slit widths at different levels or by using different neural density filters, different ratios will be attained, but in each case this ratio remains constant during control perfusion.

Reasons to Measure pH_i

Most enzymes and membrane transport mechanisms are affected quite dramatically by $[H^+]$. Since nearly all cells have transmembrane voltages in the range of -60 to -90 mV (inside negative), there is a constant tendency for the cells to accumulate H^+ to a level 10 to 30 times higher than the extracellular pH. Thus, in the steady state, most cells must use membrane mechanisms for ridding themselves of this constant acid load.

The cells of the gastric gland have particular problems in terms of regulating their pH_i. Parietal cells secrete H^+ by the action of the H^+, K^+-ATPase across the apical membrane, and the cells must get rid of this base load. As we will discuss below, this is likely accomplished by the activity of the Cl^-/HCO_3^- exchanger that resides in the basolateral membranes. The anion exchanger thereby helps to maintain pH_i constant and at the same time provide Cl^- to accompany the H^+ that is secreted across the apical membrane. Thus, the parietal cells use the same mechanisms for regulating pH_i and for generating transcellular fluxes of H^+ and Cl^-. In contrast, the chief cells and mucous neck cells that lie adjacent to the parietal cells must be able to withstand a pH of 1 in the gland lumen, so it might be expected that these cells would have well-developed mechanisms for preventing cellular acidification.

Identification of Base-Loading pH_i Regulatory Mechanisms

A variety of different transporters have been identified in the parietal cells and chief cells of gastric glands. The standard way to test for base-loading mechanisms is to acid load cells using the NH_4^+ prepulse technique[29] and then to monitor the recovery in different solutions and in the presence of different, specific inhibitors. We routinely incubate the cells with a Ringer's solution containing $20-50$ mM NH_4^+ (replacing Na^+) for $2-10$ min. When the NH_4^+ solution is changed back to the original Ringer's, the cells acidify to somewhere between pH_i 6.2 and 6.8, and recovery is then monitored.

It should be noted that although the NH_4^+ prepulse is a convenient method for acid loading cells because pH_o remains constant, the method is somewhat inconsistent because the amount of acid loading cannot be predicted before the procedure has been performed; it seems to depend on the "health" of the cells and on other factors as well. It is nearly impossible to acidify parietal cells below about pH_i 7.0 in Cl^--free solutions using the NH_4^- treatment, even with high concentrations (up to 55 mM NH_4^+) for 15 min.[30] In acid-loaded parietal cells the recovery of pH_i to control level is absolutely dependent on the presence of Na^+ and, in HEPES-buffered (HCO_3^-/CO_2-free) solutions, is nearly completely blocked by $0.5-1.0$ mM amiloride,[1,3,15,30,31] an effect that has come to be the signature of the Na^+/H^+ exchanger.[32] The rates of recovery are also unaffected by the presence or absence of Cl^-.[33] Taken together, these data indicate that there is an Na^+/H^+ exchanger in these cells. Work on isolated membranes has shown that this exchanger is on the basolateral membrane.[15] At the basal Na_i^+ of about 10 mM,[34] the Na^+/H^+ exchanger operates as a base loader.

In solutions that contain HCO_3^-/CO_2, an additional mechanism seems

[29] A. Roos, and W. F. Boron, *Physiol. Rev.* **61**, 296 (1981).
[30] M. C. Townsley, and T. E. Machen, *Am. J. Physiol.* **257**, G350 (1989).
[31] S. Muallem, D. Blissard, E. E. Cragoe, and G. Sachs, *J. Biol. Chem.* **263**, 14703 (1988).
[32] D. J. Benos, *Am. J. Physiol.* **242**, C131, (1982).
[33] M. C. Townsley, and T. E. Machen, in preparation
[34] P. A. Negulescu, A. T. Harootunian, R. Y. Tsien, and T. E. Machen, *Cell. Regul.* **1**, 259 (1990).

to be operating because recovery from an acid load is not blocked by amiloride, but it is independent on the presence of Na^+ in the solutions.[30] Such Na^+- and HCO_3^--dependent recovery that is not blocked by amiloride could, according to the known pH_i regulatory mechanisms, be due to the presence of Na^+/HCO_3^- cotransport or to Na^+-dependent anion exchange ($Na^+/H^+/Cl^-/HCO_3^-$). It is not easy to eliminate the presence of the $Na^+/H^+/Cl^-/HCO_3^-$ system, but it is relatively easy to test for the Na^+/HCO_3^- cotransporter by simply looking for recovery in HCO_3^--containing, Cl^--free solutions that is not blocked by amiloride but is blocked by the disulfonic stilbene H_2DIDS.[30] These types of experiments have led to the conclusion that the parietal cells have an Na^+/HCO_3^- cotransporter. This transporter helps the cell to recover from acidic loads, and it has been argued that it normally operates as a base loader,[30] but there is no unequivocal information on this point.

One curious aspect of all the experiments to date has been that the activity of the H^+, K^+-ATPase has never been detected. For example, in acidified parietal cells, recovery is Na^+ dependent, both in resting and stimulated cells, while the H^+/K^+ pump should be able to operate without Na^+.

Identification of Acid-Loading Mechanisms

The method of testing for mechanisms that would help prevent base loads is nearly identical to that used for looking for those that prevent acidification. Cells are alkalinized and recovery back to base line is monitored. One method for alkalinizing the cells is to treat them with a Ringer's solution containing a weak acid (e.g., acetic acid) for a few minutes, similar to the method for acid loading the cells with NH_4. When the weak acid is removed, the cell will have a pH_i that is above the control, and if there are acid-loading mechanisms, pH_i will recover back to base line. This method has been applied to the parietal cell (A. M. Paradiso, 1987, unpublished results), but the method has not gained popularity because it is easier just to alkalinize the cells by treating them with a Cl^--free solution. This treatment causes the cells to alkalinize from about pH_i 7.1 to 7.4–7.6. Recovery from this alkaline load is dependent on the presence of Cl^-, and it is blocked by 200 μM DIDS or H_2DIDS.[13,15] These changes also occur in Na^+-free solutions, so it has been concluded that the parietal cells have a Cl^-/CHO_3^- exchanger.[13,15] The anion exchanger has always been assumed to work as an acid loader, but this will, of course, depend on $[HCO_3^-]_i$, $[HCO_3^-]_o$, $[Cl^-]_i$, and $[Cl^-]_o$. With $pH_o = 7.4$, $pH_i = 7.1$, and $[Cl^-]_0 = 120$ mM, the anion exchanger will work as an acid loader as long as $[Cl^-]_i < 60$ mM.

Quantitating Rates of H^+ Flux through the pH_i Regulatory Mechanisms

All the experimental protocols mentioned to this point have yielded only qualitative information about the presence or absence of specific transporters because only rates of change of pH_i were measured. To obtain quantitative information about the rates of operation of the mechanisms, it is necessary to multiply rates of change of pH_i by the buffer capacity of the cells, i.e.,

$$J_{H+} = (\Delta pH_i / \Delta t) \beta_t \tag{1}$$

where J_{H+} is the H^+ flux (in mM/min), $\Delta pH_i / \Delta t$ is the rate of change of pH_i (in pH units/min), and β_t is the buffer capacity (in mM/pH).

An accurate determination of β is therefore crucial for an unambiguous interpretation of a variety of experiments. For example, when measuring the activity of both the Cl^-/HCO_3^- and Na^+/H^+ exchangers as a function of different pH_i, the approach has been to alter intracellular pH_i and then to measure rates of Cl^--dependent or Na^+-dependent recovery of pH_i back toward baseline.[33,35] Initial rates of change of pH_i are most informative because the gradients of Cl^- or Na^+ and pH should be relatively constant during this time. Rates of H^+ flux through the transporters at the different pH_i values can be calculated only if β_t is known at each pH_i. The cells' buffer capacity is, in turn, defined as the sum of the intrinsic (non-HCO_3^-) buffers of the cells (β_i) plus that contributed from the presence of HCO_3^-/ CO_2, $\beta_{HCO_3^-}$, which is equal to 2.3 times the cellular $[HCO_3^-]$,[29,36] i.e.,

$$\beta_t = \beta_i + \beta_{HCO_3^-} = \beta_i + 2.3[HCO_3^-] \tag{2}$$

In HCO_3^-/CO_2-containing solutions the intracellular $[HCO_3^-]$ is calculated from the Henderson–Hasselbalch equation, the known CO_2 concentration of the gassing mixture, and pH_i (measured with BCECF).

Recent experiments have shown that β_i of most cells,[27] including the parietal cell, is strongly dependent on pH_i. The general method for determining β_i at different pH_i is to incubate the cells in HCO_3^-/CO_2-free solutions, block the known pH_i regulatory mechanisms with amiloride and H_2DIDS, and then treat the cells with different $[NH_4^+]$. Determination of β_i requires the measurements of small pH_i steps induced by a defined change in $[H^+]_i$. To establish small changes of $[H^+]_i$, stepwise reductions of $[NH_4^+]_o$ are made in HCO_3^-/CO_2-free solutions.[35] It is assumed that $[NH3]_o = [NH3]_i$ (because of the rapid transmembrane permeation of NH_3), there is low NH_4^+ conductance, or low entry of NH_4^+, and pK_a 9.25. $[NH_4^+]_i$ is calculated from the Henderson–Hasselbalch equation and

[35] E. Wenzl, and T. E. Machen, *Am. J. Physiol.* **257**, G741 (1989).
[36] W. F. Boron, *Annu. Rev. Physiol.* **48**, 377 (1986).

measured pH_i. Lowering $[NH_4^+]_o$ elicits a decrease in both $[NH3]_i$ and $[NH_4^+]_i$ and the intrinsic buffer capacity (for the pH_i range over which the measurements were recorded) is calculated as

$$\beta_i = \Delta[H^+]_i / \Delta pH \qquad (3)$$

A typical experiment to determine β_i as a function of pH_i in a single parietal cell (PC) is shown in Fig. 4. Before the beginning of this trace, the parietal cell was treated for 10 min with 1 mM amiloride + 200 μM H_2DIDS. (It is difficult to solubilize amiloride and H_2DIDS in the same solution so we prepare this solution by mixing equal quantities of a solution containing 2 mM amiloride with one containing 400 μM H_2DIDS). Then, NH_4^+ was added to the inhibitor-containing solutions in the sequence 30, 20, 20, 5, and 0 mM. As shown, the NH_4^+/NH_3 caused pH_i first to increase (due to the entry of the NH_3) and then to decrease in steps as $[NH_4^+]_o$ was decreased; pH_i was stable at each concentration. When the inhibitors were washed out of the solution, pH_i recovered to above the control value (not shown). β_i was calculated for each of the NH_3/NH_4^+ and pH_i steps. A summary is shown in Fig. 5, where it can be seen that β_i was strongly dependent on pH_i. These experiments were performed at room temperature, but similar experiments at 37° gave nearly identical results.[35]

Another point regarding the activities of the anion and cation exchangers is that the pH_i measured at any point in time is the sum of the combined activities of the alkalinizing and acidifying processes. In the case where the cell has been artificially acidified using an NH_4^+ prepulse, recovery of pH_i to the control level can occur only if the combined rates of the alkalinizing mechanisms (Na^+/H^+ exchange, Na^+/HCO_3^- cotransport, and H^+, K^+-ATPase) are larger than those of the acidifying mechanisms (Cl^-/HCO_3^- exchange, metabolism, and any H^+ conductance), and it is possible that the rates of transport of the different mechanisms will change at different pH_i. Therefore, if one is attempting to compare the activities of a specific transporter under two different conditions or of two different transporters under the same conditions, it is important to measure rates of H^+ flux at the same pH_i so that the pH_i dependencies of the transporters will not become a factor.

A related issue is that a change from one steady state pH_i to another does not in itself yield unambiguous information. For example, assume that the resting parietal cell has an Na^+/H^+ exchanger and a Cl^-/HCO_3^- exchanger, and they both turn over five times/sec in a resting cell. If no other pH_i-modifying processes are occurring in this resting cell, then pH_i will remain constant (i.e., the cell is in a steady state). When the parietal cell gets stimulated by histamine, the H^+, K^+-ATPase (another base loader) becomes activated. If its turnover rate is 10 times/sec in the stimulated PC,

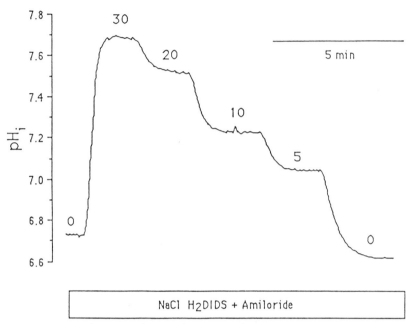

Fig. 4. An experiment showing how β_i was determined in a single parietal cell (PC) of an intact gastric gland. Amiloride (1 mM) and 200 μM H$_2$DIDS were added to the solution to block the known pH$_i$ regulatory mechanisms, and the trace begins after 10 min of this treatment. NH$_4$$^+$ was added to the solution as shown. NH$_4$$^+$ (30 μM) caused pH$_i$ to increase, and pH$_i$ decreased in steps as [NH$_4$$^+$]$_o$ was stepped down from 30, to 20, to 10, to 5, and finally to 0 mM. β_i was calculated for each of the step decreases of [NH$_4$$^+$]$_o$ as described in the text. (Taken from Ref. 27 with permission of the American Physiological Society.)

and the anion exchanger also gets activated (turnover increasing from 5 to 10 times/sec) while the Na$^+$/H$^+$ exchanger decreases its activity from 5 to 2, then pH$_i$ will still increase, and this increase will also be blocked by amiloride. Thus, an amiloride-blockade increase of pH$_i$ alone does not necessarily indicate that the Na$^+$/H$^+$ exchanger has been activated.[37] It should also be noted that many experiments involve treating cells with unphysiological solutions, e.g., the cells are acidified or alkalinized in Na$^+$-free or Cl$^-$-free solutions. These different treatments are unavoidable for obtaining information about the presence or absence of the transporters. However, these treatments can alter cellular function. As an example, Na$^+$-free solutions, which are often used to investigate Na$^+$-dependent pH$_i$ regulatory mechanisms, cause Ca$_i^{2+}$ to increase, similar to the

[37] T. E. Machen, M. C. Tonsley, E. Wenzl, P. A. Negulescu, and A. M. Paradiso, *Ann. N.Y. Acad. Sci.* **574**, 447 1989.

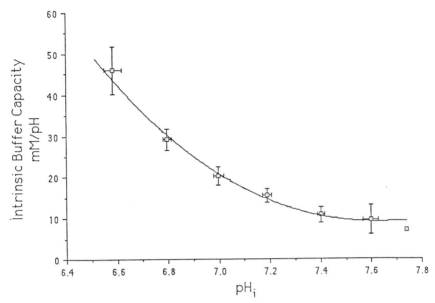

FIG. 5. Effects of pH_i on β_i in PC. Data from experiments such as those in Fig. 4 were summarized by averaging β_i values from 0.1-pH_i unit groups. Therefore, there are SEM bars on both the β_i and the pH_i measurements. (Taken from Ref. 27 with permission of the American Physiological Society.)

effect of cholinergic stimulation.[38] Further, Na^+-free and other treatments do not show how a particular transporter is operating in the normal parietal cell.

Our approach to this problem has been to treat resting and stimulated parietal cells with specific inhibitors and look for changes in pH_i immediately following their addition. Since Na^+/H^+ exchange tends to alkalinize the cell, inhibition of this transporter by amiloride should acidify the cell, and the rate of acidification reflects the activity of the exchanger before the inhibitor was added. When this rate of acidification is multiplied by the β_t at the resting $pH_i = 7.1$, we find that the Na^+/H^+ exchanger is contributing a unidirectional Na^+ influx of 2–3 mM/min in the resting parietal cell,[39] and this decreases to about 1 mM/min in cells that have been treated with maximally stimulating doses of histamine +3-isobutyl-1-methylxanthine (IBMX). Similar experiments have also been performed using H_2DIDS to inhibit the anion exchanger, which appears to increase its activity by

[38] Negulescu P. A., and T. E. Machen, *J. Membr. Biol.* in press.
[39] P. A. Negulescu, and T. E. Machen, in preparation.

fivefold with these stimulants.[40] Experiments by Muallem, Sachs, and their colleagues[31] have shown that histamine alone causes the Na^+/H^+ and Cl^-/HCO_3^- exchangers both to get activated, and we have found that carbachol also activates the Na^+/H^+ exchanger.[39] It therefore seems that there is no common path to stimulation of the parietal cell, for the stimulants do not all have the same effects on the different pH_i regulatory mechanisms. However, even these experiments are incomplete because none of them has been performed in solutions containing the physiological buffer HCO_3^-/CO_2.

Measuring Ca_i^{2+} with Fura-2

Microfluorimetry of Fura-2

The most popular Ca^{2+} indicator at present is the tetracarboxylate Fura-2. The chemical basis of this class of Ca^{2+} indicators and the general advantages of Fura-2 over its predecessor, quin2, have been described in detail by Tsien and co-workers.[2,22,33,41] Briefly, Fura-2 is distinguished by brightness, relative resistance to photodamage and bleaching, appropriate K_d for Ca^{2+} binding in the cytoplasm, and large spectral shifts upon Ca^{2+} binding. These factors contribute to a dye which can be loaded into the cytoplasm at much lower concentrations than quin2, and hence contributes less to Ca^{2+} buffering. Its availability as a membrane-permeant AM ester derivative allows it to be easily loaded into most cells.

Although Fura-2 has many advantages, there are two factors which must be considered when measuring Ca_i^{2+} in single cells with this dye. First, the optical components of the microscope will attenuate the excitation beam, even after optimization of UV transmission using special nosepieces and objectives designed for short-wavelength transmission. Thus, spectra obtained from free dye in the cuvette of a fluorimeter may be different from those obtained through the microscope. The second factor is that Fura-2 *in situ* (i.e., in the intact cell) can exhibit significantly different spectral characteristics from those observed *in vitro* (i.e., as free dye). Both these factors contribute to significant differences between calibrations based on dye in the cuvette and in cells under the microscope.

Figure 6A and B show the differences in the excitation spectra obtained from Fura-2 in a quartz cuvette and from the stage of a Zeiss IM35 inverted microscope. The same SPEX (Edison, New Jersey) fluorimeter,

---

[40] A. M. Paradiso, M. C. Townsley, E. Wenzl, and T. E. Machen, *Am. J. Physiol.* **257,** C554 (1989).

[41] R. Y. Tsien, Rink, T. J., and M. Poenie, *Cell Calcium* **6,** 145 (1985).

dye samples, and photomultiplier were used in both cases, so the differences in the spectra are due entirely to differences introduced by passing the light to, and collecting from, the microscope. The cuvette measurements (Fig. 6A) were made in a quartz cuvette with emission collected through a monochrometer set at 510 nm. For spectra obtained on the stage of the microscope, the excitation light was first passed from the fluorimeter

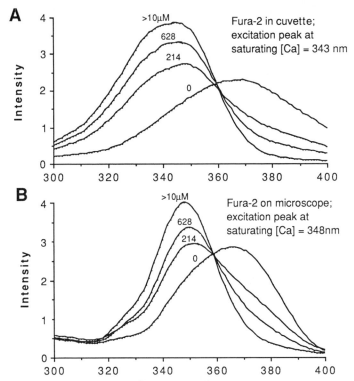

FIG. 6. Excitation spectra and Ca²⁺ sensitivity of Fura-2 *in vitro* and *in situ*. (A) *In vitro* Ca²⁺ titration of 5 μM Fura-2 free acid in a cuvette using a fluorimeter. Emission collected by a monochrometer at 510 nm. (B) Similar Ca²⁺ titration on the stage of a Zeiss IM35 fluorescence microscope. Emission light collected through a 450-nm long-pass filter. Note the attenuation at short wavelengths. See text for more details. (C) *In situ* Ca²⁺ titration of Fura-2-loaded cells under conditions identical to those in (B). Note the increased fluorescence at longer wavelengths and the decreased ratio change for a given Ca²⁺ change. All solutions contained 115 mM KCl, 15 mM NaCl, 10 mM HEPES. Various Ca²⁺ concentrations were achieved by mixing solutions containing 10 mM K₂H₂EGTA with solutions containing 10 mM K₂ CaEGTA, similar to the method used in Ref. 2. All experiments were performed at 24°. (D) Comparison of Fura-2 fluorescence responses to Ca²⁺ *in vitro* and *in situ*. Note the suppression of the fluorescence response *in vitro* on the microscope and *in situ* compared to the response in the cuvette.

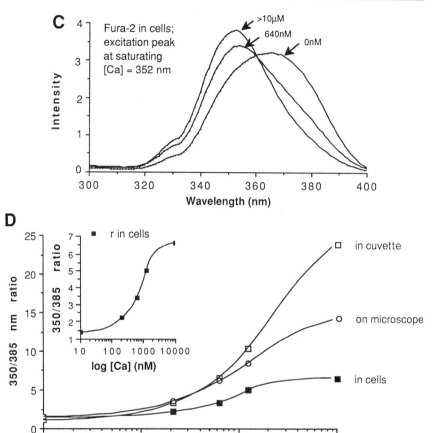

FIG. 6 *(continued)*

to the microscope through a quartz focusing lens provided by SPEX. The light then passed through a UG5 UV filter (Zeiss), was reflected via a 395 dichroic through a quartz Zeiss nosepiece to a Nikon ×40 (NA 1.3) oil immersion fluorite objective. The dye sample was placed on a glass cover-slip (type 1) identical to those on which loaded glands are placed. Emission light was filtered through a 450 long-pass filter before passing to the photomultiplier. The most obvious difference is that, despite efforts to maximize UV transmission with quartz optics, there is attenuation of short-wavelength light which shifts the excitation peak, at saturating $[Ca^{2+}]$, to longer wavelengths—an apparent red shift (Fig. 6B). The net effect of this is to reduce the detected ratio change between Ca^{2+}-free and saturating $[Ca^{2+}]$ by 30% (see Fig. 6D).

Loading of Glands with Fura-2

When unloaded glands were exposed to 5 μM Fura-2/AM dye on the microscope, an adequate signal is obtained within 10 min at room temperature. Glands were typically incubated with 5 μM Fura-1/AM for 20–30 min at room temperature to ensure adequate, uniform loading. The average intracellular concentration of Fura-2 in glands has been determined to be ~75 μM based on the uptake of dye from the loading medium.[9] Since parietal cells accumulate somewhat more Fura-2 than chief cells, this value is probably higher than 75 μM in parietal cells and lower in chief cells.

Spectral Properties of Fura-2 in Situ

Figure 6C shows a series of excitation spectra from a single Fura-2-loaded gland mounted in a perfusion chamber and placed on the stage of an inverted microscope as described above. When compared with spectra from calibration buffers (Fig. 6A and B), Fura-2 shows a red shift in gastric cells, similar to that noted by other investigators.[17,40] This shift means that increased emission is noted at longer wavelengths, which decreases the 350/385 nm ratio at any given Ca_i^{2+} and results in an underestimation of Ca_i^{2+} if calibration is based on Fura-2 behavior in Ca^{2+} buffers. This red shift can be simulated by changes in viscosity and may reduce the ratio at any Ca_i^{2+} by 15% when compared with aqueous calibration buffers.[42] The spectra from Fura-2-loaded cells also demonstrate the suppression of the total fluorescent response of Fura-2 in the cells. This can be demonstrated by dividing the ratio obtained at saturating $[Ca^{2+}]$ by that obtained at 0 $[Ca^{2+}]$. The resulting quotient is a normalized value for the total fluorescence response (see Table I). In the cuvette the Fura-2 R_{max}/R_{min} ratio is approximately 30 when saturated with Ca^{2+}. By comparison, R_{max}/R_{min} for Fura-2 in the cells was about 5 under similar conditions.

There appear to be a number of possible effects of the cytoplasm on Fura-2 fluorescence. For example, the affinity of the dye for Ca^{2+} in the cytoplasm is decreased due to increases in viscosity[42] or ionic strength,[2] and intracellular heavy metals such as manganese, zinc, iron, and copper can quench Fura-2 fluorescence.[2,19,24] In addition, Fura-2 often exhibits a smaller spectral shift upon Ca^{2+} binding. In some cases this may be due to significant autofluorescence, a problem which becomes more severe when loading is poor. In other cases, smaller detected shifts may arise from incomplete hydrolysis of the five AM esters, which yields relatively Ca^{2+}-insensitive dye.[19] The amount of Ca^{2+}-sensitive dye relative to background can be quickly determined by treating cells with ionomycin and exposing

[42] M. Poenie, J. Alderton, R. Steinhardt, and R. Y. Tsein, *Science* **233**, 886 (1986).

them to Mn^{2+}. The ionophore will allow Mn^{2+} to enter rapidly and quench the cleaved Fura-2. As shown in Fig. 7A, this technique reduced the fluorescent signal from a single parietal cell to nearly that of the unloaded cell, indicating that >95% of the intracellular dye was the free Fura-2 acid. As mentioned above, a more quantitative assessment of the quality of the loaded dye can be obtained by releasing the dye from the cell and titrating it with Ca^{2+}.[18,19] Figure 7B shows that compartmentalized dye represents a small fraction of the total intracellular Fura-2 since 10 μM digitonin releases 85% of the dye.

The effects of fluorescence measurements through the microscope and the behavior of Fura-2 in cells can be corrected by calibrating the fluorescent signal from the cells.

Calibration of Fura-2 in Situ

Calibration of Fura-2 could, in principle, be performed by equilibrating Ca_i^{2+} and Ca_o^{2+} in the presence of ionophores, similar to the technique used for BCECF and SBFI. In practice, however, it is inconvenient to routinely determine several Ca_i^{2+} levels because it is difficult to equilibrate Ca_i^{2+} and Ca_o^{2+}. Several factors make such calibration difficult, including small ion fluxes, Ca_i^{2+} regulatory mechanisms, and the fact that ionophores behave somewhat unpredictably. A simpler way to determine Ca_i^{2+} is to calculate Ca_i^{2+} according to the equation derived by Grynkiewicz et al.,[2]

$$Ca_i^{2+} = K(R - R_{min})/(R_{max} - R) \tag{4}$$

where R_{min} is the ratio of fluorescence intensities at 340 and 385 nm obtained at 0 $[Ca^{2+}]$, R_{max} is the ratio at saturating $[Ca^{2+}]$, and R is the ratio at some intermediate $[Ca^{2+}]$ level. Thus, all that is required for accurate calibration is the fluorescence intensity ratio at 0 $[Ca^{2+}]$, saturating $[Ca^{2+}]$, and the K_d. In Eq. (4) K represents $K_d(F_{min}/F_{max})$, where K_d is the dissociation constant for Fura-2 and F_{min} and F_{max} are the fluorescence intensities at 385 nm minus and plus Ca^{2+}, respectively.

These values have been determined empirically in gastric cells as follows. R_{min} was obtained by exposing the glands to a Ca^{2+}-free solution containing 10 mM EGTA at pH 8.0. The elevated pH increases the affinity of EGTA for Ca^{2+} but has negligible effects on the dye. Under these conditions Ca^{2+} leaves the cell. Nonfluorescent Ca^{2+} ionophores such as ionomycin or 4-bromo-A23187 (each at 10 μM) can be added to facilitate the process. R_{max} is obtained by exposing cells to excess $[Ca^{2+}]$ (1 mM). As mentioned above, Ca^{2+} ionophores did not reliably equilibrate Ca_i^{2+} and Ca_o^{2+} so low doses of digotonin (between 1 and 5 μM) were used. The trick is to avoid adding too much digitonin, which results in rapid release of dye

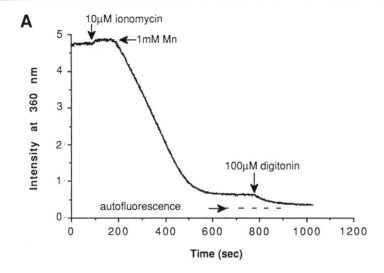

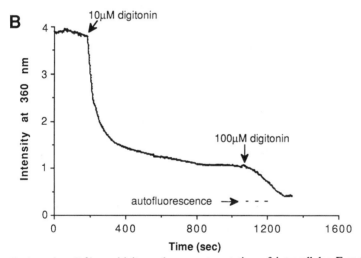

FIG. 7. Assessing Ca^{2+} sensitivity and compartmentation of intracellular Fura-2. (A) Quenching of the intracellular dye with 1 mM extracellular Mn^{2+} in the presence of 5 μM ionomycin. (B) Release of cytoplasmic trapped dye by treatment with 10 μM digitonin. These experiments were performed at room temperature.

from the cytoplasm. To determine the intracellular K_d, the intensity ratio at a known intermediate $[Ca^{2+}]$ value is necessary. Once the ratio values at 0, saturating, and an intermediate $[Ca^{2+}]$ are known, calculation of the K_d can be obtained by rearrangement of Eq. (4):

$$K_d = [Ca^{2+}](R - R_{min}/R_{max} - R)(F_{min}/F_{max}) \qquad (5)$$

where $[Ca^{2+}]$ is determined by the dissociation constant of Fura-2 in EGTA-containing solutions. Once the K_d is known, this value (along with values for R_{min} and R_{max}) can be plugged back into Eq. (4) and used to determine $[Ca^{2+}]$ at any ratio value. When this was done in gastric cells, a K_d of 350 nM was determined.[9] Subsequent calibration based on K_d, R_{min}, and R_{max} values $in\ situ$ resulted in higher basal Ca_i^{2+} values and larger changes upon addition of agonists than those calculated from analogous Fura-2 responses to $[Ca^{2+}]$ $in\ vitro$. The K_d was fairly consistent among different preparations so the value of 350 nM was used routinely. R_{min} and R_{max} were more variable and were therefore determined on each experimental day. Although the calibrations shown above were derived from spectra shown in Fig. 6, the measurements could have been made using a rotating filter wheel, since all that is required are intensities at two wavelengths.

Implicit in such a calibration is the fact that the K_d was determined at an intracellular pH of ~7.1. Although Fura-2 has a lower pK_a than the parent EGTA compound, it should be remembered that the dye is still pH sensitive, resulting in an increase in the K_d of Fura-2 as pH drops below 6.7. Therefore in experiments where large cytoplasmic acidifications are likely to occur, a correction may be necessary.[38] As a final comment on calibration, it should be noted that Fura-2 becomes less sensitive to $[Ca^{2+}]$ at levels above $1-2\ \mu M$ so that high $[Ca^{2+}]$ levels may not be accurately reported by the probe. Recently, probes with higher K_d values have been introduced[43] which are more appropriate for measuring high Ca_i^{2+} levels.

With regard to the choice of Fura-2 as a Ca^{2+}-sensitive probe, special applications may be more conveniently addressed using probes with different chromophores or affinities for Ca^{2+}.[43] For example, the short excitation wavelengths required for Fura-2 are incompatible with most confocal microscopes, whose laser emission lines are usually at longer wavelengths. Also, Fura-2 is difficult to use with the new generation of caged compounds which release or take up Ca^{2+} upon illumination with UV light[44-46]

[43] A. Minta, J. P. Y. Koa, and R. Y. Tsien, $J.\ Biol.\ Chem.$ **264,** 8171 (1989).
[44] R. Y. Tsien, and R. S. Zucker, $Biophys.\ J.$ **50,** 843 (1986).
[45] J. P. Y. Kao, A. T. Harootunian, and R. Y. Tsien, $J.\ Biol.\ Chem.$ **264,** 8179 (1989).
[46] S. R. Adams, J. P. Y. Kao, and R. Y. Tsien, $J.\ Am.\ Chem.\ Soc.$ **111,** 7957 (1989).

because fluorescent excitation of the probe would tend to photolyze the buffers. Finally, some cell types may be particularly sensitive to UV excitation. In such special cases the use of long-wavelength Ca^{2+} indicators such as Fluo-3 may obviate these problems, although one loses the advantage of excitation rationing since these probes do not shift their spectra upon binding Ca^{2+}. For the experiments we presently conduct in glands, Fura-2 remains the best choice due to the advantages listed at the beginning of this section.

Study of Ca^{2+} Metabolism in Parietal Cells with Fura-2

The Ca_i^{2+} response to carbachol in parietal cells is typical of that found in a number of epithelial cell types, including parotid, salivary, and pancreatic acinar cells. Carbachol stimulation involves responses by Ca^{2+} permeabilities and pumps in both the plasma membrane and an internal Ca^{2+} store. The physical nature of the internal Ca^{2+} store is unclear, but most evidence suggests that it is part of the endoplasmic reticulum.[47]

The activation of permeabilities in both the plasma membrane and the internal store is shown in Fig. 8. A maximal dose of carbachol (100 μM) causes a biphasic elevation in Ca_i^{2+}, the phases of which can be distinguished by their dependence on extracellular Ca^{2+}. The initial, rapid increase in Ca_i^{2+} is due to release from internal stores since it is unaffected by Ca^{2+}-free solutions. The secondary, sustained elevation of Ca_i^{2+} is due to Ca^{2+} entry across the plasma membrane because it is abolished in the absence of external Ca^{2+}. The two phases can also be distinguished by sensitivity to lanthanum ion. In parietal cells 50 μM La^{3+} had no effect on the initial Ca_i^{2+} peak but prevents the sustained plateau,[48] indicating that release of internal stores and sustained elevation of Ca_i^{2+} occurred through distinct mechanisms. These two phases can be elicited independently, suggesting that an increase in Ca_i^{2+} is not a signal for either release of the internal store or activation of the entry across the plasma membrane in the parietal cell.

The mechanisms by which Ca^{2+} permeabilities are stimulated by agonists have been the subject of intense study. With regard to the intracellular Ca^{2+} store, a variety of agonists, including carbachol, cause release of the internal Ca^{2+} store by stimulating phosphoinositide (Ptdlns) breakdown.[47] A soluble mediator, inositol trisphosphate [Ins(1,4,5)P$_3$], is generated whose receptor is also a Ca^{2+} channel located in the membrane of the internal store.[49] Although Ins(1,4,5)P$_3$-mediated release is a prevalent

[47] M. J. Berridge, and R. F. Irvine, *Nature (London)* **341**, 197 (1989).
[48] P. A. Negulescu, and T. E. Machen, *Am. J. Physiol.* **254**, C130 (1988).
[49] C. D. Ferris, R. L. Huganir, S. Supattone, S. H. Snyder, *Nature (London)* **342**, 87 (1989).

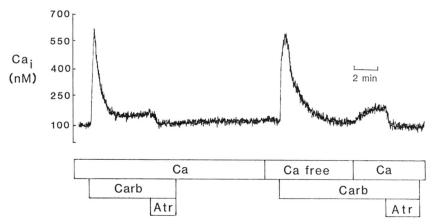

FIG. 8. Relative contributions of intracellular and extracellular Ca^{2+} to the carbachol-induced increase in Ca^{2+}. On stimulation with 100 μM carbachol (Carb), Ca_i^{2+} increased rapidly to a peak and then decreased back to a plateau level. The cholinergic antagonist atropine (Atr, 1 μM) caused Ca_i^{2+} to decrease to control levels. In Ca^{2+}-free (2 mM EGTA) solutions the peak Ca^{2+} was largely unaffected but the plateau was eliminated. (Taken from Ref. 48 with permission.)

mechanism, there are examples of agonists which do not stimulate detectable PtdIns metabolism and yet are capable of releasing the internal store. In the parietal cell, for example, histamine releases Ca^{2+} from a carbachol-sensitive internal store via H_2 receptors in a manner that appears independent of PtdIns metabolism[10] and shows some synergism with cAMP.[48] Other examples include airway epithelium, in which cAMP releases Ca^{2+} stores,[50] and platelets, in which ADP releases internal stores,[51] although it is a poor stimulant of PtdIns metabolism.[52,53]

The nature of and mediator(s) for stimulating Ca^{2+} entry across the plasma membrane in nonexcitable cells are not well established. Part of the difficulty is due to the fact that specific blockers for these channels have not been found. Indeed, the channel itself can be difficult to detect.[54] The entry pathway is distinct from voltage-gated L-type channels found in excitable tissues such as cardiac and nerve cells and hence is not sensitive to dihydropyridine blockers of those channels. Presently, the best blockers appear to be lanthanum (~100 μM) and/or nickel ($\approx$1 mM). These cations also appear reasonably impermeant, allowing one to monitor Ca_i^{2+} without

[50] J. D. McCann, R. C. Bhalla, and M. J. Welsh, *Am. J. Physiol.* **256**, L116 (1989).
[51] S. O. Sage, and T. J. Rink, *J. Biol. Chem.* **262**, 16364 (1987).
[52] J. L. Daniel, C. A. Danglemaier, M. Selak, and J. B. Smith, *FEBS Lett.* **206**, 299 (1986).
[53] J. D. Sweatt, I. A. Blair, E. J. Cragoe, and L. E. Limbird, *J. Biol. Chem.* **261**, 8660 (1986).
[54] R. Penner, G. Mathews, and E. Neher, *Nature (London)* **334**, 499 (1988).

interference from La^{3+} or Ni^{2+} binding to Fura-2. Another issue is that there appear to be a variety of mechanisms of Ca^{2+} entry. Calcium ion entry may be stimulated by a membrane receptor that is close to, or possibly part of, the channel.[55] Others have suggested that products of PI metabolism are responsible for opening these channels.[46,56] Finally, the suggestion has been made that the state of the internal store (empty or full) rather than the presence of agonist[57,58] can determine the permeability of the plasma membrane in endothelial cells.

Calcium Ion Pumping and Internal Store Refilling

The profile of the Ca_i^{2+} signal is determined as much by the activity of Ca^{2+} extruding mechanisms as by the changes in Ca^{2+} permeability. This can be seen in Fig. 8, in which the decrease of Ca_i^{2+} following the peak is due to Ca^{2+} removal from the cytoplasm. This decrease can be attributed to extrusion across the plasma membrane, rather than accumulation into the internal store, because the store remains empty in the presence of 100 μM carbachol.[59]

Calcium ion pumping across the plasma membrane could be due to either a Ca^{2+}-ATPase or an Na^+/Ca^{2+} exchanger or both. One way to determine the presence of an Na^+/Ca^{2+} exchanger would be to monitor Ca_i^{2+} when extracellular Na^+ is removed, similar to the method used to detect Na^+/H^+ exchange. Depending on the cell type, there may be two problems with this approach. First, cholinergic receptors are stimulated by some cationic replacement ions for Na^+ such as choline and tetramethylammonium. Muallem et al.[60] showed that these effects can be avoided by treating cells with the cholinergic antagonist, atropine, or by using a different replacement cation, such as N-methyl-D-glucamine (NMG). Another complication is that lowering Na_o^+ to less than 20 mM (NMG replacement) caused Ca_i^{2+} increases in both Ca^{2+}-containing and Ca^{2+}-free solutions. Thus, lowering Na_o^+ to very low levels caused release of intracellular stores. The nature of this response is unclear but it seems to be due to external, rather than internal, effects of removing external Na^+.[38] The fact that Na_o^+ above 20 mM has no effect on Ca_i^{2+} suggests that Na^+/Ca^{2+} exchange plays

[55] C. O. Benham and R. W. Tsien, *Nature (London)* **328,** 275 (1987)

[56] A. P. Morris, D. V. Gallagher, R. F. Irvine, and O. H. Peterson, *Nature (London),* **328,** 653 (1987).

[57] T. J. Hallam, R. Jacob, and J. E. Merritt, *Biochem. J.* **259,** 125 (1988).

[58] H. Takemura, A. R. Hughes, O. Thastrup, and J. W. Putney, Jr., *J. Biol. Chem.* **264,** 12266 (1989).

[59] P. A. Negulescu, and T. E. Machen, *Am. J. Physiol.* **254,** C498 (1988).

[60] S. Muallem, T. Beeker, and S. J. Pandol, *J. Membr. Biol.* **102,** 155 (1988).

no significant role in Ca_i^{2+} homeostasis and that Ca^{2+} extrusion across the plasma membrane is accomplished by a Ca^{2+}-ATPase. This Ca_i^{2+} response to low Na_o^+ may be a general phenomenon since it was first demonstrated in a variety of cultured cells.[61] Thus, one should perform appropriate controls when investigating Na^+/Ca^{2+} exchange using this approach.

Since the internal Ca^{2+} store is emptied by stimulation with 100 μM carbachol, refilling can be assessed by the magnitude of the Ca_i^{2+} response to a second stimulus. In this way, conditions required for Ca^{2+} uptake into the pool can be studied. For example, if carbachol stimulation is continued until the initial Ca^{2+} increase relaxes to either basal levels in Ca^{2+}-free solution or to plateau levels in the presence of extracellular Ca^{2+} (Ca_o^{2+}) is required for reloading. Reloading from the external space required 2–3 min and no change in Ca_i^{2+} was detected during this time. One possible explanation for this finding is that Ca^{2+} bypasses the cytoplasm on its way to the internal store. Another possibility is that Ca^{2+} uptake from the cytosol by the internal store is vigorous and regulated in such a way as to avoid generation of a Ca_i^{2+} transient during reloading.

Evidence for reloading from the cytosol is shown in Fig. 9. Cells were stimulated repeatedly but in briefly in Ca^{2+}-free solutions. The purpose of brief (10 sec) stimulations was to release the internal store yet terminate the stimulation (with atropine) before the released Ca^{2+} had all been pumped out of the cell across the plasma membrane. Under these conditions, repeated Ca_i^{2+} increases occurred in a Ca^{2+}-free solution, indicating that Ca^{2+} was being recycled from the cytosol back into the internal store. As seen in Fig. 9, following addition of atropine Ca_i^{2+} dropped at a grater rate than in the presence of carbachol. This is consistent with the additional Ca^{2+} sequestering system of the internal store now being effective. This recycling is due to two phenomena. First, the Ca^{2+} permeability of the internal store decreases immediately upon termination of the stimulus. Second, the Ca^{2+} pumping system of the internal store is very vigorous. It has a high Ca^{2+} affinity, as demonstrated by its ability to sequester Ca^{2+} at all levels of Ca_i^{2+}, and a high capacity, as shown by the rapidity of recycling ($<$30 sec under these conditions). These findings suggest that the internal store reloads from the cytosol, although it does not rule out some direct connection between the internal store and the extracellular space.

Calcium Ion Uptake during Reloading

The tight linkage between Ca^{2+} influx across the plasma membrane and uptake by the internal store makes it difficult to study Ca^{2+} entry during

[61] J. B. Smith, S. D. Dwyer, and L. Smith, J. Biol. Chem. **264**, 831 (1989).

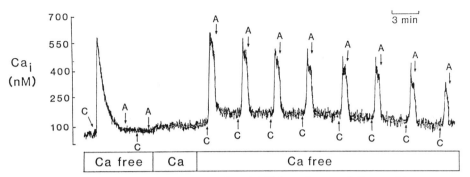

FIG. 9. Evidence for internal store reloading of Ca^{2+} from the cytosol. Repeated increases in Ca_i^{2+} can be elicited in a Ca^{2+}-free bathing solution if carbachol stimulation (C) is terminated with atropine (A) before Ca^{2+} has relaxed to baseline. Because the internal store is completely released by 100 μM carbachol, repeated increases in Ca^{2+} must be due to recycling of the released Ca^{2+}. (Taken from Ref. 59 with permission.)

reloading directly. One way to look at influx in the absence of pumping is to take advantage of the fact that Fura-2 binds all divalent cations. Therefore, experiments can be designed using divalent cations such as Mn^{2+}, Cd^{2+}, and Ba^{2+}, which may permeate Ca channels but are poor substrates for pumps. Two advantages of Mn^{2+} are that it binds Fura-2 with ~40X higher affinity than Ca^{2+},[2] increasing the probe's sensitivity to small fluxes, and that it quenches the dye, so it cannot be confused with an increase in Ca_i^{2+}. Using Mn^{2+} as a tracer for Ca^{2+} movement across the plasma membrane it has been shown that emptying the internal store is sufficient to stimulate Mn^{2+} entry (determined by the rate of Fura-2 quench) in endothelial cells and platelets.[57,62] Interestingly, emptying the internal store did not stimulate increased Mn^{2+} influx in parotid cells[63] or parietal cells (P. A. Negulescu and T. E. Machen, unpublished), suggesting that there may be different mechanisms of Ca^{2+} entry and reloading of internal stores in some cell types.

Calcium Ion Oscillations

The coordination between Ca^{2+} uptake across the plasma membrane and the internal store is best demonstrated by the phenomenon of Ca_i^{2+} oscillations (see Berridge and Galione for review[64]). These oscillations have

[62] T. J. Hallam, R. Jacob, and J. E. Merritt, *Biochem. J.* **255,** 179 (1988).
[63] J. E. Merritt, and T. J. Hallam, *J. Biol. Chem.* **263,** 6161 (1988).
[64] M. J. Berridge, and A. Galione, *FASEB J.* **2,** 3074 (1988).

been observed in a number of cell types, including lymphocytes,[65,66] mast cells,[67] endothelial cells,[68] and hepatocytes,[69] in response to a variety of agonists. Much interest is currently aimed at determining the mechanisms responsible for the generation of such oscillations and their relevance to the biological response. Their presence has led to the proposal that some nonelectrically excitable cells may respond to stimulus intensity by generating a frequency-modulated signal. Although parietal cells possess Ca^{2+} entry pathways and internal stores, they rarely exhibited any oscillatory behavior in response to carbachol. Small Ca_i^{2+} oscillations (50 nM amplitude) have been observed only at concentrations of agonist that were well below those needed to stimulate acid secretion. As carbachol doses increased, the magnitude of the Ca_i^{2+} response increased, and this correlated with increased acid secretion. Thus, our data suggest that the parietal cell interprets the strength of the cholinergic stimulus with Ca_i^{2+} signal that varies in amplitude, rather than frequency.[9]

Calcium Ion as an Intracellular Signal

In addition to correlation of Ca_i^{2+} responses with secretory responses, Fura-2 can also be used to monitor the effectiveness of Ca_i^{2+} buffers in controlling the magnitude of the Ca_i^{2+} response. One such buffer is BAPTA, which can be loaded into cells as an AM ester. Figure 10 shows that by increasing the concentration of BAPTA/AM in the loading medium, one can control the amount of Ca^{2+} buffering and attenuate the Ca^{2+} changes in response to carbachol. In parietal cells, increasing buffer correlated with greater inhibition of carbachol-stimulated acid secretion.[9] Chew and Brown[70] also showed that Ca^{2+} buffering prevented phosphorylation of proteins which may be important in mediating carbachol-stimulated acid secretion. These types of approaches can be useful in determining the importance of Ca_i^{2+} as a signal.

Calcium Ion Buffering by Fura-2

Although Fura-2 has a lower Ca^{2+} affinity than BAPTA, it could buffer Ca_i^{2+} at high enough concentrations, especially when one considers that free Ca_i^{2+} is about 100 nM in the cell and the amount of Ca^{2+} binding

[65] H. A. Wilson, D. Greenblatt, M. Phoenie, F. D. Finkelman, and R. Y. Tsien, *J. Exp. Med.* **166,** 601 (1987).
[66] R. S. Lewis, and M. D. Cahalan, *Cell. Regul.* **1,** 99 (1989).
[67] E. Neher, and W. Almers, *EMBO J.* **5,** 51 (1986).
[68] R. Jacob, J. E. Merritt, T. J. Hallam, and T. J. Rink, *Nature (London)* **335,** 40 (1988).
[69] N. M. Woods, K. S. R. Cuthbertson, and P. H. Cobbold, *Nature (London)* **319,** 600 (1986).
[70] C. S. Chew, and M. R. Brown, *Am. J. Physiol.* **256,** G99 (1989).

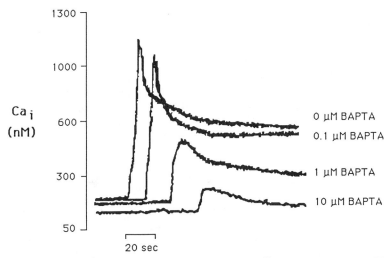

FIG. 10. Effect of various loading concentrations of the Ca^{2+} chelator BAPTA on 100 μM carbachol-stimulated Ca^{2+} increases measured with Fura-2. Traces are shifted in time for presentation purposes; the delay between the addition of the agonist and the Ca_i^{2+} increase was similar in all cases. (Taken from Ref. 9 with permission.)

indicator is about $1000\times$ higher (i.e., 100 μM), with a K_d of ~ 300 nM. At these concentrations, however, the effects of Fura-2 buffering are probably small relative to the intrinsic buffering of the cell. For example, in neutrophils, the number of cytoplasmic Ca^{2+}-binding sites corresponded to a concentration of 760 μM with a K_d of 0.55 μM.[71] This means that total cytoplasmic Ca^{2+} must be about 120 μM in order for free Ca^{2+} to be 100 nM, and the total cytoplasmic Ca^{2+} must increase by 380 μM if the cell is to achieve 1 μM free Ca^{2+}. If 100 μM Fura-2 with a K_d of 300 nM is added to the intrinsic buffer system, then the total buffer is 860 μM and a weighted K_d would be 520 nM. Under these conditions resting Ca_i^{2+} would be 84 nM and the measured Ca_i^{2+} increase would be 730 nM. In this example, then, Fura-2 would attenuate the peak free Ca_i^{2+} by 27%. Obviously, the effects would be more severe if the cell had more Fura-2 or less intrinsic Ca^{2+} buffer, thus efforts should be made to minimize loading to the levels discussed at the beginning of the chapter.

Measuring Na_o^+ with SBFI

The fluorescent Na^+-binding indicator, SBFI, is a member of a new family of fluorescent indicators for Na^+ and K^+ recently introduced by

---

[71] V. von Tscharner, D. A. Deranleau, and M. Baggiolini, *J. Biol. Chem.* **261**, 10163 (1986).

Minta and Tsien.[3] These probes have benzofuran fluorophores similar to that of Fura-2 and are therefore compatible with filter sets for Fura-2. (If possible, slightly shorter wavelengths than the typical 350/385 nm pair for Fura-2 should be used to maximize the dye's sensitivity to Na^+.) At this time there have been only two studies using SBFI to study Na^+ metabolism in cells. Harootunian et al.[72] showed that mitogens increased Na_i^+ and Na^+ influx in fibroblasts. The following discussion summarizes approaches used to study Na^+ metabolism in gastric cells using SBFI.[34]

Calibration and Spectra of SBFI

The general considerations regarding calibration of Fura-2 are especially applicable to SBFI because it behaves quite differently inside the cell. Its absolute fluorescence is dramatically increased by increases in viscosity[3] and the ratio is strongly affected by ionic strength.[34] Fortunately, on-line calibration of Na_i^+ is considerably easier than that of Fura-2 due to more effective ionophores and larger ion fluxes involved in generating millimolar changes in Na^+ concentration.

Calibration is best achieved using the monovalent cation ionophore gramicidin D (between 5 and 10 μM) which will equilibrate Na_i^+ and Na_o^+ as well as K_i^+ and K_o^+. To calibrate Na_i^+, we make up two solutions of equal ionic strength. One is Na^+ free and contains 130 mM potassium gluconate and 30 mM KCl. The other contains 130 mM sodium gluconate and 30 mM NaCl. Both solutions contain 10 mM HEPES, 1 mM $CaCl_2$, and 1 mM $MgSO_4$ and are titrated with NMG base to 7.1. By mixing these solutions together any [Na^+] between 0 and 150 mM can be obtained.

There are several spectral differences between SBFI fluorescence in solution (Fig. 11A) and in gastric cells (Fig. 11C). As in the case of Fura-2, part of the difference is due to the attenuation of short-wavelength excitation light by the microscope (Fig. 11B), which reduces the sensitivity of the probe to Na^+ (Fig. 11D). Interestingly, the red shift of fluorescence in the cells increased the sensitivity to Na^+ compared to dye on the microscope by shifting the excitation spectrum into a better detected range. Thus, Na^+ changes in the cells produced bigger 340/380 nm ratio changes in the cell than on the microscope, although the sensitivity in the cell is still attenuated with respect to dye in the cuvette (Fig. 11D). In parietal and chief cells a 10 mM Na^+ change resulted in a ~0.15 ratio unit change. This results in a resolution of approximately 2 mM in parietal cells. The Na^+ sensitivity varies among cell types. Although parietal and chief cells had

[72] A. T. Harootunian, J. P. Y. Kao, B. K. Eckert, and R. Y. Tsien, *J. Biol. Chem.* **264,** 19458 (1989).

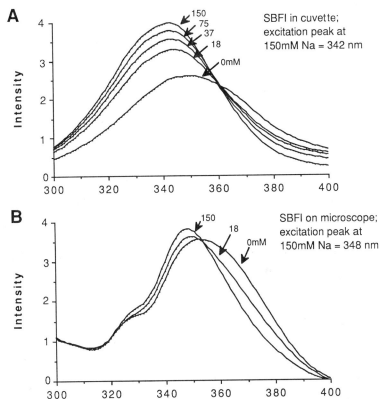

FIG. 11. Excitation spectra and Na$^+$ sensitivity of SBFI *in vitro* and *in situ*: (A) Na$^+$ titration of 5 μM SBFI in a cuvette; (B) Na$^+$ titration of 5 μM SBFI-free acid on the stage of a microscope; (C) Na$^+$ titration of SBFI-loaded glands. Gramicidin D (5 μM) was added at the time indicated. Note that the excitation peak at high Na$^+$ *in situ* is at longer wavelengths than that *in vitro* and that, *in situ*, increasing Na$_i^+$ decreases fluorescence at all wavelengths. In all experiments Na$^+$ was titrated at constant ionic strength by mixing solutions containing 150 mM Na$^+$ with solutions containing 150 mM K$^+$ as described in the text. (D) Comparison of Na$^+$ sensitivity *in vitro* and *in situ*. Note that sensitivity in the cells is greater than that on the microscope, presumably due to the red shift in cells.

approximately the same sensitivity, studies in fibroblasts indicate that the resolution can be several times better.[72]

Another difference is that the affinity for Na$^+$ *in situ* is lower than that *in vitro*: whereas in free solution the ratio begins to saturate above 50 mM, in parietal cells the calibration was linear up to 70 mM and was not saturated at 150 mM Na$^+$. Thus the K_d is higher than the 17 mM value determined *in vitro* and was not consistent in all experiments (Table I). Because of the variable K_d we generally calibrate SBFI at the end of each experiment, similar to the technique used for BCECF.

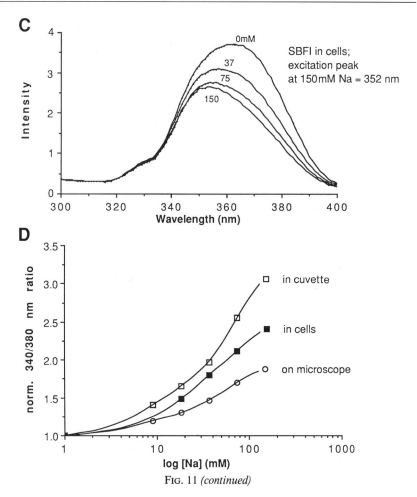

FIG. 11 *(continued)*

The most dramatic difference between SBFI spectra obtained *in vitro* and *in situ* was that the entire spectrum decreases in intensity as Na_i^+ increases (Fig. 11C). The decreases are larger at 380 nm than at 340 nm so that the ratio increases as Na_i^+ increases. This apparent quenching of fluorescence at high Na_i^+ could be related to changes in viscosity or ionic strength.

Accurate calibration *in situ* is especially important in measuring Na_i^+ because changes in steady state Na_i^+ may be small with respect to baseline. For example, upon stimulation of parietal cells with carbachol, Na_i^+ increased by 5 mM, an increase of only 50% over baseline. Na_i^+ is tightly regulated due to the Na^+ sensitivity of the Na^+ pump. Because the cell attempts to keep Na_i^+ constant, the challenge in measuring Na_i^+ is to maximize sensitivity in order not to miss changes. In contrast, Ca^{2+} increases by approximately 10-fold in response to carbachol, consistent with its role as a signal.

Selectivity in Situ

Because changes in steady state Na_i^+ on stimulation led to small changes in the SBFI ratio, we were concerned that changes in other ions, notably K_i^+, pH_i, or Ca_i^{2+}, might introduce small artifacts. The selectivity of SBFI over other ions was determined *in situ* by clamping Na_i^+ and altering the concentration of K^+ or H^+ in the presence of appropriate ionophores.[34] To determine the selectivity of SBFI for Na^+ over K^+ in parietal cells, Na_i^+ was clamped using gramicidin D and then step changes in K^+ were made to determine the independent effect of K^+ on SBFI. Following removal and readdition of Na^+ for reference, KCl was removed in steps and replaced with CsCl, which has negligible effects on SBFI (not shown). Because Cs^+ and K^+ both permeate the gramicidin pore, Cs^+ exchanges with K^+ and maintains the ionic balance of the cell. Under these conditions, removal of all the K^+ from the perfusate resulted in an "apparent" decrease of only 4 mM Na^+. In these experiments it was crucial to use Cs^+ as the cationic replacement for K^+. If NMG was substituted for Cs^+, there was a large increase in the ratio. This effect may be due to a drastic change in ionic strength since Cs^+ (as well as Cl^-) rapidly leave the cell and cannot be replaced by the impermeant NMG cation.

The effect of pH_i was more of a concern. For these experiments 10 μM nigericin (a K^+/H^+ ionophore) was added in addition to gramicidin to the high K^+ calibration medium to assure that $pH_i = pH_o$. In parietal cells, increasing pH from 7.0 to 7.5 caused a ratio increase of 0.05 units, corresponding to an "apparent" Na^+ increase of 4 mM. A total change of 1.0 pH unit (7.5–6.5) caused the SBFI fluorescence to decrease by 0.1 ratio units, which would appear as an Na_i^+ decrease of ~8 mM. Since carbachol increases pH_i by about 0.15 pH units in parietal cells, the magnitude of the carbachol-induced Na_i^+ rise may be slightly overestimated. Larger changes in pH_i would lead to even larger errors.

Although no compartmentation of the dye was apparent in gastric cells, some SBFI did accumulate in discrete regions of fibroblasts. The methods for determining compartmentation are the same as for Fura-2. Since many compartments have low pH and the SBFI ratio is decreased by acidic spaces, the ratio from the compartmented dye may be lower than the ratio of dye in the cytoplasm. Harootunian *et al.*[72] showed that adding nigericin and monensin to the cell caused the ratio of the compartmentalized dye to increase, presumably by dissipating the pH gradients in these compartments.

Measurements of Na$^+$ Fluxes Using SBFI

One way to amplify the sensitivity to changes in Na$^+$ metabolism is to measure the rate of Na$^+$ influx in the absence of Na$^+$/K$^+$ pump activity. The pump can be inactivated by either removing extracellular K$^+$ (K$_o^+$) or adding the cardiac glycoside, ouabain. The advantage of K$_o^+$ removal is that its effects are rapidly reversible, allowing one to make multiple measurements on a single cell. Ouabain washes off slowly, although the process can be accelerated by competition with high K$_o^+$ solutions. Another alternative is to use dihydroouabain, a reversible ouabain analog.

The conversion of SBFI-measured fluorescence changes into Na$^+$ fluxes requires an estimate of buffering capacity. In parietal cells, Na$^+$ buffering was estimated by comparing changes in Na$_i^+$ and pH$_i$ during recovery of cell from an acid load *via* the Na$^+$/H$^+$ exchanger. Since H$^+$ buffering is known (see Fig. 4), and the changes in pH$_i$ and Na$_i^+$ can be measured, Na$^+$ buffering can be estimated as a coefficient which makes the H$^+$ flux equal to the Na$^+$ flux. Using this technique, Na$^+$ buffering in parietal cells was found to be negligible.[34] Thus, the rate of change in Na$_i^+$ following inactivation of the Na$^+$ pump is a direct measure of net unidirectional Na$^+$ flux.

The rate at which Na$_i^+$ increases after Na$^+$/K$^+$ pump inhibition is due to Na$^+$ influx. In resting gastric cells Na$_i^+$ increased by 3.2 mM/min. Upon stimulation with carbachol, Na$^+$ influx increased by an average of 2.4-fold in both cell types. The nature of Na$^+$ entry pathways in both resting and stimulated cells is of obvious interest. The contribution of some Na$^+$-coupled ion transporters can be determined by ion substitution experiments. For example, inwardly directed NaKCl$_2$ or NaHCO$_3$ cotransport could be detected by comparing Na$^+$ influx in the presence and absence of K$^+$ and HCO$_3^-$, respectively. Potassium ion-dependent Na$^+$ influx was not detected in either resting or stimulated gastric cells and all experiments were conducted in HCO$_3^-$-free solutions (HEPES buffer), so the observed stimulation was not due to an HCO$_3^-$-dependent mechanism. If one compares BCECF-measured H$^+$ fluxes through the Na$^+$/H$^+$ exchanger with SBFI-measured Na$^+$ fluxes it appears that Na$^+$/H$^+$ exchange accounts for much of both resting Na$^+$ entry and carbachol-stimulated Na$^+$ influx. The degree to which Na$^+$/H$^+$ exchange contributes to overall Na$^+$ influx would ideally be determined directly by measuring Na$^+$ flux in amiloride-containing solutions. Unfortunately, the fluorescence of amiloride makes the measurement impossible.

Measurement of Na$^+$ Pump Activity

The activity of the Na$^+$/K$^+$ pump was determined by measuring total and ouabain-insensitive Na$^+$ efflux from Na$^+$-loaded cells (Fig. 12). Cells

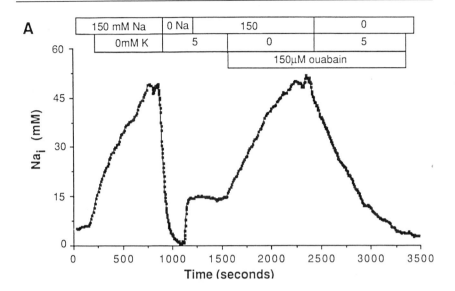

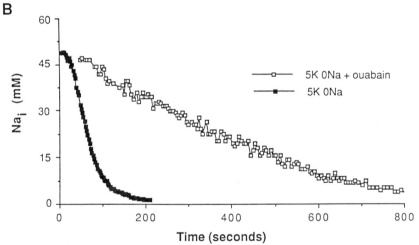

FIG. 12. Measurement of Na⁺ dependence of the Na⁺ pump. (A) Following Na⁺ loading of cells in K⁺-free medium the rate of Na⁺ efflux in Na⁺-free medium was measured in the presence and absence of 150 μM ouabain; (B) Na⁺ efflux on expanded time scale; (C) rate of Na⁺ efflux plotted as a function of Na_i^+. Rates of Na⁺ efflux were calculated from (B). Subtraction of ouabain-insensitive efflux (open squares) from total efflux (open circles) yields the activation curve of the Na⁺/K⁺ pump. *Inset:* Hill plot of Na⁺ pumping vs Na_i^+. Data averaged from seven parietal cells within a single gland and is typical of five similar experiments. (From Ref. 33.)

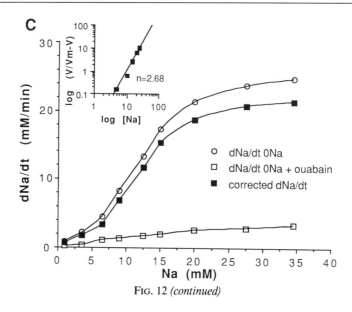

FIG. 12 *(continued)*

were Na^+ loaded by incubation in K^+-free solution (Fig. 12A). Once Na^+ had increased to about 50 mM, K_o^+ was added back and Na_o^+ was removed (NMG replacement). Under these conditions, Na^+ exits the cells through both passive mechanisms as well as the Na^+/K^+ pump. In order to account for the passive component, the experiment was repeated in ouabain. As seen in an expanded time scale (Fig. 12B), the rate of Na^+ efflux was ~sixfold slower in the presence of ouabain. When the ouabain-insensitive efflux is subtracted from the total, the activity of the pump as a function of Na_i^+ can be determined. Figure 12C shows how the rate of efflux increases as Na_i^+ increases. The pump increased its activity eight-fold between 5 and 20 mM Na_i^+. Thus, the pump is only partially activated at resting Na_i^+ and is well suited to resist changes in Na_i^+ in the case of increased Na^+ influx. This can explain why carbachol increases Na_i^+ by only 50% when the rate of Na^+ influx increases by 250%.

In summary, the initial trials of SBFI have proved it to be a useful probe for studying Na^+ metabolism. In gastric cells, the limited resolution of the probe (~2 mM) made it better suited for detecting Na^+ fluxes because larger Na_i^+ changes are measured. It remains to be seen whether its behavior in other cell types is good enough to detect submillimolar Na_i^+ changes.

Conclusion

One advantage of using fluorescent probes is that the same ratioing system can be used for measuring any of the dyes discussed above. This is particularly advantageous because there are many potential interactions among the various ions and ion transporters. For example, the effect of activation of Na^+/H^+ exchange on both pH_i and Na_i^+ can be studied. Once buffering capacities are known, net unidirectional H^+ and Na_i^+ fluxes can be compared to determine the degree to which Na^+ fluxes are due to Na^+ movement through the Na^+/H^+ exchanger. In principle, another application of parallel measurements would be to obtain stoichiometric ratios of ion transport through cotransporters and exchangers (i.e., the Na^+/HCO_3^- cotransporter or the Na^+/Ca^{2+} exchanger). Finally, the survey of various ion activities during particular experimental maneuvers can yield unexpected relationships, such as the effect of low Na_o^+ on release of Ca^{2+} stores.

Despite the general applicability of the technique, this chapter has demonstrated that the dyes should not be regarded equally. They load differently, behave differently once inside cells (Figs. 3, 6, and 11), and have different effects on cellular metabolism (see Fig. 1). With regard to ionic measurement, each probe has its own strengths and its own limitations. For example, Fura-2 exhibits a large spectral shift upon Ca^{2+} binding, but can be problematic due to compartmentation and suppression of the spectral shift in the cytoplasm. BCECF, on the other hand, has no spectral shift but is less compartmentalized and is less affected by the cytoplasmic environment. Finally, the use of each probe should be considered in the context of the cell. For example, because of the low cytoplasmic Ca_i^{2+} concentration, Ca^{2+} buffering by Fura-2 should be considered. This is not a concern with either BCECF (because of millimolar intrinsic H^+ buffering) or SBFI (because of the presence of millimolar Na^+ levels). With such considerations as guidelines, fluorescent ion indicators should continue to provide unique insights into cellular physiology for many years.

[5] Electrophysiological Techniques in the Analysis of Ion Transport across Gastric Mucosa

By S. Curci and E. Frömter

Introduction

Gastric mucosa has a complex histological structure. In the fundus region it consists of a layer of mucus-secreting surface epithelial cells (SEC) which is interrupted at short distances by the openings of gastric pits. These are shallow invaginations of the SEC layer which drain the secretion from the gastric glands. The glands which, depending on species and age, may be as long as 1 mm or more are densely packed in the subepithelial connective tissue. They run straight from bottom to top where they irrigate into the pits. In mammals, the glands are composed of enzyme-secreting chief cells (CC) and HCl-secreting parietal cells (PC) together with mucous neck cells (MNC) and endocrine cells (EC). In amphibia instead of CC and PC we find only one cell type, the oxyntic cell (OC), which secretes both enzymes and HCl. The narrow space between the glands is filled with connective tissue and blood capillaries. At the bottom the mucosa is packed on a thick layer of smooth muscle fibers embedded in connective tissue, and this layer, finally, is covered by the peritoneum.

This complex architecture causes enormous problems but also presents a great challenge to the investigator since one must identify how much which cells, which cell membranes, which tight junctions, or which fluid spaces contribute to the observable overall ion fluxes or transport properties of the organ. Ideally, by puncturing single cells, gland lumina, interstitial spaces, or lateral intercellular spaces with microelectrodes and applying the entire arsenal of electrophysiological techniques (conventional and ion-selective microelectrodes, noise analysis, impedance analysis, patch-clamp analysis, etc.) it should be possible to collect enough information to answer all questions of interest. In reality, however, it has even been difficult to measure the cell potential of SEC, and it took almost 20 years before the first reliable data on membrane potentials of OC became available. To overcome the complexity of the tissue many different attempts have been made to disrupt the system and to analyze individual parts separately. Thus single glands, single cells, or pieces of cell membrane have been isolated and analyzed, and many of these approaches have been successful, but thus far they have not yet led to an unambiguous picture of gastric transport. In the following we shall summarize the individual electrophysiological techniques that have been used to study gastric mucosa and shall briefly discuss the validity and limitations of the results obtained.

Transepithelial Electrical Measurements on Gastric Mucosa

Macroscopic Current and Voltage Measurements

By studying the transepithelial potential difference (V_t) and resistance (R_t) or, respectively, the short-circuit current (I_{sc}) as a function of ambient ion concentrations or during application of individual stimulants or inhibitors it has been possible in many epithelia to identify the major ion transport processes and the major transport properties of the vascular and contravascular cell membranes. In gastric mucosa, however, because of its complex architecture, such studies have led to ambiguous model concepts. Intact stomachs filled with bicarbonate Ringer's solution or isolated gastric mucosae (retaining usually a greater part of the muscle tissue), when bathed on either surface in identical Ringer's solutions, generate a lumen-negative V_t which varies between -10 and -60 mV, depending on the species investigated.[1] As has been found in short-circuit experiments in most species the resting state V_t is generated mainly by active Na^+ absorption and active (nonacidic) Cl^- secretion. Potassium ion and H^+ fluxes are usually negligible. On stimulation of HCl secretion with histamine, V_t, I_{sc}, and R_t decrease in most species studied: frog,[2] rat and guinea pig,[3] piglet,[4] and dog,[5] but V_t may also increase , e.g., in *Necturus*.[6] Attempts to interpret individual components of I_{sc}, e.g., the acidic and nonacidic active chloride secretion[2,7] as well as rheogenic active H^+ secretion[8-10] as reflecting independent active (rheogenic) ion pumps, have not been convincing (see Ref 11). Instead we know today that active H^+ secretion is generated

[1] R. P. Durbin, *in* "Alimentary Canal" (C. F. Code, ed.), Handbook of Physiology, Sect. 6, Vol. 2, p. 879. American Physiological Society, Washington, D.C., 1967.

[2] C. A. M. Hogben, *Am. J. Physiol.* **180**, 641 (1955).

[3] T. J. Sernka and C. A. M. Hogben, *Am. J. Physiol.* **217**, 1419 (1969).

[4] J. G. Forte and T. E. Machen, *J. Physiol.* **244**, 33 (1975).

[5] Y. J. Kuo and L. L. Shanbour, *J. Physiol.* **291**, 367 (1979).

[6] J. R. Demarest and T. E. Machen, *Am. J. Physiol.* **249**, C535 (1985).

[7] M. J. Starlinger, M. J. Hollands, P. H. Rowe, J. B. Mathiews, and W. Silen, *Am. J. Physiol.* **250**, G118 (1986).

[8] W. S. Rehm, *Am. J. Physiol.* **185**, 325 (1956).

[9] W. S. Rehm and M. E. Lefevre, *Am. J. Physiol.* **208**, 922 (1965).

[10] R. L. Shoemaker and G. Sachs, *in* "Gastric Secretion" (G. Sachs, E. Heinz, and K. J. Ullrich, eds.), p. 147. Academic Press, New York, 1972.

[11] W. W. Reenstra, J. D. Bettencourt, and J. G. Forte, *Am. J. Physiol.* **252**, G543 (1987).

by an electroneutral K^+/H^+ exchange pump[12,13] in conjunction with various electroneutral and conductive (rheogenic) ion transport mechanisms arranged in series and in parallel within the cells (for further discussion see Frömter et al.[14]).

In many epithelia, e.g., in frog skin, it has been possible to define single-cell membrane transport mechanisms by analyzing changes in V_t and R_t in response to both ion substitutions or inhibitors. In gastric mucosa, however, such attempts have been less fruitful. For example, in recent studies the presence of a basolateral HCO_3^- conductance,[15] an $NaHCO_3$ cotransporter,[16] and a rheogenic NaCl cotransporter with a coupling stoichiometry of Cl^- to $Na^+ > 2 : 1$[17] have been postulated, but the validity of these postulates and the location of the postulated transporters remained unknown.

Measurements of Current Fluctuations

More information about a tissue may be gained if instead of the total macroscopic current the microscopic current fluctuations are analyzed (noise analysis, see Ref. 18). Thus far only one study has been published in which this approach was applied to frog gastric mucosa.[19] The data obtained in the presence and absence of inhibitors (e.g., Ba^{2+}) were interpreted as supporting the existence of apical K^+ channels and Cl^- channels in stimulated frog gastric mucosa. The interpretation was not compelling, however, and it is not to be expected that this approach will ever gain the same importance for gastric physiology that it has had for the analysis of sodium transport across frog skin where the transport pattern is far less complex.

Resistance and Impedance Measurements

In many gastric mucosae, determination of the DC resistance is not as trivial as it may sound because polarization phenomena may develop

[12] G. Sachs, H. Chang, E. Rabon, R. Schackmann, M. Lewin, and G. Saccomani, *J. Biol. Chem.* **251**, 7690 (1976).

[13] J. G. Forte and H. C. Lee, *Gastroenterology* **73**, 921 (1977).

[14] E. Frömter, S. Curci, and A. H. Gitter, *in* "Epithelial Secretion of Water and Electrolytes" (J. A. Young and P. Y. D. Wong, eds.), p. 293. Springer-Verlag, Berlin, 1990.

[15] M. Schwartz, G. Carrasquer, and W. S. Rehm, *Biochem. Biophys. Acta* **819**, 187 (1985).

[16] G. Klemperer, S. Lelchuk, and S. R. Caplan, *J. Bioenerg. Biomembr.* **15**, 121 (1983).

[17] G. Carrasquer, T. C. Chu, W. S. Rehm, and M. Schwartz, *Am. J. Physiol.* **242**, G620 (1982).

[18] W. van Driessche and W. Zeiske, *Physiol Rev.* **65**, 833 (1985).

[19] W. Zeiske, T. E. Machen, and W. van Driessche, *Am. J. Physiol.* **245**, G797 (1983).

during passage of constant current which give rise to slow continuous voltage creeps, the origin of which is not always clear.[20] Neglecting such polarization phenomena and measuring the response of V_t to constant current pulses of 1- to 5-sec duration, DC resistances have been calculated which range between ~100 and 600 $\Omega \cdot cm^2$. Although these values are seemingly rather low, compared to those of tight epithelia (which range up to 80 $k\Omega \cdot cm^2$),[21] they do not allow the conclusion that gastric mucosa is a leaky or moderately leaky epithelium with low-resistance paracellular shunts. This may hold only for *Necturus* antrum,[22] in which a comparison of transepithelial resistance and cell membrane resistances suggests the presence of low-resistance tight junctions (see also discussion of cable measurements, below).

In theory more information can be obtained if, instead of the DC resistance, the impedance of a tissue is measured with alternating currents of different frequencies. In frog skin such measurements were found to fit an equivalent circuit model consisting of two membranes in series,[23] and under specific experimental situations impedance measurements allowed details of the blocking properties of Na^+ channels by amiloride to be derived.[24] Disregarding some earlier pilot studies[20] on gastric mucosa, thus far only one study has been performed in which the impedance was measured systematically during transition from resting to stimulated state.[25] In view of the complex architecture of the tissue, however, the equivalent circuit model which has been used to analyze those data was far too simple. In essence, a one-cell-type model was used consisting of two RC elements for the apical and basolateral cell membrane in series with some cable-like element representing either glandular structures, lateral spaces, or microvilli. From the stimulation-induced impedance changes, the authors concluded that the impedance properties of the SEC were negligible and that the transepithelial electrical response was largely determined by the OC. Furthermore they concluded that the lumen of the glands did not represent a major electrical resistance and that the increasing capacitance observed during stimulation mainly reflected the increasing apical surface area of the OC. However, in view of the many simplifying assumptions (neglect of different cell types, tight junctions, cell-to-cell coupling), the validity of the conclusions remains uncertain. It

[20] D. H. Noyes and W. S. Rehm, *Am. J. Physiol.* **219**, 184 (1970).
[21] J. T. Higgins, L. Cesaro, B. Gebler, and E. Frömter, *Pfluegers Arch.* **358**, 41 (1975).
[22] D. I. Soybel, S. W. Ashley, R. A. Swarm, C. D. Moore, and L. Y. Cheung, *Am. J. Physiol.* **252**, G19 (1987).
[23] P. G. Smith, *Acta Physiol. Scand.* **81**, 355 (1971).
[24] J. Warncke and B. Lindemann, *Pfluegers Arch.* **589**, S94 (1985).
[25] C. Clausen, T. E. Machen, and J. M. Diamond, *Biophys. J.* **41**, 167 (1983).

is to be hoped that the combination of impedance measurements with simultaneous intracellular recordings as applied by Kottra and Frömter[26] to *Necturus* gall bladder may provide a more reliable basis for such type of analyses.

Vibrating Probe Studies

The spatial inhomogeneity of gastric mucosa suggests that the transepithelial current flow may also be inhomogeneous or, in other words, since HCl secretion originates mainly from the pits, it may also be expected that the electrical current density differs over the pits from that over the planar SEC layer. This was tested and confirmed by Demarest *et al.*,[27] who used a vibrating electrode to scan the tissue surface. By vibrating up and down, this electrode allowed the bath potential to be measured in two definable heights above the epithelial surface which, at constant medium conductance, may serve as a measure of local current density. With amiloride present on the lumenal surface the authors demonstrated that in *Necturus* gastric mucosa, virtually all I_{sc} originated from the pits, whereas in the absence of amiloride, the surrounding SEC generated only a small Na^+ current but no Cl^- current.

Intracellular Potential and Resistance Measurements with Microelectrodes

Measurements on Intact Gastric Mucosae

Cell Potential and Voltage Divider Ratio of SEC. Mounting pieces of gastric mucosa, mucosal surface up, within an Ussing chamber and impaling the surface cells from the top with glass pipet microelectrodes is a straightforward experiment which has been performed on the stomach of various species of frog and on *Necturus* to determine SEC cell membrane potentials. Despite the apparent simplicity of the experiment, however, it has taken a number of years for reliable data on the potential difference across the apical (V_m) and basolateral (V_s) membrane of the SEC to become available. As summarized in Table I[28-36] the reported V_s values increased

[26] G. Kottra and E. Frömter, *Pfluegers Arch.* **402**, 421 (1984).
[27] J. R. Demarest, C. Scheffey, and T. E. Machen, *Am. J. Physiol.* **251**, C643 (1986).
[28] L. Villegas, *Biochim. Biophys. Acta* **64**, 359 (1962).
[29] L. Villegas, F. Michelangeli, and L. Senanes, *Biochim. Biophys. Acta* **219**, 518 (1970).
[30] R. L. Shoemaker, G. M. Makhlouf, and G. Sachs, *Am. J. Physiol.* **219**, 1056 (1970).
[31] J. G. Spenney, R. L. Shoemaker, and G. Sachs, *J. Membr. Biol.* **19**, 105 (1984).
[32] R. L. Shoemaker, *Acta Physiol. Scand. Special Suppl.*, p. **173** (1978).
[33] T. E. Machen and T. Zeuthen, *Philos. Trans. R. Soc. London, B* **299**, 559 (1982).
[34] T. Schettino and S. Curci, *Pfluegers Arch.* **403**, 331 (1985).
[35] T. Schettino and S. Curci, *Pfluegers Arch.* **383**, 99 (1980).
[36] S. Curci and T. Schettino, *Pfluegers Arch.* **401**, 152 (1984).

TABLE I
CELL MEMBRANE POTENTIALS AND VOLTAGE DIVIDER RATIOS OF SURFACE EPITHELIA
CELLS IN CONTROL CONDITIONS[a]

Species	V_s (mV)	V_m (mV)	VDR	Ref.
Rana pipiens	−49.2	−20.5	0.81	28,29
Necturus maculosus	−46.2	−29.9	1.2	30,31
	−52.8	−28.7	5.6	32
	−51.0	−34.0	4.5	33
	−57.3	−38.8	2.7	34
	−65	−40	3.5	27
Rana esculenta	−73.0	−54.5	—	35
	−66.8	−40.9	9.7	36

[a] V_s and V_m, Mean values of serosal and mucosal cell membrane potential, respectively; VDR, voltage divider ratio, defined by the voltage changes $\Delta V_m/\Delta V_s$ which are observed in response to transepithelial constant current pulses. For each species the data are listed in the order of the year of publication.

over the years, probably as a result of using smaller and, therefore, less damaging microelectrode tips.

Together with V_s usually the voltage divider ratio (VDR) was also determined. It is calculated from the simultaneously recorded response of V_s and V_t to short transepithelial constant current pulses after correction of both signals for serial fluid resistances.[37] Ideally, VDR should equal the ratio of the apical to basal cell membrane resistance (R_a/R_b). In stomach, however, the identity of VDR and R_a/R_b is not clear since the electrical field may be distorted both in the narrow connective tissue layer between the glands and in the vicinity of the gastric pit openings. As can be seen in Table I, V_s of SEC is somewhat higher in *Rana esculenta* than in *Necturus*, and the same appears to hold for VDR, although *Necturus* glands are shorter. This difference is probably not related to the just-mentioned field problems but appears to indicate a species difference in the distribution of cell membrane resistances or different keeping conditions of the animals.

Cell Potential and Voltage Divider Ratio of OC. Early attempts to puncture OC of intact frog and *Necturus* gastric mucosa were made on the same preparation as described above by simply advancing the electrode further down into the tissue and trying to identify recording sites by dye iontophoresis later.[28,32] Not unexpectedly this technique did not yield any convincing results (see Table II). Recently, however, two laboratories succeeded in dissecting off the connective tissue layer and thereby exposing the glands for direct impalement of OC with microelectrodes.[6,38]

In our laboratory the following microdissection procedure has been

[37] E. Frömter, *J. Membr. Biol.* **8,** 259 (1972).

TABLE II

CELL MEMBRANE POTENTIALS AND VOLTAGE DIVIDER RATIOS OF OXYNTIC CELLS IN
CONTROL CONDITIONS

Species	V_s (mV)	V_m (mV)	VDR	Ref.
Rana pipiens	-17.9^a	—	3.3	28,29
Necturus	-17.0^a	$+10$	8.4	32
	-45.9^b	-19.3	1.1	6
Rana esculenta	-56.9^b	-43.6	—	38
	-66.3^b	-40.7	9.4	39

[a] Deep punctures from mucosal surface.
[b] Punctures from serosal surface. All other details as in Table I.

developed. The gastric wall is mounted between the horizontal half-chambers, serosal surface up, and the muscle layer as well as the greater part of the connective tissue is removed by blunt dissection with a pair of watchmaker forceps. Then the chamber is tightened and the upper compartment is filled with solution. Under oblique inspection through a glass coverslip the area to be punctured is then further freed from connective tissue with the same forceps until individual glands "pop up" (see Fig. 1). These are then punctured under microscopic inspection by lowering a microelectrode in a perpendicular direction down onto the tissue. Usually we do not puncture the very end of the glands but advance the electrode into the gap between two glands thus impaling cells in the midregion of the glands. Here the individual tubules are more firmly fixed by the surrounding tissue so that the cells are less dislodged and less distorted before they are penetrated by the electrode. This provides better seals.

Table II summarizes cell membrane potential measurements of OC. It can be seen that the new technique yields much higher values than the earlier approaches. Comparing Tables I and II it can also be seen that in frog stomach V_s of the OC (66.3 mV) and SEC (66.8 mV) are virtually identical and the same holds for VDR of OC and SEC. There is, however, a great difference between the VDR of frog and Necturus in both cell types.

In recent years a number of microelectrode experiments have been performed on SEC and OC to determine the ionic conductances of the apical and basolateral cell membrane of these cells from changes in V_s and VDR in response to ion substitutions in the apical and basolateral fluid compartment. In these experiments K^+ conductances have been identified in the basolateral cell membrane of OC and SEC which increase under stimulation with histamine,[38] and a rheogenic $Na^+/(HCO_3^-)_n$ cotransporter has

[38] L. Debellis, S. Curci, and E. Frömter, Am. J. Physiol. 258, G631 (1990).

FIG. 1. Microphotograph of serosal surface of frog stomach before (a) and after (b) microdissection of serosal connective tissue. In (a) the muscle layer has been removed by blunt dissection. In (b) individual glands have been exposed and show up in the center of the picture.

been detected in the basolateral membrane of SEC.[39] In addition an amiloride-inhibitible Na^+ conductance has been observed in the apical cell membrane of SEC both in *Necturus*[27] and in histamine-stimulated frog stomach,[40] but no evidence was obtained for significant K^+ or Cl^- conductance in this cell membrane.

Cell-to-Cell Coupling Measurements (Two-Dimensional Cable Analysis). In fundus, and recently also in antrum mucosa of *Necturus* stomach, attempts have been made to analyze cell membrane resistances by means of two-dimensional cable analysis.[31,22] In such experiments constant current pulses are injected into a cell and the resulting potential displacement is recorded in nearby cells. If the radial current (or voltage) attenuation follows a zero-order Bessel function such measurements allow the cell-to-cell coupling resistance to be determined as well as the summed apical and basolateral cell membrane conductances from which the individual cell membrane resistances can be worked out with the help of VDR (see Ref. 37). However, due to the complex histology of gastric mucosa, the validity of those measurements is difficult to assess. In antrum mucosa which does not contain long glands the situation is less critical. By comparing the cell membrane resistances with the overall transepithelial resistance it has been concluded that antrum mucosa is a moderately leaky epithelium with a paracellular shunt resistance of 710 $\Omega \cdot cm^2$ and a lumped cellular resistance of 11.290 $\Omega \cdot cm^2$.[22]

Measurements on Isolated Gastric Glands

The above-described approach to study intact frog and *Necturus* stomach with microelectrodes is not applicable to mammalian tissue because it is usually too difficult or even impossible to keep isolated mammalian stomach preparations viable in Ussing-type chambers. To our knowledge, thus far only one attempt has been reported to puncture the SEC of dog stomach.[41] After Berglindh and Öbrink[42] had detected that gastric glands can be isolated from rabbit stomach by collagenase treatment, attempts have been made to puncture identified individual cells of these glands with microelectrodes. However, the results were disappointing. Kafoglis *et al.*[43] reported a mean value of -7 mV for V_s of PC, which is not convincing. Our own attempts to puncture cells of collagenase-isolated or microdissected gastric glands were not rewarding either.[44] The success rate of cell

[39] S. Curci, L. Debellis, and E. Frömter, *Pfluegers Arch.* **408,** 497 (1987).
[40] S. Curci, L. Debellis, and E. Frömter, unpublished.
[41] C. A. Canosa and W. S. Rehm, *Biophys. J.* **8,** 415 (1968).
[42] T. Berglindh and K. J. Öbrink, *Acta Physiol. Scand.* **96,** 150 (1976).
[43] K Kafoglis, S. J. Hersey, and J. F. White, *Am. J. Physiol.* **246,** G433 (1984).
[44] T. Schettino, M. Köhler, and E. Frömter, *Pfluegers Arch.* **405,** 58 (1985).

punctures was extremely low. The best cell potential records gave values of ~ −25 mV and except for evidence for a basolateral K^+ conductance no decent information on cell membrane properties could be obtained. In addition no difference was seen between PC identified by their autofluorescence[45] and CC.

Measurements on Cell Cultures

An alternative approach to investigate parietal cells with microelectrodes has been made by Okada and Ueda[46] by growing explanted gastric glands of newborn rats in cell culture media. Parietal cells from these cultures also showed potentials of ~ −25 mV. In our hands, the culture technique could not be well reproduced so the approach was again abandoned. To our knowledge monolayer cultures of dispersed isolated chief cells[47] or SEC[48] have not yet been investigated with microelectrodes.

Measurements on Isolated Cells

An attempt has been made to isolate cells of *Necturus* gastric mucosa and to puncture them with microelectrodes in order to determine cell potential and cell membrane resistances directly.[49] This approach, however, cannot be recommended since microelectrode punctures of single, small cells never yield useful results. There are a great number of problems with this technique: (1) the polarity of the cells is lost; (2) isolated cells are usually difficult to immobilize for puncture; (3) impalement with the microelectrode very often cause damage and introduces leak artifacts; (4) electrode solution may contaminate the cell; and (5) membrane resistances measured in series with the microelectrode resistance are subject to gross errors because the microelectrode resistance is usually unstable, particularly if the electrode tip is small and the resistance accordingly high.

Intracellular Ionic Activity Measurements

In recent years ion-selective microelectrodes have been used in a great number of tissues to determine cytoplasmic ionic activities. In gastric mucosa this approach has been applied mainly to the fundus region of frog

[45] M. Köhler and E. Frömter, *Pfluegers Arch.* **403**, 47 (1985).

[46] Y. Okada and S. Ueda, *J. Physiol.* **354**, 109 (1984).

[47] A. Ayalon, M. J. Sanders, L. P. Thomas, D. A. Amirian, and A. H. Soll, *Proc. Natl. Acad. Sci. U.S.A.* **79**, 7009 (1982).

[48] M. Rutten, D. Rattner, and W. Silen, *Am. J. Physiol.* **249**, C503, (1985).

[49] A. L. Blum, G. T. Shah, V. D. Wiebelhaus, F. T. Brennan, H. F. Helander, R. Ceballos, and G. Sachs, *Gastroenterology* **61**, 189 (1971).

and *Necturus* where SEC have been punctured. Thus far mostly single-barreled electrodes were used so that the measurements had to be analyzed by referring to cell potential measurements which were made under identical conditions but in different cells. In this way intracellular K^+ and Cl^- concentrations have been determined and possible changes in response to apical or basolateral ion substitutions have been analyzed.[5,36] These showed that the SEC of frog and *Necturus* are not able to secrete Cl^-[34,36] in contrast to what had been postulated earlier.[33] Since the microelectrode design has been recently improved[50] it now appears possible to reproducibly construct fine-tip double-barreled ion-selective electrodes, which might also be applicable to study OC function in more detail.

Patch-Clamp Experiments

The patch-clamp technique as developed 8 years ago by Neher and collaborators[51] has revolutionized electrophysiological laboratories and has brought important new insights into molecular properties of cell membrane ion transport mechanisms. With this technique tight seals are formed between a small membrane patch and the tip of a micropipet so that current flow through single ion channels can be recorded either in the cell-attached or in the excised configuration. Alternatively, if the membrane patch is ruptured in the cell-attached configuration a low-resistance access path is obtained to the cytoplasm allowing whole cell currents to be recorded. In spite of these great advantages, application of the technique to gastric mucosa has not been very successful thus far, mainly because of difficulties with seal formation. Except for some short communications[14,52] only one investigation has been published on single-channel and on whole cell recordings from parietal cells of gastric glands.[53] In this study a maxi-K^+ channel and a nonselective cation channel were observed and in another study on single cells or tubule fragments of *Necturus* gastric mucosa evidence was obtained for a cAMP-regulated K^+ channel and a Ca^{2+}-regulated K^+ channel whose exact location remained unclear, however.[54]

[50] Y. Kondo, T. Bührer, K. Seiler, E. Frömter, and W. Simon, *Pfluegers Arch.* **414**, 663 (1989).
[51] O. P. Hamill, A. Marty, E. Neher, B. Sakmann, and F. Sigworth, *Pfluegers Arch.* **391**, 85 (1981).
[52] D. D. G. Loo, J. D. Mendlein, T. Berglindh, A. H. Soll, G. Sachs, and E. M. Wright, *Fed. Proc., Fed. Am. Soc. Exp. Biol.* **44**, 643 (1985).
[53] H. Sakai, Y. Okada, M. Morii, and N. Takeguchi, *Pfluegers Arch.* **414**, 185 (1989).
[54] S. Ueda, D. D. F. Loo, and G. Sachs, *J. Membr. Biol.* **97**, 31 (1987).

Conclusion

Although virtually all different electrophysiological techniques (transepithelial potential and resistance measurements, impedance measurements, noise analysis, vibrating electrode measurements, intracellular potential and ionic activity measurements with microelectrodes, and single-channel measurements, as well as whole cell measurements with patchclamp electrodes) can be applied and have been applied to study the ion transport properties of gastric mucosa, compared to other epithelia, the amount of information which has been obtained with these techniques is still limited. This fact is largely due to the complex histological structure of the tissue, which has impeded the collection and straightforward interpretation of individual data. It is to be hoped, however, that these problems do not discourage researchers since electrophysiological analysis is so fundamental for our understanding of ion transport mechanisms, particularly in epithelia of complex histological structure. On the other hand, it is encouraging that recently a number of advances have been made, particularly in gaining access to the oxyntic cells (in amphibian stomach), which promise to open the route to a better understanding of the role of OC and of SEC at rest and during stimulation. However, further progress is badly needed.

Acknowledgment

The authors thank Dr. L. Debellis and Mr. G. Signorile for providing the microphotographs as well as Mrs. I. Harward and U. Merseburg for secretarial assistance.

[6] Gastric Glands and Cells: Preparation and in Vitro Methods

By Thomas Berglindh

Introduction

The way in which the gastric mucosa can produce hydrochloric acid in great volumes at a concentration of 160 mM has fascinated man since the days of William Beaumont's studies of Alexis St. Martins' stomach.[1] Gradually, increased knowledge of the control mechanisms emerged, but only with the advent of functioning cellular preparations and new biochemical approaches have we begun to understand the acid-secreting parietal cell.

[1] W. Beaumont, "Experiments and Observations on the Gastric Juice and the Physiology of Digestion." FP Allen, Plattsburg, New York, 1833.

The hydrogen ion pump has been identified as a unique H^+,K^+-ATPase, which even has been cloned, but perhaps surprisingly we are still far from total understanding of how the greatest ion gradient known to mammalian physiology can be generated. That the parietal cells are able to concentrate H^+ 4 million times (i.e., going from pH 7.4 to 0.8) should be compared with a potassium gradient of 30 in a normal cell. The other intriguing part is how the parietal cell and the gland lumen can withstand such high acid exposures without damage.

My interest in this area started in 1971 when I, as a Ph.D. student, was given the task by my tutor Dr. K. J. Öbrink to try to prepare isolated parietal cells. We succeeded in making isolated gastric cells, but in the process stumbled over a totally new technique, namely, how to make isolated gastric glands.[2,3,]

For almost all purposes the gland is a much better *in vitro* preparation than the cell. The gland constitutes the smallest functioning unit of the gastric mucosa and its cells have normal polarity, with tight junctions. Cell communication either by cell–cell interaction or by para- and autocrine messengers can take place. The glands are robust, have a closed off base, and probably a partially collapsed lumen.

In contrast, isolated cells, either crude or in a more purified state, have lost their polarity. Basolateral and apical membrane can mix, and the intracellular secretory channel is closed off and cannot communicate with the outer "world" as in the native cell. Cells are also much more fragile in general, and the preparation procedures normally more violent and damaging. Based on this it is easy to determine when each preparation should be used provided that parietal cell activity can be followed.

Glands. All physiological studies including receptors: Studies on electrolyte effects, as well as on permeation (where the basolateral membrane is made permeable to larger molecules by electric shock or digitonin treatment[4]), electrophysiological investigations, and fluorescent probe studies in the microscope, just to mention a few applications (for detailed references see Soll and Berglindh[5]).

Cells. Unless the isolated parietal cells are relatively pure, they cannot solve any questions better than glands do, provided that glands can be made from the species of interest. If pure, they can serve as a source for biochemical studies, be used for ion flux measurements, and lend themselves to short-term culture studies.

[2] T. Berglindh and K. J. Öbrink, *Acta Physiol. Scand.* **87,** 21A (1973).

[3] T. Berglindh and K. J. Öbrink, *Acta Physiol. Scand.* **96,** 150 (1976).

[4] D. H. Malinowska, this volume [7].

[5] A. H. Soll and T. Berglindh, *in* "Physiology of the Gastrointestinal Tract, Second Edition" (L. R. Johnson. ed.), p. 883. Raven, New York, 1987.

In the following presentation I give my personal view, and experience of almost 20 years of work on gastric glands and cells, many details of which have not been published before.

Preparation of Gastric Glands from Rabbit

Solutions

Phosphate-buffered saline (PBS): NaCl, 149.6 mM; K$_2$HPO$_4$, 3.0 mM; NaH$_2$PO$_4$, 0.64 mM; pH 7.3

Collagenase enzyme solution: NaCl, 130.0 mM; NaHCO$_3$, 12.0 mM; NaH$_2$PO$_4$, 3.0 mM; Na$_2$HPO$_4$, 3.0 mM; K$_2$HPO$_4$, 3.0 mM; MgSO$_4$; 2.0 mM; CaCl$_2$, 1.0 mM; Phenol Red, 10 mg/liter, pH 7.4

Incubation Medium: NaCl, 132.4 mM; KCl, 5.4 mM; Na$_2$HPO$_4$, 5.0 mM; NaH$_2$PO$_4$, 1.0 mM; MgSO$_4$, 1.2 mM; CaCl$_2$, 1.0 mM; Phenol Red, 10 mg/liter, pH 7.4

Animals. Male albino rabbits of New Zealand strain, 1.5–3.0 kg, are anesthetized with sodium pentobarbital (Nembutal, 30 mg/kg) through an ear vein. Open the abdomen and cannulate the aorta in a retrograde direction. Use a catheter with the biggest possible diameter. Place the catheter above the renal arterial branches. Inject 5 ml heparin (250 U/ml dissolved in PBS) with force through the cannula.

Intraarterial Perfusion. Use a high-capacity roller pump capable of delivering 1–2 liters/min, thus capable of working against a counterpressure. Keep PBS at 37° and bubble it constantly with oxygen.

Fill the pump tubing with warm PBS (cold solution will cramp the vessels). Open the thorax and clamp the thoracic aorta with a curved forceps. Start the high-pressure perfusion and continue until the stomach is exsanguinated and edematous; the liver should also turn pale. To alleviate some of the pressure you can either cannulate the portal vein or make small incissions in the liver using a scalpel.

During continued perfusion cut out the stomach and inject a lethal dose of anesthetics in the animal. Open the stomach along the lesser curvature. Empty the gastric content and rinse the mucosa with warm PBS. Discard the cardia and the antrum (antrum is easily distinguishable based on color and muscle thickness). Wash the mucosa totally free of adhering content and blot the mucosa hard against filter paper (disposable towels work fine). For gland production it does not matter and is rather an advantage if the blotting destroys surface epithelial cells. With the fingers of one hand stretch the stomach on a glass plate and gently scrape the mucosa off the muscle layer using a blunt instrument, preferably a large curved forceps. After a good perfusion this is very easily done and only small amounts of submucosa will be included. Mince the mucosa pieces into 2- to 3-mm fragments. This is most easily done on a cork plate with curved scissors. Transfer the minced pieces into a large glass Petri dish (15-cm diameter) with at least 1-cm walls. Rinse the pieces with plain collagenase enzyme solution while continuing to mince. White pieces of submucosa will float to the surface and can easily be decanted. Repeat this several times; finally decant all fluid.

Digestion of Mucosa Pieces

Bubble the enzyme solution for at least 10 min with 95% O_2, 5% CO_2 and then add 50 ml to a 100- to 150-ml round flask (flat bottom, wide neck) containing rabbit albumin (Sigma, St. Louis, MO), 1 mg/ml, 2 mg/ml glucose, and collagenase, 150 U/ml. If a proteolytic inhibitor such as TLCM (see below), is used add it at this point and let it prereact for 5 min with the collagenase. Put in a stirring bar (round or hexagonal type), which just barely spans the flat bottom. Add the minced mucosa and fill the flask with pure oxygen. Seal the flask with Parafilm and place it on a magnetic stirrer (120–150 rpm) in a 37° water bath. To minimize the need to organize different contraptions to hold the flask firmly in place we found that putting a lead weight (i.e., the lid of a lead container used to ship highly radioactive substances) on top of the neck of the flask worked very well. Closely follow the digestion process after 45 min and stop when all big pieces are gone and when a cream souplike consistency is obtained. Typically, the pH of the digestion medium will decrease slightly during the digestion (as indicated by the Phenol Red); This is quite normal and only rarely needs to be adjusted. When the digestion is finished, place the entire contents into a bigger flask and add 100–150 ml plain enzyme solution to dilute the thick digestion product. Filter through a coarse nylon cloth draped in a plastic funnel directly into new 15-ml clear plastic tubes with conical bottoms. Let the glands sediment so that they are loosely packed on the bottom. Suction off the upper solution with a fire-polished Pasteur pipet and add fresh plain enzyme solution. Resuspend gently using a 5-ml piston pipet (type Gilson) with a wide plastic tip. Repeat the solution changes several times until the upper solution is clear; then resuspend the glands in 100 ml incubation medium (with albumin and glucose, 2 mg/ml of each) and let them stand with gentle agitation at room temperature until use. Properly prepared glands should be long (>500 μm) and the rounded glandular base and protruding parietal cells should be clearly visible using interference contrast microscopy of living glands.

Important Points about the Gland Preparation

Animals. My experience is that the nonalbino rabbit is more difficult to work with in that the mucosa seems to be sticking harder to the submucosa. Glands from guinea pigs can be obtained, though the benefit of high-pressure perfusion is not as evident in that species. The guinea pig gastric mucosa and gastric wall are very thin. The rat mucosa is very tough and thus very difficult to scrape off. Some benefit is obtained from good high-pressure perfusion. Dogs likewise, have, a mucosa which is difficult to strip, but perfusion has been shown to facilitate the procedure.[6]

High-Pressure Perfusion. Vascular perfusion of the gastric arteries has several advantages. It will make the mucosa blood free and will tend to lift the mucosa off the submucosa and muscle layers as well as induce an edema and thus a mechanical separation of the glands, since the blood vessels run alongside and around the gastric glands. Taken together this will decrease gland or cell separation time, increase viability, and make the cells more responsive to secretagogues. The pressure should be as high as possible without rupturing the vessels. We have recorded pressures distal to the catheter inlet of approximately 600 mmHg.

Collagenase Digestion. The selection of proper collagenase has been given a special section (see below). Here I merely want to point out that magnetic stirring is absolutely essential; an ordinary shaker bath will not work. Also the shape of the flask is essential, i.e., a round flask with flat bottom. Depending on the quality of the perfusion and the collagenase, the time for the digestion can vary between 45 and 90 min. Never bubble solutions containing glands and albumin. The foam created can destroy the glands due to the surface tension.

Collagenase. Type I (crude) from Worthington Biochemicals (Freehold, NY) or Sigma (St. Louis, MO), typically 125–250 U/mg solid is used. Aim for a final concentration of 150 U/ml digestive solution (typically 50 ml for the use of one rabbit gastric mucosa). Over the years the specific activity has increased. This is why we now calculate by unit instead of weight. Type I collagenase always contains contaminants of proteolytic activity, the amount of which varies from batch to batch. Partial proteolytic activity is essential for a proper and fast digestion of gastric mucosa and thus types II – V cannot be recommended. However, if the balance between collagenase and proteolytic activity is incorrect, glands will not respond properly to secretagogues. Proteolytic activity, such as trypsin-like, clostripain, and neutral proteases, for each batch is normally listed on the en-

[6] T. Berglindh and D. Hansen, *Fed. Proc., Fed. Am. Soc. Exp. Biol.* **43**, 996 (1984).

closed data sheet. A good rule when ordering collagenase is to ask for batches with "normal" content of these proteolytic activities. Especially important is to look out for clostripain levels that are too high. Even with this precaution, however, one does not know if the collagenase will work properly, and at this point a potentially very tedious screening process must be undertaken. Order 100 mg of several different batches and request the company to reserve larger quantities of the same batches (this is such a common procedure that salespersons will know exactly what to do). Now take the gastric mucosa from one or possibly two rabbits (depending on how many batches will be tested in parallel), mix the minced mucosa pieces well with some digestion medium, and transfer equal amounts of them to smaller digestion flasks (Duran or equivalent, flask, flat bottom, wide neck, 50 ml). Make the total volume 25 ml and add collagenase (150 U/ml). Digest the mucosa pieces as described under Preparation of Gastric Glands from Rabbit.

Different collagenases might need different times to totally digest the mucosa into gastric glands. Although the faster the better is the rule, exceptions are not too unusual. Thus decide by visual inspection when the digestion is finished: anywhere between 45 and 90 min. *Do not forget to include the batch of collagenase now being used as a control!*

Use of Proteolytic Inhibitors. Finding *the* collagenase is very frustrating and could be a "lifetime" task. Therefore, the use of inhibitors of proteolytic activity will speed things up in that one will get acceptable results with a good, but not perfect, batch. The best I have found is N-α Tosyl-L-lysylchloromethane (TLCM; Calbiochem, Los Angeles, CA), which will inhibit clostripain activity as well as some tryptic activity.[7] Typically, 10–20 μg/150 U collagenase is added. Let TLCM prereact with the collagenase solution for 5 min before adding it to the mucosa pieces. A very striking observation is that a "good" batch will not change its behavior, whereas a "bad" batch will be significantly improved in the presence of TLCM.

Storage of Collagenase. Once the proper batch has been found, there can be other problems. Collagenase normally stored in an ordinary freezer will age rather rapidly if it is taken out to room temperature and put back in the freezer again: if this is done with a full 10-g bottle, rapid deterioration will be seen. To avoid this, 500-mg aliquots should be prepared and stored tightly sealed in a −70° freezer. However, if storage problems are experienced, TLCM will partly revive aged collagenase.

Protocol for Batch Testing. In order to get full insight into the potential

---

[7] T. Hefley, J. Cushing, and J. S. Brand, *Am. J. Physiol.* **240,** C234 (1981).

damaging properties of the new collagenase, the following protocol for aminopyrine (AP) accumulation should be run.

Control: Histamine, 1, 3, 10, 100 μM; Isobutyl methylxanthine (IMX), 10 μM; human gastrin I (HGI), 0.1, 1, 10, nM + IMX; acetylcholine (ACh), 0.1, 0.3, 1, 3 μM + IMX; ACh alone, 1 μM (add at 15 min, take at 30 min); dibutyryl-cAMP (db-cAMP), 1 mM; 2-Mercaptoethanol (BME), 10 mM; histamine, 10 μM + 2-mercaptoethanol

Incubate at 37° for 30 min according to the protocol for AP Accumulation Studies (see below).

An authentic batch testing is shown in Fig. 1, where two of the tested collagenases were fair but inferior to the reference collagenase and one batch was found to be totally unusable.

What to look for: The basal AP accumulation ratio should be 10–20. Values higher than this are normally a sign of cell destruction.

Histamine: The histamine dose–response curve must not be sluggish ($ED_{50} = 3 \mu M$), and maximum values should at least be 75, but are normally above 100.

IMX: Isobutyl methylxanthine stimulation is mostly dependent on the presence of endogenously released histamine. The higher the glandular density the stronger the stimulation. Normally the AP accumulation ratio for IMX should not be above 40–50.

ACh: Acetycholine alone will induce only a rapid transient stimulation of glandular AP accumulation. Maximum is reached after 10–15 min. Expect at least a doubling of your basal value. In dog glands and cells ACh is a very potent secretagogue with sustained stimulatory effect.

ACh plus IMX: In the presence of IMX the cholinergic response is potentiated and sustained. This is one of the strongest responses with maximum AP values of 10 times the basal. The ED_{50} should be 0.5 μM. A significant portion of the ACh plus IMX response will be blocked by H_2-receptor antagonists.

Gastrin plus IMX: Gastrin will not stimulate rabbit gastric glands by itself but depends on gastrin-induced endogenous release of histamine, the effect of which will be potentiated by IMX. The gastrin response can be totally blocked by H_2-receptor antagonists. Expect a doubling of the value for IMX alone ($ED_{50} = 1–3$ nM). In dog glands and cells a small but significant stimulation of AP accumulation is induced by gastrin itself. If you must use pentagastrin, increase the concentration 10-fold.

2-Mercaptoethanol: It appears that during the gland preparation some oxidation of sulfhydryl groups of receptors and enzymes takes place. A "bad" collagenase will make this effect worse. The presence of BME will

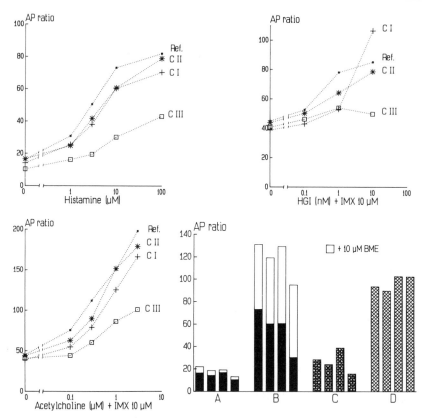

FIG. 1. Example of collagenase batch testing, with reference collagenase and three un-known collagenases (CI–CIII) run according to the batch testing protocol. Bar graph: A, Control + BME; B, histamine (10 μM) + BME; C, acetylcholine (1 μM) (15 min); and D, db-cAMP (1 mM) for reference and collagenase (CI–CIII). Abbreviations defined in text.

make the glands more responsive to most secretagogues (especially hista-mine). 2-Mercaptoethanol alone will normally increase the basal value by 20–40% and might double the histamine (10 μM) response. Do not use sulfhydryl reducing agents during the collagenase digestion since these will increase the activity of clostripain.

Preparation of Isolated Parietal Cells

Historically, preparation of isolated gastric and, in particular, parietal cells was the ultimate goal. Evidently this is a two-part task: (1) preparation of viable responsive gastric cells and (2) purification of parietal cells, since, depending on species, these constitute only 20–40% of the cell number in the gastric mucosa. In fact the latter point turned out to be the stumbling block.

Enzymatic Preparation of Gastric Cells. I had tried numerous different enzymes alone or in combination as well as the use of intermittent chelating steps to obtain a good cell preparation, but found only one technique reproducible and gentle enough for routine work.[2,8]

The starting point is exactly the same as for gland preparation, i.e., perfusion, mincing, composition of all solutions, etc., but as a first step the minced pieces are treated with Pronase.

Pronase can be obtained from a number of sources and with widely different specific activities. I would recommend Merck (Darmstadt, FRG) Pronase with an activity of around 30,000–70,000 U/g. This is, in contrast to most other Pronases, a whitish (not gray) powder. As usual, different concentrations and incubation times must be tested, but the aim should be for 0.5–1 mg/ml for 15 min. Digest with standard enzyme solution in an ordinary digestion flask; however, increase the rabbit albumin to 2 mg/ml. Fill the flask with oxygen and seal with Parafilm, run with a magnetic stirrer at 37°. After 15 min the mucosa pieces are still intact although somewhat more fluffy and there are isolated cells and some sheets of surface epithelial cells floating around. Now, wash the mucosa pieces thoroughly by letting them sediment in tubes at least three times with fresh plain enzyme solution and good resuspension. For the last wash, centrifuge at high speeds (2000 rpm) for 2 min and resuspend the mucosa pieces in regular collagenase enzyme solution, but, if possible, try to lower the collagenase amount by 25 or 50% (75–110 U/ml). Continue digestion of the pieces as usual. Very rapidly the mucosa will fall apart, almost entirely into isolated cells. The time needed for the second step will vary between 20 and 40 min. Watch carefully so that digestion time is minimized.

Dilute the digested cells three times with enzyme solution containing albumin and glucose, filter through a nylon cloth, and let the cells stand in 15-ml plastic test tubes for 5 min. With a fire-polished long-nosed Pasteur pipet remove the bottom sediment, which will contain undigested glands and larger fragments. Resuspend the cells and centrifuge at 700–1000 rpm (~200 g) for 3 min. Check carefully that most cells have been pelleted; if not, extend the centrifugation time, otherwise parietal cells will be lost. Resuspend and repeat the washing/centrifugation twice, using incubation medium with 2 mg/ml rabbit albumin and glucose.

A suspension has now been made of parietal cells, peptic cells, surface epithelial cells, and a small cell population, which are presumably endo-

[8] T. Berglindh, *Fed. Proc., Fed. Am. Soc. Exp. Biol.* **44**, 616 (1985).

crine-like cells. These can either be used directly or be purified in a single step density gradient centrifugation.

Cell Separation

In contrast to many other cell systems, it has been difficult to separate gastric cells, particularly to obtain a pure parietal cell fraction. We have tried several different density gradient methods, including Percoll and Ficoll, but with only limited success. Problems have included cell clumping, cell death, lack of reproducibility, osmotic shrinkage, and poor yields. The elutriation technique devised by Dr. Soll[5,8] will work, but it is time consuming and very difficult to obtain parietal cell fractions with a purity higher than 60%.

The Nycodenz One-Step Technique. Nycodenz (NYCO, Norway) was originally devised as an X-ray contrast chemical. It turned out, however, that it also had very interesting density properties. It is easy to obtain a density of 1.15 g/ml. It gives a low osmotic contribution, making isotonic solutions possible. It is nontoxic and easy to wash away and, not least important, it has a low viscosity, making the cells reach their isopycnic levels fast. To make 100 ml of separation medium, dissolve 27.6 g Nycodenz in 60 ml distilled water, add 1 ml KCl (540 mM), 1 ml MgSO$_4$ (120 mM), 5 ml Tris-HEPES (300 mM, pH 7.4), and 1 g bovine albumin. Adjust the pH to 7.4 with NaOH and add water to make the volume 100 ml. This medium will have a density of 1.15 g/ml. Mix this stock solution with incubation medium (containing 2 mg/ml albumin and 2 mg/ml glucose) to obtain the following densities: 1.10, 1.075, 1.05, and 1.0375 g/ml. In a 50-ml plastic centrifugation tube (with screwcap) put 10 ml 1.10 in the bottom, layer 20 ml 1.075 on top, followed by 10 ml 1.05. Resuspend 1–1.5 ml pelleted cells with 7 ml 1.0375 density medium and layer on top of the 1.05 layer. All this is done at room temperature. Centrifuge at 2500 rpm (approx 800 *g*) for 8–10 min.

Enrichment of parietal cells (80–95% pure) will be found on the 1.075 layer; the contaminants are mainly small cells. On the 1.10 layer enrichment of peptic cells with some contamination of parietal cells is obtained. At the bottom clumps and some peptic cells are seen. By increasing the density from 10.75 to 1.078 a slightly lower purity but an improved yield of parietal cells is obtained.

Parietal cells are easily distinguished using Nomarski optics or by staining with nitrotetrazolium blue, which will color the mitochondria blue. The latter technique is rapidly performed on air-dried cells on a microscope slide.[3] The separation method works for crude cells from both rabbit and dog and, in fact, with dog better than 90% purity was routinely obtained.

Typical AP accumulation ratios for these purified parietal cells are listed in Table I. The AP ratio in dog cells is 10–20 times lower than for rabbit except in response to carbachol. However, on a percentage basis, most responses are similar.

Determination of Parietal Cell Activity

A successful *in vitro* preparation is useless unless you find ways to communicate with it. In terms of the "classical" ions, Na^+, K^+, Cl^-, and Ca^{2+}, isotopes are the most common answer, but in regard to hydrogen ions (H^+) other techniques must be applied. The stimulation of H^+ formation in an isolated parietal cell would conceivably give rise to a transient change of pH in an unbuffered medium, but one must recall that for every molecule of H^+ secreted into the secretory channel an OH^- ion is released into the cytoplasm (fortunately the OH^- ion will not actually reach the cytoplasm, but will be converted to HCO_3^- by carbonic anhydrase bound to the cytoplasmic side of the apical membrane). Thus, equal amounts of acid and base will be released into the medium and hence no consistent pH change can take place. There could be a time separation in the acid and base appearance in the medium, but this could hardly be used as a routine method. Consequently there was a need for more unconventional methods to register parietal cell activity. To date, three successful approaches have been documented, i.e., oxygen consumption, CO_2 production, and weak base accumulation.

Oxygen Consumption. The parietal cell is one of the most mitochondria-rich cells of the body, a feature presumably indicative of a need for massive energy. Since acid secretion is the main task of this cell, energy utilization must be coupled to the formation and transport of H^+, and since all other cell types in a complex gastric gland have few mitochondria, any major change in oxygen consumption must originate from the parietal cell. Oxygen consumption can be measured either in a classical Warburg-

TABLE I
TYPICAL AP ACCUMULATION RATIOS IN PURIFIED RABBIT AND DOG CELLS[a]

Animal	Control	Histamine $(10 \mu M)$	Carbachol $(10 \mu M)$	Forskolin $(10 \mu M)$
Rabbit	79–169	380–1733	80–284	764–1402
Dog	9–18	55–80	80–127	85–144

[a]Based on six separate experiments with a purity of $83 \pm 6\%$ and $91 \pm 3\%$ for rabbit and dog, respectively.

type respirometer or using O_2 electrodes. Both systems have advantages and disadvantages.

Respirometer. We have used a Gilson system[3] where up to 20 flasks can be monitored simultaneously. CO_2 generated is absorbed in a center well by a filter paper wick bathed in 3 M KOH. Stimulators or inhibitors can conveniently be added from fixed side arms on the incubator flasks. Gland or cell sticking to the glass is minimized by careful siliconization prior to use. Ten milligrams (dry wt) of glands or cells in 3 ml medium can be run in each flask. Since the gas phase is huge, long-term studies can be performed with constant oxygen tension. Drawbacks of this system as follows: solutions should not contain HCO_3^-; response after secretagogue addition is somewhat sluggish; and the shaking motion is not optimal.

O_2 Electrodes. Smaller amounts of cells or glands can be used.[9] Response to change in oxygen consumption is rapid. All different kinds of media can be used. On the negative side are the following: (1) oxygen tension decreases with time and there is a build-up of CO_2, thus only short-term studies can be performed and a time-dependent secretagogue response study might be impossible; (2) it is difficult to add substances during a run; and (3) due to technical and cost considerations only a few preparations can be run in parallel.

CO_2 Generation. To measure the major metabolic end product has generally the same advantages as found in the respirometer, but is less dependent on a specific apparatus so long as the CO_2 can be trapped and analyzed. It is possible to use ^{14}C-labeled glucose[10] directly, but since the parietal cell appears to be able to utilize several different substrate types, a mixture of [^{14}C]glucose and, for example, [^{14}C]butyric or palmitic acid is recommended. An integrated device capable of delivering test substances to the cells without breaking the seal is advisable.

Weak Base Accumulation. The theory behind the use of lipid-permeable weak bases for measurement of acid formation or, more precisely, the concentration of sequestered acid in a known volume of intracellular space, has been dealt with in several other publications.[5,11] Briefly, an ideal weak base should be unprotonated and lipid soluble (high membrane penetration ability) at physiological pH. Once it enters a low pH space it should be protonated (positively charged) and lose its membrane-crossing properties. This will, according to the pH partition hypothesis,[12] lead to an accumulation which is in direct proportion to the pH (and volume) of the

[9] A. H. Soll, *J. Clin. Invest.* **61**, 370 (1978).
[10] W. D. Davidson, K. L. Klein, K. Kurokawa, and A. H. Soll, *Metabolism* **30**, 596 (1981).
[11] T. Berglindh, H. F. Helander and K. J. Öbrink, *Acta Physiol. Scand.* **97**, 401 (1976).
[12] P. A. Shore, B. B. Brodie, and C. A. M. Hogben, *J. Pharmacol.* **119**, 361 (1957).

acid space. In our search for such an "ideal" base we came across amino-pyrine (AP) (*N*,*N*-dimethyl antipyrine), which, in its unlabeled form, had been used for studies of gastric blood flow. We ordered a custom synthesis from NEN (Boston, MA) in 1973 and received a functioning product, but with rather low specific activity. The results, however, exceeded our wildest imagination. We got a basal AP accumulation which could be totally blocked by 10 m*M* NaSCN (a well known inhibitor of acid secretion). We could generate dose–response curves to histamine and other secretagogues and we could repeat the transient cholinergic stimulation seen in the oxygen consumption studies. Since then AP has been the golden standard for studies of parietal cell activity and has led to spin-offs using fluorescent weak bases for microscopy[13] and has as well been used as a principle for delivery systems of drugs to the parietal cell.[14]

Protocol for AP Accumulation Studies. The specific activity of AP is today very high, ~3.7 GBq/mmol (100 mCi/mmol). Aminopyrine is nor-mally shipped in sterile water at a concentration of 1 m*M*. To minimize internal radiation damage the concentration can be increased to 2–3 m*M* by adding high-quality unlabeled AP or by adding dimethyl sulfoxide (DMSO) to make a final concentration of 10%. Neither of these measures will affect the AP accumulation. Aminopyrine will deteriorate with time, which will give erroneously low AP ratios. This aspect has been studied in some detail[15] and there are two published thin-layer chromatography (TLC) methods showing how to check the purity of AP.[15,16] The method recommended by the manufacturer will not discriminate between genuine and metabolized AP.

Gland (or cell) concentration: In each incubation vial aim for 3–5 mg dry wt for glands and 2–3 mg dry wt for cells. Therefore start with 1–1.5 ml sedimented glands or 0.5 ml packed cells and make the volume 10 ml with incubation medium. Add 0.05 μCi AP/ml suspension.

General procedure: Prior to this 20-ml plastic scintillation vials (of the most hydrophobic type) with 1 ml medium, possibly including test sub-stances, should be prepared. Add 0.5 ml of well-stirred gland suspension, cover, and place in a 37° shaking or preferably rotating water bath (80–120 rpm). Incubate for the desired time (to reach steady state for most secretagogues like histamine, IMX, db-cAMP, forskolin, gastrin plus IMX, etc., 30 min is usually required but always prepare your own time–response curves).

[13] T. Berglindh, D. R. DiBona, S. Ito, and G. Sachs, *Am. J. Physiol.* **238**, G165 (1980).
[14] E. Fellenius, T. Berglindh, G. Sachs, L. Olbe, B. Elander, S. E. Sjöstrand, and B. Wallmark, *Nature (London)* **290**, 159 (1981).
[15] J. Sack and J. G. Spenney, *Am. J. Physiol.* **243**, G313 (1982).
[16] L. Holm-Rutili and T. Berglindh, *Am. J. Physiol.* **250**, G575 (1986).

Using a Pasteur pipet transfer the whole content of the vial to pre-weighed 1.5-ml Eppendorf test tubes (with lids). Pellet the cells or glands by centrifugation at 15,000 g for 15 sec in an Eppendorf-type centrifuge. Remove as much as possible of the supernatant and save it, put the lid on the tube with the pellet (to minimize evaporation of water), and weigh it to determine the wet weight of the pellet. To dry the pellet either put the tubes in a 60° oven for 12 hr or freeze the pellets and subsequently dry them in a freeze dryer. If the oven is used, be careful to test that the Eppendorf tubes do not lose weight due to volatile components in the plastic. If this is noted, cure the tubes for 24 hr at 60° before use. Weigh the tubes with the dry pellets and add 0.5–1.0 ml 1 M NaOH to dissolve the pellets by overnight incubation at 60°. Transfer the clear solution to scintillation vials and add 10 ml Dimilume, which is particularly suitable for alkaline solutions. Count an aliquot (200 μl) of the supernatant.

Calculation of AP accumulation: The best way to express the acid status of the parietal cell is to calculate the AP ratio, i.e., amount of AP in intracellular water (ICW)/amount AP in extracellular water (ECW). In order to do that the amount of intracellular water must be determined. Typically, this is done using an isotope labeled extracellular marker, such as inulin, combined with determination of wet and dry weights of the gland or cell pellet. Alternatively, a ^{14}C-labeled extracellular marker could be combined with tritiated water, in which case no weighing is needed since the ICW = total water (^{3}H) − ECW (^{14}C). The latter method has pre-viously given uncertain values since it involves double-isotope counting, but thanks to highly improved scintillation counters is now quite feasible. From a practical standpoint, however, it is very inconvenient to have to determine ICW for every single experiment and therefore we have devised a short cut. For isolated gastric glands, which are studied under normal conditions, we have from vast experience found that the ICW = 2 × pellet dry weight (in mg).[11] Among experiments this figure will vary only by ±10%. Exceptions to this will occur when the parietal cells are stimulated maximally with secretagogues such as db-cAMP, in which case the cells will swell, or if either nonisotonic medium or a medium containing imper-meable ions is used. A special case is the use of cytoskeletal disrupting agents, when measurements of ICW should always be done.

The ECW is, thus, total water (wet wt − dry wt) minus ICW (2 × dry wt). Use the ECW figure to subtract extracellular contribution to the total radioactivity in the pellet. Divide the pellet counts by ICW to obtain disintegrations per minute per microliter and divide that by the superna-tant counts (in dpm/μl). This is the AP ratio. To facilitate the calculations it is very easy to construct a spreadsheet, into which scintillation data can be imported directly as can data from the balance if it has a RS-232 interface.

There are two alternative methods of presenting AP accumulation data. In one, which is acceptable provided that pellet weights are exactly the same, a background vial containing glands with 10 mM NaSCN is subtracted from all pellet counts. Ten millimolar SCN$^-$ will give an AP ratio of 1, i.e., no accumulation. With this method no wet weights have to be determined.

The other method, which has appeared in the literature, is not acceptable since only the amount of AP in the pellets is presented. This "lazy man's method" does not take into account that AP actually is removed from the medium (sometimes less than 25% of the original amount remains) and thus the parietal cell activity is grossly underestimated.

What the AP Ratio Tells Us about pH in the Intracellular Channels. As presented, the AP accumulated is assumed to be evenly distributed in the intracellular water. Thus a ratio of 100 means an average pH of 3 (pK_a of AP is 5.0), a ratio of 1000 equals a pH of 2, etc.

However, the true space of acid accumulation is only a fraction of the total ICW, and varies a great deal from resting to stimulated cells, making it more or less impossible to predict or determine. However, taking this into account, we will undoubtedly have structures in the parietal cell where the pH approaches that found during maximal stimulation of the stomach, i.e., 0.8.

There is also a basic difference between how glands and isolated cells handle the accumulated AP. In glands, where parietal cells have direct contact with the lumen, AP will slowly diffuse out into the medium, lowering the maximum value. In isolated parietal cells, where the normal polarity found in the glands is lost, the secretory channel is closed off and internalized. On stimulation the acid will be trapped inside large vacuoles, which is the main reason for the very high AP ratios found in maximally stimulated purified parietal cells.

Conclusion

Isolated gastric gland preparations are today the standard for *in vitro* studies, being used for basic research and not only for aspects of acid secretion but also for questions related to the peptic cell, intrinsic factor secretion, and release of histamine. The glands are also used extensively by the pharmaceutical industry in the quest for improved therapeutic principles.

A final piece of advice from someone who has spent half a lifetime getting to know them: be gentle to the glands they will give you a lot of satisfaction back.

[7] Permeabilizing Parietal Cells

By Danuta H. Malinowska

During the last decade, a variety of methods have been developed for rendering the plasma membrane permeable, leaving intracellular organelles (e.g., mitochondria, secretory granules) and intracellular membranes (e.g., endoplasmic reticulum) intact. Irrespective of cell type, the objective of permeabilization has been to obtain access to the cell cytosol and to study mechanisms of cellular processes *in situ*. Access to the cell cytosol allows not only manipulation of intracellular conditions, but also the introduction of normally impermeable compounds. A permeable cell model is thus intermediate in complexity between the intact cell and isolated membranes or granules and enables a controlled study of intracellular mechanisms and intracellular processes, which are inaccessible in an intact system.

Methods of permeabilization have involved treatment of cells with reagents such as ATP,[1-3] lysophosphatidylcholine,[4] digitonin,[5,6] saponin,[7] filipin;[8] exposing cells to a high electric field, i.e., electric shock treatment,[9,10] and treatment of cells with hypotonic medium.[11] The two most widely used methods, electric shock treatment and the use of saponin or digitonin, differ in the size of "holes" produced in the membrane. Electric shock treatment allows equilibration of Ca^{2+}-EGTA buffers and entry of Mg-ATP and other small molecules with little release of large molecules such as the cytoplasmic enzyme, lactate dehydrogenase (M_r 130,000). In some cells, resealing occurs spontaneously after 1 hr. Treatment of cells with the detergents saponin or digitonin, thought to exert their effects by

[1] L. A. Heppel and N. Makan, *J. Supramol. Struct.* **6**. 399 (1977).

[2] B. D. Gomperts, *Nature (London)* **306,** 64 (1983).

[3] L. A. Heppel, G. A. Weisman, and I. Friedberg, *J. Membr. Biol.* **86,** 189 (1985).

[4] M. R. Miller, J. J. Castellot, Jr., and A. B. Pardee, *Exp. Cell Res.* **120,** 421 (1979).

[5] P. F. Zuurendonk and J. M. Tager, *Biochim. Biophys. Acta* **333,** 393 (1974).

[6] G. Fiskum, S. W. Craig, G. L. Decker, and A. L. Lehninger, *Proc. Natl. Acad. Sci. U.S.A.* **77,** 3430 (1980).

[7] E. G. Lapetina, S. P. Watson, and P. Cuatrecasas, *Proc. Natl. Acad. Sci. U.S.A.* **81,** 7431 (1984).

[8] H. S. Gankema, E. Laanen, A. K. Groen, and J. M. Tager, *Eur. J. Biochem.* **119,** 409 (1981).

[9] P. F. Baker and D. E. Knight, *Nature (London)* **276,** 620 (1978).

[10] D. E. Knight and P. F. Baker, *J. Membr. Biol.* **68,** 107 (1982).

[11] J. Seki, M. Lemahieu, and G. C. Mueller, *Biochim. Biophys. Acta* **378,** 333 (1975).

interaction with cholesterol in the membrane,[12] results in increases in the membrane permeability to inorganic ions, metabolites, and enzymes.[5,13] Resealing of digitonin-permeabilized cells has not been observed. Irrespective of the method used for permeabilization, it is important to remember that the cells have been artificially altered. Thus, the properties of a permeable cell model should be compared wherever possible with known properties of intact cells and isolated components.

To develop a permeable parietal cell model for studying mechanisms of acid secretion is particularly complex. The H^+,K^+-ATPase, responsible for acid secretion resides in the intracellular tubulovesicles when at rest and in the elaborated apical or secretory membrane following stimulation. Thus it is necessary to differentially permeabilize only the plasma membrane, leaving the intracellular tubulovesicles or secretory membrane intact. Acid secretion in isolated gastric glands or cells is monitored by the accumulation of the radioactive weak base, [^{14}C]aminopyrine, which measures the presence of intact acid spaces in parietal cells. Therefore, ability to accumulate weak base not only is a measure of acid secretory function, but also indicates whether the intracellular tubulovesicles or apical membrane are intact and not being permeabilized.

The first permeable parietal cell model was developed by Berglindh et al.[14] Electric shock treatment was applied to isolated rabbit gastric glands. Although some success was achieved with this technique and important new data were obtained, electron microscopy of shocked gastric glands revealed massive mitochondrial damage in parietal cells, as evidenced by swelling and vacuolation.[15] A method of permeabilizing parietal cells of gastric glands using the detergent, digitonin, was therefore developed initially only in unstimulated (resting) gastric glands.[16] Mitochondria were unaffected since they contain little cholesterol.[17] This technique was then successfully applied to gastric glands which were stimulated to secrete acid.[18,19] Both functional states of the parietal cell (resting and stimulated) were maintained after permeabilization as evidenced by measurement of acid secretion. Recently, digitonin was also used to permeabilize resting and stimulated isolated gastric cells (unpurified) as well as purified parietal

[12] T. Akiyama, S. Takagi, V. Sankawa, S. Inari, and H. Saito, *Biochemistry* **19**, 1904 (1980).

[13] W. P. Dubinsky and R. S. Cockrell, *FEBS Lett* **59**, 39 (1975).

[14] T. Berglindh, D. R. Dibona, C. S. Pace, and G. Sachs, *J. Cell Biol.* **85**, 392 (1980).

[15] D H. Malinowska and H. F. Helander, unpublished observations (1981).

[16] D. H. Malinowska, H. R. Koelz, S. J. Hersey, and G. Sachs, *Proc. Natl. Acad. Sci. U.S.A.* **78**, 5908 (1981).

[17] A. Colbean, J. Nachbaur, and P. M. Vignais, *Biochim. Biophys. Acta* **249**, 462 (1971).

[18] D. H. Malinowska, J. Cuppoletti, and G. Sachs, *Am. J. Physiol.* **245**, G573 (1983).

[19] S. J. Hersey and L. Steiner, *Am. J. Physiol.* **248**, G561 (1985).

cells, with retention of functional state.[20] A permeable, purified parietal cell model will be particularly useful in investigation of intracellular mechanisms involved in the regulation of acid secretion by the parietal cell, without interference from other cell types. Nevertheless, the permeable gastric gland model has been useful for investigation of some aspects of acid secretion,[16] as well as mechanisms of inhibition of acid secretion by a variety of inhibitors.[21]

This chapter will describe in detail the digitonin method of permeabilizing gastric parietal cells using isolated rabbit gastric glands and isolated unpurified rabbit gastric cells (which are both heterogeneous cell populations, containing mainly parietal and peptic cells) as well as purified parietal cells. Two methods of assessing the degree of permeabilization and a method to assess parietal cell functional state will also be described. Preparation of rabbit gastric glands, unpurified gastric cells, and purified parietal cells will be covered only briefly, since this is dealt with in Chapter 6.

Principle. Digitonin (M_r 1229) is a detergent which forms equimolecular complexes with free cholesterol. It increases the membrane permeability of various cell types to inorganic ions, metabolites, and enzymes.[5,13] This effect is thought to be due to the interaction of digitonin with cholesterol in the plasma membrane.[12] Mitochondria and endoplasmic reticulum are relatively unaffected because they contain little cholesterol compared to the plasma membrane.[17] However, the sites of acid secretion in the parietal cell (tubulovesicles in resting cells and the secretory membrane in stimulated cells) contain cholesterol: the molar cholesterol/phospholipid ratio is 0.48.[22] Although this value is about half of that measured in the plasma membrane, where the molar cholesterol/phospholipid ratio is 0.76,[17] great care must be taken to permeabilize only the plasma membrane.

Gastric Glands

The permeable gastric gland model, as developed for unstimulated glands[16] will be described in this section. The degree of permeabilization is assessed by measuring release of the cytoplasmic enzyme, lactate dehydrogenase (LDH). Acid secretory function is assessed by the ability of the permeable gastric glands to accumulate the weak base, [^{14}C]aminopyrine,

[20] D. H. Malinowska, unpublished data (1985, 1986).

[21] B. Wallmark, B.-M. Jaresten, H. Larsson, B. Ryberg, A. Brandstrom, and E. Fellenius, *Am. J. Physiol.* **245**, G64 (1983).

[22] J. J. Schrijen, A. Omachi, W. A. H. M. Van Groningen-Luyben, J. J. H. H. M. dePont, and S. L. Bonting, *Biochim. Biophys. Acta* **649**, 1 (1981).

on addition of ATP. In this model, prior to permeabilization, mitochondrial ATP synthesis and hydrolysis and Na^+,K^+-ATPase activity are inhibited with oligomycin and ouabain. Although not a prerequisite (see the section, Gastric Cells), this is done to abolish mitochondrial function and reduce acid secretory rates to very low levels.

Preparation of Gastric Glands. Gastric glands are prepared from rabbit gastric mucosa as described by Berglindh and Obrink.[23] Briefly, the rabbit is anesthetized and the stomach perfused under pressure with phosphate-buffered saline. The fundic mucosa is scraped off, minced, and incubated with collagenase (25 mg/ml) in enzyme medium[23] for 30–60 min. After filtering through a nylon mesh, the glands are washed several times in standard medium by resuspension and gravity sedimentation. The standard medium contains 132.4 mM NaCl, 5.4 mM KCl, 1 mM NaH_2PO_4, 5 mM Na_2HPO_4, 1.2 mM $MgSO_4$, 1 mM $CaCl_2$, 10 μg/ml Phenol Red, 2 mg/ml glucose, and 2 mg/ml rabbit albumin, pH 7.4.

Permeabilization Medium. The glands are washed three times in this medium prior to permeabilization. It resembles the cell cytosol with respect to ionic concentrations and contains 100 mM KCl, 20 mM NaCl, 1.2 mM $MgSO_4$, 5 mM Na_2HPO_4, 1 mM NaH_2PO_4, 20 mM HEPES/9.7 mM Tris (pH 7.4) and 10 μg/ml Phenol Red. The medium is Ca^{2+}, glucose and albumin free. A second medium can also be used which contains 60 mM KCl, 1.2 mM $MgSO_4$, 6 mM PO_4/12 mM Tris (pH 7.4), 20 mM HEPES/9.7 mM Tris (pH 7.4), 10 μg/ml Phenol Red, and 43 mM (tetramethylammonium)$_2SO_4$ to maintain an osmolarity of about 300 mOsm. This medium is Na^+ free, as well as Ca^{2+}, glucose, and albumin free. Intracellular Na^+ has been shown to exert inhibitory effects on acid secretion.[18,24]

Incubation and Permeabilization. Gastric glands (3 mg dry wt/ml medium) are incubated in permeabilization medium at 37° in a shaking water bath for 30 min with the H_2 blocker, cimetidine (10^{-4} M), to obtain unstimulated parietal cells and in the presence of oligomycin (10 μg/ml glandular suspension) and 10^{-4} M ouabain to inhibit mitochondrial function and Na^+,K^+-ATPase activity, respectively. Digitonin [stock: 10 mg/ml in dimethyl sulfoxide (DMSO)] is then added to the glands to give the desired final concentration. ATP (5 mM final concentration, pH 7.4) or medium is added immediately following the digitonin. Just prior to the experiment, a stock solution of 100 mM ATP is made up in permeabilization medium and the pH is adjusted to 7.4 with Tris. At timed intervals after digitonin and ATP addition, 1 ml of glands and medium is sampled

[23] T. Berglindh and K. J. Obrink, *Acta Physiol. Scand.* **96**, 150 (1976).
[24] H. R. Koelz, G. Sachs, and T. Berglindh, *Am. J. Physiol.* **241**, G431 (1981).

and rapidly spun down in an Eppendorf centrifuge. The supernatant medium is removed from the glandular pellet and kept for analysis. The glandular pellet is weighed, lyophilized, and reweighed to obtain wet and dry weights. Parallel experiments are performed for LDH release measurements and for acid secretion measurements using [^{14}C]aminopyrine.

Measurement of LDH Release. To quantitate LDH released from the gastric glands by digitonin treatment, LDH is measured in both the supernatant medium and in the glandular pellet. Lactate dehydrogenase present in the supernatant medium (i.e., LDH released) is then expressed as a percentage of total glandular LDH (i.e., $LDH_{supernatant} + LDH_{pellet}$). The LDH assay is performed directly on the supernatant medium either with or without dilution as necessary. To measure the LDH present in the glandular pellet, 5 ml H_2O is added to the lyophilized pellet, which is allowed to rehydrate for about an hour. The pellet is homogenized by pipetting up and down to release any LDH remaining in the glandular pellet into the water. After a brief centrifugation of the pellet, the LDH assay is performed on the pellet supernatant. The spectrophotometric assay using pyruvate and NADH is used to measure LDH.[25]

> Substrate/buffer: 6.2 mg sodium pyruvate, 700 mg K_2HPO_4, and 90 mg KH_2PO_4 are dissolved in 70 ml H_2O to give 0.63 mM pyruvate, 50 mM phosphate, pH 7.5
>
> NADH solution (ca. 11.3 mM): 10 mg NADH-Na$_2$ is dissolved in 1.071 ml NaHCO$_3$ (1 g/100 ml). This solution is freshly made up

Assay: At room temperature, 1 ml of substrate/buffer is mixed with 16.7 μl NADH solution in a cuvette. Sample (33.3 μl) is added to start the reaction. The OD_{340} is measured every 30 sec for 3 min. The OD_{340}/min is calculated in the linear range. After allowing for dilutions, LDH present in the medium is calculated as a percentage of the total glandular LDH. A typical experiment illustrating the effect of varying the concentration of digitonin on LDH release from unstimulated gastric glands is shown in Fig. 1 (similar results were observed with stimulated glands). As the digitonin concentration was increased, cytoplasmic LDH released from the glands also increased. Thus, at a concentration of 20 μg digitonin/ml glandular suspension, 90% of total cellular LDH was released. In parallel experiments, it is then essential to investigate acid secretory function at this concentration of digitonin.

Measurement of Acid Secretion. Acid secretion in parietal cells of gastric glands is measured as previously described[26] by the accumulation of

---

[25] H. U. Bergmeyer and E. Bernt, *in* "Methods in Enzymatic Analysis" (H. U. Bergmeyer, ed.), p. 575. Academic, New York, 1974.

[26] T. Berglindh, H. F. Helander, and K. J. Obrink, *Acta Physiol. Scand.* **97,** 401 (1976).

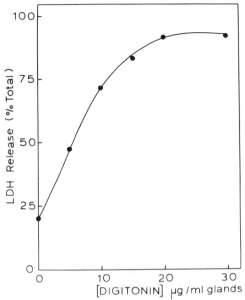

FIG. 1. Effect of digitonin on LDH release from unstimulated gastric glands. Digitonin dose dependently caused release of cytoplasmic LDH.

the radioactive weak base, [^{14}C]aminopyrine (AP). Prior to incubation, [^{14}C]AP (0.05 μCi/ml glandular suspension) is added to the glands. The incubation with cimetidine, oligomycin, and ouabain, digitonin permeabilization with and without addition of 5 mM ATP, and processing of samples is performed as described in the section *Incubation and Permeabilization*. Radioactivity present in the medium and in the glandular pellet is measured. The dried glandular pellet is dissolved in 1 M NaOH (0.5–1.0 ml) at 60°, added to 10 ml Dimilume 30 scintillation fluid (Packard, Illinois), and counted. One hundred microliters of the medium is added to 10 ml Aqueous Counting Scintillant (Amersham, Illinois) and counted. The ratio of [^{14}C]AP in intraglandular H$_2$O/[^{14}C]AP in the medium (or AP ratio) is calculated. Intraglandular H$_2$O has been calculated to be twice the dry weight of the pellet.[26] Figure 2 shows an experiment where acid secretory function (AP ratio) was measured in unstimulated gastric glands after treatment with digitonin (20 μg/ml glandular suspension) with and without 5 mM ATP. The glands were first incubated with cimetidine, oligomycin, and ouabain for 30 min, during which time the AP ratio slowly decreased from 40 to about 5. After the addition of digitonin and in the absence of ATP, the AP ratio further decreased to 2, i.e., acid secretion was abolished. However, when 5 mM ATP was added immediately following

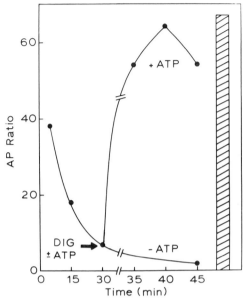

FIG. 2. Acid secretion, measured by [^{14}C]AP ratio, in unstimulated, digitonin (DIG)-permeabilized gastric glands with and without ATP. Glands were incubated with cimetidine (10^{-4} M), oligomycin (10 μg/ml glands), and ouabain (10^{-4} M) for 30 min. Digitonin (20 μg/ml glands) followed by ATP (5 mM) or medium was then added. Hatched column, control AP ratio.

the digitonin, the AP ratio increased to a maximum of about 60 after 10 min, a value similar to the AP ratio of unpermeabilized control gastric glands (hatched column). The maximum AP ratio was obtained with 5 mM ATP (data not shown).

Microscopic Appearance. The morphological appearance of digitonin-permeabilized glandular parietal cells is assessed by electron microscopy. The parietal cells responding to ATP are microscopically assessed using the metachromatic fluorescence shift of acridine orange as described by Dibona *et al.*[27] Acridine orange is a pH-sensitive fluorescent dye which accumulates in acid spaces. At low pH the green fluorescence of acridine orange at 530 nm is quenched and a red fluorescence appears at 624 nm.

Electron microscopy: Digitonin-treated gastric glands are fixed in 2% (v/v) glutaraldehyde in medium for 30–45 min and postfixed for 15 min in 1% osmium tetroxide. After dehydration, the glands are embedded in

[27] D. R. Dibona, S. Ito, T. Berglindh, and G. Sachs, *Proc. Natl. Acad. Sci. U.S.A.* **76**, 6689 (1979).

Spurr embedding medium and the sections are prepared. The nucleus, mitochondria, and tubulovesicles of unstimulated glandular parietal cells which had been incubated with oligomycin and ouabain for 30 min followed by digitonin (20 μg/ml glands) for 10 min appeared normal.[16] There was no vacuolization or swelling observed.

Fluorescent microscopy: Gastric glands are incubated with oligomycin and ouabain in the presence of 100 μM acridine orange. Digitonin (20 μg/ml glands) and ATP (5 mM) are added as described above. At timed intervals, the glands are observed with a fluorescent microscope. Without addition of ATP, only green fluorescence persisted in the parietal cells. However, addition of ATP resulted in increasing red fluorescence only in the parietal cells, indicating the development of acid spaces.[16]

Use of ATP-Regenerating Systems. In oligomycin- and ouabain-inhibited digitonin-permeabilized gastric glands, exogenously added ATP is the only available energy source for acid secretion. As ATP is utilized, the medium ADP concentration increases, which must cause a measure of inhibition of acid secretion. This has been confirmed by showing that addition of 1 mM ADP with 5 mM ATP results in 46% inhibition of the AP ratio.[20] To minimize inhibitory effects of ADP, an ATP-regenerating system can be added with the ATP. Table I shows results obtained with

TABLE I
EFFECT OF ATP-REGENERATING SYSTEMS ON
PERMEABLE RESTING GASTRIC GLANDS[a]

Agent	AP ratio
ATP (5 mM)	40
ATP (5 mM) CP (35 mM) CK (140 U/ml)	61
ATP (5 mM)	29
ATP (5 mM) PEP (2.5 mM) PK (50 U/ml)	53

[a] Conditions: 30-min pretreatment with cimetidine in the presence of oligomycin and ouabain; digitonin and ATP with and without regenerating systems were added and AP ratio was measured 10 min. later. Cimetidine, 10^{-4} M; oligomycin, 10 μg/ml; ouabain, 10^{-4} M; digitonin, 20 μg/ml glands. CP, Creatine phosphate; CK, creatine phosphokinase; PEP, phosphoenolpyruvate; PK, pyruvate kinase.

creatine phosphate : creatine phosphokinase (35 mM : 140 U/ml) and phosphoenolpyruvate : pyruvate kinase (2.5 mM : 50 U/ml) added to the permeable glands with 5 mM ATP. In both cases, the AP ratio was substantially higher in the presence of an ATP-regenerating system.

Stimulated versus Unstimulated Gastric Glands. The methods and data presented in this section have utilized an unstimulated or resting gastric gland preparation obtained by treatment of the glands with the H_2 blocker, cimetidine (10^{-4} M). A stimulated gastric gland preparation can also be used. If the glands are incubated with 10^{-3} M dibutyryl-cAMP for 30 min in the presence of oligomycin and ouabain, permeabilized with digitonin, and 5 mM ATP added as described, the AP ratio measured after 10 min with ATP was found to be double that measured in unstimulated glands as shown in Table II. Therefore, the two functional states of the parietal cell appear to be preserved after permeabilization, as reflected by AP ratios. Thus, the permeable gastric gland model may prove useful for some studies of the mechanisms involved in the regulation of acid secretion, as long as the parameters measured are parietal cell specific (i.e., [^{14}C]AP uptake), since gastric glands are a heterogeneous cell model.

Properties. The known properties of acid secretion in digitonin-permeabilized gastric glands are summarized in Table III. In some cases, the experiments could be performed only in a permeable system. For example, to investigate electrogenicity and the sole necessary energy source of the H^+ pump *in situ*, ionophores and anoxic conditions were used, respectively.[16] Such conditions and inhibitors could be used only in a system where mitochondrial function was bypassed. Permeable stimulated gastric glands retain higher rates of acid secretion compared to permeable unstimulated

TABLE II
STIMULATED VERSUS RESTING PERMEABLE
GASTRIC GLANDS[a]

Agent	AP ratio
Cimetidine (10^{-4} M)	30
dbcAMP (10^{-3} M)	60

[a] Conditions: 30-min pretreatment with cimetidine or dibutyryl cyclic AMP (dbcAMP) in the presence of oligomycin and ouabain; digitonin and ATP were added and AP ratio measured 10 min later. Oligomycin, 10 μg/ml; ouabain, 10^{-4} M; digitonin, 20 μg/ml glands; ATP, 5 mM.

TABLE III
PROPERTIES OF H^+ SECRETION IN DIGITONIN-PERMEABILIZED GASTRIC GLANDS

A. In unstimulated (resting) gastric glands[a]
  1. ATP is the sole energy source for H^+ secretion
  2. In oligomycin-inhibited glands, optimal ATP concentration for H^+ secretion is 5 mM
  3. H^+-secretion requires K^+ and Cl^-
  4. The H^+ pump is electroneutral (not electrogenic)
  5. K^+ and Cl^- conductances in the secretory membrane are low
  6. H^+ secretion is inhibited by
    a. Vanadate ($K_{0.5} = 10^{-6}$ M)[b]
    b. Omeprazole ($K_{0.5} = 2.7 \times 10^{-7}$ M)[c,d]
    c. Thiocyanate ($K_{0.5} = 8.4 \times 10^{-5}$ M)[c]
    d. Tetraethylammonium ($K_{0.5} = 2 \times 10^{-3}$ M)[e]
    e. Furosemide ($K_{0.5} = 5 \times 10^{-4}$ M)[e]
B. Prestimulation with dibutyryl-cAMP (10^{-3} M) or histamine (10^{-4} M) results in higher H^+-secretion compared to unstimulated (10^{-4} M cimetidine-treated) gastric glands[e,f]
C. Mitochondrial function is maintained and can be utilized *in situ* in the presence of mitochondrial substrates (10 mM succinate, 1 mM pyruvate) and low (1 mM) amounts of ATP or ADP to support H^+ secretion[a,e,f]

---

[a] Data from Malinowska *et al.*[16] unless otherwise stated.
[b] Data from L. D. Faller, D. H. Malinowska, E. Rabon, A. Smolka, and G. Sachs *in* "Membrane Biophysics: Structure and Function in Epithelia" p. 153. Alan R. Liss, New York, 1981.
[c] Data from Wallmark *et al.*[21]
[d] Data from E. Rabon, J. Cuppoletti, D. Malinowska, A. Smolka, H. F. Helander, J. Mendlein, and G. Sachs *J. Exp. Biol.* **106**, 119 (1983).
[e] Data from Malinowska,[20] and Malinowska *et al.*[18]
[f] Data from Hersey and Steiner.[19]

gastric glands. However, the properties of permeable stimulated gastric glands remain to be defined in greater detail. As briefly mentioned earlier, mitochondrial function in digitonin-permeabilized glands is maintained. This was originally suggested from experiments where it was found to be necessary to use oligomycin to reduce AP ratios to low levels.[16] Confirmation was obtained by Hersey and Steiner,[19] who showed that in the absence of oligomycin and in the presence of the mitochondrial substrates, succinate (10 mM) and pyruvate (1 mM), 1 mM ADP as well as 1 mM ATP supported acid secretion in digitonin-permeabilized gastric glands and these responses were inhibited by oligomycin or atractyloside. These results correlated well with unpublished observations that in the presence of oligomycin, 1 mM ATP was insufficient, but 5 mM ATP was sufficient to support acid secretion in digitonin-permeabilized gastric glands.[20]

Comments. Although the permeable gastric gland model has been a useful model for investigating some aspects of acid secretory mechanisms, it is a heterogeneous system, with peptic cells present as well as parietal cells, which curtails its use. However, the permeable gastric gland model has additionally been used to study mechanisms of pepsin secretion from the peptic cells.[28]

Gastric Cells

In order to expand and improve the permeable parietal cell model, the digitonin permeabilization technique has recently been applied to an unpurified gastric cell suspension[20] with the ultimate goal to apply it to purified parietal cells. Since the amount of material available after cell purification is small, the methodology was first worked out with an unpurified gastric cell suspension, which will be described in this section. A new method of assessing the degree of permeabilization of the cells using the change in fluorescence of ethidium bromide is described. Acid secretory function is measured by the uptake of [^{14}C]aminopyrine as described in the previous section.

Preparation of Gastric Cells. A crude, heterogeneous cell suspension is prepared from the rabbit gastric mucosa as described by Berglindh.[29] Briefly, the minced fundic mucosa is digested at 37° sequentially with Pronase (25 mg/ml) for 15 min and then with collagenase (25 mg/ml) for 30 min. The resulting cell suspension is filtered through a nylon mesh and washed several times in standard medium (for composition, see previous section) by centrifugation and resuspension. Parietal cells are purified from this crude gastric cell suspension by Nycodenz buoyant density centrifugation.[29]

Permeabilization Medium. The permeabilization medium used for cells was developed subsequent to more detailed knowledge of normal intracellular ion concentrations, as well as effects of altered intracellular ion concentrations.[18,24] The medium contains 80 mM KCl, 20 mM K$_2$SO$_4$, 5 mM Na$_2$HPO$_4$, 1 mM NaH$_2$PO$_4$, 1.2 mM MgSO$_4$, 10 mM HEPES/4.8 mM Tris, pH 7.4, 10 μg/ml Phenol Red, 10 mM disodium succinate, and 4 mM (tetramethylammonium)$_2$SO$_4$ to maintain osmolarity. Thus, the medium contains 120 mM K$^+$, 82 mM Cl$^-$, and 31 mM Na$^+$. Succinate is added as a mitochondrial substrate. The gastric cells are washed three times in this medium.

Incubation and Permeabilization. The crude gastric cells (1 mg dry

[28] S. H. Norris and S. J. Hersey, *Am. J. Physiol.* **249,** G408 (1985).
[29] T. Berglindh, *Fed. Proc., Fed. Am. Soc. Exp. Biol.* **44,** 616 (1985).

wt/ml medium) are incubated at 37° for 30 min with cimetidine (10^{-4} M) or histamine (10^{-4} M) to obtain unstimulated (resting) and stimulated parietal cells, respectively, and then permeabilized by the addition of digitonin. The degree of permeabilization is assessed fluorimetrically using ethidium bromide and acid secretion is monitored by measuring [^{14}C]AP uptake after the addition of 1 mM ATP (pH adjusted to 7.4 with Tris base), final concentration.

Ethidium Bromide Assay of Permeabilization. The method described by Gomperts[2] for mast cells is used.

Principle: Ethidium bromide (M_r 394.3) is a nuclear stain, which is normally impermeant. Upon contact with nucleic acid, it fluoresces. Its excitation wavelength is 365 nm and the fluorescence emission wavelength is 595 nm.

Assay: A baseline fluorescence trace is first obtained with 750 μl permeabilization medium containing 50 μM ethidium bromide in a cuvette maintained at 37° and provided with a magnetic stirrer. Intact cells (750 μl) are then added and the increase in fluoresence followed using a Perkin Elmer MP44 spectrofluorimeter (excitation wavelength, 365 nm; emission wavelength, 600 nm, found to give the maximum signal). After equilibration has occurred with intact cells (about 5 min), digitonin is added and the increase in fluoresence followed. Figure 3 shows an experiment using histamine-stimulated gastric cells. There was a small increase in fluorescence upon addition of cells, prior to digitonin addition, indicating that some of the cells were leaky. (Such a measurement on intact cells can be used as a viability test). Addition of 1 μg digitonin/ml cell suspension had no effect. However, as digitonin was progressively increased above 1 μg/ml cells, the fluorescence increased in steps proportional to the digitonin. With each effective digitonin addition, there was a rapid fluorescence increase followed by a slower fluorescence increase which plateaued after about 5 min. Also shown is the effect of one large dose of digitonin (10 μg/ml cells) added to the intact cells. ATP either in the absence or presence of digitonin had no permeabilizing effect, since there was no observed increase of fluorescence of ethidium bromide (data not shown). Maximum or 100% permeabilization is measured on every sample by adding large doses of digitonin (e.g., 100 μg/ml cells) at the end of the experiment and observing no further fluorescence change. Figure 4 shows the effect of digitonin on permeabilization, expressed as percentage permeabilization in 5 min. The degree of permeabilization increased with increasing digitonin concentration. Thus, at 10 μg digitonin/ml cells, the gastric cells were about 70% permeable. The dose–response curve was similar whether resting or stimulated cells were used.

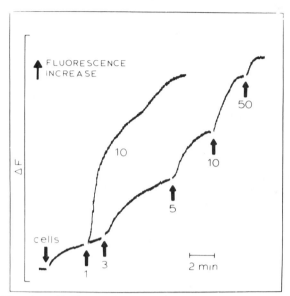

FIG. 3. Ethidium bromide fluorescence assay of digitonin permeabilization of stimulated gastric cells. Digitonin (1, 3, 5, 10, and 50 µg/ml cells) was added as indicated. Increase in fluorescence indicates increasing permeabilization. Ethidium bromide, 25 μM; excitation wavelength, 365 nm; emission wavelength, 600 nm. Cells were pretreated with 10^{-4} M histamine for 30 min.

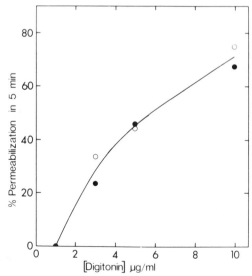

FIG. 4. Effect of digitonin on ethidium bromide uptake by gastric cells, expressed as percentage permeabilization in 5 min. The cells were pretreated with 10^{-4} M cimetidine (O) or 10^{-4} M histamine (●) for 30 min prior to permeabilization.

Comments: The ethidium bromide method of assessing the degree of permeabilization is simple and convenient. It is less time consuming and uses less cellular material than the LDH release method. It does not, however, give any indication of the size of "holes" produced by digitonin in the membrane, other than the fact that ethidium bromide of M_r 394.3 enters the cells and fluoresces on interaction with cellular nucleic acid. Although this method can be used with gastric glands, the bulky glands result in "noisy" fluorimetric traces. The ethidium bromide method is useful for rapid quantitation of permeabilization, while the LDH release method confirms that large molecules can pass through the "holes" produced.

Measurement of Acid Secretion. Uptake of [^{14}C]aminopyrine (as described under the section, Gastric Glands) is used. In these experiments, digitonin is allowed to act for 5 min prior to the addition of 1 mM ATP (pH 7.4) or medium. Figure 5 shows the effect of varying concentrations of digitonin on the AP ratio measured with and without ATP for 10 min in resting and stimulated gastric cells. This experiment was performed in parallel to the one shown in Figs. 3 and 4. In intact cells, the AP ratios after a 30-min incubation with 10^{-4} M cimetidine and 10^{-4} M histamine were 70 and 350, respectively. In the absence of ATP, increasing amounts of digitonin resulted in decreasing AP ratios. With ATP, maximum AP ratios (120, resting cells; 400, stimulated cells) were observed at digitonin concentrations of 5 and 10 μg/ml cells. The greatest difference between the AP ratio with and without ATP (ΔATP AP ratio) was, however, observed at 10 μg digitonin/ml cells. At this concentration of digitonin, the cells were 70% permeable (see Fig. 4). Higher digitonin concentrations resulted in lower AP ratios with ATP, suggesting that the tubulovesicles or secretory membrane were also being permeabilized (data not shown).

Figure 6 shows the time course of the ATP effect on acid secretion when 10 μg digitonin/ml cells is used. The AP ratio in resting cells was maintained at about 120 for 20 min with ATP, while the AP ratio in stimulated cells transiently increased to nearly 400 at 10 min and then was maintained at about 250–200. Thus the two functional states of parietal cells appear to be maintained after permeabilization.

A similar experiment, performed on purified parietal cells is shown in Fig. 7. The parietal cells were 80% pure and the ΔATP AP ratio is shown. The absolute values for the AP ratios following ATP addition were 24 for resting cells and 161 for stimulated cells. The sixfold higher ΔATP AP ratio measured in stimulated cells as compared to resting cells confirmed maintenance of the two functional states after digitonin permeabilization. Thus, the digitonin permeabilization method can also be successfully applied to purified parietal cells.

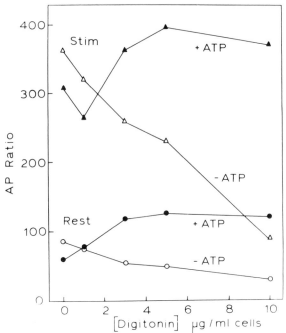

FIG. 5. Effect of digitonin with and without ATP on the AP ratio in resting and stimulated gastric cells. The cells were pretreated with cimetidine (10^{-4} M) or histamine (10^{-4} M) for 30 min, then digitonin was added for 5 min, followed by ATP (1 mM) or medium for 10 min.

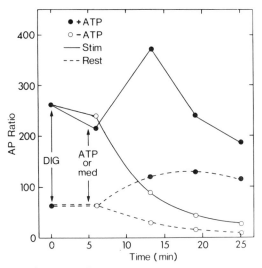

FIG. 6. Time course of the ATP effect on the AP ratio of resting and stimulated, digitonin-permeabilized gastric cells. The cells were pretreated with cimetidine (10^{-4} M) or histamine (10^{-4} M) for 30 min. Digitonin (10 μg/ml cells) was added, followed by ATP (1 mM) or medium 5 min later.

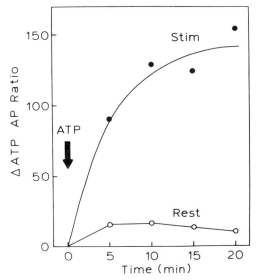

FIG. 7. Time course of the ATP effect on the AP ratio of resting (O) and stimulated (●), digitonin-permeabilized parietal cells (80% pure). ΔATP AP ratio is the difference in the AP ratio measured with and without ATP. Parietal cells were pretreated with cimetidine (10^{-4} M) or histamine (10^{-4} M) for 30 min. Digitonin (10 μg/ml cells) was then added for 5 min, followed by 1 mM ATP or medium. Absolute AP ratios with ATP were 24 and 161 for resting and stimulated cells, respectively.

Conclusions

Digitonin-permeabilized purified parietal cells, which retain their functional state, will prove to be a useful model to study cellular mechanisms regulating acid secretion as well as properties of the H^+,K^+-ATPase *in situ*, without interference from other cell types. The digitonin (or saponin) permeabilization method is being applied to an increasing variety of cell types, mainly to study mechanisms of exocytosis from secretory granules, with particular interest in Ca^{2+} and inositol 1,4,5-trisphosphate (IP_3)-mediated mechanisms.[30,31] However, some caution is advisable, particularly in light of a recent report that IP_3 caused Ca^{2+} release in saponin-permeabilized platelets, but not in platelets permeabilized by high-voltage discharge.[32] Recently, digitonin was successfully used to permeabilize the

[30] E. G. Lapetina, J. Silio, and M. Ruggiero, *J. Biol. Chem.* **260,** 7078 (1985).
[31] S. A. Lee and R. W. Holz, *J. Biol. Chem.* **261,** 17089 (1986).
[32] A. R. L. Gear and T. J. Hallam, *Fed. Proc., Fed. Am. Soc. Exp. Biol.* **45,** 225 (1986).

apical or mucosal membrane of colonic epithelial sheets, which were mounted in Ussing chambers,[33] allowing the study of the properties of the K^+ and Cl^- conductances of the basal lateral membrane. Thus, digitonin permeabilization is becoming a widely used and important technique in studies of intracellular mechanisms and membrane transport.

[33] D. Chang and D. C. Dawson, *Fed. Proc., Fed. Am. Soc. Exp. Biol.* **45**, 512 (1986).

[8] Pepsinogen Secretion in Vitro

By STEPHEN HERSEY, LAURA TANG, JAN POHL, and MELISSA MILLER

Introduction

Pepsinogen is a proenzyme, or zymogen, which is synthesized, stored, and secreted by mammalian gastric chief cells.[1] Under acid conditions (pH < 5), pepsinogen undergoes an autocatalytic cleavage of an NH_2-terminal peptide to generate the active enzyme, pepsin.[2] Pepsin belongs to the class of aspartyl proteases which includes renin, cathepsin D and E, microbial pepsins, and the human immunodeficiency virus (HIV)-1 protease.[3,4] Pepsins are distinguished from other members of this class by having an unusually low pH optimum of approximately 2.5. Although pepsinogen itself is stable to mildly alkaline conditions, the activated enzyme becomes irreversibly denatured by brief exposure (10–15 min) to pH above 7.5. Several pepsin isozymes have been identified[4] but these share the major characteristics of acid activation, low pH optimum, and alkali denaturation. Pepsin has a fairly broad substrate specificity, but preferentially attacks peptide bonds formed by aromatic residues.[5]

Based on the restricted biological distribution of pepsinogen and the unique enzymatic properties of pepsin, it has been possible to develop fairly simple procedures for the *in vitro* assay of pepsinogen secretion. The present chapter covers specific procedures used to measure pepsinogen secretion by gastric glands isolated from rabbit. However, most of the

[1] J. N. Langley, *J. Physiol. (London)* **3**, 269 (1881).
[2] T. Kageyama, *Eur. J. Biochem.* **176**, 543 (1988); M. N. G. James and A. R. Sielecki, *Nature (London)* **319**, 33 (1986).
[3] N. S. Andreeva, *Mol. Biol. (Moscow)* **19**, 185 (1985); V. Kostka (ed.), "Aspartic Proteinases and Their Inhibitors." de Gruyter, Berlin, 1985.
[4] V. Richmond, J. Tang, S. Wolf, R. E. Truco, and R. Caputto, *Biochim. Biophys. Acta* **29**, 453 (1958).
[5] J. S. Fruton, *Adv. Enzymol.* **44**, 1 (1976); J. Pohl, M. Baudys, and V. Kostka, *Anal. Biochem.* **133**, 104 (1983).

procedures may be applied, with little modification, to assays with other preparations. In addition, the general principles may be applied to the study of secretion of other enzymes or macromolecules.

Biological Preparation

Because the stomach is the major source of pepsinogen, *in vitro* preparations for studying secretion are derived from this tissue. A variety of *in vitro* models have been employed ranging from isolated, intact gastric mucosa to suspensions of purified chief cells. Depending on the experimental questions, one or more of these preparations would prove suitable. However, for routine measurement of pepsinogen secretion, the simplest and best characterized preparation is that of isolated gastric glands.[6] Although gastric glands contain multiple cell types, only the chief cells secrete pepsinogen. Therefore, it is unnecessary to further purify the chief cells in order to measure pepsinogen secretion. The procedure for obtaining gastric glands from rabbit is described and may be adapted easily to other species.

Principle

The mammalian gastric mucosa consists of numerous gastric glands containing about 50% chief cells. Several gastric glands empty into a single gastric pit, which is an infolding of the surface epithelium lined with mucus-secreting cells. A submucosal layer, consisting of connective tissue, smooth muscle, blood vessels, and nerve endings, underlies the gastric glands. The gastric glands are isolated from other tissue elements by a controlled enzymatic digestion using collagenase. During this procedure the submucosa is digested away and the surface epithelial cells are released. However, the gastric glands retain their junctional complexes and are released essentially intact. Since the intact glands are denser than isolated cells or other digestion products, they may be isolated simply by allowing them to settle in collection tubes.

Procedure

Young (2–3 kg) New Zealand White rabbits are preferred as the yield of gastric glands is near optimal. The animal is anesthetized with Nembutal (iv, 50 mg/kg) and the stomach exposed by a midline incision. With rabbits, and larger animals, the stomach is perfused retrograde under high pressure (200 mmHg) in order to create an edematous region in the submucosa. This is helpful in removing the mucosa and allowing the digestive

---

[6] T. Berglindh and K. J. Obrink, *Acta Physiol. Scand.* **96,** 150 (1976).

enzymes access to the gastric cells. With smaller animals (rats, mice, etc.) the perfusion step usually is not performed.

When perfusion is performed, a polyethylene cannula is placed in the abdominal aorta. The cannula is fitted with a three-way stopcock to allow injections and connection to a peristaltic perfusion pump. The pump should be capable of perfusion at 100 ml/min. Heparin (500 U/kg) is injected via the stopcock and allowed to circulate for 2–3 min. The perfusion is then initiated at 20–30 ml/min and, immediately, the thoracic cavity is opened and the descending aorta clamped just anterior to the diaphragm. Perfusion of the intestine is prevented by clamping the splanchnic vessels and the perfusion rate increased to 60–70 ml/min. At this point obvious perfusion of the stomach should be evident. To prevent back pressure, small cuts are made in the liver. Perfusion is continued until the surface of the stomach is cleared. This typically requires 600–800 ml and may be assisted by gentle massaging of the stomach. The perfusion solution is a phosphate-buffered saline (PBS; 0.9% NaCl, 6 mM NaPO$_4$, pH 7.2) which is prewarmed to 37° and saturated with 100% oxygen.

The stomach is removed and cut open along the lesser curvature. Any stomach contents are emptied out and the mucosa is rinsed extensively with PBS. Excess mucus or food particles can be wiped from the surface. The antral portion of the stomach is cut away and discarded. The fundic region is placed on a glass plate and the mucosal layer is scraped from the outer muscle layer using a blunt spatula. The scrapings are collected into a beaker and suspended in PBS (50 ml). The mucosal scrapings are then minced to a fine texture with scissors. This is a crucial step in the procedure since failure to mince the tissue adequately results in poor digestion and low yields. The minced tissue should have a uniform appearance with pieces less than 1 mm. The minced tissue is allowed to settle and excess PBS is decanted. The average yield of minced tissue should be 20–25 ml from a single rabbit.

The minced tissue (20–25 ml settled) is poured into a round-bottom digestion flask containing 50 ml of digestion medium. The digestion medium consists of a basic salt solution (132 mM NaCl, 5.4 mM KCl, 1.2 mM MgSO$_4$, 1.0 mM CaCl$_2$, 6 mM NaPO$_4$, pH 7.2) to which is added 2 mg/ml bovine serum albumin, 2 mg/ml glucose, 10 μg/ml Phenol Red, 1 mM pyruvate, and 0.5–1.0 mg/ml collagenase. The collagenase concentration should correspond to approximately 150 U/ml. (One unit releases 1.0 μmol of leucine from collagen in 5 hr at pH 7.4, 37°.) The flask is placed in a 37° water bath fitted with a magnetic stirrer and stirred at a moderate rate. The flask is continuously gassed with 100% oxygen and careful attention is paid to maintaining the pH at 7.2–7.4. In some procedures the digestion medium is renewed after about 15 min, but we have routinely continued the entire digestion in the original medium. Typically, the digestion requires 30–60 min, depending on the amount of tissue and the activity of the collagenase preparation. The digestion is monitored by taking samples and examining them for large pieces of tissue and the appearance of free glands. When the digestion is judged to be complete, the suspension is diluted 1:1 with an incubation medium consisting of the basic salt solution plus 1 mg/ml bovine serum albumin (BSA), 1 mg/ml glucose, 1 mM pyruvate, and 0.5 mM dithiothreitol. The diluted suspension is poured through a nylon mesh (200 μm) to remove large pieces and collected into a series of 15-ml conical, graduated centrifuge tubes. The suspension is allowed to settle for 15 min initially. During this time, the glands settle to the bottom, forming an identifiable band of 2–5 ml with a cloudy but less dense upper phase. The upper phase, which contains free cells and debris, is removed by suction and the glands resuspended in fresh incubation medium. The glands are allowed to settle for 2 additional 10-min periods with resuspension in fresh medium. At this point, the glands may be suspended at the desired final dilution in a medium appropriate for the experimental conditions. Using the gland volume obtained after a 10-min settling period, a final dilution of 1:30 (v/v) gives a reproducible and adequate suspension for pepsinogen secretion assays. A more accurate procedure is to place an aliquot (0.5 ml) of gland suspension in a tared microcentrifuge tube, centrifuge at 10,000 g for 15 sec, carefully remove the supernatant, and estimate the wet weight. With this method, a final suspension containing 10 mg wet wt/ml is satisfactory. The average yield of gastric glands from a single rabbit is 8–10 ml of settled material.

During the gland preparation procedure, there are a few additional points which should be noted. A minor, though important, point is that all solutions should be at room temperature or 37° since rapid temperature changes (particularly cooling) can result in damage to the chief cells. A second point is that the glands should be used as soon as possible after preparation but if stored for up to 3 hr should be diluted and well oxygenated. The most important point is that the yield and quality of the glands depends critically on the batch of collagenase used for the digestion. Purified collagenase does not work since it appears that contaminating proteases are essential for adequate digestion. As yet, no one has defined the exact composition of a suitable collagenase preparation. Instead, investigators must screen various batches of enzyme until a suitable preparation is found. To assist in this process, some suppliers provide a testing program in which a sample (100 mg) is provided for screening and larger amounts are placed on reserve for purchase should the batch prove suitable. It is suggested that one start with a preparation containing approximately

400 U/mg collagenase activity, 100 U/mg caseinase activity, and less than 0.1 U/mg tryptic activity.

Secretion of Pepsinogen

Principle

The secretion of pepsinogen results in the appearance of the macromolecule in the bathing medium. The medium is separated from the secreting cells and assayed for pepsinogen content. Several procedures can be used to distinguish between true secretion and nonspecific release of pepsinogen, e.g., due to cell damage.

Procedure

The routine assay of pepsinogen secretion by gastric glands is performed in a shaking waterbath at 37°. The glands are incubated at a final dilution of about 10 mg wet wt/ml. This dilution is based on the observation that 10 mg wet wt of glands contains approximately 200 μg of pepsinogen. Since, with maximal stimulation, the glands secrete 10–20% of the initial content of pepsinogen during a 30-min incubation, it may be anticipated that the pepsinogen concentration in the medium would reach 20–40 μg/ml. This allows for an assay (see below) using 25–50 μl of medium to fall within the standard range. Thus, for other tissue preparations, one should attempt to incubate a suspension containing approximately 200 μg pepsinogen/ml. The total pepsinogen content of the tissue may be estimated easily by incubating an aliquot in the presence of 0.5% Triton X-100 for 10 min at room temperature. This releases all of the pepsinogen into the medium. After centrifuging (10,000 g, 15 sec) the sample, the medium is assayed for pepsinogen. Usually, the sample must be diluted at least 1:10 in order to remain within the range of standards. Alternatively, the pepsinogen may be released by homogenization in hypotonic medium (10 mM phosphate buffer) or by repetitive freezing and thawing. Once the preparation has been characterized for average pepsinogen content, a suitable dilution can be estimated. The same procedures for estimating total pepsinogen content are used also as a normalizing factor to express the rate of pepsinogen secretion.

A variety of vessels are suitable for tissue incubation during the secretion assay and the investigator should choose one appropriate to the tissue sample. For assays using tissue similar to gastric glands or isolated cells a convenient procedure is to incubate 0.25–0.50 ml of tissue suspension in 1.5-ml microcentrifuge tubes. The tubes are oriented horizontally in a tube rack and incubated submerged in a 37° shaking water bath. A minimum shaking rate of 80–100/min should be employed to achieve adequate mixing and oxygenation. Failure to maintain adequate oxygenation leads to tissue damage and thus high basal release of pepsinogen with reduced responses to stimuli.

Using gastric glands or cells, a typical assay procedure, which may be used to screen different lots of collagenase, involves the following set-up. Pairs of microcentrifuge tubes are labeled for assay in duplicate. Test agents are prepared as 100-fold concentrated stocks and 5 μl added to the appropriate tubes. One pair of tubes receives 2.5 μl Triton X-100 and is used to estimate the total pepsinogen content. One pair of tubes is designated as an initial or zero time sample and is used to measure the pepsinogen content of the medium at the start of incubation. Additional tubes are designated for control samples (no addition or vehicle only) and test agents. Since it is known that pepsinogen secretion is stimulated by two general classes of agents, one acting via cAMP and one acting via intracellular calcium, typical test agents should include forskolin ($10^{-5} M$) or isoproterenol ($10^{-5} M$) for cAMP-mediated, and carbachol ($10^{-4} M$) or cholecystokinin octapeptide ($10^{-8} M$) for calcium-mediated stimulation.[7] After appropriate additions to the tubes, 0.5 ml of tissue suspension is added to each tube. The zero time samples are centrifuged (10,000 g, 15 sec) and the supernatant sampled for assay. The tubes containing Triton X-100 are set aside for incubation at room temperature. The remaining tubes are placed in the water bath and incubated for 20–30 min. Following the incubation period the tubes, including the Triton X-100 samples, are centrifuged (10,000 g, 15 sec) and the supernatants sampled for assay. The assay samples may be obtained by decanting supernatant into culture tubes and sampling an appropriate aliquot later, or aliquots may be pipetted directly into a set of microcentrifuge tubes for immediate assay.

The rate of pepsinogen secretion has been expressed in a variety of ways. In all cases the zero time value should be subtracted to obtain the amount of pepsinogen secreted during the incubation period. Since the rate of secretion is generally not constant with time, i.e., there is an initial rapid rate followed by a slower rate, the secretion should be expressed for the actual incubation time unless a time course is measured directly. As illustrated in Fig. 1, the time course of pepsinogen secretion may vary with different stimuli. Thus, it is essential to perform a time course experiment when comparing the efficacy of various stimuli. The amount of pepsinogen

[7] S. J. Hersey, "Physiology of the Gastrointestinal Tract, Second Edition" (L. R. Johnson, J. Christensen, M. Jackson, E. D. Jacobson, and J. H. Walsh, eds.), p. 947. Raven, New York, 1987.

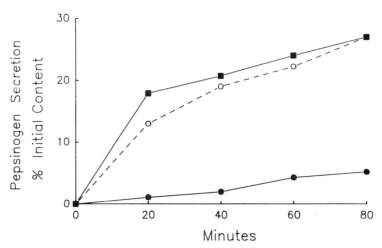

FIG. 1. Pepsinogen secretion by rabbit isolated gastric glands. Pepsinogen released into the medium is expressed as a percentage of initial gland content estimated from samples treated with Triton X-100. Time course typically shows an initial rapid phase followed by a slow continued release. Control (●); CCK (○); forskolin (■).

released may be expressed as micrograms equivalent to a standard preparation or, more usually, as a percentage of the total pepsinogen content using the Triton X-100 samples as a normalizing factor. The latter method allows for differences in pepsinogen content and amount of tissue between preparations without necessitating additional estimates, e.g., tissue weight or protein content. In some instances the stimulation of secretion is expressed as a percentage of control release; however, this manipulation of the data should be avoided unless the actual value for control release is provided.

As part of establishing a secretory assay or testing different experimental variables it is necessary to show that the appearance of pepsinogen in the medium is due to secretion rather than nonspecific release, e.g., with cell damage. Several procedures are available for distinguishing between these possibilities. The simplest test is to examine the control values. Typically, the control release is on the order of 1–4% of total pepsinogen content/30 min and is found to be relatively constant with time. If the control release is excessive, tissue damage must be suspected. A more definitive test for nonspecific release is to measure the appearance in the medium of a nonsecreted material. For this purpose, the cytosolic enzyme, lactic dehydrogenase, is a suitable and easily measured marker. Other enzymes or nonpermeant metabolic intermediates would be suitable and the choice made depending upon the ease of measurement.

Assay of Pepsinogen

Principle

A simple and selective assay for pepsinogen involves conversion of the proenzyme to the active protease, pepsin, under acid conditions and measurement of the proteolytic activity at pH 2.0.[8] The sensitivity of the assay is enhanced by reacting the trichloroacetic acid (TCA)-soluble products of proteolysis with chromogenic agents.

Procedure

A routine assay for pepsin activity uses hemoglobin as a substrate at pH 2.0. The following solutions are prepared in advance.

1% hemoglobin: Dissolve 1 g bovine hemoglobin in 100 ml 0.1 N HCl. This solution requires vigorous stirring for 1–2 hr and when decanted may still leave a residue. The solution may be aliquoted and stored frozen for several months

10% Trichloroacetic acid (TCA): A 50% (w/v) solution is prepared and dilutions made as required

0.5 N NaOH: Prepared as needed by dilution of a 2.0 N NaOH solution

The assay is performed in a constant temperature bath, usually at 37°. Aliquots of incubation supernatant (25–50 μl) containing 0.1–2.0 μg pepsinogen are placed in 1.5-ml microcentrifuge tubes. The tubes are placed in the assay bath and 450 μl of hemoglobin solution (prewarmed to assay temperature) is added at 10-sec intervals. After 10 min, the reaction is stopped by addition of 500 μl 10% TCA. The tubes are placed in a microcentrifuge and spun at 10,000 g for 3 min. One hundred microliters of the clear supernatant is transferred to a 3-ml culture tube and 100 μl of 0.5 N NaOH is added. The alkalinized supernatant is treated with a colorimetric reagent and the absorbance read in a spectrophotometer.

In earlier versions of this assay, the supernatant following TCA precipitation was read directly in a spectrophotometer at 275 nm. However, it is found that many of the agents employed for tissue incubations, e.g., peptide hormones, cyclic nucleotides, etc., absorb at this wavelength and, thus, interfere with the assay. For this reason, it is desirable to react the TCA-soluble products with a colorimetric reagent. Two such reagents have

[8] M. L. Anson and A. E. Mirsky, *J. Gen. Physiol.* **16,** 59 (1932).

proved successful in our laboratory, the Folin phenol reagent and a bicin-choninic acid reagent. Both reagents are based on the biuret reaction in which proteins or peptides convert cupric ions to cuprous ions under alkaline conditions. The reagent then reacts with the cuprous ion to gener-ate a complex which absorbs at a visible wavelength. Specific procedures for these two reagents are as follows.

Folin Phenol Reagent. A cupric ion reagent is prepared just prior to use by mixing: 0.5% $CuSO_4$, 1% sodium potassium tartrate, and 2% $NaCO_3$ dissolved in 0.1 N NaOH at a ratio of 1:1:48 (v/v/v). The individual solutions can be prepared in advance and stored for 1–2 weeks at 5° but the reagent should be prepared fresh.

One milliliter of the cupric ion reagent is added to the culture tube containing 0.2 ml of alkalinized TCA-supernatant. After 10 min at room temperature, 100 μl of Folin phenol reagent is added, the sample is vor-texed quickly, and the absorbance read in a spectrophotometer at 750 nm after 30 min, using H_2O as a reference. The Folin phenol reagent is avail-able commercially (e.g., from Sigma Chemical Co., St. Louis, MO) and usually is diluted 1:1 with H_2O prior to use in the assay.

Bicinchoninic Acid Reagent. This reagent is available commercially from Pierce (Rockford, IL) as a protein assay reagent. We have used this reagent with a slight modification of the supplier's procedure. The reagent is supplied in two parts, a cupric ion solution and an alkaline bicinchoninic acid solution. The two parts are mixed just prior to use at a ratio of 1:50 as instructed. One milliliter of the reagent is added to the culture tube con-taining 0.2 ml of the alkalinized TCA-supernatant and the sample is vor-texed quickly. The sample is incubated at 37° for 15 min and then allowed to cool to room temperature for 15 min. The sample absorbance is read in a spectrophotometer at 562 nm, using H_2O as a reference.

With either of the colorimetric procedures, the pepsinogen content of the sample is estimated from a standard curve prepared from commercially available purified pepsinogen. Typical standard curves for the two colori-metric procedures are presented in Fig. 2. The sensitivity of the assay is similar for both procedures. Under standard conditions, hemoglobin sub-strate and a 10-min incubation at 37°, the usable assay range is between 0.2 and 2.0 μg pepsinogen. The sensitivity is less using other proteins, e.g., albumin, as substrate. However, the sensitivity may be increased as much as fivefold by increasing the time of incubation with substrate to 1 hr. Once an appropriate standard curve is established, the pepsinogen content of samples may be estimated from a graph or by using a relatively simple computer program. Using the values from samples treated with Triton X-100, the values for micrograms pepsinogen may be converted to per-centage of initial content.

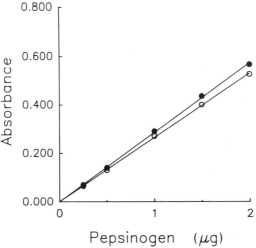

FIG. 2. Standard curves for assay of pepsin activity. Curves were prepared using purified porcine pepsinogen (Sigma Chem. Co.) assayed to contain 2800 pepsin units/mg protein. Standards were incubated at 37°, 10 min, with 1% bovine hemoglobin at pH 2.0. The TCA-soluble products were reacted with Folin phenol reagent (●) or bicinchoninic acid reagent (○).

Isolation and Identification of Pepsinogen

Under some experimental conditions, it is necessary or desirable to isolate and/or identify pepsinogens. Several detailed methods have been published describing the purification of pepsin.[9] In general, these methods involve multiple chromatographic steps, are more specific for pepsin than pepsinogen, and often result in denaturation of the enzyme. The method described here is relatively simple and identifies pepsinogen as well as pepsin in samples which are only partially purified.

Principle

Samples of incubation supernatant (or tissue homogenate supernatant) are concentrated by NH_4SO_4 precipitation and partially purified by fractionation on DEAE ion-exchange columns. An aliquot is treated with acid to convert pepsinogen to pepsin and the treated and untreated samples are compared by SDS-PAGE. Pepsinogens are identified as major protein bands in the untreated sample while pepsins are identified as slightly lower molecular weight bands which are enhanced by acid treatment. The gels may be transferred to polyvinylidene difluoride (PVDF) membranes and

[9] J. Tang, this series, Vol. 19, p 406.

submitted to amino acid composition and sequence analysis, if this is desired. Alternatively, the DEAE fractions may be purified by affinity chromatography to yield enzymatically active pepsin.

Procedure

The concentration of pepsinogen from a supernatant sample is easily accomplished by NH_4SO_4 precipitation. As shown in Fig. 3, addition of NH_4SO_4 in the range of 30–50% (w/v) achieves a complete precipitation of pepsinogen. During the precipitation care must be taken to maintain a pH above 7.0 to avoid activation of pepsinogen. The simplest procedure is to prepare a stock solution of 70% NH_4SO_4 (w/v), adjust the pH of the stock with NaOH, and add an equal volume of 70% NH_4SO_4 to the supernatant sample. The mixture is allowed to stand in the cold for approximately 1 hr and then centrifuged (10,000 g, 15 min). The supernatant is removed carefully and the precipitated proteins are redissolved in a small volume of buffer (10 mM NaPO$_4$, pH 7.4). The volume of buffer is chosen to result in a 10- to 100-fold concentration compared to the starting material.

DEAE cellulose is prepared as usual and equilibrated with buffer (10 mM NaPO$_4$, pH 7.4). Columns are poured to contain approximately 5 ml bed volume which allows a rapid elution with satisfactory fractionation. The sample (5–10% bed volume) is loaded onto the column and protein is eluted with stepwise 1.5–2.0 bed volumes of buffer (pH 7.4) containing 0.2, 0.3, 0.4, and 0.5 M NaCl. Fractions (10–15% of bed volume) are collected and assayed for pepsin activity and total protein content. Figure 4 shows a typical profile of pepsin activity recovered from a 0.8×10.0 cm ion-exchange column. The supernatant material derived from rabbit gastric glands elutes from DEAE columns as two major peaks of pepsin activity. One peak elutes with 0.3 M NaCl while a second peak requires 0.4 M NaCl. As indicated below, these peaks of activity reflect distinct proteins. Fractions containing adequate pepsin activity are pooled and reprecipitated with 35% NH_4SO_4. The precipitate is dissolved in buffer (10 mM NaPO$_4$, pH 7.4) at a volume so as to contain 0.2–0.4 μg pepsinogen/μl as estimated by the proteolytic activity. The pooled fractions are prepared for SDS-PAGE analysis or further purification of pepsin by affinity chromatography.

For SDS-PAGE analysis, an aliquot (50–100 μl) of the pooled fraction is acidified with HCl (1 N HCl; 5 μl/100 μl sample for samples in 10 mM NaPO$_4$ buffer) to pH 2.0. After 1 min, an equal volume of 1 N NaOH is added to reneutralize the sample and reduce proteolysis by the activated pepsin. Generally, it is observed that upon acidification the sample becomes turbid, but with adequate neutralization, the sample will clarify

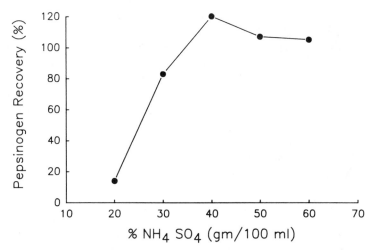

FIG. 3. Precipitation of pepsinogen by ammonium sulfate. Recovery of pepsinogen was measured as proteolytic activity after acid activation of the redissolved precipitate compared to activity measured in the original solution.

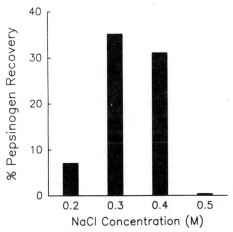

FIG. 4. Fractionation of pepsinogens by DEAE ion-exchange chromatography. Approximately 500 μg equivalent pepsin activity contained in an NH_4SO_4 precipitate from rabbit gastric glands was loaded on a DEAE column. Activity was eluted with 10 mM $NaPO_4$ buffer, pH 7.4, containing NaCl as indicated. Two major fractions of activity are recovered.

immediately. Aliquots (80 μl) of both untreated and acid-treated samples are prepared for SDS-PAGE. Twenty microliters of a concentrated SDS sample buffer is added to the samples and the mixture placed in a boiling water bath for 2 min. The concentrated sample buffer contains 313 mM Tris-HCl buffer, pH 6.8, 10% sodium dodecyl sulfate (SDS), 20% glycerol, 0.25% Bromphenol Blue, and 10% 2-mercaptoethanol. The 2-mercaptoethanol is added to the sample buffer just prior to use. The electrophoresis is performed with slab gels using the discontinuous buffer system of Laemmli.[10] We have routinely employed 1.5-mm thick, 12 × 16 cm gels with a 4% stacking gel and either a 10% or a 7.5–15% gradient gel for separation. The gels are cooled and electrophoresed at constant current (40 mA for stacking gel and 60 mA for separating gel) for approximately 2 hr, or until the tracking dye reaches the bottom of the gel. The gels are stained with Coomassie Blue R-250 according to standard methods. Figure 5 shows a typical result of SDS-PAGE analysis for pepsinogen and pepsin. It should be noted that pepsinogen stains well with Coomassie; however, pepsin stains rather poorly with this dye and quantitative comparison of the proteins cannot be made using this stain. Despite this drawback, the pepsinogen and pepsin bands can be identified by comparison of the acid-treated and untreated protein patterns.

As indicated by Fig. 5, these procedures can be used to distinguish different pepsinogens. The material eluting from the DEAE column with 0.3 M NaCl exhibits a single molecular weight band which is slightly lower than that seen with the material eluting with 0.4 M NaCl. The pepsin activity eluting with 0.3 M NaCl is tentatively identified as pepsinogen C (pepsin C; EC 3.4.23.3) while the activity eluting with 0.4 M NaCl corresponds to pepsinogen A (pepsin A; EC 3.4.23.1). Following a brief acid treatment, new protein bands appear on gel analysis corresponding to the activated pepsin molecules. Confirmation that the major protein bands represent pepsinogens can be obtained by partial sequence analysis since the NH$_2$-terminal portion of pepsinogens, the activation peptide, contains multiple basic residues and a highly conserved sequence (-Lys-Val-Pro-Leu-).[3] Alternatively, identification of pepsins may be achieved by purification on an affinity column and SDS-PAGE analysis.

In cases where amino acid composition or sequence information is required, the samples are processed through the SDS-PAGE procedure as above with the following modifications. All of the reagents employed for electrophoresis must be of high quality and, particularly, the SDS should be recrystallized from ethanol. After the gel is poured, it must be preelectrophoresed and, preferably, aged for 24 hr before running the sample.

[10] U. K. Laemmli, *Nature (London)* **227**, 680 (1970).

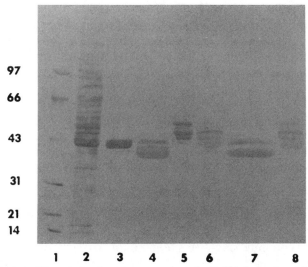

FIG. 5. SDS-PAGE analysis of pepsinogen and pepsin. Rabbit isolated gastric glands were homogenized, centrifuged at 10,000 g for 10 min, and the supernatant concentrated by NH_4SO_4 precipitation. The concentrate was fractionated by DEAE ion exchange and fractions applied to a 7.5–15% gradient gel. Following electrophoresis, proteins were transferred to PVDF membrane and stained with Coomassie. Lanes 1–8: molecular weight standards; original NH_4SO_4 precipitate; DEAE fraction eluting with 0.3 M NaCl; 0.3 M NaCl fraction after acidification; DEAE fraction eluting with 0.4 M NaCl; 0.4 M NaCl fraction after acidification; 0.3 M NaCl fraction eluted from affinity column; 0.4 M NaCl fraction eluted from affinity column. Fifteen to 20 μg of pepsinogen activity was applied to the gel for the DEAE fractions and 30–40 μg for the acidified or affinity-purified material in order to compensate for the weaker staining of pepsin relative to pepsinogen. It is noted that some unactivated pepsinogen is bound to the affinity column and elutes with the active pepsin.

Preelectrophoresis is performed using an upper (cathode) buffer containing 125 mM Tris-HCl, pH 6.8, 0.1% SDS, 1 mM glycine, and 0.1 mM sodium thioglycolate. The gel is electrophoresed for 20–30 min at 40 mA/gel. The buffer is then removed (this buffer may be saved for reuse) and the normal cathode buffer used for sample electrophoresis. These precautions are critical to prevent the proteins from being chemically modified so as to block the free NH_2 terminus for sequencing.

Following electrophoresis of the samples, the stacking gel is removed and the separating gel placed in cold transfer buffer (25 mM Tris-HCl, pH 8.6, 192 mM glycine, 0.1% SDS, and 15% methanol). The proteins are transferred to PVDF membranes using a semi-dry transfer apparatus (Hoeffer). This membrane is preferred over other transfer membranes due to the higher binding capacity and stability during sequence analysis. The electrotransfer process is fairly straightforward. Filter paper is soaked in transfer buffer and positioned on the anode surface of the apparatus. The PVDF membrane must be activated by placing it in methanol and then, without drying, into the transfer buffer. The PVDF membrane is stacked on the filter paper and the gel placed on top, followed by another layer of filter paper. Caution must be used to ensure that all layers are wet with transfer buffer and that there are no air bubbles between the layers. Also, it is helpful to presoak the gel in transfer buffer for 15–30 min prior to stacking. Transfer is complete in 90–120 min using 200 mA of current for a 12 × 12 cm gel. If transfers are run longer, they should be performed in a cold room to prevent overheating. After the transfer is complete, the membrane is stained (5 min) with Coomassie Blue R-250, destained with 70% methanol–7% acetic acid (three changes, 5 min), and air dried. The PVDF membranes may be stored indefinitely in the freezer.

`Purification of enzymatically active pepsin is achieved by affinity chromatography of the reprecipitated DEAE fractions. The affinity resin used is prepared by coupling an active site inhibitor to Sepharose.[11] The resin is washed with 10 mM NaPO$_4$ buffer, pH 3.0, and poured into a small column. The column should contain sufficient resin-binding capacity to exceed the anticipated amount of pepsin to be added. Since the binding capacity of the resins is on the order of 1–2 mg/ml, a column containing 0.5 ml resin is generally adequate.

The affinity column is rinsed extensively with 10 mM NaPO$_4$, pH 3.0, prior to sample addition. An aliquot of the DEAE fraction is acidified with HCl (1 N, 5 μl/100 μl sample), diluted 1 : 5 with pH 3.0 buffer, and immediately applied to the affinity column. The sample is rinsed into the resin with 4 bed volumes of pH 3.0 buffer. Pepsin is eluted from the column with 20 mM KPO$_4$ buffer, pH 6.0. Fractions (0.5–1.0 ml) are collected and assayed for pepsin activity; while true pepsins, i.e., pepsin A, elute at pH 6.0, some isozymes are retained on the column. These proteins may be eluted with buffer (pH 6.0) containing 10% dioxane.[11] Fractions containing pepsin activity may be pooled and concentrated by precipitation with NH$_4$SO$_4$. The concentrated material can be analyzed by SDS-PAGE or used for further enzymatic characterization. Throughout the purification procedures, care is taken to maintain the pH below 6.5 in order to avoid inactivation of the pepsin. As shown in Fig. 5, the material recovered from the affinity column migrates on SDS-PAGE identically with the protein

---

[11] J. Pohl, M. Zaoral, A. Jindra, Jr., and V. Kostka, *Anal. Biochem.* **139**, 265 (1984); The active site inhibitor, Val-D-Leu-Pro-Phe-Phe-Val-D-Leu, is available commercially and may be coupled to *N*-hydroxysuccinimide CH Sepharose 4B by standard methods. Commercially available pepstatin–Sepharose resin is less selective and binds to several aspartyl proteases besides pepsins. Also, elution from pepstatin–Sepharose requires high pH (> 8.0), resulting in denaturation of pepsin.

generated by acid treatment of the original DEAE fraction. As with the original DEAE fraction, the affinity-purified material may be transferred to PVDF membranes for sequence or composition analysis.

Acknowledgments

The authors wish to thank Ms. E. Christian for preparation of the manuscript. This work was supported, in part, by grants from the National Institutes of Health, NIDDKD, Nos. DK 14752, DK 36548.

[9] HCO₃⁻ Secretion: Stomach, Duodenum

By GUNNAR FLEMSTRÖM and HENRIK FORSSELL

Secretion of HCO_3^- by the gastric and duodenal mucosa has been shown to occur in all species tested[1] and most probably originates from the surface epithelium.[1] It can be stimulated by a variety of means, including physiological stimuli such as the presence of acid in the lumen,[2] sham feeding[3,4] or gastric distension.[5] A standing pH gradient across the mucous gel adherent to the duodenal and gastric surface and dependence of this gradient on mucosal HCO_3^- secretion has been demonstrated with pH-sensitive microelectrodes. The pH at the lumenal cell membranes thus remains near neutral in spite of acidities in the lumenal bulk solution as low as pH 2.0–3.0 in the stomach and pH 1.5–2.0 in the duodenum.[6,7] These findings strongly suggest that alkaline secretion into the mucous gel constitutes a first line of mucosal defense against lumenal acid in the stomach and may be the main mechanism in the duodenum. This chapter describes the various methods employed to study gastric and duodenal mucosal alkaline secretions in animals and humans and some of the difficulties involved in measuring the secretion and evaluating the nature

---

[1] G. Flemström, *in* "Physiology of the Gastrointestinal Tract, Second Edition" (L. R. Johnson, J. Christensen, M. Jackson, E. D. Jacobson, and J. H. Walsh, eds.), p. 1011. Raven, New York, 1987.

[2] J. R. Heylings, A. Garner, and G. Flemström, *Am. J. Physiol.* **246,** G235 (1984).

[3] H. Forssell, B. Stenquist, and L. Olbe, *Gastroenterology* **89,** 581 (1985).

[4] S. J. Konturek and P. Thor, *Am. J. Physiol.* **251,** G591 (1986).

[5] H. Forssell and L. Olbe, *Scand. J. Gastroenterol.* **22,** 627 (1987).

[6] E. M. M. Quigley and L. A. Turnberg, *Gastroenterology* **92,** 1876 (1987).

[7] G. Flemström and E. Kivilaakso, *Gastroenterology* **84,** 787 (1983).

of the transport processes. The characteristics of the secretions have recently been summarized elsewhere.[1]

Treatment of Experimental Animals and Tissues

Attention to previous handling and feeding of animals (and human volunteers) is important in studies of gastric and duodenal HCO_3^- transport. Sham feeding and feeding[3,4] as well as intralumenal acid[2,8] increase the alkaline secretion in both stomach and duodenum. The response to sham feeding is mediated via the vagal nerves whereas the rise in secretion in response to intralumenal acid is mediated by an increase in local mucosal prostaglandin synthesis as well as by humoral factors and neural reflexes. Recent experiments on isolated bullfrog duodenum and in anesthetized rats and cats indicate that mucosal prostaglandin production[8,9] and the nervous tone[10] also influence basal rates of HCO_3^- secretion. Variations in mucosal production of prostaglandins (or other mediators controlling the secretion) may affect the sensitivity to their exogenous administration.[8,9] Such variations may be induced by preoperative storing and feeding or result from experimental stress or operative trauma. Furthermore, distention has been shown to increase secretion in the stomach via local enteric neural reflexes[5] and gastric and duodenal HCO_3^- secretion in anesthetized animals is under tonic α_2-adrenergic inhibition.[10] Neural reflexes as well as circulating catecholamines contribute to the latter inhibition. It is thus important to attempt to use a constant level of anesthesia in work with anesthetized animals and to minimize trauma to or distension of the tissue in work on such animals or with isolated mucosa. Stress and sham-feeding stimuli should be avoided in conscious animals and in human volunteers. Cyclooxygenase inhibitors such as indomethacin or aspirin can be used to prevent endogenous prostaglandin synthesis.[7,8,11] Mucosal blood supply of HCO_3^- may be rate limiting for the secretion by the duodenal mucosa.[12] Changes in acid–base balance during experiments should thus represent a source of error and should be avoided or corrected.

[8] G. Flemström, A. Garner, O. Nylander, B. C. Hurst, and J. R. Heylings, *Am. J. Physiol.* **242**, G183 (1982).

[9] J. R. Heylings, S. E. Hampson, and A. Garner, *Gastroenterology* **88**, 290 (1985).

[10] C. Jönsson, O. Nylander, G. Flemström, and L. Fändriks, *Acta Physiol. Scand.* **128**, 65 (1986).

[11] K. Takeuchi, O. Furukawa, H. Tanaka, and S. Okabe, *Gastroenterology* **90**, 636 (1986).

[12] R. Schiessel, M. Starlinger, E. Kovats, W. Appel, W. Feil, and A. Simon, *in* "Mechanisms of Mucosal Protection in the Upper Gastrointestinal Tract" (A. Allen, G. Flemström, A. Garner, W. Silen, and L. A. Turnberg, eds.), p. 267. Raven, New York, 1983.

Gastric Bicarbonate Secretion

Antrum and Acid-Inhibited Fundus in Vitro

Antral and fundic mucosa from amphibia,[13,14] rabbit,[15] guinea pig,[16] and dog[17] have been mounted as a membrane in *in vitro* chambers. Chambers should be made in such a way that oxygenation of the tissue is facilitated, especially in experiments with mammalian mucosa (Fig. 1). Stable and reproducible rates of HCO_3^- (and H^+ secretion) by guinea pig antrum and fundus[16] have been obtained with chambers of the type illustrated in Fig. 2. Antral mucosa has no or very few acid-secreting (parietal) cells and is composed mainly of surface epithelial mucus cells which are morphologically indistinguishable from those of the fundic area. The antrum has thus been used as a model for studies of gastric nonparietal ion transport. In most species *in vitro*, the fundic mucosa secretes acid spontaneously. Neither histamine H_2-receptor antagonists nor the H^+,K^+-ATPase inhibitor omeprazole affect gastric HCO_3^- secretion[1] and both types of drugs have been used to inhibit fundic H^+ secretion for studies of HCO_3^- transport. Alkalinization of the unbuffered lumenal solution is titrated continuously at a predetermined level by infusion of titrant (acid) under automatic control from pH stat equipment. Basal rates of HCO_3^- secretion *in vitro* disclosed by inhibition of the acid secretion range from 0.10 to 0.50 μmol/cm²/hr in amphibian and 0.3 to 1.0 μmol/cm²/hr in mammalian preparations. Characteristics of fundic HCO_3^- secretion after use of various means to inhibit H^+ secretion have been compared in frog mucosa.[13] It was found that they were the same whether H^+ secretion had ceased spontaneously or had been inhibited by histamine H_2 antagonists or omeprazole. Furthermore, secretion by acid-inhibited fundus and spontaneously alkaline secreting antrum showed very similar properties.

Use of titration endpoints above pH 7 seems preferable in determining rates of HCO_3^- secretion. At lower pH values, measurement of surface pH with microelectrodes[7] may be used to exclude the possibility that passive diffusion of H^+ from the lumenal solution into the mucosa (back diffusion) contributes to the alkalinization. It should also be noted that levels of pH $\leq$ 2 may stimulate alkaline secretion in the stomach. Stimulation by acid in the duodenum has been observed to occur already at pH $\leq$ 5.[2,7]

[13] G. Flemström, *Am. J. Physiol.* **233**, E1 (1977).
[14] K. Takeuchi, A. Merhav, and W. Silen, *Am. J. Physiol.* **243**, G377 (1982).
[15] D. Fromm, J. H. Schwartz, R. Robertson, and R. Fuhro, *Am. J. Physiol.* **231**, 1783 (1976).
[16] H. Mattsson, K. Carlsson, and E. Carlsson, in "Mechanisms of Mucosal Protection in the Upper Gastrointestinal Tract" (A. Allen, G. Flemström, A. Garner, W. Silen, and L. A. Turnberg, eds.), p. 141. Raven, New York, 1983.
[17] Y. J. Kuo, L. L. Shanbour, and T. A. Miller, *Dig. Dis. Sci.* **12**, 1121 (1983).

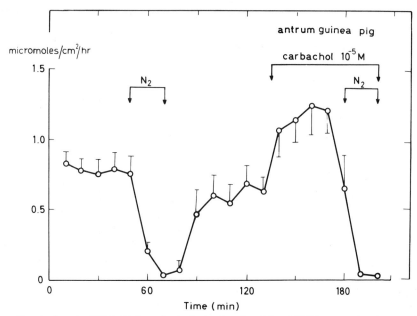

FIG. 1. Anoxia (N_2) inhibits carbachol-stimulated and basal HCO_3^- secretion by antral mucosa isolated from the guinea pig and mounted in an *in vitro* chamber as illustrated in Fig. 2, indicating dependence of secretion on tissue metabolism. Means $\pm$ SE, $n = 6$. (Reproduced from Mattsson *et al.*[16] with permission.)

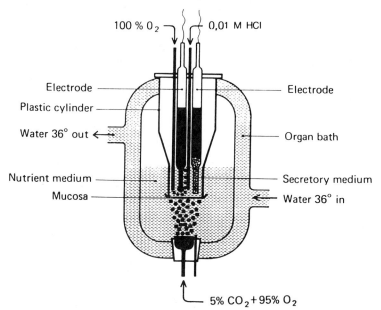

FIG. 2. Chamber for studies of secretion by mammalian gastric mucosa *in vitro* constructed to facilitate oxygenation of the tissue. (H. Mattsson with kind permission.)

Most studies on isolated gastric mucosa are performed with an unbuffered solution on the lumenal side. The nutrient (submucosal) side is bathed with an HCO$_3^-$-containing solution gassed with 95% O$_2$ and 5% CO$_2$. It is thus necessary to consider the possibility that passive permeation of HCO$_3^-$ or CO$_2$ affects the lumenal alkalinization. HCO$_3^-$ would alkalinize and CO$_2$ might acidify the lumenal side. Nutrient side HCO$_3^-$/CO$_2$ removal with HEPES/O$_2$ replacement does not affect the basal secretion or HCO$_3^-$ in fundic mucosa from the European frog *Rana temporaria*.[13] With fundic mucosa from the American bullfrog *Rana catesbeiana* both independence and some dependence of lumenal alkalinization on nutrient HCO$_3^-$/CO$_2$ have been reported.[14] In bullfrog antrum, removal of nutrient HCO$_3^-$/CO$_2$ results in a 20 to 40% decrease in alkalinization, and in isolated rabbit antrum, complete cessation of lumenal alkalinization has been observed.[15] Alkalinization by both frog and mammalian fundic and antral mucosa *in vitro* is, however, abolished by metabolic inhibitors such as anoxia, 2,4-dinitrophenol, or cyanide (cf. Fig. 1), indicating that secretion depends on tissue metabolism. Finally, it should be stressed that agents which desquamate the epithelium (e.g., ethanol or aspirin) or increase paracellular pathways (EDTA) induce passive permeation of HCO$_3^-$. The lumenal alkalinization thus induced in a damaged (leaky) mucosa must be distinguished from the metabolic-dependent secretion of HCO$_3^-$ by intact epithelium. The "leaky" mucosa is characterized by a markedly lower transepithelial electrical resistance and potential difference.[1,18] Use of nutrient buffers other than HCO$_3^-$/Co$_2^-$ have been used *in vitro* in distinguishing between metabolic-dependent transport and passive leakage of HCO$_3^-$.[1] In whole animals with such a mucosa, considerable amounts of serum albumin and glucose appear in the gastric lumen.[18]

Even if potent inhibitors of H$^+$ secretions are used to disclose gastric fundic HCO$_3^-$ secretion, there may be a residual, small and therefore masked H$^+$ secretion. An increase in this acid secretion may cause an apparent inhibition of the net alkaline (HCO$_3^-$ minus H$^+$) secretion. Comparison between antral and fundic epithelium or simultaneous determination of acid and alkaline secretions (see *in vivo*) could be used in elucidating this problem.

Measurement of Gastric HCO$_3^-$ Secretion in Vivo

Grossman[19] demonstrated titratable HCO$_3^-$ in the juice from canine antral pouches, and Hollander[20] reported the occurrence of HCO$_3^-$ in juice

---

[18] H. W. Davenport, *N. Engl. J. Med.* **276,** 1307 (1967).

[19] M. I. Grossman, *Proc. 21st Int. Congr. Physiol. Sci., Buenos Aires,* p. 226 (1959).

[20] H. F. Hollander, *Arch. Intern. Med.* **92,** 107 (1954).

from fundic pouches in dogs where acid secretion had been decreased by vagotomy and antrectomy. Previous studies also involve demonstrations of HCO_3^- in samples of gastric juice from achlorhydric patients and in healthy volunteers during spontaneous H^+ secretory rest. More recently, potent inhibition of acid secretion by use of histamine H_2 antagonists or omeprazole has permitted studies of some properties of alkaline fundic secretion *in vivo*. Methodology enabling concomitant determination of gastric H^+ and HCO_3^- secretions has also been developed. This is valuable especially since low lumenal pH itself may stimulate the HCO_3^- secretion. It is also likely that there is a further interrelation between HCO_3^- and H^+ secretion in that HCO_3^- produced by the parietal cells during the acid secretory process and released interstitially is used by the HCO_3^--secreting surface epithelial cells.

Acid and bicarbonate interact intragastrically with consequent release of CO_2 and water. In one type of methodology, used for simultaneous determination of these secretions in guinea pigs,[21] humans,[3,22,23] and cats,[24] the concentration of free HCO_3^- in gastric instillates or perfusates is calculated from pH and pCO_2 using the Hendersson–Hasselbalch equation. The concentration of CO_2, calculated from its solubility coefficient and pCO_2, is added to provide the total HCO_3^- concentration (Eq. 1):

$$HCO_3^- \text{ (total)} = S \times pCO_2 \times 10^{pH-pK_a} + S \times pCO_2 \qquad (1)$$

It should be noted that both the solubility coefficients (S) for CO_2 (0.033 mM/mmHg at 37° in isotonic saline) and the dissociation constant (pK_a) for carbonic acid used in the Hendersson–Hasselbalch equation (6.31 in isotonic saline at 37°) depend on the ion strength and temperature of the solutions.[25,26] For example, the solubility coefficient in isotonic saline increases to 0.039 and 0.052 mM/mmHg if temperature is decreased to 30 and 20°, respectively. A prerequisite for calculating bicarbonate and acid from intragastric pH and pCO_2 is that CO_2 released intragastrically originates from the reaction between secreted bicarbonate acid. Diffusion of CO_2 out from or into the lumen is thus a possible source of error. Inhibition of acid secretion decreases the formation of Co_2 from secreted HCO_3^- and has been used in some studies to minimize the possible loss by diffu-

[21] A. Garner and G. Flemström, *Am. J. Physiol.* **234,** E535 (1978).
[22] W. D. W. Rees, D. Botham, and L. A. Turnberg, *Dig. Dis. Sci.* **27,** 961 (1982).
[23] C. Johansson, A. Aly, E. Nilsson, and G. Flemström, *Adv. Prostaglandin Thromboxane Leukotriene Res.* **12,** 395 (1983).
[24] L. Fändriks and L. Stage, *Acta Physiol. Scand.* **128,** 563 (1986).
[25] D. D. van Slyke, J. Sendroy, A. B. Hastings, and J. M. Neill, *J. Biol. Chem.* **78,** 765 (1928).
[26] A. B. Hastings and J. Sendroy, *J. Biol. Chem.* **65,** 445 (1925).

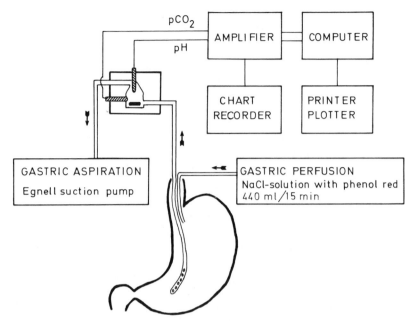

FIG. 3. Method for continuous measurement of gastric pH and pCO$_2$ in gastric perfusate in humans with computerized calculation of the H$^+$ and HCO$_3^-$ secretions.

sion from the lumen. Furthermore, artificially imposed CO$_2$ gradients are maintained in the stomach for prolonged periods and quantitative recovery of instilled exogenous HCO$_3^-$ has been demonstrated in most studies. One practical problem which should be noted is the possible loss of CO$_2$ from extragastric connections into the air. These should these be made from materials impermeable to CO$_2$.

Methodology for continuous computerized determination of gastric H$^+$ and HCO$_3^-$ secretions from pH and pCO$_2$ in humans[3,27] is illustrated in Fig. 3. Similar methodologies have been developed more recently for use in anesthetized cats.[24] To minimize possible escape of CO$_2$ from the perfusate into the mucosa the stomach is rapidly perfused. The pCO$_2$ and pH in the effluent are measured continuously and calculations are made every 30 sec [Eq. (1)] and stored by a computer. The infusion port is located just below the cardia of the stomach and the outlet is connected to a suction pump (Egnell, Trollhättan, Sweden) producing a negative pressure of 15–20 kPa once a second. The intragastric volume is determined by use of a marker (Phenol Red, 5 mg/liter). The measuring chamber (Fig. 4), con-

[27] H. Forssell and L. Olbe, *Scand. J. Gastroenterol.* **20,** 767 (1985).

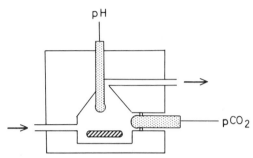

Fig. 4. The electrode-containing chamber employed in the method illustrated in Fig. 3. A 12-mm magnetic stirrer in the chamber permits thorough mixing of the gastric effluent.

taining one pH and pCO_2 electrode, has its outlet at the top, preventing air bubbles intermingled with the gastric effluent from disturbing the measurements of pH and pCO_2. A small magnetic stirrer ensures thorough mixing of the effluent. Both salivary and pancreaticobiliary secretions contain HCO_3^-. Contamination of gastric perfusates by these secretions is simple to avoid in experiments in animals but must be carefully controlled in humans. Measurement of effluent concentrations of amylase and bilirubin have been used for this purpose.[3] Isoamylase determination would be an optimal method for differentiation between salivary and pancreatic origin of the possible contamination.

Another interesting technique for simultaneous determination of gastric H^+ and HCO_3^- secretions has been developed by Feldman[28,29] for use in human subjects. The technique is based on the two-component model of gastric secretion,[20] assuming that HCO_3^- is secreted at a fixed concentration and its amount thus directly related to the volume of nonparietal secretion. Alkaline secretion dilutes and neutralizes gastric acid, leading to a reduction in acidity and osmolality. The HCO_3^- output is thus calculated from measurements of gastric juice volume, osmolality, and H^+ concentration. It is assumed in these calculations that there is a fixed relation between the osmolality of plasma and that of gastric parietal and nonparietal secretions. A further prerequisite for the method is a low permeability of the gastric mucosa to water. High permeability would result in movement of water along its chemical gradient from the lumen into the mucosa, increasing lumenal osmolality and leading to an underestimate of HCO_3^- secretion.

Values of gastric HCO_3^- concentration ($\sim$100 mM) and output ($\sim$2.6

[28] M. Feldman and C. C. Barnett, *J. Clin. Invest.* **72,** 295 (1983).
[29] M. Feldman, *Am. J. Physiol.* **248,** 6188 (1985).

mmol/hr) estimated from osmolality are 5- to 10-fold greater that those determined from pCO_2 and pH[3,22,23] and those observed previously in healthy volunteers during spontaneous H^+ secretory rest.[30] The fact that measurements have been made in acid-secreting subjects and after inhibition or arrest of acid secretion, respectively, may suggest that stimulation of HCO_3^- secretion by lumenal acid accounts for part of the difference. Maximal stimulation by (instilled) lumenal acid observed in the stomach in experimental animals is, however, only two- to threefold. Another explanation for the discrepancy is that the hypotonicity of gastric juice arises not only from interaction between secreted HCO_3^- and H^+ but also from the processes for transport of HCl and water within the gastric tubules.[31] This would result in an overestimate of HCO_3^- secretion measured by osmolality. It should be noted, however, that gastric HCO_3^- secretion calculated from osmolality and that calculated from pH and pCO_2 exhibit qualitatively very similar sensitivities to stimulants and inhibitors.

Measurement of Duodenal Mucosal HCO_3^- Secretion

Preparations used included mucosa isolated from bullfrogs[32,33] or rats[34] and mounted in *in vitro* chambers, segments of duodenum cannulated *in situ* in anesthetized animals.[7,8,11] chronic duodenal loops or pouches in conscious animals,[4,35] and segments of duodenum isolated between balloons in human volunteers.[36] The secretion of HCO_3^- has been measured by continuous titration of lumenal perfusates or by titration of the HCO_3^- concentration in duodenal effluents. In designing studies of duodenal mucosal HCO_3^- secretion, it should be noted that there is a gradient in secretory rate along the duodenum with proximal segments secreting at greater rates. This has been observed in the rat, dog, and human, all of which contain Brunner's glands in the proximal duodenum.[1] Very similar findings have also been made in the bullfrog, a species devoid of these glands.[33] Furthermore, secretion in proximal and more distal duodenum in the rat shows very similar sensitivity to stimulation by prostaglandins and inhibition by acetazolamide.[1] This suggests that the surface epithelium is a main source of HCO_3^- also in mammalian proximal duodenum.

---

[30] M. Kristensen, *Scand. J. Gastroenterol.* **10**(Suppl. 32), 1 (1975).
[31] W. S. Rehm, C. F. Butler, S. G. Spangler, and S. S. Sanders, *J. Theor. Biol.* **27**, 443 (1970).
[32] G. Flemström, *Am. J. Physiol.* **239**, G198 (1980).
[33] J. N. L. Simson, A. Merhav, and W. Silen, *Am. J. Physiol.* **240**, G401 (1981).
[34] S. Bridén, G. Flemström, and E. Kivilaakso, *Gastroenterology* **88**, 295 (1985).
[35] J. I. Isenberg, B. Smedfors, and C. Johansson, *Gastroenterology* **88**, 303 (1985).
[36] J. I. Isenberg, D. L. Hogan, M. A. Koss, and J. A. Selling, *Gastroenterology* **91**, 370 (1986).

Studies of Duodenum in Vitro

Proximal duodenum isolated from the American bullfrog and mounted as a tube[32] (Fig. 5) or a flat sheet[33] have been used in several studies. Use of tube preparations should minimize edge damage while flat preparations allow determination of electrical short-circuit current and resistance. Both types of preparations alkalinize the lumenal perfusate at a greater rate ($\sim$1.0 μmol/cm^2/hr) than does gastric antrum or fundus (0.1–0.4 μmol/cm^2/hr) from the same species. Secretion of HCO_3^- by the bullfrog duodenum is abolished by inhibitors of tissue metabolism and stimulated by a variety of agents, including dibutyryl cyclic AMP, phosphodiesterase inhibitors, and prostaglandins. Transport involves neutral Cl^-/HCO_3^- exchange at the lumenal membrane stimulated by some gastrointestinal hormones and inhibited by furosemide, and electrogenic transport of HCO_3^- stimulated by cyclic AMP and prostaglandins. Cl^-/HCO_3^- exchange as well as a conductance pathway with equal affinity for HCO_3^- and Cl^- have recently been identified in apical vesicles from guinea pig duodenal enterocytes.[37] Furthermore, HCO_3^- secretion by rat duodenum *in vivo* is stimulated by prostaglandin E_2 and glucagon and inhibited by furosemide in a manner very similar to that observed in bullfrog preparations *in vitro*. These findings strongly suggest that transport by bullfrog duodenum *in vitro* and mammalian duodenum *in vivo* occur by similar processes. Duodenum isolated from the rat has also been used for *in vitro* studies of the secretion.[34] This mammalian preparation, however, has much smaller basal rates of secretion and a lower sensitivity to stimulants and inhibitors than does rat duodenum *in vivo* or isolated frog duodenum. This may reflect insufficient oxygenation and/or supply of nutrient HCO_3^- to the relatively thick mammalian preparation.

Duodenal Mucosal HCO_3^- Secretion in Vivo

Segments of duodenum with intact blood supplies and devoid of pancreatic and biliary secretions have been perfused with saline circulated by means of a gas lift or a pump.[7,8,11] The rate of lumenal alkalinization was measured by continuous titration using a pH stat or by determination of the HCO_3^- concentration in effluents. Conscious animal models include rats with transplanted open duodenal loops[35] and dogs with pouches made of loops of proximal or distal duodenum.[4] A chamber suitable for *in situ* titration of HCO_3^- secretion by segments of duodenum in anesthetized rats, guinea pigs, cats, and rabbits is illustrated in Fig. 5.

---

[37] C. D. A. Brown, C. R. Dunk, and L. A. Turnberg, *Am. J. Physiol.* **257**, G661 (1989).

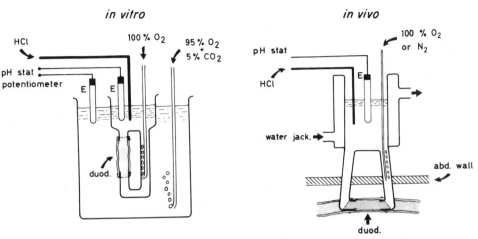

FIG. 5. Diagram of glass chambers used to perfuse segments of bullfrog duodenum *in vitro* and mammalian duodenum *in situ*. Mucosal fluid (4 ml with the frog preparation, 5 ml in rats, and 20 ml in cats and rabbits) is circulated rapidly by a gas lift (100% O$_2$ or N$_2$). The rate of HCO$_3^-$ secretion is determined by titration with HCl controlled by a pH stat system.

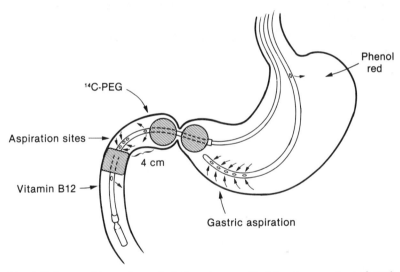

FIG. 6. Tube assembly used to occlude the pylorus and isolate a 4-cm segment of proximal duodenum in humans for studies of mucosa HCO$_3^-$ secretion. Two pediatric endotracheal cuffs were fluoroscopically positioned to straddle the pylorus while a third cylindrical balloon was positioned 4 cm beyond. A double-lumen nasogastric tube was also placed so that its tip was in the middle of the gastric antrum. Three nonabsorbed markers were infused: phenol red into the gastric lumen, [14C]PEG into the isolated test segment, and vitamin B$_{12}$ just beyond the test segment. (Reproduced from Isenberg *et al.*[36] with permission.)

The release of CO_2 or decrease in osmolality when acid solutions are instilled into the duodenum have been used to estimate duodenal HCO_3^- secretion in some previous studies. Such studies by Harmon and collaborator[38] provided the first clear evidence that the duodenal mucosa devoid of Brunner's glands and pancreaticobiliary secretion transports HCO_3^- at high rates. It should be observed, however, that the high permeability of the duodenal mucosa to CO_2 and water is likely to result in some underestimate of the HCO_3^- production when measured by such techniques.

A method enabling measurement of duodenal mucosal secretion in human patients has been reported recently.[36,39] Short segments (4 cm) of either the most proximal duodenum, including the duodenal bulb, or the distal duodenum (10–40 cm beyond the pylorus) have been isolated between balloons as illustrated in Fig. 6. The isolated segment is perfused with saline and HCO_3^- measured in the effluent. Nonabsorbed markers (Phenol Red, vitamin B_{12} and [^{14}C]polyethylene glycol) are used for determination of possible leakage out from or into the segment. Trypsin is measured in the effluent to further exclude any contamination with pancreaticobiliary secretion. Balloons may cause some distention of the duodenal wall and stimulation of gastric alkaline secretion has been observed during distention of the stomach.[5] The stimulation of duodenal mucosal secretion by lumenal acid, prostaglandins, and vasoactive intestinal polypeptide and the inhibition by indomethacin observed in humans are, however, very similar to the effects observed in acute and chronic duodenal segments in experimental animals.

[38] J. W. Harmon, M. Woods, and N. J. Gurll, *Am. J. Physiol.* **235,** E692 (1978).
[39] L. Knutson and G. Flemström, *Gut* **30,** 1708 (1989).

[10] Isolation of H^+,K^+-ATPase-Containing Membranes from the Gastric Oxyntic Cell

By W. W. REENSTRA and J. G. FORTE

Introduction

Hydrochloric acid secretion by the oxyntic cell of the mammalian stomach has been shown to be dependent on the activity of the H^+,K^+-ATPase (EC 3.6.1.36).[1,2] At rest the oxyntic cell is characterized by an apical membrane with short microvilli, and a profusion of intracellular membranes at the apical pole.[3,4] These membranes, the tubulovesicles, contain the H^+,K^+-ATPase.[5] Secretagogue-induced stimulation of the oxyntic cell results in a 5- to 10-fold increase in the surface area of the apical membrane, elongation of the microvilli, and a marked decrease in the number of tubulovesicles.[2,6] Stimulation also causes a translocation of the H^+,K^+-ATPase into the apical membrane and an increase in the rate of oxygen consumption.[5,7,8] This increased O_2 consumption is blocked by specific inhibitors of the H^+,K^+-ATPase.[9] Thus stimulation causes both a massive fusion of the tubulovesicles with the apical membrane, the membrane fusion hypothesis,[10] and an activation of the H^+,K^+-ATPase.

Microsomal vesicles isolated from the stomachs of slaughterhouse animals have been shown to contain an electroneutral ATP-driven H^+/K^+ exchange pump.[11] However, as these membranes do not have pathways for K^+ and Cl^- entry, pH gradients cannot be generated unless the K^+ ionophore valinomycin is added in order to allow K^+ access to the interior of

[1] J. G. Forte and H. C. Lee, *Gastroenterology*, **73**, 921 (1977).
[2] G. Sachs, J. Kaunitz, J. Mendlein, and B. Wallmark, *in* "Handbook of Physiology Section 6" (J. G. Forte, ed.), Vol. 3, p. 229. American Physiological Society, Washington, D.C., 1989.
[3] G. M. Forte, T. E. Machen, and J. G. Forte, *Gastroenterology* **73**, 941 (1977).
[4] H. F. Helander, *Int. Rev. Cytol.* **70**, 217, (1981).
[5] A. H. Smolka, H. F. Helander, and G. Sachs, *Am. J. Physiol.* **245**, G589 (1983).
[6] J. A. Black, T. M. Forte, and J. G. Forte, *Gastroenterology* **81**, 509 (1981).
[7] D. K. Hanzel, T. Urushidani, and J. G. Forte, *Am. J. Physiol.* **256**, G1082 (1989).
[8] T. Berglindh, H. F. Helander, and K. J. Obrink, *Acta Physiol. Scand.* **97**, 401 (1976).
[9] B. Wallmark, C. Briving, J. Fryklund, K. Munson, R. Jackson, J. Mendlein, E. Rabon, and G. Sachs, *J. Biol. Chem.* **262**, 2077 (1987).
[10] J. G. Forte, J. A. Black, T. M. Forte, T. E. Machen, and J. M. Wolosin, *Am. J. Physiol.* **241**, G349 (1981).
[11] G. Sachs, H. H. Chang, E. Rabon, R. Schackmann, M. Lewin, and G. Saccomani, *J. Biol. Chem.* **251**, 7690 (1976).

METHODS IN ENZYMOLOGY, VOL. 192

the vesicle.[12] For this reason these vesicles cannot be representative of the stimulated apical membrane; they are currently thought to be derived from the tubulovesicles.[13] In addition, the absence of parallel pathways for the entry of K^+ and Cl^- has been suggested to be the reason why the H^+,K^+-ATPase is inactive in the resting oxyntic cell.[10,14]

These observations allow two independent criteria to be set for isolated apical membrane vesicles from the stimulated oxyntic cell. (1) Stimulated apical vesicles must contain the H^+,K^+-ATPase; (2) in addition, they should also have pathways for K^+ and Cl^- flux so that ATP-driven accumulation of hydrochloric acid can be achieved without the addition of K^+ ionophores. These criteria have been used to isolate and characterize apical membrane vesicles from stimulated rabbit oxyntic cells. The properties of these stimulated apical membrane vesicles (SA vesicles) are contrasted with tubulovesicles, isolated from the microsomal fraction of resting oxyntic cells. Methods are described for preparing both SA vesicles and tubulovesicles from gastric mucosa *in vivo* and from gastric glands.[15]

Methods

Solutions

Phosphate-buffered saline (PBS): 150 mM NaCl, 5 mM phosphate, pH 7.4

Hypotonic homogenizing buffer (MSEP): 125 mM mannitol, 40 mM sucrose, 1 mM ethelenediaminetetraacetic acid (EDTA), 5 mM PIPES-Tris, pH 6.7

Isotonic suspending medium (SM): 300 mM sucrose, 0.2 mM EDTA, 5 mM Tris-HCl, pH 7.3

Sedative cocktail: 60 mg/ml ketamine hydrochloride (Ketaset; Bristol Laboratories, Syracuse, NY), 6 mg/ml xylazine (Rompun; Miles Laboratories, Shawnee, KA), 1.2 mg/ml acepromazine maleate (Aveco, Fort Dodge, IA).

Incubation buffer: 132 mM NaCl, 5.4 mM KCl, 5 mM Na_2HPO_4, 1 mM NaH_2PO_4, 1.2 mM $MgSO_4$, 1 mM $CaCl_2$, 5 mM sodium acetate, 11 mM glucose, 1 mg/ml bovine serum albumin (BSA), 25 mM HEPES, pH 7.4

---

[12] H. C. Lee and J. G. Forte, *Biochim. Biophys. Acta* **508**, 339 (1978).

[13] J. A. Black, T. M. Forte, and J. G. Forte, *Anat. Rec.* **196**, 163 (1980).

[14] S. J. Hersey, L. Steiner, S. Matheravidathu, and G. Sachs, *Am. J. Physiol.* **254**, G856 (1988).

[15] T. Berghlindh, this volume [6].

Vesicles Preparation from Rabbit Gastric Mucosa

Adult New Zealand White rabbits (1.5 to 2 kg) of either sex are fasted overnight with free access to water. They are fed standard rabbit chow *ad libitum* for 20 min before being sedated with 1.0 ml/kg of the cocktail (sc) and given (1) the H_1-receptor antagonist chlorpheneramine maleate (0.3 ml of 15 mg/ml) and (2) histamine (0.2 ml of 100 mM, pH 7), both sc. After 10 min the sedated rabbit is given two 0.2-ml doses of histamine (iv), separated by 10 min. (Additional cocktail can be given if necessary.) Five minutes after the final histamine injection the rabbit is anesthetized with 2.0 ml of 50 mg/ml sodium pentobarbital (Nembutal; Abbott Laboratories, Chicago, IL), iv. Resting rabbits are treated in the same manner except that (1) they are not fed, (2) they are not given chlorpheneramine, and (3) the histamine is replaced by equal volumes of 150 mg/ml (pH 7) cimetidine, an H_2-receptor antagonist. All subsequent steps are identical for both preparations. For studies where it is necessary to know the rate of acid secretion a fistula can be made in the stomach and gastric juice collected with an inserted cannula.[16] With this surgical procedure the rate of acid secretion can be determined by titrating the collected juice, and the tissue can be taken at a defined physiological state for subsequent membrane fractionation.[17]

The stomach is removed 2 min after the injection of pentobarbital, cut open, the contents removed, and washed with ice-cold PBS. (All subsequent steps are performed with ice-cold solutions.) The fundic region is cut out and placed on a glass plate that has been chilled on ice. Any remaining stomach contents and mucus are removed by blotting with filter paper and the mucosal tissue is separated from the muscularis by scraping with a glass microscope slide. The scraped tissue is collected in a plastic beaker, weighed, covered with MSEP, and thoroughly minced with scissors. The minced tissue is allowed to settle and the supernatant decanted, additional MSEP is added, and the process repeated until the supernatant is clear. (This removes contaminating mucus and connective tissue that has been isolated with the epithelial cells.) MSEP is added (25 ml/g of scraped tissue) and the tissue triturated in a Potter–Elvehjem homogenizer by 12 passes with a tight-fitting Teflon pestle at 200 rpm.

The homogenate (Hom) is spun at 80 g_{max} for 10 min. The supernatant is decanted and saved while the pellet is resuspended in MSEP, rehomogenized, and spun as before. The pellet from the second low-speed spin (P_0) is retained and the two supernatants are combined and spun at 4300 g_{max} for 10 min. The pellet from the 4300 g spin (P_1) is retained and the

---

[16] B. P. Curwain and N. C. Turner, *J. Physiol. (London)* **331**, 431 (1981).
[17] J. G. Forte and D. J. Keeling, *FASEB J.* **2**, 1275A (1988).

supernatant respun at 14,000 g_{max} for 10 min. The supernatant from the 14,600 g spin is recentrifuged at 48,000 g_{max} for 90 min, while the pellet (P_2) is retained. The pellet from the 48,000 g spin (P_3) and the supernatant (S_3) are retained. All pellets are resuspended in SM and assayed, as is S_3. The apical membrane-rich P_1 and the tubulovesicle-rich P_3 fractions can be further purified by density gradient centrifugation.

Density Gradient Centrifugation of P_1 and P_3

Apical membranes in P_1 are separated from the other components (largely mitochondria and nuclei) by centrifugation on a step gradient built of Ficoll 400 (Sigma, St. Louis, MO) in SM. P_1 is resuspended at a final concentration of 12% Ficoll in SM and layered over a cushion of 18% Ficoll in SM. SM is layered above the 12% layer. The gradient is spun at 500 g_{max} for 30 min and then the speed is increased to 137,000 g_{max} for 90 min. (The initial low-speed spin removes many dense particles that at high speeds tend to drag SA vesicles into the 18% layer.) Membranes at the 0–12% and the 12–18% interfaces are removed by syringe, diluted fivefold with SM, and pelleted at 143,000 g_{max} for 45 min. These pellets, $P_1 12$ and $P_1 18$, respectively, are suspended in SM. Membranes can be stored at $-80°$ for 3 weeks without loss of activity. Both fractions show H^+,K^+-ATPase activity, with $P_1 18$ having a higher yield and specific activity; henceforth, apical membranes or SA vesicles will refer to the $P_1 18$ fraction.

Tubulovesicles can be purified from P_3 by either of two methods. In one method P_3, suspended in SM, is placed on the top of a cushion of 5% Ficoll (in SM) and spun for 2 hr at 135,000 g_{max}. Membranes ($P_3 5$) are removed from the interface with a syringe, diluted fivefold with SM, and pelleted. Pellets are resuspended in SM and can be stored at $-80°$. Additional material can be collected by increasing the percentage of Ficoll; however, membranes that float on 10 and 16% Ficoll have lower specific activities for K^+-stimulated ATPase and capacities for pH gradient formation than membranes that float on 5% Ficoll.[18,19] An alternative procedure is to use a step gradient of sucrose (20, 27, and 33%) in 0.2 mM EDTA and 5 mM Tris-HCl, pH 7.3. P_3 suspended in SM is placed on top of the sucrose gradient and spun for 4 hr at 135,000 g_{max}. Fractions are isolated as described for the Ficoll gradient but can be stored directly without pelleting and resuspension in SM. Henceforth, tubulovesicles will refer to either the 5% Ficoll fraction ($P_3 5$) or to the 20% sucrose fraction.

[18] B. H. Hirst and J. G. Forte, *Biochem. J.* **231,** 641 (1985).
[19] W. W. Reenstra, unpublished observations.

Membrane Preparation from Gastric Glands

Both tubulovesicles and SA vesicles can also be made from gastric glands.[20] Despite the rather limited amount of material in these preparations gastric glands offer several experimental advantages over the whole animal preparation. For example, it is possible to conveniently study the time course and dose–response characteristics for stimulation of the oxyntic cell in one preparation and to perform labeling studies on oxyntic cell membrane proteins. Gastric glands are prepared as described by Berglindh.[15] Glands that have settled at unit gravity are washed in incubation buffer, settled again, and then resuspended at a 20 to 1 dilution. Appropriate stimulants (or inhibitors) are added (e.g., histamine or stimulators of adelyalate cyclase or, for resting glands, cimetidine), and the glands are incubated under continuous oxygenation and shaking for 40 min. The pH is monitored and readjusted every 10 min. After incubation the glands are spun down, washed, and resuspended in ice-cold MSEP at twice the concentration of the incubation. The glands are homogenized in a Potter-Elvehjem homogenizer with an extremely tight-fitting Teflon pestle turning at 200 rpm. Fractionation is done as described for scraped mucosa except that P_0 is spun at 40 g_{max} for 5 min. Due to the small amount of material density gradient centrifugation of P_1 is done by suspending the P_1 pellet in 18% Ficoll (in SM) and overlaying with SM. After spinning for 2 hr at 135,000 g_{max} the purified apical membranes ($P_1$18) have floated to the interface between the SM and the 18% Ficoll while the other membranes have been pelleted. $P_1$18 is collected from the interface and Ficoll is removed as described for the mucosal preparation.

Assay of H^+,K^+-ATPase Activity

Hydrolysis of ATP can be measured by the liberation of inorganic phosphate. Fractions are incubated in 1.0 ml of 37° reaction buffer, containing 1 mM MgSO$_4$, 1 mM ATP, 100 μM ouabain, 10 mM Na-PIPES, pH 6.8. Reactions are run in the presence or the absence of 150 mM KCl for 10 min. Reactions are quenched with 1.0 ml of ice-cold 14% trichloroacetic acid and inorganic phosphate determined by extraction of the phosphomolybdate complex into butyl acetate.[21]

Several technical problems complicate the use of this assay for the determination of H^+,K^+-ATPase activity in whole homogenate and crude cell fractions. Crude fractions are rich in both mitochondrial (F-type) and

[20] T. Urishidani and J. G. Forte, *Am. J. Physiol.* **252**, G458 (1987).
[21] H. Sanui, *Anal. Biochem.* **60**, 489 (1974).

vacuolar (V-type) ATPases,[22] thus the background ATPase activity is high. Although the activity of these ATPases is not directly affected by K[+], so that in principle the K[+]-stimulated ATPase could be used as a marker for the H[+],K[+]-ATPase, their activities can be affected by membrane potentials formed as a consequence of the addition of K[+] and K[+] ionophores. The addition of specific inhibitors can help to alleviate the problem of high background, but care must be taken as most inhibitors are not absolutely specific. For example oligomycin, while reasonably specific for the F-type ATPase, will also inhibit P-type ATPases at the concentrations required for total inhibition of the F-type ATPase.[23] A second feature that makes routine assaying of the ATPase activity difficult is the fact that the sites for ATP binding and for activation by K[+] are on opposite sides of the membrane.[24] Thus for vesicular preparations the access of K[+] to the activation sites must be addressed. In tubulovesicles, but not in apical membranes, the addition of a K[+] ionophore such as valinomycin or nigericin markedly increases the K[+]-stimulated activity.[25] However, as these agents are inhibitory at high concentrations special attention must be paid to the concentrations employed, making their application to the routine assay of crude fractions difficult. A third problem arises from the fact that in some H[+],K[+]-ATPase-containing fractions, notably apical membranes, a large fraction of the H[+],K[+]-ATPase activity is cryptic, so that the addition of a small amount of a detergent such as octylglucoside increases K[+]-stimulated H[+],K[+]-ATPase activity, even in the presence of ionophores.[18] Unfortunately at high concentrations these detergents inhibit the H[+],K[+]-ATPase, making quantification of H[+],K[+]-ATPase activity in the presence of detergents difficult.

Many of the above problems can be circumvented by assaying the p-nitrophenylphosphatase (pNPPase) activity of the H[+],K[+]-ATPase. Studies have shown that the K[+]-stimulated pNPPase activity is a partial reaction of the H[+],K[+]-ATPase[9,26] and that this activity is less susceptible to the problems described above for the ATPase activity. First, pNPP is not hydrolyzed by the F- and V-type ATPases, and hydrolysis by the Na[+],K[+]-ATPase can be inhibited by ouabain without inhibiting the H[+],K[+]-ATPase.[27] In addition, the site where K[+] activates pNPP hydrolysis appears be

[22] P. Pederson and E. Carafoli, *TIBS* **12**, 146 (1987).
[23] J. R. Sachs, *J. Physiol. (London)* **302**, 219 (1980).
[24] H. B. Steward, B. Wallmark, and G. Sachs, *J. Biol. Chem.* **256**, 2682 (1981).
[25] H. C. Lee, H. Briebart, M. Berman, and J. G. Forte, *Biochim. Biophys. Acta* **553**, 107 (1979).
[26] J. G. Forte, A. L. Ganser, R. C. Beesley, and T. M. Forte, *Gastroenterology* **69**, 175 (1975).
[27] J. M. Crothers, W. W. Reenstra, and J. G. Forte, *Am. J. Physiol.* (in press).

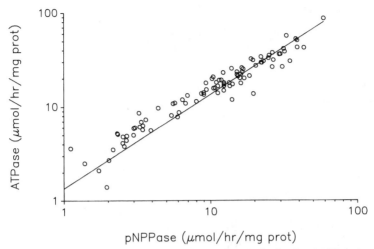

FIG. 1. Correlation between K⁺-stimulated pNPPase and K⁺-stimulated ATPase in microsomal vesicles prepared from rabbit stomachs of varying age. Potassium ion-stimulated H⁺,K⁺-ATPase and pNPPase activities were measured as described in the text with 1 μM nigericin added to the ATPase assay. The membrane fractions assayed included the P_3 fraction and the sucrose density-purified subfractions of P_3. Stomachs were taken from resting rabbits aged 3 to 60 days.[28] Over this period of ontogeny the wide variation in specific activity of the membrane fractions is due to the synthesis of the H⁺,K⁺-ATPase as the animals mature. Even with the variation specific activity that spans two orders of magnitude there is a high degree of correlation between K⁺-pNPPase and K⁺-ATPase activities. The linear regression has a slope of 0.75 ± 0.02 and an intercept of −0.4 ± 3.3.

on the same side of the membrane as the pNPP-binding site, so that K⁺ ionophores are not required.

In our standard assay the K⁺-stimulated hydrolysis of pNPP is measured by incubating membrane fractions for 10 min at 37° in 1.0 ml of 5 mM MgSO₄, 5 mM Na₂pNPP (Calbiochem, Los Angeles, CA), 100 μM ouabain, 10 mM PIPES/Tris, pH 7.2, either with or without 10 mM KCl. Reactions are stopped with 1.5 ml of NaOH, protein precipitated, and the optical density of p-nitrophenolate read at 410 nm. Potassium ion-stimulated activities of the fractions are determined from the difference in optical density with and without K⁺. In order to demonstrate the correlation between the pNPPase and the ATPase activities of the H⁺,K⁺-ATPase purified fractions from P_3 with specific activities that differed by 100-fold were assayed for both activities. As shown in Fig. 1[28] there is an excellent correlation ($R^2 = 0.92$) between the K⁺-stimulated pNPPase and ATPase activities, establishing that K⁺-stimulated pNPP activity is directly proportional to H⁺,K⁺-ATPase activity.

[28] J. M. Crothers, W. W. Reenstra, and J. G. Forte, unpublished observations.

Distribution of H^+,K^+-ATPase Activity

As shown in Table I, for resting preparations most of the H^+,K^+-ATPase activity is found in P_3, with relatively little activity seen in the heavier fractions. In contrast, for stimulated preparations the majority of the activity is seen in P_1 and P_0, with little activity in P_3. Thus stimulation results in a dramatic redistribution of total activity from P_3 to heavier fractions, particularly P_1. This procedure gives a clear separation of the two fractions that contain the H^+,K^+-ATPase as the centrifugal forces have been adjusted to give relatively little activity in the P_2 fraction. It should be noted that in stimulated and resting preparations, a large fraction of the total activity is in the P_0 fraction. Rehomogenization of the first pellet helps to release some of this activity into P_1 but more vigorous homogenization appears to break the apical membrane into smaller membrane vesicles so that the most of the activity in stimulated preparations is recovered in P_3 (see below). For stimulated mucosa the increase in specific activity of $P_1$18 over Hom is only 5-fold, while in resting preparations the specific activity of $P_3$5 is usually more than 10-fold greater than the specific activity of Hom. The difference is largely reflective of the presence of a large number of additional proteins in the apical membrane fraction. It also should be noted that the total activity in $P_1$18 and $P_1$12 is only $17 \pm 5\%$ of the activity in P_1. Most of the K^+-stimulated pNPPase activity from P_1 sediments with the mitochrondria-rich pellet, P_1P. For the six stimulated preparations in Table I, P_1P had 83% of the total activity in P_1 but the mean specific activity was not significantly different from that of the homogenate.

We have used the ratio of the total K^+-stimulated pNPPase activity in P_1 to the total activity in P_3 (P_1/P_3 ratio) as a biochemical index for the degree of stimulation of the mucosa at the time of sacrifice.[18,20] Typical values for the P_1/P_3 ratio obtained from stimulated and resting rabbits and from stimulated and resting gastric glands are given in Table II. Also, as shown in Fig. 2, there is a good correlation between the rate of acid secretion at the time of sacrifice and the P_1/P_3 ratio. In the study shown an H_2-receptor antagonist SKF-93479[29] was given to rabbits after they had been maximally stimulated with histamine. Rates of acid secretion were measured at various times by collecting the gastric juice from rabbits with cannulated stomachs. The indicated P_1/P_3 ratios ratios are for the same animals sacrificed at the indicated times after the administration of SKF-93479.[17] The data in Table II also show that in glands there is a good

[29] R. C. Blakemore, T. H. Brown, G. J. Duran, C. R. Ganellin, M. E. Parsons, A. C. Rasmussen, and D. A. Rawlings, *Br. J. Pharmacol.* **74**, 200P (1981).

TABLE I
DISTRIBUTION AND SPECIFIC ACTIVITY OF pNPPASE

A. Distribution of total K^+-pNPPase activity among crude cell fractions
Percentage of total recovered activity[a]

	P_0	P_1	P_2	P_3	S_3
Resting	24 ± 4	11 ± 1	5 ± 1	55 ± 1	5 ± 2
Stimulated	37 ± 4	38 ± 5	6 ± 1	15 ± 1	4 ± 1

B. Specific activity of pNPPase (μmol/hr/mg)

	Hom	P_1	$P_1 18$	P_3	$P_3 5$
Resting	1.9 ± 0.3	1.0 ± 0.1	3.5 ± 0.6	11.8 ± 3.3	24 ± 9
Stimulated	2.4 ± 0.4	3.1 ± 0.5	12.0 ± 1.9	6.8 ± 1.2	ND^b

[a] The sum of the recovery activity was 78% of the total homogenate activity.
[b] ND, Not determined.

correlation between acid production, measured with the aminopyrine (AP) ratio,[15] and the P_1/P_3 ratio. Maximal values for the P_1/P_3 ratio in glands are always smaller than those that can be achieved in live animals. We suspect that this reflects a decreased level of stimulation of the oxyntic cell in gastric glands.

Alternative procedures for the isolation of apical membranes from

TABLE II
P_1, P_3 RATIO FROM STIMULATED AND RESTING MUCOSA
AND GLANDS

	P_1/P_3	n	AP ratio[a]
Gastric mucosa			
Stimulated	2.8 ± 0.3	30	
Resting	0.18 ± 0.02	4	
Gastric glands[b]			
Stimulated			
10^{-4} M Histamine	0.46 ± 0.08	4	79 ± 23
10^{-4} M Histamine +	1.02 ± 0.19	11	157 ± 21
10^{-5} M forskolin			
Resting			
10^{-4} M Cimetidine	0.17 ± 0.07	4	11.6 ± 1.7

[a] AP, Aminopyrine.
[b] Data from Ref. 20.

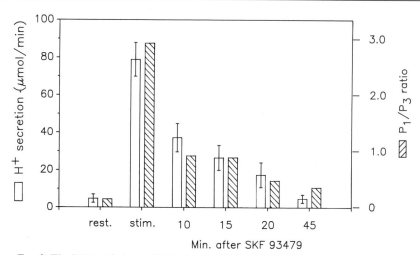

FIG. 2. The P_1/P_3 ratio is a valid biochemical index of parietal cell stimulation. Cannulas were placed in the stomach of sedated rabbits, gastric juice was collected, and the rates of acid secretion were determined. Fasting rabbits (rest.) were treated with the H_2-receptor antagonist SKF 93479. Stimulated rabbits were treated with histamine and either sacrificed during maximal secretion (stim.) or given SKF 93479 and sacrificed at the indicated time after the drug. Membrane fractions (P_1 and P_3) were prepared from rabbits sacrificed at the indicated times and the P_1/P_3 ratio was determined as the ratio of total K^+-stimulated pNPPase activity in P_1 to the total activity in P_3. The data show a close correspondence of P_1/P_3 ratios with the HCl secretory rate.

rabbit and rat oxyntic cells[30,31] have homogenized in an isotonic homogenizing medium at pestle speeds 5 to 15 times that used in this procedure. In the reported cases membranes with SA vesicle-like characteristics are isolated along with microsomal membranes in a high-speed microsomal fraction. From such preparations it is not possible to use the convenience of differential centrifugation to separate apical membranes from microsomes or to use the P_1/P_3 ratio as an indicator of oxyntic cell stimulation. However, an elaborate centrifugation scheme with a D_2O gradient for the separation of H^+,K^+-ATPase-rich vesicles with either "resting" or "stimulated" characteristics has been reported.[31]

In paired studies the total K^+-stimulated ATPase activity per gram of mucosal tissue in the homogenate is decreased by stimulation. This is largely due to the cryptic activity in P_1. When corrections are made for the cryptic activity, the total activities of resting and stimulated homogenates

[30] J. Cuppoletti and G. Sachs, *J. Biol. Chem.* **259,** 14952 (1984).
[31] E. Rabon, W. B. Im, and G. Sachs, this series, Vol. 157, p. 649.

are not significantly different. In addition, no alteration in the kinetic properties of the H^+,K^+-ATPase have been observed on stimulation. Studies in our laboratory have failed to find any difference between the activity of the H^+,K^+-ATPase in SA vesicles and in tubulovesicles[18] or in the ability of the enzyme to form a phosphoenzyme.[19] Thus we must conclude that stimulation does not alter the activity of the H^+,K^+-ATPase per se.

Assay of pH Gradient Formation

We have standardly assayed the formation of acid-interior pH gradients by H^+,K^+-ATPase-containing vesicles with acridine orange, a fluorescent amine.[12] Typical assay conditions are 150 mM sucrose, 75 mM KCl, 250 μM MgSO$_4$, 250 μM ATP, 5 mM PIPES/Tris, pH 7.0, and 1 μM acridine orange. Membranes are added to a final concentration of between 10 and 30 μg/ml and the decrease in fluorescence (excitation at 493 nm, emission at 530 nm) resulting from the self-quenching of the accumulated acridine orange is monitored. The assay can also be carried out by using the change in absorbance of acridine orange at 492 nm.[32] Typical results for the fluorescence assay are shown in Fig. 3, where pH gradient formation by SA vesicles and tubulovesicles is compared. For the tubulovesicles a significant rate of acridine orange fluorescence quenching is seen only after the addition of valinomycin.[12] In contrast, SA vesicles show fluorescence quenching that is independent of valinomycin.[33] While tubulovesicles require high concentrations of both K^+ and Cl^- for pH gradient formation, the rate of pH gradient formation by SA vesicles is unaffected by reducing either K^+ or Cl^- to 5 mM, provided the other ion is maintained at 75 mM.[19] The most straightforward explanation for this result is that the maximal rates of the ion transport pathways in SA vesicles are much greater than the rate of ATP-driven proton pumping. Characterization of the ion transport pathways has shown that the conductances for both K^+ and Cl^- are greater in SA vesicles than in tubulovesicles. Both Ba^{2+}[34] and triethylamine (TEA)[35] have been shown to inhibit K^+ entry. At 75 mM Cl^- the Cl^- channel appears to be insensitive to the Cl^- channel blockers. DPC (N-phenylanthranilic acid) and NPPB [5-nitro-2-(3-phenylpropylamino-benzoate]. However, at 75 mM K^+ and 1 mM Cl^- both DPC and NPPB inhibit H^+ uptake with K_I values of 500 and 50 μM, respectively.[19]

[32] E. Rabon, H. H. Chang, and G. Sachs, *Biochemistry* **17**, 3345 (1978).
[33] J. M. Wolosin and J. G. Forte, *FEBS Lett.* **125**, 208 (1981).
[34] J. M. Wolosin and J. G. Forte, *J. Membr. Biol.* **83**, 261, (1985).
[35] J. Cuppoletti and D. H. Malinowska, *Gastroenterology,* **94**, A82 (1988).

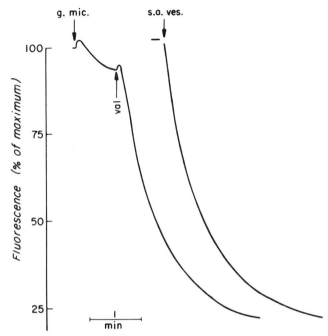

FIG. 3. pH gradient formation by gastric vesicles prepared from resting or stimulated stomach. Either tubulovesicles (g. mic.) or stimulated apical membranes (s.a. ves.) were added at 20 μg/ml to proton uptake medium containing 75 mM KCl, 150 mM sucrose, 250 μM Mg-ATP, 1 μM acridine orange, and 5 mM PIPES-Tris, pH 7.0. For tubulovesicles the addition of 1 μM valinomycin (val) is required in order to achieve optimal rates of quenching of the acridine orange fluorescence, whereas for the stimulated apical membranes the quenching of acridine orange fluorescence is independent of valinomycn.

Size and Orientation of SA Vesicles

Stimulated apical vesicles can also be differentiated from tubulovesicles on the basis of size.[34] Stimulated apical vesicles have a larger mean diameter, 0.29 μm, than that of tubulovesicles, 0.10 μm. The size distribution of SA vesicles is also quite skewed, with many vesicles having diameters three times the mean; in contrast the variance in size is much smaller for tubulovesicles. Electron microscopy has shown that many SA vesicles appear to be multilamellar and multivesicular.[10,36] This may account in part for the presence of cryptic ATPase activity in these vesicles. Both electron microscopic and biochemical studies have shown that f-actin is

[36] J. M. Wolosin and J. G. Forte, *J. Biol. Chem.* **256,** 3149 (1981).

present inside SA vesicles.[37] Since the f-actin is most likely a remnant of the microfilaments present on the cytoplasmic side of the apical microvilli, this would imply that some fraction of these vesicles are oriented with the cytoplasmic face inside. This orientation would also be consistent with cryptic H^+,K^+-ATPase activity. Studies that have combined the ability of the detergent octylglucoside to expose cryptic H^+,K^+-ATPase activity with the ability of trypsin to inactivate H^+,K^+-ATPases with exposed cytoplasmic domains have led to the conclusion that for SA vesicles only 30 to 50% of the H^+,K^+-ATPase molecules are oriented with cytoplasmic domains exposed to the external surface of the vesicle.[18] For isolated tubulovesicles virtually all of the H^+,K^+-ATPase is oriented with the cytoplasmic domain on the external surface of the vesicle.

Proteins in SA Vesicles and Tubulovesicles

SDS-PAGE gels of isolated apical membranes ($P_1 18$) and the tubulo-vesicle-rich fraction P_3 are shown in Fig. 4A. These gels show that stimulation causes the transfer of a 94-kDa protein, known to be the catalytic subunit of the H^+,K^+-ATPase,[2] from P_3 to $P_1 18$. This is in agreement with the redistribution of H^+,K^+-ATPase activity. Figure 4A also shows that many of the proteins isolated in stimulated $P_1 18$ are also present in resting $P_1 18$. Two such proteins are actin (M_r 45, 000) and an 80-kDa phosphoprotein whose level of phosphorylation increases with stimulation.[38,39] As shown in Fig. 4B these proteins remain with the H^+,K^+-ATPase in $P_1 18$ after sonication and recentrifugation. Immunocytochemistry has been used to demonstrate that both the 80-kDa protein and actin are present in the apical membrane of both the stimulated and the resting oxyntic cell.[7] These observations suggest that the $P_1 18$ fraction from resting preparations contains resting apical membranes, i.e., apical membranes that lack the H^+,K^+-ATPase. Also shown in Fig. 4B is a gel of purified tubulovesicles. The second band seen between 60 and 80 kDa has been shown to be a glycoprotein that is associated with the catalytic subunit of the H^+,K^+-ATPase in a one-to-one stoichiometry.[40] This protein may represent a subunit for the H^+,K^+-ATPase like the β-subunit of the Na^+,K^+-ATPase (see [1] in Vol. 156 of this series).

[37] J. M. Wolosin, C. Okamoto, T. M. Forte, and J. G. Forte, *Biochim. Biophys. Acta* **761**, (1983).

[38] T. Urushidani, D. K. Hanzel, and J. G. Forte, *Biochim. Biophys. Acta.* **930**, 209 (1987).

[39] T. Urushidani, D. K. Hanzel, and J. G. Forte. *Am. J. Physiol.* **256**, 1070 (1989).

[40] C. T. Okamoto, J. M. Karpilow, A. Smolka, and J. G. Forte, *Biochim. Biophys. Acta* **1087**, 360 (1990).

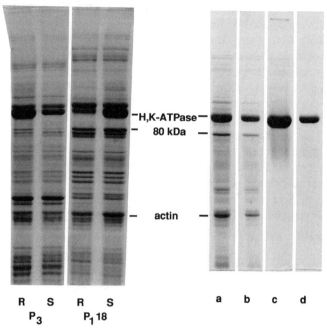

R   S   R   S   a   b   c   d
  P₃   P₁18

FIG. 4. Proteins of membrane fractions from stimulated and resting gastric mucosas revealed by SDS-PAGE. (A) Coomassie Blue-stained gel of tubulovesicle-rich fraction, P_3, and apical membrane-rich fraction, $P_1 18$, from both resting (R) and stimulated (S) gastric glands. Note the major stimulation-related redistribution of the 94-kDa band (H^+,K^+-ATPase) from P_3 to $P_1 18$, and the lack of change in the 80 kDa and actin bands of $P_1 18$. (B) Proteins in highly purified preparations of SA vesicles and tubulovesicles. Highly purified apical membranes (lanes a and b) were prepared by sonicating $P_1 18$ membranes prepared as described in the methods, and resedimenting the membranes through a Ficoll density gradient. Highly purified tubulovesicles (lanes c and d) were obtained by first sedimenting P_3 on a sucrose step gradient, as described in the text. The fraction that layered on 20% sucrose was collected, washed, and then resedimented on the same gradient; the 20% layer was collected. Lanes a and c, 50 μg protein; lanes b and d, 25 μg protein. Note the enrichment of the 80-kDa band and actin (45 kDa) in the apical membranes and their complete absence in the purified tubulovesicles.

Conclusions

Immunological studies have shown that stimulation of the gastric oxyntic cell results in a redistribution of the proton pump, H^+,K^+-ATPase, from cytoplasmic tubulovesicles to the apical plasma membrane. The methods described here allow for the biochemical isolation of both of these H^+,K^+-ATPase-containing membranes from *in vivo* preparations and from isolated gastric glands. The conditions of homogenization allow the

H^+,K^+-ATPase in the stimulated apical membrane to be separated from the H^+,K^+-ATPase in the tubulovesicles by differential centrifugation. This permits a biochemical assessment of the degree of oxyntic cell stimulation from the distribution of H^+,K^+-ATPase activity in the crude fractions. Density gradient centrifugation procedures give further purification of apical membrane vesicles and tubulovesicles from their respective crude membrane fractions. Characterization of these two membranes shows that the activity of the H^+,K^+-ATPase per se is not altered by stimulation. However, the permeability to K^+ and Cl^- of the membrane containing the H^+,K^+-ATPase is altered by stimulation. This alteration in the specific ion permeabilities of the H^+,K^+-ATPase-containing membranes appears to be both the major mode of regulation of the H^+,K^+-ATPase activity and a important component in the regulation of HCl secretion.[10,14]

[11] Peptic Granules from the Stomach

By BRIAN E. PEERCE

Introduction

Stimulus–secretion coupling is thought to involve five stages: (1) stimulation, (2) Ca^{2+} influx, (3) granule translocation, (4) fusion of granule membrane and cell membrane, and (5) membrane recovery (exocytosis). Recent studies have focused on the problem from the isolated cell level to determine the mechanisms involved in secretogogue stimulation of secretion (for review, see Ref. 1). These studies have led to the conclusion that the secretory granule may have an active role in steps 4 and 5 of the secretory process. Active participation by the granule in step 5 has long been recognized[2]; however, a role in granule membrane/cell membrane fusion has only recently been appreciated.[3,4] Granule membrane/cell membrane fusion has recently been suggested to involve ion gradients across the granule membrane as well as cell membrane potential.[5] The nature of these ion gradients and how they are generated is best studied at the level of the isolated secretory granule. Essential to these studies is a

[1] R. D. Burgoyne, *Biochim. Biophys. Acta* **799**, 201 (1984).

[2] W. W. Douglas, *Br. J. Pharmacol.* **34**, 475 (1968).

[3] H. B. Pollard, C. J. Pazoles, C. E. Creutz, and O. Zinder, *Int. Rev. Cytol.* **58**, 159 (1976).

[4] F. S. Cohen, J. Zimmersberg, and A. Finkelstein, *J. Gen. Physiol.* **75**, 251 (1980).

[5] W. A. Kachadorian, J. Muller, and A. Finkelstein, *J. Cell Biol.* **91**, 584 (1981).

stable secretory granule preparation whose *in vivo* behavior approximates its behavior in the cell.

Secretory granules may be subdivided into preproteolytic enzyme-containing granules (zymogen) and nonzymogen-containing granules. As a class, the nonzymogen-containing granules have been more extensively studied (for reviews, see Refs. 6 and 7). This may be due to the inherent instability of the isolated zymogen granule and the presence of multiple cell types in tissues containing zymogen granules. Pepsinogen granules in the gastric chief cell have been recognized since the 1880s.[8] However, attempts to isolate these granules have been unsuccessful until recently.[9] The granules are very osmotic sensitive, pH sensitive, and Cl^- sensitive, having a half-life of approximately 20 hr in basic sulfate buffer at 4°. Pepsinogen granules vary in size between 0.5 and 1 μm, with an average size of approximately 0.8 μm.[10] Better than 90% of the granule contents is the inactive proteolytic enzyme, pepsinogen.

This chapter describes a method which has been developed to isolate gastric chief cell pepsinogen granules in hypertonic, basic, Cl^--free medium using a differential centrifugation step for initial purification, followed by a sucrose linear gradient centrifugation for the final purification step. The method results in approximately eightfold purified pepsinogen granules which retain low-speed pelletable pepsinogen for 3 days following isolation.

Materials and Methods

Isolation of Pepsinogen Granules

Five White Norfolk rabbits are sacrificed by decapitation, their stomachs excised and washed with distilled water, and placed in ice-cold isolation buffer containing 270 mM sucrose, 200 μM EGTA, 100 μM phenyl-methylsulfonyl fluoride (PMSF), and 40 mM Tris-SO$_4$, pH 8. PMSF was made as a methanolic solution and then added to 1 liter of isolation buffer. The mucosa was wiped free of mucus, scraped into isolation buffer, and homogenized by 10 strokes of a glass–Teflon homogenizer set at 2500 rpm. The homogenate was then passed through three layers of cheesecloth

[6] H. B. Pollard, C. E. Creutz, and C. J. Pazoles, *Methods Cell Biol.* **23**, 313 (1981).

[7] J. D. Castle, A. M. Castle, and W. L. Hubbell, *Methods Cell Biol.* **23**, 335 (1981).

[8] R. R. Bensley, *in* "Special Cytology" (E. V. Cowdry, ed.), Vol 1, p. 199. Harper (Hoeber), New York, 1932.

[9] B. E. Peerce, A. Smolka, and G. Sachs, *J. Biol. Chem.* **259**, 9255 (1984).

[10] H. F. Helander, *Cell Tissue Res.* **189**, 287 (1978).

to remove coarse debris and the filtrate was rehomogenized with 20 strokes of a tight-fitting ground glass–glass homogenizer set at 250 rpm. This second homogenization step was found to be necessary to improve cell rupture and increase granule yield without having any effect on granule integrity. The homogenate was then diluted to 500 mol with isolation buffer and centrifuged for 10 min at 120 g. The supernatant from this spin was then centrifuged at 700 g for 10 min. The supernatant from the second 7000 g min^{-1} spin was then diluted with 2 M sucrose to a final density of 1.12 g/ml and centrifuged for 10 min at 5000 g. The pellets were collected and resuspended in 1 M sucrose + 10 mM Tris-SO$_4$ by gentle agitation and centrifuged again. The advantage of this procedure was found to be increased granule stability and a reduction of mitochondrial contamination. Increasing the medium to 1.13 g/ml prior to the initial centrifugation step made mucus removal very difficult. The final granule pellet was then resuspended in 1 M sucrose + 40 mM Tris-SO$_4$ and layered onto a linear gradient of 38 to 50% sucrose containing 100 μM EGTA and 40 mM Tris-SO$_4$ and centrifuged for 4.5 hr at 108,000 g in an AH-627 Sorvall swinging bucket rotor. Fractions (1.5 ml) were collected by displacement with 60% sucrose, and the sucrose concentrations in the gradient fractions were measured with an Abbe refractometer (Bausch and Lomb, Rochester, NY). The fraction corresponding to 48 to 49.5% sucrose was then diluted threefold and centrifuged for 10 min at 3500 g. The pellet was resuspended in 500 mM sucrose and centrifuged again. The supernatants were then centrifuged for 10 min at 3500 g. The final pellets were then resuspended in 500 mM sucrose and used immediately.

Granule Morphology

Electron microscopy granules were resuspended in a solution containing 4% glutaraldehyde, 4% paraformaldehyde, 100 μM MgSO$_4$, and 100 mM cacodylate buffer, pH 7.2. Following an overnight fixation, the granules were washed in 100 mM cacodylate buffer and postfixed in 2% OsO$_4$ buffered with 100 mM cacodylate, pH 7.2. After 1 hr the granules were washed three times in the cacodylate buffer and then embedded in 2% agar buffered with 100 mM cacodylate. The agar was found to be particularly useful for maintaining granule stability during the dehydration steps. Following dehydration through a series of graded alcohols, the granules were embedded in Epon, stained with lead citrate and uranyl acetate, and thin sectioned using an LKB ultramicrotome (LKB, Piscataway, NJ). The sections were examined on a Phillips 400 electron microscope.

Enzyme Assays

Pepsin was assayed according to the method of Anson and Mirsky using denatured hemoglobin as substrate.[11] ATPase activity was measured as described by Fujita *et al.*[12] and H^+,K^+-ATPase as described by Saccomani *et al.*[13] Succinate dehydrogenase (SDH) and cytochrome *c* oxidase activities were measured according to the methods of King[14] and Cooperstein and Lazarow,[15] respectively. Acid phosphatase was measured at pH 5 and alkaline phosphatase at pH 10 according to Torriani with *p*-nitrophenyl phosphate as substrate.[16] Protein was measured according to the method of Lowry *et al.*[17] with bovine serum albumin (BSA) as standard.

Results

Characterization of Fractions

Table I shows the distribution of enzyme activities following the initial phase of the isolation procedure summarized in Fig. 1. Following the differential centrifugations, one major and two minor sources of contamination remain in the crude granule pellet, P_3. The major contaminant is mitochondrial, as determined by the activities of succinate dehydrogenase and cytochrome *c* oxidase associated with P_3. The minor contaminants are lysosomal and plasma membrane, as judged by the acid phosphatase and ouabain-sensitive Na^+,K^+-ATPase activities, respectively. Table I also illustrates the advantage of dilution of S_2 with 2 *M* sucrose. Of the total SDH activity found in S_2, 85.7% remains in the 5000 *g*/10 min supernatant, while 60% of the peptic activity associated with S_1 is recovered in P_3.

Figure 2 shows the elution profile following the linear gradient centrifugation step. The mitochondrial contamination equilibrates between 38 and 40% sucrose, or near the start of the linear gradient. The intact peptic granules will enter 48% sucrose, but will not enter 50%. There is some breakdown of granules during the centrifugation; however, this accounts for less than 20% of the peptic activity applied to the gradient.

[11] M. L. Amson, and A. E. Mirsky, *J. Gen. Physiol.* **16,** 59 (1932).
[12] M. Fujita, H. Matsui, K. Nagamo, and M. Nakao, *Biochim. Biophys. Acta* **233,** 404 (1971).
[13] G. Saccomani, H. B. Stewart, D. Shaw, M. Lewin, and G. Sachs, *Biochim, Biophys. Acta* **465,** 311 (1977).
[14] T. E. King, this series, Vol. 10, p. 322.
[15] S. J. Cooperstein, and A. Lazarow, *J. Biol. Chem.* **188,** 665 (1951).
[16] A. Torriani, this series, Vol. 12B, p. 212.
[17] D. H. Lowry, N. J. Rosebrough, A. L. Farr, and R. J. Randall, *J. Biol. Chem.* **193,** 265, (1951).

TABLE I
DIFFERENTIAL CENTRIFUGATION

Marker enzyme[a]	S_1[b]	S_2[c]	S_3[d]	P_3[e]	Percentage recovery[f]
Protein (mg/ml)	30.1 ± 0.8 ($n = 5$)	8.2 ± 0.2 ($n = 8$)	7.23 ± 0.17 ($n = 8$)	4.47 ± 0.38 ($n = 8$)	82.1
Alkaline phosphatase (μmol/mg/hr)	0.5 ± 0.1 ($n = 4$)	1.35 ± 0.27 ($n = 4$)	1.48 ± 0.13 ($n = 4$)	0 ($n = 4$)	96.1
Acid phosphatase (μmol/mg/hr)	2.06 ± 0.1 ($n = 4$)	7.58 ± 0.48 ($n = 4$)	6.9 ± 0.74 ($n = 4$)	1.58 ± 0.2 ($n = 4$)	86.5
Na^+,K^+-ATPase (μmol/mg/hr)	5.8 ± 0.4 ($n = 4$)	14.2 ± 0.9 ($n = 4$)	13.9 ± 0.8 ($n = 4$)	1.57 ± 0.8 ($n = 4$)	84.2
H^+,K^+-ATPase (μmol/mg/hr)	3.1 ± 0.6 ($n = 4$)	7.52 ± 0.2 ($n = 4$)	6.41 ± 0.2 ($n = 4$)	0 ($n = 4$)	75.1
Cytochrome c oxidase (μmol/mg/hr)	10.1 ± 0.5 ($n = 2$)	28.56 ± 0.3 ($n = 2$)	25.7 ± 0.3 ($n = 2$)	1.59 ± 0.1 ($n = 2$)	81.1
Succinate dehydrogenase (μmol/mg/hr)	168.5 ± 0.8 ($n = 10$)	41.5 ± 0.13 ($n = 10$)	281 ± 17.9 ($n = 10$)	59 ± 11.6 ($n = 10$)	85.8
RNA (μg/mg protein)	28.6 ± 0.5 ($n = 3$)	4.1 ± 0.9 ($n = 3$)	1.3 ± 0.1 ($n = 3$)	0 ($n = 3$)	18.9
Mg^{2+}-ATPase (μmol/mg/hr)	2.13 ± 0.2 ($n = 4$)	5.22 ± 1.1 ($n = 4$)	2.67 ± 0.1 ($n = 4$)	6.58 ± 0.2 ($n = 4$)	83.5
Pepsin (Mu/mg protein)	163.4 ± 0.17 ($n = 5$)	683 ± 0.24 ($n = 10$)	312 ± 20.4 ($n = 10$)	414.8 ± 8.2 ($n = 10$)	62.3

[a] Enzyme activities were determined as described in Materials and Methods. All values are the means ±SE for the enzyme activities of the various fractious.
[b] S_1, 120 g for 10 min (supernatant).
[c] S_2, 700 g for 10 min (supernatant).
[d] S_3, 5000 g for 10 min (supernatant).
[e] P_3, 5000 g for 10 min (pellet).
[f] Percentage recovery is calculataed on the basis of the activity of S_1 versus the activities found in $P_3 + S_3$.

The distribution of marker enzymes following the linear gradient centrifugation is shown in Table II. Fraction I contains the bulk of the plasma membrane contamination and 10% of the recoverable peptic activity. The lysosomal contamination appears to cosediment with the SDH and cytochrome c oxidase activities in fraction II, which corresponds to 38 to 40% sucrose. Fraction III contains only 0.2% of the total SDH activity and 0.5% of the cytochrome c oxidase activity of S_2. There is no measurable plasma membrane contamination in this fraction as judged by the absence of Na^+,K^+-ATPase activity associated with this fraction. In addition there is

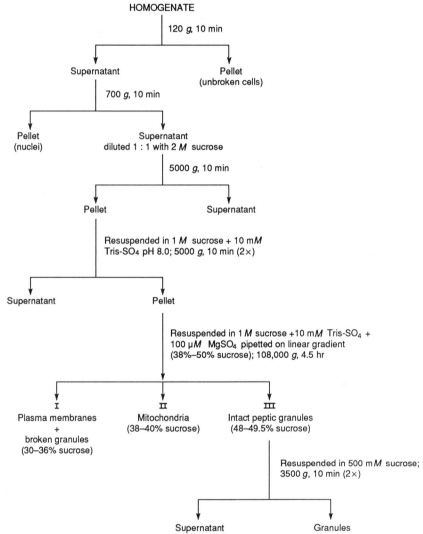

HOMOGENATE

120 g, 10 min

Supernatant   Pellet
                                               (unbroken cells)
700 g, 10 min

Pellet   Supernatant
(nuclei)   diluted 1 : 1 with 2 M sucrose

                     5000 g, 10 min

              Pellet   Supernatant

Resuspended in 1 M sucrose + 10 mM
Tris-SO$_4$ pH 8.0; 5000 g, 10 min (2×)

Supernatant   Pellet

                Resuspended in 1 M sucrose +10 mM Tris-SO$_4$ +
                100 μM MgSO$_4$ pipetted on linear gradient
                (38%–50% sucrose); 108,000 g, 4.5 hr

I   II   III
Plasma membranes   Mitochondria   Intact peptic granules
+   (38–40% sucrose)   (48–49.5% sucrose)
broken granules
(30–36% sucrose)
                                          Resuspended in 500 mM sucrose;
                                          3500 g, 10 min (2×)

                              Supernatant   Granules

FIG. 1. Summary of the method of isolation of peptic granules.

no measurable RNA in fraction III and only 0.5% of the acid phosphatase activity of P_3 is recovered in fraction III.

Fraction III is marked by the copurification of two enzyme activities. Peptic activity is 8.4-fold enriched in the fraction and accounts for approximately 90% of the total protein. The second enzyme activity is an Mg^{2+}-stimulated ATPase activity which is 3.4-fold enriched over the activity seen

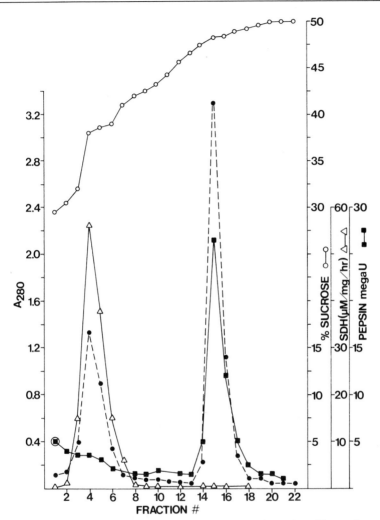

FIG. 2. Elution profile of fractions from the linear sucrose gradient. The crude granule pellet, P$_3$, was placed on a 38 to 50% linear gradient of sucrose and centrifuged for 4.5 hr at 108,000 *g*. Following the centrifugation, 1.5-ml fractions were collected and assayed as described in Materials and Methods. SDH, Succinate dehydrogenase.

in S$_2$. The lower enrichment of the ATPase may be the result of the presence of other ATPases with basal Mg^{2+}-stimulated activities in S$_2$ lowering the apparent enrichment of the granule-associated ATPase activity.

Figure 3 is an electron micrograph of fraction III taken from the linear

TABLE II
Linear Gradient Centrifugation

Marker enzyme[a]	P_3	I[b]	II[c]	III[d]	Percentage recovery[e]	Enrichment[f]
Protein (mg/ml)	4.47 ± 0.38 (n = 8)	0.4 ± 0.07 (n = 8)	0.53 ± 0.1 (n = 8)	4.1 ± 0.3 (n = 8)	104.2	2.27
Alkaline phosphatase (μmol/mg/hr)	0 (n = 4)	0 (n = 4)	0 (n = 4)	0 (n = 4)	—	0
Acid phosphatase (μmol/mg/hr)	1.58 ± 0.2 (n = 4)	0.27 ± 0.03 (n = 4)	0.72 ± 0.1 (n = 4)	0.1 ± 0.04 (n = 4)	26.2	.15
Na$^+$,K$^+$-ATPase (μmol/mg/hr)	1.57 ± 0.15 (n = 4)	1.52 ± 0.1 (n = 4)	0.26 ± 0.07 (n = 4)	0 (n = 4)	37.8	0
H$^+$,K$^+$-ATPase (μmol/mg/hr)	0 (n = 4)	0 (n = 4)	0 (n = 4)	0 (n = 4)	—	0
Cytochrome c oxidase (μmol/mg/hr)	1.59 ± 0.1 (n = 2)	0 (n = 2)	0.44 ± 0.05 (n = 2)	0.16 ± 0.01 (n = 2)	26.8	.10
Succinate dehydrogenase (μmol/mg/hr)	59.14 ± 11.6 (n = 10)	0 (n = 10)	56.0 ± 1.0 (n = 10)	0.74 ± 0.15 (n = 10)	63.1	.05
RNA (μg/mg protein)	0 (n = 3)	0 (n = 3)	0 (n = 3)	0 (n = 3)	—	0
Mg^{2+}-ATPase (μmol/mg/hr)	6.58 ± 0.2 (n = 4)	1.4 ± 0.8 (n = 4)	6.2 ± 0.5 (n = 4)	20.4 ± 0.1 (n = 4)	273.7	3.38
Pepsin (MU/mg/protein)	414.8 ± 8.2 (n = 10)	11.5 ± 1.3 (n = 10)	6.98 + 0.13 (n = 10)	1406 + 6.8 (n = 10)	84.7	8.43

[a] Enzyme activities were determined as described in Materials and Methods. Values are the means ±SE for the enzyme activities of the indicated fractions.
[b] Fraction I: 30–36% sucrose.
[c] Fraction II: 38–40% sucrose.
[d] Fraction III: 48–49.5% sucrose.
[e] Recoveries are calculated on the basis of P_3 versus the total activity of all fractions.
[f] Enrichment is calculated on the basis of the activity in fraction III divided by that measured in S_1 (Table I).

172

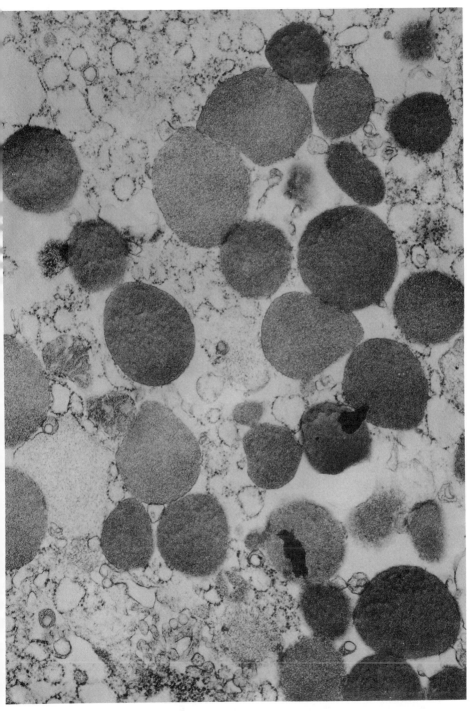

FIG. 3. Electron micrograph of fraction III. The fraction sedimenting between 48 and 49.5% sucrose was prepared for electron microscopy as described in Materials and Methods. (×16,600)

gradient and prepared as detailed in the Materials and Methods section. The micrograph demonstrates the absence of recognizable intact mitochrondria in agreement with the marker enzyme studies of this fraction. In addition to the intact granules, microsomal-like material appears to be present despite the absence of RNA or H^+,K^+-ATPase activity in this fraction. To determine the source of this material, the various fractions separated during the isolation procedure were tested for their reactivity with a monoclonal antibody which has been demonstrated to be specific for the H^+,K^+-ATPase isolated from hog mucosa and to cross-react with the rabbit enzyme.[18] The results of these studies demonstrated the absence of any contaminating parietal cell apical membranes. Although the source of this material remains unknown, it is possible that these membrane forms represent granules which have lysed during the fixation procedure, similar to the results seen with mast cell granules[19] and serotonin granules.[20]

Methodological Variation

At the initial stage of the differential centrifugation procedure, it is possible to improve the purification prior to the linear sucrose gradient centrifugation step. The method is a variation of the method of Castle et al. for the isolation of rat parotid secretory granules.[21] The method utilizes a low-speed centrifugation, 5000 g for 20 min, and 40% sucrose plus 40 mM Tris-SO$_4$, pH 8. The granules are layered directly on the 40% sucrose, and 30% sucrose plus 40 mM Tris-SO$_4$, pH 8 is placed on top of the granules. Lighter contamination floats to the 30% sucrose, while heavy contamination forms a pellet. The pepsinogen granules remain at the 40–30% sucrose interface and may be collected by aspiration. This variation avoids the problems associated with pelleting the mechanically labile granules and granule aggregation, but does tend to dilute the sample prior to linear gradient centrifugation. If the sample volume is too large, a 5000 g/10 min centrifugation may still be necessary to pellet the granules prior to the linear gradient centrifugation.

Further Comments

Once isolated, continued washing in 0.5 M sucrose plus 40 mM Tris-SO$_4$, pH 8 followed by low-speed centrifugation results in decreased yield

[18] A. Smolka, H. F. Helander, and G. Sachs, *Am. J. Physiol.* **245**, G589 (1983).
[19] P. G. Kruger, D. Lagunoff, and H. Wan, *Exp. Cell Res.* **129**, 83 (1980).
[20] S. E. Carty, R. G. Johnson, and A. Scarpa, *J. Biol. Chem.* **256**, 11244 (1981).
[21] J. D. Castle, J. D. Jamieson, and G. E. Palade, *J. Cell Biol.* **64**, 182 (1975).

of pepsinogen with little or no change in purification of the associated ATPase activity. It is thought that the granules are labile to continued pelleting by centrifugation, and it is not recommended to wash and pellet the isolated granules more than two or three times. If further washes are desired, the methodological variation using 40% sucrose discussed above is recommended. This method avoids the loss of granules due to pelleting while providing a method of washing the final granule fraction as well as concentrating this fraction.

Conclusion

A method is described which isolates a four- to eightfold purified pepsinogen granule fraction from the gastric chief cell. The degree of purification is similar to other zymogen granule preparations using differential centrifugation[22] or Percoll gradient centrifugation.[23] Like other zymogen granule preparations, this preparation has limited stability even in low ionic strength and hypertonic buffers. Electron micrographs suggest two possible sources of contamination in the pepsinogen granule fraction: the empty vesicles, which may be the source of the associated ATPase activity, and a microsomal-like material. An alternative explanation is that during preparation of the granule fraction for electron microscopy, a percentage of the granules lysed, and the empty vesicles and the aggregated material originate from the granules. This interpretation is supported by the marker enzyme studies indicating no measureable RNA or H^+,K^+-ATPase and the immunological studies. Whatever the source of this material, the pepsinogen granule fraction is suitable for studies on ion pathways, provided an assay unique to the granules is used. Release of pepsinogen from the granules as assayed by loss of pelletable pepsinogen is such an assay.

[22] S. S. Rothman, *Biochim. Biophys. Acta* **241**, 567 (1971).
[23] R. C. DeLisle, I. Schultz, T. Tyrakowski, W. Haase, and U. Hopfer, *Am. J. Physiol.* **246**, G411 (1984).

[12] Isolation and Primary Culture of Endocrine Cells from Canine Gastric Mucosa

By TADATAKA YAMADA

Introduction

Somatostatin and gastrin are two polypeptide hormones that function to regulate acid secretion. They are localized within discrete endocrine cells in the gastric mucosa. While there are many intact systems that have been utilized to study the mechanisms controlling the release of these two hormones, the results obtained have been somewhat difficult to interpret for the following reasons: (1) both hormones may respond to a common stimulus, yet somatostatin inhibits gastrin release and gastrin stimulates somatostatin release, thus the release of one hormone may influence the release of the other; (2) the release of both hormones may be regulated by a variety of neural, hormonal, interstitial (paracrine), and lumenal factors, thus any stimulatory or inhibitory influence on hormone secretion cannot be confirmed as being direct or indirect; and (3) the cellular basis for the secretion of the two hormones at the level of the receptor or postreceptor signal transduction cannot be determined in an intact tissue. To complicate the issue further, within the gut, somatostatin is present in neural elements as well as in endocrine/paracrine cells. Thus, the source of somatostatin released from an intact tissue cannot be identified with ease. For these reasons, the development of methods to isolate somatostatin- and gastrin-secreting endocrine cells has greatly facilitated the efforts to elucidate the mechanisms governing their regulation.[1-10] Furthermore, these

[1] A. H. Soll, T. Yamada, J. Park, and L. P. Thomas, *Am. J. Physiol.* **247,** G558 (1984).

[2] T. Yamada, A. H. Soll, J. Park, and J. Elashoff, *Am. J. Physiol.* **247,** G567 (1984).

[3] A. H. Soll, D. A. Amirian, L. P. Thomas, J. Park, J. D. Elashoff, M. A. Beaven, and T. Yamada, *Am. J. Physiol.* **247,** G715 (1984).

[4] A. H. Soll, D. A. Amirian, J. Park, J. D. Elashoff, and T. Yamada, *Am. J. Physiol.* **248,** G569 (1985).

[5] T. Yamada, J. Park, A. Soll, K. Sugano, T. Chiba, and A. Todisco, *in* "Regulatory Peptides in Digestive, Nervous and Endocrine Systems" (M. J. M. Lewin and S. Bonfils, eds.), INSERM Symp. p. 19. Elsevier, Paris, 1985.

[6] K. Sugano, J. Park, A. Soll, and T. Yamada, *Am. J. Physiol.* **250,** G686 (1986).

[7] M. Matsumoto, J. Park, K. Sugano, and T. Yamada, *Am. J. Physiol.* **252,** G315 (1987).

[8] T. Chiba, J. Park, and T. Yamada, *in* "Somatostatin" (S. Reichlin, ed.), p. 229. Plenum, New York, 1987.

[9] A. Todisco, J. Park, E. Lezoche, H. Debas, Y. Tache, and T. Yamada, *Gastroenterology* **92,** 919 (1987).

[10] K. Sugano, J. Park, A. H. Soll, and T. Yamada, *J. Clin. Invest.* **79,** 935 (1987).

techniques have proved to be widely applicable toward isolation of other endocrine cells from the gastrointestinal tract.[11-13]

Principles

For the initial studies, canine gastric mucosa was selected as the tissue source because of the extensively available correlative information on the physiology of gastrointestinal hormones in dogs. Furthermore, techniques for dispersion and isolation of canine gastric parietal cells had been described already by Soll.[14] This method coupled the principles of enzyme digestion of mucosal tissues with a modification of the method of Amsterdam and Jamieson,[15] who noted, in their isolation of pancreatic exocrine cells, that chelation of calcium promoted disruption of cell junctions without cytotoxic effects. Thus, cells could be dispersed from mucosa in a functional state without excessive damage from prolonged exposure to proteolytic enzymes. With a minimum of additional effort these techniques were modified to isolate gastric endocrine cells.

For further purification of the cells of interest, counterflow elutriation offered advantages over other approaches because it permitted high-yield rapid enrichment of viable and functional cells. The physical theory of cell elutration is discussed in detail elsewhere.[16] Briefly, cells are separated by equilibration of three forces: centrifugal force, buoyant force, and fluid drag force. Since the magnitude of the first two forces varies with the third power of the diameter but only the first power of the density, the primary determinant for cell separation is size, although density is a minor factor not to be excluded. Initially, the freshly elutriated endocrine cells were used for study but unlike parietal cells, which function well in this state, they exhibited a high level of basal activity and did not respond reproducibly to any stimuli. Culturing the cells for 40 hr on a collagen matrix eliminated these problems and, in addition, enhanced endocrine cell enrichment.

Instruments and Materials

1. Beckman J6M refrigerated centrifuge with JE-6B or JE-10X elutriator rotors (Beckman Instruments, Palo Alto, CA)
2. Cole-Parmer Masterflex peristaltic pumps (1–100 rpm and 6–600 rpm) with 7015-20 pumpheads (Cole-Parmer, Chicago, IL)

[11] D. L. Barber, A. M. J. Buchan, J. Walsh, and A. H. Soll, *Am. J. Physiol.* **250**, G374 (1986).
[12] D. L. Barber, A. M. J. Buchan, J. Walsh, and A. H. Soll, *Am. J. Physiol.* **250**, G385 (1986).
[13] D. L. Barber, J. H. Walsh, and A. H. Soll, *Gastroenterology* **91**, 627 (1986).
[14] A. H. Soll, *J. Clin. Invest.* **61**, 370 (1978).
[15] A. Amsterdam and J. D. Jamieson, *J. Cell Biol.* **63**, 1037 (1974).
[16] P. C. Keng, C. K. N. Li, and K. T. Wheeler, *Cell Biophys.* **3**, 41 (1981).

3. Nylon mesh (63 and 240 threads/in.; Naz-Dar/KC, Troy, MI)
4. Tissue culture plates (24-mm multiwell plates; Flow Laboratories, McLean, VA)
5. Rat tail collagen (prepared by the method of Bornstein[17])
6. Collagenase (type I; Sigma, St. Louis, MO)
7. Hanks' balanced salt solution (HBSS), Earle's balanced salt solution (EBSS), basal medium Eagle (BME), Dubecco's modified Eagle's/ Ham's F-12 media (DME/HF-12, 1:1 mixture; Irvine Scientific, Santa, Ana, CA)
8. Other routine chemicals obtained through standard commercial sources

Tissue Digestion and Cell Dispersion

Mongrel dogs are conditioned in the university laboratory animal facility for 1 week prior to study. Following an overnight fast, the animals are anesthetized with intravenously injected sodium pentobarbital and their stomachs quickly excised. After washing with ice-chilled HBSS containing 0.1% bovine serum albumin (BSA), the stomach is opened and divided into antral and fundic (body) segments, which are then each divided into two roughly equal halves. The fundus serves as a source for somatostatin-containing D cells and the antrum provides both D cells and gastrin-containing G cells. The fundic mucosa is bluntly dissected free from the submucosal tissues with relative ease in a 15-cm Petri dish in iced HBSS/0.1% BSA. In contrast, the antral mucosa is more tightly anchored to the submucosal layers, thus the tissue must be pinned to a corkboard and an incision made through the mucosa before the two layers can be separated gently but tenaciously at their junction by blunt dissection. The mucosa may require light scraping with a glass microscope slide to remove any mucus that is present. The separated mucosal sheets are then placed in a beaker on ice and finely minced with a sterilized scalpel (#21 blade). After being divided into equal portions of approximately 10–12 g each, the minced tissue is placed into a 250-ml screw-top Corning flask (Corning, NY) containing 50 ml of BME with 0.22% $NaHCO_3$, 10 mM HEPES, 2 mM glutamine, 0.1% BSA, and 350 mg/liter crude collagenase (buffer A). The mixture is gassed thoroughly with 95% air/5% CO_2, then the flask is tightly capped and incubated for 15 min at 37° in a shaking water bath set at 150 oscillations/min. At the end of the incubation, the medium is decanted and the cells are washed for 30 sec with 50 ml of Ca^{2+}/Mg^{2+}-free EBSS containing 0.22% $NaHCO_3$, 10 mM HEPES, and 2 mM EDTA

[17] M. B. Bornstein, *Lab. Invest.* **7**, 134 (1958).

(buffer B), then incubated for 10 min at 37° in the same buffer supplemented with only 1 mM EDTA and, in addition, BME, amino acids, 2 mM glutamine, and 0.1% BSA. After washing with buffer B the cells are incubated sequentially for 15, 60, and 30 min in buffer A. The results of a typical antral G cell dispersion are depicted in Fig. 1. The highest yields for endocrine cells are usually obtained in the final two collagenase digestion periods, thus these cells are utilized for further processing.

The dispersed cells are pooled, diluted with an equal volume of ice-chilled HBSS/0.1% BSA, then filtered through relatively coarse nylon mesh (63 threads/in.) and centrifuged at 50g for 5 min in 50 ml plastic centrifuge tubes. The cell pellets are washed in HBSS/0.1% BSA and centrifuged as before three additional times before filtering again through a finer (240 threads/in.) nylon mesh. At this point the cells should be clearly dispersed with no evidence of clumping. The average yield approximates 10^8 cells/g of mucosa and viability, as determined by Trypan Blue exclusion, generally exceeds 95%.

Cell Elutriation

The choice of rotor utilized for elutriation is dictated by the number of cells to be separated. The small (JE-6B) and large (JE-10X) Beckman

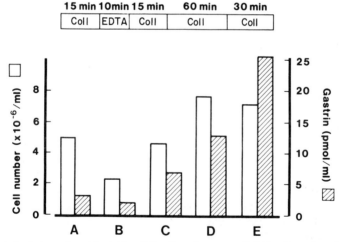

FIG. 1. Isolation of enriched G cell fractions by sequential digestion of canine antral mucosa. Cell number and gastrin concentration from a representative elutriation are depicted. (A) Collagenase (Coll) digestion (15 min); (B) EDTA treatment (10 min); (C) collagenase digestion (15 min); (D) collagenase digestion (60 min); (E) collagenase digestion (30 min).

elutriator rotors are functionally equal except that the latter can handle a 5- to 10-fold greater load of cells in a single run. The number of cells obtained from one dog antrum is usually sufficiently limited to permit efficient elutriation with the small rotor in a few runs, but for the more numerous fundic cells the larger rotor is optimal if one does not wish to work long into the night.

From this stage, it is important to prevent bacterial contamination of the cells, thus sterile techniques are applied throughout media preparation and cell separation and the elutriator rotor and tubing are sterilized by circulating 70% ethanol through the system. Roughly $2-4 \times 10^8$ cells are suspended in 40 ml of HBSS/0.1% BSA and loaded into the JE-6B rotor via a three-way stopcock connected to the lower speed peristaltic pump (1–100 rpm). The elutriation speed is set at 2800 rpm and the flow rate of the pump is set at 21.5 ml/min. After a 150-ml wash with HBSS/0.1% BSA to clear the debris, bacteria, and erythrocytes and other small cells, the elutriator speed is decreased to 2100 rpm. The effluent (150 ml) is collected in three 50-ml conical centrifuge tubes in a small portable laminar flow hood. Then the rotor speed is decreased to 2000 rpm and the flow rate is increased successively to 32.3, 44.5, and 103.5 ml/min with 150 ml of effluent collected at each step. These methods can be adapted to the larger elutriator rotor (JE-10X) with the use of the higher speed peristaltic pump (6–600 rpm). Approximately 2×10^9 cells suspended in 80 ml of HBSS/0.1% BSA are washed at a flow rate of 35 ml/min with the rotor speed set at 1200 rpm. After a 500-ml wash, the flow rate is increased to 75 ml/min and the next 500 ml of effluent is collected for the D cell-enriched fractions.

Figure 2 provides an example of a typical elutriation of antral mucosal cells. There is approximately 85% recovery of the 3×10^8 cells applied and the fraction with the largest number of G cells are 4,5 and 6,7. In the antrum, D cells appear to be widely distributed across all of the elutriated fractions but the greatest concentrations are found in 2,3. In the fundus, however, D cells are mostly concentrated in the 2,3 fraction and are reasonably well separated from other cellular elements (Fig. 3). It is difficult to assess enrichment of endocrine cells at this point on the basis of extractable immunoreactivity per unit tissue mass because of the inevitable loss of peptide by spontaneous release or cellular damage during the isolation procedure. A more accurate reflection of the degree of purification is obtained by immunohistochemistry. The endocrine cells compose roughly 3–5% of the cell population in the most enriched elutriator fractions while they compose a miniscule proportion, perhaps less than 0.1%, of the total gastric mucosal cells.

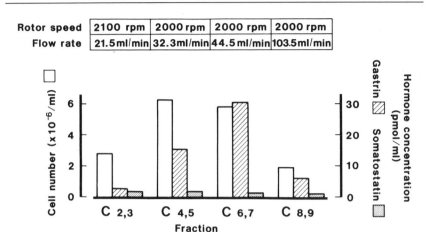

FIG. 2. Gastrin-containing G cell enrichment from antral mucosa by counterflow elutriation. Cell numbers as well as gastrin and somatostatin concentration from a typical elutriation are depicted. Rotor speed and flow rate are noted at the top of the figure.

Cell Culture

The freshly elutriated cells are centrifuged in 50-ml aliquots at 1500 rpm in a Beckman TJ-6 centrifuge, then resuspended in 10 ml of DME/HF-12 medium supplemented with gentamicin (100 μg/ml) and 10% dog serum (inactivated by treating for 30 min at 56°). After the cell concentration is determined in a Coulter counter, additional medium is applied to adjust the cell count to $2-3 \times 10^6$/ml. The cells are now ready for plating into tissue culture wells.

The tissue culture plates must be prepared with collagen gel beds approximately 3 hr prior to plating cells. The collagen gel mixture consists of rat tail collagen (2.0 g/liter), BME (1.5×), and NaOH (0.02 N) and is kept ice chilled until 1 ml is poured into each tissue culture well. The collagen is polymerized and gels are formed by incubating the plates for 30 min at 37°, then 2 ml of DME/HF-12 medium supplemented with gentamicin (100 μg/ml) and 10% dog serum is added to each well and the plates are equilibrated for 2 hr at 37° in a humidified atmosphere of 95% air/5% CO_2. In a laminar flow hood, the equilibration medium is carefully aspirated, and 1 ml of elutriated cells is poured into each well. After the cells are incubated for 40 hr at 37° in 95% air/5% CO_2, they adhere to the collagen support as single cells or form islands of grouped cell monolayers (Fig. 4). On electron microscopy the cells retain their ultrastructure and their polarity (Fig. 5). A curious and fortuitous, yet unexplained, phenomenon that results from culturing the elutriated cells for 40 hr is endocrine cell enrich-

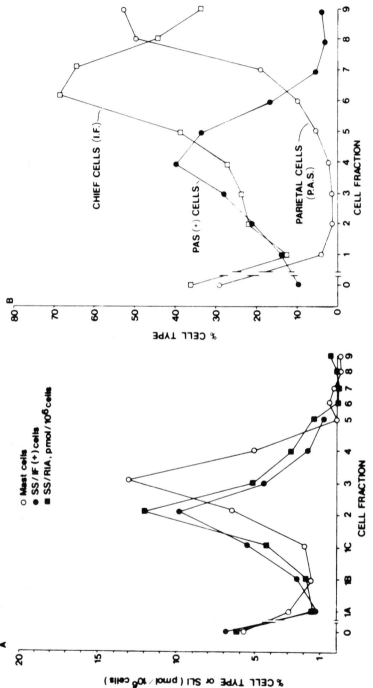

FIG. 3. Somatostatin-containing D cell enrichment from fundic mucosa by counterflow elutriation. Somatostatin (SS) distribution in (A) was determined by immunofluorescence (IF) and radioimmunoassay (RIA). Distribution of chief cells by IF, mucous cells by periodic acid–Schiff stain [PAS(+)], and parietal cells (also by PAS) is depicted in (B). (From Ref. 1.)

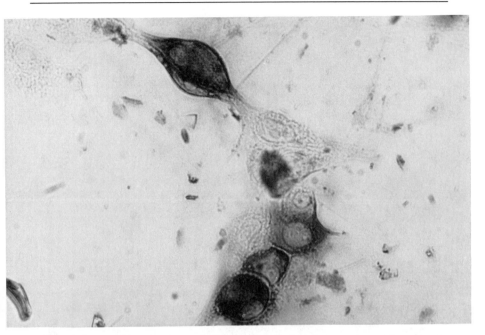

FIG. 4. Immunohistochemistry of enriched G cells in primary culture. Mucosal cells enriched with G cells by counterflow elutriation were cultured in plastic culture wells on a thin collagen film for 2 days. Media were carefully aspirated and washed with Earle's balanced salt solution twice, then the cells were fixed in Bouin's fixative. After washing, fixed cells were stained by the peroxidase–antiperoxidase method using diaminobenzidine as a substrate and antibody 5135 as a primary antibody. (From Ref. 10.)

ment (Table I). D cells are enriched from approximately 5% of the total population following elutriation to 70% after culture, while G cells are enriched from 2–3% to 20–30%.

Although G cells attach to the collagen gels as well as D cells, they exhibit a relatively high apparent level of background peptide release, even when the cultures are kept metabolically inactive by lowering the incubation temperature to 4°. There are several possible explanations for this phenomenon,[10] but it can be avoided almost entirely by using a thin collagen film in place of the gel as a cell support. This method is less efficient than the collagen gel in supporting cell attachment, as reflected in the smaller amount of gastrin harvested (collagen bed: 94 ± 8 pmol/well vs collagen coat: 21 ± 3 pmol/well, mean ± SE, $n = 9$). However, basal gastrin levels are sharply reduced, thus an improved signal-to-noise ratio can be obtained with this method in gastrin release studies (Fig. 6). To prepare the collagen films, 1 ml of a solution containing 3.3 g/liter rat tail collagen is poured into each tissue culture well and incubated at 25° for 15 min.

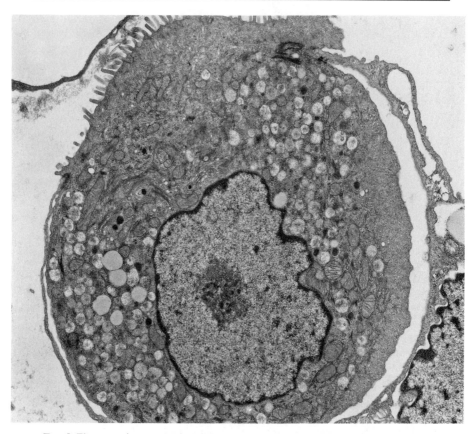

FIG. 5. Electron micrograph of a cultured G cell taken at a magnification of ×10,400. As with G cells *in situ*, the cultured cells have maintained some polarity, with an apical brush border, a central nucleus, and a supranuclear Golgi apparatus. The secretory granules are basally located, roughly of equal size, and filled with light flocculent material. (From K. Sugano, J. Park, W. O. Dobbins, and Yamada, *Am. J. Physiol.* **253**, G62 (1987).

Then, the collagen solution is aspirated and the plates are inverted over lids into which one drop of concentrated NH_4OH has been placed. After 15 min, the NH_4OH is aspirated, the lids are wiped clean and replaced, and the plates are dried overnight at 25°. Prior to pouring the cells, each culture well is equilibrated in 1 ml of supplemented DME/HF-12 medium for at least 1 hr.

Peptide Release Studies

In preparing the tissue culture plates for peptide release studies, they are first gently agitated to remove loosely attached cells. Then the medium is

TABLE I
CELL TYPES PRESENT AT INITIAL PLATING OF FUNDIC MUCOSAL CELLS
AND AFTER 48 HR OF CULTURE[a]

Stain	Percentage cells at initial plating	Percentage cells after 48 hr of culture
SLI	4.8 ± 0.9 (5)	69.8 ± 6.5 (5)
PGLI	8.9 ± 1.5 (4)	6.5 (2)
GLI	5.1 ± 1.4 (4)	2.8 (2)
5-HTLI	4.8 ± 1.7 (4)	0.0 (2)
Toluidine Blue	20.2 ± 6.5 (5)	5.8 ± 5.3 (3)

[a] Values are means ± SE; number of preparations is indicated in parentheses. Cells in the elutriator fractions at the time of plating and after 48 hr in culture on a bed of collagen were studied as follows: peroxidase–antiperoxidase staining was performed using antibodies to somatostatin-like immunoreactivity (SLI), pepsinogen-like immunoreactivity (PGLI), glucagon-like immunoreactivity (GLI), and serotonin-like immunoreactivity (5-HTLI); mast cells were detected by the presence of metachromasic granules on Toluidine Blue staining. (From Ref. 1.)

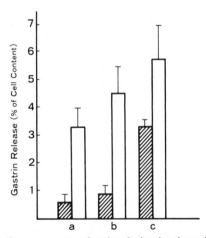

FIG. 6. Effects of a collagen support on basal and stimulated gastrin release. Hatched bars and open bars represent the data obtained from the culture wells coated with a collagen film and those with a collagen bed, respectively, under the following conditions: (a) basal gastrin release at 4°, (b) basal gastrin release at 37°, (c) bombesin (10^{-9} M)-stimulated gastrin release at 37°. At 4° bombesin (10^{-4} M) produced no stimulation of gastrin release over basal values as shown in (a). All experiments were performed with 2 hr of incubation and data represent mean ± SE ($n = 24$). (From Ref. 10.)

aspirated with a sterile Pasteur pipet and quickly replaced with 1 ml of EBSS supplemented with 10 mM HEPES and either 0.1% gelatin (for D cells) or 0.1% BSA (for G cells). The different additives are used because BSA appears to stimulate somatostatin release somewhat while gelatin may interfere with the radioimmunoassay for gastrin. In rapid succession the EBSS is aspirated and the cells are washed two more times with additional 1-ml aliquots of supplemented EBSS. Finally, 1 ml of supplemented EBSS containing the specific test agent to be studied is added to each of the wells, taking care to adjust the pH of the medium to 7.4 with dilute HCl or NaOH. The standard test plates are incubated at 37° in 95% air/5% CO_2 but a control plate should be incubated at 4°. Furthermore, in each plate incubated at 37°, two wells should be reserved as unstimulated control cultures. For general studies a 2-hr incubation is optimal inasmuch as peptide release increases linearly during that interval and a large increment over values obtained with the 4° cultures can be demonstrated (Fig. 7).

At the end of the 2-hr incubation period, the medium is aspirated, placed into a 1.5-ml conical centrifuge tube, and microfuged for 1 min. The supernatant is stored at −20° until asay for immunoreactive peptide. Since the concentration of endocrine cells in any given preparation may vary, it is inappropriate to express the response of the cell cultures to various treatments as absolute peptide concentrations in the medium. The differences between preparations may obscure the data and make comparisons between different experiments difficult to analyze. Thus, the most meaningful expression of the results can be obtained if peptide release is measured as a function of total cell peptide content. To obtain this value for D cells, the collagen gels are lifted off the tissue culture wells, placed into test tubes, and extracted in acetic acid at a final concentration of 3%. The G cells are lifted off the collagen-coated wells with 1 ml of NaOH and the mixture is placed into test tubes. The test tubes are placed into a boiling water bath for 10 min, vortexed, and centrifuged at 1000 g for 10 min. The supernatants are stored at −20° until time of assay.

Comments

The only experiments described in this chapter are peptide release studies, but it is obvious that these gut endocrine cells can be utilized for a wide range of pharmacological, physiological, and biochemical experiments with little or no adaptation of the described methodology. However, the investigator who intends to utilize these isolated cells must be willing to make a major commitment of his laboratory resources toward such an effort. The preparations generally require the attentions of more than a single invesigator and must be carefully planned well in advance. A typical

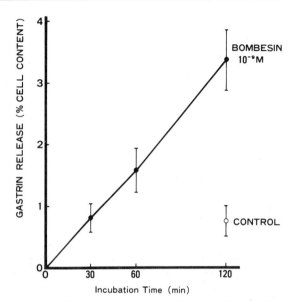

FIG. 7. Time course of bombesin-induced gastrin release from cultured antral G cells. Cultured cells were incubated with bombesin (10^{-9} M) and the media from two wells were harvested at the time indicated. The data (mean $\pm$ SE) from three separate dog preparations were used for data calculation. (From Ref. 10.)

week's schedule might include media and glassware preparation on Monday, cell isolation, elutriation, and culture on Tuesday, release studies and cell extractions on Thursday, and radioimmunoassays on Friday. Furthermore, as anyone who has experimented with primary cell cultures is well aware, the ability to obtain reproducible results with uncontaminated cells week after week depends on the dedicated effort of one individual who is uncompromisingly committed to the preparations.

Acknowledgments

The techniques described in this chapter were made possible by the generous willingness of Dr. Andrew Soll to share information and laboratory resources in a close collaborative effort that has continued over 7 years. Doctor Kentaro Sugano, Dr. John DelValle, and Mrs. Jung Park have made important contributions in this effort. This research was supported by NIH Grants DK33500 and DK34306 and by funds from the Michigan Gastrointestinal Peptide Research Center (DK34933). I am grateful to Lori Ennis for typing this manuscript.

[13] Isolation of Pancreatic Islets and Primary Culture of the Intact Microorgans or of Dispersed Islet Cells

By CLAES B. WOLLHEIM, PAOLO MEDA, and PHILIPPE A. HALBAN

Anatomical Distribution and Cellular Organization of Pancreatic Islets

In most animals, the cells of the endocrine pancreas are grouped into numerous (about one million in man, 2000–3000 in most laboratory rodents) and small (20–800 μm in diameter) clusters called islets of Langerhans.[1,2] These clusters are scattered throughout the exocrine portion of the pancreas and together comprise only about 1% of the volume of the adult gland. About one-third of the islets of Langerhans are found in the lower portion of the pancreas head, the so-called duodenal pancreas, which in laboratory rodents is vascularized by branches of the superior mesenteric artery and is drained by the main ventral pancreatic duct. The remaining islets are found in the upper part of the head, the body, and the tail of the gland, the so-called splenic pancreas, which is vascularized by branches of the coeliac trunk and is drained by the main dorsal pancreatic duct.[3]

In both pancreatic regions, the islets of Langerhans are composed of at least four different types of endocrine cells, each making a separate hormone.[4] The topographical and numerical relationships of these cells are characteristic of a given species. In laboratory rodents, the insulin-producing β cells form the central core of the islet, whereas the glucagon-producing α cells, the somatostatin-producing δ cells and the pancreatic polypeptide-producing PP cells are arranged in one or two layers at the islet periphery. Thus, each islet consists of a homocellular central region formed almost exclusively by β cells and a heterocellular peripheral region in which β, α, δ and PP cells are intermixed.[5] While the proportions of insulin (60–80% of the islet cells) and somatostatin-containing cells (2–5% of the islet cells) are similar throughout the pancreas, those of α and PP cells

[1] B. W. Volk and K. F. Wellmann, "The Diabetic Pancreas" (B. W. Volk and E. R. Arquilla, eds.), p. 117. Plenum, New York and London, 1985.
[2] J. R. Henderson, *Lancet* 2, 469 (1969).
[3] L. Orci, *Diabetes* 31, 538 (1982).
[4] L.-I. Larsson, F. Sundler, and R. Hakanson, *Diabetologia* 12, 211 (1976).
[5] L. Orci and R. H. Unger, *Lancet* 2, 1243 (1975).

differ in the islets from the splenic and the duodenal pancreas.[6] Whereas in the former region α and PP cells represent 15 and 0.5% of the islet cells, respectively, in the latter region these proportions are reversed, with α cells representing 1% and PP cells 14% of the islet cells.[6]

Besides the endocrine cells, large numbers of capillaries lined by fenestrated endothelial cells are the most conspicuous component of pancreatic islets.[1,7] These vessels form an intricate network which is closely apposed to islet cells from which it is separated by a basal lamina and, usually, by only minute amounts of connective tissue.

Isolation Methods for Obtaining Islets from the Adult Pancreas

Islets of Langerhans can be isolated mechanically or by the combination of a mechanical and an enzymatic approach.

Microdissection

For microdissection, the pancreas is gently stretched over a black-stained Sylgard (Dow Corning, Midland, MI)-coated dish and secured to it with small dissecting pins. The preparation is continuously bathed in a Krebs-Ringer-bicarbonate (10 mM) buffer containing 5 mM HEPES and viewed under a dissecting binocular microscope ($\times$ 15). In control animals, islets of Langerhans are seen as ovoid or round whitish structures which can be freed by cutting and tearing the surrounding pink–yellow exocrine tissue, using two 30-gauge hypodermic needles. The method has been mostly used to isolate mouse islets but can be adapted to several other species as well.[8,9]

The main advantage of microdissection is to avoid the use of enzymes which may damage islet cells or alter their functioning. Thus, differences between microdissected and collagenase-isolated islets have been reported.[8,10] Furthermore, the technique is unique in providing a way to isolate islet tissue when enzymatic procedures cannot be applied, e.g., after chemical or physical fixation of the pancreas.[8] The main disadvantage of microdissection is that only a limited number of islets, most often incompletely cleaned of the surrounding acinar cells and connective tissue, can

[6] L. Orci, D. Baetens, M. Ravazzola, Y. Stefan, and F. Malaisse-Lagae, *Life Sci.* **19**, 1811 (1976).

[7] O. Ohtani, T. Ushiki, H. Kanazawa, and T. Fujita, *Arch. Histol. Jpn.* **49**, 45 (1986).

[8] J. H. Nielsen and A. Lernmark, "Cell Separation: Methods and Selected Applications" (T. G. Pretlow and T. P. Pretlow, eds.), p. 99. Academic Press, New York, 1983.

[9] C. Hellerstrom, *Acta Endocrinol.* **45**, 122 (1964).

[10] J. C. Henquin and H. P. Meissner, *Experientia* **40**, 1043 (1984).

be isolated in any reasonable period of time. Furthermore, microdissection is quite laborious when small or medium-size islets are required (large islets are usually selected for dissection) or when a modification of the appearance of the endocrine or exocrine tissue complicates the identification of the islets within the pancreas. Thus, microdissection is hardly applicable for the isolation of islets from animals with insulin-dependent diabetes or in which β cells have been degranulated by chronic secretagogue stimulation. Under the latter conditions, islets appear pale gray and a vital staining of their capillary network is required to allow for their detection within the pancreas.[8,11]

Collagenase Digestion of the Pancreas

The method of choice for obtaining relatively large numbers of functional islets from a variety of animal species is to digest the pancreas with collagenase. A number of methods based on collagenase digestion have been published. All are aimed at obtaining the maximal number of intact islets, free of surrounding exocrine tissue. These methods can be subdivided into those which involve cutting the pancreas into small pieces before exposure to collagenase and those which depend upon distention of the pancreas by infusing collagenase into the pancreatic duct. Methods using distention through a blood vessel rather than the duct are less successful. The digestion procedure is followed in all cases by a separation procedure allowing for purification of the isolated islets from exocrine tissue. These procedures will be considered separately.

Digestion of Small Pieces of Pancreas. This method is adapted from the original procedure of Lacy and Kostianovsky.[12] It is suitable for the isolation of islets from rat and mouse pancreas. It has not been found to be successful for the isolation of islets from the pancreas of larger animals, these organs being, typically, more fibrous.

In rodents the pancreas is removed immediately after the animal is killed or anesthetized. Several pancreases can be combined in Krebs–Ringer–bicarbonate buffer containing 10 mM HEPES (KRB-HEPES), 2.8 mM glucose (for recipe, see Appendix). The pancreas is minced with scissors to obtain pieces measuring approximately 1 mm. The tissue is transferred to 15-ml conical centrifugation tubes and then washed in KRB-HEPES to remove free cells and fat (two centrifugations of 1 min each at 500 g). It is convenient for this and the following steps to use a plastic (polycarbonate is preferable) screw-cap centrifuge tube with calibrations indicating volume. Note however, that for any procedure involving

[11] V. Bonnevie-Nielsen, L. T. Skovgaard, and A. Lernmark, *Endocrinology* 112, 1049 (1983).
[12] P. E. Lacy and M. Kostianovsky, *Diabetes* 16, 35 (1967).

preparation of islet or islet cells it is essential that any graduations are indicated on the outside of the tube rather than being embossed on the inner face (both islets and cells adhere to these irregular embossed graduations). The volume of packed tissue is noted and an equal volume of KRB-HEPES added. The tissue is now ready for the addition of collagenase. The amount of enzyme used will depend upon its specific activity and its commercial source/purity. We have tested collagenase from three sources, and found batches suitable for islet isolation from each. A given collagenase type from a given supplier will not, however, always prove useful, since activity/properties change from lot to lot. We recommend that one of the following be tried:

1. Sigma Chemical Company (St. Louis, MO), collagenase type V or type XI: One unit (Mandl unit) is defined as the amount which liberates peptides from collagen equivalent to 1.0 μmol of leucine/5 hr at pH 7.4 and 37°. Typical activity is as follows: Type V, 200 U/mg; type XI, 1200 U/mg.

2. Serva Fine Biochemicals, Inc. (Westburg, NY), collagenase from *Clostridium histolyticum*, 0.6–0.8 U/mg, lyophilized research grade: One unit catalyzes the hydrolysis of 1 μmol 4-phenylazobenzyloxycarbonyl-L-prolyl-L-leucylglycyl-L-prolyl-D-arginine/min at 25°, pH 7.4. The activity of this collagenase is equivalent to approximately 500 Mandl units/mg.

3. Cooper Biomedical (Worthington) (Malvern, PA): collagenase type CLS IV, specific activity 125–180 Mandl units/mg.

As a rough guide, we suggest that initially digestion be attempted using 500 Mandl units/ml total volume of tissue plus KRB-HEPES. Collagenase should be added directly to the test tube containing the tissue. The combined volume of tissue and buffer should not exceed 5 ml (i.e., split the tissue into several tubes rather than overfilling one tube). If glass, rather than plastic, tubes are used, they should be siliconized. Each tube is now stoppered and shaken vigorously by hand while immersed in a 37° water bath. Several laboratories use a wrist-action shaker (we use model 75 from Burrell Co., Pittsburgh, PA); shake on position 6, which makes it possible to standardize the shaking and to digest many tubes at once. The digestion time will depend on the collagenase used and its concentration. The first signs of digestion will be the gradual disappearance of visible pieces of pancreas and the accompanying cloudiness and discoloration of the solution. Digestion (with shaking) should be allowed to continue until few large pieces of tissue are left. When the tube is inverted, the suspension should coat the walls of the tube and appear to have a milky consistency and to contain very small particles. To become accustomed to judging the digestion end-point, it is recommended that aliquots are taken repeatedly as from 4 min of digestion for observation under a dissection microscope (epi- rather than transillumination) or by phase-contrast microscopy. Islets appear as spheroid, opaque objects [of 50- to 500-μm diameter (Fig. 1)] against a sandy background of acinar cell debris. Collagenase digestion of a normal rat pancreas typically yields from 200 to a maximum of 700 islets, i.e., only ~10% of the total available islet population of a pancreas. If two rat pancreases were being digested in a total volume of 8 ml, and 100 μl was taken as an aliquot to monitor digestion, the most one could hope for would be 20 islets.

Although there are published methods to differentiate the isolated islets from contaminating lymph nodes and exocrine acini under the dissecting microscope, in our hands these have not proved useful. Thus, it is strongly recommended that the investigator take a sample of what is regarded as being islet and exocrine material and have it examined by electron microscopy.

If the pancreas is underdigested, there will be many islets left either trapped by exocrine and connective tissue or surrounded by clusters of exocrine cells. If overdigested, the islets will no longer show their normal, relatively smooth surface (Fig. 1). Overdigestion will also result in complete disruption of some islets, with an associated decrease of islet yield.

At the end of the digestion, the tissue is washed three times with KRB-HEPES containing 0.5% bovine serum albumin (BSA). The first wash is followed by centrifugation at 150 g for 30 sec, the second and third at 80 g for 30 sec, respectively. The final pellet is resuspended in KRB-HEPES-BSA (for handpicking) or in Histopaque (for gradient centrifugation); both methods are described below. Note that if the isolated islets are left in a vessel (whether plastic or glass) without protein in the buffer/medium, they will often stick to its walls.

Digestion of the Whole Rat Pancreas. Digestion in a shaking water bath: This method is an adaptation of that of Noel *et al.*[13] The rat (optimal weight 170–200 g) is anesthetized (phenobarbital or ether) and a laparotomy performed. The common bile duct is ligated at its exit into the duodenum. The duct is then cannulated (1-in. length, 20- or 22-gauge Teflon catheter) at its exit from the liver. The pancreas may then be distended by injecting 15 ml TC199 culture medium (with Hanks' salts) containing 10% newborn calf serum (NBCS) and collagenase. The type of collagenase used is as for digestion of pancreas pieces (see previous section). Again, the concentration for optimal digestion must be established by the investigator for each collagenase lot. As a guide, a useful concentra-

[13] J. Noel, A. Rabinovitch, L. Olson, G. Kyrinkides, J. Miller, and D. H. Mintz, *Metabolism* **31**, 184 (1982).

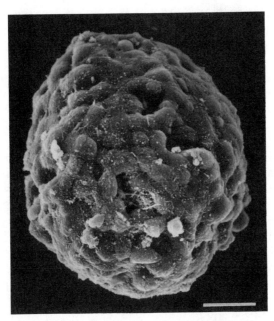

FIG. 1. Intact rat islet isolated by collagenase digestion and seen by scanning electron microscopy. (Bar = 25 μm.)

tion for initial screening is typically 500 Mandl units/ml. Some of the collagenase solution will inevitably leak out of the pancreas as it is distended. Excessive leaking and only modest, or localized, distention of the pancreas indicates improper cannulation of the duct. The catheter may be held in place (and leaking around the catheter prevented) by gentle pressure using small forceps. Once distended, the pancreas is dissected out of the animal and transferred to a 25 cm^2/50 ml capacity plastic tissue culture flask. Up to five pancreases may be combined in one flask, which is kept on ice until all the tissue has been added to it. The flask is now tightly closed and placed in a shaking water bath at 37°. Digestion proceeds with shaking (120 strokes/min). The approximate digestion time will be 15–25 min, depending on the collagenase used. As for digestion of pancreas pieces, the end-point of digestion must be established by the investigator by taking aliquots and examining them under the dissecting or phase-contrast microscope. At the end of digestion the collagenase solution will contain a fine suspension of tissue with some much larger fragments of undigested tissue and blood vessels. The color of the solution changes from red/orange at the start of the incubation to red/brown at its termination. This change, when using the recommended culture medium, is both pronounced and abrupt.

The digested tissue is transferred to a 50-ml capacity conical plastic tube and washed twice in TC199/NBCS (1-min centrifugation at 500 g). The pellet is resuspended in TC199/NBCS and vortexed vigorously before filtration through a 40-mesh wire filter. This will remove large fragments of undigested tissue. The material remaining in the filter is transferred back into the conical tube and suspended to TC199/NBCS. After vortexing, it is once again passed over the filter and centrifuged at 500 g for 1 min. The pellet is then taken for purification of islets from exocrine tissue, as described below.

Static digestion: In many laboratories (including our own) the pancreas is digested, following distention with collagenase, under static conditions rather than in a shaking water bath.[14] The digestion time may be somewhat longer under such conditions. Ideally rats weighing 180–220 g should be used. Sprague-Dawley and Wistar (or Wistar-Lewis) provide good yields of islet; some other strains may prove less good. The procedure is as follows: distend the pancreas (same method as above) with 6 ml Hanks' balanced salts solution (HBSS-HEPES; see Appendix for recipe) supplemented with 5 mM CaCl$_2$ and 15 mg type I collagenase (Sigma) (for this purpose weigh out 15 mg collagenase in one tube and, just before injection, dissolve in 6 ml HBSS-HEPES supplemented with 30 μl of 1 M CaCl$_2$). It is critical that the rat be bled by cutting its heart just before the pancreas is distended (this dramatically improves islet yield). The pancreas is carefully removed from the rat to ensure that it does not leak extensively (some loss of collagenase solution is inevitable), any fat is dissected away, and the organ is then transferred to a 50-ml plastic conical tube. The tissue is kept on ice until all pancreases in the series have been distended and transferred to their individual tubes. The pancreas is (are) washed 2× with HBSS-HEPES prewarmed to 37° and then incubated at 37° in 10 ml HBSS-HEPES. Note that collagenase is not added to the external medium during digestion and that for these operations BSA is not present either. At the end of digestion (typically 20 min) add ice-cold HBSS-HEPES-BSA (0.35 g/100 ml) and swirl the tube to disrupt the tissue. Allow the tissue to settle for 1 min on ice. Aspirate the supernatant. Add 8 ml HBSS-HEPES-BSA and pass 6× through a 14-gauge (trocar) needle. Any large fragments of tissue should be removed to prevent clogging of the needle. Centrifuge for 10 sec at 50 g, aspirate the supernatant, and replace with 8 ml HBSS-HEPES-BSA. Repeat two aspirations through the 14-gauge needle. Centrifuge again and resuspend in 8 ml HBSS-HEPES-BSA. Pass through a strainer (~40 mesh; we use a nylon tea strainer for this purpose!) and collect in a 100-ml

---

[14] R. Sutton, M. Peters, P. McShane, D. W. R. Grey, and P. J. Morris, *Transplantation* **42,** 689 (1986).

siliconized glass beaker. Wash the tube and strainer extensively with HBSS-HEPES-BSA. Centrifuge at 500 g for 10 sec. Aspirate the supernatant, replenish with HBSS-HEPES-BSA and centrifuge at 350 g for 10 sec. Aspirate as much supernatant as possible. The preparation is now ready for density gradient separation of islets from exocrine tissue (see the section Methods Based on Density, below). After washing, the islets should ideally be handpicked (see the section, Handpicking, below) before use.

Application of existing methods to pancreas of larger animals or man: The collagenase distention method (with either shaking or static conditions for digestion, and various other modifications appropriate to each animal species) has been used with success on the pancreas of, notably, dog,[13] pig,[15] and man.[16] Despite recent advances, including the development of automated techniques for the mass isolation of human islets,[17] there is, to our knowledge, no published method which reliably and reproducibly provides large quantities of clean, functional, human islets.

Purification Methods to Separate Islets from Exocrine Tissue

Methods Based on Size

We have developed a method for islet purification which depends upon the relative sizes of islets and collagenase-digested exocrine tissue.[18] This method is based on filtration through meshes of varying sizes and is useful only when the islets obtained by collagenase digestion are clearly of a different size than acini. This is often the case for rat islets but not necessarily for those of other animal species. In addition, some lots of collagenase produce a fine granular suspension of exocrine cells which adhere to islets and clog up the filters. We now recommend the use of Ficoll or Histopaque gradients over the filtration method, despite the disadvantages noted below.

Methods Based on Density

Several methods for separating islets from exocrine tissue based on their relative densities have been described. These methods have depended on the use of a gradient of varying concentrations of sucrose, Percoll, Ficoll, Histopaque, BSA, or metrizamide. The best separations are achieved using Ficoll or Histopaque. The method of choice for rat islets used on a routine

---

[15] C. Ricordi, E. H. Finke, and P. E. Lacy, *Diabetes* **35,** 649 (1986).
[16] D. W. R. Gray, P. McShane, A. Grant, and P. J. Morris, *Diabetes* **33,** 1055 (1984).
[17] C. Ricordi, E. H. Finke, B. J. Olack, and D. W. Scharp, *Diabetes* **37,** 413 (1986).
[18] R. E. Offord and P. A Halban, *Biochem. Biophys. Res. Commun.* **82,** 1091 (1978).

basis in our laboratory depends on gradients of Histopaque 1077 obtained from Sigma (this is a mixture of Ficoll 400 and sodium diatrizoate). The pellets obtained from the collagenase digestion, filtration, and washing of the rat pancreas (the product of one pancreas per conical tube) are resuspended in 10 ml Histopaque and well mixed. It is important that the suspension is homogeneous. The suspension is then overlayered with 10 ml HBSS-HEPES-BSA (0.35 g/100 ml), and the tube centrifuged in a swing-out rotor for 20 min at 900 g and 15°. Acceleration and deceleration must be slow to prevent any disturbance to the interface. The islets are collected from the interface and washed twice in HBSS-HEPES-BSA (1 min at 400 g followed by 30 sec at 250 g). One or two extra 15-sec centrifugation steps at 40 g will ensure that the islets recovered in the final pellet are essentially free of fine debris. The entire procedure provides remarkably clean islets but is dependent on rigorous experimental procedure. In particular, it is critical that (1) the HBSS-HEPES-BSA is prepared at the correct osmolarity and pH and (2) that the pancreas digest pellet contain essentially no supernatant before the Histopaque is added (to avoid dilution). Changing the ratio of tissue to Histopaque may also reduce the efficiency of this purification step.

It should be noted that although the combination of the pancreas distention/static digestion technique with purification on Histopaque gradients provides large numbers of quite pure islets (typically 500–800 islets/rat pancreas), a period of tissue culture is required before they can be used for studies. We recommend 18–24 hr in Dulbecco's modified Eagle's medium (DMEM) containing 8.3 mM glucose and 10% NBCS.

Handpicking

The least traumatic separation method from the point of view of the islets is to handpick them under a dissection microscope. This is, however, the most traumatic for the investigator, being time consuming and stressful on the eyes and wrist! It also yields fewer clean islets, most probably because many of them go unrecognized and since only large islets are usually handpicked. The method is certainly not applicable to the purification of tens of thousands of islets by one operator.

For handpicking, a good epiillumination, as provided by a fiber-optic light source, which is bright but nonheating, is required. The islets (suspended in KRB-HEPES-BSA) should be observed against a black background. A useful way to achieve this is to spray black paint on the outer surface of the bottom of a glass Petri dish. It is also important to avoid using old, scratched dishes since the scratched surface will not provide a good background for picking. The black surface often supplied with dissecting microscopes is far from ideal since it typically has some reflective areas which closely resemble islets. The islets may be conveniently picked with a siliconized glass Pasteur pipet with a drawn-out tip, an Eppendorf or Gilson pipet using a yellow tip, or with a braking pipet (25-μl constriction pipet).

Isolation of Islet Cells

Numerous techniques have been devised to prepare dispersed islet cells. Most of these methods rely on a sequential chemical, mechanical, and enzymatic treatment of collagenase-isolated islets.[19-29] The following is a description of the procedure we have adopted for preparing single rat islet cells for culture, with some indications of the variations in use in other laboratories.

Chemical Treatment

Collagenase-isolated rat islets (2000–6000) are collected into a siliconized glass tube containing 1 ml KRB-HEPES supplemented with 0.5% (w/v) BSA and 200 mg/dl of glucose. This medium is changed twice to wash the isolated islets which are allowed to sediment by gravity for a few minutes between each of these short (2–3 min) rinsing periods. The islets are then exposed for 1 min to 5 ml of a KRB-HEPES-BSA prepared without adding Ca^{2+} and are finally incubated for 15 min at room temperature in 1 ml of a similar Ca^{2+}-poor KRB supplemented now with 3 mM EGTA.

As in most other tissues, calcium removal decreases islet cell adhesion,

[19] A. Lernmark, *Diabetologia* **10**, 431 (1974).
[20] J. Ono, R. Takaki, and M. Fukuma, *Endocrinol. Jpn.* **24**, 265 (1977).
[21] P. Meda, E. L. Hooghe-Peters, and L. Orci, *Diabetes* **29**, 497 (1980).
[22] H. Kromann, M. Christy, J. Egeberg, A. Lernmark, J. Nerup, and H. Richter-Olesen, *Med. Biol.* **58**, 322 (1980).
[23] D. G. Pipeleers and M. A. Pipeleers-Marichal, *Diabetologia* **20**, 654 (1981).
[24] T. Dyrberg, S. Baekkeskov, and A. Lernmark, *J. Cell Biol.* **94**, 472 (1982).
[25] J. O. Nielsen and A. Lernmark, *in* "Cell Separation: Methods and Selected Applications" (T. G. Pretlow and T. P. Pretlow, eds.), Vol. 2, p. 99. Academic Press, New York, 1983.
[26] A. Lernmark, T. Dyrberg, J. H. Nielsen, and the Hagedorn Study Group, *in* "Methods in Diabetes Research" (J. Larner and S. L. Pohl, eds.), Vol. 1C, p. 259. Wiley, New York, 1984.
[27] D. G. Pipeleers, *In* "Methods in Diabetes Research" (J. Larner and S. L. Pohl, eds.), Vol. 1B, p. 185. Wiley, New York, 1984.
[28] K.-D. Kohnert and B. Hehmke, *J. Biochem. Biophys. Methods* **12**, 81 (1986).
[29] P. A. Halban, C. B. Wollheim, B. Blondel, P. Meda, E. N. Niesor, and D. H. Mintz, *Endocrinology* **111**, 86 (1982).

probably by perturbing both Ca^{2+}-dependent cell adhesion molecules and intercellular junctions. Indeed, microscopic examination shows that after this chemical treatment, cells within islets have lost their compactness and polygonal shape, have become round, and remain in contact only at focal points across enlarged intercellular spaces.

Although calcium removal is the first step in most islet dispersion procedures in use today, a number of variations have been reported concerning the duration of the islet incubation in the Ca^{2+}-free medium (from a rapid washing up to 1 hr), the type (EGTA or EDTA) and concentration (1–3 mM) of the calcium chelator used, and the temperature (ambient to 37°) at which the incubations are performed.[19-29] In the absence of complete comparative studies of these different recipes, it would appear conservative to shorten as much as possible the exposure of the islets to the Ca^{2+}-free medium and to the chelator. However, in our hands, removal of Ca^{2+} for up to 1 hr did not obviously decrease the yield of viable β cells, as judged by their ability to attach to dishes and retain immunostainable insulin in culture.

Mechanical Treatment

At the end of the incubation in the Ca^{2+}-free and EGTA-containing medium, the islets are aspirated through a disposable polypropylene tip using a 1-ml pipet and are immediately redispensed by placing the pipet tip against the inside wall of a siliconized glass tube. This procedure is repeated three times using a large pipet tip (Gilson blue model, internal diameter of about 0.75 mm) and, then, three other times with a smaller pipet tip (Gilson yellow model, internal diameter of about 0.40 mm). After the first three passages through the large tip, the islets maintain their normal shape and size but separate from each other. By contrast, as soon as the preparation is aspirated through the small tip, the islets disrupt and the surrounding medium becomes cloudy due to the dense suspension of dispersed islet cells. At this stage, microscopic examination shows that the preparation consists of a mixture of single cells and clumps comprising 2 to 30 cells. Alternatively, the same dispersion is achieved by aspirating the islets first through an 18-gauge needle (three times), then a 20-gauge needle (three times), and, eventually, through a 25-gauge needle (three times). We have found this latter procedure more reliable.

Whereas most of the procedures developed to disperse islet cells include a mechanical treatment of the islets, the way (syringe needle, siliconized glass Pasteur pipet, polypropylene pipet tip, adapted micromixer), the duration, and the time at which this treatment is performed (only once after the chemical treatment, throughout the enzymatic treatment, etc.)

vary from one procedure to the next.[19-29] The heterogeneous islet cell preparation obtained after mechanical dispersion of the islets is already suitable for a number of physiological, biochemical, and morphological studies but not for acute experiments requiring only single cells or for the establishment of long-term islet cell cultures. Indeed, plating islet cell clumps leads almost inevitably to a rapid contamination of the cultures by fibroblasts and, rarely, to formation of endocrine monolayers.

In principle, a chemical–mechanical treatment avoiding the exposure of the islets to enzymes appears ideal. In practice, however, the preparations obtained without enzymatic treatment are usually unsatisfactory. Thus, the yield of living cells is limited and the cell population is rather heterogeneous, both because it always contains a mixture of single cells and clumps and because it is contaminated by other cell types, mostly fibroblasts, which are presumably trapped within the clumps. Thus, in most protocols, a further enzymatic treatment of the dispersed cells is performed.

Enzymatic Treatment

When a homogeneous preparation of single cells is required, the islets are dispersed as described above after replacement of the Ca^{2+}-free and EGTA-containing medium by 0.9 ml of a similar medium supplemented with 0.1% (w/v) trypsin (type 1 : 250; Difco, Detroit, MI).

The dispersed cell suspension is then transferred into a 25-ml siliconized glass spinner flask (Bellco, Vineland, NJ) and diluted with 10 ml of the Ca^{2+}-free and trypsin-containing medium. After a 3- to 5-min incubation at room temperature, under continuous rotation at about 80 rpm, the enzymatic treatment is stopped by adding to the cell suspension 10 ml of cold (4°) KRB (if cells are to be used in acute experiments) or RPMI 1640 medium (if cells are to be cultured). At this stage, microscopic examination shows that the preparation consists almost exclusively of fully dissociated, round cells (Figs. 2 and 3).

As for the other steps of the dispersion protocol, a number of differences exist between the various procedures in use, with respect to the enzymatic treatment of isolated islets.[1-10] Thus, several enzymes (trypsin, DNase, dispase, Pronase) have been used at various concentrations (0.02–5 mg/ml for trypsin; 1–5 μg/ml for DNase; 2–5 mg/ml for dispase; 1 mg/ml for Pronase), for different durations (3–30 min), and at various temperatures (ambient to 37°).[1-10] Furthermore, while in some protocols enzymes are used after incubation of the islets under conditions causing Ca^{2+} removal, in other protocols the Ca^{2+} removal and the enzymatic treatment are performed simultaneously.[19-29] The advantages/disadvan-

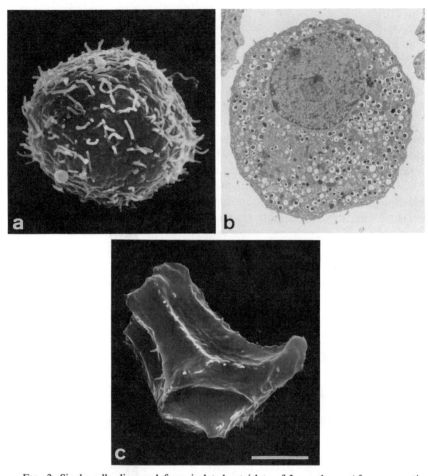

Fig. 2. Single cells dispersed from isolated rat islets of Langerhans. After enzymatic dispersion, islet cells have a globular shape and numerous microvilli scattered at their surface (a). The presence of typical secretory granules in the cytoplasm allows for the identification of most of the isolated islet cells as insulin-producing β cells (b). The appearance of enzymatically dispersed cells (a and b) contrasts markedly with that of cells dissociated mechanically from fixed islets and which have a polygonal shape with essentially smooth sides (c). (Bar = 5 μm.)

tages of each procedure remain to be established since few laboratories have compared different procedures on the same islet preparation and since some of the steps of the dispersion protocols are hardly amenable to a careful quantitative comparison (e.g., the activity of enzymes from a given manufacturer varies from batch to batch). Each laboratory should therefore perform careful quality control studies on its own dispersed cells.

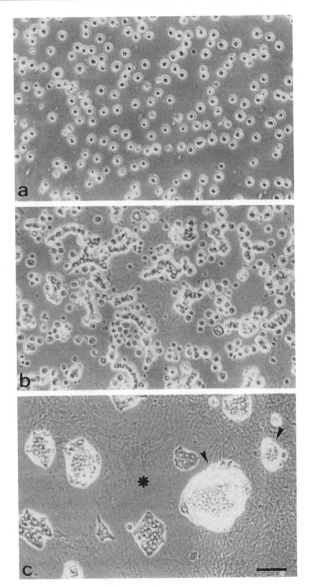

FIG. 3. Pattern of reaggregation of islet cells in static culture. Fully isolated cells obtained by mechanical and enzymatic (trypsin) dispersion of islets were plated (a). After 12 hr of culture, the islet cells reaggregated in small linear aggregates (b). After 24 hr of culture, larger clusters of islet cells are found attached to the culture dish (c). On a substratum of osmotically disrupted fibroblasts (asterisk), some islet cell clusters flatten to form monolayers while others (arrowheads) retain a more clumpy tridimensional organization (c). (Bar = 60 μm.)

Collection of Undamaged Cells

Although islet cells are ready to use immediately after dispersion, a separation of damaged from undamaged cells may be desired or required for some experiments. This can be achieved by layering the dispersed cells over either a Percoll medium at a 1.045 density,[21,28] or a 4% solution of BSA,[19,27] and by centrifuging the gradient for 5–10 min at 500 or 50 g, respectively. The intact cells, which penetrate the gradient, are collected at the bottom of the tube and are rinsed a few times in culture medium before use.

Preparation for Culture

In the absence of expensive sterile rooms or special equipment, it is virtually impossible to perform the procedure summarized above under strictly sterile conditions. Thus, if the dispersed cells are to be cultured, additional processing is required before plating. One milliliter of the dispersed cell preparation is layered on top of 12 ml of sterile RPMI 1640 medium supplemented with 10% fetal calf serum, in a sterile 15-ml conical tube. The cells are pelleted by a 10-min centrifugation at 80 g, the supernatant is carefully removed and the cell pellet is gently resuspended, by two or three aspirations through a sterile siliconized glass Pasteur pipet, in 100 μl of sterile culture medium. After transfer of this 100 μl of suspension into a second tube and addition of 12 ml of sterile medium, the cells are centrifuged again. This procedure is repeated three times and the cells are then suspended in 1 ml of the final medium to be used for culture. If this medium is supplemented with the regular antibiotics (penicillin and streptomycin), this protocol consistently ensures long-term cultures without obvious contamination.

Viability and Yield

The classic Trypan Blue exclusion test remains the most convenient and fastest way to assess the viability of dispersed islet cells. To perform this test, 10 μl of islet cell suspension is incubated for 10 min at 37° in 100 μl of 0.2% Trypan Blue freshly prepared in albumin-free KRB. A 10-μl aliquot is then taken and introduced into a hemocytometer for the evaluation of cell number and proportion of Trypan Blue-stained cells. While it should be recognized that lack of Trypan Blue stain is not a definitive index of cell viability, comparative tests have shown that the sensitivity of this simple method is similar to that of more laborious tests.[22] The viability of single islet cells usually appears better following isolation by enzymatic treatment than following a chemical–mechanical dispersion. However, it is still unclear whether this difference is real or only apparent, i.e, due to the fact that enzymes eliminate all altered cells, including those which, being only slightly damaged, are pelleted with the healthy cells in the absence of enzymes.

In our hands, 6 hr is required for three persons to isolate about 3000–5000 islets (from six adult rat pancreases) and to plate the $2-4 \times 10^6$ cells which can be dispersed from these microorgans. A high proportion of these cells are single, exclude Trypan Blue, attach to culture dishes and maintain, in short-term cultures, a normal ultrastructure and the ability to release insulin in response to a variety of secretagogues.

Separation of Islet Cell Types

The suspension of islet cells obtained by the methods described above will consist of a mixture of the four major endocrine cell types (α, β, δ, and PP cells). There will also inevitably be some contaminating exocrine cells, fibroblasts, and ductal and vascular cells. Finally, there will be a finite number of dead cells. For many purposes such a mixture will be unacceptable. Of the methods described for separating out the various cell populations, the most promising is the use of fluorescence-activated flow cytometry (the "cell sorter"). The method has been best described, and most successfully employed, by Pipeleers and co-workers and will therefore not be described again here in detail.[30] We shall, however, go over the essential steps of the method, and its limitations based upon our own experience.

The cell suspension is first subjected to filtration (to remove large aggregates) and to a centrifugation through Percoll (density 1.04 g/ml) for 6 min at 300 g to separate living cells (in the pellet) from dead cells and debris. The pellet is then resuspended in a physiological buffer and kept at low ambient glucose (2.8 mM) throughout the subsequent isolation and purification steps. The suspension is then submitted to autofluorescence-activated cytometry using an excitation wavelength set at 488 nm and an emission recorded at 510–550 nm. In this way, the observed cell fluorescence represents predominantly that of FAD. β cells can be separated from non-β cells based on two parameters: (1) increased forward-angle light scatter and (2) increased FAD fluorescence. Populations with the highest light scatter and FAD fluorescence consist mostly of β cells. After β cells have been sorted from non-β cells, the latter cells are submitted to a second passage through the cell sorter. Before this, the cells are exposed to 20 mM glucose at 37° for 15 min. Excitation in the cell sorter at 351–363 nm, with emission recorded at 400–470 nm, allows for measurement of cellular NAD(P)H. A major population with low fluorescence corresponds to α cells.

---

[30] D. G. Pipeleers, P. A. In't Veld, M. Van de Winkel, E. Maes, F. C. Schuit, and W. Gepts, *Endocrinology,* **117,** 806 (1985).

Another method for purifying β cells away from non-β cells is to use an anti-β cell antibody. This method has been described by Alejandro et al.,[31] using a mouse monoclonal antibody (R2D6) directed against a ganglioside found on the surface of islet β cells but not on non-β cells. Dispersed rat islet cells were incubated in the presence of R2D6 ascites (diluted by 1 : 100) in the cold, followed by exposure to a second fluorescein-labeled antibody directed against mouse IgM. β cells which bind the R2D6 antibodies were in this way fluorescently tagged and, accordingly, were sorted in a flow cytometer.

We have used these two methods for islet cell purification. Although both methods do indeed produce substantial purification of a given islet cell type, they also suffer from certain limitations. First, the cell preparation must be in the form of a single cell suspension since large aggregates will not be sorted appropriately, unless the cytometer gates are modified accordingly. Furthermore, in order to obtain a highly purified subpopulation of cells, the sorting gates must be set in a very stringent fashion to avoid any overlap in the parameters of other cell types. This results in a dramatic reduction in yield. In our hands, a 50% yield of highly purified cells is an unusually high value. This issue becomes particularly important if one is obliged to pass cells through the cytometer twice (as for obtaining purified α cells). Starting from one rat pancreas, which provides 500–1000 clean isolated islets, each islet can be expected to yield 1000 living single cells. Therefore, it follows that under optimal conditions, each pancreas will yield 500,000–1,000,000 single islet cells, of which only about 15% will be α cells. The first passage through the flow cytometer which will separate β from non-β cells based on FAD fluorescence, will allow for recovery of 50% of the α cells, i.e., about 40,000–80,000 cells. The second sorting, based on NAD(P)H fluorescence following an incubation at high glucose (during which some cells aggregate and are therefore lost for sorting purposes) will again yield only 50% recovery, i.e., about 20,000–40,000 cells. Such a yield severely limits the nature of subsequent experiments.

Primary Culture

Intact Islets

Many different methods for the *in vitro* maintenance of nondissociated mammalian islets have been described since the first report by Moskalewski in 1965.[32] The islets are either isolated by collagenase digestion or by

[31] R. Alejandro, F. L. Schienfold, S. V. Hajek, M. Pierce, R. Paul, and D. H. Mintz, *J. Clin. Invest.* **70**, 41 (1982).
[32] S. Moskalewski, *Gen. Comp. Endocrinol.* **5**, 342 (1965).

microdissection. Two different procedures have been used in which the microorgan is either allowed to attach to a substratum or not.

Culture Conditions Promoting Islet Attachment. The attachment of islets to glass coverslips or on conventional tissue culture plastic Petri dishes is thought to be facilitated by collagenase digestion,[33] probably because this digestion loosens or destroys the islet connective capsule. The culture medium (most often TC199 or RPMI 1640) is supplemented with antibiotics (100 U/ml penicillin G and 100 μg/ml streptomycin) and 10% NBCS or 10% fetal calf serum (FCS). Antibiotics should also be included in all buffers during the isolation procedure. The choice of glucose concentration of the medium will depend on the experimental design. Best results in terms of insulin biosynthesis and secretion have usually been seen in islets maintained at glucose concentrations between 6 and 11 mM (see below). In general, medium changes are not required for short-term (5–7 days) culture of the islets. If the cultured islets are to be used in experiments requiring nonattached tissue, they can be detached from the substratum either by the use of a rubber policeman or by adding trypsin (0.1%).

Culture of Islets in Suspension. The maintenance of free-floating islets under conditions not allowing attachment facilitates islet harvesting but makes medium changes more cumbersome. We have cultured collagenase-isolated islets from rats, mice, or Chinese hamsters by suspending 200–400 islets in 4–5 ml of DMEM, RPMI 1640, or T199 medium containing 8.3 mM glucose, 100 U/ml penicillin, 100–200 μg/ml streptomycin, and 10% heat-inactivated NBCS (or FCS).[34] Glutamine is added only to culture media devoid of this amino acid. As the islets are kept in a tissue culture incubator with a humidified atmosphere gassed with 5% CO_2/95% air, it is necessary to include in the culture medium the appropriate concentration of $NaHCO_3$ The islets are placed in Petri dishes designed for bacterial culture (e.g., No. 1007; Falcon, Oxnard, CA) to which they do not adhere.

The culture of islets in suspension has the advantage of not favoring fibroblast outgrowth. The usual concentration of calf serum (10%) can be reduced (to 1%) and, for shorter culture periods, be replaced by bovine serum albumin. Islets of *ob/ob* mice have also been successfully maintained in chemically defined medium in the absence of serum for 7 days.[35] The culture of intact islets has at least one major drawback. Most islets, and particularly the larger ones, tend to develop central necrosis in less

---

[33] A. Andersson and C. Hellerström, *Diabetes* **21** (Suppl. 2), 546 (1972).

[34] C. B. Wollheim, M. Kikuchi, A. E. Renold, and G. W. G. Sharp, *J. Clin. Invest.* **60**, 1165 (1977).

[35] A. Buitrago, E. Gylfe, B. Hellman, L.-A. Idahl, and M. Johansson, *Diabetologia* **11**, 535 (1975).

than 24 hr, probably because of insufficient oxygenation. This is prevented by culturing the islets at 20°, a condition which, however, is not suitable for secretion studies. In general, the islets are kept at 37°, but good survival has also been reported at 8°.

A variation of islet culture involving long-term (up to 36 days) perifusion of intact rat islets which retain good secretory function has been described.[36] In this preparation, the islets lodge on the filter of a perifusion chamber, show no fibroblastic proliferation, and display best secretory responses when exposed to variable, rather than to a constant, glucose concentration.

Monolayer Culture of Islets

Nondispersed Islets. For many purposes, it is necessary to culture islet cells in monolayers rather than as intact islets. This can be achieved by plating a single cell suspension (see the section, Dispersed Islet Cells, below) or by the two following methods.

The first method involves the partial trypsinization of collagenase isolated rat islets.[37] Islets (900–1200) are resuspended in 3 ml of Ca^{2+}- and Mg^{2+}-free phosphate-buffered saline containing 0.25% trypsin. The islets are then disrupted by gentle pipetting at 37° for 2 min. The resulting small cell clumps (10–25 cells) are plated in plastic Petri dishes suitable for mammalian culture, each receiving 0.9 ml of CMLR 1066 supplemented with antibiotics, vitamins, 2 mM glutamine, 5.6 mM glucose, and 15% fetal calf serum. The cell clumps attach within 4 to 5 days and few cells remain free floating. As in other monolayer cultures fibroblast overgrowth is seen with time.

The other method reported to produce monolayers of nondissociated islets employs an inhibitor of cyclic nucleotide phosphodiesterases, 3-isobutyl-1-methylxanthine (IBMX). The rat islets are plated into plastic culture dishes in TC199 medium containing antibiotics, 10% fetal calf serum, and 5 mM glucose. After 2 days in the presence of 0.1 mM IBMX, the islets start attaching and spreading to form small portions of epithelioid cell monolayers around the islet. Again fibroblast overgrowth becomes with time a major problem.[38] In both methods, the extent of monolayer formed is much smaller and fibroblast contamination occurs much faster than when one starts with a single cell suspension, as described in the next section.

[36] P. E. Lacy, E. H. Finke, S. Conant, and S. Naber, *Diabetes* **25**, 484 (1976).
[37] M. Kostianovsky, M. L. McDaniel, M. F. Still, R. C. Codilla, and P. E. Lacy, *Diabetologia* **10**, 337 (1974).
[38] H. Ohgaware, R. Carroll, C. Hofmann, C. Takahashi, M. Kikuchi, A. Labrecque, Y. Hirata, and D. F. Steiner, *Proc. Natl. Acad. Sci. U.S.A.* **75**, 1897 (1978).

Dispersed Islet Cells. Monolayer culture on nontreated surfaces: For obtaining successful primary cultures of adult islet cells in monolayer, it is essential to start the cultures with a suspension of single islet cells (see above). These cells are dispensed into plastic Petri dishes suitable for mammalian tissue culture. The use of DMEM containing 10% fetal calf serum has been found to optimize monolayer formation. The cells initially settle and adhere to the bottom of the dish, while remaining spherical (Fig. 3). During the next 1–2 days the cells start spreading to form monolayers (Fig. 3), the extent of which is dependent on several factors, including the seeding cell concentration. Monolayer formation is improved by seeding the cells as 25-μl droplets and at a concentration of 5–10 $\times$ 10^5 cells/ml. This can be done by dispersing many drops into a 35-mm Petri dish. It is essential to keep the culture environment fully humidified to avoid any loss of such a minute volume due to evaporation. At densities lower than 2 $\times$ 10^5 cells/ml, many cells will remain single and will not spread on the dish. For some purposes (i.e., screening antibody binding by solid-phase colorimetric assay — ELISA) the same procedure can be used to seed the cells in 96-well plates.

The regulation of insulin release from adult islet cells in monolayer culture appears similar to that of the intact islets. There is a problem with overgrowth of fibroblasts even when highly purified islets are used as starting tissue. Such growth can be at least partially controlled using one of the methods described below.

Plating of cells on osmotically disrupted fibroblasts and chemically treated surfaces: Attempts have been made to promote the attachment, survival, and function of islet cells in culture, by plating on a variety of substrata chosen to reconstitute the physiological environment of β cells more closely than the usual culture plastic. Thus, islet cells have been grown on an extracellular matrix,[39,40] on collagen,[41] on feeder layers of irradiated fibroblasts,[42,43] and on osmotically disrupted fibroblasts[44] (Figs. 3 and 5). In the latter approach, neonatal pancreatic fibroblasts are grown at confluence in regular culture dishes and are disrupted by a 3-min exposure to sterile bidistilled water. After removal of all fluid, the dishes, which are still coated with fibroblast membranes and matrix, are stored at

[39] C. H. Thivolet, P. Chatelain, H. Nicoloso, A. Durand, and J. Bertrand, *Exp. Cell Res.* **159**, 313 (1985).

[40] N. Kaiser, A. P. Corcos, A. Tur-Sinai, Y. Ariav, and E. Cerasi, *Endocrinology* **123**, 834 (1988).

[41] R. Montesano, P. Mouron, M. Amherdt, and L. Orci, *J. Cell Biol.* **97**, 935 (1983).

[42] A. Rabinovitch, T. Russel, and D. H. Mintz, *Diabetes* **28**, 1108 (1979).

[43] L. E. Malick, A. Tompa, C. Kuszynski, P. Pour, and R. Langenbach, *In Vitro* **17**, 947 (1981).

[44] P. Meda, E. Hooghe-Peters, and L. Orci, *Diabetes* **29**, 497 (1980).

4° under sterile conditions. At the time of plating, islet cells are simply seeded at the desired density after the dishes have been rapidly rinsed with culture medium. This procedure accelerates and promotes the attachment of fully dispersed islet cells irrespective of plating density.[44,45] Thus, osmotically disrupted fibroblasts permit the formation of extensive monolayers, when islet cells are plated at high density (Fig. 5), as well as the culture of single islet cells, when plating is performed at a density low enough to prevent the spontaneous reaggregation of islet cells.[43,44] Adhesion of islet cells to inert surfaces (plastic or glass) is also favored by pretreating such surfaces with polylysine or polyornithine.[45]

Suspension culture: When cells dispersed from adult rat islets are cultured in suspension, at a sufficiently high density, they aggregate into three-dimensional aggregates rather than forming monolayers.[46,47] Under certain conditions, the cellular composition and organization of these aggregates is analogous to that of native islets, whereas their average size is approximately half that of an adult rat islet. Thus, reaggregation of adult islet cells *in vitro* might become a powerful tool for studying the factors responsible for islet cell recognition/adhesion. The method is as follows: Adult rat islets are trypsinized using the method described above. The dispersed cells are then suspended at a concentration of ~5 × 10^5 cells/ml in DMEM medium supplemented with 10% fetal calf serum, 8.3 mM glucose. A lower cell density will result in smaller aggregates and a longer reaggregation period, whereas a higher density favors the formation of large aggregates which ultimately fuse to form cell masses approximately 5× the size of native islets. It is critical that the cell suspension be completely free of extraneous material (ducts, cell debris, dust particles, etc.) which could interfere with the reaggregation process. Single islet cells are seeded in a volume of 1.2–3 ml into 35-mm nonadhering plastic Petri dishes suitable for bacterial culture (i.e., Falcon type 1007). Cells start adhering to each other within a few hours. By 24 hr there are numerous small aggregates which assume a typical spheroid appearance (Fig. 4). We usually study the aggregates after 6 days in culture.

Primary Cultures of Newborn and Fetal Islet Cells

These methods take advantage of the fact that fetal and neonatal pancreatic endocrine cells have a higher, albeit limited, proliferative capac-

[45] G. C. Weir, P. A. Halban, P. Meda, C. B. Wollheim, L. Orci, and A. E. Renold, *Metabolism* 33, 447 (1984).
[46] P. A. Halban, S. L. Powers, K. L. George, and S. Bonner-Weir, *Diabetes* 35 (Suppl. 1), 45a (1986).
[47] D. W. Hopcroft, D. R. Mason, and R. S. Scott, *In Vitro Cell. Dev. Biol.* 21, 421 (1985).

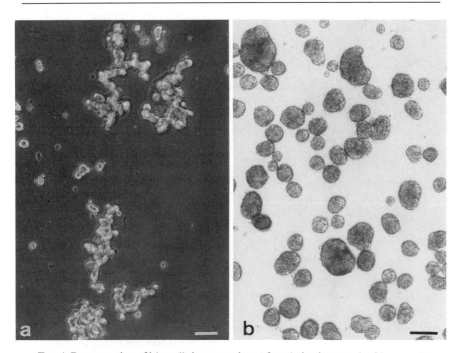

FIG. 4. Reaggregation of islet cells in suspension. After plating into regular 35-mm culture dishes (2 ml of medium), most dispersed islet cells reaggregate spontaneously without attaching to the plastic. After 24 hr of static culture, islet cells in suspension form linear or branched assemblies (a). After 5 days of culture, tridimensional clumps called pseudoislets are observed (b). [(a) Bar = 50 μm; (b) bar = 150 μm.]

ity than adult cells, as well as the fact that young pancreas displays a somewhat larger (~2–3%) proportion of islet tissue. We shall not consider methods in which explants of the whole fetal pancreas are cultured, but focus on the more widespread preparation of monolayer cell cultures.[48]

Usually 30–100 1- to 3-day-old pups are decapitated and pinned on their sides to a board in groups of 10. After swabbing with 70% ethanol, the animals are placed in a laminar flow hood and an incision is made in their left flank over the spleen. By pulling out the spleen, the pancreas can be freed from the stomach and then excised. The pancreases are first collected in glass beakers containing phosphate-buffered saline (PBS) and kept on ice, transferred to a glass Petri dish, trimmed free from remaining fragments of spleen and fat, and then transferred to another dish with the same buffer. The glands are then cut with iris scissors to pieces of about 1 mm which are transferred to a 50-ml Erlenmeyer flask containing a Ca^{2+}- and Mg^{2+}-free PBS. The pooled pancreatic tissue is washed for 10 min by

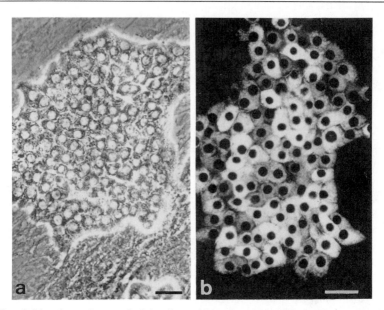

Fig. 5. Monolayer clusters of adult primary islet cells cultured for 1 week. In most cells, a normal content of secretory granules is revealed by the dense appearance of cytoplasm under phase contrast (a) and by the intense immunofluorescent staining for insulin (b). [(a and b) Bar = 30 μm.]

gentle stirring at room temperature. After removal of the supernatant, 10 ml of the divalent cation-free PBS containing 5.6 mM glucose, 1 mg/ml trypsin, and 0.2 mg/ml collagenase (Worthington CLS IV from Cooper Biomedical, Malvern, PA) at pH 7.6 is added. The tissue is dissociated by gentle stirring in the enzyme-containing PBS for 10 min at 37° in an incubator. The tissue is then allowed to settle and the supernatant containing the isolated cells is collected and transferred to 50-ml glass tubes. To stop the action of the enzymes, 15 ml of cooled TC199 medium containing 10% NBCS is added and the cells kept on ice until the final centrifugation. A new aliquot of the enzyme mixture is added to the pancreatic pieces and the procedure is repeated 8–10 times until virtually no undigested tissue remains. The first two supernatants are discarded, as they usually contain mainly cell debris. The remaining pooled supernatants from 8 to 10 successive incubations are washed twice by centrifugation (10 min at 150 g), then made up to the appropriate final volume for plating.

As fibroblastoid cells attach more rapidly than epithelioid ones, the cell suspension is first placed in 10-cm-diameter plastic tissue culture Petri dishes (usually one dish/20 ml of medium/25 pancreases). After about 16 hr, aliquots of supernatant cells (usually the equivalent of 2.5 pancreases) are transferred to 35-mm-diameter dishes. The cells are left to settle for 48 hr, by which time clusters of endocrine cells of varying size (10–100 cells/cluster) are formed (Fig. 6). The medium (TC199 with antibiotics and 10% NBCS) is changed 48 hr after the first plating and its glucose concentration decreased from 16.7 to 5.6 mM. This is done to reduce the rate of insulin secretion and to allow the cells to become well granulated.[48]

These cell cultures contain well-differentiated pancreatic endocrine cells of the four major types, i.e., β cells (insulin), α cells (glucagon), δ cells (somatostatin), and PP cells (pancreatic polypeptide), as well as varying amounts of fibroblasts. Exocrine cells do not survive in culture.[49]

Similar methods for mouse and hamster pancreases have been described. The method is equally applicable to fetal, rather than newborn, rat pancreas. In this case the fetuses are removed from 1-day pregnant rats by uterotomy. Thereafter, the procedure is as described above, except that fewer treatments with the trypsin-collagenase mixture are necessary before the glands are entirely digested.

It is of interest that fetal and neonatal pancreas can be partially dissociated with collagenase. If such a preparation is plated, islets will form on a lawn of fibroblast cells (for details see Refs. 50 and 51).

Growth and Replication

Adult Islets

β cells can probably be formed by both differentiation of stem cells or by division of existing β cells. It is well known that adult islet cells have a limited potential for mitotic division, with a mitotic index estimated *in vivo* around 0.5%, i.e., six- to sevenfold lower than that seen in fetal pancreas. While the differentiation and proliferation of stem cells has been studied only to a limited extent, the proliferation of β cells has been examined in adult islets in culture. Total DNA content of isolated adult rat islets was only modestly increased after culture for 2–3 weeks but was further increased in the presence of human growth hormone. This action was species specific, as it was not seen in adult mouse islets under the same conditions.[52]

[48] C. B. Wollheim, B. Blondel, P. A. Trueheart, A. E. Renold, and G. W. G. Sharp, *J. Biol. Chem.* **250**, 1354 (1975).
[49] L. Orci, A. A. Like, M. Amherdt, B. Blondel, A. Kanazawa, E. B. Marliss, A. E. Lambert, C. B. Wollheim, and A. E. Renold, *J. Ultrastruct. Res.* **43**, 270 (1973).
[50] C. Hellerström, N. J. Lewis, H. Borg, R. Johnson, and N. Freinkel, *Diabetes* **28**, 769 (1979).
[51] R. C. McEvoy, *in* "Methods in Diabetes Research" (J. Larner and S. Pohl, eds.), Vol. 1A, p. 227. Laboratory Methods, Wiley, New York, 1984.
[52] J. H. Nielsen, *Endocrinology* **110**, 600 (1982).

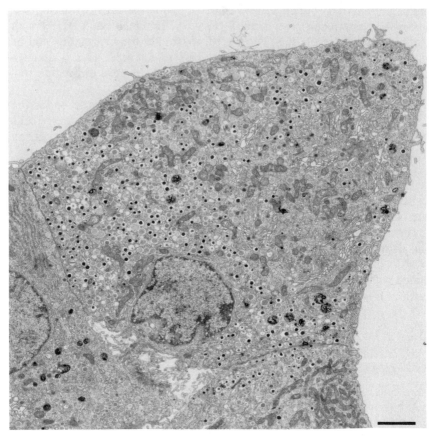

FIG. 6. Cluster of islet cells from a 1-week culture, showing the normal ultrastructural organization of three β cells. (Bar = 2 μm.)

In another study, DNA content and DNA synthesis were compared following a 7-day exposure of mouse islets to nicotinamide, an inhibitor of poly(ADP-ribose) synthetase.[53] DNA synthesis was measured by adding 1 μCi/ml of [*methyl*-³H]thymidine to the culture medium during the last 18 hr of culture. At the end of the experiment, batches of 50 islets were washed with Hanks' balanced salt solution containing 10 mM nonradioactive thymidine. The islets were then disrupted by sonication and the homogenates treated with 5% (w/v) trichloroacetic acid. The precipitate was then filtered on glass microfiber filters to separate labeled DNA from nonincorporated thymidine. The filters were dried and then counted by

[53] S. Sandler and A. Andersson, *Diabetologia* **29**, 199 (1986).

liquid scintillation spectrometry. Total DNA in islets was measured by a modification of the fluorimetric method of Kissane and Robins.[54] The study showed that nicotinamide increased DNA synthesis but not the total islet content of DNA.[53] Although some polyploidy has been reported in islets, the islet DNA content is generally considered a better index of changes in cell number than [³H]thymidine incorporation.[55] The latter method is subject to variations in the activity of thymidine kinase and of the other enzymes of thymidine metabolism as well as in the length of the S phase. A similar discrepancy between [³H]thymidine incorporation and total DNA was seen in mouse islets exposed for 5 days to various glucose concentrations. For example, 16.7 mM glucose stimulated thymidine incorporation without changing total islet DNA. In other studies, theophylline and IBMX have been shown to increase islet DNA synthesis. A final problem inherent to studies employing intact islets is that the proliferating cell population must be identified in order to establish whether β cells are indeed involved. This has not always been done.

Monolayer Cultures of Newborn Islet Cells

The method to correlate the degree of labeling with [³H]thymidine, as established by autoradiography (Fig. 7), with β cell identification, as established by aldehyde-thionine staining, in pancreatic monolayer cultures of neonatal rat pancreas has been described in detail by Rabinovitch *et al.*[56] and will only be outlined here. [³H]Thymidine (10 μCi/ml) is added to the culture medium during the last 16–20 hr of culture. The culture is terminated by washing the cells three times with PBS and then fixing them with Bouin's solution (75 ml formaldehyde and 25 ml picric acid plus 2 ml of glacial acetic acid). The cultures are rinsed extensively in tap water and are left to dry at room temperature. They are then stained with aldehyde-thionine and can thereafter be kept for weeks in the dark before autoradiography. The culture dishes are exposed 3–10 days to a photographic emulsion (Ilford L4). Finally, the [³H]thymidine labeling index is established by counting a given number of aldehyde-thionine-positive cells in each dish and scoring the frequency of labeled nuclei in these cells.

β cell proliferation in monolayers can be confirmed by estimating the β cell mitotic index. To this end 1 μg/ml N-desacetyl-N-methylcolchicine (Colcemid; GIBCO, Grand Island, NY) is added to the cells for the last 18 hr of culture to arrest division in metaphase. After culture, mitoses are better visualized by short swelling of the cells in hypotonic solution (e.g.,

[54] J. M. Kissane and E. Robins, *J. Biol. Chem.* **233**, 184 (1958).
[55] I. Swenne, *Diabetes* **31**, 754 (1982).
[56] A. Rabinovitch, B. Blondel, T. Murray, and D. H. Mintz, *J. Clin. Invest.* **66**, 1065 (1980).

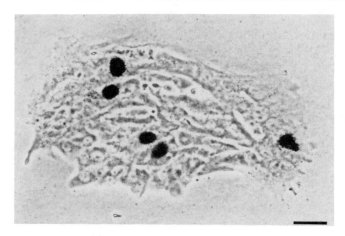

FIG. 7. Light microscopy autoradiograph of a cluster of newborn islet cells cultured for 1 day in the presence of [³H]thymidine. Autoradiographic grains are concentrated over the nucleus of five islet cells which had entered the S phase of the cell cycle, possibly in preparation for mitosis. (Bar = 30 μm.)

0.075 M KCl). Thereafter, the cultures are fixed and stained with aldehyde-thionine as above and the mitotic index is established by scoring the cells showing mitotic figures.

The two methods usually reveal a similar pattern, although the [³H]thymidine index is higher than the mitotic index.[55] Using these methods, it has been shown that glucose, tolbutamide, growth hormone, and multiplication stimulation activity (MSA, the rat equivalent of insulin-like growth factor II) can promote β cell replication.

Glucose has also been reported to increase the number of β cells entering the cell cycle in experiments on rat fetal islets in culture. The islet cells were synchronized with hydroxyurea and then pulse labeled with [³H]thymidine during culture at different glucose concentrations. By this alternative method it was confirmed that only a very limited fraction of β cells is capable of proliferating.[55]

Technical Problems Encountered in the Culture of Islet Cells

Overgrowth of Fibroblasts

A problem in all types of cultures, with the exception of free-floating islets, is the overgrowth of fibroblasts. This growth is favored by serum which, however, cannot be easily suppressed as it is necessary for preservation of a satisfactory secretory response of islet cells. The following methods have been found useful for limiting fibroblast growth.

Decantation. The simplest, but least efficient, way of reducing fibroblast contamination is decantation based upon the differential adhesive properties of fibroblasts and epithelioid cells. This decantation is used routinely in neonatal monolayer culture preparation (see above). It should not be performed later than 16 hr after plating to avoid loss of too many endocrine cells already attached to the dish. A second decantation can be carried out 6 hr later. In that case, it must be verified that the clusters of endocrine cells have not yet attached. By measuring hormone contents before and after decantation we found that up to one-third of the endocrine cells may be lost in this way.

Cystine-Free Media. Culture in cystine-free medium in the presence or absence of serum results in fibroblast death in neonatal rat pancreatic monolayers.[56,57] We have found that the following treatment is most effective. At the first medium change (48 hr after plating) the cells are exposed for 24–48 hr to Dulbecco's modified Eagle's medium prepared without L-cystine and without serum. The cultures are then returned to complete normal medium. Large endocrine clusters resist this treatment better than small clusters. It is clear, however, that if the exposure to the cystine-free medium is prolonged, islet cells become damaged and die. The cystine-free medium is relatively efficient, even in the presence of serum, especially in cultures with moderate fibroblast contamination, such as monolayers established by plating dispersed fragments of isolated adult islets. In this preparation, however, we found that plating the cells in cystine-free medium and maintaining them for 2–3 days in this medium inhibited glucose-stimulated insulin release, as assessed during a short-term incubation in KRB-HEPES on the third or fourth day. This inhibition was not reversible by maintaining the cells for 24 hr in normal medium prior to glucose challenge. The reason for this loss of responsiveness is not clear, but changes in glucose metabolism may be involved, as the cells responded normally to phorbol esters (C. B. Wollheim, unpublished observations).

Treatment with Iodoacetate. Pünter *et al.* first described in preliminary experiments that the Na^+ salt of iodoacetic acid preferentially eliminates fibroblasts from monolayer cultures.[58] We have added 2 μg/ml of this agent in complete medium to islet cell cultures 72 hr after plating and left it for a 15-hr period. Thereafter, the cultures were washed carefully and fresh medium added. In some cases, this treatment eliminates fibroblasts completely. Unfortunately, the effectiveness varies from one experiment to another. Rabinovitch and co-workers have described the use of 2.5 μg/ml iodoacetate for 3–5 hr as being usually efficient, and recommend removal

---

[57] W. L. Chick, A. A. Like, and V. Lauris, *Endocrinology* **96,** 637 (1975).
[58] J. Pünter, F. Wengenmayer, K. Engelbart, and H. Rolly, *In Vitro* **18,** 291 (1982).

of the drug as soon as signs of fibroblast necrosis start appearing. The treatment may be repeated a second time 2–3 days later.[59] Iodoacetate does not perturb insulin biosynthesis and secretion. It is more effective in removing fibroblasts than another sulfhydryl (SH) reagent, thiomerosal, which appears more toxic for islet cells.[60]

D-valine. It can be used instead of L-valine in the culture medium, but we have no experience with this method.[61]

IBMX. The phosphodiesterase inhibitor IBMX has also been proposed to eliminate fibroblasts. However, when intact islets of Langerhans are cultured in the presence of IBMX (see above) fibroblast growth was still observed.[38]

Glucose Concentration of the Culture Medium

Best results for the maintenance of β cell secretory function have been obtained at glucose concentrations between 6 and 11 mM. However, it is extremely difficult to give general guidelines in this context, as the results depend on several factors, such as the species from which the islets are isolated, the type of serum used (fetal or newborn calf, human, etc.), and the culture medium. In neonatal monolayer cultures, best results were obtained by plating the cells in the presence of 16.7 mM glucose and then decreasing the concentration to 5.6 mM after 48 hr, which allows β cells to regranulate.[48] In cultures of free-floating adult islets, 8.3 and 2.8 mM glucose yielded similar islet contents of insulin over a 6-day culture period. However, there was a selective and reversible inhibition of glucose-stimulated insulin release at 2.8 mM glucose after both 24 hr and 6 days of culture.[62] This situation somewhat resembles that observed in vivo after starvation, in that islet glucose metabolism is reduced by low glucose concentrations. High glucose can be replaced by L-leucine or 2-ketoisocaproic acid for maintaining insulin mRNA levels, biosynthesis, and secretion in cultured mouse islets.[63] Whether prolonged exposure of islet cells to a high (i.e., 16.7 mM) glucose concentration is deleterious to β cell function is still controversial.

[59] J. A. Romanus, A. Rabinovitch, and M. M. Rechler, Diabetes 34, 696 (1985).
[60] J. T. Braaten, M. J. Lee, A. Schenk, and D. H. Mintz, Biochem. Biophys. Res. Commun. 61, 476 (1974).
[61] S. F. Gilbert and B. R. Migeon, Cell 5, 11(1975).
[62] E. G. Siegel, C. B. Wollheim, G. Janjic, G. Ribes, and G. W. G. Sharp, Diabetes 32, 993 (1982).
[63] M. Welsh, J. Brunstedt, and C. Hellerström, Diabetes 35, 228 (1986).

Analytical Methods

Hormone Release

Monolayer Cultures. Monolayer cultures of either newborn[48] or adult islet[45] cells have been used to study the release of their constituent hormones. For this purpose, the dishes are washed 3 × with KRB-HEPES-BSA containing 2.8 mM glucose and are then preincubated for 30 min at 37° in KRB-HEPES-BSA, 2.8 mM glucose. This allows the cells to reach a stable basal hormone release pattern during the subsequent incubation period. The dishes are washed once with KRB-HEPES-BSA (glucose, 2.8 mM) and then incubated for 15 min to 1 hr at 37°, using KRB-HEPES-BSA (1 ml for a 35-mm dish) containing the selected test substances for stimulating or inhibiting release. If the release of glucagon is to be measured, it is normal practice to include the protease inhibitor aprotinin at a concentration of 250 kallikrein inhibitory units/ml. Care should be taken to ensure that the pH of the buffer is always adjusted to 7.4 regardless of the test substances added. At the end of the incubation period, the dishes are placed on ice and the medium is transferred to test tubes, which are then centrifuged in the cold for 10 min at 150 g to pellet any cell which may have detached during the incubation. In order to extract hormones from the cells, 1 ml of acid ethanol (for recipe, see Appendix) is added to each dish once the incubation buffer has been removed. The dishes are sealed with Parafilm and left at 2–4° for 24 hr. The acid ethanol is then transferred to tubes which are tightly capped and kept at −20°. There is no need to scrape the cells from the dishes since all insulin is rapidly solubilized in the acid ethanol.

Intact Islets. The approach for studying hormone release from intact, adult islets is similar to that for monolayers. The major difference is in the handling of the tissue. Islets freshly isolated or maintained in tissue culture may be used. The islets are washed three times in KRB-HEPES-BSA, 2.8 mM glucose (by centrifugation at 500 g for 1 min). They are then preincubated for 15–30 min at 37° in the same buffer. After one wash, the islets are dispensed in batches of 10 into 12 × 75 mm siliconized glass test tubes using a dissecting microscope and an Eppendorf or Gilson pipet with a yellow tip (the 10 islets should be picked in 25–50 μl buffer). The test substances are then added to the tubes in 1 ml KRB-HEPES-BSA and the islets are incubated for 15 to 60 min at 37°. The tubes are placed on ice and, after gentle mixing, by handtapping of the tubes the samples are centrifuged for 10 min at 150 g in the cold. The supernatant is removed under the dissecting microscope to ensure that no islets are collected, and stored as described above. Alternatively, after centrifugation, 800 μl buffer is carefully removed without allowing the pipet tip to approach the islet pellet. In this way, the use of the microscope is avoided. The islets are then extracted with 1 ml acid ethanol (as described for monolayer cultures) for the measurement of hormone content.

Islet Perifusion. For the assessment of dynamic insulin secretion, isolated islets are usually perifused (superfused). If the islets have been cultured under free-floating conditions they can be picked directly from the culture dish with the aid of an Eppendorf pipet (usually in a volume of 50 μl) by placing the dish under a dissecting microscope. The number of islets placed in the perifusion chamber should be adapted to the flow rate to allow adequate detection of basal hormone release. We have used 40 to 100 islets/chamber and the chambers had a volume of either 70 or 700 μl. When the larger chambers are used,[64] a flow rate of up to 2.5 ml/min can be applied, while the smaller chambers are used at a flow rate of 1 – 1.4 ml/min.[34] The buffer is either a normal KRB medium, in which case continuous gassing with 95% oxygen/5% CO_2 is necessary, or, to avoid gassing, the same KRB-HEPES buffer as used for static incubations can be employed. In general the islets are perifused for 30 – 45 min under basal conditions (2.8 mM glucose) and then exposed to the stimuli (e.g., 16.7 mM glucose) for 15 – 45 min. Perifusion experiments of up to 4-hr duration have been reported but, in general, experiments should be limited to 3 hr in view of the simple composition of the buffer. Perifusion of islets with culture medium has made it possible to continue the studies for several weeks.[36] Care must be taken to control the temperature of the perifusion buffer, as insulin secretion from isolated islets is extremely temperature sensitive. We have described a system where both the perifusion chamber, the tubing, and the rotating oxygenator drums, serving as buffer reservoirs, are immersed in a thermostatted water bath.[34] Alternative approaches can be used, including the use of Millipore (Bedford, MA) chambers. The medium is collected in fraction collectors and assayed for immunoreactive insulin. It is usually not necessary to centrifuge the samples, as cells which might detach from the islets are retained in the chamber by the filter (usually an 8-μm filter). This method of perifusion has been applied not only to the measurement of insulin secretion from islets of different species but also to the assessment of efflux rates of a number of metabolites and radioisotopes.

Reverse Hemolytic Plaque Assay for Studying Insulin Release. This method provides a way to visualize insulin release from individual β cells and β cell clusters.[65,66] The dispersed islet cells are mixed in a KRB-

[64] M. Kikuchi, A. Rabinovitch, W. G. Blackard, and A. E. Renold, *Diabetes* 23, 550 (1974).
[65] D. Salomon and P. Meda, *Exp. Cell Res.* 162, 597 (1986).
[66] D. Bosco, L. Orci, and P. Meda, *Exp. Cell Res.* 184, 72 (1989).

HEPES, 0.1% (w/v) BSA, with 4% packed sheep red blood cells (Behring-werke AG, Marburg, West Germany) to which 0.5 mg/ml protein A is attached using 0.1 mg/ml chromium chloride. Aliquots of the mixed cell suspension are then introduced into 50-μl Cunningham's glass chambers coated with 0.1 mg/ml poly-L-lysine (M_r 150,000–300,000) and placed for 45 min at 37°. After rinsing of the chambers, the cells are incubated in the presence of an anti-insulin serum (we use a polyclonal serum,[67] at a dilution of 1:50), under the experimental conditions to be tested. Subsequently, the chambers are rinsed with an excess of KRB-HEPES and incubated again for 60 min at 37° with guinea pig complement (Behring-werke Ag, Marburg, FRG), diluted 1:10–1:50. Alternatively, complement can be added during the secretion test, i.e., together with the anti-insulin serum. In both cases, cell viability is then tested by the Trypan Blue exclusion test before the chambers are fixed in Bouin's solution and immunostained for insulin.[65,66] Finally, the chambers are sealed and stored in the dark at 4°. Under such conditions, they can be examined repeatedly for months. Insulin release is detected by the appearance of hemolytic plaques (Fig. 8) induced, around secreting β cells, by the lysis of red blood cells which bear insulin–anti-insulin complexes bound to protein A.[65] Quantification of these plaques around Trypan Blue-unstained and immunohistochemically identified β cells allows for detection and quantification of insulin release on a per cell basis,[65,66] thus providing a rapid and sensitive method to study β cell function under a variety of conditions. This assay can be combined with other techniques, such as microelectrode[65] or patch-clamp electrophysiology[68] and autoradiography,[69] to study simultaneously on the same β cell insulin secretion and other relevant parameters. It can also be used to study repeatedly a population of the very same cells during successive secretion tests. Since the method is noninvasive, it should also be of interest in the screening for insulin-secreting clones in mixed cultures.[70] Furthermore, using appropriate antibodies against glucagon, somatostatin, and pancreatic polypeptide, the method should be applicable also to study the secretion of α, δ, and PP islet cells.

Hormone Biosynthesis

Insulin biosynthesis is most commonly assessed by monitoring the incorporation of radioactively labeled amino acids into insulin and its precursors, using either monolayer cultures or intact islets. The handling of

[67] P. H. Wright and W. J. Malaisse, *Diabetologia* 2, 178 (1966).
[68] B. Soria, M. Chauson, and P. Meda, *Diabetologia* 32, 543A (1989).
[69] P. Meda and D. Bosco, *Diabetologia* 32, 516A (1989).
[70] M. A. Feldman and J. I. Chou, *In Vitro* 19, 171 (1983).

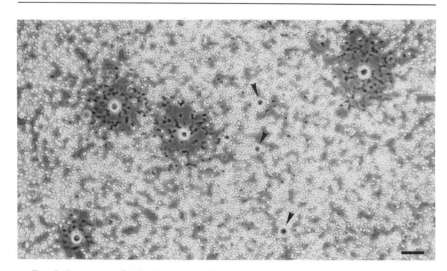

FIG. 8. Reverse hemolytic plaque assay for insulin secretion. Glucose (16.7 mM) stimulation induced insulin release from four single β cells, as judged by the formation of hemolytic plaques, which are seen as round areas devoid of refringent red blood cells and containing dark erythrocyte ghosts. The arrowheads point to three other islet cells which did not induce hemolytic plaques following glucose challenge. (Bar = 30 μm.)

the cultured cells or islets is as described above for studying hormone release. A typical protocol for studying both biosynthetic rates and proinsulin-to-insulin conversion would be as follows. Incubate islets (20 islets/ 12 × 75 mm tube) or monolayer cultures (35-mm Petri dish) for 5 min at 37° in 1 ml KRB-HEPES-BSA containing 100 μCi/ml [³H]leucine (specific radioactivity 120–160 Ci/mmol). If the effects of glucose on insulin synthesis are to be investigated, then different concentrations of the sugar will be tested. The cultures or islets are placed on ice and washed three times in KRB-HEPES-BSA or culture medium, as described above, to remove unincorporated [³H]leucine. Some tubes (dishes) are kept for analysis of incorporation of label into cells during the 5-min pulse. The others are incubated at 37° for a chase period which extends typically for up to 3 hr.[71] The conversion of labeled proinsulin to insulin and the release of newly synthesized, as well as stored insulin can be followed during the chase incubation.[71] At the end of the chase period, or at fixed times during the chase, the incubation buffer or the culture medium is removed, centrifuged (150 g, 10 min), and stored for analysis. The cells or islets are

[71] C. J. Rhodes and P. A. Halban, *J. Cell Biol.* **105**, 145 (1987).

extracted by sonication in an aqueous buffer, such as PBS, containing 0.1% BSA. Radioactively labeled proinsulin and insulin are then measured by immunoprecipitation[72] or by high-pressure liquid chromatography (HPLC).[73]

Morphology of Cultured β Cells

Morphology is instrumental in identifying β cells among other islet cell types and fibroblasts and in assessing the maintenance of the differentiated β cell phenotype in culture.

Primary islet cells are distinguished from fibroblasts by their morphology, their rate and pattern of adhesion, and their growth. Under phase contrast, living islet cells appear as small (about 12 μm in diameter), round (when single), or polygonal cells (within clusters) filled with dense cytoplasmic granules and which adhere rather slowly to culture dishes, usually without much spreading. Thus, most islet cells retain a globular shape after attachment to the culture dishes (Figs. 2 and 3), until they merge with neighboring cells to form monolayer clusters or tridimensional clumps (Figs. 3 and 5). By contrast, fibroblasts have a larger and more elongated shape and, at least at low density, show a very prominent cytoskeleton network. Fibroblasts adhere rapidly to most culture substrata and usually flatten readily after attachment. Furthermore, while primary islet cells proliferate very slowly if at all under most culture conditions, fibroblasts divide much faster and rapidly overgrow the islet cells. Thus, cells proliferating actively in primary islet cell cultures are almost always fibroblasts.

While it is usually easy to distinguish fibroblasts and islet cells by mere observation of the cultures, it is virtually impossible to identify the different types of islet cells under phase contrast. Thus, a positive identification of β cells requires the demonstration of their characteristic cytoplasmic content of β secretory granules by either conventional electron microscopy (Fig. 6) or immunostaining, using a specific serum against insulin (Fig. 5).

Ultrastructural analysis and immunolabeling also provide useful indications about the maintenance of the differentiated phenotype of primary β cells in culture, a sensitive indicator of which is a normal content of

[72] P. A. Halban, C. B. Wollheim, B. Blondel, and A. E. Renold, *Biochem. Pharmacol.* **29**, 2625 (1980).
[73] D. Gross, A. Skvorak, G. Hendrick, G. Weir, L. Villa-Komaroff, and P. Halban, *FEBS Lett.* **241**, 205 (1988).

typical β granules. Dedifferentiation of β cells *in vitro* is most often associated with the loss of their characteristic granularity.

Appendix

1. Krebs-Ringer-bicarbonate-HEPES (KRB-HEPES) used for islet isolation and short-term incubation studies:

Stock solution A: NaCl (34.7 g), KCl (1.77 g), KH_2PO_4 (0.808 g), $MgSO_4 \cdot 7H_2O$ (1.463 g). Make up to 1 liter with distilled H_2O, then add $CaCl_2 \cdot 2H_2O$ (1.87 g)
Stock solution B: $NaHCO_3$ (2.625 g), NaCl (5.347 g). Make up to 1 liter with distilled H_2O

These stock solutions may be kept for up to 2 weeks at $2-4°$.
Prepare KRB-HEPES *fresh* each day:

Solution A: 20 ml
Solution B: 16 ml
HEPES solution (1 M stock):   1 ml
Distilled H_2O: 58 ml

Equilibrate by bubbling through air for 10 min. Add bovine serum albumin if required. Adjust pH to 7.4 (prewarm solution to temperature desired for the experimental procedure) with 1 M NaOH.

2. Hanks'–HEPES buffered salts solution (HBSS-HEPES) used for islet isolation and Histopaque gradients:

NaCl: 8 g
KCl: 0.4 g
$MgSO_4 \cdot 7H_2O$:   0.2 g
$Na_2HPO_4 \cdot 2H_2O$:   0.06 g
KH_2PO_4:   0.06 g
$NaHCO_3$:   0.35 g
HEPES: 2.38 g
D-Glucose $\cdot H_2O$:   0.4 g

Make up to 1 liter with distilled H_2O, then add $CaCl_2 \cdot 2H_2O$ (0.185 g). Equilibrate by bubbling through air for 20 min. Adjust pH to 7.2 with 1 M NaOH at room temperature.

For all steps after digestion of the pancreas with collagenase and the subsequent washes of digested tissue at $37°$, add bovine serum albumin (0.35 g/100 ml). The Hanks' solution can be conveniently stored as a $4\times$ stock solution (without glucose or albumin).

Acid ethanol:

Ethanol (absolute):   750 ml
Concentrated HCl: 15 ml
Distilled H$_2$O: 235 ml

Keep in a tightly closed bottle at $2-4°$.

Acknowledgments

We thank Dr. Benigna Blondel for invaluable help in the preparation of this manuscript. This work was supported by Grants from the Swiss National Science Foundation Nos. 32-25665.88 (C.B.W.), 31-26625.89 (P.M.), 31-9394.88 (P.A.H.), Grant No. 187384 from the Juvenile Diabetes Foundation International (P.M.), a grant from the Greenwall Foundation (P.A.H.), and Grant No. DK-35292 from the National Institutes of Health (P.A.H.)

[14] Establishment and Culture of Insulin-Secreting β Cell Lines

By Claes B. Wollheim, Paolo Meda, and Philippe A. Halban

In order to study the molecular mechanism of insulin production a very large number of highly purified pancreatic β cells is required. The most successful methods for isolating islets from large mammals, including man (see [13], this volume) can provide up to 200,000 islets/pancreas. The procedures are, however, time consuming and costly. This number of islets consists of some 5×10^8 cells, a modest number for detailed biochemical analysis. Only approximately 75% are β cells, and there is furthermore a significant contamination with exocrine cells. These major obstacles to the study of β cell function have been overcome in recent years by the use of insulin-producing cell lines. The aim of this chapter is to describe the various types of lines available, as well as their functional characteristics.

Lines Derived from Naturally Occurring or Induced Insulinomas

Naturally Occurring Insulinomas

Only a limited number of cell lines have been derived from spontaneously occurring insulinomas of either animal or human sources. Typically, the tumor is dissected out of the pancreas and then digested with collagenase with (or without) the addition of trypsin. The tumor cells are then allowed to grow in tissue culture. The major problem encountered in virtually all attempts to obtain a permanent insulin-producing line in this way is that the cells dedifferentiate (as manifested by a dramatic decrease in cellular insulin content) rather rapidly. There appears to be an inverse relationship between proliferative capacity and differentiated status. There is no reproducible method available for selecting cells with maintained, differentiated function, although approaches including the use of extracellular matrices[1,2] have been tried. It is, however, possible on occasion to maintain or restore, at least in part, differentiated function by inducing tumor growth during the *in vivo* passage of the cells.[3,4] The general view remains, however, that obtaining a well-differentiated line with useful *in vitro* proliferative capacity depends largely upon chance. Indeed, to our knowledge no human cell line fulfilling these criteria on a long-term basis exists. Despite these reservations, various groups have used insulinoma cells of both human[5-7] and animal[3,8-10] origin for their studies.

Induced Tumors

Insulinomas have been induced in the rat by either radiation[11] or by the combination of streptozotocin and nicotinamide.[12,13] The tumor cells have then been used to establish cell lines. It must again be stressed that, as with

[1] C. H. Thivolet, P. Chatelain, P. Nicoloso, A. Durand, and J. Bertrand, *Exp. Cell Res.* **159**, 313 (1985).
[2] R. Muschel, G. Khoury, and L. M. Reid, *Mol. Cell. Biol.* **6**, 337 (1986).
[3] P. A. Rae, C. C. Yip, and B. P. Schimmer, *Can. J. Physiol. Pharmacol.* **57**, 819 (1979).
[4] P. R. Flatt, M. G. DeSilva, S. K. Swanston-Flatt, C. J. Powell, and V. Marks, *J. Endocrinol.* **118**, 429 (1988).
[5] W. L. Chick, V. Lauris, J. S. Soeldner, M. H. Tan, and M. Grinsberg, *Metabolism* **22**, 1217 (1973).
[6] K. Adcock, M. Austin, W. C. Duckworth, S. S. Solomon, and L. R. Murrell, *Diabetologia* **11**, 527 (1975).
[7] C. H. Thivolet, A. Demidem, M. Haftek, A. Durand, and J. Bertrand, *Diabetes* **37**, 1279 (1988).
[8] A. F. Gazdar, W. L. Chick, H. K. Oie, H. L. Sims, D. L. King, G. C. Weir, and V. Lauris, *Proc. Natl. Acad. Sci. U.S.A.* **77**, 3519 (1980).
[9] C. A. Carrington, E. D. Rubery, E. C. Pearson, and C. N. Hales, *J. Endocrinol.* **109**, 193, (1986).
[10] G. A. Praz, P. A. Halban, C. B. Wollheim, B. Blondel, A. J. Strauss, and A. E. Renold, *Biochem. J.* **210**, 345 (1983).
[11] W. L. Chick, W. Shields, R. N. Chute, A. A. Like, V. Lauris, and K. C. Kitchen, *Proc. Natl. Acad. Sci. U.S.A.* **74**, 628 (1977).
[12] W. L. Chick, M. C. Appel, G. C. Weir, A. A. Like, V. Lauris, J. G. Porter, and R. N. Chute, *Endocrinology* **107**, 954 (1980).
[13] P. Masiello, C. B. Wollheim, B. Blondel, and A. E. Renold, *Diabetologia* **24**, 30 (1983).

the use of spontaneously occurring insulinomas (see above), there is no guarantee that a given tumor will provide a useful line. The main, but not the only,[12] exception to this rule has been an X-ray-induced rat insulinoma.[8] This tumor can be maintained by *in vivo* passage (serial transplantation) in inbred (NEDH) rats.[11] Several tumors derived from the original one now exist. They differ in their rate of growth and, typically, those which grow the slowest display the highest degree of differentiation. There has also been a suggestion that growing the tumor under the kidney capsule, rather than subcutaneously (the more conventional method), may further favor differentiation.[14]

The first cell lines derived from this insulinoma were the RIN cells. These lines were established by Chick and associates by growing isolated tumor cells in culture.[8] These investigators deserve credit for their perseverance since their initial attempts were unsuccessful. The RIN cells have indeed provided many groups with a useful source of insulin-producing cells for a variety of purposes. The most commonly used subline is RINm5F. This line was initially selected since it had a relatively high insulin content, with nondetectable amounts of glucagon and somatostatin.[8,15]

Culture and Characteristics of RINm5F Cells and Derived Sublines

These cells are grown in RPMI 1640 (11 mM glucose) medium supplemented with 10% fetal calf serum (FCS). Typically, cells are seeded into plastic culture vessels at a concentration of 2×10^5 cells/ml. The medium is changed after 3 and 6 days, and the cells trypsinized after 1 week. For this purpose the cells are rinsed with Ca^+/Mg^{2+}-free phosphate-buffered saline and then exposed to 0.025% (w/v) trypsin (1:250; Difco, Detroit, MI) and 0.27 mM EDTA in the same buffer for 1–3 min at 37°. The cells are dislodged from the culture surface by tapping the side of the vessel, and are then washed in medium before being reseeded. In order to maintain the functional status of RIN cells, it is essential that they be exposed to no more than 15–20 tissue culture transfers. Beyond this number there is an ever-increasing chance of the cells suddenly and unpredictably showing a dramatic loss of insulin content and responsiveness to secretagogues. It is unclear what causes this undesirable loss of function, but this phenomenon is by no means unique to RIN cells. Because of this limitation in passage number, it is important to have an adequate reserve of cells in storage. For

[14] M. Hoenig and G. W. G. Sharp, *Endocrinology* **119**, 2502 (1986).
[15] H. K. Oie, A. F. Gazdar, J. D. Minna, G. C. Weir, and S. B. Baylin, *Endocrinology* **112**, 1070 (1983).

this purpose, cells are suspended in phosphate-buffered saline or complete culture medium (10% FCS), with the addition of 10% dimethyl sulfoxide (DMSO) ($4-6 \times 10^7$ cells/ml) and the temperature then progressively decreased before storage in liquid nitrogen. The cells are thawed by rapidly raising the temperature to 37°, and then transferring them to RPMI 1640 medium supplemented with 20% FCS (the seeding density should be $\sim 10^6$ cells/ml). As soon as the cells have attached to the culture surface, the FCS concentration may be decreased to 10%. When handled as described, the subclone of RINm5F used in our laboratories (RINm5F-2A) has maintained apparently unchanged function for over 7 years.[10,16] This subline was established by cloning RINm5F cells using the limiting dilution method (i.e., by allowing colony formation from a single cell).

For RIN-r cells, another RIN cell clone, culture in chemically defined medium in the absence of serum was found to sustain cell proliferation.[17]

Lins Derived from Tumors in Transgenic Mice

Transgenic mice expressing simian virus 40 (SV40) T antigen uniquely in the pancreatic β-cell have been derived by injection into a mouse embryonic pronucleus of a fusion gene construct consisting of the structural region of the SV40 T antigen gene driven by the insulin promoter.[18] These animals develop insulinomas and a cell line (B-TC cells) has been derived from such an insulinoma.[19] An alternative approach has been to use a gene construct allowing for random tissue expression of T antigen in the transgenic mice. In this case an insulinoma was also found (although not all mice had such tumors) and a cell line again prepared (IgSV195 cells).[20] Such mouse β cell lines have been less extensively studied than the RIN (see above) or HIT (see below) cell lines. This is due in part to the fact that they have only recently been described. They may be grown in Dulbecco's modified Eagle's medium (DMEM) containing 10% FCS and 25 mM glucose, using the same basic culture techniques described for RIN cells.

[16] C. B. Wollheim, M. J. Dunne, B. Peter-Riesch, R. Bruzzone, T. Pozzan, and O. H. Petersen, *EMBO J.* **7**, 2443 (1988).

[17] H. K. W. Fong, W. L. Chick, and G. H. Sato, *Diabetes* **30**, 1022 (1981).

[18] D. Hanahan, *Nature (London)* **315**, 115 (1985).

[19] S. Efrat, S. Linde, H. Kofold, D. Spector, M. Delannoy, S. Grant, D. Hanahan, and S. Baekkeskov, *Proc. Natl. Acad. Sci. U.S.A.* **85**, 9037, 1988.

[20] A. Gilligan, L. Jewett, D. Simon, I. Damjanov, F. M. Matschinsky, H. Weik, C. Pinkert, and B. B. Knowles, *Diabetes* **38**, 1056 (1989).

Lines Obtained by Viral Transformation of β Cells

Primary cultures of pancreatic endocrine cells (see [13], this volume) have been transformed with SV40 and clonal cell lines obtained.[21,22] One line in particular, HIT cells, derived from SV40 infection and transformation of Syrian hamster β-cells,[22] has proved useful for studies on β cell function. HIT cells (and in particular the T15 subclone), are maintained and grown in tissue culture in much the same way as RIN cells (see above) with the following differences. The culture medium originally recommended was F-12 containing 15% horse serum and 2.5% FCS, with the addition of 10 μg/ml glutathione, 0.1 μM selenous acid.[22] We now use RPMI 1640 (11 mM glucose) supplemented with 10% FCS, and again containing 0.1 μM selenous acid. As for the other cell lines described, HIT cells display a marked phenotypic alteration with time in culture,[23] and we therefore strongly recommend that an adequate stock of early passage number cells be stored in liquid nitrogen and that the cells not be exposed to more than 10–20 tissue culture passages.

Transfection of Secretory Cells with the Insulin Gene

Transformed secretory cells have been transfected with the insulin gene driven by a viral promoter.[24-26] In order to facilitate the selection of stable transfectants, we transfect cells with a construct (DOL vector) carrying the insulin gene driven by the murine leukemia virus long terminal repeat, colinear with the neomycin resistance gene.[25] The transfected cells are grown in G418 (geneticin) at a concentration toxic (175 μg/ml) to cells not expressing the neomycin resistance gene. Of these G418-resistant clones, approximately 30% were found to produce detectable levels of immunoreactive insulin.

The transfection of noninsulin-producing cells with the insulin gene is, typically, useful for studying the mechanism and regulation of insulin expression. It furthermore allows for the expression of mutant insulin

[21] E. J. Niesor, C. B. Wollheim, D. H. Mintz, B. Blondel, A. E. Renold, and R. Weil, *Biochem. J.* **178,** 559 (1979).

[22] R. F. Santerre, R. A. Cook, R. M. D. Crisel, J. D. Sharp, R. J. Schmidt, D. C. Williams, and C. P. Wilson, *Proc. Natl. Acad. Sci. U.S.A.* **78,** 4339 (1981).

[23] H.-J. Zhang, T. F. Walseth, and R. P. Robertson, *Diabetes* **38,** 44 (1989).

[24] H.-P. Moore, M. D. Walker, F. Lee, and R. B. Kelly, *Cell* **35,** 531 (1983).

[25] D. J. Gross, P. A. Halban, C. R. Kahn, G. C. Weir, and L. Villa-Komaroff, *Proc. Natl. Acad. Sci. U.S.A.* **86,** 4107 (1989).

[26] S. K. Powell, L. Orci, C. S. Craik, and H.-P. H. Moore, *J. Cell Biol.* **106,** 1843 (1988).

genes.[25-27] The cell line most frequently used for such studies is the AtT20 (mouse pituitary corticotroph) transformed line. These cells may be grown in F-10 medium supplemented with 15% horse serum and 2.5% FCS. We have found that cell growth is favored by raising the glucose concentration to 25 mM. Other cell types have been used for transfection studies, and these include HIT cells.[28]

Other Approaches to Obtaining Insulin-Producing Cell Lines

There have been attempts to obtain cell lines from cultures of β cells of neonatal[29] or fetal[30] origin. This approach depends upon the random and spontaneous outgrowth of cells displaying peculiarly elevated proliferative capacity (possibly protodifferentiated stem cells). No stable cell line has yet become available using this approach.

Analytical Methods for Studying Insulin-Producing Cell Lines

Morphology of β Cell Clones

The identification of insulin-containing cells in a clone or line cannot be made by the mere observation of living cultures since permanent cells vary in shape and size and, usually, do not contain sufficient amounts of secretory granules for a consistent detection under phase contrast. Granules are much less abundant in the β cell clones presently available than in control primary β cells and in the tumors from which permanent β cell lines have been derived (Fig. 1). Conventional electron microscopy is also of limited value since the secretory granules of most cloned β cells vary in morphology and are often rather different from typical β granules (Fig. 3).

[27] D. J. Gross, L. Villa-Komaroff, C. R. Kahn, G. C. Weir, and P. A. Halban, *J. Biol. Chem.* **264,** 21486 (1989).

[28] G. Gold, M. D. Walker, D. L. Edwards, and G. M. Grodsky, *Diabetes* **37,** 1509 (1988).

[29] K. W. Ng, P. R. Gummer, B. L. Grills, V. P. Michelangeli, and M. E. Dunlop, *J. Endocrinol.* **113,** 3 (1987).

[30] I. Matsuba, M. Narimiya, H. Yamada, Y. Ikeda, T. Tanese, M. Abe, and H. Ishikawa, *Jikeikai Med. J.* **28,** 257 (1981).

FIG. 1. Identification of insulin-containing cells in a transplantable islet cell tumor, induced by X-ray irradiation of NEDH rats. Immunostaining of the tumor reveals the presence of numerous insulin-containing cells which occur in clusters or as scattered single cells among abundant connective tissue (a). At the electron microscope level, most of these cells contain numerous typical β secretory granules (b) and show an ultrastructural organization analogous to that of normal β cells. [(a) Bar = 100 μm; (b) bar = 1 μm.]

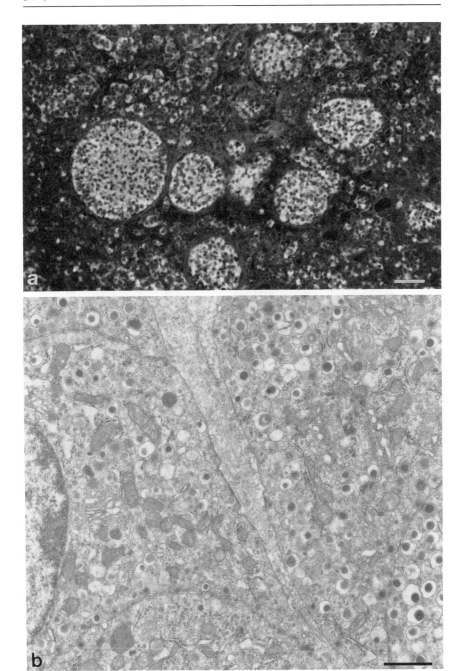

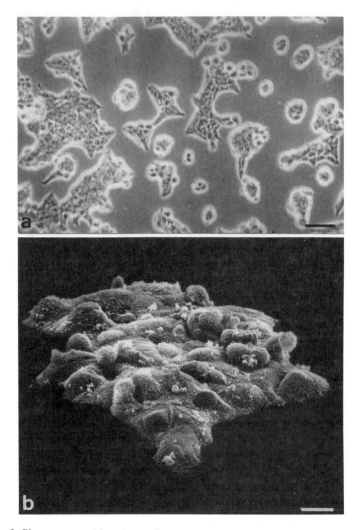

Fɪɢ. 2. Phase-contrast (a) and scanning electron microscopy (b) showing characteristic clusters of insulin-producing cells from the RIN-5F clone, under conventional culture conditions. [(a) Bar = 100 μm; (b) bar = 20 μm.]

Furthermore, since these few granules are heterogeneously distributed among cells, a large number of sections should be screened to obtain a fair evaluation of the actual granularity of the whole cell population. Immunolabeling for insulin is instrumental in addressing this question morphologically. However, due to the poor granularity of β cell clones, this approach

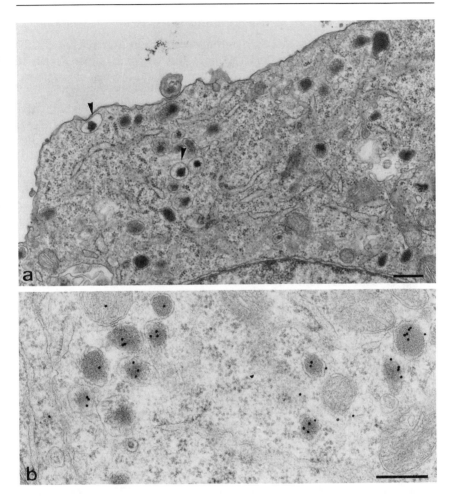

FIG. 3. Ultrastructural organization of an RIN-5F cell. (a) Secretory granules are scarce in these cells and rarely (arrowheads) show the characteristic clear halo of typical β granules. (b) Insulin is immunolocalized by gold particles in a population of small, halo-devoid secretory granules of an RIN-5F cell. (Bars = 0.5 μm.)

inconsistently labels only a few cells at the light microscope level and, thus, should be performed at the ultrastructural level (Fig. 3).

The growth rate of cloned β cells is usually higher than that of primary β cells. Thus, periodic morphological control shows readily the progressive growth of cloned β cells in clusters (Fig. 2) and a high frequency of mitosis. The growth rate of β cell clones varies quite extensively, depending on the culture substratum,[1,2] which also appears able to modify cell secretion.

Biochemistry

Short-Term Insulin Release. Immunoreactive insulin release can be studied from insulin-producing cell lines just as from native β cells in monolayer cultures. The procedures are similar and the reader is referred to Chapter [13], this volume. Note, however, that the cellular insulin content of insulin-producing cell lines is usually less than 1% that of native β cells and that fractional insulin release is often much higher (basal release of 10% of cellular insulin content/hr is not unusual). A typical protocol[10] for RINm5F cells would be as follows (for buffer recipes, see Appendix, Chapter [13], this volume). RIN cells are seeded into 24-well test plates at a density of 2×10^5 cells/ml in 1 ml RPMI 1640, 10% fetal calf serum. The medium is changed after 3 days and insulin release studied 1 day later. The cells are washed three times with 1 ml KRB-HEPES-BSA, 2.8 mM glucose, and then preincubated in this buffer for 15 min at 37°. Insulin release is then followed by incubating the cells in 1 ml KRB-HEPES-BSA, with selected addition of secretagogues, for 30–60 min at 37°. The incubation buffer is taken and centrifuged at 2–4° for 10 min at 150 g to pellet any cells which may have detached from the wells during the incubation. This supernatant is kept for radioimmunoassay of insulin. Acid ethanol (1 ml) is added to the cells to extract cellular insulin (leave at 2–4° overnight, then decant the acid ethanol and store it at −20° in stoppered tubes). Cellular DNA may be measured after extracting insulin in this way.

Insulin secretion can also be measured from cells in suspension by performing either static incubations or perifusions. In both cases the cells are detached from the culture vessel with trypsin as for routine passaging, washed, and transferred to a 100-ml volume spinner culture flask. We use RPMI 1640 medium with the same composition as for culture, except for the use of 1% newborn calf serum to minimize cell clumping. The usual cell concentration is around $1–2 \times 10^6$ cells/ml. The spinner culture is performed at 37° for 3 hr, a time considered sufficient to allow the recovery from trypsin treatment of cell surface structures.

For static incubation the cells are then washed and resuspended in an appropriate volume of KRB-HEPES-BSA buffer and distributed into siliconized 3-ml glass or plastic tubes. The concentration must be adapted to the cellular content of insulin, but for RINm5F cells we use approximately 0.5×10^6 cells/ml and an incubation volume of 1 ml. Usually the cells are preincubated for 15–30 min under basal conditions followed by centrifugation at room temperature (150 g, 5 min). The supernatant is discarded. Alternatively only 0.8 ml is removed and kept at −20° for measurement of immunoreactive insulin. The cells are then resuspended in the various test solutions and reincubated at 37° for 10–30 min. The incubation is stopped by chilling the tubes, followed by centrifugation in the cold, decantation, and addition of 1 ml of acid ethanol to the pellet for the assessment of cellular hormone content. The partial removal of the preincubation medium is less perturbing to the cells, but requires the correction of the secretion during the incubation by subtraction of the hormone remaining in the tube from the preincubation.[31]

For perifusion, the cells are resuspended in KRB-HEPES-BSA to yield about $2-10 \times 6$ cells/ml. They are placed in the perifusion chamber (chamber volume 700 μl). Usually the cells are perifused for 30–45 min under basal conditions followed by a 15- to 30-min stimulation period. The basal buffer is then reintroduced to verify that no drift in baseline secretion has occurred.[32] This procedure is very similar to that described for the perifusion of isolated islets.[33] It should be stressed that it is essential to keep the temperature of the incubations and perifusions at 37° for optimal secretory responsiveness. Alternative methods, including the perifusion of cells in small BioGel P 2 columns[34] (Bio-Rad, Richmond, CA) or perifusion of cells attached to glass coverslips,[35] have been reported.

Insulin Biosynthesis. Proinsulin biosynthesis and the conversion of proinsulin to insulin may be studied in cell lines in much the same way as in native islet cells (see [13], this volume). The major difference lies in the amount of proinsulin synthesized per unit time relative to total, noninsulin-related, proteins. For native β cells this value can reach ~50% over a short pulse-label period (i.e., not exceeding 30 min) at high glucose. Such is not the case even for the best differentiated cell lines, where values of 1% or less would be typical. This means that cell extracts can normally not be analyzed directly by reversed-phase high-pressure liquid chromatography (HPLC). Rather, it is customary to first immunoprecipitate radioactively labeled products using anti-insulin serum and Protein A–Sepharose as described in detail in Ref. 36. The precipitated products are then displaced from the immune complex and analyzed by HPLC.[25,27,37]

---

[31] S. Ullrich and C. B. Wollheim, *Mol. Pharmacol.* **28**, 100 (1985).
[32] C. B. Wollheim, S. Ullrich, and T. Pozzan, *FEBS Lett.* **177**, 17 (1984).
[33] M. Kikuchi, A. Rabinovitch, W. G. Blackard, and A. E. Renold, *Diabetes* **23**, 550 (1974).
[34] P. Knudsen, H. Kofod, A. Lernmark, and C. J. Hedeskov, *Endocrinol. Metab.* **8**, E338 (1983).
[35] R. S. Hill and A. E. Boyd III, *Diabetes* **34**, 115 (1985).
[36] P. A. Halban and C. B. Wollheim, *J. Biol. Chem.* **255**, 6003 (1980).
[37] D. Gross, A. Skvorak, G. Hendrick, G. C. Weir, L. Villa-Komaroff, and P. Halban, *FEBS Lett.* **241**, 205 (1988).

Differentiated Function of β Cell Lines

Despite the obvious advantages of having an unlimited source of insulin-producing cells, there are a number of limitations to the use of such lines as a valid model for native β cell function. It is clearly beyond the scope of this chapter to list all of the similarities and differences found between the various lines available and their native β cell counterparts. The individual investigator will always be obliged to make the appropriate comparison using his or her particular line and analytical approaches. In general terms, however, it is perhaps useful to note the following. The native β cell from most adult animals (but not from ruminants) responds to a rise in ambient glucose with a marked increase in insulin release. This stimulation displays an apparent K_m of 7–8 mM glucose, and reaches a maximum at ~15 mM. Such is not the case for any known insulin-producing cell line. Thus, the RIN cell responds (if at all) with increased secretion only when glucose is raised from 0 to 2 mM.[10] The HIT cell originally responded in the glucose range of 2–7 mM[22] but seems to have changed its characteristics in many laboratories where a pronounced shift of the glucose dose–response curve to the left has been reported.[38] Furthermore, it has become necessary to preincubate HIT cells in the absence of glucose (e.g., 30 min) in order to observe a subsequent stimulation of insulin release.[38]

Another striking difference between β cell lines and native β cells is their insulin content. The highest values observed (250–1000 ng/10^6 cells) in RIN and HIT cells correspond to ~0.5–2% that of native cells. At the other extreme, some less-well-differentiated lines have been studied and insulin cell contents as low as 0.01% of a native β cell have been reported. We have found that monitoring cellular insulin content at regular intervals is a useful way of ensuring maintained differentiated function. Although least marked in RIN cells (and most pronounced in HIT and B-TC cells), all lines studied by us have shown sudden and precipitous declines in insulin content as a function of tissue culture passage number.

One possible use of β cell lines is as a target for measuring anti-islet cell antibodies in the serum of diabetic patients.[39] Here again it is important to note that there are considerable differences in the surface features of these cells compared with native cells[40] and, as for insulin content and sensitivity to secretagogues, these features also change with time in culture.[39]

---

[38] M. D. Meglasson, C. D. Manning, H. Najafi, and F. M. Matschinsky, *Diabetes* **36**, 477 (1987).

[39] J. W. Thomas, V. J. Virta, and L. J. Nell, *J. Immunol.* **138**, 2896 (1987).

[40] P. A. Halban, S. L. Powers, K. L. George, and S. Bonner-Weir, *Endocrinology* **123**, 113 (1988).

Finally, even apparently well-defined lines (such as the RIN lines) display remarkable plasticity in differentiated function. Indeed, there are RINm5F sublines which produce more glucagon than insulin, despite their having been selected initially[8] for their preferential insulin production. Attempts to improve the differentiated status of RIN cells by exposing them to sodium butyrate have met with limited success. Thus, some RIN cells respond to butyrate with increased levels of insulin production[41] and with better expression of typical β cell surface antigens[42] whereas other sublines do not.[43] It would appear as though only poorly differentiated RIN cell sublines will respond to any extent to this particular agent.[43]

Acknowledgments

We thank Dr. Benigna Blondel for invaluable help in the preparation of this manuscript. This work was supported by Grants from the Swiss National Science Foundation Nos. 32.25665.88 (C.B.W.), 31-26625.89 (P.M.), 31-9394.88 (P.A.H.), and Grant No. 187384 from the Juvenile Diabetes Foundation International (P.M.), a grant from the Greenwall Foundation (P.A.H.), and Grant No. DK-35292 from the National Institutes of Health (P.A.H.).

[41] J. Philippe, D. J. Drucker, W. L. Chick, and J. F. Habener, *Mol. Cell. Biol.* **7**, 560 (1987).
[42] R. K. Bartholomensz, I. L. Campbell, and L. C. Harrison, *Endocrinology* **124**, 2680 (1989).
[43] S. L. Gardner, F. Dotta, R. C. Nayak, K. L. George, G. S. Eisenbarth, and P. A. Halban, *Diabetes Res.* **12**, 93 (1989).

[15] Membrane Potential Measurements in Pancreatic β Cells with Intracellular Microelectrodes

By Hans Peter Meissner

Introduction

In 1968 it was reported for the first time that glucose and other insulin secretagogues induce membrane depolarization and trigger electrical activity in pancreatic β cells.[1] In early electrophysiological studies the membrane potential of β cells was found to be low.[1,2] Later, with the improvement of the microelectrode technique more negative membrane potentials were measured, similar to those known from nerve or muscle.[3,4]

[1] P. M. Dean and E. K. Matthews, *Nature (London)* **219**, 389 (1968).
[2] P. M. Dean and E. K. Matthews, *J. Physiol. (London)* **210**, 255 (1970).

It was further shown that the electrical activity of β cells is closely related to insulin release[2,3,5-7] and it became evident that this electrical activity is one of the first and crucial steps in the cascade of events leading finally to insulin secretion by exocytosis.

Elucidation of the ionic mechanisms underlying the membrane potential changes induced by different test conditions markedly increased our knowledge of the regulation of insulin release. All these data have been reviewed extensively.[8-11] However, many questions on the process of insulin secretion are still unanswered and further electrophysiological experiments may, therefore, be helpful in clarifying the complexity of the stimulus–secretion coupling in pancreatic β cells.

In the following, a detailed description of the method for membrane potential measurements in β cells will be given. It has been developed in the Department of Physiology of the University of Saarland in Homburg/Saar and was first described in 1974.[3] In this chapter only the conventional microelectrode technique will be presented. The patch-clamp technique, which was introduced to islet research in 1984,[12-14] will be described by Petersen *et al.* in Chapter [20] of this volume.

Perifusion System

The perifusion system developed for membrane potential measurements in β cells[3] is shown schematically in Fig. 1. The perifusion chamber which contains the preparation consists mainly of three parts which are squeezed together by screws like a sandwich (Fig. 2). In the upper part of the chamber, built of translucent Perspex, a small rectangular recess having a volume of about 1 ml is milled. On the back of this recess there is a bore hole in which the reference electrode can be inserted. On the right side a small channel ends through which the perifusing solutions enter the tissue bath. The front of the recess is closed by a thin Perspex pane.

[3] H. P. Meissner and H. Schmelz, *Pfluegers Arch.* **351,** 195 (1974).
[4] H. P. Meissner, *J. Physiol. (Paris)* **72,** 757 (1976).
[5] H. P. Meissner and I. J. Atwater, *Horm. Metab. Res.* **8,** 11 (1976).
[6] P. M. Dean, E. K. Matthews, and Y. Sakamoto, *J. Physiol (London)* **246,** 459 (1975).
[7] I. Atwater, E. Rojas, and A. Scott, *J. Physiol. (London)* **291,** 57P (1979).
[8] J. C. Henquin and H. P. Meissner, *Experientia* **40,** 1043 (1984).
[9] E. K. Matthews, *in* "The Electrophysiology of the Secretory Cell" (A. M. Poisner and J. M. Trifaro, eds.), p. 93. Elsevier, New York, 1985.
[10] S. Ozawa and O. Sand, *Physiol. Rev.* **66,** 887 (1986).
[11] O. H. Petersen and I. Findlay, *Physiol. Rev.* **67,** 1054 (1987).
[12] F. M. Ashcroft, D. E. Harrison, and S. J. H. Ashcroft, *Nature (London)* **312,** 446 (1984).
[13] D. L. Cook and C. N. Hales, *Nature (London)* **311,** 271 (1984).
[14] D. L. Cook, M. Ikeuchi, and F. Y. Fujimoto, *Nature (London)* **311,** 269 (1984).

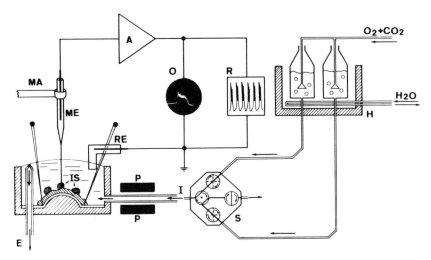

FIG. 1. Schematic presentation of the perifusion system. H, Heater to warm up the perifusing solutions; S, stopcock for rapid change of the perifusing solutions; I, inflow to the tissue bath; P, Peltier elements and heat exchanger to regulate the temperature in the tissue bath to 37°; IS, isolated islets in the tissue bath; E, outflow tube; ME, microelectrode with an Ag–AgCl wire; RE, reference electrode connected to ground; MA, holder of the micromanipulator to fix the microelectrode; A, amplifier; O, oscilloscope; R, ink recorder.

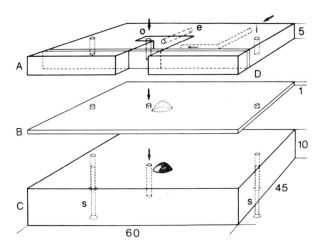

FIG. 2. Perifusion chamber. A, Upper part with the recess for the tissue bath; e, bore hole for the reference electrode; o, outflow tubing; i, channel for the inflowing solution; B, middle part consisting of soft plastic material; C, lower part with the prominent half-sphere; s, screws to combine the different parts; D, Perspex pane to close the front of the tissue bath. Dimensions are given in millimeters.

The lower part of the chamber, also made of Perspex, possesses a small, prominent half-sphere which projects into the tissue bath after assembling of the chamber. The middle part is made of a soft plastic material, 1 mm thick, into which the prominent half-sphere of the lower part is impressed by heating. When the different parts are assembled the tissue bath is sealed with petroleum jelly. The whole chamber is mounted on a metal column, the height of which can be changed.

The temperature in the chamber is regulated by two Peltier elements placed on both sides of a heat exchanger. The heat exchanger, made of stainless steel, is located on the inflow side of the perifusion chamber (Fig. 1). The temperature is electronically controlled via a thermistor inserted in the stainless steel block.

By using a special stopcock[15] (Fig. 1, S) the different perifusing solutions can easily be changed. The solutions are continuously gassed with a mixture of 95% O_2 and 5% CO_2. To avoid air bubbles in the perifusion chamber, the bursting of which usually dislodge the microelectrode from the cell, the perifusing solutions are preheated to about 40° and then cooled by the Peltier elements so that the temperature in the chamber is 37°.

The outflow of the solutions in the tissue bath is sucked through an overflow pipe by an evacuating pump.

The flow rate of the perifusing solutions is about 3.5 ml/min, thus, the total exchange of the fluid in the tissue bath requires approximately 17 sec. From the stopcock to the entry of the tissue bath the solution needs about 10 sec.

To prevent vibrations of the preparation and the microelectrode the chamber with the micromanipulator is mounted on a heavy iron plate which is additionally loaded with bricks of lead. The metal plate is placed on four tennis balls held by metal rings on an especially fortified table. For further shock absorption the table is set in boxes filled with sand. In another setup, a Barry-Isolair table (Barry Controls GmbH, Raunheim, FRG) was successfully used for damping of vibrations. This table utilizes air bolsters to absorb vibrations.

Animals and Preparation

For the experiments, NMRI mice, mostly female, were used. The animals, weighing 25–30 g, had free access to a standard pellet diet and tap water until death. Two hours before sacrifice 20 mg/kg pilocarpine was injected intraperitoneally. Such treatment facilitates visualization and

---

[15] K. H. Kilb and R. Stämpfli, *Naunyn-Schmiedebergs Arch. Pharmacol.* **285**, 293 (1974).

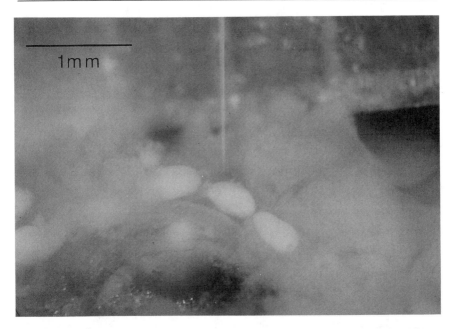

FIG. 3. Isolated islets in the perifusion chamber. In the center are three isolated islets attached to the exocrine tissue with a small part of their basis. Above the centrally located islet, a microelectrode is placed. On the left side, additional, not-yet isolated islets can be seen.

preparation of the islets without affecting the membrane potential changes brought about by physiological stimuli of β cells.[16] The animals were killed by decapitation. A small piece of the corpus of the pancreas was then excised and immediately transferred to the perifusion chamber. This piece of pancreas is spread out over the spherical vaulted bottom of the tissue bath and fixed with fine pins to the elastic bottom of the chamber. The preparation is continuously perifused at 37° with a Krebs-Ringer-bicarbonate buffer.

Using obliquely incident light and a dark background the islets can be easily distinguished with a binocular preparation microscope (Zeiss, Oberkochen, FRG) as spherical and ovoid whitish cell accumulations which are distinctly delimited from the exocrine part of the pancreas. Their number is highest along the bigger blood vessels. Sometimes they are interconnected by small tissue bridges and form complexes of islets. Some islets reach a diameter of about 300 μm (Fig. 3).

The mouse islets consist of 80 to 90% β cells and only 10 to 20% are

[16] J. C. Henquin and H. P. Meissner, *Am. J. Physiol.* **240**, E 245 (1981).

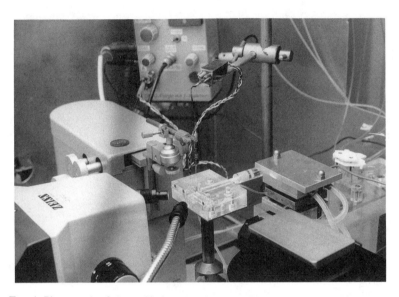

Fig. 4. Photograph of the perifusion chamber and the surrounding equipment. *Center*: The perifusion chamber with the reference electrode at the back. *Left*: The microscope with light fibers and the micromanipulator with a microelectrode. *Right*: The stopcock and the heat exchanger with the Peltier elements. Above the chamber, the input of the amplifier connected by a wire with the microelectrode may be seen. In the background the amplifier is visible.

other cells, namely, α, δ, and PP cells,[17,18] which surround the centrally located β cells like a mantle.

To illuminate the preparation fiberglass optics are used (Fiber-Optic AG, Spreitenbach, Switzerland). The dissection microscope is mounted so that the view is to the front of the perifusion chamber (Fig. 4).

As described by Hellerström[19] and Dean and Matthews[2] several islets are gently freed from the surrounding exocrine tissue with the aid of fine watchmaker forceps (Dumont, Switzerland) and trabecular scissors under microscopic control. The dissected islets remain attached to the exocrine tissue with only a small part of their bases (Fig. 3). When the thin capsule of connective tissue, which preserves the structural integrity of the islets, is injured during dissection, successful impalement with a microelectrode becomes difficult. We have almost never succeeded in measuring electrical activity of β cells in injured islets.

[17] G. J. Patent and M. Alfert, *Acta Anat.* **66,** 504 (1967).
[18] H. Ferner and H. Kern, *in* "Handbuch des Diabetes Mellitus" (E. F. Pfeiffer, ed.), p. 41. J. F. Lehmann, München, 1969.
[19] C. Hellerström, *Acta Endocrinol.* **45,** 122 (1964).

Microelectrodes and Electrical Equipment

Microelectrodes are used for intracellular recording of the membrane potential. They are made from omega dot capillaries with a filament (Frederick Haer and Company, Brunswick, Maine). The outer diameter of these capillaries is 2 mm, the inner 1 mm. The microelectrodes are pulled with a vertical conventional electrode puller. Best measurements of membrane potential are obtained with electrodes having a long and flexible tip. The microelectrodes are filled with a filtered 2 M potassium citrate solution adjusted to a pH of 7.2–7.4 with citric acid.

As the diameter of β cells is only 10 to 14 μm, very fine electrodes are necessary for impalement. Only with microelectrodes having a tip resistance between 100 and 300 MΩ is it possible to impale β cells successfully and to remain in a cell for 1–2 hr.

The microelectrode is electrically connected through an Ag–AgCl wire with the input of an FET amplifier (input resistance > 10^{11} Ω, grid current < 10^{-12} A, amplification factor 50) with capacity compensation. The output potential of the amplifier is continuously recorded on tape and monitored on an oscilloscope and two ink recorders (W+W Scientific Instruments Inc., Basel, Switzerland and Gould Inc., Cleveland, Ohio). The reference electrode is connected to ground (Fig. 1). It consists of a small sintered Ag–AgCl plate in a small Perspex reservoir filled with saturated KCl. This reservoir is connected to the tissue bath by a small tube filled with KCl-agar.

In order to avoid electrical interferences the perifusion chamber, the stopcock, the micromanipulator, and the input of the amplifier are electrically screened by a Faraday cage.

Solutions

As basic solution a Krebs-Ringer-bicarbonate buffer is used, having the following ionic composition: NaCl, 122 mM; CaCl$_2$, 2.6 mM; MgCl$_2$, 1.2 mM; and NaHCO$_3$, 21 mM.[20] It is gassed with a mixture of 95% O$_2$ and 5% CO$_2$ and the pH is 7.4. No albumin is present. When glucose is added to the medium, no correction is made to maintain osmolarity.

In early experiments a Krebs-Henseleit solution in the modification of Dean and Matthews[2] was used. This solution had the following composition (mM): NaCl, 103; KCl, 4.7; CaCl$_2$, 2.56; MgCl$_2$, 1.13; NaHCO$_3$, 25; NaH$_2$PO$_4$, 1.15; glucose, 2.8; sodium pyruvate, 4.9; sodium fumarate, 2.7; and sodium glutamate, 4.9. However, pyruvate is able to potentiate the

---

[20] H. P. Meissner and W. Schmeer, *in* "The Mechanism of Gated Calcium Transport across Biological Membranes" (S. T. Ohnishi and M. Endo, eds.), p. 157. Academic Press, New York, 1981.

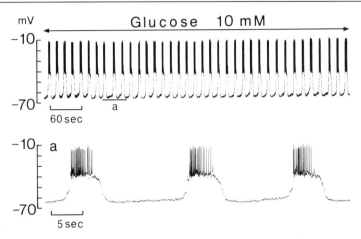

FIG. 5. Regular glucose-induced electrical activity of a single β cell. The upper part of the recording shows a train of slow waves in the presence of 10 mM glucose. Record (a) shows the three slow waves marked by a bar in the upper part at an expanded time scale. Note the regularity of glucose-induced electrical activity.

glucose-induced electrical activity,[21] therefore in later experiments pyruvate, glutamate, and fumarate were omitted from the medium. Also, in later experiments NaH_2PO_4 was not added to the solution.

Impalement and Membrane Potential of β Cells

For impalement of a β cell with a microelectrode a Leitz (Wetzlar, FRG) micromanipulator is used. The microelectrode approaches the isolated islet from above (Figs. 3 and 4). After penetration of the superficial parts of the islet the microelectrode is inserted into a β cell by slightly knocking with a finger tip on the plate on which the perifusion chamber and the micromanipulator are mounted.

β cells are identified by the typical electrical activity they exhibit in the presence of a stimulatory glucose concentration.[3,4] For this reason it is necessary to start the experiments with a stimulatory glucose concentration (10 or 15 mM). As shown in Fig. 5, the electrical activity consists of the succession of slow waves with bursts of rapid spikes superimposed on the plateau level and of silent repolarization phases (intervals).

A criterion for an acceptable measurement of membrane potential in

[21] H. P. Meissner and J. C. Henquin, in "Diabetes 1982" (E. N. Mngola, ed.), p. 353. Excerpta Medica, Amsterdam, 1983.

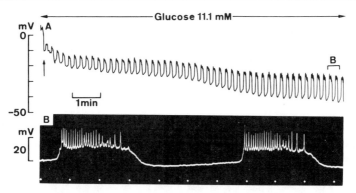

FIG. 6. Effect of sealing on the membrane potential of a single β cell perifused in the presence of 11.1 mM glucose. The ink recording (A) shows electrical activity over a period of about 10 min. At the beginning of (A), marked by the arrow, the electrode was impaled into the β cell. (A) was recorded by a chart writer of low-frequency response; therefore, the amplitude of spikes is reduced. Oscilloscope record (B), obtained at the time marked in (A), shows two slow waves with spikes of full amplitude. The interval between two points in (B) corresponds to 2 sec. Note the gradual increase of membrane potential combined with an augmentation of the amplitude of slow waves in record (A).

the presence of 10 or 15 mM glucose is that the plateau potential be not less negative than −30 mV and the amplitude of the slow waves not smaller than 15 mV.

In most experiments the high membrane potential values are not immediately achieved after the impalement of the electrode into the β cell. There is rather a gradual polarization of the membrane lasting several minutes until the membrane reaches a stable potential (Fig. 6). This behavior may be due to sealing of the membrane around the electrode tip.

As also seen in Fig. 5, the glucose-induced burst pattern of electrical activity is very regular and the duration of successive slow waves and intervals remains fairly constant. However, this remarkable regularity of slow waves and intervals does not characterize the response of all cells stimulated by constant concentration of glucose. On the contrary, fluctuations of these electrical events were evident in certain cells (Fig. 7), where relatively regular variations in the duration of slow waves and/or intervals can be seen.[22]

A further criterion for an acceptable measurement of the membrane potential is also the prompt response of a single cell to a change of glucose concentration. Thus, the duration of slow waves lengthens, whereas the

[22] J. C. Henquin, H. P. Meissner, and W. Schmeer, *Pfluegers Arch.* **393**, 322 (1982).

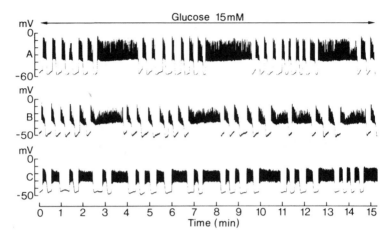

FIG. 7. Cyclic variations of glucose-induced electrical activity of single β cells. These records were obtained from three different cells. The electrical activity was induced by 15 mM glucose.

intervals shorten when the glucose concentration is raised (Fig. 8).[3,4,23] At glucose concentrations above 20 mM the β cell membrane remains persistently depolarized at the plateau potential and exhibits a continuous spike activity (Fig. 8C).

If the intensity of the electrical activity is expressed as a fraction of plateau phase (i.e., the fraction of time spent at the plateau level with spike activity), its increase with the concentration of glucose is characterized by a sigmoid relationship (Fig. 9),[3,23] which markedly resembles the sigmoid relationship between glucose concentration and insulin release.[24]

At a nonstimulatory glucose concentration, i.e., a glucose concentration that also does not stimulate insulin release, the membrane of β cells is polarized and the potential is stable. In a glucose-free medium the resting potential usually ranges between -60 and -70 mV.[16] Higher values up to -90 mV are occasionally recorded. In one series of our experiments a mean value $\pm$SEM of -63.4 ± 1.2 mV ($n = 25$) was found for the resting potential in the absence of glucose.[16]

As shown in Fig. 10, there is no significant difference of membrane potentials measured in a glucose-free medium or in the presence of 3 mM glucose.[23] However, when glucose concentration is raised beyond 3 mM the membrane depolarizes. At a critical level of about -50 mV (threshold potential) the β cell becomes electrically active and the potential oscillates

[23] H. P. Meissner and M. Preissler, in "Treatment of Early Diabetes" (R. A. Camarini-Davalos and B. Hanover, eds.), p. 97. Plenum, New York, 1979.
[24] S. H. J. Ashcroft, J. M. Bassett, and P. J. Randle, *Diabetes* **21**, (Suppl. 2), 538 (1972).

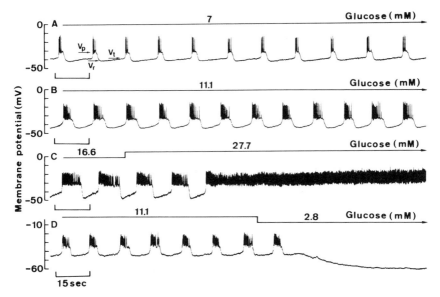

FIG. 8. Effect of variation of glucose concentration on the electrical activity of single β cells. (A) and (B) were obtained from the same cell, (C) and (D) are records from two other cells. The arrows in (A) mark the threshold potential (V_t), the plateau potential (V_p), and the repolarization potential (V_r). Note that there is no difference of V_t, V_p, and V_r in (A) and (B) despite the variation of the glucose concentration. In (C) and (D) the glucose concentration was altered at the step of the trace on top of each record. The time scale is the same for all records. (From Ref. 23.)

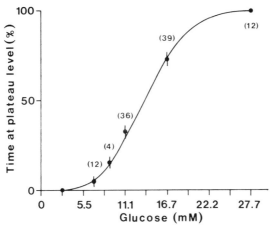

FIG. 9. Relationship between glucose concentration (abscissa) and fraction of plateau phase (ordinate). The latter was obtained from perifusion periods of several minutes by measuring the total time during which the membrane was depolarized at the plateau potential. One hundred percent time at plateau level corresponds to continuous spike activity. Results are given as mean ± SEM; number of cells in parentheses. (From Ref. 23.)

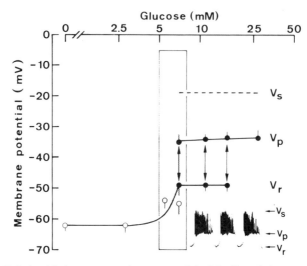

FIG. 10. Relationship between membrane potential of β cells and glucose concentration. Values are given as mean ± SEM, and were obtained from 4 (5.5 mM glucose) up to 29 (2.8 mM glucose) cells. The unfilled circles correspond to membrane potential values at which no electrical activity was observed. The hatched column indicates the range of glucose concentrations at which cells in an electrically active or silent state could be found. The inset with the three slow waves clarifies the potential levels used in the plot. V_s, Reversal potential of the spikes. (From Ref. 23.)

between the repolarization potential and the plateau potential from which the spike activity originates. The potential levels of the electrical activity are more or less independent of the glucose concentration; only at glucose concentrations beyond 20 mM does the membrane remain depolarized at the plateau level.

Acknowledgments

I thank Prof. Meves (Department of Physiology, University of Saarland, Homburg/Saar) for reading the manuscript and for support. I am grateful to W. Schmeer for many years of skillful assistance. I am also grateful to the Deutsche Forschungsgemeinschaft and to Prof. Stämpfli for their support of my work during my stay in Homburg/Saar.

[16] Stimulation of Secretion by Secretagogues

By Stephen A. Wank, Robert T. Jensen, and Jerry D. Gardner

Introduction

Stimulation of pancreatic secretion is controlled by both neural and hormonal factors. All physiologically relevant neurotransmitters and gastrointestinal hormones are either acetylcholine or peptides. The most physiologically relevant neurotransmitters that regulate pancreatic secretion are acetylcholine and vasoactive intestinal peptide (VIP), and in some species, gastrin-releasing peptide. The most physiologically relevant gastrointestinal hormones that stimulate pancreatic secretion are cholecystokinin (CCK) and secretin. Some peptides such as somatostatin and pancreatic polypeptide have an inhibitory action on pancreatic secretion *in vivo*; however, these pancreatic islet peptides have no direct activity on pancreatic acinar cell secretion *in vitro*.[1] Numerous other agents of less physiologic relevance are able to stimulate pancreatic secretion and have been summarized extensively.[2]

Several *in vitro* pancreatic preparations have been used for studying secretagogue-stimulated pancreatic secretion. A detailed discussion of these preparations as well as their advantages and disadvantages is presented elsewhere in this volume.[2a] A preparation of dispersed pancreatic acini is most desirable for the study of secretagogue-stimulated pancreatic enzyme secretion because this preparation contains at least 96% acinar cells, demonstrates high levels of enzyme secretion, remains responsive for up to 5 hr *in vitro*, allows multiple identical sampling, and can be used for studies investigating secretagogue interaction with specific receptors and postreceptor biochemical events.

Secretagogue stimulation of pancreatic acinar cells results in the release of a number of digestive enzymes, including the serine proteases (trypsinogen, chymotrypsinogen, proelastase, and kallikreinogen), exopeptidases (procarboxypeptidases A and B), lipases, (prophospholipase A_2, lipase, and colipase) and α-amylase. Amylase, unlike the other digestive enzymes secreted by the pancreas, is not a proenzyme requiring enzymatic activa-

---

[1] J. A. Williams and I. D. Goldfine, *in* "The Exocrine Pancreas" (V. L. W. Go, J. D. Gardner, F. P. Brooks, E. Lebenthal, E. P. DiMagno, and G. A. Scheele, eds.), p. 347. Raven, New York, 1986.

[2] J. D. Gardner and R. T. Jensen, *in* "The Exocrine Pancreas" (V. L. W. Go, J. D. Gardner, F. P. Brooks, E. Lebenthal, E. P. DiMagno, and G. A. Scheele, eds.), p. 109. Raven, New York, 1986.

[2a] D. Menozzi, R. T. Jensen, and J. D. Gardner, this volume [18].

tion and does not require cofactors for its activity. The measurement of amylase activity in biological fluids as an important aid in the diagnosis of pancreatic disease has spawned a number of convenient commercial assays. The immediate activity of amylase and the availability of convenient assays have made measuring amylase secretion the most popular method of assessing secretagogue stimulation of enzyme secretion from pancreatic acinar cells.

Measurement of Amylase Secretion

Early methods for determining amylase activity such as those based on physical changes in the substrate (viscosity or turbidity), the generation of reducing sugars (saccharogenic), or the amyloclastic starch–iodine reaction are either too complex, insensitive, or inaccurate to be of practical value for measuring amylase activity from multiple samples taken from the incubation medium during the stimulation of pancreatic acinar cells.[3-5] Recently three commercially available amyloclastic assays, Phadebas, Cibacrom, and Procion Yellow, have been developed.[3-5] They each utilize an insoluble, cross-linked starch polymer conjugated to either a blue or yellow dye. α-Amylase catalyzes the hydrolysis of the 1–4 glycan bonds, resulting in the liberation of soluble starch–dye components. The starch polymer is separated from the soluble starch–dye components by centrifugation. The absorbance of the dye in the supernatant is proportional to the α-amylase activity in the sample. These tests have a linear relation between enzyme concentration and hydrolysis of starch polymer under conditions of substrate saturation with a correlation coefficient of 0.95[3,4] and a sensitivity of about 35 Somogyi units/liter for serum.[3]

A later method developed for the study of secretagogue stimulation of dispersed pancreatic acinar cells is based upon monitoring the discharge of radiolabeled secretory proteins.[6] Isolated acinar cells from pancreas are pulse–chase labeled with [^{3}H]leucine (60 μCi/ml) for 10 min at 37° and diluted 10-fold with chase solution containing 4.0 mM nonradioactive leucine. After a chase period of 1 hr at 37°, the acinar cells are then stimulated with secretagogues and sampled. A 1-ml sample is separated from the medium by layering over a 1-ml discontinuous gradient of solution containing 4% bovine plasma albumin. The separated cell pellets and medium are then precipitated and washed with 0.5 N perchloric acid prior to liquid scintillation counting. This assay gives results which are in close

[3] M. Ceska, K. Birath, and B. Brown, *Clin. Chim. Acta* **26**, 437 (1968).
[4] O. H. Jung, *Clin. Chim. Acta* **100**, 7 (1980).
[5] H. Rinderknecht, P. Wilding, and B. J. Haverback, *Experientia* **23**, 805 (1967).

agreement with methods used to assay amylase secretion. Although this assay is dependent on cellular events preceding packaging into storage granules and involves the use of relatively expensive radiochemicals, it is more sensitive than the amylase assay.[6]

Most recently, a reverse hemolytic plaque assay has been developed that allows direct visualization of amylase secretion by single pancreatic acinar cells as well as pancreatic acini. In this assay, sheep red blood cells coated with protein A are mixed with pancreatic acini or acinar cells and allowed to attach to Cunningham chambers.[7] Anti-amylase serum (1:2) is then added with a secretagogue at variable concentrations for different periods of time at 37°. During this incubation amylase is secreted and bound by anti-amylase. The amylase–anti-amylase complex is then bound to protein A on the surface of the red blood cells. After rinsing the chambers with appropriate incubation solution, guinea pig serum (1:10) is added as a source of complement, which causes lysis of the surrounding red blood cells that have the amylase–anti-amylase coupled on their surface. The response of amylase secretion can then be quantitated by the percentage and size of the hemolytic plaques surrounding the secreting pancreatic acini or acinar cells. This assay has the advantage over standard amylase assays of assessing visually the functional heterogeneity of the response in amylase secretion among individual pancreatic acini and acinar cells. It also offers the potential, through the use of various antisera, to assess the possible heterogeneity of zymogens among pancreatic acini and whether they are secreted in a parallel manner.[7]

Experimental Design for Studying Secretagogue Stimulation of Dispersed Pancreatic Acini by the Amyloclastic Method Using the Phadebas Amylase Test

The amyloclastic method using the Phadebas amylase test was chosen among other assays for measuring the secretagogue stimulation of pancreatic acinar cells because of the immediate enzymatic activity of amylase, the high sensitivity, reproducibility and accuracy of the test over a wide range of enzyme concentrations, and the ease and availability of the assay.

1. Prepare pancreatic acini as described[2a] and dilute acini with incubation solution as follows: guinea pig, one pancreas in 200 ml; rat, one pancreas in 300 ml; mouse, one pancreas in 100 ml

---

[6] A. Amsterdam and J. D. Jamieson, *J. Cell Biol.* **63,** 1057 (1974).
[7] D. Bosco, M. Chenson, R. Bruzzone, and P. Meda, *Am. J. Physiol.* **254,** G664 (1988).

2. Have available the following:

Polypropylene tubes (17 × 100 mm) (Falcoa, #2059; Becton Dickin-
son, Lincoln Park, NJ)
Polystyrene tubes (17 × 100 mm) (Falcon #2017)
Microcentrifuge tubes (0.4 ml) (PGC Scientific, Gaithersburg, MD)
Screw-cap conical centrifuge tubes (50 ml) (Corning, Corning, NY)
Phadebas amylase test tablets (Pharmacia, Piscataway, NJ)
Spectrophotometer (capable of reading at 620 nm)
Microcentrifuge
Centrifuge capable of holding tube (17 × 100 mm) and obtaining
1500 g
Erlenmeyer flask (250 ml)
Vortex mixer
Magnetic stir plate
O_2 (100%)
Dubnoff shaking water bath
Nyosil oil (William F. Nye, Inc., New Bedford, MA)
NaOH (0.5 N)
Lysing solution: NaH_2PO_4 (0.01 M, pH 7.8), sodium dodecyl sulfate
(0.1%), bovine serum albumin (0.1%), $CaCl_2$ (100 mM)
Amylase solution: NaH_2PO_4 (0.02 M, pH 7.0), NaCl (0.05 M), NaN_3
(0.02%)
Secretagogues under investigation

Methods

Secretagogue Stimulation of Amylase Secretion

1. Dispense 1 ml of dispersed pancreatic acini (held in uniform suspension by swirling in a 250-ml Erlenmeyer flask) into Falcon #2059 incubation tubes in triplicate. This should be performed with each set of tubes
containing the experimental secretagogues and a control tube with no
secretagogue. The control will be used to determine the basal secretion
during the incubation.

2. Vortex each incubation tube, gas with 100% O_2, cap each tube, and
incubate at 37° for the desired duration (usually 30 min for stoichiometric
dose–response studies) in a shaking water bath.

3. After dispensing acini into one of the three sets of tubes in the
triplicate series and before starting another set, dispense 300 μl from the
cell suspension in the 250-ml Erlenmeyer flask into three 400-μl microfuge
tubes (prefilled with 50 μl Nyosil oil) and immediately spin for 15 sec.
These will serve as the zero time tubes.

4. At various times during a kinetic experiment or at the end of a 30-min incubation for dose–response studies, vortex and sample 300 μl from each incubation tube into a 400-μl microfuge tube prefilled with 50 μl of Nyosil and centrifuge for 15 sec.

5. At any time during the incubation, dilute 500 μl of acinar cell suspension into 4.5 ml of lysing solution in each of six 50-ml conical centrifuge tubes and vortex. These media tubes will be used to assay for the total amylase contained in a sample of the cell suspension.

Measurement of Amylase by the Phadebas Amylase Test

1. Suspend the Phadebas amylase test tablets for a final concentration of one tablet in 8 ml of amylase solution (the number of tablets and total volume will depend on the number of samples to be assayed) using a magnetic stirrer.

2. Dispense 200 μl of the lysing solution into each amylase tube (Falcon #2017) corresponding to each experimental condition, and into each of the zero tubes and 4.5 ml into each of the six media tubes.

3. Sample 50 μl of supernatant from each microfuge tube and 150 μl from each media tube into their corresponding amylase tubes.

4. Dispense 2 ml of the above suspension of Phadebas amylase test reagent (held in uniform suspension by continuous stirring) into each amylase tube and incubate at 37° in a shaking water bath for 10–30 min, depending on the amount of amylase activity in the samples. (Incubating too long will necessitate further dilution of the samples to bring them into the linear range of the concentration–absorbance curve).

5. Stop the reaction by adding 500 μl 0.5 N NaOH.

6. Dilute each tube with 5–10 ml of H_2O, depending on the intensity of blue color. (Add the same volume of water to each tube.)

7. Centrifuge each tube at 1500 g for 5 min.

8. Measure the absorbance of the sample supernatants at 620 nm using a reagent blank set at zero absorbance.

9. The percentage of amylase activity in the acini at the beginning of the incubation that is released into the extracellular medium during the incubation is calculated according to the formula:

$$\text{Percentage total amylase secretion} = \left[\frac{OD_{sample} - OD_{zero}}{(OD_{media})(3.33) - OD_{zero}} \right] 100$$

Discussion

Using the dye-coupld starch polymer method to study secretagogue-stimulated amylase release from pancreatic acini is versatile and has been applied to the analysis of the kinetics of enzyme secretion, stoichiometric

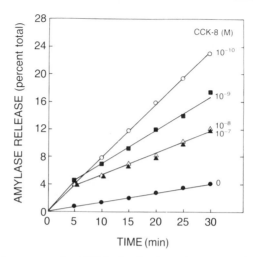

FIG. 1. The time course for CCK-8-stimulated amylase release from dispersed acini prepared from guinea pig pancreas. Acini were suspended in standard incubation solution containing the indicated concentrations of CCK-8. The percentage of total amylase released into incubation medium was determined at 5-min intervals during a 30-min incubation at 37°.

dose–response relationships, and desensitization of enzyme secretion through the use of sequential incubations and potentiation of stimulated secretion by combining two secretagogues in the same incubation.

An example of a kinetic study of secretagogue-stimulated amylase release from dispersed pancreatic acini incubated with CCK-8 is shown in Fig. 1.[8] Dispersed pancreatic acini were suspended in standard incubation solution and the indicated concentrations of CCK-8. At 0.1 nM CCK-8, amylase secretion is maximal and constant for 30 min, while at higher concentrations of CCK-8 (up to 10 mM) the rate of secretion is maximal for the first 5 min and then decreases during the remainder of the incubation in direct relation to the increase in the concentration of CCK-8.

An example of a stoichiometric dose–response study of secretagogue-stimulated amylase secretion is shown in Fig. 2 for CCK-8, bombesin, carbachol, vasoactive intestinal peptide (VIP), substance P, and calcitonin gene-related peptide (CGRP).[9] Each secretagogue varies in potency and efficacy. For secretagogue mediated by cellular cyclic AMP (VIP and

[8] S. R. Peikin, A. J. Rottman, S. Batzri, and J. D. Gardner, *Am. J. Physiol.* **235**, E743 (1978).
[9] R. T. Jensen, D. H. Coy, Z. A. Saeed, P. Heinz-Erian, S. Mantey, and J. D. Gardner, *Ann. N.Y. Acad. Sci.* **547**, 138 (1988).

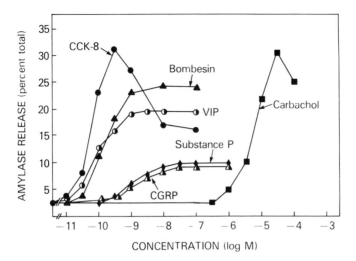

CONCENTRATION (log M)

FIG. 2. Dose–response curves for the secretagogue-induced increases in amylase secretion from dispersed acini from guinea pig pancreas. Acini were incubated with the secretagogues specified for 30 min at 37°. Amylase secretion is expressed as the percentage of total cellular amylase released into the medium during the incubation. VIP, Vasoactive intestinal peptide; CGRP, calcitonin gene-related peptide.

CGRP),[10,11] the configuration of the dose–response curve is monophasic. For secretagogues whose stimulation of enzyme secretion is mediated by cellular calcium,[12-16] the configuration of the dose–response curve is monophasic (bombesin and substance P) or biphasic (CCK-8 and carbachol).

In pancreatic acinar cells, two forms of desensitization have been described; homologous, in which exposure to a secretagogue attenuates the subsequent response only to secretagogues interacting with the same specific class of receptors, and heterologous, in which exposure to one secretagogue attenuates the subsequent response to secretagogues that interact with other classes of receptors. In a typical desensitization experiment,

[10] J. D. Gardner and M. J. Jackson, *J. Physiol. (London)* **270**, 439 (1977).
[11] Z.-C. Zhou, M. L. Villanueva, M. Noguchi, S. W. Jones, J. D. Gardner, and R. T. Jensen, *Am. J. Physiol.* **251**, G391 (1986).
[12] R. T. Jensen, G. F. Lemp, and J. D. Gardner, *Proc. Natl. Acad. Sci. U.S.A.* **77**, 2079 (1980).
[13] R. T. Jensen and J. D. Gardner, *Fed. Proc., Fed. Am. Soc. Exp. Biol.* **40**, 2486 (1981).
[14] R. T. Jensen, T. Moody, C. Pert, J. E. Rivier, and J. D. Gardner, *Proc. Natl. Acad. Sci. U.S.A.* **75**, 6139 (1978).
[15] J. D. Gardner, T. P. Conlon, H. L. Klaeveman, T. D. Adams, and M. A. Ondetti, *J. Clin. Invest.* **56**, 366 (1975).
[16] R. T. Jensen and J. D. Gardner, *Proc. Natl. Acad. Sci. U.S.A.* **76**, 5679 (1979).

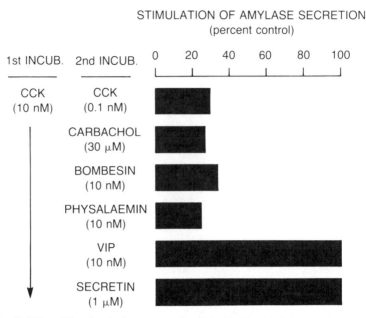

FIG. 3. Effect of first incubating pancreatic acini with 10 nM cholecystokinin (CCK) on subsequent secretagogue-stimulated amylase release. Pancreatic acini were first incubated with 10 nM CCK for 60 min at 37°, washed, and then reincubated with the secretagogues specified during a second incubation for 30 min at 37°. Amylase release was measured during the second incubation. Amylase release is expressed as a percentage of the amylase released from acini first incubated with no additions. VIP, Vasoactive intestinal peptide.

pancreatic acini are first incubated with an agonist, washed, and reincubated with the same or various different agonists during a second incubation and the response measured. Figure 3 illustrates an example of results from sequential incubations of the same acini with either the same secretagogue, CCK, or different secretagogues, carbachol, bombesin, and physalaemin (an analog of substance P), applied to the study of homologous and heterologous desensitization of amylase secretion from dispersed pancreatic acini, respectively.[17] Cholecystokinin causes homologous desensitization with a 70% reduction in secretin stimulated by CCK and heterologous desensitization with a 65–70% reduction in secretion stimulated by carbachol, bombesin, and physalaemin. Interestingly, the heterologous desensitization induced by CCK is limited to secretagogues that utilize a calcium-mediated pathway[17] and does not alter the stimulation of enzyme secretion caused by vasoactive intestinal peptide or secretin that utilize a cyclic AMP-mediated pathway.

[17] S. Abdelmoumene and J. D. Gardner, *Am. J. Physiol.* **239,** 6272 (1980).

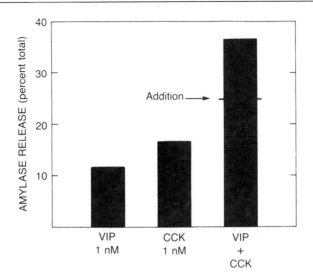

Fig. 4. Stimulation of amylase release from pancreatic acini incubated with vasoactive intestinal peptide (VIP) alone, cholecystokinin (CCK) alone, and VIP plus CCK together for 30 min at 37°. Amylase release is expressed as the percentage of total cellular amylase released into the medium during the incubation.

Receptors on the surface of pancreatic cells, having been occupied by a specific secretagogue, activate one of two distinct pathways comprising a series of intracellular events that ultimately results in stimulation of enzyme secretion. The major intracellular mediator for one of these pathways is calcium and for the other is cyclic AMP. When two secretagogues that are both mediated by cellular calcium or both by cellular cyclic AMP are combined in the same incubation, the stimulation in enzyme secretion is less than or equal to the sum of the secretion stimulated by either secretagogue acting alone.[18] However, when a secretagogue that is mediated by cellular calcium is combined with a secretagogue that increases cellular cyclic AMP, the stimulation of enzyme secretion is greater than the sum of secretion stimulated by each secretagogue acting alone and is termed "potentiation."[18] An example for the potentiation of enzyme secretion is illustrated in Fig. 4 for enzyme secretion stimulated by 1 nM VIP, a secretagogue that increases cyclic AMP, 1 nM CCK, a secretagogue that increases cellular calcium, and the combination of the two secretagogues.[8] The increase in the secretion by the combination of VIP and CCK is greater than the sum of the increase caused by each secretagogue acting alone.

[18] J. D. Gardner and R. T. Jensen, in "Physiology of the Gastrointestinal Tract" (L. R. Johnson, ed.), p. 831, New York, Raven, 1987.

[17] Pancreatic Secretion: *In Vivo*, Perfused Gland, and Isolated Duct Studies

By R. M. CASE and B. E. ARGENT

Variations in Pancreatic Structure and Function

The exocrine pancreas consists of two functional units, acini and ducts. Quantitatively, the acini dominate; duct cells have been calculated to occupy only 14% by volume of the gland in humans, 4% in guinea pig, and 2% in the rat.[1] The ductal tree begins with the centroacinar cells which line each acinus and which connect to the smallest elements of the true ductal system, the intercalated ducts. These open into intralobular ducts, which run within the pancreatic lobules, and which in turn join larger interlobular ducts.[2,3] The final division of the ductal tree is usually the main pancreatic duct. However, in the rat, mouse, and hamster a variable number of interlobular ducts open into the bile duct forming a common bile–pancreatic duct. In terms of experimental studies, the implications of this arrangement are clear. To collect pure pancreatic juice from a cannula inserted into the common duct, it is first necessary to ligate the duct beyond the limit of pancreatic tissue so as to prevent contamination with bile. In chronic animal experiments, cannulation of the bile duct at this point is obviously necessary in order to convey bile to the duodenum.[4,5]

Cats and dogs often have accessory ducts which open into the duodenum close to the main pancreatic duct. Rather than attempt to cannulate the accessory duct, it is usual to ligate it, and thereby direct all the secretion through the main duct, or ignore it. In most species the main pancreatic duct, or bile-pancreatic duct, empties into the second part of the duodenum. However, in the rabbit and guinea pig it enters the third part of the duodenum, a long way from the pylorus.

It is generally agreed that acini secrete digestive enzymes and a variable (usually small) quantity of chloride-rich fluid in response to cholecystokinin-pancreozymin (CCK) and vagal stimulation (acting via acetylcho-

[1] S. Githens, *J. Pediatr. Gastroenterol. Nutr.* **7**, 486 (1988).

[2] R. M. Case and B. E. Argent, *in* "The Exocrine Pancreas: Biology, Pathobiology, and Diseases" (V. L. W. Go, J. D. Gardner, F. P. Brooks, E. Lebenthal, E. P. DiMagno, and G. A. Scheele, eds.), p. 213. Raven, New York, 1986.

[3] R. M. Case and B. E. Argent, *in* "Handbook of Physiology: The Gastrointestinal System III" (S. G. Schultz, J. G. Forte, and B. B. Rauner, eds.), p. 383. Oxford University Press, New York, 1989.

[4] F. J. Haberich, T. Bozkurt, and W. Reschke, *Z. Gastroenterol.* **18**, 427 (1980).

[5] S. Ormai, M. Sasvári, and E. Endröczi, *Scand, J. Gastroenterol.* **21**, 509 (1986).

line), and that the ducts secret a bicarbonate-rich fluid in response to secretin. However, there are wide species variations in the volume of spontaneous secretion, the sensitivity to hormones and neurotransmitters, and the maximum bicarbonate concentration in pancreatic juice.[3,6] A knowledge of these variations, which are summarized in Table I, is essential for designing and interpreting both *in vivo* and *in vitro* experiments on the pancreas.

In Vivo Studies

Today, *in vivo* studies are performed for three reasons.

1. To study physiological control mechanisms: Studies on the intact animal will always be necessary to probe the interaction between nerves, hormones, and paracrine agents in the regulation of pancreatic secretion.[7,8] Most such studies are performed in conscious dogs or, to a lesser extent, cats and humans. Chronic animal experiments demand a permanent, satisfactory arrangement for collecting pancreatic juice. The simplest way to achieve this would be to create a fistula between the external body surface and duodenum. However, this is unsatisfactory because secreted pancreatic juice comes into contact with the duodenum. As a result, pancreatic proteases are activated, skin erosion occurs, and the flow of pancreatic juice becomes continuous.[9] The solution to this problem is to install a wide cannula into the duodenal wall opposite the main pancreatic duct through which pancreatic juice can be collected temporarily, by cannulating the main duct, and returned to the duodenum.[10,11] In humans, either the duodenal contents are aspirated[12] or the pancreatic duct is cannulated directly using endoscopic techniques.[13]

[6] R. M. Case, *in* "Experimental Pancreatitis" (G. Glazer and J. H. C. Ranson, eds.), p. 100. Ballière Tindall, London, 1988.

[7] Z. Itoh, R. Honda, and K. Hiwatashi, *Am. J. Physiol.* **238,** G332 (1980).

[8] M. Singer, *in* "The Exocrine Pancreas: Biology, Pathobiology, and Diseases" (V. L. W. Go, J. D. Gardner, F. P. Brooks, E. Lebenthal, E. P. DiMagno, and G. A. Scheele, eds.), p. 315. Raven, New York, 1986.

[9] R. A. Gregory, "Secretory Mechanisms of the Gastro-Intestinal Tract." Arnold, London, 1962.

[10] J. E. Thomas and J. O. Crider, *Am. J. Physiol.* **131,** 349 (1940).

[11] J. E. Thomas, "The External Secretion of the Pancreas." Thomas, Springfield, Illinois, 1950.

[12] E. P. DiMagno, *in* "The Exocrine Pancreas: Biology, Pathobiology, and Diseases" (V. L. W. Go, J. D. Gardner, F. P. Brooks, E. Lebenthal, E. P. DiMagno, and G. A. Scheele, eds.), p. 193. Raven, New York, 1986.

[13] S. Domschke, W. Domschke, W. Rosch, S. J. Konturek, E. Wunsch, and L. Demling, *Gastroenterology* **70,** 533 (1976).

TABLE I
SPECIES-DEPENDENT PATTERNS OF ELECTROLYTE SECRETION[a,b]

Species	Stimulus	Volume	Maximum [HCO₃⁻] (mM)
Dog (1,2), cat (3),	Spontaneous	0(+)	—
human (4)	+Secretin	+++++	145
	+CCK	+	60
	+Vagus	+	?
Rat (5)	Spontaneous	+	25
	+Secretin	++	70
	+CCK	+++	30
	+Vagus	++	?
Rabbit (6)	Spontaneous	++	60
	+Secretin	+++	130
	+CCK	++	110
	+Vagus (carbachol)	+++	120
Pig (7,8)	Spontaneous	+	?
	+Secretin	+++++	160
	+CCK	++	35
	+Vagus	++++	150
Guinea pig (9,10)	Spontaneous	+	95
	+Secretin	+++++	150
	+CCK	+++	140
	+Vagus	+++	120
Hamster (11)	Spontaneous	+	60
	+Secretin	++++	140
	+CCK	+	40
	+Vagus (carbachol)	++	80

[a] This table gives an idea of the response to stimuli given alone: potentiation often occurs when stimuli are given together. The references are to key papers which illustrate most of the features. Except for Refs. 1, 2, 4, and 8 (see below), all data were obtained from studies on anesthetized animals: quantitative differences may occur in conscious animals, especially in the rat, in which secretion is increased fivefold in conscious animals (see Ref. 5). CCK, Cholecystokinin-pancreozymin.

[b] Key to references: (1) W. M. Hart and J. E. Thomas, Gastroenterology 4, 409 (1945); (2) H. T. Debas and M. I. Grossman, Digestion 9, 469 (1973); (3) R. M. Case, A. A. Harper, and T. Scratcherd, J. Physiol. 201, 335 (1969); (4) S. Domschke, W. Domschke, W. Rosch, S. J. Konturek, E. Wunsch, and L. Demling, Gastroenterology 70, 533 (1976); (5) W. A. Sewell and J. A. Young, J. Physiol. 252, 379 (1975); (6) F. Seow and J. A. Young, Proc. Aust. Physiol. Pharmacol. Soc. 17, 199P (1986); and K. T. F. P. Seow, R. M. Case, and J. A. Young Pancreas, in press; (7) J. C. D. Hickson, J. Physiol. 206, 275, 299 (1970); (8) S. L. Jensen, J. F. Rehfeld, J. J. Holst, O. V. Nielsen, J. Fahrenkrug, and O. B. Schaffalitsky de Muckadell, Acta Physiol. Scand. 111, 225 (1981); (9) J. S. Davison and V. Dickson, in "Secretion: Mechanisms and Control" (R. M. Case, J. M. Lingard, and J. A. Young, eds.), p. 225. Manchester University Press, Manchester, 1984; (10) P. J. Padfield, A. Garner, and R. M. Case, Pancreas 4, 204 (1989); (11) A. E. Ali, S. C. B. Rutishauser, and R. M. Case, Pancreas 5, 314 (1990).

2. To assay the effects of newly discovered peptides, drugs, etc.: Although *in vitro* studies can provide much information, they cannot predict the influence of secondary phenomena such as changes in blood flow (e.g., Ref. 14). Usually anesthetized animals are used for this purpose. In this case the pancreatic duct can be cannulated permanently either from the duodenum or, more easily, through an incision in the pancreatic duct at the point where it passes obliquely through the duodenal wall.

3. To study pancreatic growth and adaption, and models of pancreatic disease: Most studies on growth and dietary adaptation[15] and experimental pancreatitis[16,17] are carried out on rats, largely for reasons of economy, while studies on experimental cancer often use hamsters.[18]

Acinar Cell Atrophy

As mentioned above, the majority of pancreatic bicarbonate secretion is derived from the ducts. Because duct cells comprise such a small proportion of the gland, studying their function is difficult. One way around this problem is to selectively destroy the acinar tissue. This can be achieved in two ways.

1. By duct ligation: In many species, including the mouse, rat, guinea pig, rabbit, and dog, ligation of the main duct causes atrophy of the acinar cells, but not the ductal cells or islets of Langerhans. Following duct ligation in the rat, for example, the acini essentially disappear within 72 hr and cuboidal duct cells proliferate so that by the fifth day ductlike structures form the bulk of the lobular structure.[19]

2. By feeding a copper-deficient diet: As first noticed by Müller,[20] rats fed a copper-free diet develop a noninflammatory acinar cell atrophy in the absence of changes in ductal tissue. Addition to the diet of a copper-chelat-

[14] R. M. Case and T. Scratcherd, *J. Physiol. (London)* **226**, 393 (1972).
[15] U. R. Fölsch, *Clin. Gastroenterol.* **13**, 679 (1984).
[16] G. Adler, H. F. Kern, and G. A. Scheele, *in* "The Exocrine Pancreas: Biology, Pathobiology, and Diseases" (V. L. W. Go, J. D. Gardner, F. P. Brooks, E. Lebenthal, E. P. DiMagno, and G. A. Scheele, eds.), p. 407. Raven, New York, 1986.
[17] M. Steer, *in* "Experimental Pancreatitis" (G. Glazer and J. H. C. Ranson, eds.), p. 207, Baillière Tindall, London, 1988.
[18] D. S. Longnecker, *in* "The Exocrine Pancreas: Biology, Pathobiology, and Diseases" (V. L. W. Go, J. D. Gardner, F. P. Brooks, E. Lebenthal, E. P. DiMagno, and G. A. Scheele, eds.), p. 443, Raven, New York, 1986.
[19] A. W. Pound and N. I. Walker, *Br. J. Exp. Pathol.* **62**, 547 (1981).
[20] H. B. Müller, *Virchows Arch. A: Pathol. Anat.* **350**, 353 (1970).

ing agent, such as D-penicillamine[21] or triethylenetetramine,[22] accelerates the process. To produce atrophy, young, male rats (125–125 g) are allowed free access to distilled water and are fed a copper-deficient diet (about 20 g daily for 6–10 weeks) containing no more than 3 mg kg^{-1} copper and between 10 and 30 mg kg^{-1} zinc. This diet can be purchased from SDS, Ltd. (Witham, Essex, England). Provided that the copper and zinc contents are as specified, it is unnecessary to add chelating agents. However, it is important to prevent coprophagy by housing the rats in wire-bottomed cages, and to ensure that cage fittings are not made from copper-containing alloys.[22]

The secretory behavior of such atrophic glands in anesthetized rats is as would be predicted from Table I, i.e., the fluid secretory response to secretin is preserved while that to CCK is greatly attenuated.[21,22] Unfortunately attempts to apply this method to other laboratory species have so far failed (e.g., the hamster is not affected; U. R. Fölsch, personal communication) so that it is unlikely to provide useful information about comparative aspects of regulation of duct cell function. However, it is a very convenient method for providing uncontaminated ductal tissue for biophysical studies (see below).

Ductal Perfusion

The main pancreatic duct modifies the electrolyte composition of the primary secretions produced by the small ducts and acini. One way of studying the properties of the main pancreatic duct *in vivo* is to perfuse the duct lumen in anesthetized animals. In this method, first described in the cat,[23] the main duct is cannulated both at the tail of the gland and at the point where it traverses the duodenal wall. The duct is then perfused with an appropriate fluid (artificial "pancreatic juice") from tail to head using a motor-driven syringe. This technique is particularly suited to those species like the cat where there is no spontaneous secretion, so that perfusion fluids are not contaminated with endogenous secretions. However, with the inclusion of a volume marker in the perfusion fluid, it has also been used in other species such as the rabbit.[24] Localization of the duct at the tail of the gland is often difficult and requires microdissection. It is therefore tempting to thread a narrow flexible tube through the duct from the duodenal end and withdraw it from the tail. However, such treatment dramatically

[21] U. R. Fölsch and W. Creutzfeldt, *Gastroenterology* **73**, 1053 (1977).
[22] P. A. Smith, J. P. Sunter, and R. M. Case, *Digestion* **23**, 16 (1982).
[23] R. M. Case, A. A. Harper, and T. Scratcherd, *J. Physiol. (London)* **201**, 335 (1969).
[24] H. A. Reber, C. J. Wolf, and S. P. Lee, *Surg. Forum* **20**, 382 (1969).

changes the permeability of the ductal epithelium and should therefore be avoided.[25]

Ductal perfusion was used first to demonstrate that significant passive fluxes of bicarbonate and chloride occur across the ductal epithelium and that such fluxes are responsible for the flow rate-dependent changes in anion composition of pancreatic juice (i.e., the decrease in bicarbonate and reciprocal increase in chloride which occur as secretory rate declines).[23] As well as being used to characterize the physiological properties of the ductal epithelium, this technique has also been used to study the effect on ductal permeability of factors implicated in the pathogenesis of pancreatitis.[26]

Isolated Gland Studies

The first *in vitro* preparation of the pancreas (essentially, pancreatic slices) was used in the 1930s to study pancreatic metabolism. Similar preparations have subsequently been used with huge success to study the cellular mechanisms and control of pancreatic enzyme secretion. More recently there has been a move toward using dissociated acini and single acinar cells for such studies. These *in vitro* techniques fall beyond the scope of this chapter and are described elsewhere in this volume.

Pancreatic fluid secretion clearly cannot be studied in slices or dissociated acini because ductal integrity must be maintained in order to collect and measure the secretory product. To achieve this, isolated whole gland preparations are required. Such preparations have two major uses.

1. To study the function of the gland in the absence of interference from other organs which may complicate interpretation of the data: In this way it has been observed, for example, that the inhibitory effect of prostaglandin E on fluid secretion in the anesthetized cat is indirect (caused by reflex vasoconstriction), as the same substance stimulates fluid secretion in the perfused cat pancreas.[14] A number of peptides also have different effects on fluid secretion *in vivo* and *in vitro*, perhaps for the same reason.

2. To study the cellular mechanisms responsible for fluid secretion by observing the effects of ion substitution and transport inhibitors: Although direct studies of duct tissue are now possible (see below), isolated gland preparations have provided a great deal of useful information about pancreatic electrolyte secretion and its cellular control.[2,3]

[25] R. M. Case and T. Scratcherd, *Biochim. Biophys. Acta* **219,** (1970).
[26] H. A. Reber, G. Adler, and K. R. Wedgwood, *in* "The Exocrine Pancreas: Biology, Pathobiology, and Diseases" (V. L. W. Go, J. D. Gardner, F. P. Brooks, E. Lebenthal, E. P. DiMagno, and G. A. Scheele, eds.), p. 255. Raven, New York, 1986.

Although methods for isolating and perfusing the dog pancreas were described in the 1920s[27,28] and perfused rat preparations were first used to study insulin secretion in 1947,[29] not until the 1960s were glands isolated from the rabbit and cat used to study pancreatic exocrine secretion. The techniques used in these two species are quite different.

The Isolated Rabbit Pancreas

The pancreas in rabbit lies between the descending and ascending limbs of the first intestinal loop. Taking advantage of the extreme thinness of this gland, Rothman and Brooks[30] devised a method in which the gland was suspended in a bath of physiological salt solution. New Zealand White rabbits are fasted overnight and anesthetized with urethane. To isolate the pancreas, the attachments of other intestinal segments to the first intestinal loop are ligated and severed and its distal end cut away from the remaining intestine and from the mesentery medial to the portal vein. To prevent damage to the gland, a segment of rectum is left attached. The pancreatic duct which, as mentioned above, enters the third part of the duodenum is then cannulated with a fine polyethylene tube. Finally, the intestinal loop containing the pancreas is removed from the animal, attached by hooks to a Plexiglas frame, and the whole place in a heated chamber containing a physiological salt solution (Fig. 1).

This preparation has been used in a number of laboratories for studies of pancreatic fluid secretion,[30-34] epithelial permeability,[35] stimulus–secretion coupling,[36] and the "endocrine" secretion of pancreatic enzymes.[37] It is quick and easy to set up. However, it has one major disadvantage: its rate of spontaneous secretion is five- to sixfold greater than that of the gland *in vivo* and, consequently, it is almost insensitive to secretin stimulation.[30]

[27] B. P. Babkin and E. H. Starling, *J. Physiol. (London)* **61**, 245 (1926).
[28] B. Goldstein, *Z. Gesamte Exp. Med.* **61**, 649 (1928).
[29] E. Anderson and J. A. Long, *Endocrinology* **40**, 92 (1947).
[30] S. S. Rothman and F. P. Brooks, *Am. J. Physiol.* **208**, 1171 (1965).
[31] K. A. Hubel, *Am. J. Physiol.* **212**, 101 (1967).
[32] A. S. Ridderstap, *Pfluegers Arch.* **311**, 199 (1969).
[33] C. H. Swanson and A. K. Solomon, *J. Gen. Physiol.* **62**, 407 (1973).
[34] C. R. Caflisch, S. Solomon, and W. R. Galey, *Pfluegers Arch.* **380**, 121 (1979).
[35] J. W. C. M. Jansen, J. J. H. H. M. de Pont, and S. L. Bonting, *Biochim. Biophys. Acta* **551**, 95 (1979).
[36] V. V. A. M. Schreurs, H. G. P. Swarts, J. J. H. H. M. de Pont, and S. L. Bonting, *Biochim. Biophys. Acta* **404**, 257 (1975).
[37] L. D. Isenman and S. S. Rothman, *Proc. Natl. Acad. Sci. U.S.A.* **74**, 4068 (1977).

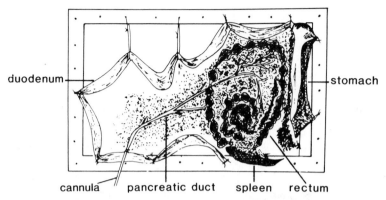

duodenum — stomach

cannula   pancreatic duct   spleen   rectum

FIG. 1. Schematic diagram of the isolated rabbit pancreas and attached duodenum mounted on a Plexiglas frame, which is then suspended in a chamber filled with a physiological salt solution. The stippled area represents pancreatic tissue. (Courtesy of Professor J. J. H. H. M. de Pont.)

Vascular Perfused Gland Preparations

As an alternative to the rabbit gland preparation, Case et al.[38] developed a more conventional preparation for studies of electrolyte secretion in which the cat pancreas was surgically isolated and perfused through its arterial supply with physiological salt solutions. The pancreas receives its blood supply from a number of sources: as an example the supply to the cat gland is illustrated in Fig. 2. Therefore considerable surgery is required in isolating the gland prior to perfusion. Although the principles involved in such surgery are similar in all species, the details will clearly differ according to minor anatomical variations. In addition to the cat[38] details of gland isolation and perfusion have been described for the following species: dog,[39-41] rat,[42] pig,[43] guinea pig,[44] and Syrian golden hamster.[45] Recent

[38] R. M. Case, A. A. Harper, and T. Scratcherd, *J. Physiol. (London)* **196,** 133 (1968).
[39] G. L. Nardi, J. M. Greep, D. A. Chambers, C. McCrae, and D. B. Skinner, *Ann. Surg.* **158,** 830 (1963).
[40] J. Hermon-Taylor, *Gastroenterology* **55,** 488 (1968).
[41] D. Augier, J. P. Boucard, J. P. Pascal, A. Ribet, and N. Vaysse, *J. Physiol. (London)* **221,** 55 (1972).
[42] T. Kanno, *J. Physiol. (London)* **226,** 353 (1972).
[43] S. L. Jensen, J. Fahrenkrug, J. J. Holst, C. Kühl, O. V. Nielsen, and O. B. Schaffalitzky de Muckadell, *Am. J. Physiol.* **235,** E381 (1978).
[44] T. Matsumoto and T. Kanno, *Peptides* **5,** 285 (1984).
[45] R. H. Bell, S. Place, P. McCullough, M. B. Ray, and D. H. Rogers, *Int. J. Pancreatol.* **1,** 71 (1986).

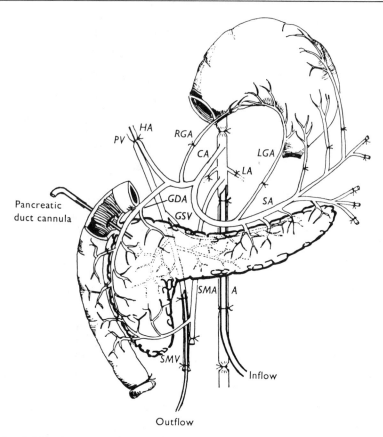

FIG. 2. Vascular supply of the cat pancreas. For clarity, the stomach is shown separated from the duodenum, displaced anteriorly and turned through 180°. A, Aorta; CA, celiac axis; GDA, gastroduodenal artery; GSV, gastrosplenic vein; HA, hepatic artery; LA, lumbar artery; LGA, left gastric artery; PV, portal vein; RGA, right gastric artery; SA, splenic artery; SMA, superior mesenteric artery; SMV, superior mesenteric vein. (Reproduced with permission from R. M. Case, A. A. Harper, and T. Scratcherd, *J. Physiol.* **196**, 133, 1968.)

reviews of the techniques involved have been published for the cat[46] and rat[47] and will not be repeated here.

In addition to studies on the mechanisms and cellular control of pancreatic electrolyte secretion,[2,3] perfused glands can be used to study a

[46] T. Scratcherd, *in* "The Exocrine Pancreas: Biology, Pathobiology, and Diseases" (V. L. W. Go, J. D. Gardner, F. P. Brooks, E. Lebenthal, E. P. DiMagno, and G. A. Scheele, eds.), p. 245. Raven, New York, 1986.

[47] T. Kanno *in* "In Vitro Methods for Studying Secretion" (A. M. Poisner and J. M. Trifaró, eds.) p. 45. Elsevier Biomedical, Amsterdam, 1987.

variety of functions, including control of enzyme secretion,[48,49] the "endo-crine" secretion of pancreatic enzymes,[50,51] amino acid transport into the gland,[52] determination of the neurotransmitters involved in regulation,[48,53] assaying secretin,[54] tissue metabolism,[55] hemodynamics and vasomotor phenomena,[41,56,57] production of circulatory shock factors,[58] and experimental pancreatitis.[59,60]

The type of solution used as a perfusion fluid is determined to some extent by the needs of the experiment, by convenience, and by cost. For most studies a simple bicarbonate-buffered physiological salt solution is adequate: in the case of ion substitution studies it is essential. With such solutions it is necessary to add a small amount of albumin (0.1%) to prevent binding of stimulatory peptides to the glassware of the perfusion circuit. The low viscosity of these simple salt solutions allows rapid perfusion at low perfusion pressures (<50 mmHg) and hence provision of adequate oxygenation even in the absence of oxygen carriers. So long as perfusion is sufficiently rapid (about 0.5 ml g^{-1} min^{-1}), addition of erythrocytes makes no difference to oxygen consumption or secretion.[55] Of course, use of salt solutions rapidly leads to the development of simple edema. However, there is no evidence that this is deleterious to secretory function. Thus, using the secretory response to secretin as a criterion, the saline-perfused cat pancreas is viable for 9 hr or more.

In studies of vasomotor phenomena and experimental pancreatitis, the use of autologous blood as perfusate is usually preferred. In such cases the perfusate is recirculated and therefore it is important to exclude the duodenum from the perfusion circuit so as to prevent contamination of the blood with humoral products which could subsequently affect pancreatic function. Unfortunately, such a precaution is not usually taken by those working with blood-perfused preparations.[40,41,59] The use of emulsified fluoro-

[48] B. E. Argent, R. M. Case, and T. Scratcherd, *J. Physiol. (London)* **216**, 611 (1971).
[49] B. E. Argent, R. M. Case, and T. Scratcherd, *J. Physiol. (London)* **230**, 575 (1973).
[50] A. Saito and T. Kanno, *Jpn. J. Physiol.* **23**, 477 (1973).
[51] R. J. L. Anderson, J. M. Braganza, and R. M. Case, *Pancreas* **5**, 394 (1990).
[52] G. E. Mann and P. S. R. Norman, *Biochim. Biophys. Acta* **778**, 618 (1984).
[53] J. J. Holst, O. B. Schaffalitsky de Muckadell, and J. Fahrenkrug, *Acta Physiol. Scand.* **105**, 33 (1979).
[54] T. Scratcherd, R. M. Case, and P. A. Smith, *Scand. J. Gastroenterol.* **10**, 821 (1975).
[55] J. J. Holst, S. L. Jensen, O. V. Nielsen, and T. W. Schwartz, *Acta Physiol. Scand.* **109**, 7 (1980).
[56] R. N. Saunders and C. A. Moser, *Arch. Int. Pharmacodyn. Ther.* **197**, 86 (1972).
[57] E. E. Elisha, D. Hutson, and T. Scratcherd, *J. Physiol. (London)* **351**, 77 (1984).
[58] W. W. Ferguson, T. M. Glenn, and A. M. Lefer, *Am. J. Physiol.* **222**, 450 (1972).
[59] P. Saharia, S. Margolis, G. D. Zuidema, and J. L. Cameron, *Surgery* **82**, 60 (1977).
[60] S. S. Hong, R. M. Case, and K. H. Kim, *Pancreas* **3**, 450 (1988).

carbons as oxygen carriers in place of autologous blood has been found successful over a period of 4 hr.[61,62] However, because salt solutions can provide sufficient oxygen, if edema formation is considered a problem, a simpler alternative to autologous blood is to add dextran of M_r 60,000–80,000 to the perfusate as a 6% (v/w) solution

Isolated Pancreatic Ducts

Although isolated gland preparations have contributed greatly to our understanding of pancreatic electrolyte secretion (as well as other aspects of pancreatic physiology), they do not permit the precise biophysical studies which are necessary for a complete description of ion transport by the ductal epithelium. For this purpose it is necessary to isolate either ductal epithelial cells[63–66] or intact pancreatic ducts.[66–81] For most transport studies the latter approach is preferred since the morphological polarity of the epithelium is retained.

[61] K. Kowalewski and A. Kolodej, *Surg. Gynecol. Obstet.* **146,** 375 (1978).

[62] V. P. O'Malley, D. M. Keyes, and R. G. Postier, *J. Surg. Res.* **40,** 210 (1986).

[63] I. Schulz, K. Heil, S. Miltinović, W. Haase, D. Terreros, and G. Rumrich, *in* "Cell Populations, Methodological Surveys, (B)" (E. Reid, ed.), Vol. 9, p. 127. Ellis Harwood, Chichester, 1979.

[64] I. Schulz, K. Heil, A. Kribben, G. Sachs, and W. Haase, *in* "Biology of Normal and Cancerous Exocrine Pancreatic Cells" (A. Ribet, L. Pradayrol, and C. Susini, eds.), p. 3. Elsevier/North-Holland Biomedical, Amsterdam, 1980.

[65] M-S. Tsao and W. P. Duguid, *Exp. Cell Res.* **168,** 365 (1987).

[66] M. A. Gray, J. R. Greenwell, and B. E. Argent, *J. Membr. Biol.* **105,** 131 (1988).

[67] V. Wizemann, A.-L. Christian, J. Wiechmann, and I. Schulz, *Pfluegers Arch.* **347,** 39 (1974).

[68] U. R. Fölsch, H. Fischer, H.-D. Söling, and W. Creutzfeldt, *Digestion* **20,** 277 (1980).

[69] S. Githens, D. R. G. Holmquist, J. F. Whelan, and J. R. Ruby, *J. Cell Biol.* **85,** 122 (1980).

[70] S. Githens, D. R. G. Holmquist, J. F. Whelan, and J. R. Ruby, *In Vitro* **16,** 797 (1980).

[71] S. Githens, D. R. G. Holmquist, J. F. Whelan, and J. R. Ruby, *Cancer* **47,** 1505 (1981).

[72] S. Githens and J. F. Whelan, *J. Tissue Cult. Methods* **8,** 97 (1983).

[73] B. E. Argent, S. Arkle, M. J. Cullen, and R. Green, *Q. J. Exp. Physiol.* **71,** 633 (1986).

[74] S. Arkle, C. M. Lee, M. J. Cullen, and B. E. Argent, *Q. J. Exp. Physiol.* **71,** 249 (1986).

[75] S. Githens, J. J. Finley, C. L. Patke, J. A. Schexnayder, K. B. Fallon, and J. R. Ruby, *Pancreas* **2,** 427 (1987).

[76] M. E. Madden and M. P. Sarras, *J. Histochem. Cytochem.* **35,** 1365 (1987).

[77] S. R. Hootman and C. D. Logsdon, *In Vitro* **24,** 566 (1988).

[78] I. Novak and R. Greger, *Pfluegers Arch.* **411,** 58 (1988).

[79] I. Novak and R. Greger, *Pfluegers Arch.* **411,** 546 (1988).

[80] E. L. Stuenkel, T. E. Machen, and J. A. Williams, *Am. J. Physiol.* **254,** G925 (1988).

[81] S. Githens, J. A. Schexnayder, K. Desai, and C. L. Patke, *In Vitro* **25,** 679 (1989).

Small interlobular, and intralobular, ducts have been isolated from the pancreas of the rat,[66,68–71,73–76,78–81] cat, [67] hamster,[71,72,75] and guinea pig.[77] The techniques employed fall into two categories: (1) dissociation of the gland with enzymes and mechanical shearing, following by isolation of ducts by either manual selection, centrifugation or microdissection,[66,68–77,80,81] and (2) microdissection without prior tissue dissociation.[67,78,79] Unfortunately, the yield of ducts obtained by microdissection alone is low, and may be a limiting factor for biochemical studies. Prior dissociation of the gland with enzymes gives a much higher yield, although it is our experience, and the experience of others,[69] that this approach produces ducts in which the epithelial cells are morphologically poorly preserved. We suspect the damage largely results from the action of digestive enzymes, particularly lipases, which are difficult to inhibit pharmacologically, released from acinar cells during the prolonged dissociation periods employed by most workers. Furthermore, ducts isolated in this way always contain adherent acinar tissue.[69–72]

Isolation of Small Pancreatic Ducts

We have successfully combined a high yield of ducts, with excellent morphological preservation of the ductal epithelium, by using glands taken from copper-deficient rats.[73,74] As a starting point for duct isolation this preparation has two advantages. First, the proportion of duct cells in the gland is markedly increased and, second, the content of potentially harmful digestive enzymes is markedly reduced.

Copper-deficient rats, produced as described above, are killed by cervical dislocation and exsanguinated. The pancreas is then immediately excised, trimmed free of fat and mesentery, and 2 ml of HEPES-buffered Dulbecco's modification of Eagle's medium (HEPES-DMEM), containing 100 U collagenase ml^{-1} and 400 U hyaluronidase ml^{-1}, injected into the interstitium of the gland. The chromatographically pure collagenase (CLSPA 4000 U mg^{-1}) is obtained from Worthington (Freehold, NJ) and bovine hyaluronidase (type IV, 810 NF U mg^{-1}) from Sigma (Poole, Dorset, England). All solutions are sterilized by filtration through 0.2-μm filters.

The bloated gland is then transferred to a small glass flask, containing another 2 ml of HEPES-DMEM plus enzymes, and coarsely chopped with scissors. After top gassing with 100% O_2, the flask is incubated at 37° for 40 min in a shaking water bath (80 cpm). At the end of this period as much fluid as possible is removed and 3 ml of fresh digestion buffer added to the flask. The tissue is then mechanically disrupted by five passes through a glass Pasteur pipet (internal tip diameter 1.2 mm), top gassed with 100%

O_2, and incubated for another 40 min at 37° as described above. Finally, the tissue suspension is passed 10 times through a Pasteur pipet and then centrifuged at 300 g for 3 min. After removing the supernatant, together with any floating fatty tissue, the pellet is resuspended in 3 ml HEPES-DMEM and recentrifuged. This washing procedure is repeated twice before the tissue pellet is finally resuspended in 5–10 ml HEPES-DMEM containing 3% (w/v) bovine serume albumin (BSA) (pH 7.4).

A sample of the tissue suspension, which contains remnants of acinar tissue, blood vessels, nerves, islets of Langerhans, and pancreatic ducts, is placed on a glass slide and viewed under transmitted light at a total magnification of either ×40 or ×100. Interlobular ducts are microdissected using sharpened stainless steel needles and then transferred, using a micropipet, to a storage well positioned at the edge of the slide. This well contains a small drop of HEPES-DMEM plus 3% (w/v) BSA covered by light paraffin oil. Micropipets are pulled by hand from thin-walled glass tubing (1.2-mm i.d., 1.6-mm o.d.), and have fire-polished tips with internal diameters of between 50 and 150 μm. Twenty to 50 interlobular ducts (25–100 μm in diameter; up to 2 mm in length) can be isolated from each gland. Microdissection of the smaller intercalated ducts (there are no distinctive intralobular ducts in the rat pancreas[82]) is technically difficult but, conveniently, they are often obtained as branches of the interlobular ducts.

In addition to morphology, simple biochemical criteria such as O_2 consumption,[74] adenine nucleotide concentrations,[74] and incorporation of amino acids into proteins[68] have been employed to show that these isolated ducts are viable. Importantly, they increase their cyclic AMP content when stimulated by secretin and vasoactive intestinal peptide (VIP), whereas CCK, glucagon, bovine pancreatic polypeptide, and carbamycholine have no effect.[68,74] These findings indicate that the isolated ducts retain secretin and VIP receptors, and that these receptors are functionally coupled to adenylate cyclase. Secretin is probably the most important physiological regulator of pancreatic bicarbonate secretion and there is very good evidence from intact gland studies that this peptide uses cyclic AMP as an intracellular messenger.[2,3]

Culture of Isolated Pancreatic Ducts

Ducts isolated from the rat and hamster pancreas can be cultured in agarose or collagen gels for up to 20 weeks.[1,70–73,75] We find it more convenient to culture ducts for short periods (2–3 days) on polycarbonate filter rafts which float on the growth medium.[73] This has the advantage that

---

[82] H. Takahashi, *Arch. Histol. Jpn.* **47,** 387 (1984).

individual ducts can easily be removed from the filters using micropipets. The growth medium consists of McCoy's 5A (Iwakata and Grace modification) supplemented with 10% (v/v) fetal calf serum, 2 mM glutamine, 20 mU ml^{-1} insulin, and 20 μg ml^{-1} dexamethasone. Cultures are incubated at 37° in an atmosphere of 5% CO_2 in air.

Maintaining the ducts in culture results in a marked dilatation of their lumens, a flattening of the epithelium against the surrounding connective tissue layer, and an overall swelling of the ducts.[1,70-73,75] These morphological changes are caused by sealing of the ducts during the early stages of culture, followed by fluid secretion into the closed lumenal space. The speed with which this occurs probably reflects the morphological and biochemical preservation of the freshly isolated ducts. It takes 2–4 days for ducts isolated from the normal rat and hamster pancreas,[70-72,75] but occurs within 8 hr for ducts isolated from the copper-deficient rat gland.[73] If swollen ducts are punctured the lumen collapses and the epithelium returns to its normal dimensions. Whether the cut ends of the ducts simply stick together, or whether sealing is associated with cell growth is unknown.

Use of Isolated Ducts to Study Ductal Bicarbonate Transport

Recently, our understanding of the cellular mechanism by which duct cells secrete bicarbonate ions has been advanced by application of microelectrode, microperfusion, patch-clamp electrode, and fluorescent probe techniques to isolated pancreatic ducts.[66,78-80,83-85] It is also possible to measure directly fluid secretion from cultured interlobular ducts, using micropuncture technology.[73] For these experiments cultured ducts are immobilized in a Perspex tissue bath (volume, 5 ml) using a suction pipet, illuminated by a fiber optic light source, and viewed at a magnification of ×100 using a stereomicroscope. Suction pipet are pulled from microelectrode glass (outside diameter 1 mm, Clarke Electromedical, Pangbourne, England) and their tips, which measure about 80 μm in diameter, fire polished on a microforge. The ducts are then micropunctured using a collection pipet filled with light paraffin oil stained with Oil Red O. Collection pipets are also made from microelectrode glass and their tips, which measure about 6 μm in diameter, are beveled for ease of penetration.

[83] M. A. Gray, J. R. Greenwell, and B. E. Argent, in "Cellular and Molecular Basis of Cystic Fibrosis" (G. Mastella and P. M. Quinton, eds.), p. 205. San Francisco Press, San Francisco, 1988.
[84] M. A. Gray, J. R. Greenwell, and B. E. Argent, in "Epithelial Secretion of Electrolytes and Water" (J. A. Young and P. Y. D. Wong, eds.), p. 253. Springer-Verlag, Heidelberg, 1990.
[85] I. Novak, in "pH Homeostasis: Mechanisms and Control" (D. Häussinger, ed.), p. 447. Academic Press, New York, 1988.

Successful micropuncture is confirmed by the injection of a small drop of colored oil into the lumen of the duct. The lumenal fluid is then aspirated, causing the duct to collapse, and ejected to waste in the bath. Following repuncture along the same entry track, fluid collection is started by application of a very small subatmospheric pressure to the pipet. The length of the collection period is usually 1 hr, during which time the duct is perifused with a Krebs-Ringer buffer at 37°.

At the end of the experiment, the subatmospheric pressure on the collection pipet is relieved and the tip of the pipet carefully removed from the duct and then rapidly withdrawn from the tissue bath. Collected fluid is then immediately ejected onto a siliconized cavity microscope slide under oil, and the volume calculated from the drop diameter. Depending on the size of the duct and the stimulus protocol, the volume of secreted fluid varied between 0.1 and 7 nl. After measuring the dimensions of the duct, the epithelial volume is calculated and secretion rates finally expressed as nl hr^{-1}nl^{-1} duct epithelium.

Figure 3 shows some results obtained using this technique. Future

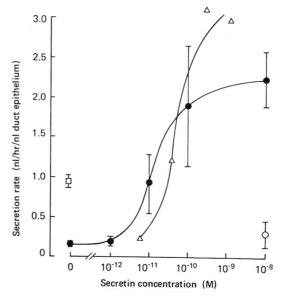

FIG. 3. The effects of secretin (●), dibutyryl cyclic AMP (□), and replacing extracellular bicarbonate ions with HEPES (○) on fluid secretion from cultured rat interlobular ducts. For comparison, the dose–response curve for secretin-stimulated fluid secretion from the perfused rat pancreas is also shown, (△). Dibutyryl cyclic AMP concentration was 2×10^{-4} M. All data points are mean ± SEM. (Drawn from data in B. E. Argent, S. Arkle, M. J. Cullen, and R. Green, $Q. J. Exp. Physiol.$ **71,** 633, 1986.)

studies of this type offer the possibility of establishing unequivocally which hormones and neurotransmitters stimulate the duct cells, how these agents interact, and the true composition of ductal fluid.

Acknowledgments

Recent work has been supported by grants from the Medical Research Council (United Kingdom) and the Cystic Fibrosis Research Trust (United Kingdom).

[18] Dispersed Pancreatic Acinar Cells and Pancreatic Acini

By DELIA MENOZZI, ROBERT T. JENSEN, and JERRY D. GARDNER

Introduction

Initially the secretory function of the pancreas *in vitro* was studied using slices or fragments of the gland. An advance in the study of pancreatic function *in vitro* was the development of a technique for preparing dispersed pancreatic acinar cells.[1-3] This preparation consisted of a homogeneous suspension of dispersed single acinar cells. A few years later a modification of the technique for preparing dispersed cells led to a method for preparing dispersed pancreatic acini.[4-6] Dispersed acini represent clumps of pancreatic acinar cells containing 10–20 cells/clump.

Preparation of Pancreatic Acinar Cells

Dispersed pancreatic acinar cells are prepared with a minor modification of the technique developed by Amsterdam and Jamieson.[1-3] This preparation contains at least 96% acinar cells and these cells retain their responsiveness to pancreatic secretagogues for up to 5 hr *in vitro*. Dispersed acinar cells have been prepared using pancreas from rats, mice, guinea pigs, or rabbits.

1. Prepare the following stock solution in advance and store at 4°:

NaCl (100 mM)
KCl (6 mM)

[1] A. Amsterdam and J. D. Jamieson, *Proc. Natl. Acad. Sci. U.S.A.* **69,** 3028 (1972).
[2] A. Amsterdam and J. D. Jamieson, *J. Cell Biol.* **63,** 1037 (1974).
[3] A. Amsterdam and J. D. Jamieson, *J. Cell Biol.* **63,** 1057 (1974).

KH_2PO_4 (2.5 mM)
HEPES (24.5 mM; pH 7.4)
Sodium pyruvate (5 mM)
Sodium glutamate (5 mM)
Sodium fumarate (5 mM)
Glucose (14 mM)
Glutamine (2 mM)
Soybean trypsin inhibitor (0.1 mg/ml)
Eagle's basal amino acid mixture ([100×; 1% (v/v)]
Essential vitamin mixture [100×; 1% (v/v)]

2. Prepare two separate wash solutions:

Wash solution 1: Stock solution plus $CaCl_2$ (0.1 mM), $MgCl_2$ (1.2 mM)
Wash solution 2: Stock solution plus $CaCl_2$ (1 mM), $MgCl_2$ (1.2 mM), albumin [4% (w/v)]

3. Prepare three separate digestion solutions:

Digestion solution 1: Stock solution plus $CaCl_2$ (0.1 mM), $MgCl_2$ (1.2 mM), type I collagenase (0.75 mg/ml; Sigma, St. Louis, MO); type I-S hyaluronidase (1.5 mg/ml; Sigma)
Digestion solution 2: Stock solution plus EGTA (2 mM)
Digestion solution 3: Same as digestion solution 1 except use collagenase (1.25 mg/ml) and hyaluronidase (2 mg/ml)

4. Prepare incubation solution: Stock solution plus $CaCl_2$ (0.5 mM), $MgCl_2$ (1.2 mM), and albumin [1% (w/v)]

The animal is sacrificed by CO_2 narcosis and the pancreas is removed, trimmed of fat and mesentery, and pinned to a wax tray. Five milliliters of digestion solution 1 is injected into the tissue using a syringe and a 25-gauge needle. The pancreas and the digestion solution are transferred to a siliconized 25-ml Erlenmeyer flask, gassed for 30 sec with 100% O_2, capped, and incubated in a Dubnoff metabolic shaking incubator (160 oscillations/min) at 37° for 15 min. After 15 min, the digestion solution is decanted and replaced with 8 ml of digestion solution 2. The flask is gassed with 100% O_2, capped, and returned to the incubator for 5 min at 37°. At the end of the 5-min incubation the digestion solution is decanted and the pancreas is washed twice with 10 ml of wash solution 1. The washes are discarded. Five milliliters of digestion solution 3 is added to the flask. The flask is gassed with 100% O_2, capped, and returned to the incubator at 37° for 20 min. After 20 min the pancreas is dispersed by pipetting up and down five times through a Pasteur pipet and then passing it in and out of a

TABLE I
USES OF PANCREATIC ACINAR CELLS

Function measured	Reference
Amylase secretion	7–9
Protein secretion	3
Receptor binding	10
Cellular cAMP	11
Cellular cGMP	12,13
Calcium transport	14
Hydrolysis of polyphosphoinositides	15
Na^+ and K^+ content	9

syringe bearing a 19-gauge needle five times. The resulting cell suspension is passed through nylon mesh (Spectramesh, 70 μm; Spectrum Medical Industries, Los Angeles, CA) to remove large clumps of nondispersed tissue. To separate the acinar cells from the digestion solution and cell debris, the cell suspension is gently layered into two 15-conical centrifuge tubes each containing 6 ml of wash solution 2. The tubes are centrifuged at 50 g for 5 min and the supernatant is discarded. The cell pellets are combined and washed twice with 6 ml of wash solution 2 by centrifugation at 50 g for 5 min and resuspension. After the second wash, the supernatant is removed and discarded and the acinar cells are suspended in an appropriate volume of incubation solution.

Uses of Pancreatic Acinar Cells

Table I lists the published uses of pancreatic acinar cells.[7-15] Acinar cells can be permeabilized by electrical or chemical treatment; however, the preparation and applications of permeabilized acinar cells is reviewed in Chapter [19] in this volume.

[4] S. R. Peikin, A. J. Rottman, S. Batzri, and J. D. Gardner, *Am. J. Physiol.* **235**, 743 (1978).
[5] J. A. Williams, M. Korc, and R. L. Dormer, *Am. J. Physiol.* **235**, E517 (1978).
[6] A. Amsterdam, T. E. Solomon, J. D. Jamieson, *Methods Cell Biol.* **20**, 361 (1978).
[7] J. D. Gardner and M. J. Jackson, *J. Physiol. (London)* **270**, 439 (1977).
[8] D. E. Chandler and J. A. Williams, *J. Membr. Biol.* **32**, 201 (1977).
[9] J. A. Williams, P. Cary, and B. Moffat, *Am. J. Physiol.* **231**, 1562 (1976).
[10] J. P. Christophe, T. P. Conlon, and J. D. Gardner, *J. Biol. Chem.* **251**, 4629 (1976).
[11] J. D. Gardner, T. P. Conlon, and T. D. Adams, *Gastroenterology* **70**, 29 (1976).
[12] J. P. Christophe, E. K. Frandsen, T. P. Conlon, G. Krishna, and J. D. Gardner, *J. Biol. Chem.* **251**, 4640 (1976).
[13] R. N. Lopatin and J. D. Gardner, *Biochim. Biophys. Acta* **543**, 465 (1978).
[14] I. Schultz, *Am. J. Physiol.* **239**, G335 (1980).
[15] J. W. Putney, Jr., G. M. Burgess, S. P. Halenda, J. S. McKinney, and R. P. Rubin, *Biochem. J.* **212**, 483 (1983).

Preparation of Dispersed Pancreatic Acini

The procedure for preparing dispersed pancreatic acini is a minor modification of the technique reported by Peikin et al.[4] Others[5,6] have also reported methods for preparing dispersed acini. These techniques[4-6] when used with rat pancreas result in acini that show a greater than fivefold increase in enzyme secretion in response to a maximally effective concentration of cholecystokinin. Recently a technique has been reported for preparing acini using a short digestion period.[16] Unfortunately this rapid method for preparing acini gives rise to cells that show a greater than fivefold increase in enzyme secretion only in response to a maximally effective concentration of a cholecystokinin analog. Freshly prepared dispersed acini can be used for at least 5 hr *in vitro* and have been very useful to investigate the mechanism of action of various secretagogues. Dispersed acini have been prepared using pancreas from mice, rats, guinea pigs, or rabbits.

1. Prepare the following stock solution:

    NaCl (100 mM)
    KCl (6 mM)
    Glucose (14 mM)
    KH$_2$PO$_4$ (2.5 mM)
    HEPES (24.5 mM; pH 7.4)
    MgCl$_2$ (1.2 mM)
    Sodium pyruvate (5 mM)
    Sodium fumarate (5 mM)
    Sodium glutamate (5 mM)
    Glutamine, (2 mM)
    Soybean trypsin inhibitor (0.1 mg/ml)
    Eagle's basal amino acid mixture [100×; 1% (v/v)]
    Essential vitamin mixture [100×; 1% (v/v)]

2. Prepare two separate wash solutions;

    Wash solution 1: Stock solution plus CaCl$_2$ (2 mM) and albumin [4% (w/v)]
    Wash solution 2: Stock solution plus CaCl$_2$ (2 mM)

3. Prepare digestion solution: Stock solution plus collagenase (564 U/ mg, 0.04 mg/ml type CLSPA; Worthington Biochemical Corp., Freehold, NJ), albumin [0.2% (w/v)], and CaCl$_2$ (2 mM)

4. Prepare incubation solution: Stock solution plus CaCl$_2$ (0.5 mM) and albumin [1% (w/v)]

[16] R. Bruzzone, P. A. Halban, A. Gjinovci, and E. Trimble, *Biochem. J.* **226,** 621 (1985).

The animal is sacrificed by CO_2 narcosis and the pancreas is removed, trimmed of fat and mesentery, and pinned to a way tray. Five milliliters of digestion solution is injected into the tissue using a syringe and a 25-gauge needle. The pancreas and the digestion solution are transferred to a siliconized 25-ml Erlenmeyer flask, gassed for 30 sec with 100% O_2, capped, and incubated in a Dubnoff metabolic shaking incubator (160 oscillations/min) at 37° for 10 min. After 10 min of incubation, the digestion solution is decanted and replaced with 5 ml of fresh digestion solution. The flask is gassed, capped, and placed in the incubator for 10 min at 37°. After the second 10-min incubation, the digestion solution is again replaced with 5 ml of fresh digestion solution and the flask is gassed, capped, and placed in the incubator for 10 min at 37°. After the third 10-min incubation, the digestion solution is decanted and replaced with 5 ml of fresh digestion solution. The flask is gassed and capped, and the tissue is disrupted by shaking the flask vigorously by hand for 15 sec. After discarding the duct system and large tissue fragments, the suspension of acini is gently layered into two 15-ml conical centrifuge tubes, each containing 6 ml of wash solution. The tubes are centrifuged at 200 g for 5 sec and the supernatant is discarded. The cell pellets are combined and any large clumps of tissue are removed with a Pasteur pipet. The acini are washed twice with wash solution by centrifugation (200 g for 5 sec) and resuspended. After the last wash, the supernatant is discarded and the acini are suspended in an appropriate volume of incubation solution.

Uses of Dispersed Pancreatic Acini

Table II lists the published uses of dispersed pancreatic acini.[17-40]

[17] J. D. Gardner and R. T. Jensen, in "Physiology of the Gastrointestinal Tract" (L. R. Johnson, ed.), p. 831. Raven, New York, 1981.

[18] J. D. Gardner and R. T. Jensen, in "Physiology of the Gastrointestinal Tract" (L. R. Johnson, ed.), 2nd Ed., p. 1109. Raven, New York, 1987.

[19] S. R. Hootman and J. A. Williams, in "Physiology of the Gastrointestinal Tract" (L. R. Johnson, ed.), 2nd Ed. p. 1129. Raven, New York, 1987

[20] J. A. Williams, A. C. Bailey, and E. Roach, Am. J. Physiol. **254**, G521 (1988).

[21] R. S. Izzo, C. Pellecchia, and M. Praisman, Am. J. Physiol. **255**, G738 (1988).

[22] S. A. Rosenzweig, L. J. Miller, and J. D. Jamieson, J. Cell Biol. **96**, 1288 (1983).

[23] C. D. Logsdon and J. A. Williams, Biochem. J. **223**, 893 (1984).

[24] S. Muallem, Annu. Rev. Physiol. **51**, 83 (1989).

[25] G. R. Gunther and J. D. Jamieson, Nature (London) **280**, 318 (1979).

[26] J. D. Gardner and A. J. Rottman, Biochim. Biophys. Acta **627**, 230 (1980).

[27] D. B. Burnham and J. A. Williams, Am. J. Physiol. **246**, G500 (1984).

[28] M. Noguchi, H. Adachi, J. D. Gardner, and R. T. Jensen, Am. J. Physiol. **248**, G692 (1985).

[29] F. S. Gorelick, J. A. Cohn, S. D. Freedman, N. G. Delahunt, J. M. Gershouni, and J. D. Jamieson, J. Cell Biol. **97**, 1294 (1983).

TABLE II
USES OF DISPERSED PANCREATIC ACINI

Function measured	Reference
Amylase secretion	17-19
Receptor binding	17,18
Ligand internalization	20-23
Cytosolic calcium	24
$^{45}Ca^{2+}$ transport	24
Cellular cAMP	18,19
Cellular cGMP	25,26
Hydrolysis of polyphosphoinositides	19
Formation of diacylglycerol	19
Protein kinase C activity	27,28
Calmodulin-dependent protein kinase activity	27,29
cAMP-Dependent protein kinase activity	27,30,31
cGMP-Dependent protein kinase activity	30
Phosphorylation of cellular protein	32,33
Prostaglandins and leukotrienes	19
Ion channels and membrane potential	34
Cytosolic pH	35
Glucose uptake	36,37
Amino acid uptake	37,38
[^{3}H]Thymidine incorporation into DNA	39,40

[30] R. T. Jensen and J. D. Gardner, *Gastroenterology* 75, 806 (1978).
[31] O. Holian, C. T. Bombeck, and L. M. Nyhus, *Biochem. Biophys. Res. Commun.* 95, 553 (1980).
[32] D. B. Burnham, H.-D, Söling, and J. A. Williams, *Am. J. Physiol.* 254, G130 (1988).
[33] C. K. Sung and J. A. Williams, *Diabetes* 39, 544 (1989).
[34] O. H. Petersen and D. V. Gallacher, *Annu. Rev. Physiol.* 50, 65 (1988).
[35] K. S. Carter, L. R. Rutledge, M. L. Steer, and W. Silen, *Am. J. Physiol.* 253, G690 (1987).
[36] M. Korc, J. A. Williams, and I. D. Goldfine, *J. Biol. Chem.* 254, 7624 (1979).
[37] H. Sankaran, I. D. Goldfine, A. Bailey, V. Licko, and J. A. Williams, *Am. J. Physiol.* 242, G250 (1982).
[38] Y. Iwamoto and J. A. Williams, *Am. J. Physiol.* 238, G440 (1980).
[39] C. D. Logsdon and J. A. Williams, *Am. J. Physiol.* 244, G675 (1983).
[40] C. D. Logsdon and J. A. Williams, *Am. J. Physiol.* 250, G440 (1986).

Dispersed Pancreatic Acini in Culture

The procedure for preparing dispersed pancreatic acini in culture is an aseptic modification of the techniques reported previously for preparing fresh dispersed pancreatic acini. This method allows one to study long-term actions of various agents.

Short-term primary cultures are prepared according to a modification[41] of the procedure described by Lodgson and Williams.[39]

The following culture medium is prepared in advance and stored at 4°.

1 : 1 mixture of Ham's F-12 and Dulbecco's modified minimal essential medium (MEM)
HEPES (10 mM; pH 7.4)
Albumin (1.0 mg/ml)
Soybean trypsin inhibitor (0.1 mg/ml)
Ascorbic acid (0.1 mM)
Epidermal growth factor (EGF) (2 nM)
ITS: Insulin (5 μg/ml), transferrin (5 μg/ml), selenium (5 ng/ml) (Collaborative Research, Lexington, MA)
Penicillin (100 U/ml)
Streptomycin (100 μg/ml)
Amphotericin B (25 μg/ml)

Acini from one pancreas are suspended in 80–120 ml of culture medium and 10-ml aliquots of the suspension are placed in disposable plastic tissue culture flasks and incubated for up to 48 hr at 37° in a humidified atmosphere of 5% CO_2 in air. Agents to be tested are added at appropriate times during the culture period. Pancreatic acini can be maintained in a differentiated state in suspension culture for up to 48 hr. The acini retain their ability to secrete enzymes, although cultured acini are less sensitive and responsive to agonists than freshly prepared acini. Pancreatic acini in short-term culture can be used to measure the same functions that have been measured using freshly prepared acini but have the advantage of permitting one to examine actions that require 1 to 2 days to became maximal.[41,42]

Long-term primary cultures (up to 14 days) have been described for pancreatic acini from mouse.[40] A monolayer culture of pancreatic acini has recently been described.[40] In this condition pancreatic acini first undergo a period of adaptation during which differentiation occurs. Next, the acinar

---

[41] S. R. Hootman, M. E. Brown, J. A. Williams, and C. D. Logsdon, *Am. J. Physiol.* **251**, G75 (1986).
[42] C. K. Sung, S. R. Hootman, E. L. Stuenkel, C. Kuroiwa, and J. A. Williams, *Am. J. Physiol.* **254**, G242 (1988).

cells spread and divide to form a confluent monolayer on collagen gel. These cultured cells retain their ability to respond to cholecystokinin in a trophic manner, but lack the normal content of secretory enzymes and do not show stimulation of enzyme secretion with pancreatic secretagogues.

Perifusion of Pancreatic Acini

Pancreatic acini, prepared as described earlier in this chapter, are placed in a small chamber and immobilized using filters[43] or gel.[44,45] Incubation solution is continuously passed over the acini and the effluent is collected using a fraction collector. This procedure provides the advantage that a population of pancreatic acini can be subjected to a continuous flow of material *in vitro*, thereby both supplying fresh modifier and removing the products of the response, as is likely to happen *in vivo*. Moreover the outflow is collected continuously into fractions that are then analyzed, thereby permitting constant monitoring of the dynamics of the system. Despite the potential of this system in elucidating the kinetics of pancreatic enzyme secretion, this procedure has not been characterized extensively and used to study pancreatic acinar cell functions.

This method has been used to measure amylase secretion,[43–45] $^{45}Ca^{2+}$ efflux,[44] and cellular cAMP.[44]

Strengths and Drawbacks of Various Preparations

During the last decade dispersed pancreatic acini have been used much more extensively than acinar cells, primarily because acini are easier to prepare and show a much greater increase in enzyme secretion in response to secretagogues than do acinar cells. Table III compares the strengths and drawbacks of the various preparations.

[43] K. Imamura, H. Wakasugi, H. Shinozaki, and H. Ibayashi, *Jpn. J. Physiol.* **33**, 687 (1983).
[44] E. K. Frandsen, *in* "Biology of Normal and Cancerous Exocrine Pancreatic Cells" (A. Ribet, L. Pradayrol, and C. Susini, eds.), p. 27. Elsevier/North-Holland Biochemical, Amsterdam, 1980.
[45] P. Singh, I. Asada, A. Owlia, T. J. Collins, J. C. Thompson, *Am. J. Physiol.* **254**, G217 (1988).

TABLE III
STRENGTHS AND DRAWBACKS OF PANCREATIC PREPARATIONS

Preparation	Strengths	Drawbacks
Acinar cells	Population of cells can be purified to homogeneity and studied in absence of other cell types	Relatively difficult and time consuming to prepare
	Can count the number of cells	Cannot measure those functions that reflect the polarity of the acinar cell
	Can take multiple representative samples	Stimulation of enzyme secretion is poor
	Can examine simultaneously the effects of different agents under identical conditions	
Dispersed acini	Do not expose tissue to potentially deleterious actions of chelators of divalent cations	Cannot count the number of cells
	Secrete substantially more enzyme	Cannot measure those functions that reflect polarity of the acinar cells
	Can take multiple representative samples	
	Can examine simultaneously the effects of different agents under identical conditions	
Dispersed acini in short-term culture	Same as freshly prepared acini	Same as freshly prepared acini
	Can study longer term actions of appropriate agents	Less sensitive and responsive to secretagogues then freshly prepared acini
Perifusion of pancreatic acini	Can monitor continuously the kinetics of the biological response	Cannot count the number of cells
		Less convenient and limited sampling
		Long time required to reach equilibrium
		Cannot examine simultaneously the effects of different agents under identical conditions

[19] Permeabilizing Cells: Some Methods and Applications for the Study of Intracellular Processes

By IRENE SCHULZ

Introduction

Physiological function of many cell types is regulated by hormones and neurotransmitters that are recognized by the cell via specific receptors. External signals are translated by the cell into intracellular signals. This involves the action of different membrane-bound proteins and lipids as well as cytosolic messenger molecules which exert their action on intracellular membranes and/or the plasma membrane by opening ion channels or activating enzymes. Although the stimulus–response coupling can be very similar in different cell types and involves similar signal transduction pathways, the physiological response of the cell depends on the nature of target proteins which finally lead to specific cell answers, such as contraction, growth, metabolism, or secretion. Gaining access to intracellular sites which are involved in signal transduction pathways offers the possibility to study intracellular processes by controlled manipulation of the cell interior. Successful techniques to bypass the limiting plasma membrane include permeabilization of the cell barrier, leaving intracellular organelles intact. This allows one to introduce substances into the cell which do not pass the plasma membrane. Thus the role of second and third messengers of hormones, such as nucleotides, peptides, or sugars, as well as ionic requirements can be directly studied in the context of whole cellular function. Techniques for permeabilization of cells have many, sometimes severe, limitations. Ideally they should allow incorporation of substances into the cell without destroying its integrity. Often it is a balancing act to find the right conditions which allow perfect permeabilization of the plasma membrane without damage of intracellular organelles.

In this chapter some methods of controlled cell permeabilization are reviewed which have proved useful for studying many aspects of exocytotic secretion. They include permeabilization by high voltage discharge,[1] treatment with mild nonionic detergents, such as saponin or digitonin,[2-6] and

[1] D. E. Knight and P. F. Baker, *J. Membr. Biol.* **68**, 107 (1982).
[2] H. Wakasugi, T. Kimura, W. Haase, A. Kribben, R. Kaufmann, and I. Schulz, *J. Membr. Biol.* **65**, 205 (1982).
[3] D. Knight and M. Scrutton, *Biochem J.* **234**, 497 (1986).
[4] J. Brooks and S. Treml, *J. Neurochem.* **40**, 468 (1983).
[5] L. Dunn and R. Holz, *J. Biol. Chem.* **258**, 4983 (1983).
[6] S. Wilson and N. Kirshner, *J. Biol. Chem.* **258**, 4994 (1983).

TABLE I
PORE SIZE AND MOLECULAR MASSES OF PERMEANT SOLUTES IN CELLS PERMEABILIZED
BY DIFFERENT METHODS

Method	Pore diameter (nm)	Permeant molecule	(kDa)	References
High-voltage electric discharge	~2	Ions, nucleotides	Up to 1	1
Digitonin (~10 μg/ml)	8–10	Enzymes	Up to 200	28,29,31
Saponin (~45 μg/ml)	8–10	Enzymes	Up to 200	
α-Toxin (20–50 μg/ml)	2–3	Ions, nucleotides	Up to 1	48
Streptolysin O (0.1–1 μg/ml)	>15	Enzymes, immunoglobulins	>200	56

pore-forming bacterial toxins, such as α-toxin from *Staphylococcus aureus* and streptolysin O from β-hemolytic streptococci.[7,8] Other methods use osmotic shock, brief freeze/thaw,[9] treatments of cells with Sendai virus,[10,11] ATP,[12-16] or Ca^{2+} chelators EGTA[17] or EDTA.[18] Washing of cells in a nominal Ca^{2+}-free medium (i.e., without addition of calcium and without Ca^{2+} chelators) has been also employed to induce leakiness in pancreatic cells.[19] In a single cell the whole cell patch-clamp technique, which records conductances over the entire plasma membrane, allows the introduction of substances into the cell over the patch pipet.[20]

[7] G. Ahnert-Hilger, S. Bhakdi, and M. Gratzl, *J. Biol. Chem.* **260,** 12730 (1985).
[8] G. Ahnert-Hilger and M. Gratzl, *TIPS* **9,** 195 (June, 1988).
[9] R. A. Nichols, W. C. Wu, J. W. Haycock, and P. Greengard, *J. Neurochem.* **52** (2), 521 (1989).
[10] C. C. Impraim, K. A. Foster, K. J. Micklem, and C. A. Pasternak, *Biochem. J.* **186,** 847 (1980).
[11] C. A. Pasternak and K. J. Micklem, *J. Membr. Biol.* **14,** 293 (1973).
[12] J. P. Bennet, S. Cockcroft, and B. D. Gomperts, *J. Physiol. (London)* **317,** 335 (1981).
[13] G. A. Weisman, B. K. De, and R. S. Pritchard, *J. Cell Physiol.* **138** (2), 375 (1989).
[14] T. H. Steinberg, A. S. Newman, J. A. Swanson, and S. C. Silverstein, *J. Biol. Chem.* **262** (18), 8884 (1987).
[15] F. Di Virgilio, V. Bronte, D. Collavo, and P. Zanovello, *J. Immunol.* **143** (6), 1955 (1989).
[16] P. E. R. Tatham and M. Lindau, *J. Gen Physiol.* **95,** 459 (1990).
[17] G. B. McClellan and S. Winegrad, *J. Gen. Physiol.* **72,** 737 (1978).
[18] S. Knasmueller, A. Szakmary, and A. Wottawa, *Mutat. Res.* **216** (4), 189 (1989).
[19] H. Streb and I. Schulz, *Am. J. Physiol.* **245,** G347 (1983).
[20] E. Neher, *J. Physiol.* **395,** 193 (1988).

Electropermeabilization

Electropermeabilization, often also referred to as "electrical break-down," "electropermeation," "electroporation," or "high voltage electric discharge," has been known about for two decades already.[21,22] However, it was some time before its use for making cells leaky and gaining access to intracellular sites was appreciated by cell biologists.[23,24] Electrical break-down of the plasma membrane is achieved when cells are exposed to a field pulse of high intensity (kV/cm range) and short duration (nsec to μsec range).

The basic principles of electropermeabilization have been discussed extensively elsewhere[1,25] and are only briefly outlined here. If a cell is exposed to an electric field, the magnitude of the potential difference imposed across the cell membrane will depend on the intensity of the field (E, in V/cm) and the radius (a) of the cell and will vary around the cell's circumference. With reference to Fig. 1 the voltage (V) at point G (V_G) across the membrane of a spherical cell is given by the simplified equation

$$V_G = C \times a \times E_G \times \cos \vartheta$$

C depends on the relative electrical conductivities of the extracellular fluid, the cytosol, and the membrane and the size of the cell. For membrane thicknesses of 5–10 nm it does not differ significantly from a value of 1.5.[1,25] From the equation it can be seen that the potential difference imposed across the membrane will be a maximum at the points A and B ($\cos \vartheta = \pm 1$) and will decrease around the cell to positions D and F where the electric field has no influence on the membrane ($\cos \vartheta = 0$). The breakdown voltage V_G is therefore first reached in field direction ($E = E_G$). It is reached in membrane sites oriented at a certain angle to the field direction if field strength E is higher than E_G (Fig. 1). Breakdown of the membrane occurs if the critical breakdown voltage V_G (~ 1 V) is reached by further increase in the external field strength. It is assumed that the breakdown causes the formation of transmembrane pores (filled with con-ductive solution) which leads to immediate discharge of the membrane. From the exponential influx and efflux of marker substances in bovine adrenal medullary cells an effective pore size of 2 nm was calculated to be formed by exposure to an electric field of 2 kV/cm and duration of 200 μsec.[1] For models how pore formation accurs, the reader is referred to the review by Zimmerman.[25]

[21] U. Zimmermann, J. Schulz, and G. Pilwat, *Biophys. J.* **13**, 1005 (1973).
[22] U. Zimmermann, G Pilwat, and F. Riemann, *Biochim. Biophys. Acta* **375**, 209 (1975).
[23] P. F. Baker and D. E. Knight, *Nature (London)* **276**, 620 (1978).
[24] P. F. Baker and D. E. Knight, *J. Physiol. (Paris)* **76**, 497 (1980).
[25] U. Zimmermann, *Rev. Physiol. Biochem. Pharmacol.* **105**, 175 (1986).

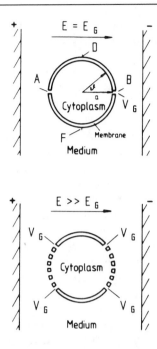

$$V_G = 1.5 \times a \times E_G \times \cos \vartheta$$

FIG. 1. Schematic diagram of a cell exposed to high electric field strengths. The membrane potential V_m, which is built up across the membrane in response to the external field, is highest at sites A and B oriented in field direction. For membrane sites D and F it is zero. The breakdown voltage V_G is therefore first reached in field direction $E = E_G$. If higher field strength $(E > E_G)$ is applied, breakdown of the membrane occurs, as indicated by the formation of transmembrane pores at membrane sites oriented at a certain angle ϑ to the field direction. (From Ref. 25 with small modifications.)

Chemical Permeabilization

Digitonin and Saponin

Chemical permeabilization of cells can be achieved by treatment of cells with mild nonionic detergents, such as digitonin or saponin. These plant glycosides permeabilize membranes by complexing with membrane cholesterol and other unconjugated β-hydroxysterols.[26] Some authors feel that permeabilization by digitonin or saponin is difficult to control and that these detergents might damage intracellular membranes and even lead to release of intracellular organelles.[8] However, the degree of damage

[26] R. A. Mooney, this series, Vol. 159, p. 193.

depends very much on the detergent concentration. Study of the effects of both saponin and digitonin on the electrical conductance of black lipid membranes and on the surface pressure of lipid monofilms have shown that both agents induced channel-like fluctuations in planar bilayers made either of diphytanoylphosphatidylcholine (DPhPC) or of DPhPC and cholesterol (2/1, w/w). In cholesterol-free bilayers the amount needed to induce an increase in conductance was 0.3–1 mg/ml for saponin and about 0.2 mg/ml for digitonin. In contrast, in cholesterol-containing bilayers the concentration needed to induce pores was about 10 μg/ml for both saponin and digitonin.[27] Electron micrographs of mixtures of cholesterol and saponin showed hexagonal arrangements of holes with a diameter of ~ 8 nm.[28,29] Digitonin–cholesterol complexes showed spherical micelles as well as rigid tubular structures in electron micrographs.[30] By electron microscopy on Rous sarcoma virus and cell membranes, pits of 8–10 nm were observed, following treatment with saponin or digitonin.[28,29,31] The number of these holes increased with increasing saponin concentrations,[31] and it was suggested that this selective action of glycosides results from interaction of their lipophilic heads with cholesterol forming a ring with a central hydrophilic hole in a micellar arrangement.[29] It therefore may be expected that the properties of cholesterol-rich membranes, such as plasma membranes, are primarily affected,[32-34] whereas cholesterol-poor membranes, such as those of endoplasmic reticulum and mitochondria,[33] should be much less affected. These micellular pores allow passage of molecules ranging up to molecular weights of at least 200,000. Thus, the cytosolic enzyme lactate dehydrogenase (M_r 134,000) is easily released from digitonin- and saponin-treated cells. A study in hepatocytes showed that the amount of protein retained by these cells drops sharply at low digitonin concentrations and remains constant at ~ 50% of proteins retained between 200 and 900 μg/ml of digitonin. Above concentrations of 1000 μg/ml digitonin, additional cell protein is lost.[35] At these high concentrations, digitonin can solubilize membrane phospholipids and proteins

[27] H. Gögelein and A. Hüby, *Biochim. Biophys. Acta* **773**, 32 (1984).
[28] A. D. Bangham and R. W. Horne, *Nature (London)* **196**, 952 (1962).
[29] A. M. Glauert, J. T. Dingle, and J. A. Lucy, *Nature (London)* **196**, 953 (1962).
[30] A. Martonosi, *Biochim. Biophys. Acta* **150**, 694 (1968).
[31] R. R. Dourmashkin, R. M. Dougherty, and R. J. C. Harris, *Nature (London)* **194**, 1116 (1962).
[32] M. P. Blaustein, R. W. Ratzlaff, N. C. Kendrick, and E. S. Schweitzer, *J. Gen. Physiol.* **72**, 15 (1978).
[33] E. D. Korn, *Science* **153**, 1491 (1966).
[34] D. Thines-Sempoux, A. Amar-Costesec, H. Beaufay, and J. Berthet, *J. Cell Biol.* **43**, 189 (1969).
[35] P. H. Weigel, D. A. Ray, and J. A. Oka, *Anal. Biochem.* **133**, 437 (1983).

like a typical nonionic detergent whereas between 200 to 900 μg/ml only cytosolic proteins are released.[35] Even at digitonin concentrations as high as 3000 μg/ml neither lysosomes nor mitochondria release their soluble enzymatic content. This indicates that intracellular membranes that lack cholesterol are not permeabilized by digitonin.[36] Saponin, however, has been used to permeabilize membranes of the endoplasmic reticulum (ER). Whereas at saponin concentrations lower than 200 μg/ml intracellular membranes do not seem to be affected[37-39] at higher concentrations selective release of secretory proteins from the ER occurred[40] and 50% of pulse ^{35}S-labeled albumin was released.[41] Even at 5 mg/ml less than 10% of the membrane of the endoplasmic reticulum was solubilized, as judged by the degree of release of a membrane-bound enzyme specific for this organelle.[41] To decide whether the observed effects of Ca^{2+} on pancreatic protein release in the presence of saponin require functioning components of the secretory machinery in the cell or could be due to rupture of zymogen granules in the cells without involvement of fusion, we have tested amylase release from isolated zymogen granules (ZG) at different saponin or digitonin concentrations in the presence and absence of Ca^{2+}. Saponin concentrations tested up to 500 μg/ml and digitonin concentrations tested up to 400 μg/ml had no effect on amylase release. Test substances which stimulate enzyme secretion from permeabilized pancreatic acini, such as cAMP (2 mM) plus the phorbol ester 12-O-tetradecanoyl phorbol 13-acetate (TPA, 10^{-7} M) as the most effective stimulants, had no significant effects on amylase release from isolated zymogen granules with or without digitonin and in the presence or absence of Ca^{2+}.

Pore-Forming Bacterial Toxins

Two pore-forming proteins, α-toxin secreted from *Staphylococcus aureus* and streptolysin O from β-hemolytic streptococci, have been successfully used to permeabilize secretory cells.

Staphylococcal α-toxin causes lysis of erythrocytes[43] and creates transmembrane pores in the plasma membrane of different cell types.[44] The

[36] J. Mackall, M. Meredith, and M. D. Lane, *Anal. Biochem.* **95**, 270 (1979).
[37] A. Colbeau, J. Nachbauer, and P. M. Vignais, *Biochim. Biophys. Acta* **249**, 462 (1971).
[38] R. Henning, H. D. Kaulen, and W. Stoffel, *Hoppe-Seyler's Z. Physiol. Chem.* **351**, 1191 (1970).
[39] S. Fleischer and M. Kervina, this series. Vol. 31, p. 147.
[40] G. J. Strous and P. van Kerkhof, *Biochem. J.* **257**, 159 (1989).
[41] M. Wassler, I. Jonasson, R. Persson, and E. Fries, *Biochem. J.* **247** (2), 407 (1987).
[42] T. Kimura, K. Imamura, L. Eckhardt, and I. Schulz, *Am. J. Physiol.* **250**, G698 (1986).
[43] S. Bhakdi and J. Tranum-Jensen, *Philos. Trans. R. Soc. London B* **306**, 311 (1984).
[44] S. Bhakdi and J. Tranum-Jensen, *Rev. Physiol. Pharmacol.* **107**, 147 (1987).

native form of the toxin assembles into characteristic ring structures in the target membrane. It becomes intimately associated with the lipid bilayer[45-47] and creates stable transmembrane pores with a diameter of 2–3 nm,[48-50] allowing permeation of smaller molecules up to 1000 Da, such as ions or nucleotides.[8] In a marker release study conducted with resealed human erythrocyte ghosts[48] it was found that liberation of the small marker sucrose paralleled the extent of hemolysis and was complete at an α-toxin concentration of 35 μg/ml with human and 7 μg/ml with sheep erythrocyte ghosts. Release of inulin (effective molecular diameter of 3 nm) occurred with some retardation. Inulin release was complete at 70 μg/ml α-toxin concentration.[48] Myoglobin (effective diameter 4 nm) was not released from resealed human and sheep erythrocyte ghosts at maximal α-toxin concentrations tested (70 μg/ml).

The property of staphylococcal α-toxin to form pores and to allow permeation of small molecules has been utilized to study contractility in smooth muscle cells,[51] enzymatic activities of internal hepatocyte organelles,[52] secretion from rat pheochromocytoma (PC 12) cells,[7,53] from adrenal medullary chromaffin cells,[54] and from rat insulinoma (RINA 2) cells.[55]

Membrane lesions, which are produced by *streptococcal* membranolysins, streptolysin S, and streptolysin O, are heterogeneous in size and structure.[56] Both toxins act on eukaryotic cell membranes, but apparently in different ways. Studies with erythrocytes have shown that streptolysin S lyses cells by a colloid-osmotic process[57,58] while streptolysin O is thought to produce larger primary lesions, resulting in noncolloid-osmotic lysis.[59]

[45] J. P. Arbuthnott, J. H. Freer, and B. Billcliff, *J. Gen. Microbiol.* **75**, 309 (1973).
[46] J. H. Freer, J. P. Arbuthnott, and A. W. Bernheimer, *J. Bacteriol.* **95**, 1153 (1968).
[47] J. H. Freer, J. P. Arbuthnott, and B. Billcliff, *J. Gen. Microbiol.* **75**, 321 (1973).
[48] R. Füssle, S. Bhakdi, A. Sziegoleit, J. Tranum-Jensen, T. Kranz, and H. J. Wellensiek, *J. Cell Biol.* **91**, 83 (1981).
[49] A. W. Bernheimer, K. S. Kim, C. C. Remsen, J. Antanavage, and S. W. Watson, *Infect. Immun.* **6**, 636 (1972).
[50] M. Thelestam and R. Möllby, *Biochim. Biophys. Acta* **557**, 156 (1979).
[51] P. Cassidy, P. E. Hoar, and W. G. L. Kerrick, *Biophys. J.* **21**, 44a (1978).
[52] B. F. McEwen and W. J. Arion, *J. Cell Biol.* **100**, 1922 (1985).
[53] G. Ahnert-Hilger, S. Bhakdi, and M. Gratzl, *Neurosci. Lett.* **58**, 107 (1985).
[54] M.-F. Bader, D. Thiersé, D. Aunis, G. Ahnert-Hilger, and M. Gratzl, *J. Biol. Chem.* **261**, 5777 (1986).
[55] K. J. Föhr, J. Scott, G. Ahnert-Hilger, and M. Gratzl, *Biochem. J.* **262**, 83 (1989).
[56] L. Buckingham and J. L. Duncan, *Biochim. Biophys. Acta* **729**, 115 (1983).
[57] J. L. Duncan and L. Mason, *Infect. Immun.* **14**, 77 (1975).
[58] W. Hryniewicz, S. Szmigielski, and M. Janiak, *Zentralbl. Bakteriol. Mikrobiol. Hyg. Abt. 1, Orig. A:* **238**, 201 (1977).
[59] J. L. Duncan, *Infect Immun.* **9**, 1022 (1974).

Escape of labeled marker molecules of various sizes from resealed sheep erythrocytes ghosts treated with the toxins for 30 min allowed estimation of the sizes of the channels formed.[56] Streptolysin S formed lesions ranging in size from 0.9 to 4.5 nm in diameter, whereas lesions produced by streptolysin O was dependent on the amount of toxin presented to the membrane. After treatment with less than 25 hemolytic units of streptolysin O, little or no release of molecules larger than 7.2 nm in diameter (M_r 68,000) was observed. After treatment with 300–500 hemolytic units of toxin lesions > 12.8 nm were seen.[56] Others have reported streptolysin O pores with an approximately 10-fold larger diameter (30–35 μm) than α-toxin channels.[60] They are large enough for the free passage of molecules, such as enzymes or immunoglobulin.

Experimental Procedure

Experimental Buffers

In experiments where intact pancreatic acinar cells were used, the experimental buffer had an ion composition similar to that of the extracellular fluid and contained the following:

Buffer A: 130 mM NaCl (or 145 mM NaCl); 4.7 mM KCl; 1.2 mM KH_2PO_4; 2 mM $MgCl_2$; 1 mM $CaCl_2$; 15 mM glucose. The solution was buffered with 10 or 20 mM HEPES adjusted with NaOH to pH 7.4 and also contained 0.1 g/liter trypsin inhibitor and 2 g/liter bovine serum albumin (BSA)

In experiments with permeabilized acinar cells ions were adapted to the intracellular milieu and contained the following:

Buffer B: 130 mM KCl (120 or 145 mM KCl); 1.2 mM KH_2PO_4; 2 mM $MgSO_4$ (or 1 or 2 mM $MgCl_2$); 15 mM glucose; 20 mM (or 25 mM) HEPES (pH 7 adjusted with Tris); 0.1 g/liter trypsini inhibitor, and 2 g/liter BSA. Free Ca^{2+} concentrations of 10^{-8} and 10^{-7} mol/liter were adjusted with 1 mmol/liter ethyleneglycol-bis(β-aminoethylether)N,N'-tetraacetic acid (EGTA) and with varying total Ca^{2+} concentrations and 10^{-6} and 10^{-5} mol/liter free $[Ca^{2+}]$ was adjusted with 10^4 mol/liter nitrilotriacetic acid (NTA), as had been described.[42] Free Ca^{2+} concentrations were measured with a Ca^{2+}- selective electrode[19] and was found to differ not more

[60] S. Bhakdi, J. Tranum-Jensen, and A. Sziegoleit, *Infect. Immun.* **47,** 52 (1985).

than 10% from the calculated free Ca^{2+} concentration. For experiments in which secretion of proteins was studied, the buffer contained 10^{-4} mmol/liter $CaCl_2$ and no EGTA. This is a Ca^{2+} concentration higher than used by other authors[1,61] and was chosen since higher effects of secretagogues, such as carbachol or cAMP, on protein release from pancreatic acinar cells were obtained at 10^{-4} mol/liter as compared to 10^{-5} mol/liter $[Ca^{2+}]$ (Fig. 6a and b). In experiments in which Ca^{2+} uptake into permeabilized pancreatic acinar cells was monitored with a Ca^{2+} macroelectrode, the standard incubation buffer also contained Mg-ATP (5 or 10 mmol/liter) and an ATP-regenerating system (creatinine phosphate; 10 mmol/liter; creatinine kinase; 8 or 10 U/ml), potassium pyruvate, 5 mmol/liter; potassium succinate, 5 mmol/liter; antimycin A, 0.01 mmol/liter; oligomycin, 0.005 mmol/liter; and NaN_3; 10 mmol/liter.[19,62]

Isolation of Pancreatic Acinar Cells and Acini

Pancreatic acini were isolated by digestion of pancreatic tissue with collagenase according to standard methods.[2,63] Briefly, adult male Wistar rats (200–500 g), which had been fasted overnight, were stunned and killed by cervical dislocation. The pancreases of two or three rats were removed for each experiment, trimmed free of fat, lymph nodes, and connective tissue, and finely minced. The pancreatic fragments were then incubated in 20 ml of minimum essential medium (MEM) plus Earle's salt without L-glutamine and L-leucine (GIBCO Europe, Karlsruhe, FRG), containing 15 mmol/liter glucose, 0.1 g/liter trypsin inhibitor and 2 g/liter BSA (pH 7.4 with HEPES/NaOH). When [3H]-labeled protein release was to be measured, [3H]leucine (75 Ci, 2.775 MBq) was added. The supernatant was decanted after 30 min and replaced with 10 ml of buffer A containing in addition 1500 U collagenase (Sigma type III). The supernatant was again decanted after a further 15 min, the tissue segments briefly rinsed with fresh buffer A and incubated with a further 10 ml buffer as above containing 2250 U collagenase for 30 to 40 min. To obtain isolated acinar cells a 5-min incubation in Ca^{2+}-free buffer containing 1 mmol/liter ethylenediaminetetraacetic acid (EDTA) was included between the first and second collagenase digestion steps. Following the second incubation

[61] D. E. Knight and E. Koh, *Cell Calcium* 5, 401 (1984).
[62] F. Thévenod, M. Dehlinger-Kremer, T. P. Kemmer, A.-L. Christian, B. V. L. Potter, and I. Schulz, *J. Membr. Biol.* 109, 173 (1989).
[63] A. Amsterdam and J. D. Jamieson, *Proc. Natl. Acad. Sci. U.S.A.* 69, 3028 (1972).

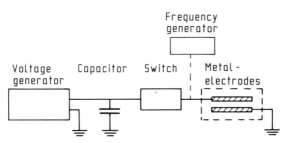

FIG. 2. Experimental set-up for electrical breakdown of cell membranes. The cell suspension is placed between two stainless steel electrodes 1 cm apart. The electrodes are connected to a capacitor via a switch. The capacitor is charged by a voltage generator to a certain voltage (up to 20 kV/cm). When the switch is closed, the capacitor is exponentially discharged through the cell suspension (the resultant electric field being 2 kV/cm, time constant of exponential delay 200 μsec). (From Ref. 25 with small modifications.)

with collagenase, the tissue was washed with fresh buffer A, centrifuged at 50–100 g, and the acini dispersed by pipetting through 5-ml plastic pipet tips, the ends of which had been cut and then fire polished in a flame. The dispersed acini were filtered twice through a single layer of gauze and then layered over 25 ml of buffer A containing 40 g/liter BSA and centrifuged for 5 min at 50–100 g. The pellet was resuspended and washed twice more with buffer A, the final pellet being taken up in 60 ml buffer A as before but containing 10^{-3} mol/liter Ca^{2+} if experiments were to be performed with intact cells or 10^{-4} mol/liter Ca^{2+} if permeabilized cells were to be used.

Permeabilization of Cells by High-Voltage Electric Discharge

Immediately prior to the experiment, a 5-ml aliquot of the acinar suspension was taken and centrifuged for 5 min at 50–100 g and the pellet resuspended in 1 ml of experimental buffer B (see above for composition) and kept on ice. The cell suspension was placed in a small plastic cuvette (dimensions 1.2 × 1.5 × 1.2 cm). Two stainless steel electrodes were inserted through the cover of the cuvette so that they were positioned flat against the sides of the cuvette, 1 cm apart. A series of high-voltage electric shocks were then delivered to the cells from a specially constructed 2-μF capacitor which was discharged through the cell suspension with a time constant of 200 μsec in KCl, the resultant electric field being 2 kV/cm (Fig. 2). After permeabilization, the cells were removed and added to incubation vials containing 5 ml of cold experimental buffer (final volume 6 ml). The final protein concentration averaged 3.36 ± 0.18 mg/ml ($n = 20$).

Assessment of Leakiness

Leakiness of permeabilized cells can be assessed by counting the number of cells which had taken up Trypan Blue,[42,64] by flow cytometry, and by the requirement for exogenous NADPH for oxygen consumption.[65] Another way to assess the permeability of the plasma membrane to high-molecular-weight substances is to estimate the release of cytosolic enzyme lactate dehydrogenase (LDH) from permeabilized cells. Furthermore, comparison of the rates of release of LDH with granular proteins of secretory cells allows one to decide if exocytosis rather than leakage of proteins from damaged cells had occurred. As had been shown by Knight and Baker[1] challenge of protein release with Ca^{2+} from permeabilized adrenal medullary cells had no effect on the rate of release of LDH but increased that of dopamine-β-hydroxylase (DβH) and catecholamine. As a percentage of their total cellular contents 7.5% of total DβH was released in association with 15% of the total catecholamine. The same results were also found in intact cells.[1] Since the molecular mass of both DβH (290,000) and adrenaline (330) differ by a factor of ~ 900 the rate of appearance of the catecholamine should be ~ 30 times faster than that of DβH if diffusion from broken granules and not exocytosis had occurred.[1] Accessibility of low-molecular-weight extracellular solutes to the interior of the cell has been also determined by measuring the influxes of trace amounts of ^{45}Ca-labeled EGTA, ^{51}Cr-labeled EDTA, or of [^{4}C]ATP or by measuring the effluxes of ^{86}Rb-labeled or 3-*O*-methyl-D-[^{14}C]glucose.[1]

In a study on protein release from electropermeabilized pancreatic acini we have found the optimal conditions by challenging protein release with a combination of TPA (10^{-7} mol/liter) and cAMP (2×10^{-3} mol/liter) at varying conditions of applied voltage and number of discharges.[10] As illustrated in Fig. 3, increasing either parameter (i.e., voltage or number of pulses) results in a corresponding increase in basal secretion as compared to intact cells; at the highest voltage tested (2 kV, 10 discharges), basal protein release was 11.3% of total after 30 min. Conversely, stimulated protein secretion was sharply attenuated by increasing the number of discharges. The maximal stimulated protein secretion was observed at 1 kV and 6 discharges ($73 \pm 11.7\%$ above basal; basal $= 7.9 \pm 0.5\%$ of total, $n = 14$), whereas at 2 kV for 10 discharges protein secretion was reduced to only 10.1% above basal. Consequently, in further experiments the parameters of electrical breakdown employed were six discharges of 1 kV. Secretion from electrically permeabilized acini in response to TPA

[64] C. M. Fuller, L. Eckhardt, and I. Schulz, *Pfluegers Arch.* **413**, 385 (1989).
[65] S. Grinstein and W. Furuya, *J. Biol. Chem.* **263** (4), 1779 (1988).

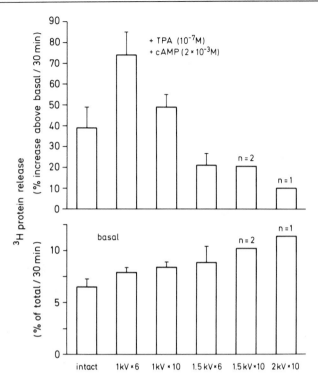

FIG. 3. Effect of varying conditions of high-voltage electric discharge on basal and stimulated secretion (evoked by TPA, 10^{-7} mol/liter, = cAMP, 2×10^{-3} mol/liter) from dispersed
pancreatic acini. Bars represent the mean ± SEM of at least three experiments. Experiments
on "intact" cells were performed in the presence of buffer A (see the section, Experimental
Buffers) and those on cells exposed to electric shocks were performed in the presence of buffer
B in the presence of 10^{-4} mol/liter Ca^{2+}. After permeabilization the cells were removed and
added to 5 ml of cold buffer B. For determination of ^3H-labeled protein release [^3H]leucine-
prelabeled cells (see the section, Isolation of Pancreatic Acinar Cells and Acini) were incubated in buffer B gassed with O_2 at 37° for 30 min in a shaking water bath. Samples of 200 μl
were taken at intervals throughout the incubation period, were centrifuged at 1100 rpm in an
Eppendorf microfuge for 15 sec, and the ^3H-labeled protein release was determined from the
supernatant following precipitation with ice-cold trichloroacetic acid (TCA). Total ^3H-labeled
protein in cell suspensions was precipitated with TCA, the samples were centrifuged at
3000 rpm for 15 min, the TCA supernatants were discarded, and the pellets were dissolved in
0.2 M NaOH. Radioactivity was counted and ^3H-labeled protein release in the supernatant
was expressed as a percentage of the total ^3H-labeled protein present in the cell suspension.
(From Ref. 64.)

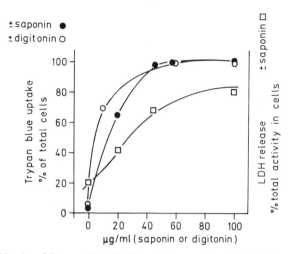

FIG. 4. Uptake of Trypan Blue (0.5%) and release of lactate dehydrogenase (LDH) in saponin- and digitonin-treated acini at different detergent concentrations. Acini were isolated as described in the text and incubated in buffer B with or without detergents for 100 min. Lactate dehydrogenase activity was determined with an LDH Merckotest (No. 3339) from Merck (Darmstadt, FRG) in the supernatant and the pellet of cells which had been centrifuged for 1 min at 14,000 g in an Eppendorf microfuge. Lactate dehydrogenase release was expressed as percentage of total activity present in cells before adding saponin.

alone was $56.3 \pm 13.2\%$ above basal/30 min (basal = $6.7 \pm 0.8\%$ total/ 30 min, $n = 5$) while that evoked by cAMP alone was $25.2 \pm 7.6\%$ above basal/30 min (basal = $8.4 \pm 1\%$ total/30 min, $n = 4$).

Permeabilization of Cells by Digitonin or by Saponin

Pancreatic acini were preincubated in buffer B (see above) with digitonin (10 μg/ml) or with saponin (45 μg/ml) and were incubated in the same solution during the experiment.

The leakiness of cells were estimated by the release of lactate dehydrogenase, by uptake of Trypan Blue and by the optimal effects of ATP and of cAMP, which do not permeate intact cells.[2,42] As shown in Fig. 4, ~60–70% of cells treated with saponin and digitonin took up Trypan Blue at 20 and 10 μg/ml, respectively. At detergent concentrations, at which nearly 100% of cells were leaky, LDH release was only ~60–70% (Fig. 4). In cells pretreated with saponin, $^{45}Ca^{2+}$ uptake increased rapidly after addition of ATP (Fig. 5). Maximal uptake was obtained at 45 μg/ml of saponin, whereas at 20 μg saponin/ml $^{45}Ca^{2+}$ uptake was slower and still increased within the period observed. At 60 and 100 μg saponin/ml, however, ATP-

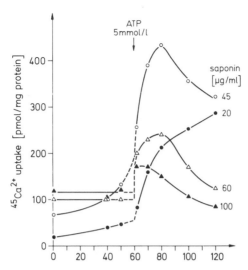

FIG. 5. Effects of saponin at different concentrations on ATP-promoted $^{45}Ca^{2+}$ uptake in isolated pancreatic acini. Acini were isolated as described above and incubated in buffer B with a total calcium concentration of 10^{-6} – 10^{-5} mol/liter and in the presence of saponin. At time 0 $^{45}Ca^{2+}$ (2 μCi/ml) and 60 min later ATP (5×10^{-3} mmol/liter) were added. Acini were separated from the incubation medium by a Millipore filtration method. (From Ref. 2.)

promoted $^{45}Ca^{2+}$ was smaller as compared to 45 μg/ml. Concentrations of saponin of 45 μg/ml and of digitonin of 10 μg/ml therefore appeared to be most useful in permeabilizing cells while leaving intracellular mechanisms, such as ATP-dependent $^{45}Ca^{2+}$ uptake, apparently intact (Fig. 5). Therefore these detergent concentrations have also been used for studies on protein release as stimulated by secretagogues, such as carbachol or the second messenger cAMP, at clamped Ca^{2+} concentrations of 10^{-8} and 10^{-7} mol/liter adjusted with EGTA and at 10^{-6} and 10^{-5} mol/liter adjusted with NTA. Figure 6 shows the effect of different free Ca^{2+} concentrations on basal and carbachol-stimulated (Fig. 6a) and on cAMP-stimulated protein release (Fig. 6b) in "intact" and in saponin (45 μg/ml)-treated acini. Whereas intact acini did not respond to increasing Ca^{2+} concentrations with ^{3}H-labeled protein release, ^{3}H-labeled protein release increased with increasing Ca^{2+} concentrations in leaky acini and was about twofold at 10^{-4} M free Ca^{2+} concentration. In the presence of carbachol and the absence of Ca^{2+}, ^{3}H-labeled protein release was higher than without carbachol, and further increase occurred at free Ca^{2+} concentrations $> 10^{-6}$ mol/liter in both intact and saponin-treated cells. Cyclic AMP (2×10^{-3} mmol/liter), which had no effect in intact cells (not shown), increased protein release with increasing Ca^{2+} concentrations, reaching

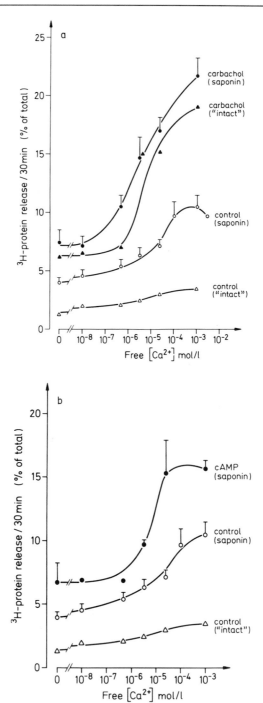

maximal protein secretion at 10^{-5} mol/liter (Fig. 6b). The phorbol ester TPA, which activates protein kinase C, mimicking the effect of diacylglycerol, a second messenger in different cells,[66,67] increases enzyme secretion from the exocrine pancreas.[42] As shown in Fig. 7 under conditions of incubation with either 10^{-7} mol/liter TPA or 5×10^{-6} mol/liter CCh, both intact cells and cells treated with digitonin (10 μg/ml) respond to approximately the same extent; in response to TPA, ^{3}H-labeled protein release was increased by 44.7 ± 9.5% above basal (basal = 6.3 ± 0.8% of total, $n = 11$) in intact cells and by 22.5 ± 7.3% above basal (basal = 10.6 ± 1.7% of total, $n = 3$) in digitonin-treated acini. The response to 5×10^{-6} mol/liter CCh was also similar in both intact and leaky cells, secretion being increased by 107.5 ± 19.7% above basal (basal = 4.5 ± 0.4% of total, $n = 8$) in intact cells and by 78.8 ± 9.8% (basal = 8.7 ± 0.7% of total, $n = 17$) in leaky cells. When intact acini were incubated with TPA and 2×10^{-3} mol/liter cAMP, secretion was not any greater than that observed in the presence of 10^{-7} mol/liter TPA alone (39.2 ± 10.4% above basal; basal = 6.5 ± 0.8% of total, n = 8). In contrast, under conditions of combined stimulation with TPA and cAMP in leaky cells, secretion was markedly increased by 111.2 ± 14.8% above basal (basal = 9.7 ± 0.6% of total protein, $n = 14$) (Fig. 7). This indicates that in permeabilized cells the muscarinic receptor was preserved and that the intracellular machinery leading to protein release was stimulated to similar extents in both intact and leaky cells if the secretagogue, such as TPA, could pass the plasma membrane. However, a second messenger of hormone action, such as cAMP, which does not permeate the cell membrane, was effective only in permeabilized cells.

[66] M. Castagna, M. Takai, M. Kaibuchi, K. Sano, U. Kikkawa, and Y. Nishizuka, *J. Biol. Chem.* **257**, 7847 (1982).
[67] Y. Nishizuka, *Nature (London)* **308**, 693 (1984).

FIG. 6. (a) Effect of different free Ca^{2+} concentrations with or without carbachol (10^{-5} mol/liter) on ^{3}H-labeled protein release from saponin-treated (45 μg/ml) and "intact" acini. Acini were preincubated for 40 min in the presence of 45 μg/ml saponin in buffer B and were incubated in the same solution during the experiment. Intact acini were incubated in buffer A. Free Ca^{2+} concentrations were adjusted with EGTA and NTA (see the section, Experimental Buffers); ^{3}H-labeled protein was determined as described in Fig. 3. Mean values ± SE from 4 to 14 separate experiments and from 1 experiment for "intact acini." (From Ref. 42.) (b) Effect of different Ca^{2+} concentrations on cAMP (2 mmol/liter)-stimulated ^{3}H-labeled protein release from saponin (45 μg/ml)-treated pancreatic acini. Conditions were similar as described for Figs. 3 and 6a. Mean values ± SE from one experiment for intact acini and at least three experiments for saponin-treated acini.

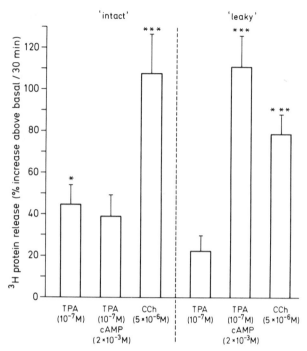

FIG. 7. ^{3}H-Labeled protein release from intact and digitonin (10 μg/ml)-permeabilized acini after 30 min of incubation with secretagogues. Experiments in intact cells were performed in the presence of 10^{-3} mol/liter Ca^{2+} in buffer A and those in leaky acini in the presence of 10^{-4} mol/liter Ca^{2+} in buffer B. Digitonin was added to the incubation buffer immediately following the addition of cells to the buffer. Bars are the means of at least three experiments $\pm$ SE. $*,p < 0.05$; $***,p < 0.001$. (From Ref. 64.)

Permeabilization of Cells by α-Toxin and Streptolysin O

Rat insulinoma cells (RINA2) and pheochromocytoma (PC 12) cells have been permeabilized with α-toxin in the following way:

Before permeabilization, the cells were washed in a buffer containing 150 mmol/liter KCl, 5 mmol/liter NaNO$_3$, 20 mmol/liter 3-(N-morpholino)propane sulfonic acid (MOPS), pH 7.2, and 1 mmol/liter EGTA. Cells were then incubated with 300 hemolytic units of α-toxin/10^{-7} cells for 10 min on ice followed by 30 min at 30°. Other instructions report the use of 20–50 μg/ml of α-toxin, usually dialyzed against buffered potassium glutamate solution corresponding to an α-toxin to total cell protein ratio of 0.1–0.3 (w/w). Following permeabilization the cells are washed three times with the same buffer but without EGTA.[55] Streptolysin O lyses PC12 cells at concentrations between 0.1 and 1 μg/ml after activation with 4 mmol/liter dithiothreitol. Similarly as with α-toxin, the experimental protocol involves binding of the toxin to the extracellular side of the plasma membrane in the cold followed by warming to 30° which triggers pore formation.[8,44] For detailed permeabilization procedures the reader is referred to the review by Bhakdi and Tranum-Jansen and the literature cited therein.[43,44]

The Choice of Anions

In adrenal medullary cells it has been found that Cl^- inhibits Ca^{2+}-evoked secretion of both catecholamine and dopamine-β-hydroxylase (DβH). Therefore glutamate was used in studies of Ca^{2+}-dependent catecholamine release from bovine adrenal medullary cells after exposure to intense electric fields.[1] The order of effectiveness at inhibiting calcium-dependent catecholamine release was SCN^-> Br^-> Cl^-> acetate>glutamate.[1]

In a study on the ionic dependence of enzyme secretion from permeabilized rat pancreatic acini we have shown that secretion stimulated by carbachol or by TPA plus cAMP was significantly reduced when 130 mmol/liter Cl^- in the buffer was replaced by I^-, NO_3^-, SCN^-, or cyclamate. Stimulated secretion could be increased with increasing concentrations of Cl^- in the medium, where Cl^- was substituted for by an equimolar amount of cyclamate.[64] The role of Cl^- in both permeabilized adrenal medulla and permeabilized pancreatic cells appears to be discrepant and could be related to the occurrence of regulated Cl^- channel and a Cl^-/OH^- exchanger in the zymogen granule membrane of pancreatic acinar cells involved in the event of exocytosis.[68,69] This is further supported by the observation that the anion exchange and Cl^--conductance blocker 4,4'-diisothiocyantostilbene-2,2'-disulfonic acid (DIDS) inhibited both enzyme secretion in permeabilized pancreas cells and Cl^- conductance in isolated pancreatic zymogen granules whereas it had no effect on permeabilized adrenal medullary cells. It therefore seems to be necessary to find optimal ionic requirements for each cell under study. Discrepant data, such as absence of cAMP-dependent enzyme secretion from permeabilized pancreatic acinar cells in glutamate buffer[61] and cAMP-dependent enzyme release in Cl^- buffer,[42,64] (see also Fig. 6b) could have its reason in different conditions used.

[68] R. C. DeLisle and U. Hopfer, *Am. J. Physiol.* **250**, G489 (1986).
[69] C M. Fuller, H. H. Deetjen, A. Piiper, and I. Schulz, *Pfluegers Arch.* **415**, 29 (1989).

The Choice of Cations

In permeabilized adrenal medulla cells no significant difference of Ca^{2+}-evoked protein release could be detected when cells were stimulated in K^+, Na^+, or sucrose buffer. However, in permeabilized pancreatic cells neither Na^+ nor $NMDG^+$ could adequately replace K^+ in the incubation buffer. Secretion evoked by carbachol was reduced by approximately 60% in the presence of NaCl buffer and by almost 80% when K^+ was replaced by $NMDG^+$. However, Rb^+ was as good as K^+ in supporting stimulated secretion.[64] Studies on isolated zymogen granules indicated the presence of a K^+ conductance pathway in the granular membrane[70] that might be involved in secretion.[64]

The Role of Nucleotides

ATP. In ATP-permeabilized adrenal medullary cells Mg-ATP is required for Ca^{2+}-dependent exocytosis. Removal of Mg-ATP from the incubation buffer abolished Ca^{2+}-evoked secretion, whereas readmission of Mg-ATP restored the sensitivity to Ca^{2+}.[1] ATP in the absence of Mg^{2+} was unable to support Ca^{2+}-dependent exocytosis. Raising the free Mg^{2+} concentration above 2 mmol/liter in the presence of 5 mmol/liter Mg-ATP reduced the maximum extent of exocytosis.[1] In the absence of ATP, Mg^{2+} alone had no effect on secretion.[1] Other nucleotides, such as GTP, CTP, UTP, or ITP, were about 20% as effective as ATP, whereas the ATP analogs AMP-PNP or ATP-γ-S and cyclic nucleotides were without effect in adrenal medullary cells.[1,71] It thus suggests that in this type of cell ATP hydrolysis is required for exocytosis. In other cells, such as platelets, a wider range of nucleoside triphosphates is effective, CTP being nearly as effective as ATP.[71] In PC12 cells exocytosis is exclusively triggered by Ca^{2+} and does not require the presence of Mg^{2+}-ATP.[7,53,72,73] ATP is also not required for exocytosis from permeabilized mast cells.[74]

Similarly in permeabilized pancreatic acinar cells the presence of ATP in the incubation buffer was not necessary for stimulating protein release, although the presence of ATP (1–5 mmol/liter) enhanced protein secretion somewhat by 10–20% (T. Kimura and I. Schulz, unpublished). The nature of the role for ATP in the secretory response does not seem to be clear. It could be required for phosphorylation of specific proteins involved in the stimulus-secretion pathway. ATP could also be necessary for energy provision, in the event of exocytosis, for example in fusion of zymogen

[70] K. W. Gasser, J. DiDomenico, and U. Hopfer, *Am. J. Physiol.* **254**, G93 (1988).
[71] D. E. Knight V. Niggli, and M. C. Scrutton, *Eur. J. Biochem.* **143**, 437 (1984).
[72] G. Ahnert-Hilger and M. Gratzl, *J. Neurochem.* **49**, 764 (1987).
[73] G. Ahnert-Hilger, M. Bräutigam, and M. Gratzl, *Biochemistry* **26**, 7842 (1987).
[74] P. E. Tatham and B. D. Gomperts, *Biosci. Rep.* **9** (1), 99 (1989).

granules with the lumenal plasma membrane. However, concentrations of ATP usually necessary for protein phosphorylation are in the micromolar range, whereas electropermeabilized adrenal medullary cells require millimolar levels of ATP.[1] Recently it has been shown that ATP has a modulatory function in the regulation of Cl^- channels in the zymogen granular membrane of pancreatic acinar cells. Whereas ATP at 1 μmol/liter reduced Cl^- conductance in isolated pancreatic zymogen granules, it stimulated Cl^- conductance above 50 μmol/liter.[75]

GTP. A role of GTP-binding proteins in exocytosis has been postulated for many cell types. Whereas in permeabilized platelets,[76] mast cells,[77] and adrenal chromaffin cells[78] GTP and GTP analogs act synergistically with calcium, in other systems, such as permeabilized rabbit neutrophils[79] or insulinoma RINm5F cells,[80] GTP analogs both potentiate the response to calcium and activate exocytotic secretion independently of calcium. The effect of GTP-γ-S on catecholamine release from α-toxin-permeabilized adrenal chromaffin cells did not occur when ATP was absent during stimulation. In contrast, guanine nucleotides could synergize with calcium to trigger secretion in the absence of ATP in mast cells.[77] ATP enhanced, however, the effective affinity for Ca^{2+} and GTP analogs in the exocytotic process without altering the maximum extent of secretion.[7] These results present evidence for at least two sites of action of GTP in stimulus-secretion coupling: one site in the plasma membrane is a GTP-binding protein, which transduces receptor signals to activation of phospholipase C[81]; the second site is concerned with the exocytotic mechanisms[79] and could be located in the granular membrane.

We have recently shown the presence of nine small molecular weight GTP-training proteins (20–30 kDa), of three G_i-like proteins (G_{i1}, G_{i2}, and G_{i3}) as well as G_0 to be present in the membrane of pancreatic zymogen granules (ZG) (S. Schnefel, A. Pröfrock, K.-D. Hinsch, and I. Schulz, *Eur. J. Biochem.*, submitted, 1990). Furthermore, the GTP analog GTPγS (10^{-4} mol/l) evoked enzyme secretion from digitonin-permeabilized pancreatic acinar cells and shifted the maximal stimulatory cholecystokinin-octapeptide (CCK-8) concentration from 10^{-9} mol/l to 10^{-11} mol/l. The Cl^- conductance in ZG isolated from CCK-8-prestimulated pancreatic acini was increased to the same level as could be increased by direct addition of

[75] F. Thévenod, K. Gasser, A. Goldsmith, and U. Hopfer, *J. Gen. Physiol.* **94**, 32a (1989).
[76] R. J. Haslam and M. L. Davidson, *FEBS Lett.* **174**, 90 (1984).
[77] T. W. Howell, S. Cockcroft, and B. D. Gomperts, *J. Cell Biol.* **105**, 191 (1987).
[78] M.-F. Bader, J.-M. Sontag, D. Thiersé, and D. Aunis, *J. Biol. Chem.* **264**, 16426 (1989).
[79] M. M. Barrowman, S. Cockcroft, and B. D. Gomperts, *Nature (London)* **319**, 504 (1986).
[80] L. Vallar, T. J. Biden, and C. B. Wollheim, *J. Biol. Chem.* **262**, 5049 (1987).
[81] S. Cockcroft and B. D. Gomperts, *Nature (London)* **314**, 534 (1985).

GTPγS (10^{-5} mol/l) to ZG from unstimulated acini (A. Piiper, L. Eckhardt, and I. Schulz, *Eur. J. Biochem.*, submitted, 1990).

Summary

The techniques described allow controlled permeabilization of plasma membranes from different types of cells for gaining access to the cell interior and enables one to control intracellular events. Most common techniques are electropermeabilization, permeabilization with mild nonionic detergents such as saponin and digitonin and by pore-forming toxins, such as α-toxin and streptolysin O. Whereas electropermeabilization and α-toxin create small pores of ~2 nm, digitonin, saponin, and streptolysin O form bigger holes and therefore also allow the introduction of large molecules, such as enzymes and immunoglobulins. A disadvantage of the latter methods is the loss of cytosolic constituents which might be necessary for signal-transduction pathways in the cell.[82] In secretory cells the main requirement for exocytosis appears to be Ca^{2+}, which brings about the full response comparable to hormone effects in some cells (platelets),[71] adrenal medullary cells,[1] but not in all cells (pancreatic acinar cells). The nucleotide, anion, and cation requirements are different for different cell types and are probably intimately related to the cell-specific mechanisms involved in exocytosis such as regulation of ion channels and ion carriers, or the involvement of nucleotide-binding proteins. Since permeabilized cells are preparations intermediate between intact cells and isolated organelles, they offer great opportunities for the advancement of our understanding of the mechanisms involved in stimulus–response coupling.

[82] T. Sarafian, D. Aunis, and M.-F. Bader, *J. Biol. Chem.* **262**, 16671 (1987).

[20] Electrophysiology of Pancreatic Acinar Cells

By O. H. Petersen, M. Wakui, Y. Osipchuk, D. Yule,
and D. V. Gallacher

Introduction

The most important pancreatic acinar secretagogues, acetylcholine (ACh) and cholecystokinin (CCK), evoke both enzyme and Cl^--rich fluid secretion.[1] These processes are Ca^{2+} dependent[2] and associated with an increase in the cytosolic free Ca^{2+} concentration ($[Ca^{2+}]_i$)[3] mediated by the

[1] O. H. Petersen and N. Ueda, *J. Physiol. (London)* **264**, 819 (1977).
[2] N. Ueda and O. H. Petersen, *Pfluegers Arch.* **370**, 179 (1977).
[3] D. I. Yule and D. V. Gallacher, *FEBS Lett.* **239**, 358 (1988).

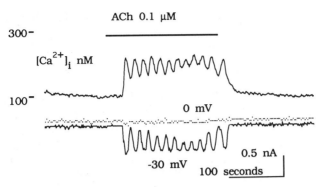

FIG. 1. Mouse pancreatic acinar cell. Simultaneous recording of $[Ca^{2+}]_i$ with dual-excitation microfluorimetry using Fura-2 (top trace) and Ca^{2+}-dependent Cl^- inward current (at -30 mV) in the patch-clamp whole cell recording configuration (bottom trace). ACh, Acetylcholine. [Adapted from Y. V. Osipchuk, M. Wakui, D. I. Yule, D. V. Gallacher, and O. H. Petersen, *EMBO J.* **9**, 697 (1990).]

internal messenger inositol(1,4,5)trisphosphate.[4] The increase in $[Ca^{2+}]_i$ activates ion channels and the resulting ionic fluxes are important for the fluid secretion.[5-8] The secretagogue-evoked increase in $[Ca^{2+}]_i$ is also involved in triggering enzyme secretion.[9,10]

Stimulant-Evoked Increase in $[Ca^{2+}]_i$

In resting pancreatic acinar cells $[Ca^{2+}]_i$ is about $100-150$ nM and maximal doses of ACh or CCK evoke an increase to about 1 μM (for references see the recent review by Muallem[10]). In single acinar cells challenged with submaximal concentrations of ACh, oscillations in $[Ca^{2+}]_i$ are frequently observed[3] (Fig. 1). In the presence of normal (millimolar) Ca^{2+} concentrations in the extracellular solution such oscillations in $[Ca^{2+}]_i$ continue as long as the stimulus is maintained but quickly disappear when the secretagogue is withdrawn (Fig. 1). In the absence of extracellular Ca^{2+}, a sustained submaximal ACh stimulus gives rise to only a few pulses of intracellular Ca^{2+} increase and a sustained rise in $[Ca^{2+}]_i$ does not occur until Ca^{2+} is added to the bath solution.[3] Removal of external Ca^{2+} during sustained ACh stimulation immediately evokes a fall in $[Ca^{2+}]_i$ to the

[4] H. Streb, R. Irvine, M. J. Berridge, and I. Schulz, *Nature (London)* **306**, 67 (1983).
[5] O. H. Petersen and Y. Maruyama, *Nature (London)* **307**, 693 (1984).
[6] O. H. Petersen, *Am. J. Physiol.* **251**, G1 (1986).
[7] O. H. Petersen, *Physiol. Rev.* **67**, 1054 (1987).
[8] O. H. Petersen and D. V. Gallacher, *Annu. Rev. Physiol.* **50**, 65 (1988).
[9] I. Schulz and H. H. Stolz, *Annu. Rev. Physiol.* **42**, 127 (1980).
[10] S. Muallem, *Annu. Rev. Physiol.* **51**, 83 (1989).

resting (prestimulation) level and this effect is quickly reversible upon readmission of Ca^{2+} to the bath.[3] It is therefore clear that the initial stimulant-evoked increase in $[Ca^{2+}]_i$ is due to release of Ca^{2+} from an internal store whereas the sustained increase is acutely dependent on Ca^{2+} influx from the extracellular space.

It is now generally accepted that the initial stimulant-evoked intracellular Ca^{2+} release is mediated by inositol(1,4,5)trisphosphate [Ins(1,4,5)P$_3$] generated by breakdown of phosphatidylinositol(4,5)bisphosphate (PIP$_2$)[4,11] and in permeabilized pancreatic acinar cells Ins(1,4,5)P$_3$ evokes Ca^{2+} release from a nonmitochondrial store.[4] PIP$_2$ breakdown to Ins(1,4,5)P$_3$ and 1,2-diacylglycerol (DG) is due to the receptor-mediated activation of phospholipase C via a GTP-binding protein.[11]

The mechanism by which Ca^{2+} influx during the sustained phase of stimulation is regulated has not yet been fully clarified. It is known that Ins(1,4,5)P$_3$ is very rapidly phosphorylated to Ins(1,3,4,5)P$_4$ via the Ca^{2+}-dependent Ins(1,4,5)P$_3$-3 kinase[12] and that very rapid formation of Ins(1,3,4,5)P$_4$ occurs following activation of either ACh or CCK receptors, but as far as the pancreatic acinar cells are concerned there is still no direct evidence for a role of Ins(1,3,4,5)P$_4$ in regulating Ca^{2+} influx.[13] In internal cell perfusion studies on the not entirely dissimilar lacrimal acinar cells, in which $[Ca^{2+}]_i$ is monitored by measurement of Ca^{2+}-dependent K^+ current (see below), it has been shown that Ins(1,4,5)P$_3$ evokes only a transient internal Ca^{2+} release whereas a combination of Ins(1,4,5)P$_3$ and Ins(1,3,4,5)P$_4$ induces a sustained Ca^{2+} mobilization that is acutely dependent on external Ca^{2+}.[13-16] Ins(1,3,4,5)P$_4$ can enhance Ins(1,4,5)P$_3$-evoked internal Ca^{2+} release[15] (in the absence of external Ca^{2+}) and while the action of Ins(1,4,5)P$_3$ is acutely reversible the effects of Ins(1,3,4,5)P$_4$ are long lasting and continue for 5–10 min following its removal.[16] Figure 2 shows a very schematic model providing an undoubtedly much oversimplified view of how these two inositol polyphosphates may regulate Ca^{2+} handling in the lacrimal acinar cells.

Internal perfusion studies employing single pancreatic acinar cells have so far revealed very long-lasting effects of intracellular Ins(1,4,5)P$_3$ stimulation that are independent of external Ca^{2+},[17] suggesting very large internal Ca^{2+} stores. Similar effects were also obtained with ACh stimulation.[17]

[11] M. J. Berridge and R. F. Irvine, *Nature (London)* **341**, 197 (1989).
[12] T. J. Biden and C. B. Wollheim, *J. Biol. Chem.* **261**, 11931 (1986).
[13] O. H. Petersen, *Cell Calcium* **10**, 375 (1989).
[14] A. P. Morris, D. V. Gallacher, R. F. Irvine, and O. H. Petersen, *Nature (London)* **330**, 653 (1987).
[15] L. Changya, D. V. Gallacher, R. F. Irvine, B. V. L. Potter, and O. H. Petersen, *J. Membr. Biol.* **109**, 85 (1989).
[16] L. Changya, D. V. Gallacher, R. F. Irvine, and O. H. Petersen, *FEBS Letts.* **251**, 43 (1989).
[17] M. Wakui, B. V. L. Potter, and O. H. Petersen, *Nature (London)* **339**, 317 (1989).

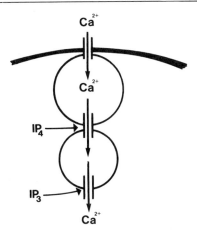

FIG. 2. Simple model concept showing that whereas Ins(1,4,5)P$_3$ (IP$_3$) opens channels, allowing Ca^{2+} to move from within stores to the cytosol, Ins(1,3,4,5)P$_4$ (IP$_4$) connects different Ca^{2+} stores, thereby also establishing a connection to the extracellular compartment. (Adapted from Changya et al.[16])

Since the effects of Ins(1,4,5)P$_3$ could also be obtained with the nonmetabolizable analog inositol(1,4,5)trisphosphorothioate [Ins(1,4,5)PS$_3$][17] it would appear that in the internally perfused mouse pancreatic acinar cells Ins(1,4,5)P$_3$ by itself is much more effective in causing prolonged internal Ca^{2+} release than in the internally perfused mouse lacrimal acinar cells. This result may, however, not be representative of the intact mouse pancreatic acinar cells where a submaximal ACh dose evokes only a few pulses of intracellular Ca^{2+} release in the absence of external Ca^{2+},[3] in spite of the fact that the internal store is very far from empty since a larger dose can mobilize much more Ca^{2+}. Connections between Ca^{2+} pools that under some circumstances require the continued presence of Ins(1,3,4,5)P$_4$ (Fig. 2) may in other experimental situations be permanently activated.[16]

The existence of at least two different nonmitochondrial Ca^{2+} pools, one sensitive and the other insensitive to Ins(1,4,5)P$_3$, is now clearly established,[18] but the exact location of these pools as well as the nature and regulation of possible connections between them remain obscure.

Ion Channels

The first patch-clamp studies of ionic pores in pancreatic acinar cell membranes revealed the presence of Ca^{2+}-activated nonselective cation channels in the basolateral acinar cells membranes.[5] These studies are

[18] F. Thevenod, M. Dehlinger-Kremer, T. P. Kemmer, A. L. Christian, B. V. L. Potter, and I. Schulz, J. Membr. Biol. 109, 173 (1989).

historically important as they were the first to demonstrate directly intracellular messenger-mediated opening of ion channels evoked by receptor activation, but the nonselective cation channel is probably not normally activated in the intact cells when physiological doses of secretagogues are used, since it is blocked by intracellular ATP present in millimolar concentration.[19]

Potassium ion-selective channels activated by internal Ca^{2+} have been directly demonstrated in the basolateral membranes of human, pig, as well as guinea pig pancreatic acinar cells.[7,8,19] Two types of channels (low and high conductance) are present, but they have similar Ca^{2+} and voltage dependence.[7,8,19] In the mouse and rat pancreatic acinar cells these channel types have not been found.[5]

Early microelectrode studies on mouse pancreatic acinar cells revealed that ACh evokes a large increase in membrane Cl^- conductance[20] and the selectivity of these ACh-opened Cl^- channels was studied in some detail.[21] Later patch-clamp studies on mouse acinar cells fully confirmed these findings.[17] Figure 1 shows that the time course of the ACh-evoked oscillatory increase in Cl^- current exactly reflects the time course of the changes in $[Ca^{2+}]_i$. Indeed, the ACh-evoked increase in Cl^- current can be abolished by filling up the internally perfused acinar cell with a solution containing a high concentration of a Ca^{2+} chelator[17] and is therefore clearly a Ca^{2+}-dependent current. Unfortunately there is no single-channel evidence concerning these Cl^- channels in the pancreatic acini, but in the rat lacrimal acinar cells a similar current has been found and in these cells single-channel studies indicate a very low unit conductance of about $1-2$ pS.[22]

The Cl^- channels have not been localized with certainty to basolateral or lumenal membranes. In analogy with what is known about the salivary glands it is, however, tempting to suggest that they may be specifically concentrated in the lumenal membranes.[6,23]

Calcium Ion Oscillations Studied by Measuring Ca^{2+}-Dependent Cl^- Current

Figure 1 shows ACh-evoked oscillations in $[Ca^{2+}]_i$ and Ca^{2+}-dependent Cl^- current and indicates that it may be possible simply to monitor $[Ca^{2+}]_i$

[19] K. Suzuki and O. H. Petersen, *Am. J. Physiol.* **255**, G275 (1988).
[20] O. H. Petersen, "The Electrophysiology of Gland Cells." Academic Press, New York and London, 1980.
[21] O. H. Petersen and H. G. Philpott, *J. Physiol. (London)* **306**, 481 (1980).
[22] A. Marty, Y. P. Tan, and A. Trautmann, *J. Physiol. (London)* **357**, 293 (1984).
[23] K. R. Lau and R. M. Case, *Pfluegers Arch.* **411**, 670 (1988).

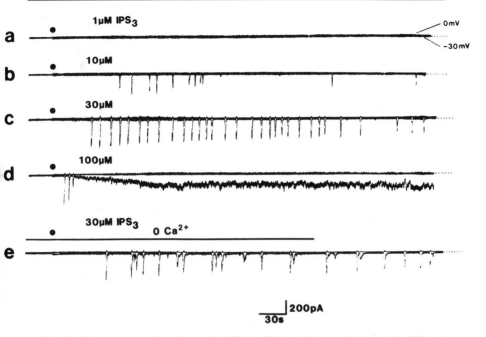

FIG. 3. Mouse pancreatic acinar cell. Effects of intracellular perfusion with different concentrations of Ins(1,4,5)PS$_3$ (IPS$_3$) on Ca^{2+}-dependent Cl$^-$ current (downward deflection) measured at -30 mV membrane potential in patch-clamp whole cell recording configuration. (Adapted from Wakui *et al.*[17])

near the plasma membrane by measuring this current. The Ca^{2+} oscillations evoked by various hormones and neurotransmitters in many different cell types have been thought to reflect oscillations in Ins(1,4,5)P$_3$ concentration.[24] Although it is with present technology impossible to measure Ins(1,4,5)P$_3$ levels in single cells with high time resolution, we have shown that internal perfusion of single pancreatic acinar cells with constant levels of Ins(1,4,5)P$_3$ or the nonmetabolizable analog Ins(1,4,5)PS$_3$ gives rise to discrete pulses of Ca^{2+}-dependent Cl$^-$ current (Fig. 3), indicating that fluctuations in the Ins(1,4,5)P$_3$ level are unnecessary for the generation of pulsatile internal Ca^{2+} release.[17]

The currently most attractive hypothesis explaining stimulant-evoked Ca^{2+} oscillations is based on the existence of at least two separate intracellular nonmitochondrial Ca^{2+} pools, one sensitive and the other insensitive to Ins(1,4,5)P$_3$. An initial release of Ca^{2+} from the Ins(1,4,5)P$_3$-sensitive pool could act as a primer for further Ca^{2+} release from Ins(1,4,5)P$_3$-insen-

---

[24] T. J. Rink and R. Jacob, *Trends Neurosci.* **12**, 43 (1989).

sitive pools, producing a spike that could spread throughout a cell, but although there is evidence for Ca^{2+}-induced Ca^{2+} release in the electrically excitable muscle, nerve, and chromaffin cells the position is unclear in electrically nonexcitable cells, such as, for example, exocrine acinar cells.

By combined and simultaneous recordings of changes in $[Ca^{2+}]_i$ near the plasma membrane (assessed by measurement of Ca^{2+}-dependent Cl^- current in the patch-clamp whole cell recording configuration) as well as in the cytoplasm as a whole (assessed by measuring fluorescence of Fura-2 by photon counting over the surface of a single cell) the stimulant-evoked Ca^{2+} oscillations in pancreatic acinar cells (Fig. 1) have been further characterized. The surface membrane receptors for ACh and CCK have been stimulated or direct activation of G proteins has been achieved through internal application of the nonhydrolyzable GTP analog GTP-γ-S. $Ins(1,4,5)P_3$, $Ins(1,4,5)PS_3$, or Ca^{2+} have also been infused into the cells. All these procedures evoke pulsatile Ca^{2+} mobilization, but the pattern of Ca^{2+} spike generation varies considerably. The Ca^{2+} spikes observed during both $Ins(1,4,5)P_3$ and Ca^{2+} infusion (Fig. 4) are short and more easily detected near the plasma membrane (Ca^{2+}-dependent Cl^- current) than in the cell at large (microfluorimetry). The ACh-evoked Ca^{2+} pulses can in some cases be equally brief, but are mostly broader (Fig. 1). The Ca^{2+} spikes induced by CCK or GTP-γ-S are generally of somewhat longer duration. Caffeine, a drug long known to enhance Ca^{2+}-induced Ca^{2+} release, potentiates the action of stimulants so that, for example, brief Ca^{2+} pulses evoked by $Ins(1,4,5)P_3$ become longer and larger and are seen also in the microfluorimetric traces.[25] These data indicate that Ca^{2+}-induced Ca^{2+} release plays an important role in the generation of pulsatile internal Ca^{2+} release, but also suggest that surface receptor activation may influence the pattern of Ca^{2+} pulse generation by factors additional to $Ins(1,4,5)P_3$ formation.

Importance of Ionic Currents for Acinar Fluid Secretion

The most popular model for exocrine acinar fluid secretion is based on Ca^{2+} activation of K^+ channels in the basolateral and Cl^- channels in the lumenal membrane (Fig. 5). The K^+ channel activation is essential in order to allow Cl^- uptake via a $Na^+/K^+/2Cl^-$ cotransporter and in the steady state Na^+ and K^+ recirculate via cotransporter, K^+ channel, and Na^+/K^+ pump.[5-8]. The Cl^- channels allow Cl^- to escape into the lumen, at the

[25] Y. V. Osipchuk, M. Wakui, D. I. Yule, D. V. Gallacher, and O. H. Petersen, *EMBO J.* **9,** 697 (1990).

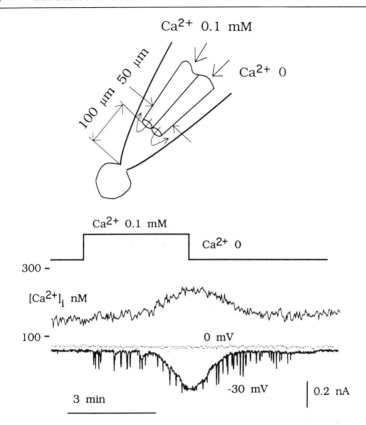

FIG. 4. Mouse pancreatic acinar cell. Two plastic tubes with opening diameter of about 20 μm were used to exchange solution at the tip of the patch pipet. Application of 200 mmHg pressure to the tube containing 0.1 mM Ca^{2+} resulted in a slow rise of [Ca^{2+}]$_i$ and Ca^{2+}-induced Cl$^-$ current and in repetitive short-lasting spikes of Ca^{2+}-dependent Cl$^-$ current. Application of pressure to the tube filled with Ca^{2+} free solution reversed the effect. (Adapted from Osipchuk et al.[25])

same time generating lumen negativity attracting Na$^+$ via the paracellular pathway.[5-8] In all salivary and lacrimal gland acinar cells so far studied such Ca^{2+}-dependent Cl$^-$ and K$^+$ currents have been found.[6] In the pancreatic acinar cells, however, the situation is complicated since in some species there are K$^+$-selective channels and in others (mouse and rat) not.[6-8] On the other hand large Ca^{2+}-dependent Cl$^-$ currents have been found only in mouse and rat pancreatic acinar cells.[17] It would appear that acinar cells from the pancreas are less complete with regard to Ca^{2+}-activated currents than other exocrine glands and therefore one might expect them

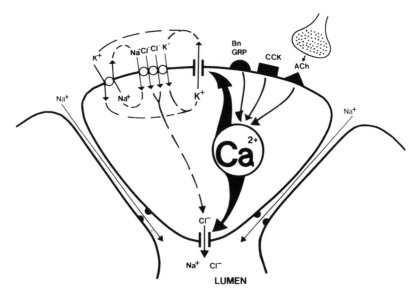

FIG. 5. Transport model for pancreatic acinar cell based on general transport model for exocrine acinar cells as described by Petersen and Maruyama[5] and Petersen.[6] Bn, Bombesin; GRP, gastrin-releasing peptide, ACh, acetylcholine; CCK, cholecystokinin.

not to be able to produce large quantities of fluid rapidly. This is indeed the case. Salivary glands may easily produce 500 μl/g gland wt/min during maximal parasympathetic stimulation,[26] whereas the rat pancreas can secrete only about 6 μl/g gland wt/min during optimal stimulation.[27]

[26] O. H. Petersen and J. H. Poulsen, *Acta Physiol. Scand.* **73**, 93 (1968).
[27] W. A. Sewell and J. A. Young, *J. Physiol.* **252**, 379 (1975).

[21] Characterization of a Membrane Potassium Ion Conductance in Intestinal Secretory Cells Using Whole Cell Patch-Clamp and Calcium Ion-Sensitive Dye Techniques

By Michael E. Duffey, Daniel C. Devor, Zahur Ahmed, and Steven M. Simasko

Introduction

Cell membrane K$^+$ conductances are important to mammalian cell function. One of the most significant roles of K$^+$ conductances is to cause the resting membrane electrical potential difference, which is generated by the electrodiffusion of K$^+$ from the cell. Regulation of this transport pathway will affect membrane potential. Thus, control of this pathway is a convenient means by which cells can regulate potential-sensitive events.

This form of control is important in the process of Cl$^-$ secretion by epithelial cells, such as those found in intestinal crypts or the trachea. During Cl$^-$ secretion, the exit of Cl$^-$ across the apical cell membrane depolarizes the cells and decreases the driving force for Cl$^-$ exit. Secretion is sustained by the opening of basolateral membrane K$^+$ channels, which results in hyperpolarization of the apical membrane potential.[1] This chapter describes a strategy for characterizing the changes in membrane K$^+$ conductance that result from exposure of isolated cells of the human secretory cell line T84 to a cholinergic agonist, and the role of intracellular Ca^{2+} as a mediator of that process.

Isolated Cell Studies

Membrane conductances are difficult to study during secretion by intact epithelial tissues because secreting cells are not easily isolated from other transporting cells and endogenous paracrine and neural elements. This difficulty is greatly reduced if a homogeneous secretory cell line can be studied instead. The clonal cell line T84 originated from a lung metastasis of a human colonic carcinoma,[2] and secretes Cl$^-$ in response to various gastrointestinal hormones and neurotransmitters when grown as a

---

[1] P. L. Smith and R. A. Frizzell, *J. Membr. Biol.* 77, 187 (1984).
[2] H. Murakami and H. Masui, *Proc. Natl. Acad. Sci. U.S.A.* 77, 3463 (1980).

monolayer.[3,4] Clonal cells are convenient models when their transport properties parallel those of the primary tissue. However, caution must be observed before conclusions derived from clonal cell studies are applied to normal cell functions since signal transduction may be altered in transformed cells.

T84 cells grown in a monolayer secrete Cl^- when exposed to the cholinergic agonist, carbamylcholine (carbachol).[4] The apical membrane of these cells appears to contain a Cl^- transport pathway that is normally open and not affected by carbachol. Thus, secretion may not involve activation of apical membrane Cl^- channels. Rather, a basolateral membrane K^+ pathway is apparently activated during stimulation by carbachol. It has been proposed that the opening of these K^+ channels, with a resulting membrane hyperpolarization, is the primary mechanism of Cl^- secretion.[4] However, this hypothesis has not been proved since it was based on the results of isotopic Cl^- and Rb^+ (substituted for K^+) flux measurements, which do not distinguish between conductive and carrier-mediated membrane pathways. In addition, the signal transduction steps responsible for activation of the membrane K^+ pathway are not known, although carbachol causes a rise in intracellular Ca^{2+} [4] and Ca^{2+}-activated K^+ channels exist in other epithelial cells.[5]

We have used whole cell patch-clamp techniques[6] and the fluorescent Ca^{2+} indicator Fura-2[7] to characterize membrane conductance changes that occur when isolated T84 cells are exposed to carbachol. Isolated cells were used to simplify the interpretation of electrical responses that would otherwise be affected by cell-to-cell interactions present in epithelial monolayers (i.e., via gap junctions).

For these studies, T84 cells are grown in Dulbecco's modified Eagle's medium (DMEM) and Ham's F-12 (1:1) supplemented with 20 mM HEPES, 10 mM glucose, 5% newborn calf serum, 50 μg/ml penicillin, 50 μg/ml streptomycin, and 100 μg/ml neomycin. The cells are incubated in a humidified atmosphere of 5% CO_2 at 37°. When a confluent monolayer is seen (7–10 days) subculture (1:3 dilution) is required, using 0.05% trypsin and 0.50 mM EDTA to remove the cells from the culture dish surface.

[3] K. Dharmsathaphorn, J. A. McRoberts, K. G. Mandel, L. D. Tisdale, and H. Masui, *Am. J. Physiol.* **246,** G204 (1984).

[4] K. Dharmsathaphorn and S. J. Pandol, *J. Clin. Invest.* **77,** 348 (1986).

[5] N. K. Wills and A. Zweifach, *Biochem. Biophys. Acta* **906,** 1 (1987).

[6] D. C. Devor, S. M. Simasko, and M. E. Duffey, *Am. J. Physiol.* **258,** C318 (1990).

[7] D. C. Devor, Z. Ahmed, and M. E. Duffey, *J. Gen. Physiol.* **94,** (Abstract), 9A (1989).

Membrane Potential and Current Measurements

Patch-clamp measurement of membrane potential is done on cells which have been plated onto glass coverslips. To assure cell stability, cells are plated 1 day before use. Coverslips containing cells are placed in a small Plexiglas chamber (volume approximately 0.5 ml) mounted on the stage of an inverted phase-contrast microscope contained within a Faraday cage. Measurements are made on single, isolated, rounded cells which appear to have a smooth surface and no apparent blebs. After formation of a high-resistance seal ~ 10 GΩ) between the electrode and cell, suction is applied so that the cell membrane under the electrode ruptures to achieve the whole cell configuration.[8] Membrane potential (and current, see below) are referenced to the bath by an Ag/AgCl pellet. Liquid junction potential is reduced in those experiments in which Cl$^-$ in the bath is decreased by using an Hg/HgCl electrode connected to the solution by a 1 M KCl-agar bridge.

TW150 patch pipets are fabricated from 1.5-mm-diameter borosilicate glass (World Precision Instruments, Sarasota, FL) using a two-stage micropipet puller, and coated with silicon polymer (Sylgard; Dow Corning, Midland, MI) to reduce pipet electrical noise. The pipets are then polished on a microforge before use. The composition of the pipet filling solution is important since the cell interior will be quickly dialyzed by the pipet solution in the whole cell patch configuration (see below).[8] EGTA and Ca^{2+} are added to maintain a known free Ca^{2+} concentration of 100 nM.[9] In our experiments, ATP and GTP are added to help maintain cell viability and as possible substrates for membrane signal transduction. The pipet solution is slightly hypotonic to the bath solution (280 vs 300 mOsm) to facilitate patch seal formation.

The standard cell bathing solution contains (in mM): 145 NaCl, 5 KCl, 1 CaCl$_2$, 1 MgCl$_2$, 10 HEPES, and 10 glucose; pH 7.4. The pipet solution contains (in mM): 130 KCl, 5 NaCl, 4 MgCl$_2$, 0.12 CaCl$_2$, 10 HEPES, 2 ATP, 0.5 GTP, and 0.2 EGTA; pH 7.4. Measurements are made at room temperature (22–25°).

The effect of carbachol on membrane potential in an isolated T84 cell is shown in Fig. 1. Membrane potential stabilizes in this cell in approximately 60 sec at −35 to −40 mV after formation of a whole cell patch. Carbachol (100 μM) exposure causes a rapid hyperpolarization to approxi-

[8] O. P. Hamill, A. Marty, E. Neher, B. Sakmann, and F. J. Sigworth, *Pfluegers Arch.* **391**, 85 (1981).
[9] A. E. Martell and R. M. Smith, "Critical Stability Constants." Plenum, New York, 1974.

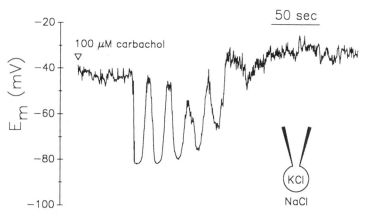

FIG. 1. Membrane potential (E_m) recorded in the whole cell patch-clamp configuration (*inset*) as a function of time for a cell exposed to 100 μM carbachol in standard bath (145 mM Na$^+$) and pipet (130 mM K$^+$) solutions (see text). Carbachol exposure (∇) continued for the duration of the recording. Capacitive transients (see text) have been filtered out for clarity. [Modified with permission from D. C. Devor, S. M. Simasko, and M. E. Duffey, *Am. J. Physiol.* **258,** C318 (1990).]

mately -75 mV, followed by a series of oscillations of the potential and a decline to near-baseline levels. The rapid hyperpolarization observed in this cell could have been caused by a variety of changes in membrane conductance(s). However, the most likely cause is an increase in membrane K$^+$ conductance since the potential approaches the K$^+$ equilibrium potential ($E_{K+} = -83$ mV). Likewise, the depolarizing phase of the response could be caused by a decrease in K$^+$ conductance. On the other hand, a decrease in membrane Cl$^-$ conductance, followed by an increase, would produce the same result.

The underlying ionic causes of this potential change can be isolated by measuring membrane current during voltage clamp. In our protocol, cells are voltage clamped to 0 mV. Under our standard conditions, this results in a large outwardly directed electrochemical driving force for K$^+$ and a large inwardly directed driving force for Na$^+$, with little driving force for Cl$^-$ (1.7 mV inward). Thus, a K$^+$ current can be recognized as an outward current flowing from the cell to the bath. Voltage pulses (30 mV, 0.2 Hz, 30-msec duration) are applied by a stimulator to the cell membrane after formation of the whole cell patch; membrane capacitive transients are used to monitor integrity of the whole cell configuration; membrane resistance is measured to monitor viability of the electrode–membrane seal. Typically, currents are filtered at 3 kHz.

The effects of continuous exposure of two cells to 100 μM carbachol on membrane current are shown in Fig. 2. Carbachol exposure is begun approximately 90 sec after formation of a whole cell patch to allow sufficient time for cell dialysis. After a delay of 45–70 sec the current rises rapidly to a peak and then oscillates as it returns to zero. This current response is consistent with a K$^+$ current flowing out of the cell. Cell input resistance during this response fluctuates as membrane current oscillates, as shown in Fig. 3, where the effects of +30 mV, 30-msec voltage pulses on membrane current are seen. As membrane current rises to a peak (B) during carbachol exposure the current response to the voltage pulse increases relative to the control response (A), then decreases as current declines (C) during an oscillation. This result demonstrates that carbachol causes a change in membrane K$^+$ conductance, which oscillates.

It is difficult to determine other electrophysiological properties (reversal potential, rectification, etc.) of this conductance by standard current–

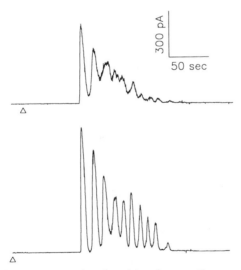

FIG. 2. Membrane current as a function of time for two cells exposed to 100 μM carbachol in standard bath (145 mM Na$^+$) and pipet (130 mM K$^+$) solutions and voltage clamped to 0 mV. Carbachol exposure ($\triangle$) continued for the duration of the recording. In this and all subsequent figures an outward current is represented as an upward deflection from baseline. [Modified with permission from D. C. Devor, S. M. Simasko, and M. E. Duffey, *Am. J. Physiol.* **258**, C318 (1990).]

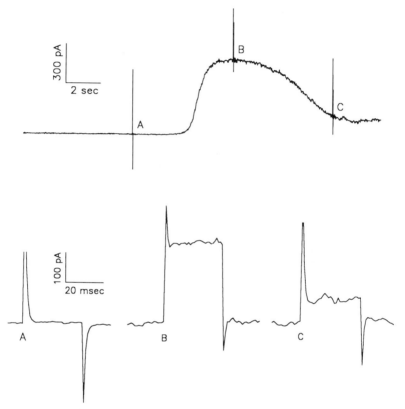

FIG. 3. Changes in membrane conductance induced by 100 μM carbachol. Top trace shows current response without filtering capacitive transients resulting from applied voltage pulses. Bottom traces are expanded views of the corresponding lettered pulses in top trace.

voltage techniques because of its oscillatory nature. However, with symmetric KCl solutions (130 mM) in the bath and pipet, current responses are not observed from cells voltage clamped at 0 mV and exposed to carbachol.[6] This suggests that the current reverses at 0 mV, a reversal potential that is predicted if carbachol activates a K$^+$ conductance under these conditions.

The results shown in Figs. 1–3 demonstrate that carbachol induces a K$^+$ conductance in these cells. However, a Cl$^-$ conductance must also exist during Cl$^-$ secretion. An apical membrane Cl$^-$ influx has previously been demonstrated in T84 cells grown as a monolayer under basal conditions.

This flux was not altered during cholinergic stimulation of Cl$^-$ secretion.[4] In order to directly determine whether carbachol activates a Cl$^-$ conductance two experiments are useful. First, K$^+$ and Cl$^-$ currents can be separated by voltage clamping cells to E_{Cl^-} or E_{K^+}, respectively. For example, at E_{Cl^-} the driving force for Cl$^-$ movement will be zero so that Cl$^-$ current will be eliminated and only K$^+$ current will be present under our standard conditions. Similarly, only Cl$^-$ current will be visible when the membrane is clamped to E_{K^+}. Thus, if T84 cells are alternately voltage clamped to 0 mV (E_{Cl^-}) and then to -76 mV ($\sim E_{K^+}$) for 1-sec intervals, a K$^+$ current may be observed at 0 mV and a Cl$^-$ current at -76 mV. During such an experiment, carbachol induces an oscillating K$^+$ current at 0 mV.[6] However, while these cells are clamped to -76 mV no inward current is induced by carbachol in 84% of the cells tested ($n = 14$) (an increased inward current would be expected if a Cl$^-$ conductance is activated). In a second type of experiment, K$^+$ current can be eliminated by replacing all the K$^+$ in the bath and pipet with Na$^+$; Na$^+$ or Cl$^-$ currents can then be identified in response to an applied voltage. When T84 cells are voltage clamped to either $+60$ or -60 mV carbachol does not induce a current in 89% of the cells tested ($n = 20$).[6] Thus, in both protocols the dominant response to carbachol is a K$^+$ current.

Carbachol does not usually activate a Cl$^-$ conductance in T84 cells, suggesting that a Cl$^-$ conductance must exist under basal conditions if cholinergic agonists stimulate active Cl$^-$ secretion. Prior to carbachol exposure, voltage pulses to E_{K^+} from E_{Cl^-} result in a current of less than 5 pA. This current is likely to result from a basal Cl$^-$ conductance, as well as resting K$^+$ and nonselective cation conductances. Existence of such a basal Cl$^-$ conductance can be demonstrated using patch-clamp techniques in current-clamp mode. Membrane potential can be measured before, and after, the driving force for Cl$^-$ movement across the cell membrane is altered. A change in membrane potential corresponding to a change in gradient would demonstrate the existence of a Cl$^-$ conductance. Conditions are optimized by replacing all K$^+$ in the bath and pipet with Na$^+$ so that K$^+$ currents will not be seen. This manipulation results in a depolarization of membrane potential to -4 ± 1 mV ($n = 12$), a value near that predicted for a membrane that is a perfect Cl$^-$ electrode (-2 mV).[6] When the bath solution is then changed to a solution in which 140 mM Cl$^-$ is replaced by gluconate, membrane potential reverses to a positive potential (21 ± 2 mV), as predicted for Cl$^-$ leaving the cell down its concentration gradient. This effect of a low bath Cl$^-$ concentration is reversible. This shift is less than that predicted for a perfect Cl$^-$ electrode but demonstrates that a Cl$^-$ conductance is present in unstimulated cells.

Pharmacological Properties of Membrane K^+ Conductance

Our results suggest that carbachol induces a K^+ conductance. This K^+ conductance can be pharmacologically characterized by use of the classic K^+ channel blockers Ba^{2+}, tetraethylammonium ion (TEA), and Cs^+. Each blocker can be applied before switching to a solution of the blocker plus carbachol (100 μM) while the cell is voltage clamped to 0 mV, so that the blocker reaches the desired concentration before carbachol reaches the cells. Ba^{2+} (2 mM), applied in this way, has no effect on the carbachol-induced K^+ current.[6] On the other hand, when TEA (30 mM) is in the bath only 27% ($n = 15$) of the cells respond to carbachol, although the average peak current of the responding cells is no different from that of control cells. Since one would predict that the peak current should be reduced if this compound were blocking K^+ channels, TEA must have another effect, such as blocking muscarinic receptors.[10]

This technique can also be used to determine the ability of the K^+ channel blocker, Cs^+, to block the K^+ conductance in T84 cells. First, a cell can be bathed in symmetric KCl solutions with CsCl added to the bath solution. The cell can then be voltage clamped to positive or negative voltages. The predicted effect of this manipulation is that Cs^+ will be driven into the K^+ channels by the imposed electrical gradient when the cell is clamped to negative voltages, but Cs^+ will have no affect at positive voltages. Second, the experiment can be repeated after Cs^+ is added to the pipet but not the bath; if Cs^+ blocks K^+ channels in this configuration rectification would be observed when the cell is clamped to positive voltages. Rectification is not observed under either of these conditions in T84 cells when 10 mM CsCl and ± 60 mV voltages are applied.[6] Thus, Cs^+ does not block the carbachol-induced K^+ channels from the intracellular or extracellular side of the channels.

These results demonstrate that the carbachol-induced K^+ conductance in T84 cells cannot be blocked by several classical K^+ channel blockers. This is consistent with the findings that these inhibitors, as well as 4-aminopyridine and apamine, do not block carbachol-induced basolateral membrane Rb^+ efflux from intact T84 monolayers.[4] Thus, the K^+ conductance induced by carbachol appears to be novel since these blockers are effective in a variety of other epithelial tissues.[5]

Effects of Intracellular Ca^{2+} on Membrane Current

Cholinergic agonists are known to cause an elevation of intracellular Ca^{2+} in many cell types,[11,12] including T84 cells.[4] Since certain types of

[10] J. Kehoe, *J. Physiol.* **225**, 115 (1972).

membrane K$^+$ channels are known to be activated by intracellular Ca^{2+} it is important to determine whether the K$^+$ current response to carbachol is Ca^{2+} sensitive. Our standard pipet solution contains 200 μM EGTA in equilibrium with enough Ca^{2+} to produce a solution Ca^{2+} concentration of 100 nM. When a whole cell patch is formed the cell interior is quickly dialyzed by this solution. The significance of intracellular Ca^{2+} to the current response can be determined by removal of Ca^{2+} from the pipet and the addition of 4 mM EGTA so that intracellular stores of Ca^{2+} are likely to be depleted. Under these conditions, the carbachol-induced current should be eliminated if the release of Ca^{2+} from intracellular stores is necessary for the response. This maneuver does eliminate the K$^+$ current induced by carbachol in T84 cells (Fig. 4, bottom). Next, the Ca^{2+} buffering capacity of the pipet solution can be altered by adding EGTA (2 mM) in the pipet while maintaining the free Ca^{2+} concentration at 100 nM. Under these conditions, a K$^+$ current is induced by carbachol, but no oscillations occur (Fig. 4, middle). Thus, since the K$^+$ conductance oscillations are dependent on intracellular Ca^{2+} buffering, they are likely to be dependent on the mechanisms that regulate intracellular free Ca^{2+}. Finally, the role of extracellular Ca^{2+} in the K$^+$ current response can be examined in cells bathed in the standard solutions made with no Ca^{2+} and 1 mM EGTA added to chelate trace Ca^{2+} contamination. Under these conditions, carbachol produces a response in T84 cells that is similar to cells bathed in solutions that contain Ca^{2+} (Fig. 4, top). Thus, the K$^+$ current response does not depend on the influx of Ca^{2+} from the bath.

Intracellular Ca^{2+} Concentration

The results of these studies with EGTA provide evidence for the involvement of intracellular Ca^{2+} in the carbachol-induced K$^+$ conductance. It is important, however, to determine whether intracellular Ca^{2+} oscillations underlie the K$^+$ conductance oscillations. Although intracellular Ca^{2+} has been measured and shown to rise during carbachol stimulation of T84 cells,[4] these studies have been done on cells in a confluent monolayer, where oscillations of Ca^{2+} have not been seen.

Measurement of Ca^{2+} in single isolated cells can be performed on cells which had been plated onto coverslips the previous day. Cells are loaded with the permeant form of the fluorescent Ca^{2+} indicator, Fura-2/AM,[13] by

---

[11] J. K. Foskett, P. J. Gunter-Smith, J. E. Meluin, and R. J. Turner, *Proc. Natl. Acad. Sci. U.S.A.* **86,** 167 (1989).

[12] D. L. Ochs, J. I. Korenbrot, and J. A. Williams, *Am. J. Physiol.* **249,** G389 (1985).

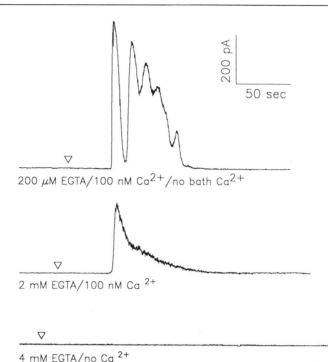

200 pA

50 sec

∇

200 μM EGTA/100 nM Ca^{2+}/no bath Ca^{2+}

∇

2 mM EGTA/100 nM Ca $^{2+}$

∇

4 mM EGTA/no Ca $^{2+}$

FIG. 4. Effects of manipulation of intracellular and extracellular Ca^{2+} on membrane current in response to 100 μM carbachol (∇) applied for the duration of the recording. Cells were voltage clamped to 0 mV. Bath and pipet contained standard solutions, except as noted. The top recording was made after bath Ca^{2+} was reduced to zero and in the presence of 1 mM EGTA. The center and bottom recordings were made in the presence of normal bath Ca^{2+}, but with 2 mM EGTA and 100 nM free Ca^{2+} in the pipet (center), or 4 mM EGTA and no Ca^{2+} in the pipet (bottom). [Reproduced with permission from D. C. Devor, S. M. Simasko, and M. E. Duffey, *Am. J. Physiol.* **258**, C318 (1990).]

incubation of the cells for 20 min in the standard bath solution in the absence of Ca^{2+} and in the presence of 5 μM Fura-2/AM. The cells are then incubated for 1 hr in the standard growth media to allow the intracellular dye to deesterify into its impermeant form.[13] Both the loading of Fura-2/AM and the subsequent incubation in growth media are done at room temperature to reduce the possibility of dye inclusion into intracellular organelles.[14]

[13] G. Grynkiewicz, M. Poanic, and R. Y. Tsien, *J. Biol. Chem.* **260**, 3440 (1985).
[14] M. Poenie, J. Alderton, R. Steinhart, and R. Y. Tsien, *Science* **233**, 886 (1986).

Coverslips with Fura-2-loaded cells are then mounted onto the bottom of a Plexiglas chamber (approximately 0.3-ml volume) which is then mounted onto the stage of an inverted microscope equipped for epifluorescence at ×40 magnification. The chamber is continuously perfused with electrolyte solution (6–7 ml/min) using a pump and aspirator, such that a complete solution exchange is accomplished in less than 20 sec. All measurements are made at room temperature.

For Ca^{2+} measurements cells are exposed to UV light from a 75-W xenon lamp which is alternately passed through 340- and 380-nm interference filters mounted on a wheel. The rotation speed of this wheel (normally 12,000 rpm) is controlled by an AC motor with optical sensors that indicate which filter is being used. The UV light exposure time is controlled by a Uniblitz (Vincent Associates, Rochester, NY) shutter mounted between the lamp housing and the filters. Emitted fluorescence from the cell specimen is collected by a photomultiplier tube after it is passed through a 510-nm bandpass filter. A pinhole turret mounted between the specimen and the filter is used to collect light from only a selected portion of the field (~ 1.5× a cell diameter).

The photomultiplier tube current produced during the 340- and 380-nm excitation is filtered at 1 kHz using an 8-pole lowpass filter, and sampled at a rate of 8 μsec/point using an LSI 11/23+ computer. The time interval between data collected at 340- and 380-nm excitation wavelengths is 2.5 msec. Intracellular Ca^{2+} is determined from each of three ratios, then the average of the three ratios is used to produce one data point. This process is repeated at either 0.5- or 1-sec intervals. Intracellular Ca^{2+} is calculated using the equation

$$[Ca^{2+}]_i = K_D F_0 (R - R_{max})/F_s (R - R_{max})$$

where K_D is the apparent equilibrium dissociation constant for Fura-2/AM (225 nM)[15]; the R terms are the fluorescence ratios calculated using the 340- and 380-nm excitation wavelengths; F_0/F_s is the fluorescence ratio obtained at zero and saturating Ca^{2+} levels using the 380-nm excitation wavelength; R_{max} and R_{min} are the F_{340}/F_{380} ratios obtained at saturating (maximum) and zero (minimum) Ca^{2+} concentrations, respectively.[13]

Calibration of the Ca^{2+} fluorescence signal is performed by a microcapillary method. A solution is made that simulates intracellular fluid (in mM): 140 KCl, 10 NaCl, 2 MgCl$_2$, 10 HEPES, pH 7.3, to which 50 μM Fura-2 (acid form) is added. The solution is made either Ca^{2+} free (2 mM EGTA) or enough Ca^{2+} is added to saturate the Fura-2 (2.5 mM CaCl$_2$,

---

[15] J. A. Connor, *Proc. Natl. Acad. Sci. U.S.A.* **83**, 6179 (1986).

2 mM EGTA). This solution is then put into rectangular quartz microcapillary tubes (optical path length 20 μm) and mounted on the stage of the microscope. F_{340}/F_{380} ratios are determined for these solutions as outlined above and R_{max}, R_{min}, and F_0/F_s are calculated. Alternatively, the Fura-2 fluorescence can be calibrated *in vivo*[13]: After each experiment the standard bathing solutions of Fura-2-loaded cells are replaced with a solution very like the composition of the intracellular milieu plus ionomycin (1–2 μM), which allows equilibration of intra- and extracellular Ca^{2+}. The free Ca^{2+} concentration in the bathing solution is controlled by varying CaEGTA/EGTA and can be determined by potentiometric titration of free and total EGTA concentrations by $CaCl_2$ and $CdSO_4$, respectively.[16]

The effect of 100 μM carbachol on intracellular Ca^{2+} in one T84 cell is illustrated in the top trace of Fig. 5A. The control Ca^{2+} concentration in this cell is 60 nM, which is essentially the same as the average of 58 $\pm$ 7 nM in 44 cells.[7] As can be seen, carbachol induces an initial rapid rise in intracellular Ca^{2+}, which averages 173 $\pm$ 16 nM in 44 cells, followed by a series of oscillations. These carbachol-induced oscillations are observed in 68% of the responding cells. The classic muscarinic antagonist atropine (10 μM) applied during a response causes the response to immediately terminate. This Ca^{2+} response pattern is very similar to that of the K^+ current seen in Fig. 2, suggesting that intracellular Ca^{2+} is the underlying factor that regulates membrane K^+ conductance during carbachol exposure.

The oscillatory Ca^{2+} pattern produced by carbachol exposure is quite variable from cell to cell. These patterns range from Ca^{2+} oscillations around a plateau value, to an initial spike of Ca^{2+} followed by a series of small oscillations as Ca^{2+} returns to baseline. While individual cells show a great deal of variability in their response, the response of a given cell is highly reproducible, as illustrated in Fig. 5. The bottom trace of Fig. 5A shows the response of the same cell as in the top trace, but during a second carbachol exposure 25 min after the first. This second response is very similar to the first, a pattern of reproducibility described as the "Ca^{2+} fingerprint" of the cell.[17]

Figure 4 shows that the K^+ conductance response to carbachol does not depend on bath Ca^{2+}. If the status of intracellular Ca^{2+} determines the state

[16] D. A. Williams, K. E. Fogarty, R. Y. Tsien, and F. S. Fay, *Nature (London)* **313**, 558 (1985).
[17] M. Prentki, M. J. Glennon, A. P. Thomas, R. L. Morris, F. M. Matschinski, and B. E. Corkey, *J. Biol. Chem.* **236**, 11044 (1988).

A   **B**

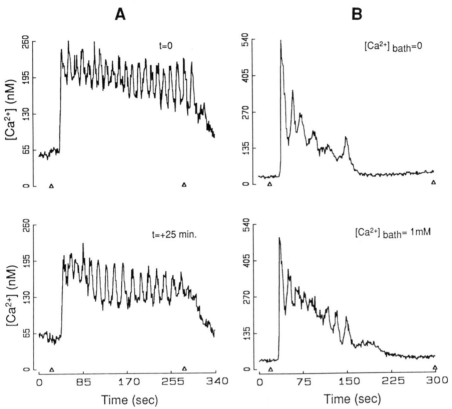

FIG. 5. Free intracellular Ca²⁺ concentration during carbachol exposure (100 μM; applied between △ symbols). The traces of (A) show responses from a cell at time = 0 (top) and 25 min later (bottom). The traces of (B) show responses from a cell at time = 0 in the absence of bath Ca²⁺ (top) and 25 min later in the presence of bath Ca²⁺ (bottom).

of this conductance then Ca²⁺ oscillations also may not depend on bath Ca²⁺. The role of bath Ca²⁺ can be determined by measurement of intracellular Ca²⁺ after removing Ca²⁺ from the bathing solution. The top trace of Fig. 5B shows that under these Ca²⁺-free conditions, 100 μM carbachol produces a rise in intracellular Ca²⁺ and oscillations that are quantitatively similar to those observed in the same cell 25 min after addition of 1 mM Ca²⁺ to the bathing solution (bottom trace of Fig. 5B). These results further support the hypothesis that release and/or uptake of Ca²⁺ from intracellular stores modulates the carbachol-induced changes in membrane K⁺ conductance.

Conclusions

Our experimental approach has characterized the increase and oscillations in membrane K^+ conductance that occur when T84 cells are exposed to the cholinergic agonist, carbachol. These findings in isolated T84 cells are consistent with the proposed model for muscarinic agonist-induced Cl^- secretion by an epithelium,[4] in which an agonist-induced increase in basolateral membrane K^+ conductance hyperpolarizes the cells and causes secretion by increasing the driving force for Cl^- exit across the apical membrane. In addition, we show that intracellular Ca^{2+} is the likely mediator of this process since a qualitatively similar pattern of Ca^{2+} oscillations is also observed. However, our measurements have not demonstrated a concurrence between oscillations of membrane K^+ current and intracellular Ca^{2+}. This issue would be best addressed using simultaneous whole cell patch-clamp and Fura-2 fluorescence techniques.

There are several advantages to the whole cell patch-clamp techniques used in these studies. First, changes in cell membrane conductance can be readily visualized (Fig. 3); these changes are not easily seen using conventional microelectrodes because of their low seal resistance. Second, intracellular solute composition can be precisely controlled by dialysis of the cell interior by the patch pipet solution. For example, this advantage was used to study the effect of Cs^+ on the membrane K^+ conductance, as described above. Furthermore, intracellular Ca^{2+} buffering capacity was altered to expose the role of intracellular Ca^{2+} in the current response to carbachol.

Care must be used in interpretation of results from dialyzed cells. This is illustrated in our studies by the difference seen between the duration of the carbachol-induced K^+ current oscillations and Ca^{2+} oscillations. In every cell studied, the K^+ current returned to baseline in the continued presence of carbachol (approximately 90 sec). On the other hand, Ca^{2+} oscillations usually continued for the entire period of carbachol exposure (300 sec or more). The reason for this difference is not known. It is unlikely that muscarinic receptor desensitization[18] is responsible for the short duration of the K^+ current response because such desensitization would also affect the Ca^{2+} response. This difference is most likely due to K^+ channel "rundown," in which cell dialysis during whole cell patch clamping results in the loss of cellular substrates necessary for normal K^+ channel activation.[19] A technique that could be used to avoid rundown is the "perforated

[18] R. G. Siman and W. Klein, *Brain Res.* **262**, 99 (1983).
[19] A. Marty, R. Horn, Y. P. Tan, and J. Zimmerberg, *in* "Secretion and Its Control" (G. S. Oxford and C. M. Armstrong, eds.), p. 97. Rockefeller Univ. Press, New York, 1989.

patch" method. In this method, cells are patched using a pipet which contains a nonselective pore-forming agent, like nystatin.[20] In this configuration, an electrical pathway is made through the cell membrane but minimal dialysis of the cellular compartment occurs.

The location of the membrane channels that underlie the carbachol-induced K$^+$ conductance remain to be determined since these sites cannot be identified using whole cell patch-clamp techniques and isolated cells. To answer this question, measurements of single K$^+$ channel activity at identified loci are preferable.[21] These measurements could be made on the basolateral membrane of inverted T84 monolayers or enzymatically isolated intestinal crypts.[22] Such measurements could provide information on channel gating characteristics (Ca^{2+} activated vs ligand gated) and additional electrophysiological properties (e.g., single-channel conductance, voltage dependence, and selectivity).

Using Fura-2 fluorescence ratios for the measurement of intracellular Ca^{2+} has several advantages since fluorescence intensity is dependent on illumination intensity, dye concentration, quantum efficiency, emission collection efficiency, and effective cell thickness in the optical beam. With the exception of quantum efficiency, all other factors are poorly quantified or are often variable. The ratio of fluorescences at two excitation wavelengths cancels out the effects of most, or all, of these variables. In addition, suitably chosen wavelengths can increase the signal-to-noise ratio, as in the case of Fura-2 at 340 and 380 nm. Our method provides a dynamic measurement of whole cell Ca^{2+} which is best used where temporal resolution is necessary, but provides little information on the spatial distribution of cytosolic Ca^{2+}. Digital imaging techniques are preferable for spatial resolution of cell Ca^{2+},[11,15] but at the expense of temporal resolution because of presently available technology.

The significance of intracellular Ca^{2+} oscillations in isolated cells to intact monolayers of T84 cells or intestinal crypts is not known. It has been suggested that a frequency-modulated signaling scheme could allow for great variation in signal strength while maintaining a low level of intracellular Ca^{2+}. A correlation between the frequency of Ca^{2+} oscillations and carbachol concentration has been observed in hepatocytes.[23] On the other

[20] R. Horn and A. Marty, *J. Gen. Physiol.* **92,** 145 (1988).

[21] M. Diener, W. Rummel, P. Mestres, and B. Lindemann, *J. Membr. Biol.* **108,** 21 (1989).

[22] D. D. F. Loo and J. D. Kaunitz, *J. Membr. Biol.* **110,** 19 (1989).

[23] N. W. Woods, K. S. R. Cuthbertson, and P. H. Cobbold, *Cell Calcium* **8,** 79 (1987).

hand, oscillations may be part of a scheme in which the cell responds to a Ca^{2+} trigger, followed by mechanisms designed to maintain a low intracellular Ca^{2+} concentration. Recent reviews of this topic are available.[24,25]

[24] M. J. Berridge, P. H. Cobbold, and K. S. R. Cuthbertson, *Philos. Trans. R. Soc. London, B* **320**, 325 (1988).
[25] M. J. Berridge and A. Galione, *FASEB J.* **2**, 3074 (1988).

[22] Isolation of Intestinal Epithelial Cells and Evaluation of Transport Functions

By GEORGE A. KIMMICH

Introduction

A great deal of solute "traffic" occurs across the intestinal epithelium in relation to the vital role intestinal tissue plays in nutrient absorption and in salt and water transfer. These solute fluxes involve transport across two separate plasma membrane boundaries and often involve a metabolic or chemical energy input at one of the two boundaries. The energy-dependent steps allow transfer against a gradient of chemical or electrochemical potential and provide the energetic basis for vectorial transfer of solutes from intestinal lumen to the circulatory system.

In addition to its usual absorptive function for salt and water, the intestine also has the capacity to undergo a startling functional transformation and become a secretory organ for salt and water. This capability can be physiologically useful in helping maintain fluidity of the intestinal contents to facilitate digestion in order to allow complete nutrient absorption and to improve peristaltic movement of the lumenal contents through the intestinal tract. In certain disease states, however, including cholera and various diarrheal diseases, intestinal secretion goes beyond the bounds of a regulated activity and can lead to serious clinical problems involving severe dehydration and death due to collapse of the peripheral vasculature.

No matter which functional mode is considered, in general, it is the function of the mucosal epithelial cells lining the intestinal villi and the intervillous crypt regions which are responsible for the transport capability of the tissue. A thorough understanding of intestinal transport events therefore implies an understanding of the transport properties of these cells and the regulation of cellular transport events.

Many experimental approaches aimed at resolving intestinal transport events have utilized a variety of intact tissue preparations (loops, sacs, sheets, slices, etc.). These preparations are morphologically complex (villi, crypts, lamina propria, connective and muscle tissue) and contain a variety of cell types which are unrelated in function to those transport events which are characteristic of the tissue. The use of isolated epithelial cells circumvents some of the complexity otherwise associated with interpretation of transport properties observed for the intact tissue preparations.

Isolation of Villous Epithelial Cells

General Considerations

It is easy to scrape the intestinal mucosal epithelium free from underlying submucosal elements using the edge of a glass slide, ruler, or similar object.[1] The epithelial gemisch produced by this mechanical procedure includes intact cells, cell clumps and considerable cell debris, lamina propria, and connective tissue fragments. Intuitively, one could conclude that this material might be an ideal starting point for subsequent isolation of a functional cell population by differential centrifugation. This ideal cannot be realized, however, due to excessive amounts of mucus released by the scraping procedure which creates a gelatinous mass that prevents easy separation of the smaller tissue elements.[2] Extensive proteolysis also occurs, presumably due to release of various proteolytic enzymes from the damaged cells and tissue. For these reasons, scraping or severe abrading of the tissue must be avoided in any procedure for which harvesting of functional epithelial cells is an ultimate aim.

On the other hand, a variety of methods can be utilized to release epithelial cells from intestinal tissue without involving a scraping process. These methods include vibration of an intact intestine everted over a metal rod, the end of which is inserted into a Vibra-Mix motor (A.G. Chemap, Zurich, Switzerland). By varying the amplitude and interval of vibration cells can be released from different loci on the intestinal villi.[3] Longer, more vigorous vibration intervals remove cells progressively further down the villus axis toward the crypt region. Crypt cells themselves are not released unless the intestinal wall is distended with air pressure.[4]

The vibration procedure is a variation of other approaches involving simple rotation of the everted intestine at varying speeds either with or without mechanical pressure applied to the exposed mucosal surface.[5]

[1] F. Dickens and H. Weil-Malherbe, *Biochem. J.* **35,** 7 (1941).
[2] J. W. Porteus and B. Clark, *Biochem. J.* **96,** 159 (1965).
[3] D. W. Harrison and H. L. Webster, *Exp. Cell Res.* **55,** 257 (1969).
[4] H. L. Webster and D. D. Harrison, *Exp. Cell Res.* **56,** 245 (1969).
[5] F. S. Sjostrand, *J. Ultrastruct. Res.* **22,** 424 (1968).

Additional variations include the use of various chelating agents (EDTA, citrate)[6] or proteolytic enzymes (trypsin, collagenase)[7] in the medium bathing the tissue while applying mechanical pressure with the fingers.

All of the above procedures produce isolated cells and cell aggregates in differing yields. However, evaluation of metabolic and/or transport properties of the cell populations shows that their functional capability is not maintained for significant intervals.[6,8,9] In our own experience, use of chelating agents either alone or in combination with proteolytic enzymes produces a cell population with particularly unstable characteristics. It is not unusual to observe extensive autolysis and loss of functional capability within 20 min when either chelators or proteases have been employed during the preparation. The following procedure for preparation of avian intestinal cells represents an alternative in which the cells exhibit stable transport and metabolism for at least 2 hr following isolation.[10,11] It is important to note that the same procedure when applied to rat intestinal tissue does not produce a useful cell population. Cell yields are low and the functional properties of the isolated cells are very poor. This represents the collective experience of our own laboratory as well as the oral communications from numerous investigators throughout the world. Although we have not evaluated application of the cell isolation procedure to other commonly used laboratory animals extensively, it does appear that cells with acceptable properties can be released from rabbit tissue (G. A. Kimmich, unpublished results).

Isolation of Chicken Intestinal Cells

Sacrifice a 4- to 8-week-old chicken by rapid decapitation. Open the abdominal cavity immediately and rapidly excise an intestinal segment beginning at the gizzard and extending for approximately four lengths of the pancreas. Pancreatic tissue occupies the space between the first duodenal loop. The pancreas is a convenient yardstick which varies in length with the age of the chicken in proportion to the length of the developing intestine. For a 6-week-old chicken it will be approximately 2 in. in length so that the intestinal segment removed will approximate 8 in.

[6] B. K. Stern, *Gastroenterology* **51,** 855 (1966).
[7] B. K. Stern and R. W. Reilly, *Nature (London)* **205,** 563 (1965).
[8] B. K. Stern and W. E. Jensen, *Nature (London)* **209,** 789 (1966).
[9] W. G. J. Iemhoff, J. W. O. Van Den Berg, A. M. DePyper, and W. C. Hulsmann, *Biochim. Biophys. Acta* **215,** 229 (1970).
[10] G. A. Kimmich, *Biochemistry* **9,** 3659 (1970).
[11] G. A. Kimmich, *in* "Methods in Membrane Biology" (E. D. Korn, ed.), Vol. 5, p. 51. Plenum, New York, 1975.

Rinse out the intestinal contents by flushing 50 ml of 0.9% ice-cold saline through the segment with the aid of a syringe. Slit the intestine lengthwise to make a flat sheet of tissue, and cut the sheet transversely into several segments about 2 in. in length.

Incubate the short tissue segments for 30 min at 37° in 10 ml of isolation medium with gentle oscillation (approximately 100 cpm).

Isolation medium: 125 mM NaCl, 20 mM HEPES-Tris (pH 7.4), 10 mM mannose, 2.5 mM glutamine, 0.5 mM β-hydroxybutyrate, 3.0 mM K_2HPO_4, 1.0 mM $MgCl_2$, 1.0 mM $CaCl_2$, 1 mg/ml bovine serum albumin (BSA), 1 mg/ml hyaluronidase (Sigma, type I)

Glutamine and β-hydroxybutyrate are included in the medium because of work with perfused tissue which indicates that intact intestine metabolizes these two compounds preferentially over other serum constituents.[12] Mannose is rapidly glycolyzed by isolated intestinal cells but is not transported by the Na^+-dependent sugar transport system. Together these three fuel molecules allow the isolated cells to establish and maintain more stable ion and sugar gradients than in their absence. After incubation, some cells and cell aggregates may have detached from the tissue. Much more extensive detachment can be induced by mixing the suspension of segments with the tip of a plastic pipet. Sometimes it is necessary to "pin" a segment down on the floor of the incubation beaker with the pipet tip and "sweep" the segment across the floor of the beaker a few times. By this procedure large numbers of cells can be released, as indicated by an increasing degree of opacity of the isolation medium.

At this point, pour the contents of the isolation beaker through a section of nylon stocking material in order to filter off the intestinal segments and to remove mucus which will cling to the mesh. Collect the filtrate containing the cell population in a clean plastic beaker. It is important that all procedures be carried out in plasticware in order to avoid fragmentation of the cell population, which can occur when cells are exposed to the microscopically jagged surfaces associated with glass surfaces.

Hyaluronidase is removed from the suspension medium by centrifugation of the cells at low speed (100–150 g) and resuspension in isolation medium without hyaluronidase. The cell pellet is very loosely packed by this procedure such that complete removal would require several wash steps. However, one or two washes produces a population with acceptable functional characteristics. Because the cell pellet aggregates readily due to residual mucus it is necessary to resuspend the pellets by gently pulling the

---

[12] H. G. Windmueller and A. E. Spaeth, *J. Biol. Chem.* **253**, 69 (1977).

cells and medium in and out of a plastic pipet a few times until a visually homogeneous suspension is obtained.

If the suspension is examined microscopically at this point, it will be observed to consist of a few isolated cells and cell doublets, but primarily of cell clumps containing a few to 10 or 20 cells each. The individual cells will appear rounded morphologically, as should be expected once the restraint represented by neighbor cells has been released. Careful observation will reveal the brush border "mustache" over a portion of the cell membrane representing the original lumenal surface. Despite the lack of fully isolated individual cells, the resuspension can be sampled readily and reproducibly with a narrow-bore micropipet. Replicate samples, carefully taken, will contain uniform amounts of total protein.

Each chick intestine will yield 1 to 2 ml of loosely packed cells or about 50 mg of cell protein. Total cell protein can be determined spectrophotometrically by incubating an aliquot of the suspension with Biuret reagent in comparison to BSA standards.[13] For most purposes we usually prepare cell suspensions containing 15–20 mg protein/ml and store the suspensions on ice until they are used experimentally. For best results, the cells should be used within 1 hr of the time of preparation. Because the cells settle and aggregate rapidly, it is always necessary to swirl the suspension vigorously a few times in order to ensure uniform sampling of the stock suspension.

It is not advisable to try and achieve further separation of the cell population. This can be accomplished by vigorous stirring or refluxing in and out of a pipet, but only at the expense of the total number of intact cells and concomitant loss of functional properties. Because of the nature of the cell suspension it is difficult to count cells or to determine vital dye staining accurately so that it is better to rely on protein or DNA determinations for standardizing cell numbers. Stability of metabolic and functional capabilities (e.g., ion or organic solute transport) proves to be a more reliable index of viability than do staining techniques.

Evaluation of Transport

A variety of metabolic parameters can be evaluated readily for the cell population[10]: (1) glucose utilization, (2) lactate production, (3) oxygen consumption, (4) CO_2 production from various substrates, and (5) adenine nucleotide ratios or energy charge. The magnitude and stability of each of these parameters individually and collectively convey information relating to biochemical function of the isolated cells. However, they do not convey

---

[13] A. G. Gornall, C. S. Bardawill, and M. M. David, *J. Biol. Chem.* **177**, 751 (1949).

much information regarding the integrity of the plasma membrane for the cell population. Indeed, cell-free extracts of many tissues can catalyze similar metabolic activities. For this reason, it seems likely that the magnitude and stability of solute gradients maintained by the cell population convey more meaningful information about the quality of the cell preparation. For energy-dependent transport events which create transmembrane concentration gradients of solutes, assessment of gradient magnitude and stability provides experimental insight not only to the integrity of the cellular plasma membrane, but also to the set of metabolic events responsible for establishing the energetic forces driving the transport systems and the integration between metabolic and transport events.

Accumulation of α-Methylglucoside (α-MG)

Because of a marked tendency for the cell population to clump and form large aggregates when incubated in a quiescent environment, it is necessary to perform experimental work under conditions in which the suspension is continually shaken at an adequate rate. This provides sufficient homogeneity for reproducible sampling as well as appropriate oxygenation of the medium to sustain the rather high aerobic metabolism of the epithelial cells. In our experience, the surface-to-volume ratio of the incubation vessel exhibits a rather narrow optimum. This relates in part to the necessity for avoiding too great a depth of incubation medium such that the shaking action of the incubator bath can be low enough to allow convenient sampling without being so slow that cells congregate in the lower levels of the suspension, leaving a lower cell density near the surface. Oscillating shaker baths are better than rotary because the latter tend to create higher cell densities in the periphery of the incubation vessel. We have found that a 4-ml total incubation volume in a 50-ml beaker oscillated at 100 cpm provides an ideal set of conditions when the cell density is between 3 to 5 mg protein/ml. In order to conserve isotope and limit the number of cells required we have also used a 2.2- to 2.5-ml volume in 30-ml beakers.

An appropriate format for assessing transport capability for several kinds of solutes can be illustrated by evaluating the uptake of α-[^{14}C]methylglucoside as follows:

Incubation Vessel Preparation. Prepare a set of 30-ml plastic beakers such that each beaker contains 2.2 ml of the same medium used for cell isolation, but without hyaluronidase. In addition, each beaker should contain 100 μM α-methylglucoside and 0.1–0.2 μCi/ml of α-[^{14}C]MG. One of these beakers will have no further additions and will be used to assess α-MG uptake under control conditions. Another will contain 200 μM

phlorizin, which is a potent inhibitor of Na^+-dependent sugar transport in these cells. Other beakers will contain whatever agents one might wish to test in terms of possible action on the sugar transport capability of the cells. For instance, one might wish to test several concentrations of ouabain because of its inhibitory action on Na^+ transport in order to assess the effect of dissipating the $\Delta\mu_{Na^+}$ on the capability of the cells for establishing sugar gradients. In this case, the set of incubation beakers might include the following:

> All beakers: 2.2 ml incubation medium + 0.25 μCi α-[^{14}C]MG ($\sim$0.1 μCi/ml)
> Beaker 1: No addition
> Beaker 2: + 200 μM phlorizin
> Beaker 3: +0.5 μM ouabain
> Beaker 4: +1.0 μM ouabain
> Beaker 5: +2.0 μM ouabain
> Beaker 6: +5.0 μM ouabain

The beaker set should be incubated at 37° with gentle shaking (100 cpm) in order to thermally equilibrate the suspension medium. A stock cell suspension ($\sim$20 mg cell protein/ml) should be thermally equilibrated at the same time.

Cellular sugar uptake is initiated by transferring a 0.1-ml sample of cell suspension to each of the experimental beakers at time zero. At intervals after the cell transfer, 100-μl samples of the incubation medium are taken with the aid of an automatic micropipet. It is important that the shaker continue operating during the sampling procedure in order to achieve uniform samples of the cell suspension. Even a short interruption of the shaking motion can lead to cell aggregation and varying amounts of cell protein from sample to sample with resultant data scatter. In order to study sugar uptake during formation of a steady state sugar gradient, an appropriate sampling schedule might be 1, 5, 10, 20, 40, and 60 min. For the suggested experimental format (surface-to-volume ratio) it is difficult to take more than six samples per incubation beaker before the medium becomes too shallow for easy sampling.

Each 100-μl sample should be diluted into 4 ml of ice-cold incubation medium (without radioactivity) *immediately* after being taken. It is convenient in terms of later sample processing to have the diluent for each experimental aliquot in 5-ml polyethylene scintillation minivials. The diluted samples should be centrifuged as soon as possible using a rapidly accelerating tabletop centrifuge. If the six incubations are run in parallel, each set of six samples for a given time point can be processed while waiting to take the sample set for the next time point. After centrifugation for 30 sec the supernatant is removed by aspiration and 4.0 ml of fresh ice-cold wash medium is added to each tube. After a second centrifugation and supernatant aspiration the cell pellets are ready for extraction and measurement of the amount of isotope accumulated. Using an Autocrit (with modified head) (Clay Adams, Parsippany, NJ) or an MHCT II centrifuge, both wash steps can be completed in less than 3 min. The carryover of extracellular radioactivity following the two washes can be evaluated by incubating the cells with impermeant [^{14}C]polyethylene glycol. For the recommended amount of cellular protein this carryover is so small (~0.02 μl or 4 CPM at 0.1 μCi/ml) that it can be ignored when calculating cellular uptake of sugar.

Chilling of the diluted cell aliquots is sufficient to prevent subsequent sugar fluxes without the necessity of including any transport inhibitors in the wash medium. This can be evaluated by maintaining diluted samples taken under comparable conditions on ice for varying intervals of time prior to centrifugation and comparing the cellular content of isotope as a function of storage time. For α-MG only 5% of the accumulated sugar is lost to the wash medium even after 20 min of sample storage time. Because sample processing is typically complete in 2–3 min, no correction is required. This should be evaluated for each solute whose transport is examined, however, in order to ascertain if corrections due to solute loss during sample processing are necessary.

Once the washed cell pellets are prepared they can be extracted for scintillation counting simply by dissolving each pellet in a 5-ml volume of Liquiscint (National Diagnostics, Manville, NJ). If polyethylene minivials are utilized for processing the cell samples, it is only necessary to add the Liquiscint directly to the minivial for a few minutes prior to counting each vial. For the conditions described, the cell pellets do not introduce any quenching or chemiluminescent properties, so no corrections for those events are necessary.

Sample Calculation. If a 10-μl sample is taken from each incubation beaker *prior* to adding cells, the count rate determined for each one (CPM) can provide a measure of either the volume specific activity or the picomolar specific activity for the α-[^{14}C]MG in each experimental run.

Volume specific activity (VSA) = (CPM/10) × D, where D is the dilution factor, due to adding cell suspension without isotope to the incubation medium (0.96 for the conditions described earlier). The units for VSA are simply CPM/μl for the final incubation suspension.

VSA is useful because it can provide a quick assessment of intracellular volume/mg of cell protein. For instance, in the experimental case which includes phlorizin, concentrative accumulation of α-MG is inhibited so that the final steady state uptake of sugar simply represents the point at which α-MG has equilibrated between the extracellular water and that part of the cellular water to which sugar has access. If the *steady state CPM in the presence of phlorizin* is divided by VSA and the result is divided by milligrams cell protein in each experimental aliquot one obtains the microliters of cell water per milligram cellular protein for the cell population. We routinely find a value of between 2.5 and 3 μl/mg cell protein by this procedure. By obtaining a value for the number of cells in each sample this can be converted to a volume per cell basis. However, because of cell aggregation in the suspension, mentioned earlier, it is difficult to determine an accurate value for cell number and we routinely express transport data in terms of uptake per milligram cell protein.

Picomolar specific activity (PSA) = VSA/C, where C is the final concentration of α-MG in the experimental beakers (in pmol/μl). The units of PSA are CPM per picomole. If CPM for any experimental sample is divided by PSA the result gives a value for the *amount* (in picomoles) of α-MG accumulated by the cells in that sample over the time interval allowed before sampling. These values can be useful in determining the rate of uptake of a particular solute under different experimental conditions. They are best employed in short-term unidirectional influx experiments in which uptake is linear with time so that initial transport rates can be compared. Such initial rates can provide insight to the magnitude of the driving forces acting on the transport system and allow kinetic analysis of the function of the system for various experimental conditions.

Solute Gradients. Assuming that phlorizin-inhibited cells only allow equilibration of sugar between extracellular and intracellular water the *steady state* counts observed with phlorizin (CPM$_{Pz}$) can also provide an approximation for gradient-forming capacity for any of the other conditions evaluated. It is only necessary to divide CPM for a particular sample by CPM$_{Pz}$ to calculate the steady state sugar distribution ratio achieved in that sample. By this procedure we find that chick intestinal cells can routinely establish an *apparent* steady state sugar gradient of about 70- to 90-fold under control conditions when extracellular α-MG is 100 μM. Approximately 30 min is required to achieve the steady state. Stability of the sugar gradient during subsequent incubation provides a measure of the stability of those driving forces acting to create the sugar gradient (Na$^+$ gradient and membrane potential).

Sugar gradients calculated by the method described above are "apparent" because the procedure assumes that the extracellular sugar concentration has not changed during the experimental interval. For the conditions described where the cellular volume represents about 0.25% of the total incubation volume (2.0 mg cell protein $\times$ 3 μl cell water/mg protein in a 2.3-ml volume), a 70-fold accumulation of solute relative to the *initial* extracellular concentration will decrease the extracellular concentration by

TABLE I

SUGAR ACCUMULATION RATIOS FOR INTESTINAL
EPITHELIAL CELLS INCUBATED AT DIFFERENT
OUABAIN CONCENTRATIONS

Ouabain concentration (μM)	Sugar accumulation ratio[a]
0	81
0.5	56
1.0	36
2.0	17
5.0	3
0 + 200 μM Phlorizin	1

[a] Defined after a 40-min incubation as the ratio of α-[^{14}C]MG counts in an aliquot of cells for the indicated ouabain concentration to that observed in the presence of 200 μM phlorizin.

approximately 15–20%. This depletion of medium solute can be detected by counting the isotope present in an aliquot of the supernatant above the cell pellet following the first centrifugation of the diluted cell samples. When the decrease in extracellular concentration is taken into account the actual accumulation ratio for α-MG established by the isolated chick cells typically is about 70/(1 − 0.15) or ~80-fold.[14] Table I shows the actual accumulation ratios observed for α-MG in the experiment described above for different ouabain concentrations.

Unidirectional Influx. As mentioned earlier, short-term experiments similar to those described above can be used to determine the *rate* of accumulation of solute by the cell suspension.[15] In this situation it is usually necessary to take aliquots of the suspension about every 10–15 sec for a total of 40–60 sec, by which time appreciable backflux begins to occur such that linearity of influx is lost. The data provide information regarding the capacity of the transport system for the particular conditions employed. For any given solute, entry usually occurs by more than one route so that a procedure is needed to identify that part of the total flux which occurs by the route of interest. In the case of α-MG it is convenient to use phlorizin sensitivity as the means of identifying influx via the Na$^+$-dependent sugar carrier. Table II shows the influx of α-MG at an extracellular concentration of 100 μM and at four different phlorizin concentrations in comparison to influx observed without phlorizin or without

[14] G. A. Kimmich and J. Randles, *Am. J. Physiol.* **237**, C56 (1978).
[15] J. Randles and G. A. Kimmich, *Am. J. Physiol.* **234**, C64 (1978).

TABLE II
UNIDIRECTIONAL INFLUX OF 100 μM α-[^{14}C]MG
INTO ISOLATED CHICKEN INTESTINAL CELLS
(EFFECT OF PHLORIZIN AND LACK OF Na$^+$)

Incubation conditions	Unidirectional influx[a] of μ-MG (pmol/min/mg protein)
Control	424
+0.5 μM Phlorizin	358
+1.0 μM Phlorizin	299
+10.0 μM Phlorizin	169
+200 μM Phlorizin	25
No Na^{+b}	22

[a] The influx was determined from five samples taken over a 36-sec interval in each case. Linear uptake is maintained for this interval. Influx = $CPM_t/(PSA \cdot t \cdot mg$ protein) where t = time of sampling in minutes.
[b] NaCl in the incubation medium was replaced by 125 mM tetramethylammonium chloride.

Na$^+$. Note that about 95% of the total α-MG influx is blocked by phlorizin or by lack of sodium.

Sodium Ion-Independent Sugar Transport

Intestinal epithelial cells accumulate various sugars via a concentrative Na$^+$-dependent sugar transport system localized in the brush border boundary of the cell. Accumulated sugar escapes from the cell to the circulatory systems via a facilitated diffusion transport system localized in the basolateral boundary. α-MG satisfies only the Na$^+$-dependent transport system so that it represents the sugar of choice for evaluating function of this system.[16] On the other hand, 2-deoxyglucose (2-DOG) selectively satisfies the Na$^+$-independent carrier so that unidirectional influx of 2-DOG represents a means of studying basolateral transfer of sugar.[17] Because 2-DOG can be phosphorylated by hexokinase it is not possible to use longer term steady state experiments with 2-DOG in order to determine equilibration time or cellular volumes.

Certain nonmetabolized sugars such as 3-O-methylglucose (3-OMG) are transported by both of the intestinal sugar transport systems. At the

[16] G. A. Kimmich and J. Randles, *Am. J. Physiol.* **241,** C227 (1981).
[17] G. A. Kimmich and J. Randles, *J. Membr. Biol.* **27,** 353 (1976).

TABLE III

EFFECT OF VARIOUS INHIBITORS OF INTESTINAL SEROSAL SUGAR TRANSFER (Na$^+$ INDEPENDENT) ON ACCUMULATION OF 3-[^{14}C]OMG BY ISOLATED INTESTINAL CELLS (Na$^+$ DEPENDENT)

Agent[a]	Concentration (μM)	Steady state 3-OMG accumulation ratio[b]
—	—	10
Dihydroquercitin	100	15
Theophylline	5	25
Hesperetin	100	28
Naringenin	100	32
Phloretin	100	35
Apigenin	100	43
Kaempferol	100	48
Cytochalasin B	100	65

[a] Dihydroquercitin, hesperetin, and naringenin are flavanones. Apigenin and kaempferol are flavones. Phloretin is an analog of the flavanones in which a heterocyclic ring has been opened.

[b] For 100 μM 3-[^{14}C]OMG.

steady state, cells transporting 3-OMG accumulate the sugar against a concentration gradient at the brush border and lose it across the basolateral boundary via Na$^+$-independent transfer. The magnitude of the steady state 3-OMG gradient maintained reflects the function of both transport events.[14] Agents which selectively interfere with the passive facilitated diffusional transfer system will allow the cells to establish better concentration gradients of 3-OMG than control cells can maintain. This can be demonstrated using the same techniques already described, but with 3-[^{14}C]OMG as the test sugar. Phloretin,[15] theophylline,[15] cytochalasin B,[14] and various flavones or flavanones[18] can all be used to interfere with passive sugar efflux as shown in Table III. The degree of gradient enhancement is proportional to the degree of inhibition of the Na$^+$-independent transport system for agents which act solely on the facilitated diffusion carrier.

ATP-Depleted Epithelial Cells

Isolated intestinal cells can be depleted of more than 90% of their ATP content by a brief incubation with 20 to 80 μM rotenone.[19] Because the ATP pool in these cells turns over several times per minute, most of the

[18] G. A. Kimmich and J. Randles, *Membr. Biochem.* **1**, 221 (1978).
[19] C. Carter-Su and G. A. Kimmich, *Am. J. Physiol.* **237**, C57 (1979).

depletion occurs in the first minute after rotenone addition. If ouabain is then added to prevent possible slow turnover of the Na^+,K^+-ATPase due to the residual ATP, it is possible to dissipate the $\Delta\mu_{Na^+}$ which is ordinarily established due to monovalent ion transport.

The ATP-depleted cells can be utilized experimentally as if they are giant membrane vesicles in order to demonstrate the role for various driving forces acting to energize concentrative transport of sugar or other solutes. They offer significant advantages over conventional plasma membrane vesicles because a more favorable surface-to-volume ratio allows much longer intervals over which influx remains linear with consequent greater ease of flux measurement and better reliability of the experimental values determined. Gradients of chemical potential or electrical potential or both can be experimentally imposed across the plasma membrane of the ATP-depleted cells and the role of each can be determined in terms related to energization of the Na^+-dependent transport system.[19,20] Unidirectional influx for the solute of interest as well as transient gradient forming capability due to the imposed driving force can each be used for providing insight to the transport mechanism.

Imposed Na^+ Gradients of Defined Magnitude $(\Delta\mu_{Na^+})$

In order to selectively study effects of a change in Na^+ gradient, it is necessary to prevent changes in diffusion potentials $(\Delta\psi)$ due to diffusional flux of the imposed Na^+ gradient.[21] This can be accomplished best by the inclusion of high concentrations of highly permeant ions during the ATP depletion procedure and subsequent incubation with Na^+ in order to "short circuit" the system and "clamp" the potential near zero. For this purpose, we have found that replacing all or part of the NaCl during ATP depletion with tetramethylammonium (TMA$^+$) nitrate is effective. Cells which have been equilibrated with tetramethylammonium (sodium) nitrate can be transferred to an incubation medium with 125 mM sodium nitrate in order to create the experimental Na^+ gradient. Sometimes in order to ensure that the potential remains constant it is necessary to include potassium nitrate plus valinomycin in both the preincubation and the incubation medium in place of part of the tetramethylammonium nitrate. The high permeability of both NO_3^- and K^+ (with valinomycin) relative to Na^+ provides an especially stable $\Delta\psi$ when Na^+ is added back to the system.[22]

[20] C. Carter-Su and G. A. Kimmich, *Am. J. Physiol.* **238**, C73 (1980).
[21] G. A. Kimmich and J. Randles, *Biochim. Biophys. Acta* **596**, 439 (1980).
[22] G. A. Kimmich, J. Randles, D. Restrepo, and M. Montrose, *Am. J. Physiol.* **248**, C399 (1985).

Imposed Membrane Potentials of Defined Magnitude ($\Delta\psi$)

If cells are ATP depleted in a medium containing 50 mM potassium gluconate with valinomycin and 100 mM sodium gluconate and then introduced to an incubation medium with 100 mM NaNO$_3$ and 50 mM tetramethylammonium nitrate, it is possible to create a diffusion potential due to the imposed gradients of permeant ions but with no imposed Na$^+$ gradient. By varying the amount of extracellular NO$_3^-$ (replaced with slowly permeant gluconate) or TMA$^+$ (replaced with K$^+$), it is possible to manipulate the $\Delta\psi$ over a range of more than 60 mV and to determine the effect on a given Na$^+$-dependent solute transport system.[23,24]

Imposed Electrochemical Gradients of Defined Magnitude ($\Delta\tilde{\mu}_{Na^+}$)

By using a combination of the above conditions, gradients of both $\Delta\psi$ and $\Delta\mu_{Na^+}$ can be created. A useful approach is to ATP deplete the cells in potassium gluconate and to transfer them to an incubation medium with NaNO$_3$. A range of $\Delta\tilde{\mu}_{Na^+}$ for different cell populations is constructed by varying the amounts of Na$^+$, K$^+$, and/or NO$_3^-$ in the two working media. Because the ATP-depleted cells cannot establish either element of a $\Delta\tilde{\mu}_{Na^+}$ metabolically, the imposed gradients will dissipate due to diffusional ion fluxes. Nevertheless, the degree to which an imposed gradient is effective in energizing a solute transport system can be assessed by examining the magnitude of "overshoot" or transient gradient-forming capacity the cells exhibit for the solute. Table IV shows the peak accumulation ratio observed for 100 μM α-MG when the various gradients described above were imposed on isolated intestinal cells. In each case, the accumulation ratio was defined as the CPM for the peak sample divided by CPM for the sample at eventual steady state (i.e., fully dissipated gradients).

Measurement of Intracellular Na$^+$ and K$^+$

The procedure described for monitoring α-MG accumulation can be modified slightly in order to provide values for intracellular Na$^+$ and K$^+$ content for different experimental conditions. In this case, instead of dissolving the cell pellets in Liquiscint for scintillation counting, they are extracted in 3% PCA and an aliquot of the supernatant, after centrifugation of the denatured cellular protein, is used for flame photometry. The

[23] G. A. Kimmich, J. Randles, D. Restrepo, and M. Montrose, *Ann. N.Y. Acad. Sci.* **456**, 63 (1985).
[24] D. Restrepo and G. A. Kimmich, *J. Membr. Biol.* **87**, 159 (1985).

TABLE IV
SUGAR TRANSPORT IN ATP-DEPLETED
INTESTINAL CELLS DRIVEN BY AN
EXPERIMENTALLY IMPOSED Na$^+$ GRADIENT,
MEMBRANE POTENTIAL, OR ELECTROCHEMICAL
POTENTIAL GRADIENT FOR Na$^+$

Driving force	Peak sugar gradient
$\Delta\mu_{Na^+}$	11.0
$\Delta\psi$	7.4
$\Delta\tilde{\mu}_{Na^+}$	19.6
$\Delta\tilde{\mu}_{Na^+}$ + phlorizin	1.0

amount of monovalent ions determined in this manner can be expressed as an intracellular concentration by utilizing a value for cell volume and assuming that each ion is uniformly distributed in the cell water. Concentrations calculated in this manner are not highly reliable, however, because of ion binding to cellular anionic constituents and possible nonuniform distribution among organelles. Both phenomena create an intracellular ion activity which is considerably lower than values determined on the basis of ion amount per volume measurements. This is particularly true for Na$^+$, where the activity measured with ion-selective electrodes is usually about half the calculated concentration.[25] Nevertheless, determination of the amount of intracellular ion under different experimental conditions can give qualitative information regarding changes in the magnitude of ion gradients maintained by the cells.

Measurement of Membrane Potentials

The unidirectional influx or rate of uptake of a lipophilic cation by any cell population should be proportional to the magnitude of the membrane potential maintained by those cells. If the influx is studied as the membrane potential is varied experimentally, one can construct a calibration curve for the relationship between flux and $\Delta\psi$. This calibration curve can be used to determine the membrane potential for situations in which it has not been experimentally created.[22,23]

A useful lipophilic ion to be used for this purpose is [^{14}C]tetraphenyl-phosphonium (TPP$^+$). Gradients of potassium gluconate imposed on ATP-depleted cells can be used to create potentials of defined magnitude. Valinomycin is used to enhance K$^+$ permeability and this, in conjunction

[25] J. O'Doherty, J. F. Garcia-Daz, and W. McD. Armstrong, *Science* **203**, 1349 (1979).

with the low permeability of gluconate, allows formation of a K^+ equilibrium potential, the magnitude of which can be calculated from the Nernst equation[22]:

$$\Delta\psi = (RT/F) \ln ([K^+]_o/[K^+]_i)$$

in which T is temperature, F is the Faraday constant, and R is the ideal gas law constant. A useful procedure is to ATP deplete the cells in a medium containing 150 mM potassium gluconate with 20 mM TMA-HEPES buffer and 1 mM CaSO$_4$. No other salt constituents are necessary. In order to initiate the influx measurement the cells are diluted 1 : 10 into 150 mM tetramethylammonium gluconate, 20 mM HEPES, and 1 mM CaSO$_4$ or into media in which various mixtures of tetramethylammonium gluconate and potassium gluconate are used in order to vary the extracellular K^+ concentration. Valinomycin and 1 μM [^{14}C]TPP$^+$ are included in the incubation medium and four to five samples are taken during the first minute of incubation in a manner similar to that already described for monitoring unidirectional influx of any solute. All influx values are expressed *relative* to the value determined when no K^+ gradient is imposed which is arbitrarily assigned a value of 1.0. This is defined as the flux at a diffusion potential of 0 mV. Relative fluxes for several different K^+ gradients (i.e., potentials) are given in Table V.

The use of relative fluxes instead of absolute values avoids the necessity of having to determine an independent value for the permeability constant for TPP$^+$. For intestinal epithelial cells, the TPP$^+$ flux vs $\Delta\psi$ relationship is the same as that predicted by the Goldman flux equation.[23,24] This calibration curve can then be used to determine an unknown potential. Only two flux measurements must be made. In one, the TPP$^+$ influx is determined

TABLE V
[^{14}C]TPP$^+$ INFLUX INTO INTESTINAL CELLS WITH
DIFFERENT POTASSIUM GLUCONATE GRADIENTS

$[K^+]_i/[K^+]_o$	$\Delta\psi$ (mV)	Relative TPP$^+$ influx[a]
1	0	1.0
2	−18	1.4
5	−42	2.0
10	−60	2.6

[a]All flux values are given as a multiple of the flux observed when no K^+ gradient ($\Delta\psi$) was imposed which was assigned an arbitrary value of 1.0.

TABLE VI
EFFECT OF Na⁺-DEPENDENT SUGAR TRANSPORT ON [¹⁴C]TPP⁺ INFLUX IN ISOLATED
INTESTINAL CELLS

Incubation conditions	Relative TPP⁺ influx	$\Delta\psi$ (mV)
Control	2.2	−55
+ 10 mM α-MG	1.9	−45
+ Rotenone and ouabain	1.0	0

for the particular experimental conditions for which you wish to determine a $\Delta\psi$. This influx is expressed *relative* to (i.e., as a multiple or fraction of) the influx measured in cells which are ATP depleted and incubated sufficiently long to dissipate ion gradients. This case is defined as $\Delta\psi = 0$ just as in the case described above in which the calibration relationship was defined relative to a situation with no imposed K⁺ gradient. It is now only necessary to pick the $\Delta\psi$ off the calibration curve for the particular flux ratio determined by the two flux measurements ($J_{\psi=x}/J_{\psi=0}$). Table VI shows the TPP⁺ influx ratio and corresponding $\Delta\psi$ determined for intestinal cells incubated under standard conditions and with 10 mM α-MG added. Actively accumulated sugars are known to partially depolarize the membrane potential due to electrogenic function of the Na⁺-dependent sugar transport system. Note that the added sugar causes about a 10-mV depolarization of the membrane potential measured in this manner, just as reported in experiments in which $\Delta\psi$ was directly monitored with microelectrodes.[26]

Summary

Epithelial cells can be isolated from the small intestine of chickens by a procedure involving hyaluronidase treatment of the intact tissue. The isolated cells retain a high degree of functional activity as assessed by the formation of 70-fold gradients of α-MG. Stability of the sugar gradients reflects maintenance of stable electrochemical Na⁺ gradients across the plasma membrane. The cells can be used to evaluate the properties of Na⁺-dependent sugar transport, Na⁺-independent sugar transport, ion transport, metabolism, membrane potentials, and the integration of these events, all of which are important to achieving a stable sugar gradient.

[26] R. C. Rose and S. G. Schultz, *J. Gen. Physiol.* **57,** 639 (1971).

[23] Isolation of Enterocyte Membranes

By AUSTIN K. MIRCHEFF and EMILE J. J. M. VAN CORVEN

Isolated brush border and basolateral membrane vesicles have been used in numerous studies of the individual transport mechanisms which underlie the small intestine's ability to absorb nutrients and electrolytes. These studies were usually conceived in preparative terms. That is, one began with the premise that it was possible to obtain a subcellular fraction which represented a sample of the membrane population of interest, then proceeded to use this fraction to verify the presence of and to characterize specific transport processes. This general approach seemed well justified, since preceding work with intact preparations had generated a number of more or less detailed predictions. Thus, Hopfer's original study of glucose transport in isolated brush border membrane vesicles[1] verified the major predictions of the sodium gradient hypothesis for glucose absorption.

In the years intervening since Hopfer and co-workers first used isolated brush border membrane preparations to delineate absorptive mechanisms, it has become clear that a strictly preparative approach is subject to the potential limitation that the transport system being characterized might, at least in part, be associated with contaminants rather than with the plasma membranes of primary interest to the goals of the study. Two different factors make it especially imperative that one consider this possibility in designing studies with intestinal basolateral membrane fractions. First, the basolateral membranes are relatively inaccessable to direct transport measurements. This means that one knows relatively little, *a priori*, about the transport mechanisms that will be found in these membranes. It also means that it will be difficult to validate in the intact preparations the conclusions drawn from vesicle studies. Second, the basolateral membrane vesicle population has been shown repeatedly to overlap endoplasmic reticulum- and Golgi-derived vesicle populations with respect to the physical properties which are most commonly detected by membrane separation procedures. Two examples are presented below to illustrate the uncertainty which can result from the possible presence of cytoplasmic membrane contaminants in basolateral membrane fractions.

The presence of an ATP-driven Ca^{2+} pump in intestinal basolateral membranes is predicted by analogy to the Ussing model for Na^+ absorption. ATP-dependent Ca^{2+} transport activity has been detected in basola-

---

[1] U. Hopfer, K. Nelson, J. Perrotto, and K. J. Isselbacher, *J. Biol. Chem.* **248**, 25 (1973).

teral membrane-containing fractions[2-4] and, in consonance with the model, pump specific activity in the isolated vesicle preparation increases when vitamin D is administered to vitamin D-deficient animals. However, it is already well known that ATP-driven Ca^{2+} pumps are primarily localized to cytoplasmic membranes in a number of different cell types.[5] It was not known how the pumps might be partitioned between the basolateral membranes and the cytoplasmic membranes which are present in typical intestinal basolateral membrane fractions. Recent studies, based on the separation procedures outline below, lead to the conclusion that most of the recoverable Ca^{2+} transport activity in the basolateral membrane fractions, as in the intestinal cell as a whole, is associated with various populations of cytoplasmic membranes.[6] It will now be necessary to determine whether vitamin D regulates both the cytoplasmic and the basolateral Ca^{2+} pumps, and in the future it will be particularly important to investigate the role of the cytoplasmic Ca^{2+} pumps in the cell's mechanisms for regulating cytosolic Ca^{2+} activity and using Ca^{2+} as a messenger while simultaneously generating an absorptive flux of Ca^{2+}.

The second example derives from attempts to use isolated membrane vesicles to survey amino acid transport pathways. A partially purified basolateral membrane fraction was shown to possess the Na^+-independent, equilibrating transporter for neutral, L-α-amino acids which would be predicted by the sodium gradient hypothesis, and characterization of this transporter indicated that it was similar, if not identical, to the classical system L.[7] Surprisingly, the basolateral membrane-containing fraction also exhibited Na^+-stimulated amino acid transport activity which was mediated by systems distinct from those of the brush border membrane. However, it has never been determined whether these transporters are associated with the basolateral membranes or with the endoplasmic reticulum and Golgi membranes which are also present in the fraction. Preliminary results with proximal tubular cells support the latter interpretation[8] and raise the intriguing possibility that the Na^+-stimulated transporters might shuttle between cytoplasmic sites and the basolateral membranes, depending on the cell's metabolic requirements.

[2] W. E. J. M. Ghijsen, M. D. deJong, and C. H. van Os, *Biochim. Biophys. Acta* **689,** 327 (1982).
[3] B. Hildmann, A. Schmidt, and H. Murer, *J. Membr. Biol.* **65,** 55 (1982).
[4] H. H. Nellans nd T. E. Popovitch, *J. Biol. Chem.* **256,** 9932 (1981).
[5] A. K. Mircheff and C. H. van Os, *in* "The Precorneal Tear Film" (F. J. Holly, ed.), p. 392. Dry Eye Institute, Lubbock, Texas, 1986.
[6] E. J. J. M. van Corven, C. H. van Os, and A. K. Mircheff, *Biochim. Biophys. Acta* **861,** 267 (1986).
[7] A. K. Mircheff, C. H. van Os, and E. M. Wright, *J. Membr. Biol.* **52,** 83 (1980).
[8] C. B. Hensley and A. K. Mircheff, *Fed. Proc. Fed. Am. Soc. Exp. Biol.* **44,** 1742 (1985).

It is important, for the sake of both rigor and depth of understanding, that one be aware of the potential for contamination in basolateral and brush border membrane fractions. This chapter will begin by describing methods for comprehensive analytical subcellular fractionation of enterocytes; implicit in this description is a summary of the separation properties of the cytoplasmic membrane populations which are likely to overlap the basolateral membranes with respect to sedimentation coefficient and equilibrium density. It will then describe how the separation properties of the plasma membrane populations are exploited by some of the current methods for rapid isolation of partially purified brush border and basolateral membrane fractions.

Tissue Preparation

The methods described below can be used with either isolated enterocytes or mucosal scrapings as a starting point. Procedures for enterocyte isolation are described in Chapter [22] of this volume. If mucosal scrapings are to be used, the intestinal segments should be thoroughly washed with physiological saline containing 1 mM dithiothreitol[9,10] to reduce the likelihood that mucus will cause excessive membrane aggregation.

Isolation and Density Gradient Media

Most of the procedures described below employ histidine-imidazole-buffered sorbitol solutions which contain 0.5 mM NaEDTA. The histadine-imidazole buffer is prepared as a fivefold concentrated stock solution; this is done by mixing 25 mM histidine and 25 mM imidazole to a final pH of 7.5. NaEDTA is prepared as a 100 mM stock solution, with pH adjusted to 7.5 with NaOH. The basic isolation buffer used for tissue disruption, differential sedimentation, and sample resuspension and storage contains 5% D-sorbitol, 5 mM histidine-imidazole buffer, and 0.5 mM NaEDTA. All media also contain 0.2 mM phenylmethylsulfonyl fluoride and 9 μg/ml aprotinin as protease inhibitors. Although the sorbitol concentrations in the solutions used in intestinal membrane isolation studies are most commonly expressed in units of weight/volume, it is most convenient to prepare the most highly concentrated solutions in a weight/weight basis. Thus, the appropriate weight of sorbitol, the appropriate volumes of the histidine-imidazole and NaEDTA stock solutions, and sufficient water

[9] M. M. Weiser, *J. Biol. Chem.* **248**, 2536 (1978).
[10] A. K. Mircheff, D. J. Ahnen, A. Islam, N. A. Santiago, and G. M. Gray, *J. Membr. Biol.* **83**, 95 (1985).

to achieve the appropriate final weight are placed in a large beaker; the beaker should be sealed with a double thickness of plastic film held in place with a rubber band. After standing overnight at room temperature, the thick paste will have dissolved almost completely; the solution should be thoroughly mixed before use. Table I presents final weights for 1 liter volumes of the most commonly used sorbitol density gradient media.

Comprehensive Subcellular Fractionation

Cell Disruption

The strategy behind this analytical approach is to delineate membrane populations by examining the distributions of enzymatic markers following a consistent sequence of physical separation procedures. It is frequently useful to employ a cell disruption procedure which causes complete vesiculation of the brush border membranes; this obviates the need to account for both a brush border vesicle population and a brush border fragment population. This degree of disruption can be accomplished with a nitrogen cavitation bomb,[11] a Tissumiser (Tekmar Instruments, Cincinnati, OH),[6,10] or a Polytron (Brinkman Instruments, Westbury, NY). If the volume of the initial homogenate can be kept to roughly 30 ml, it will be possible to avoid having to concentrate membranes by high-speed centrifugation prior to performing equilibrium density gradient centrifugation. In practical terms, this strategy can save 90 min or more of preparation time.

Differential Sedimentation

The homogenate is centrifuged at 1000 g for 10 min. The initial pellet is resuspended in 20 ml isolation buffer, homogenized, and again centrifuged at 1000 g for 10 min. The pooled supernatants, collectively designated S_0, are then subjected to equilibrium density gradient centrifugation. The low-speed pellet, designated P_0, is resuspended in isolation buffer and set aside for biochemical marker determinations.

Differential sedimentation is also employed to concentrate membranes and to separate them from soluble components and the density gradient medium after density gradient centrifugation. The fractions are diluted with 0.8 vol of isolation buffer and thoroughly mixed. If a high-performance rotor is available, the membranes can then be harvested in high yield by centrifugation at 250,000 g for 60 min. Decreases in the maximal centrifugal force available can be compensated for by increasing either the dilution factor or the centrifugation time.

[11] A. K. Mircheff, S. D. Hanna, M. W. Walling, and E. M. Wright, *Prep. Biochem.* **9**, 133 (1979).

TABLE I
FINAL WEIGHTS OF DENSITY GRADIENT MEDIA

Sorbitol concentration (% w/v)	Final weight/liter (g)
35.0	1114.9
55.0	1182.6
65.0	1213.9
70.0	1227.1
80.0	1259.6
87.4	1278.5

Density Gradient Centrifugation: Zonal Rotor Method

General procedures for the use of zonal rotors are provided in Vol. 172 of this series. This chapter will provide specific details for use of the Z-60 zonal rotor (Beckman, Fullerton, CA) to isolate membranes from rat duodenal enterocytes.[6,10] Note that the zonal rotor is loaded and unloaded at 2000 rpm and that a rotating seal designed for this purpose is available from the manufacturer.

It is possible to generate reproducible hyperbolic gradients by using a piston to maintain a constant fluid volume in the mixing chamber of a standard two-chamber gradient maker. A peristaltic pump is used to drive fluid from the gradient maker into the rotor. Five lengths of Tygon tubing should be joined, with T connecters, into an H-shaped configuration, with the peristaltic pump placed at the center of the H. One of the branches proximal to the pump, tubing line A, is connected to the gradient maker. The other proximal branch, tubing line B, will be used to introduce the sample, overlays, and cushions. One of the branches distal to the peristaltic pump, tubing line C, leads to the rotating seal and zonal rotor. The other distal branch, tubing line D, is used for bleeding off air bubbles and excess fluid. Flow through the various branches is controlled by clamping the tubing with hemostats.

In preparation for loading the zonal rotor, the reservoir chamber of the two-chamber gradient maker is loaded with 250 ml 65% sorbitol, and the channel between the reservoir and mixing chambers is primed with the 65% sorbitol solution. One hundred and eighty milliliters of 30% sorbitol is placed in the mixing chamber. Since the piston will be lowered to the surface of the 30% sorbitol solution, it must be provided with a small hole, for venting the displaced air, and a sturdy rod, which will be used to hold the piston in place. If the lower surface of the piston has a concave contour, it will be possible to remove all the air from the chamber without displacing any of the 30% sorbitol solution.

After the piston has been positioned, an additional 25 ml of 30% sorbitol is placed over it. This extra fluid is used to prime tubing line A. After line A has been purged of all air bubbles, the vent is plugged. Tubing lines B and C are then primed with isolation buffer, which will serve as an overlay for the density gradient. Once the tubing lines have been completely primed, 10 ml of isolation buffer is pumped into the rotor. The overlay of isolation buffer is followed by 50 ml of S_0, which is followed by the hyperbolic gradient of 30% to approximately 65% sorbitol. Mixing and pumping are stopped when the reservoir chamber has been emptied and a few bubbles of air have entered the mixing chamber. A cushion of 25 ml 80% sorbitol will be delivered via tubing line B. Note that line B was initially primed with isolation buffer, which must now be replaced by fluid of an appropriate density. This can be done most conveniently by allowing a small volume of fluid to flow out of the mixing chamber via lines A and B. Loading of the 80% sorbitol cushion is stopped when all air has been displaced from the rotor and fluid has begun to emerge through the center line of the rotating seal.

After the rotor has been capped and a vacuum has been established in the ultracentrifuge chamber, the rotor is accelerated to 57,500 rpm. The run continues for 60 min at this speed, after which the rotor is decelerated back to the loading speed. The density gradient will displaced from the rotor by an 80% sucrose solution, which is delivered via line B; lines B and C should be primed with this solution before the rotating seal is placed back onto the rotor. Fractions of 14.5 ml each are collected into polycarbonate centrifuge bottles suitable for the Beckman Ti 50.2, Sorvall (Wilmington, DE) TFT 50.38, or equivalent rotor. The bottle containing fraction 1 is filled to capacity with isolation buffer and thoroughly mixed with a small paddle attached to a stirring motor. Subsequent tubes are balanced against the first tube by addition of the appropriate volumes of isolation buffer. The supernatants resulting from high-speed centrifugation of the density gradient fractions are pooled, the total volume measured, and an aliquot set aside for biochemical marker determinations.

Density Gradient Centrifugation: Swinging Bucket Rotors

The procedures outlined above can be scaled down so that the same density gradient design can be employed with conventional swinging bucket rotors. Although vertical rotors have not yet been employed for these separations, we are aware of no reason why they could not also be used. Several modifications or the procedures outlined in the previous section are appropriate for the small gradient marker used with centrifuge tubes for swinging bucket rotors. For example, we have found that a soft rubber stopper suffices to fix the volume in the mixing chamber of the gradient maker. If a 20-gauge syringe needle has been passed through this stopper, it can be inserted without increasing the air pressure in the chamber; the stopper will reseal once the needle has been withdrawn.

Although gravity can be used to drive fluid from the gradient maker to the centrifuge tube, one will have better control of the flow rate if one uses a peristaltic pump. A single tubing line usually suffices for delivering the gradient. Since the gradient medium will be delivered to the centrifuge tube in the low- to high-density sequence, the outflow from the peristaltic pump should always be to the bottom of the centrifuge tube. A rigid, e.g., glass or polyethylene, tube should be attached to the outflow line. If a Buchler (Fort Lee, NJ) Auto-Densi Flow device is available, it can be used both to hold the outflow tube and, after the gradient is complete, withdraw it with minimal disruption of the gradient. The gradient can be prepared in advance and kept in the cold room until the sample, S_0, is ready. S_0 and an overlay of isolation buffer are applied manually to the top of the density gradient. Centrifugation at 26,000 rpm for 5 hr, or an equivalent force–time combination, reproduces the same marker distribution patterns as would be obtained with the zonal rotor method.

Marker Density Distributions

The individual density gradient fractions should be assayed for a battery of biochemical markers, including Na^+,K^+-ATPase, acid phosphatase, sucrase or maltase, NADPH-cytochrome c reductase, succinate dehydrogenase, galactosyltransferase, and protein. References and additional technical information for these assays are provided in Vol. 172 of this series. It is convenient to divide the density gradient fractions into several *density windows* on the basis of salient features of the marker distribution patterns. When enterocytes have been disrupted with the Tissumiser and fractionated with the zonal rotor method described above,[6] *density window I*, fractions 1–6, will be relatively free of membranes. *Window II*, fractions 7–11, will be dominated by membranes derived from the endoplasmic reticulum and trans elements of the Golgi complex; it will be relatively free of basolateral membrane contamination. *Window III*, fractions 12–17, will be the major locus of the basolateral membranes, although it will also contain substantial amounts of material derived from the endoplasmic reticulum and Golgi complex. *Window IV*, fractions 18–20, has not been analyzed in detail. *Window V*, fractions 21–24, will contain overlapping peaks of mitochondria and brush border membrane vesicles.

Separation of Brush Border Vesicles from Mitochondria

As described in detail in Vol. 172 of this series, the brush border membrane vesicles can be separated from the mitochondria, which also equilibrate in density window V, by differential rate sedimentation through a column of 17.5% sorbitol.

Phase Partitioning Analysis of Windows II and III

Additional separation procedures are necessary to resolve the microsomal membrane populations which equilibrate in density windows II and III. Countercurrent distribution, a liquid chromatographic procedure which separates membranes on the basis of differences in their partitioning in aqueous polymer two-phase systems, has been used in several analytical subcellular fractionation studies of intestinal epithelium cells.[6,10,12] Detailed descriptions of the necessary apparatus and instructions for preparing phase systems are provided in Vol. 172 of this series.

Golgi- and endoplasmic reticulum-derived populations from density window II can probably be best separated from each other in a phase system which contains 5% dextran T-500 and 3.5% polyethylene glycol 8000, 5% sorbitol, 8.3 mM imidazole, and 10 μM EDTA, pH 7.6.[12] The endoplasmic reticulum-derived membranes have a low partition coefficient and will remain near the sample loading zone of the countercurrent distribution run. The Golgi-derived populations have relatively high partition coefficients in this two-phase system, and in a 120-step countercurrent distribution run, will migrate to fractions 60–100.

Phase partitioning is of limited utility for resolving the basolateral membranes from window III when differential and density gradient centrifugation have been performed as described above. The difficulty is illustrated by the countercurrent distribution analysis depicted in Fig. 1. This analysis was performed in a standard 5% dextran–3.5% polyethylene glycol phase system, with pH 6.5. It is clear from the marker distribution patterns that density window III contains elements of at least three distinct membrane populations; one population is apparent from the overlapping peaks of markers centered between partitioning fractions 21 and 30. The other two populations are apparent from the overlapping, but slightly skewed, peaks of galactosyltransferase and Na$^+$,K$^+$-ATPase centered between partitioning fractions 51 and 90. Thus, window III contains at least one Golgi-derived population whose partitioning characteristics are similar, but not identical to, those of the basolateral membranes. Fractions

[12] D. J. Ahnen, A. K. Mircheff, N. A. Santiago, C. Yoshioka, and G. M. Gray, *J. Biol. Chem.* **258**, 5960 (1983).

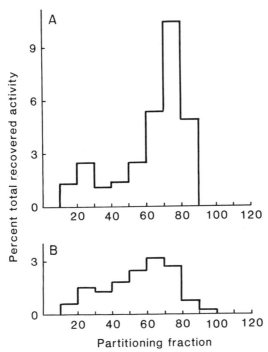

FIG. 1. Distributions of Na$^+$,K$^+$-ATPase (A) and galactosyltransferase (B) after counter-current distribution analysis of a basolateral membrane-containing density window. Density window III, obtained by the zonal rotor method described in text, was analyzed in a 5% dextran–3.5% polyethylene glycol two-phase system. Skewing between the major peaks indicates that the markers are primarily associated with two populations which are distinct but which have similar partitioning distributions.

81–90 represent a sample of basolateral membrane vesicles relatively free of Golgi contamination, although this sample would represent a relatively small portion of the total basolateral membrane population. The overlap between the Golgi and basolateral membrane populations and the fact that Na$^+$,K$^+$-ATPase is a native constituent of both membrane populations make it impossible to identify a good sample of the Golgi-derived population. A second limitation which affects the utility of phase partitioning when the goal of a study is to obtain membranes for transport experiments is that the long separation times required lead to deterioration of labile transport systems, most notably ATP-dependent Ca^{2+} pumps. These limitations can be circumvented by the second density gradient centrifugation procedure outlined below.

Second Density Gradient Centrifugation

The advantage of this procedure is that it is rapid and requires only conventional centrifugation equipment. It separates the major basolateral membrane and Golgi populations which overlap in density window III by exploiting the fact that the relative equilibrium densities of these populations are sensitive to the density, i.e., the osmolarity, of the medium in which they are applied to the gradient.

A hyperbolic gradient of 30 to 55% sorbitol is prepared in a 38-ml-capacity centrifuge tube by passing 23 ml 55% sorbitol through a mixing chamber initially loaded with 16 ml of 30% sorbitol. An additional 3 ml of 55% sorbitol, followed by 3 ml of 65% sorbitol, are then delivered directly to the centrifuge tube. A 4.00-g aliquot of density window III in 5% sorbitol is brought to a final sorbitol concentration of approximately 55% by addition of 5.42 g of an 87.4% sorbitol stock solution. This is delivered to the centrifuge tube just above the 55% sorbitol cushion. The resulting gradient is centrifuged at 26,000 rpm for 4 hr in a swinging bucket rotor. Following centrifugation, a Buchler Auto-Densi Flow device is used to collect a series of 6-ml fractions. Membranes are harvested from these fractions by high-speed centrifugation after dilution with isolation buffer or other appropriate medium. Fractions 3 and 4 contain roughly 50% of the recovered Na^+,K^+-ATPase activity, enriched a further 2.2-fold with respect to galactosyltransferase; this corresponds to a 4.6-fold overall enrichment of Na^+,K^+-ATPase with respect to galactosyltransrerase.[6]

The density gradient centrifugation procedure described in this section can also be used to resolve the major Golgi and endoplasmic reticulum populations present in density window II of the first density gradient. Fraction 3 will contain an endoplasmic reticulum sample which is apparently well purified with respect to galactosyltransferase, while fraction 5 will contain the major galactosyltransferase-containing membrane population. Fraction 4 will contain a mixture of at least four distinct endoplasmic reticulum and Golgi populations, including one population in which galactosyltransferase is enriched nearly 30-fold with respect to the initial homogenate.

Four-Dimensional Isolation of Highly Purified Basolateral Membranes

Highly purified samples of the basolateral membrane population can be obtained by following the sequence of differential sedimentation, equilibrium density gradient centrifugation after sample loading in 5% sorbitol, and density gradient centrifugation after sample loading in 55% sorbitol described in the above sections, then using countercurrent distribution to subfractionate fractions 3 and 4 from the second density gradient separation of density window III. Countercurrent distribution is performed in the pH 6.5 phase system described above. The resulting marker distribution patterns are depicted in Fig. 2; the distribution patterns for fraction 5 are also included for comparison. Countercurrent distribution resolves basolateral membranes, marked by Na^+,K^+-ATPase, from trans Golgi-derived membranes, marked by galactosyltransferase, so that the samples in partitioning fractions 51–70 contain Na^+,K^+-ATPase enriched 13- to 26-fold with respect to galactosyltransferase. Note that this countercurrent distribution analysis would have separated the Golgi-derived and basolateral membrane-derived populations in fractions III,3 and III,4 even without the second density gradient centrifugation procedure. The countercurrent distribution analysis of fraction III,5 emphasizes the necessity of the second density gradient centrifugation step, since it delineates the Golgi-derived population which overlaps the partitioning distribution of the basolateral membranes.

Rapid Isolation of Partially Purified Basolateral Membranes

Procedures which incorporate at least three different separation methods are necessary for obtaining highly purified basolateral membrane samples and for performing detailed analyses of the subcellular distributions of transport activities, receptors, or enzymes which might be present in both basolateral and cytoplasmic membrane populations. Thus, one can combine separation methods based on sedimentation coefficient, equilibrium density after sample loading in 5% sorbitol, and equilibrium density after sample loading in 55% sorbitol to isolate samples of the mitochondria and the brush border, basolateral, Golgi, and endoplasmic reticulum membrane populations. Separations based on partitioning in aqueous polymer two-phase systems can then be used to verify the homogeneity of the samples obtained with the first three separation methods, and to survey the full multiplicity of membrane populations present in the initial homogenate.

A rapid, preparative-type procedure is available for isolating a partially purified basolateral membrane fraction without swinging bucket or high-performance rotors. This procedure is based on differential and density gradient centrifugation, so it should be clear that the final "basolateral membrane" fraction also contains samples of the membrane populations which occupy the same density window as the basolateral population.

After cell disruption and low-speed centrifugation have been performed with one of the methods described above, the pooled low-speed supernatants, S_0, are centrifuged at 95,000 g for 20 min. The resulting pellet is resuspended in isolation buffer (~10 ml/intestine) with 100 strokes of a

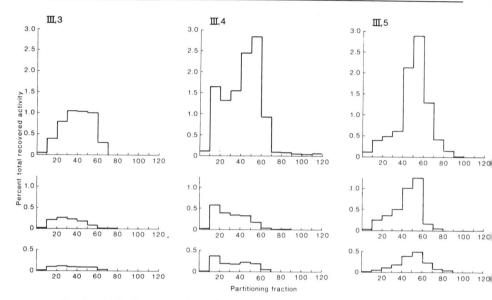

FIG. 2. Distributions of markers after countercurrent distribution analyses of fractions 3, 4, and 5 obtained after density window III from the zonal rotor method was subfractionated by the second density gradient procedure described in text. *Upper*: Na$^+$,K$^+$-ATPase. *Middle*: galactosyltransferase. *Lower*: protein. These analyses confirm that the galactosyltransferase-rich population whose partitioning distribution overlaps that of the basolateral membranes is separated from the basolateral membranes by the second density gradient centrifugation procedure. Fractions 51–70 from III,3 and III,4 represent highly purified samples of the basolateral membrane population.

loose-fitting Dounce apparatus (Dounce, Wheaton, NJ). The volume of this fraction is measured, and 1.4 vol of 65% sorbitol is added. After thorough mixing, the sample is distributed to centrifuge tubes; typically, 20-ml aliquots would be placed in 26-ml-capacity centrifuge tubes, and an overlay of 4 ml 25% sorbitol is added to each tube; the samples are centrifuged at 95,000 *g* for 60 min. The band collecting at the 25%/40% sorbitol interface will represent a basolateral membrane sample containing Na$^+$,K$^+$-ATPase enriched 10-fold with respect to the initial homogenate. This fraction is useful for detailed characterization of functions which are known to be localized to the basolateral membranes. It should be recalled, however, that the fraction is quite heterogeneous and that a further 1.8- to 2.0-fold enrichment of the basolateral membranes would be achieved by application of a third dimension separation procedure such as free-flow electrophoresis or countercurrent distribution.

Rapid Isolation of Brush Border Membranes: Divalent Cation Precipitation Method

Several modifications of the divalent cation precipitation procedure initially described by Schmitz et al.[13] are available. These procedures exploit the fact that a number of membrane populations, other than brush border vesicles, aggregate in the presence of 10 mEq/liter concentrations of Ca^{2+} or Mg^{2+}. Important advantages shared by these procedures are that they are rapid and that they yield tightly sealed vesicles which exhibit marked overshoots for Na^+- or H^+-coupled transport processes. The possible multiplicity of distinct membrane populations in such preparations has not been studied extensively, but it is noteworthy that phase partitioning analysis showed that the homologous preparations obtained from renal cortex contained elements of four distinct membrane populations; the maltase-rich brush border vesicle population accounted for only 50% of the total protein in the sample.[14] The identities of the other populations in the sample have not yet been determined.

Isolation of Brush Border Membranes Free of Divalent Cations

A variation of the original separation strategy employed by Miller and Crane[15] and by Forstner et al.[16] provides an alternative method for rapid isolation of brush border membrane vesicles. Intestinal cells are disrupted by a method which is gentle enough to leave the brush border surfaces intact. This is accomplished by either of two methods: (1) resuspension of tissue in isolation buffer followed by homogenization with 100 strokes of a loose-fitting Dounce homogenizer[17] or (2) resuspension in 5 mM EDTA followed by homogenization in a Waring blendor.[15,16] The resulting brush border fragments are quite massive, so they can be separated from mitochondria and from microsomal populations by centrifugation at 450 g for 10 min. Up to eight cycles of resuspension, homogenization, and centrifugation might be required[18] to completely remove the microsomal populations which would, if present, equilibrate at the same density as brush

---

[13] J. Schmitz, H. Preiser, D. Maestracci, B. K. Ghosh, J. J. Cerda, and R. K. Crane, *Biochim. Biophys. Acta* **323**, 98 (1973).

[14] A. K. Mircheff, H. E. Ives, V. J. Yee, and D. J. Warnock, *Am. J. Physiol.* **246**, F853 (1984).

[15] D. Miller and R. K. Crane, *Anal. Biochem.* **2**, 284 (1961).

[16] G. G. Forstner, S. M. Sabesin, and K. J. Isselbacher, *Biochem. J.* **106**, 381 (1968).

[17] M. Fujita, H. Ohta, K. Kawai, H. Matsui, and M. Nakao, *Biochim. Biophys. Acta* **274**, 336 (1972).

[18] A. K. Mircheff and E. M. Wright, *J. Membr. Biol.* **28**, 309 (1976).

border membrane vesicles. The brush border fragments are vesiculated for use in transport studies by application of more vigorous mechanical disruption techniques, such as a motor-driven glass-Teflon homogenizer in 12% sorbitol. The final membrane vesicles are separated from the nuclear material which had sedimented with the intact brush border fragments by a 2-hr centrifugation on a gradient of 25 to 65% sorbitol.

[24] Established Intestinal Cell Lines as Model Systems for Electrolyte Transport Studies

By KIERTISIN DHARMSATHAPHORN and JAMES L. MADARA

Introduction

This chapter will deal only with established intestinal cell lines, and the introductory remarks will emphasize the advantages and disadvantages of continuous cell lines. Primary culture of intestinal epithelial cells still has a great many obstacles to overcome before it can serve as a useful model system. The difficulties of primary culture result mainly from the relatively short life span of the mature intestinal epithelial cells, the contamination by microorganisms normally existing in the gastrointestinal tract, and the heterogeneity of the isolated cells. In contrast, studies of native isolated intestinal epithelium or whole animals are readily performed and have been used to identify pertinent physiological or pharmacological processes. However, native intestinal epithelium is exceedingly complex, containing many types of cells. Furthermore, the influence of various peptides and other regulatory agents normally present in the intestine also complicates the precise study of electrolyte transport mechanisms. To circumvent some of the difficulties encountered with the above approaches, one may use homogeneous monolayer cultures established from epithelial cell lines to study the mechanisms involved in an electrolyte transport process as well as its regulatory controls. The simple approach introduced by Dr. Ussing and Zerahn,[1] radionuclide uptake and efflux techniques, and other newer techniques can be readily applied to such cultured epithelial monolayers. As a matter of fact, interpretation of the results obtained from cultured epithelia is quite straightforward due to the homogeneity of the cells. Cultured cell lines serve well as model systems to study receptor-mediated regulation of transport starting from receptor binding to the activation of

[1] H. H. Ussing and K. Zerahn, *Acta Physiol. Scand.* **23**, 110 (1951).

secondary messengers and, finally, to the regulation of electrolyte transport pathways. The disadvantages of cultured cells arise only if an investigator has unrealistically high hopes that these cells will serve as normal intestinal epithelium. Regardless of the source, cultured cell lines must be "abnormal" in some aspects since they are abnormal at least in their growth regulation and perhaps in other ways as well. Cells grown in culture may express a function not normally expressed or, more often, fail to express many functions they should express. Therefore, caution must be exercised in extrapolating conclusions drawn from cultured epithelial cells to the parent tissue. With these reservations in mind, it is likely that cultured cell lines may provide some insights into the function of normal human cells just as do native epithelia derived from animals. Investigations at cellular or molecular levels are likely to benefit most from such cell culture studies. As a matter of fact, the potential pitfalls of cultured cells may be turned into advantages. For example, mutants of cultured cells lacking a specific function can be selected for study. The value of mutant cells[2-5] is analogous to that provided by patients with rare genetic diseases who facilitate the elucidation of the genetic defect. An investigator hoping to study a specific physiological function might thus find it ideal to study both cell lines in which that function is well preserved and mutant cell lines in which that function is missing.

Availability of Cultured Cell Lines Derived from the Intestine

Most of the intestinal cell lines currently available were established initially for cancer research and they are derived from cancer cells. Stem cells from rat intestine have also been established successfully.[6] Human cancer cell lines are readily established in nude mice and grown in cell culture dishes since they propagate without the normal constraint of aging. In nude mice, simple inoculation with the cancer cell line usually is sufficient to promote subsequent propagation of the cell line, provided that the cells do not secrete harmful substances which kill the mice.[7] In culture

[2] J. A. McRoberts, C. T. Tran, and M. H. Saier, Jr., *J. Biol. Chem.* **258**, 12320 (1983).
[3] J. J. Gargus, I. L. Miller, C. W. Slayman, and E. A. Aldelberg, *Proc. Natl. Acad. Sci. U.S.A.* **75**, 5589 (1978).
[4] J. J. Gargus and C. W. Slayman, *J. Membr. Biol.* **52**, 245 (1980).
[5] D. W. Jayme, E. A. Adelberg, and C. W. Slayman, *Proc. Natl. Acad. Sci. U.S.A.* **78**, 1057 (1981).
[6] A. Quaroni, J. Wands, R. L. Trelstad, and K. J. Isselbacher, *J. Cell Biol.* **80**, 248 (1979).
[7] L. M. Reid, J. Holland, C. Jones, B. Wolf, G. Niwayama, R. Williams, N. O. Kaplan, and G. Sato, *in* "Proceedings of the Symposium on the use of Athymic (Nude) Mice in Cancer Research" (D. Itouchens and A. Overjerea, eds.), p. 107. Fischer, New York, 1978.

dishes, the culture media must contain the appropriate peptides and other essential growth factors. Serum usually serves this purpose well. However, because a well-defined environment is desirable for studies of regulatory control, development of a serum-free medium should be sought.[8,9]

The list of gastrointestinal cell lines has expanded rapidly. Most of them are derived from the colon (>70 lines). However, cells from the stomach, small intestine, biliary tract, and pancreas are also available. The main sources, the American Type Culture Collection and National Institute of General Medical Sciences, update their catalogs yearly. Many newer cell lines that are not available through these sources can be identified by doing a Medline search and usually the cells can be obtained from individual investigators. To date, very few gastrointestinal cell lines have been utilized as model systems for electrolyte transport physiology. Therefore, one may find little physiological information concerning gastrointestinal cell lines, as compared to renal and nonepithelial cell lines.[10-16] To date, only two colonic cell lines, T_{84} and $CaCO_2$, have been utilized for extensive physiologic studies. Another line, the HT29, which appears to exhibit less well-differentiated electrolyte transport functions, has been used mainly for the study of vasoactive intestinal peptide (VIP) receptors and differentiation requirements.[17,18] The physiologically better differentiated T_{84}[19-31] and

[8] D. Barnes, J. van der Bosch, H. Masui, K. Miyazaki, and G. Sato, this series, Vol. 79, p. 368.
[9] H. Murakami and H. Masui, Proc. Natl. Acad. Sci. U.S.A. 77, 3464 (1980).
[10] J. S. Handler, F. M. Perkins, and J. P. Johnson, Am. J. Physiol. 238, F1 (1980).
[11] D. S. Misfeldt, S. T. Hamamoto, and D. R. Pitelka, Proc. Natl. Acad. Sci. U.S.A. 73, 1212 (1976).
[12] N. L. Simmons, J. Membr. Biol. 59, 105 (1981).
[13] E. Stefani and M. Cereijido, J. Membr. Biol. 73, 177 (1983).
[14] P. Geck, C. Pietrzyk, B. C. Burckhardt, B. Pfeiffer, and E. Heinz, Biochim. Biophys. Acta 600, 432 (1980).
[15] E. K. Hoffmann, L. O. Simonsen, and I. H. Lambert, J. Membr. Biol. 78, 211 (1984).
[16] B. Sarkadi, R. Cheung, E. Mack, S. Grinstein, E. W. Gelfand, and A. Rothstein, Am. J. Physiol. 248, C480 (1985).
[17] M. Laburthe, M. Rousset, C. Boissard, G. Chevalier, A. Zweibaum, and G. Rosselin, Proc. Natl. Acad. Sci. U.S.A. 75, 2772 (1978).
[18] A. Zweibaum, M. Pinto, G. Chevalier, E. Dussaulx, N. Triadou, B. Lacroix, K. Haffen, J.-L. Brun, and M. Rousset, J. Cell. Physiol. 122, 21 (1985).
[19] K. Dharmsathaphorn, K. G. Mandel, J. A. McRoberts, L. D. Tisdale, and H. Masui, Am. J. Physiol. 246, G204 (1984).
[20] K. Dharmsathaphorn, J. A. McRoberts, H. Masui, and K. G. Mandel, J. Clin. Invest. 75, 462 (1985).
[21] K. G. Mandel, J. A. McRoberts, G. Beuerlein, E. S. Foster, and K. Dharmsathaphorn, Am. J. Physiol. 250, C486 (1986).
[22] C. A, Cartwright, J. A. McRoberts, K. G. Mandel, and K. Dharmsathaphorn, J. Clin. Invest. 76, 1828 (1985).
[23] J. A. McRoberts, G. Beuerlein, and K. Dharmsathaphorn, J. Biol. Chem. 260, 14163 (1985).

CaCO$_2$ lines[32,33] are able to form occluding junctions with high resistance, allowing them to exhibit vectorial electrolyte transport properties. The authors' experience derives mainly from the T$_{84}$ cells which secrete Cl$^-$ in response to a variety of hormones or neurotransmitters (Table I). This chapter, in essence, summarizes our experience with this cell line. It is hoped, however, that the methodologies may also be applicable to other cultured cells.

Methods

Growth and Maintenance of T$_{84}$ Cells

T$_{84}$ cells are grown as monolayers in a 1 : 1 mixture of Dulbecco-Vogt modified Eagle's (DME) medium and Ham's F-12 (F-12) medium supplemented with 15 mM Na$^+$-HEPES buffer, pH 7.5, 1.2 g NaHCO$_3$, 40 mg/liter penicillin, 8 mg/liter ampicillin, 90 mg/liter of streptomycin, and 5% (v/v) newborn calf serum. A more detailed description of the medium is given in the Appendix. A serum-free medium has also been developed.[9] The growth requirement of the T$_{84}$ cells is not overly rigid, as attested to by its ability to grow in a variety of serum-containing media (K. Dharmsathaphorn and J. Madara, personal observations). The differences, as compared to other cultured cells, are mainly the slow growth exhibited by T$_{84}$ cells (doubling time approximately 2.5 to 3 days) and the ability of these cells to attach quite firmly to the culture plate. Because of the latter characteristics, T$_{84}$ cells require a relatively vigorous trypsinization for subculturing. Confluent monolayers are subcultured by trypsinization with 0.1% trypsin and 0.9 mM EDTA in Ca^{2+} and Mg^{2+}-free phosphate-buffered saline. This process usually takes at least 15–30 min and the cells

---

[24] K. G. Mandel, K. Dharmsathaphorn, and J. A. McRoberts, *J. Biol. Chem.* **261**, 704 (1986).

[25] K. Dharmsathaphorn, P. Huott, C. A. Cartwright, J. A. McRoberts, K. G. Mandel, and G. Buerlin, *Am. J. Physiol.* **250**, G806 (1986).

[26] A. Weymer, P. Huott, J. A. McRoberts, and K. Dharmsathaphorn, *J. Clin. Invest.* **76**, 1828 (1985).

[27] K. Dharmsathaphorn and S. Pandol, *J. Clin. Invest.* **77**, 348 (1986).

[28] J. Madara and K. Dharmsathaphorn, *J. Cell Biol.* **101**, 2124 (1985).

[29] P. Huott, K. Barrett, S. Wasserman, and K. Dharmsathaphorn, *Gastroenterology* **90**, A201 (abstract) (1986).

[30] W. Liu, P. Huott, R. A. Giannella, and K. Dharmsathaphorn, *Gastroenterology* **90**, A201 (abstract) (1986).

[31] H. Ammon and K. Dharmsathaphorn, *Clin. Res. 90*, A201 (abstract) (1986).

[32] E. Grasset, M. Pinto, E. Dussaulz, A. Zweibaum, and J.-F. Desjeux, *Am. J. Physiol.* **247**, C260 (1984).

[33] E. Grasset, J. Bernabeu, and M. Pinto, *Am. J. Physiol.* **248**, C410 (1985).

TABLE I
AGENTS AFFECTING Cl⁻ SECRETION ACROSS T$_{84}$
CELL MONOLAYERS[a]

Agent	Mechanism[b]
Secretagogues	
VIP	cAMP (*1–7*)
PGE$_1$	cAMP (*8*)
Adenosine	cAMP (*9*)
Cholera toxin	cAMP (*4*)
Carbachol	Ca^{2+} (*10*)
Histamine	Ca^{2+} (*9*)
A23187	Ca^{2+} (*1,3–6*)
Heat-stable toxin of *E. coli*	cGMP (*11*)
Bile salts	Ca^{2+} (*1,12*)
Fatty acids	Pending investigation (*1*)
β-Adrenergic agonists	Pending investigation (*1*)
Antisecretagogues	
Somatostatin	Pending investigation (*1*)
Verapamil	Ca^{2+} (*1*)

[a] T$_{84}$ cells do not respond to bomesin, neurotensin, substance P, serotonin, met-enkephalin, and α-adrnergic agonists (*1*).

[b] Key to references: (*1*) K. Dharmsathaphorn, K. G. Mandel, J. A. McRoberts, L. D. Tisdale, and H. Masui, *Am. J. Physiol.* **246**, G204 (1984); (*2*) K. Dharmsathaphorn, J. A. McRoberts, H. Masui, and K. G. Mandel, *J. Clin. Invest.* **75**, 462 (1985); (*3*) K. G. Mandel, J. A. McRoberts, G. Beuerlein, E. S. Foster, and K. Dharmsathaphorn, *Am. J. Physiol.* **250**, C486 (1986); (*4*) C. A. Cartwright, J. A. McRoberts, K. G. Mandel, and K. Dharmsathaphorn, *J. Clin. Invest.* **76**, 1828 (1985); (*5*) J. A. McRobrts, G. Beuerlein, and K. Dharmsathaphorn, *J. Biol. Chem.* **260**, 14163 (1985); (*6*) K. G. Mandel, K. Dharmsathaphorn, and J. A. McRoberts, *J. Biol. Chem.* **261**, 704 (1986); (*7*) K. Dharmsathaphorn, P. Huott, C. A. Cartwright, J. A. McRoberts, K. G. Mandel, and G. Beuerlein, *Am. J. Physiol.* **250**, G806 (1986); (*8*) A. Weymer, P. Huott, J. A. McRoberts, and K. Dharmsathaphorn, *J. Clin. Invest.* **76**, 1828 (1985); (*9*) P. Huot, K. Barrett, W. Wasserman, and K. Dharmsathaphorn, *Gastroenterology* **90**, A201 (abstract) (1986); (*10*) K. Dharmsathaphorn and S. Pandol. *J. Clin. Invest.* **77**, 348 (1986); (*11*) W. Liu, P. Huott, R. A. Giannella, and K. Dharmsathaphorn, *Gastroenterology* **90**, A201 (abstract) (1986); (*12*) H. Ammon and K. Dharmsathaphorn, *Clin. Res.* in press (abstract).

appear to tolerate it well. For physiological studies, 10^6 cells are plated on a permeable support (2-cm^2 surface area) and maintained for 7 days prior to use. The supports are similar to the "filter-bottom dish" developed by Handler *et al.*,[34] consisting of a rat tail collagen-coated polycarbonate filter (Nuclepore, 5-μm pore size; Nuclepore Corporation, Pleasanton, CA) glued to one open end of a Lexan ring (Fig. 1). By laying the ring assembly on top of a layer of glass beads, the monolayers are suspended over the bottom of a 100-mm culture dish to permit "bottom feeding." We found that this method allows the cells to become better polarized and achieve the appearance of a tall columnar epithelium (Fig. 2). Moreover, for the determination of the sidedness of transport phenomenon and for analysis of vectoral transport, such permeable supports are a necessity. Crude rat tail collagen is solubilized by dissolving 1 g rat tail tendons in 100 ml 1% acetic acid. Further steps in preparation of rat tail collagen and procedures for collagen-coating Nuclepore filters follow those described by Cereijido *et al.*,[35] and are outlined in the Appendix.

Morphological Studies

Morphological techniques applied to intact tissues are also generally applicable to T_{84} cells. We have found some minor variations to be useful, however.[28] For embedment in epoxy resins, thin sectioning and subsequent ultrastructural analysis, we find standard fixation in 2.5% glutaraldehyde, 2% formaldehyde in 0.1 M sodium cacodylate buffer, pH 7.4 to be useful but often not ideal. The high osmolarity of this fixative may induce striking cellular shrinking. This may be partially obviated by fixation in 2% glutaraldehyde in the above buffer alone. In either case, after subsequent routine postfixation in osmium,[28] it is critical to *en bloc* stain with a 2% uranyl acetate solution for 30 min before dehydration. This procedure helps us to maintain a more uniform cytoplasmic electron density for subsequent ultrastructural study. After dehydration, if monolayers on filters are exposed to propylene oxide, they roll in jelly-roll fashion, and this may facilitate sampling of a larger area of cells in 1-μm sections. However, for thin-section studies, one may wish to avoid propylene oxide altogether, since it appears to us that Nuclepore filters may be partially soluble in propylene oxide and this material may occasionally appear as electron-luscent areas within the cytoplasm of cells. For freeze-fracture analysis of

---

[34] J. S. Handler, R. E. Steele, M. K. Sahib, J. B. Wade, A. S. Preston, N. L. Lawson, and J. P. Johnson, *Proc. Natl. Acad. Sci. U.S.A.* **76**, 4151 (1979).
[35] M. Cereijido, E. S. Robbins, W. J. Dolan, C. A. Rotunno, and D. D. Sabatini, *J. Cell Biol.* **77**, 853 (1978).

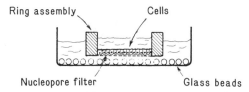

Fig. 1. T_{84} cells grown on permeable supports. T_{84} cells are obtained by trypsinization of cells grown on 100-mm culture dishes. Approximately 1.5×10^6 cells are plated on collagen-coated glutaraldehyde-cross-linked Nuclepore filters glued to Lexan rings with a surface area of 2 cm². The ring assemblies, which serve as a component of the Ussing chamber, are suspended over the bottom of a dish by laying them on top of a layer of glass beads to provide "bottom feeding." The cells are kept for at least 5 days in culture, at which point in time columnar epithelial structure and transepithelial resistance become fully developed (see Figs. 2 and 5).

tight junctions, the above combined glutaraldehyde-formaldehyde fixative is preferable to glutaraldehyde alone. For reasons which are not clear to us, it appears that the combined fixative yields a greater percentage of fractured junctions while the single fixative yields a greater percentage of cross-fractured cells. After fixation, epithelial sheets may be removed from the supporting filter by gentle scraping with a rubber policeman. If properly performed, the cells should lift over the edge of the rubber policeman in continuous sheets. Such sheets form redundant folds, when free in solution and, after glycerin infiltration, may be mounted in a pellet-like fashion on freeze-fracture disks for subsequent routine preparation.[28]

The T_{84} cell line is one of the very few cell lines which exhibits a remarkable structural resemblance to the native intestinal cell for which it serves as a model for transport studies: the colonic crypt cell.[28] In serum-containing media, T_{84} cells form tall columnar epithelial monolayers when grown on permeable supports with the basolateral surface attached to the support and the apical surface exposed (Fig. 2). the intercellular tight or occluding junction develops within 18 hr after plating at the above density and continues to mature both structurally and functionally. A peak, stable resistance is reached within 5–6 days (Figs. 3–5). When grown in a culture dish, the cells appear flatter with less distinction between the apical and basolateral surface and the tight junctions are less uniform. In the serum-free medium developed by Murakami and Masui,[9] the cells grow more rapidly, forming a "glandlike" structure rather than a monolayer and are less well differentiated.[9] However, the methodologies for electrolyte transport studies described below require confluent T_{84} cell monolayers and thus, at present, a serum-free medium applicable for growing T_{84} cell monolayers to be used for transport studies is not available. Hopefully, a suitable serum-free medium can be developed to replace the serum-con-

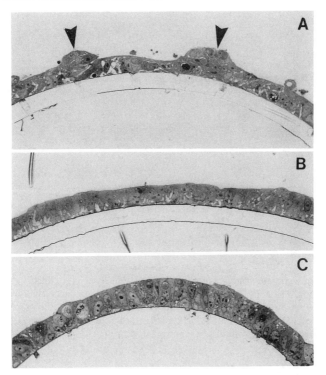

FIG. 2. Secretions (1 μm) of T_{84} epithelial monolayers 18 hr (A), 2 days (B), and 5 days (C) after plating 10^6 cells on collagen-coated Nuclepore filters. The monolayers were grown as described in Fig. 1. At 18 hr, although confluent by light microscopy, true monolayers are not formed. Rather, in occasional foci cells are multilayered. Thin sections of such multilayered areas revealed beltlike occluding junctions associated with only the most superficial cells. Progressive differentiation toward true confluent monolayers composed of taller, polarized cells occurs from 2 to 5 days. (~×200) [From J. L. Madara and K. Dharmsathaphorn, *J. Cell. Biol.* **101**, 2124 (1985) with permission from the publisher.]

taining media in the future but until this time serum is required to obtain confluent monolayers.

Transepithelial Electrolyte Transport Studies (Ussing Chamber Studies)

General Considerations

The Ussing chamber, which measures active transepithelial transport of electrolytes,[1] has been modified for cultured cell monolayers. Figure 6 provides the diagrammatic representation of the modified Ussing chamber.

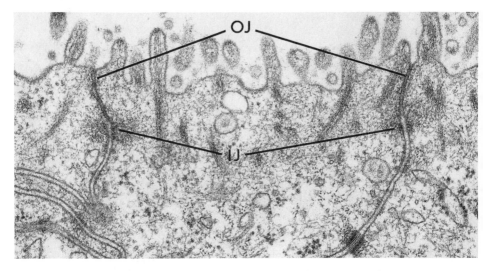

FIG. 3. Thin-section appearance of occluding junctions from T_{84} epithelial monolayers 5 days after plating. The appearance of occluding junctions became relatively uniform in thin sections by 5 days. Junctional organization was also characterized by the ordered relationships between occluding junctions (OJ) and underlying intermediate junctions (IJ). ($\times32,000$) [From J. L. Madara and K. Dharmsathaphorn, *J. Cell. Biol.* **101**, 2124 (1985) with permission from the publisher.]

The principle underlying the Ussing chamber method has been provided in other chapters in this volume and therefore will not be repeated here. Due to the fragility of cultured epithelial monolayers, some minor modifications of the Ussing chamber must be made in order to preserve the monolayers' integrity during the study.[19] These modifications revolve around two considerations: (1) minimizing the turbulence created by the airlift system, and (2) avoiding edge damage. Turbulence has been reduced by using a smaller airlift. Edge damage is completely avoided by our modifications; this is possible because the filter on which the monolayer grows is not clamped into the chamber, but rather is premounted on a Lexan ring which serves as a component of the Ussing chamber itself. The ring assembly (Nuclepore filter glued to the Lexan ring) allows the cells to grow to cover the edge before the entire ring assembly is inserted into the chamber. No pressure is exerted directly on the monolayer and hence edge damage is avoided. These modifications permit the monolayers to remain intact with a constant resistance for at least 2.5 hr and probably much longer.

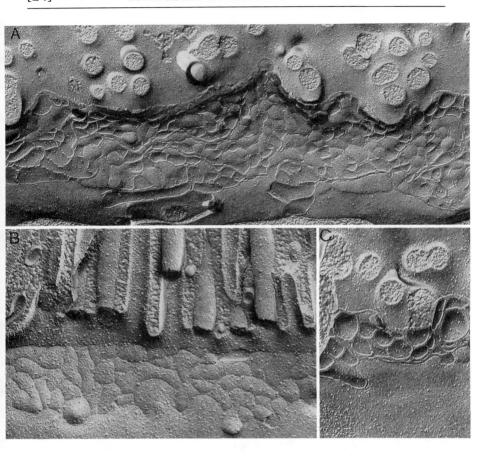

FIG. 4. Freeze-fracture images of occluding junctions from T_{84} epithelial monolayers 5 days after plating. (A) Occluding junctions which display high strand counts throughout their length, rather than focally, constitute a minor junctional population at this time. Although presumably of extremely high resistance, overall such junctions would contribute insignificantly to measurements of monolayer resistance. (B and C) The major population of junctions at this time has substantially fewer strands than that shown in (A) and are uniform and regularly composed of three to five grooves or strands. Such junctions dictate the high resistance of these monolayers. ($\times 64,000$) [From J. L. Madara and K. Dharmsathaphorn, *J. Cell. Biol.* **101**, 2124 (1985) with permission from the publisher.]

Isotopic Flux Experiments

Description of a receptor-mediated electrolyte transport phenomenon is usually the initial objective. The procedure for transepithelial flux experiments follow those of isolated intestine.[36] In most of our studies, both

[36] H. J. Binder and C. L. Rawlins, *Am. J. Physiol.* **225**, 1232 (1973).

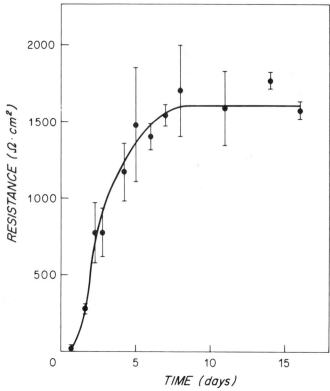

FIG. 5. Progressive rise in T_{84} epithelial monolayer resistance to passive ion flow with time. Even though confluent by light microscopy 18 hr after plating, monolayers have minimal resistance at this time. Resistance progressively rises in the ensuing 5-day period to stabilize at high resistance values of approximately 1500 $\Omega \cdot cm^2$. [From J. L. Madara and K. Dharmsathaphorn, *J. Cell. Biol.* **101**, 2124 (1985) with permission from the publisher.]

reservoirs contained identical volumes of oxygenated Ringer's solution (pH 7.4, at 37°) containing 140 mM Na$^+$, 5.2 mM K$^+$, 1.2 mM Ca^{2+}, 1.2 mM Mg^{2+}, 119.8 mM Cl$^-$, 25 mM HCO$_3^-$, 0.4 mM H$_2$PO$_4^-$, 2.4 mM HPO$_4^{2-}$, and 10 mM glucose. Any balanced salt solution containing nutrients can be used. Depending upon the study purpose, the solutions can be isotonic, hypertonic, or hypotonic and a diffusion potential can be established. The potential difference (PD) across the cell monolayers is measured by calomel electrodes in 3 M KCl. Throughout the experiment, except for 5–10 sec at intervals while the potential difference is being recorded, this potential is nullified by short circuiting with an automatic voltage clamp and Ag:AgCl$_2$ electrodes. Tissue conductances are calculated from the potential difference and the short-circuit current according

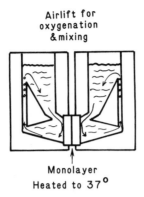

Airlift for
oxygenation
&mixing

Monolayer
Heated to 37°

FIG. 6. Diagram of modified Ussing chamber. The ring assembly with monolayers attached (shown in Fig. 1) together with another identical ring without the monolayers are inserted between two fluid-filled reservoirs. The monolayers are exposed to both reservoirs, which effectively separate the two. [From K. Dharmsathaphorn, K. G. Mandel, J. A. McRoberts, L. D. Tisdale, and H. Masui, *Am. J. Physiol.* **246**, G204 (1984) with permission from the publisher.]

to Ohm's law. Unidirectional $^{22}Na^+$ and $^{36}Cl^-$ (or $^{42}K^+$ and $^{36}Cl^-$) fluxes can be carried out simultaneously in monolayer pairs with similar conductance. Unidirectional flux rates of $^{22}Na^+$ and $^{36}Cl^-$ are stable for at least 2.5 hr and, under basal conditions, vary directly with the conductance. Figure 7 shows the results of Na^+ and Cl^- fluxes from a typical paired experiment performed on such a pair of monolayers. Isotopic flux experiments allow one to determine the rates of unidirectional and net ion movements. In the T_{84} cells, we have observed Cl^- secretion in response to a variety of secretagogues. However, we have not observed any changes in Na^+ or K^+ movements.[20-22, 25-27, 29-31] Therefore, the changes in the short-circuit current across the T_{84} monolayers reflect the changes in net Cl^- secretion. The data derived from the Ussing chamber thus are instrumental in defining this Cl^- secretory phenomenon and served as the basis for subsequent studies of the transport pathways. A number of receptor-mediated Cl^- secretory mechanisms in the T_{84} cells have been identified (Table I), as has a synergistic Cl^- secretory effect between Ca^{2+} and cAMP or Ca^{2+} and cGMP.[22,27,29,30]

Other Applications of the Ussing Chamber

Besides transepithelial isotopic fluxes, other Ussing chamber methodologies described for isolated intestine are applicable to cultured monolayers. When a specific inhibitor is available, it can be applied to either fluid

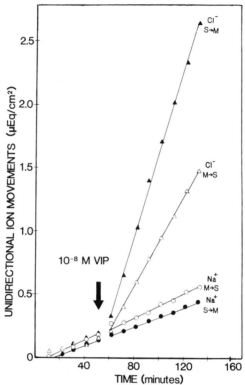

FIG. 7. Vasoactive intestinal peptide (VIP) stimulates unidirectional transport of Cl^- across T_{84} cell monolayers while having no effect on unidirectional Na^+ transport. This figure depicts the results for a representative pair of T_{84} monolayers grown on the permeable support; the monolayers in this experiment had an initial conductance of 0.5 mS/cm^2 and ion flux was determined as described in the text. VIP (10^{-8} M) was added 52 min after the addition of $^{22}Na^+$ and $^{36}Cl^-$ to the chambers. Samples were obtained at 10-min intervals. M → S represent mucosal to serosal flux or ion flux from the apical reservoir to the basolateral reservoir and S → M represents flux in the opposite direction. Chloride ion fluxes in both directions increased with the addition of VIP, with the increase in S → M flux being greater, resulting in a net Cl^- secretion. [From K. Dharmsathaphorn, J. A. McRoberts, H. Masui, and K. G. Mandel. *J. Clin. Invest.* **75**, 462 (1985) with permission from the publisher.]

reservoir bathing a cell monolayer in the Ussing chamber. This type of study may provide clues as to which transport pathway is involved in a given transport process and as to which side (apical versus basolateral) the transport pathway is located. Similarly, information regarding a secondary messenger involved in a receptor-mediated phenomenon can be obtained if an inhibitor is available. Studies utilizing specific inhibitors are relatively easy to do and may provide guidance for direct studies which are more difficult. The sidedness of a transport pathway or a membrane receptor can be easily determined in cells cultured on permeable supports because both the basolateral membranes and the apical membrane of the cultured epithelial monolayers are accessible. Cultured cell lines may also be helpful for identifying a drug or an antibody which specifically blocks a transport pathway. This is particularly the case if the cell line examined displays only one transport process. For example, agents that inhibit Cl^- secretion when administered to the apical membrane of the T_{84} cells must either block the Cl^- channel or interfere with an intracellular mechanism crucial for the Cl^- secretory process.[24-26]

Radionuclide Uptake and Efflux Studies in Whole Cells

General Considerations

The purpose of these techniques are two-fold: (1) to verify the presence of a given transport pathway which may be an ATPase pump, a carrier, or an ionic channel, and (2) to identify and study influences on transport rate of both activators or inhibitors of the transport system of interest. The principle underlying these techniques is analogous to those utilized for isolated cells and plasma membrane vesicles. As a matter of fact, most techniques pioneered by a membrane physiologists can be applied to cultured cells. Radionuclide uptake and efflux studies are carried out using two assay conditions which complement one another: (1) steady state and (2) nonsteady state.

The steady state conditions are those in which there is no electrochemical gradient imposed and all the transport pathways in the cell are functional. Steady state uptake and efflux conditions are analogous to the situation in the Ussing chamber discussed earlier. These studies are carried out under "physiological" conditions which in our studies equate with a bath composed of modified Ringer's solution or normal growth medium. The results of these studies performed on monolayers can thus be extrapolated to transport processes exhibited by monolayers mounted under similar conditions in the Ussing chamber. At basal state, there should be no net ion movement into or out of the cell. However, once a transport mechanism is activated, secondary events usually follow to maintain cellular homeostasis. This results in simultaneous movement of ions through several transport pathways, changes in membrane potential, and changes in cell volume. The complexity of this response to stimulation of a transport process makes interpretation of steady state results difficult. Moreover, the initial rate of ion transport under a steady state condition is also more difficult to measure. For these reasons, the nonsteady state conditions discussed below are often carried out to complement the steady state studies. These two complementary methods usually explain the mechanism involved in a transepithelial transport phenomenon.

Nonsteady state assay conditions for ion uptake and efflux follow the same principles as those of plasma membrane vesicles or whole cells. An abrupt change in the extracellular media may be used to set up a desirable electrochemical gradient driving ion movements either in the uptake or efflux direction. Various blockers, e.g., ouabain, bumetanide, and barium, can be used to inhibit secondary movements of ions which may confound interpretation of the results, leaving only the transport pathway of interest in operation. Blockage of transport pathways not under investigation can also be achieved by adjusting the composition of extracellular (and intracellular) ion, so as to deplete the substances these pathways recognize. Nonsteady state assays usually are carried out by varying the extracellular ionic composition utilizing cells in which the intracellular ionic composition is known.[24] Varying the extracellular ionic content is more practical because active transport mechanisms in living cells make the intracellular ionic contents more difficult to alter and less predictable. However, direct measurements of intracellular ionic composition can be obtained and one should not hesitate to alter the preincubation procedure to achieve the intracellular ionic composition needed for the study. Table II shows the intracellular concentrations of ions obtained after various preincubation procedures used in our studies.

Radionuclide uptake or efflux assays are carried out either in monolayers grown on permeable supports or monolayers attached to culture dishes (Fig. 8). Monolayers grown on permeable supports are most useful for study of the sideness of a transport event. The cultured epithelial cells with well-developed tight junctions obtained by growing monolayers on permeable supports allow for the segregation of the apical from basolateral membrane transport processes. This segregation is possible because the monolayers permit separate and specific access to either the apical or basolateral surfaces. Therefore, one may utilize these cultured monolayers to obtain information analogous to that provided by studies of purified apical and basolateral membrane vesicles. In addition, efflux of ion across the apical and basolateral surfaces can be measured separately and simultaneously using monolayers grown on permeable supports. This is done by preloading the monolayers with the isotope and mounting them between two fluid-filled chambers (Ussing chamber). Because growing monolayers on permeable supports requires more labor, when feasible, uptake and efflux studies are carried out in monolayers grown on culture dishes. Monolayers grown on culture dishes are ideal when the transport pathway of interest is on the apical surface, e.g., Cl$^-$ channels of the T$_{84}$ cells.[24]

TABLE II

CELLULAR IONIC COMPOSITION UNDER DIFFERENT INCUBATION CONDITIONS[a,b]

Incubation condition	Cell volume (μl/mg)	Ion (tracer)	Media concentration (mM)	Intracellular concentration (mM)
Culture medium	10.7 ± 0.2	K^+ ($^{86}Rb^+$)	4.6	84.0 ± 2.6
		Na^+ ($^{22}Na^+$)	165	25.0 ± 1.1
		Cl^- ($^{36}Cl^-$)	126	29.4 ± 2.9
KCl buffer (for 90 min)	7.6 ± 0.3	Rb^+ ($^{86}Rb^+$)	140	137.0 ± 6.0
		Na^+ ($^{22}Na^+$)	0	1.7 ± 0.2
		Cl^- ($^{36}Cl^-$)	140	97.4 ± 7.7
KCl buffer (for 90 min) followed by sucrose–ouabain buffer (for another 90 min)	6.6 ± 0.1	Rb^+ ($^{86}Rb^+$)	0	71.3 ± 2.8
		Na^+ ($^{22}Na^+$)	0	0.7 ± 0.2
		Cl^- ($^{36}Cl^-$)	0	6.4 ± 0.3

[a] From K. G. Mandel, K. Dharmsathaphorn, and J. A. McRoberts, *J. Biol. Chem.* **261,** 704 (1986).
[b] Cell volume determinations R. F. Kletzien, M. W. Pariza, J. E. Becker, and V. R. Potter, *Anal. Biochem.* **68,** 537 (1975) were made in glucose-free media following a 90-min incubation in buffers containing 2, 5, or 10 mM 3-O-methylglucose. The intracellular levels of the respective ions were determined after a 90-min incubation in the buffers containing the indicated tracers, or on cells preloaded in the previous step with the given isotope. Values are the mean ±95% confidence limits (cell volume) or SD (ion concentrations) of triplicate determinations. The composition of culture medium is listed in the appendix. KCl buffer contains 140 mM KCl, 10 mM Tris-HEPES, and 1 mM $MgSO_4$. Sucrose–ouabain buffer contains 241 mM sucrose and 0.5 mM ouabain.

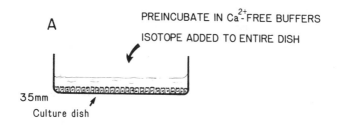

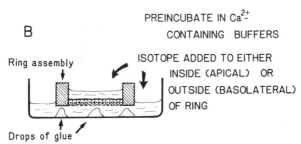

FIG. 8. Radionuclide uptake or efflux studies. The studies can be carried out with monolayers attached to impermeable supports, such as a culture dish (A), or monolayers attached to permeable supports, i.e., the ring assembly for the Ussing chamber (B). The latter provides information regarding the sidedness of the transport process.

Access to the basolateral surface can also be achieved by preincubating the monolayers in a Ca^{2+}-free medium to disrupt the tight junctions. This approach has been used to study the basolaterally localized K^+ channels of the T_{84} cells.[23]

Uptake studies are carried out by adding isotope to the monolayer, then washing away the isotope at time intervals and counting the intracellular radioactivity. Efflux studies are carried out after preincubating the monolayers with radioisotope. The efflux process is initiated by adding medium (without the isotope) to the monolayer, washing the medium off at appropriate time intervals, and counting the remaining intracellular radioactivity. The wash procedure for both uptake and efflux involves the use of $MgCl_2$- or $MgSO_4$-sucrose solution for reasons indicated below. Uptake or efflux is terminated by washing the monolayer rapidly two or three times with 2 ml ice-cold Mg^{2+}-sucrose wash buffer (100 mM $MgCl_2$, 137 mM sucrose, if $^{36}Cl^-$ flux is not being measured, or 100 mM $MgSO_4$, 137 mM sucrose for $^{36}Cl^-$ flux). For uptake studies, this is followed by an incubation on ice in this wash buffer for 3 to 5 min. We found that the wash procedure removes approximately 80% of the extracellular bound $^{22}Na^+$, $^{86}Rb^+$, or $^{36}Cl^-$ while retaining more than 90% of the intracellular radioisotope.[23,24,37,38] Effluxes are terminated without the incubation on ice because specific activity in the extracellular fluid is small and thus relatively little radioactivity is bound to the cells.[2,23,24] The intracellular radioactivity to be subsequently counted is extracted either by leaving the cells overnight in water or by adding 0.5 N NaOH. For uptake studies using confluent monolayers grown on permeable supports, Ca^{2+}-containing buffers are used and isotope is added to either the apical surface or basolateral surface. At appropriate time intervals, the uptake is terminated by the same Mg^{2+}-sucrose wash procedure and intracellular radioactivity is measured as described above. Efflux of $^{86}Rb^+$ is usually carried out after mounting the preloaded monolayers in Ussing chambers by sampling the bathing medium at intervals. As far as tracer isotopes are concerned, $^{22}Na^+$, $^{36}Cl^-$, and $^{42}K^+$ are appropriate isotopes for Na^+, Cl^-, and K^+, respectively. $^{86}Rb^+$, which has a longer half-life as compared to $^{42}K^+$, has been utilized as a tracer for K^+ after we have demonstrated its suitability for this purpose (Table III). It should be pointed out that $^{42}K^+$ is a better choice for a less extensive study. This is especially true for transepithelial flux experiments when more than one K^+ transport pathway is involved because of the complexity of verifying the suitability of $^{86}Rb^+$ as a tracer.

[37] J. A. McRoberts, S. Erlinger, M. J. Rindler, and M. H. Saier, Jr., *J. Biol. Chem.* **257**, 2260 (1982).
[38] M. J. Rindler, J. A. McRoberts, and M. H. Saier, Jr., *J. Biol. Chem.* **257**, 2254 (1982).

TABLE III
COMPARISON OF $^{86}Rb^+$ AND $^{42}K^+$ AS TRACERS FOR Rb^+ AND K^+ [a,b]

Cation	Tracer	Initial rate of uptake [nmol/(min · mg)]		Net rate of efflux (hrs^{-1})	
		VIP stimulated (ΔV)	A23187 stimulated (ΔV)	VIP stimulated (Δk)	A23187 stimulated (Δk)
K^+	$^{86}Rb^+$	10.5 ± 1.9	7.4 ± 1.5	0.41 ± 0.10	0.81 ± 0.15
K^+	$^{42}K^+$	4.4 ± 1.8	6.9 ± 1.6	0.31 ± 0.06	0.94 ± 0.14
Rb^+	$^{86}Rb^+$	6.6 ± 1.8	6.9 ± 1.6	0.32 ± 0.06	0.81 ± 0.14

[a] The validity of $^{86}Rb^+$ as a tracer for K^+ was tested by comparing the levels of VIP- and A23187-stimulated uptake and efflux when $^{86}Rb^+$ was used as a tracer for Rb^+ and when $^{86}Rb^+$ or $^{42}K^+$ were used as tracers for K^+. The increase in the rate of uptake (ΔV) and the increase in the apparent rate of efflux (Δk) in the presence of activators over the solvent controls are shown. The detailed methodologies are described in Ref. 23. The apparent rate of A23187-stimulated uptake and efflux did not differ significantly under any of the three conditions. VIP-stimulated K^+ uptake measured with $^{42}K^+$ approximated the results measured with $^{86}Rb^+$. It was only when $^{86}Rb^+$ was used as a tracer for K^+ that the apparent rate of uptake was significantly greater. The efflux results did not differ significantly from one another. Taken together, the results suggest that the A23187-stimulated mechanism of cation transport does not discriminate significantly between Rb^+ or K^+, but the VIP-stimulated mechanism may preferentially transport Rb^+ over K^+. Overall, $^{86}Rb^+$ should be a suitable tracer for qualitative studies.

[b] From J. A. McRoberts, G. Beuerlein, and K. Dharmsathaphorn, *J. Biol. Chem.* **260**, 14163 (1985).

In the Cl^--secreting T_{84} cell, we have verified the presence of an $Na^+/K^+/Cl^-$ cotransport pathway, an Na^+,K^+-ATPase, and two types of K^+ channels on the basolateral membrane as well as a Cl^- channel on the apical membrane. All four transport pathways participate in the Cl^- secretory mechanism. An Na^+/H^+ antiport has also been recognized, although its role has yet to be defined. The discussion to follow will be limited to these transport pathways. However, by employing similar strategies, proper experimental conditions could theoretically be created to study any known transport pathway.

$Na^+/K^+/Cl^-$ Cotransport Pathway. This pathway is most easily quantitated as the initial rate of bumetanide-sensitive $^{22}Na^+$, $^{86}Rb^+$, or $^{36}Cl^-$ uptake into Na^+-depleted cells. Exclusion of Na^+, K^+, or Cl^- has similar effects as the use of bumetanide. The preincubation protocol involves two steps. First, the monolayer is incubated in isotonic KCl buffer (140 mM KCl, 10 mM Tris-HEPES, 1 mM MgSO$_4$) for 90 min. This is followed by a second preincubation in isotonic sucrose-ouabain buffer (241 mM sucrose, 0.5 mM ouabain) for another 90 min. This preincubation protocol reduces

the intracellular Na^+ to less than 1 mM (Table II) and inhibits both the Na^+,K^+-ATPase and the Na^+/H^+ antiporter by depleting the intracellular substrate, Na^+. Intracellular K^+ and Cl^- concentrations are reduced to approximately 71 and 6 mM, respectively (Table II). The presence of ouabain further assures that the Na^+,K^+-ATPase pump is inactivated. Because the effective gradient driving net ion movement across the plasma membrane is the difference between $[Na^+][K^+][Cl^-]^2$ on the two sides, any uptake medium with a relatively higher $[Na^+]K^+][Cl^-]^2$ will be able to drive uptake of Na^+,K^+, and Cl^- into the cell. The only requirement is the simultaneous presence of all three ions. In our studies, uptake buffers contained 7 to 35 mM Na^+, 7 to 35 mM Rb^+ (or K^+), and 35 to 140 mM Cl^-. The use of bumetanide or exclusion of any one of the three ions involved allows us to verify the presence of an $Na^+/K^+/Cl^-$ cotransport pathway (Fig. 9). Uptakes under our study conditions are linear for 90 to 180 sec and extrapolation of the initial rates of uptake suggests that Na^+, Rb^+, and Cl^- are transported with a stoichiometry of 1 : 1 : 2 (Fig. 10). Bumetanide-sensitive $Na^+/K^+/Cl^-$ cotransport can also be quantitated in the efflux direction. This is most readily demonstrated using $^{86}Rb^+$ (or $^{42}K^+$) because the cells can be loaded to high specific activity by the Na^+,K^+-ATPase. A steady state accumulation of $^{86}Rb^+$ is obtained in 3 to 4 hr with intracellular Rb^+ concentration in the range of 90 to 110 mM. Both steady state and nonsteady state efflux of $^{86}Rb^+$ can be assessed by simply washing the monolayers and incubating them in the appropriate medium without the radioactive tracer. $^{86}Rb^+$ efflux studies in the presence or absence of 0.1 mM bumetanide will estimate the efflux via the $Na^+/K^+/Cl^-$ cotransport pathway. The rate constant of Rb^+ efflux can be estimated from the time course of efflux. Long time intervals (4 to 6 hr) are needed in order to obtain first order kinetics for steady state studies because $^{86}Rb^+$ is reaccumulated by the Na^+,K^+-ATPase. Under a nonsteady state experimental condition when the Na^+,K^+-ATPase is inhibited, all or most of the $^{86}Rb^+$ will eventually leave the cell and the data can be plotted directly on semilogarithmic plots to obtain first order kinetics. Bumetanide inhibits the rate of $^{86}Rb^+$ efflux and increasing the $t_{1/2}$ from about 2 hr to about 6 hr.[23]

The sidedness of bumetanide-sensitive $Na^+/K^+/Cl^-$ cotransport can be demonstrated easily by using cells grown on permeable supports. Radioactive uptake buffers are added to only one side of the ring (apical or basolateral side). Using this technique, we have shown that bumetanide-sensitive uptake of $^{22}Na^+$, $^{86}Rb^+$, and $^{36}Cl^-$ occurs across the basolateral membrane but not the apical membrane of T_{84} cells.[20]

Potassium Ion Channels. Potassium ion channels can be easily demonstrated using either steady state or nonsteady state $^{86}Rb^+$ efflux. Because

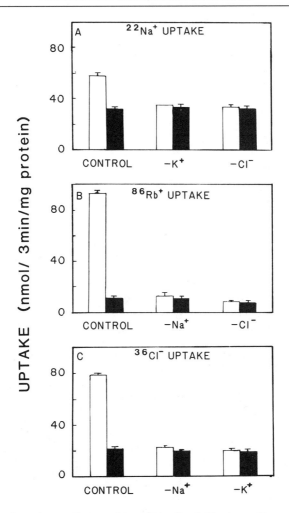

FIG. 9. Interdependence of bumetanide-sensitive ^{22}Na$^+$, ^{86}Rb$^+$, and ^{36}Cl$^-$ uptake, suggesting the existence of an Na$^+$/K$^+$/Cl$^-$ cotransport pathway. Open bars represent uptake in the absence of bumetanide, and solid bars represent uptake in the presence of 0.2 mM bumetanide. T$_{84}$ monolayers attached to 35-mm culture dishes were preincubated as described in the text. In the control experiments, uptake buffer contained 35 mM sodium gluconate, 35 mM potassium gluconate, 70 mM N-methylglucamine chloride, 10 mM Tris-SO$_4$, pH 7.5, and 1.2 mM MgSO$_4$. In the ion substitution experiments, when Na$^+$, K$^+$, or Cl$^-$ were excluded from the uptake buffers, sucrose was added to maintain isotonicity. Experiments depicted in each panel were carried out on different days with different sets of confluent monolayers. The interdependence of Na$^+$, K$^+$, and Cl$^-$ uptakes and their sensitivity to bumetanide inhibition are demonstrated and suggest the presence of a bumetanide-sensitive Na$^+$/K$^+$/Cl$^-$ cotransport pathway. [From K. Dharmsathaphorn, J. A. McRoberts, H. Masui, and K. G. Mandel, *J. Clin. Invest.* **75,** 462 (1985) with permission from the publisher.]

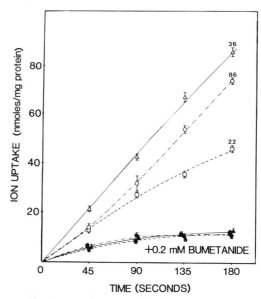

Fig. 10. Time course of bumetanide-sensitive $^{22}Na^+$, $^{86}Rb^+$, and $^{36}Cl^-$ uptake by T_{84} cell monolayers to determine the stoichiometry of the $Na^+/K^+/Cl^-$ cotransport pathway. A single set of confluent monolayers was preincubated as described in the text with the final preincubation buffer containing both 0.5 mM ouabain and 0.5 mM amiloride. Uptake buffer contained 70 mM choline chloride, 35 mM $NaNO_3$, 35 mM $RbNO_3$, 10 mM Tris-NO_3, pH 7.5, and 1.2 mM $Mg(NO_3)$, with trace amounts of either $^{86}Rb^+$ (O, ●), $^{22}Na^+$ (□, ■), or $^{36}Cl^-$ (△, ▲). Uptakes were carried out on the presence (●, ■, ▲) or absence (O, □, △) of 0.2 mM bumetanide for the given time intervals and terminated using $MgSO_4$–sucrose wash, as described in the text. Each point represents the mean ± SD of triplicate determinations. Zero time values (1–2 nmol/mg protein) have been subtracted from all data. Extrapolation of the initial rate of the bumetanide-sensitive $^{22}Na^+$, $^{86}Rb^+$, and $^{36}Cl^-$ uptake suggests that Na^+, Rb^+, and Cl^- are transported with a stoichiometry of 1:1:2. [From K. Dharmsathaphorn, J. A. McRoberts, H. Masui, and K. G. Mandel, *J. Clin. Invest.* **75**, 462 (1985) with permission from the publisher.]

the monolayers can be loaded with $^{86}Rb^+$ to a high specific activity, the steady state $^{86}Rb^+$ efflux rate can be easily measured using preloaded monolayers mounted in the Ussing chamber by counting the increased radioactivity in the bathing medium at time intervals. Addition of either VIP or A23187 (in the presence of Ca^{2+}) dramatically stimulates the rate of $^{86}Rb^+$ efflux from $^{86}Rb^+$-loaded T_{84} monolayers mounted in the Ussing chamber (Fig. 11).[21] Similar results are obtained from $^{86}Rb^+$-loaded cells grown on culture dishes (Fig. 12).[23] Under steady state conditions in the Ussing chamber, VIP, PGE_1, adenosine, A23187, carbachol, or histamine increase the rate of $^{86}Rb^+$ efflux across the basolateral membrane by one- to threefold, but had little or no effect on $^{86}Rb^+$ efflux across the apical

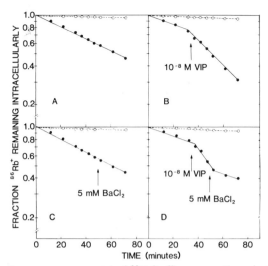

FIG. 11. Simultaneous measurement of $^{86}Rb^+$ efflux across the apical and basolateral membrane of T_{84} monolayers mounted in the Ussing chamber: Evidence for a basolaterally localized K^+ efflux pathway stimulated by VIP and blocked by barium. For this set of experiments, monolayers were grown on collagen-coated Nuclepore filters, under conditions identical to those used in the isotopic flux studies described in Fig. 7. Prior to use, the monolayers were equilibrated overnight with $^{86}Rb^+$ by adding 0.35 μCi/ml $^{86}Rb^+$ to the culture medium (~4.3 mM K^+) and sterile filtering. The preincubated monolayers were washed rapidly in two changes of Ringer's solution, mounted in the Ussing chambers, and voltage clamped. The washing and mounting procedure took approximately 15 sec. Serial samples were taken simultaneously from the serosal and mucosal reservoirs (every 5 or 10 min for this study) with buffer replacement. VIP (10 nM) or 5 mM $BaCl_2$ was added as indicated to both mucosal and serosal reservoirs. The data were corrected for sampling volume (0.5 ml of a 5-ml reservoir) and previous sampling loss due to buffer replacement. The results for serosal and mucosal $^{86}Rb^+$ efflux were analyzed independently using the closed three-compartment system with the assumption that $^{86}Rb^+$ acts as a perfect tracer for K^+. The rate of return flux back into the cell from either compartment was negligibly small as compared to the rate of $^{86}Rb^+$ efflux [K. G. Mandel, J. A. McRoberts, G. Beuerlein, E. S. Foster, and K. Dharmsathaphorn, *Am. J. Physiol.* **250**, C486 (1986)]. Therefore, the differential equations describing the three-compartment system [Solomon, *Adv. Biol. Med. Phys.* **3**, 65 (1953)] can be reduced to those describing simple decay. The apparent rate constants for serosal (k_s) and mucosa (k_m) efflux were determined for each sampling time using the relationships: $k_s = (\Delta Q/t) [1/(P_0\text{-}R\text{-}Q)]$ and $k_m = (\Delta R/t) [1/(P_0\text{-}R\text{-}Q)]$, where P_0 is the initial amount of $^{86}Rb^+$ in the cell, and R and Q are the amount of isotope in the mucosal and serosal compartments, respectively. P_0 was determined experimentally as the total amount of $^{86}Rb^+$ in replicate tissues and as the total amount in the experimental tissues at the end of the experiment plus the amount in the bathing medium. The intracellular K^+ of the monolayers was estimated to be approximately 639 nmol at the start of the experiment (derived from $^{86}Rb^+$ counts of the monolayers as compared to $^{86}Rb^+$ counts in the culture medium with $[K^+]$ of 4.3 mM). Extracellular K^+ in the bathing reservoir was 26 μmol (5 ml of solution with $[K^+]$ of 5.2 mM). The results of serosal efflux ($\bullet$) and mucosal efflux ($\circ$) are plotted as $1 - [Q/(P_0\text{-}R\text{-}Q)]$ for serosal efflux and $1 - [R/(P_0\text{-}R\text{-}Q)]$ for mucosal efflux. The negative slope of the line over time represents the rate constant for $^{86}Rb^+$ efflux. VIP (10^{-8} M) stimulatd $^{86}Rb^+$ efflux across the basolateral membrane but has no effect on $^{86}Rb^+$ efflux across the apical membrane. $BaCl_2$, which by itself has no effect on $^{86}Rb^+$ efflux, effectively abolished VIP-stimulated efflux.

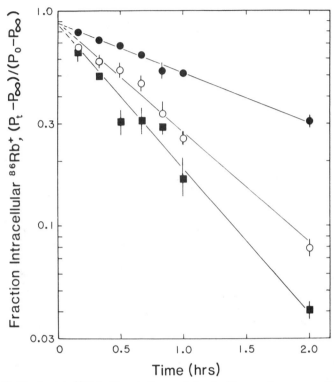

Time (hrs)

Fig. 12. Steady state ^{86}Rb$^+$ efflux studies carried out with monolayers on culture dishes: Evidence for a VIP- and A23187-stimulated K$^+$ efflux pathway. Confluent monolayer cultures of T$_{84}$ cells on replicate 35-mm culture dishes were preloaded with ^{86}Rb$^+$ as described [J. A. McRoberts, G. Beuerlein, and K. Dharmasathaphorn, *J. Biol. Chem.* **260**, 14163 (1985)]. Efflux was initiated by rapidly washing the monolayer three times with 2 ml ^{86}Rb$^+$-free loading buffer without an activator (●), with 10^{-7} M VIP (○), or with 10^{-6} M A23187 plus 1.2 mM CaCl$_2$ (■). At the indicated times, effluxes were terminated using the MgCl$_2$ wash procedure and processed for counting as described in the text. The data are presented on a log scale of $(P_t - P_\infty)/(P_0 - P_\infty)$. The initial amount of loaded Rb$^+$ (P_0) was 706 ± 15 nmol/dish (0.74 mg protein). The value for the new cellular steady state ^{86}Rb$^+$ (P_∞) was determined after 5 hr in ^{86}Rb$^+$-free loading buffer and in each case was 160 ± 5 nmol/dish. P_t is the amount of intracellular ^{86}Rb$^+$ at time t. Using the 3-O-methyl-D-glucose method of volume determination [J. Kletzien, *Anal. Biochem.* **68**, 537 (1975)], the initial steady state concentration of intracellular Rb$^+$ was 103 mM. Each point represents the mean of duplicate determinations and error bars the range. Lines are the best least-squares fit to all determinations. The results showed that this method detects both VIP- and A23187-stimulated ^{86}Rb$^+$ efflux similar to the results obtained with monolayers mounted in the Ussing chamber. [From J. A. McRoberts, G. Beuerlein, and K. Dharmsathaphorn, *J. Biol. Chem.* **260**, 14163 (1985) with permission from the publisher.]

membrane. The increase in the rate of ^{86}Rb$^+$ efflux across the basolateral membrane induced by VIP, PGE$_1$, or adenosine (cAMP-mediated) is more sensitive to Ba^{2+}, while those induced by A23187, carbachol, or histamine (Ca^{2+} mediated) are less sensitive.[21,26,27,29] Furthermore, cAMP- and Ca^{2+}-mediated responses are additive, suggesting the presence of two different K$^+$ channels.[22,27] Despite the simplicity of the experiments, interpretation of the steady state results is complicated by the activities of the Na$^+$,K$^+$-ATPase and the Na$^+$/K$^+$/Cl$^-$ cotransport pathway, both of which transport K$^+$ simultaneously. To determine whether the K$^+$ channel is directly activated by these agents, the Na$^+$,K$^+$-ATPase and the Na$^+$/K$^+$/Cl$^-$ cotransport pathways are inhibited by incubating the preloaded cells in isotonic sucrose or choline chloride buffer containing ouabain and bumetanide. Both VIP and A23187 stimulate ^{86}Rb$^+$ efflux into K$^+$-free NaCl buffer containing ouabain and bumetanide (Fig. 13).[23] The results suggest that the opening of K$^+$ channels is a direct action of VIP and A23187 independent of the activities of the Na$^+$,K$^+$-ATPase and the Na$^+$/K$^+$/Cl$^-$ cotransport pathway. Finally, the K$^+$ channels can also be assayed in the uptake direction using the same preincubation protocol. Both VIP and A23187 stimulate the initial uptake rate of ^{86}Rb$^+$ and ^{42}K$^+$.[23]

Chloride Ion Channel. Chloride ion channel activity, similar to that of K$^+$ channel, can also be demonstrated under steady state as well as nonsteady state conditions. The Cl$^-$ channel, when opened, allows bidirectional movement of Cl$^-$ across the plasma membrane, thus the electrochemical gradient across this membrane determines the direction of net flux. Because of a relatively low intracellular Cl$^-$ concentration and a relatively low specific activity of ^{36}Cl$^-$ available commercially, Cl$^-$ efflux studies (Fig. 14)[24] require a large, prohibitively expensive amount of ^{36}Cl$^-$ to adequately load the cell. Since at the initiation of uptake studies no intracellular ^{36}Cl$^-$ exists, this approach yields clearer results, requires smaller amounts of ^{36}Cl$^-$, and thus is preferred. These methods are useful in screening agents which may influence Cl$^-$ channel activity, since the assays are carried out under physiological conditions. For example, we have demonstrated under steady state conditions that addition of VIP, but not A23187, to T$_{84}$ cells stimulate the rate of ^{36}Cl$^-$ efflux as well as uptake.[24] However, the increased initial rate of uptake under these conditions is barely detectable. This difficulty in detecting changes results from the high rate of Cl$^-$/Cl$^-$ exchange between intra- and extracellular compartments paired with the relatively low intracellular Cl$^-$ concentration. Furthermore, the results are complicated by the reduction in cellular Cl$^-$ concentration (from about 40 to 25 mM) as well as the decrease in cell volume (about 20–35%) elicited by VIP. Nonsteady state studies, therefore, become important adjuncts in this instance. In these nonsteady state

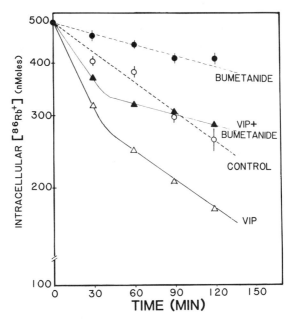

FIG. 13. Nonsteady state $^{86}Rb^+$ efflux studies carried out with monolayers on culture dishes: Evidence for the direct stimulation of the K^+ efflux pathway by VIP. Confluent monolayer cultures of T_{84} cells were loaded to a steady state with $^{86}Rb^+$ as described in text. The loading buffer consisted of 135 mM NaCl, 10 mM Tris-Cl, pH 7.5, 10 mM glucose, 5 mM KCl, 1.2 mM MgCl$_2$, and 0.5 μCi/ml $^{86}Rb^+$. The monolayers were then treated for 1 hr with 0.5 mM ouabain in 140 mM choline chloride buffer. Efflux was initiated by replacing this buffer with 140 mM NaCl buffer containing 0.5 mM ouabain, with (●, ▲) or without (○, △) 0.1 mM bumetanide and with (●, ○) or without (▲, △) 10^{-7} M VIP. In the first 30 min, $^{86}Rb^+$ efflux stimulatad by VIP occurs despite the inhibition of the Na$^+$,K$^+$-ATPase and Na$^+$/K$^+$/Cl$^-$ cotransport activity. The results indicate that VIP directly activates the $^{86}Rb^+$ efflux pathway.

studies, $^{36}Cl^-$ uptake associated with opening of the Cl^- channel is facilitated by preincubating the cells in such a way as to produce both a low intracellular Cl^- concentration and a positive intracellular electrical potential. These conditions are achieved by using the same Na$^+$ depletion protocol described earlier for the study of Na$^+$/K$^+$/Cl$^-$ cotransport. Monolayers, either on plastic or permeable supports, are first preincubated in KCl buffer (140 mM KCl, 10 mM Tris-HEPES, 1 mM MgSO$_4$) followed by preincubation in sucrose-ouabain buffer (241 mM sucrose, 0.5 mM ouabain). This preincubation procedure reduces intracellular Cl^- to about 6 mM. Besides lowering intracellular Cl^- concentration, the preincubation procedure produces a positive intracellular potential (normally about -40 mV). Therefore, $^{36}Cl^-$ uptake is facilitated and the increased uptake can be

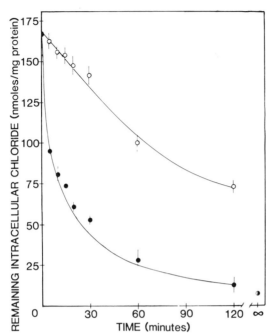

FIG. 14. $^{36}Cl^-$ efflux studies carried out with monolayers on culture dishes: Acceleration of the rate of $^{36}Cl^-$ efflux from T_{84} cells by VIP. T_{84} monolayers were loaded with $^{36}Cl^-$ by a 3-hr incubation at room temperature in KCl buffer (140 mM KCl, 10mM Tris-SO$_4$, pH 7.5, 1 mM MgSO$_4$), washed twice with 2 ml sucrose buffer (241 mM sucrose, 10 mM Tris-SO$_4$, pH 7.5, 1 mM MgSO$_4$), and incubated for 1 hr in 1 ml sucrose buffer containing 0.5 mM ouabain and 0.1 mM bumetanide. Efflux was then initiated by aspiration of the sucrose–ouabain buffer and addition of 1 ml of a buffer containing 120 mM sodium gluconate, 10 mM potassium gluconate, 10 mM NaCl, 20 mM HEPES-Tris, pH 7.5, 1 mM MgSO$_4$, 0.5 mM ouabain, and 0.1 mM bumetanide. The loss of intracellular isotope was measured in the presence (●) or absence (○) of 10^{-7} M VIP. At indicated times, efflux was terminated using the Mg^{2+}-sucrose wash procedure described in the text. The results are presented as the mean remaining intracellular $^{36}Cl^- \pm$ SD of triplicate monolayers versus time. $^{36}Cl^-$ efflux from T_{84} monolayers was accelerated by VIP. The results suggest that VIP opens a Cl$^-$ efflux pathway in the T_{84} cells. [From K. G. Mandel, K. Dharmasthaphorn, and J. A. McRoberts, *J. Biol. Chem.* **261,** 704 (1986) with permission of the publisher.]

readily detected when the Cl$^-$ channel is opened. This method for measuring the activity of Cl$^-$ channels also eliminates interference with changes in cell volume and membrane potential.

$^{36}Cl^-$ uptake studies are initiated by the addition of an uptake buffer containing 140 mM sodium or potassium gluconate with $^{36}Cl^-$ (2 to 6 mM final Cl$^-$ concentration) to the monolayers preincubated as described. A two- to fivefold stimulation in the initial rate of $^{36}Cl^-$ uptake is observed after prestimulation with VIP or PGE$_1$ (Fig. 15).[24,26,27] The accelerated

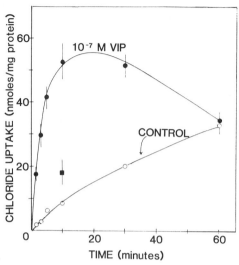

FIG. 15. $^{36}Cl^-$ uptake studies carried out with monolayers on culture dishes: Stimulation of the initial rate of Cl^- uptake by VIP. Confluent monolayers grown on 35-mm dishes were washed twice and incubated in 1 ml KCl buffer (140 mM KCl, 10 mM Tris-SO_4, pH 7.5, 1 mM $MgSO_4$) for 2 hr at room temperature. Cells were then washed with three changes of sucrose buffer (241 mM sucrose, 10 mM Tris-SO_4) and incubated in 1 ml of this buffer containing 0.5 mM ouabain and 0.1 mM bumetanide with or without 10^{-7} M VIP. After a 5-min preincubation, this buffer was aspirated and replaced with 0.7 ml sodium gluconate uptake buffer (140 mM sodium gluconate, 10 mM Tris-SO_4, pH 7.5, 1 mM $MgSO_4$, 0.5 mM ouabain, 0.1 mM bumetanide, and 1.0 μCi/ml $^{36}Cl^-$). The final $^{36}Cl^-$ concentration was approximately 2 mM with or without 10^{-7} M VIP. Uptakes were terminated and intracellular isotope counted as described in the text. The results shown represent the mean $\pm$ SD of triplicate measurements. $^{36}Cl^-$ uptake was accelerated by VIP. The results suggest that the VIP-regulated pathway functions in both efflux and uptake directions. [From K. G. Mandel, K. Dharmsathaphorn, and J. A. McRoberts, *J. Biol. Chem.* **261,** 704 (1986) with permission from the publisher.]

$^{36}Cl^-$ uptake is not inhibited by bumetanide or 4,4'-diisothiocyanostilbene-2,2'-disulfonate (DIDS), but is inhibited by *n*-phenyl-anthranilic acid, a putative Cl^- channel blocker.[24] Using monolayers grown on permeable supports, we were able to demonstrate that the secretagogue-stimulated $^{36}Cl^-$ uptake pathway was localized to the apical membrane of T_{84} cells. Subsequently, this pathway in the T_{84} cell has been shown by the patch-clamp technique to be a Cl^- channel (R. A. Frizzell, personal communication). Therefore, we believe that these radionuclide uptake methods, which are relatively simple, are useful for quantitation of the activity of Cl^- channels in cultured epithelial cells.

Na⁺/H⁺ Exchange and Other Na⁺ Transport Pathways

More potent and specific amiloride analogs for inhibition of the Na^+/H^+ and Na^+/Ca^{2+} exchange carriers and the Na^+ channel have recently become available. The amiloride analogs facilitate the investigation of these Na^+ transport pathways. 5-N-Ethyl-5-N-isopropylamiloride is a potent inhibitor for the Na^+/H^+ exchange carrier. 2,4-Dichlorophenylmethylguanidinoamiloride and benzamil are more specific for the Na^+/Ca^{2+} exchange pathway and the Na^+ channel, respectively. Appropriate experimental conditions can be designed to assess the role of these Na^+ transport pathways by kinetic analysis of Na^+ uptakes and effluxes.[38-40] An Na^+-depleted condition similar to that used for the studies of $Na^+/K^+/Cl^-$ cotransport and the Cl^- channel is suitable for studying the Na^+/H^+ exchange pathway.[40] Na^+/Ca^{2+} exchange can be demonstrated using Na^+-loaded monolayers, either as diuretic-sensitive $^{22}Na^+$ efflux or diuretic sensitive $^{45}Ca^{2+}$ uptake. Preincubation and assay under steady state conditions measure primarily the accelerative Na^+/Na^+ exchange,[38] which may represent transport via the Na^+ channel. Net Na^+ transport via an Na^+ channel should be electrogenic.

An Na^+/H^+ exchange pathway has been identified in the T_{84} cells. Sodium ion uptake via this pathway is sensitive to amiloride and has been shown to be modulated by pH gradients (Fig. 16). To our surprise, this transport carrier seems to be localized predominantly to the basolateral membrane. Amiloride-sensitive $^{22}Na^+$ uptake across the basolateral membrane is approximately ninefold higher than apical membrane uptake (13.6 ± 3.6 nmol/3 min per monolayer versus 1.6 ± 2.8 nmol/3 min per monolayer, respectively). The role of the Na^+/H^+ exchange pathway in the T_{84} cells remains to be elucidated.

Na⁺,K⁺-ATPase and Study of Driving Force

In spite of the suitability of the culture cell model for the study of transmembrane electrochemical driving force, our experience with such studies is limited. Understanding of the driving forces requires knowledge of the membrane potential, the free intracellular concentrations of all major ions, and a good estimation of cell volume, preferably with detailed time course values. The importance of the Na^+,K^+-ATPase pump to the Cl^- secretory process is attested to by the ability of ouabain to inhibit Cl^- secretion when this inhibitor is applied to the basolateral surface. We have

[39] M. J.Rindler, M. Taub, and M. H. Saier, Jr., *J. Biol. Chem.* **254**, 11431 (1979).
[40] M. J. Rindler and M. H. Saier, Jr., *J. Biol. Chem.* **256**, 10810 (1981).

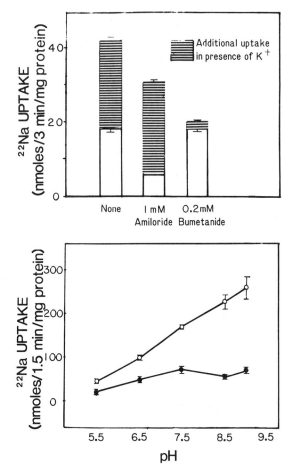

FIG. 16. $^{22}Na^+$ uptake studies: Evidence for an amiloride-sensitive Na^+/H^+ exchange pathway. *Top:* The effect of amiloride and bumetanide on $^{22}Na^+$ uptake in the presence and absence of K^+ and evidence for an amiloride-sensitive $^{22}Na^+$ transport pathway in addition to the $Na^+/K^+/Cl^-$ cotransport system. In this experiment, T_{84} monolayers were preincubated in KCl buffer followed by sucrose–ouabain buffer and assayed for 3 min in 14 mM NaCl buffer (clear bar) with no K^+ or 14 mM NaCl plus 35 mM KCl buffer (solid bar). N-Methylglucamine chloride was added to maintain isotonicity. $^{22}Na^+$ uptake in potassium-free medium, as shown by the clear bars, was inhibited by 1 mM amiloride, but not by 0.2 mM bumetanide. The solid bars represent additional uptake in the presence of K^+. This additional uptake was not inhibited by amiloride but was sensitive to bumetanide. Valves shown represent the mean ± SD of three determinations. *Bottom:* The sensitivity of this amiloride-sensitive pathway to H^+ gradient. The T_{84} cells were preincubated in 14 mM NaCl, 126 mM choline chloride, 10 mM Tris-PO$_4$, pH 7.5, and 0.5 mM ouabain for 2 hr. Uptakes were carried out in the same buffer with trace amounts of $^{22}Na^+$ with (●) or without (○) 30 μM amiloride at the indicated pH. Values shown represent the mean ± SE of three determinations.

also detected the opening of the Cl^- and K^+ channels as well as uptake of Cl^- via $Na^+/K^+/Cl^-$ cotransport. However, the changes in cell volume and membrane potential complicate the clear separation of primary and secondary contribution of these pathways to Cl^- secretion. Preliminary studies at steady states using isotopic techniques provide some useful insights. As expected, intracellular Cl^- and K^+ concentrations at the basal state are above their electrochemical equilibrium. Upon stimulation by VIP, which increases intracellular cAMP, the apically localized Cl^- channels as well as the basolaterally localized K^+ channels open. Since intracellular K^+ and Cl^- are above their electrochemical equilibrium, these two ions flow out of the cell. Indeed, at the new steady state level after VIP treatment there is reduction of intracellular Cl^- concentration together with reduction in cell volume.[24] Because efflux of K^+ would hyperpolarize the cell while efflux of Cl^- would depolarize the cell, changes in membrane potential which may regulate the opening and closing of ion channels become an important issue. Another important factor is cell volume regulation. With the exit of K^+ and Cl^-, one might expect a reduction in cell volume. Indeed, we have observed a 20 to 35% reduction in cell volume after VIP stimulation.[24] It is possible that the decrease in cell volume caused by K^+ and Cl^- loss might stimulate Na^+, K^+, and Cl^- uptake via the $Na^+/K^+/Cl^-$ cotransport pathway, analogous to volume regulatory increases described in other cells. Further investigation of the driving force should use other techniques which can provide the time courses of membrane potential, free intracellular ionic concentrations, and cell volume during activation by cGMP, cAMP, and Ca^{2+}. Measurements of membrane potential may be achieved with microelectrodes, ion-selective microelectrodes could be used to measure the concentration of free intracellular ions, and measurements of cell volume could be performed using morphometric techniques. These complementary methods should contribute valuable insights into the driving force for Cl^- secretion, including how Cl^- is maintained above its electrochemical equilibrium and the role of cell volume in the Cl^- secretory process.

Other Techniques

All other methodologies developed to study electrolyte transport in epithelial cells can be readily applied to cultured cells. Actually, the application to cultured cells may be easier. Among the techniques currently available, patch clamp and microelectrodes have already been applied successfully by other investigators. Conventional methods for plasma membrane preparations[41-43] can be applied to purify the apical and baso-lateral membrane vesicles. Because intact monolayers duplicate many ad-vantages of purified plasma membrane vesicles, we believe that membrane vesicles from homogeneous culture cells are most useful as starting mate-rial for biochemical or molecular characterization of the transporter pro-tein. The fact that some cell lines have only one transport function, e.g., Cl^- secretion by T_{84} cells, may facilitate the selection of monoclonal anti-bodies or specific blockers. Antibodies or agents that inhibit Cl^- secretion when applied to the apical surface either block the Cl^- channel itself or inhibit an intracellular mechanism which is intimately involved in the Cl^- secretory process. Finally, mutant cells defective for a given transport pathway may be selected to aid the study of that pathway and its relation to others.[2-5]

Regarding receptor-mediated mechanisms, standard techniques for measurement of cAMP, cGMP, free cytosolic Ca^{2+}, inositol triphosphate, and diacylglycerol can be easily applied to the monolayer sheet. In our experience, the correlation between receptor binding and biological re-sponses is straightforward in monolayers and easier to interpret than in animal studies. This may be due to the fact that culture cells are homoge-neous and free of peptides or neurotransmitter contaminants.

Problems Encountered

It is fortunate that the problems encountered by us are relatively few. Routine cell culture problems, such as contamination and other mishaps, arise when less care is given to the procedure involved. Theoretically, mutation may occur and the cells may change, therefore we have taken care to freeze the cells at intervals.[44]

Conclusions

Cultured cell lines may serve as model systems for the study of electro-lyte transport and its regulatory mechanisms. This chapter reviews our experience with a colonic cell line, T_{84}, which has the capacity to secrete

[41] E. M. Wright, C. H. van Os, and A. K. Mircheff, *Biochim. Biophys. Acta* **597**, 112 (1980).
[42] A. K. Mircheff, C. H. van Os, and E. M. Wright, *J. Membr. Biol.* **52**, 83 (1980).
[43] M. Kessler, O. Acuto, C. Storelli, H. Murer, M. Muller, and G. Semenza, *Biochim. Biophys. Acta* **506**, 136 (1978).
[44] P. Masur, *Am. J. Physiol.* **247**, C125 (1984).

Cl⁻ in response to regulatory agents. Morphologically, the T_{84} cell line retains many characteristics of the major population of cells found in intestinal crypts which appear to be the Cl⁻ secretory cells of the intestinal epithelium. Functionally, the cells possess many receptor-mediated regulations for Cl⁻ secretion similar to normal colon. The uniformity of cultured cells facilitates the identification of transport pathways involved in the Cl⁻ secretory process and the cascade of activation by regulatory agents.

Identification of transport pathways involved in receptor-mediated Cl⁻ secretion has been carried out using two complementary methods: (1) the Ussing chamber, which identifies transepithelial transport phenomena, and (2) radionuclide uptake and efflux studies, which identify transport pathways and detect their activation by regulatory agents. Other electrophysiological methods, e.g., microelectrodes and patch-clamp techniques, have been successfully applied to the cell line by other investigators. The study of intracellular amplifying mechanisms, e.g., cAMP, cGMP, free cytosolic calcium, inositol trisphosphate, diacylglycerol, and protein phosphorylation, is facilitated by the homogeneity of the cultured cells. Future studies of the secondary messengers in conjunction with identification of the transport pathway activated may explain the interaction between calcium and cAMP or cGMP, whose actions appear to be synergistic in this cell line. Finally, we have also identified other methodologies that may be applicable to culture cells. We predict that purified brush border and basolateral membranes will be useful for molecular characterization of the transport pathways. Development of monoclonal antibodies, which are critical for the application of molecular biology and molecular genetic techniques to the study of transport pathways, may be aided by the availability of culture cell lines which retain only one specific function. In addition, mutants defective in a specific transport pathway can be developed and selected for physiological studies. Hopefully, the methodologies presented here may also be applicable to other cultured epithelial systems.

Appendix

Solutions for T_{84} Cell Culture
1. HEPES stock solution: 1.5 M HEPES (357.4 g/liter); adjust pH to 7.4 with NaOH
2. Antibiotic stock solution: 1.0 g ampicillin, 4.8 g penicillin, 10.8 g streptomycin; adjust volume to 80 ml with water

3. Culture medium:

	Weight or volume needed for desired volumes		
	1 liter	10 liters	20 liters
F-12 (#430-1700; GIBCO, Grand Island, NY):	5.37 g	53.67 g	107.33 g
DME (GIBCO #430-2100):	6.78 g	67.84 g	135.67 g
NaHCO$_3$:	1.20 g	12.00 g	24.00 g
1.5 M HEPES stock solution:	10 ml	100 ml	200 ml
Antibiotic stock solution:	2 ml	20 ml	40 ml

Add newborn calf serum (Irvine Scientific, Santa Ana, CA), 5% (v/v), prior to use

4. Ca^{2+}- and Mg^{2+}-free phosphate-buffered saline (PBS) solution (137 mM NaCl, 2.7 mM KCl, 1.5 mM KH$_2$PO$_4$, 8 mM NaH$_2$PO$_4$, 0.015 g/liter Phenol Red):

Compound	M_r	Weight (g) needed for desired volumes			
		1 liter	6 liters	10 liters	20 liters
NaCl	58.44	8.00	48.00	80.0	160.0
KCl	74.55	0.20	1.20	2.0	4.0
KH$_2$PO$_4$	136.09	0.20	1.20	2.0	4.0
NaH$_2$PO$_4$	141.96	1.136	6.81	11.35	22.71
Phenol Red	—	0.015	0.09	0.15	0.30

5. Trypsin solution: 1.0 g trypsin (#103140; ICN Biochemicals, Irvine, CA), 0.3 g Na$_2$EDTA · 2H$_2$O; adjust volume to 1 liter with Ca^{2+} and Mg^{2+}-fre PBS solution above

Note: All solutions are sterilized by filtration or by autoclave.

Preparation of Rat Tail Collagen

1. Tails from rats are stored frozen.
2. Strip skin from tails. Grasp the end tailbone with a hemostat and twist it to break connection from adjoining bone. Pull the tail bone away sharply. This yields a tiny bone fragment with long, white, glistening tendons attached. The tendons are almost pure collagen. Cut off tendons from the bone fragment, mince them into short lengths with scissors, and place in a weighing boat.
3. Pull many of these tendon strands and weigh them (get at least 1 g).

4. Soak the tendon strands in 70% ethanol for a few minutes. This dehydrates them and also sterilizes them.

5. Solubilize 1 g of tendons in 100 ml 1% (v/v) acetic acid. The acid solution should be ice cold. Stir the mixture slowly in the cold room overnight (4°). By the following morning, the solution should be viscous. There will be a fair amount of undissolved tendon in the solution which is to be discarded.

6. Pour the collagen solution into centrifuge tubes. Centrifuge at 2000 g for 30 min at 4°. Decant the supernatant and discard the pelleted material.

7. Pour the supernatant containing the collagen into dialysis tubing (Spectropor #1). Dialyze the collagen solution overnight (in the cold room, 4°) against two changes of distilled water. Use 10 vol of water.

8. Store the rat tail collagen solution in small aliquots (100 ml or so) at 4°.

Procedure for Coating Nuclepore Filters

1. Use a forcep to dip the Nuclepore filter (5-μm pore size, Nuclepore Corporation #110613) into the rat tail collagen solution to completely coat it. This is most easily done in a 100-mm Petri dish on ice. *Keep the collagen cold!*

2. Drain excess collagen off filter by lifting it up with a forcep and scrap one side slowly over the edge of a glass slide. Place the "scrapped" side *down* on a glass or plastic plate. *Do not use a metal plate.*

3. Suspend the glass or plastic plate with the coated filters attached over vapor from a 2 M NH$_4$OH solution for 1 hr in a sealed plastic container. This precipitates the collagen to the filter surface.

4. Remove the plate with filters attached from the NH$_4$OH container and allow the filters to air dry for about 30 min.

5. Immerse the plate in a container of 4% (w/v) glutaraldehyde in water and allow to fix for 1 hr.

6. Remove the plate from the glutaraldehyde solution. The glutaraldehyde solution can be saved in the refrigerator and reused several times. Wash the plate with filters attached in running water for about 2 hr to remove the glutaraldehyde.

7. Air dry the filters overnight, then peel each of them off the plate and glue the filter onto the open end of a Lexan ring. Use silicone rubber adhesive (General Electric, RTV118) as glue. The side of the filter that was facing up will form the surface to plate cells.

8. Leave the entire ring assembly in ethanol to sterilize it before use.

Procedure for Trypsinization, Subculturing, and Plating T_{84} Cells

1. Aspirate media from confluent monolayers on culture dish(es) or flask(s).

2. Wash the culture dish with Ca^{2+}- and Mg^{2+}-free phosphate-buffered saline (PBS) solution. Use approximately 10 ml, then aspirate and discard the PBS solution.

3. Add 5 ml of trypsin solution (0.1% trypsin and 0.9 mM EDTA in PBS). Let stand in the hood for 5 min.

4. Aspirate and discard all but approximately 0.5 ml of the trypsin solution. Then return the culture dish(es) or flask(s) to the incubator. Cells will detach in 15–45 min, depending mainly on the degree of confluency of the monolayer and the activity of trypsin.

5. When cells slide upon tilting or shaking the plate, shake them off gently. Add 9 ml medium to the culture dish. The serum in the medium will inactivate the trypsin. Draw cell suspension up and down in a pipet to break up clumps and get an even suspension.

6. For subculturing, we generally split one confluent dish into four dishes and have a final medium volume of 10 ml in each 100-mm culture dish. This is done by adding 2.5 ml cell suspension from step 5 to 7.5 ml of medium. Since doubling time is 2.5 to 3 days, these new monolayers should reach confluency in about 1 week. Confluent monolayers in each 100-mm culture dish contain approximately 10 to 20×10^6 cells.

7. For plating cells on the Lexan ring assembly, the cell suspension obtained in step 5 is counted in a hemocytometer or Coulter counter to obtain a value for the number of cells per milliliter. Plating density for the ring assembly is 1.5×10^6 cells/ml. If the suspension obtained in step 5 is too dilute, centrifuge the suspension at 1000 g for 2 min, aspirate off the supernatant, and resuspend in the appropriate volume of medium for the final desired concentration of 1.5×10^6 cells/ml. One milliliter of this suspension is added to the center well of each ring. About 6 to 7 of these ring assemblies can be laid in a 100-mm culture plate containing approximately 80 glass beads (3-mm size; American Scientific #G6000-1) and 10 ml of culture medium.

Procedure for Feeding T_{84} Cells

We generally feed stock plates (100 mm) Monday and Friday by simply aspirating off the old medium and adding 10 ml fresh medium.

Procedure for Freezing T_{84} Cells

T_{84} cells can be frozen and stored in liquid nitrogen. Freezing medium consists of the culture medium supplemented with 10% serum and 5%

DMSO. Cells are frozen at a density of 5–10 million/ml/tube. We freeze the cells using a BF-5 biological freezer from Union Carbide. The freezing protocol is 30 min at 4° at level C, 30 min at level C over the liquid N_2 tank, and finally 90 min at level A over the liquid N_2. The tubes are then attached to the canes and immersed in liquid nitrogen. Viability upon thawing rapidly at 37° should be >50%.

Acknowledgments

The T_{84} cell line was originally established by Dr. Hideo Masui and colleagues at the University of California, San Diego, California. Most of the methods described are modifications of commonly used methods established by Dr. Gordon Sato, Dr. Hans Ussing, Dr. Joseph Handler, Dr. Marcelino Cereijido, and Dr. Milton Saier. Dr. Kenneth Mandel and Dr. James McRoberts are the two key associates who designed and carried out most of the experiments. Mr. Greg Beuerlein provided technical assitance. Ms. Bambi Beuerlein and Ms. Carol Gaul typed and edited the chapter.

This work was supported by Grants R01 AM28305 and R01 AM35932 from the National Institues of Health, a grant from the University of California Cancer Research Coordinating Committee, a grant from The Burroughs Wellcome Fund and a grant from the National Foundation for Ileitis and Colitis, Inc. Dr. K. Dharmsathaphorn is a recipient of a Research Career Development Aware, AM01146, from the National Institues of Health and an American Gastroenterological Association/Glaxo Research Scholar Award. Dr. James Madara is a recipient of an American Gastroenterological/Ross Research Scholar Award.

[25] Sodium Chloride Transport Pathways in Intestinal Membrane Vesicles

By ULRICH HOPFER

Overview of Models of Intestinal NaCl Transport

The classical tools for investigating epithelial electrolyte transport consist mainly of measurements of radioactive tracer and electrical fluxes across intact epithelia or of short-term tracer uptake or release by epithelial cells. Major, new biochemical and cell biological approaches were developed in the 1970s and include localization of key transport enzymes, i.e., transport ATPases, and cell fractionation with separate isolation of membrane vesicles from the lumenal and basolateral regions of epithelial cells. The membrane vesicle systems provide opportunities to investigate molecular events associated with the transport of Na^+ and Cl^- across these membranes.[1]

---

[1] U. Hopfer, K. Nelson, J. Perrotto, and K. J. Isselbacher, *J. Biol. Chem.* **248**, 25 (1973).

METHODS IN ENZYMOLOGY, VOL. 192

Several different modes of Na^+ and Cl^- transport can be distinguished in the intestine.[2,3] One major criterion is the overall flux of electrical charges associated with transepithelial NaCl flow which can be easily measured in terms of the short-circuit current; the transport is termed electrogenic if under short-circuited conditions the "active" net flow of either Na^+ or Cl^- across the epithelial cell layer is associated with a corresponding electrical flux. Other criteria are obligatory involvement of other solutes and the direction of NaCl transport. The following modes have been defined: (1) Sodium–nutrient cotransport (e.g., sodium–glucose cotransport) which, under physiological conditions, results in nutrient-dependent Na^+ absorption, (2) electrically neutral NaCl absorption, which is typically found in the small intestine of mammals and birds, but is also present in the colon of Na^+-replete rats, (3) electrogenic Na^+ absorption, which is typically present in the most distal portion of the gut and regulated by mineralocorticoids, (4) K^+-dependent, electrogenic NaCl absorption, which is typically found in the intestine of carnivorous fish, and (5) NaCl secretion. This chapter focuses on modes 2 and 5, i.e., electrically neutral NaCl absorption and NaCl secretion. Figure 1 describes the current model for electrically neutral NaCl absorption and Fig. 2 the one for NaCl secretion. Although the overall picture of salt movements across the epithelium appears well supported by experimental findings, molecular information about transporters involved and regulation of their activity is still missing or grossly incomplete. Therefore, well-established methods that exist for isolation of intestinal plasma membranes as well as assays for ATP-independent Na^+ and Cl^- transport across intestinal plasma membranes will be summarized, with particular emphasis on methods in use in the author's laboratory.

Figure 1 illustrates that NaCl absorption comes about by the presence of (1) Cl^-/HCO_3^- and (2) Na^+/H^+ exchangers in the lumenal plasma membrane in tandem with primary active Na^+ transport via the (3) Na^+,K^+-ATPase in the basolateral plasma membrane. The (4) Cl^- exit pathway in the basolateral plasma membrane is not yet well investigated in mammalian enterocytes. In the steady state, the cell must recycle the K^+ that is pumped into the cell by the Na^+,K^+-ATPase in exchange for Na^+. This extra K^+ leaves the cell through (5) K^+ channels. While the existence of Na^+/H^+ exchange,[4-14] Na^+,K^+-ATPase,[15-17] as well as K^+ channels[18,19]

[2] U. Hopfer and C. M. Liedtke, *Annu. Rev. Physiol.* **49,** 51 (1987).

[3] M. Field, M. C. Rao, and E. B. Chang, *N. Engl. J. Med.* **321,** 800, 879 (1989).

[4] H. Murer, U. Hopfer, and R. Kinne, *Biochem. J.* **154,** 597 (1976).

[5] R. Knickelbein, P. S. Aronson, W. Atherton, and J. W. Dobbins, *Am. J. Physiol.* **245,** G504 (1983).

[6] H. J. Binder, G. Stange, H. Murer, B. Stieger, and H. P. Hauri, *Am. J. Physiol.* **251,** G382 (1986).

BLOOD   LUMEN

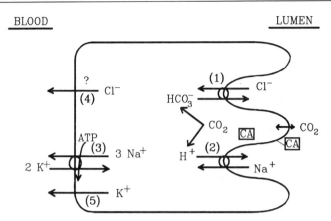

FIG. 1. Model of intestinal NaCl absorption from the lumenal to the blood-side compartment. Overall, active transepithelial absorption is a result of (1) Cl^-/HCO_3^- and (2) Na^+/H^+ exchanges at the brush border membrane; primary active movement of Na^+ out of the cell by (3) Na^+,K^+-ATPase and passive exit of (4) Cl^- and (5) K^+ at the basolateral plasma membrane. The nature of (4) Cl^- exit is not well understood. Carbonic anhydrase (CA) serves to accelerate equilibration of CO_2 with HCO_3^- and protons.

is generally accepted, the nature of the Cl^- transporters at both plasma membranes and the types of transport reactions are still debated. For example, at the brush border membrane electrically neutral Cl^- transport has been demonstrated; however, it is experimentally difficult to distinguish between Cl^-/HCO_3^- exchange, Cl^-/OH^- exchange, and HCl cotransport.

Na^+/H^+ exchange activity is prominent in brush border membranes,

[7] I. W. Booth, G. Strange, H. Murer, T. R. Fenton, and P. J. Milla, *Lancet* **1**, 1066 (1985).

[8] E. S. Foster, P. K. Dudeja, and T. A. Brasitus, *Am. J. Physiol.* **250**, G781 (1986).

[9] R. Fuchs, J. Graft, and M. Peterlik, *Biochem. J.* **230**, 441 (1985).

[10] J. N. Howard and G. A. Ahearn, *J. Exp. Biol.* **135**, 65 (1988).

[11] J. G. Kleinman, J. M. Harig, J. A. Barry, and K. Ramaswamy, *Am. J. Physiol.* **255**, G206 (1988).

[12] K. Kikuchi, N. N. Abumrad, and F. K. Ghishan, *Gastroenterology* **95**, 388 (1988).

[13] Y. Miyamoto, D. F. Balkovetz, V. Ganapathy, T. Iwatsubo, M. Hanano, and F. Leibach, *J. Pharmacol. Exp. Ther.* **245**, 823 (1988).

[14] K. Ramaswamy, J. M. Harig, J. G. Kleinman, M. S. Harris, and J. A. Barry, *Biochim. Biophys. Acta* **981**, 193 (1989).

[15] C. Stirling, *J. Cell Biol.* **53**, 704 (1972).

[16] V. Harms and E. Wright, *J. Membr. Biol.* **53**, 119 (1980).

[17] P. B. Vengesa and U. Hopfer, *J. Histochem. Cytochem.* **27**, 1231 (1979).

[18] E. Grasset, P. Gunter-Smith, and S. G. Schultz, *J. Membr. Biol.* **71**, 89 (1983).

[19] J. Costantin, S. Alcalen, A. de Souza Otero, W. P. Dubinsky, and S. G. Schultz, *Proc. Natl. Acad. Sci. U.S.A.* **86**, 5212 (1989).

but has also been described in basolateral plasma membrane vesicles.[20-22] The physiological role of basolateral Na^+/H^+ exchange is assumed to be a proton pump transporting protons out of the cell into blood. This proton pumping at the basolateral side accompanies HCO_3^- secretion at the lumenal pole, indicating the importance of Na^+/H^+ exchange in basolateral membranes for overall HCO_3^- secretion from blood to lumen. An alternative view regards this Na^+/H^+ exchange as a compensating reaction for HCO_3^- secretion at the lumenal pole, thus maintaining cellular pH homeostasis. In some species, e.g., rodents, basolateral Na^+/H^+ exchange may also play a role for concentrative Cl^- uptake (see next paragraph).

Figure 2 illustrates the steps involved in NaCl secretion: Secondary active Cl^- uptake (6) into the cell at the basolateral pole, recycling of Na^+ through the Na^+,K^+-ATPase (3), and of K^+ through K^+ channels (5) at the basolateral pole, and exit of Cl^- via Cl^- channels (7) in the lumenal plasma membrane. The secondary active Cl^- uptake across the basolateral plasma membrane probably occurs by loop diuretic-sensitive $Na^+/K^+/2Cl^-$ co-transport in humans and rabbits and the combination of Na^+/H^+ and Cl^-/HCO_3^- exchangers in rats. However, little information about these basolateral transporters has been obtained through flux studies in isolated plasma membrane vesicles. The lumenal Cl^- channel is thought to reside mainly in crypt cells[3]; however, it can also be detected in fully differentiated enterocytes and brush border membrane vesicles.[23-26]

Based on intestinal physiology and pathology, the transporter distribution is probably not homogeneous along the longitudinal axis. The same consideration applies to the crypt–villus axis. Knickelbein et al.[21] have actually demonstrated that Na^+/H^+ exchange exists in both brush border and basolateral membranes of villus cells of rabbit ileum, but only in basolateral membranes of crypt cells.

Isolation of Brush Border Membranes

Highly purified membrane vesicles from the brush border region can be isolated with relative ease. Two major types of preparations are commonly

[20] F. Barros, P. Dominguez, G. Velasco, and P. S. Lazo, Biochem. Biophys. Res. Commun. 134, 827 (1986).

[21] R. G. Knickelbein, P. S. Aronson, and J. W. Dobbins, J. Clin. Invest. 82, 2158 (1988).

[22] A. J. Moe, J. A. Hollywood, and M. J. Jackson, Comp. Biochem. Physiol. 93, 845 (1989).

[23] F. Giraldez, F. V. Sepulveda, and D. N. Sheppard, J. Physiol. (London) 395, 597 (1988).

[23a] F. Giraldez, K. Y. Murray, F. V. Sepulveda, and D. N. Sheppard, J. Physiol. 416, 517 (1989).

[24] C. P. Stewart and L. A. Turnberg, Am. J. Physiol. 257, G334 (1989).

[25] G. W. Forsyth and S. E. Gabriel, J. Membr. Biol. 107, 137 (1989).

[26] U. Hopfer, Int. Congr. Physiol. Sci. 31st, S1023 (abstract) (1989).

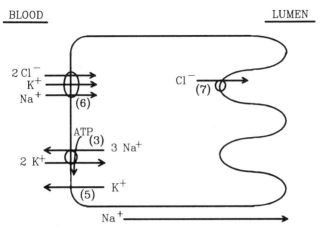

FIG. 2. Model of intestinal NaCl secretion from the blood side to the lumen. Overall, transepithelial active secretion is achieved by concentrative Cl⁻ uptake at the basolateral plasma membrane via (6) $Na^+/K^+/2Cl^-$ cotransport. This uptake is driven by the physiological Na^+ gradient (extracellular > intracellular). In the steady state, Na^+ is extruded back across the basolateral plasma membrane by (3) Na^+,K^+-ATPase and K^+ leaves through specific (5) K^+ channels. Chloride ion is secreted through (7) Cl⁻ channels in the brush border membrane. Charge compensation for the Cl⁻ movement into the lumen is mainly by Na^+ as the major extracellular cation moving through intercellular space and junction.

employed. The first one involves thorough homogenization of mucosal scrapings, aggregation of nonbrush border membranes and particulates by divalent cations, and subsequent differential centrifugation (see the section, Aggregation Method).[27,28] The other method is based on an initial isolation of intact brush borders[29,30] and subsequent depolymerization of the cytoskeletal material associated with the brush border membrane (see the section, Depolymerization Method).[31]

The quantity and quality of isolated membrane preparations are usually assessed on the basis of yield, enrichment, and specific activity of marker enzymes, such as sucrase or alkaline phosphatase for the brush border membrane and Na^+,K^+-ATPase for the basolateral membrane. With the aggregation method, brush border membranes can be prepared with 25- to 30-fold enrichments in sucrase relative to a homogenate of mucosal scrapings at a yield of 30 to 50%. The depolymerization method

[27] J. Schmitz, H. Preiser, D. Maestracci, B. K. Ghosh, J. J. Cerda, and R. K. Crane, *Biochim. Biophys. Acta* **323**, 98 (1973).

[28] H. Hauser, K. Howell, R. M. C. Dawson, and D. E. Boyer, *Biochim. Biophys. Acta.* **602**, 567 (1980).

[29] G. G. Forstner, S. M. Sabesin, and K. J. Isselbacher, *Biochem. J.* **106**, 381 (1968).

[30] A. Eichholz and R. K. Crane, this series, Vol. 31 p. 123 (1974).

[31] U. Hopfer, T. D. Crowe, and B. Tandler, *Anal. Biochem.* **131**, 447 (1983).

can yield enrichments of 2 to 4 times that of the aggregation method (i.e., 60- to 120-fold enrichments of brush border marker enzymes) at a yield of 40 to 70%. The membranes prepared by the two methods are not identical, e.g., because of differences in associated cytoskeletal material. In addition, more cytosolic contaminants are trapped by the aggregation method when vesicles are formed during the initial homogenization. While the vesicle orientation with both preparations is predominantly right side out based on electron microscopy and detergent activation of marker enzymes, the depolymerization method provides access to the cytosolic side at the intermediate stage of isolated brush borders because the cytoskeleton prevents resealing of the membrane.

Aggregation Method

This methodology is based on the report by Schmitz et al.[27] with many modifications by others.[32] The following protocol provides a brief outline of the essential steps in this method. All procedures are carried out on ice or with ice-cold solutions.

1. The small intestine is removed as intact tube from the animal, rinsed with cold buffered saline, and inverted over a thin rod. Excess mucus is removed by wiping with paper tissues.

2. The mucosa is scraped with glass slides, and the scrapings are suspended in about 20× their volume in a low ionic strength buffer containing 0.1 M mannitol, 10 mM Tris-HEPES, pH 7.4 The scrapings are then thoroughly homogenized with a high-speed blender (e.g., a Sorvall Omnimixer (Dupont Instruments, Wilmington, DE) at full speed for 5 min).

3. Either 10 mM (final concentration) $MgCl_2$ or $CaCl_2$ is added to the homogenate to aggregate nonbrush border particulates during a period of 10–15 min.

4. The aggregated material is removed by low-speed centrifugation at 6000 g for 10 min.

5. The brush border membranes are collected by medium-speed centrifugation at 45,000 g for 30 min.

Steps 3–5 of this protocol can be repeated to increase the purification of the brush border membrane. The efficiency of the aggregation is dependent not only on the divalent cation, but also on the protein concentration. The aggregation method has been successfully used with many animal species. Although both Ca^{2+} and Mg^{2+} aggregate nonbrush border material, the results with these two different cations are not identical.[28,33]

[32] H. Murer, P. Gmaj, B. Stieger, and B. Hagenbuch, this series Vol. 172, p. 346 (1989).
[33] H. Aubry, A. R. Merrill, and P. A. Proulx, Biochim. Biophys. Acta **856**, 610 (1986).

Depolymerization Method

This method is based on an initial isolation of brush borders as intact "organelles" from small intestinal scrapings and subsequent separation of cytoskeletal material from the membrane. Procedures for the preparation of brush borders of many different species have previously been described.[30] A modification for rat small intestine is provided below.

1. Small intestinal scrapings are obtained as described above for the aggregation method.

2. The intestinal scrapings from 1 rat (about 0.5 g) are suspended in 0.5 liter of homogenization buffer (5 mM EGTA, 1 mM HEPES adjusted to pH 7.4 with Tris hydroxide) and homogenized in a blender for about 10 sec. The time must be optimized for each blender and volume, based on an optimal recovery of intact brush borders: if shearing is insufficient, brush borders are not liberated from the cells; if shearing is too much, the brush borders are broken into small pieces that cannot be isolated by low-speed centrifugation. The effectiveness of the homogenization can be quantitated by the yield of marker enzyme(s) in the purified brush border fraction obtained in step 5.

3. The homogenate is passed through a Nitex (Tetko Inc., Elmsford, NY) (30 μm) filter and subsequently a thin layer of Pyrex glass wool (Corning Glass Works, Corning, NY) to remove debris and nuclei. Both filter and glass wool are prewashed with homogenization buffer.

4. The filtered homogenate is centrifuged at 1000 g for 15 min to collect the intact brush borders.

5. The 1000 g, low-speed pellet is resuspended in the homogenization buffer by careful trituration and centrifuged again at 1000 g for 10 min. The new low-speed pellet is resuspended in a small volume of homogenization buffer and an aliquot examined by light microscopy for purity. At least 90% of visible particles should be brush borders. If necessary, the fourth step can be repeated to achieve greater purity.

The resulting brush borders are stable for several hours, particularly if Ca^{2+} concentrations are kept low and the ionic strength is above 0.03 at pH 7.4.

The cytoskeleton can be separated from the membranes by a cycle of depolymerization and repolymerization of the microfilaments that form the core of brush borders. Shearing of the brush borders after depolymerization of the microfilaments results in vesiculation of the membrane, preventing reassembly of the original filaments. The de- and repolymerization can be manipulated using the conditions for G and F actin transformation. The depolymerization can be achieved by high concentrations of chaotropic ions.[31] The method described below is based on the lability of microfilaments to high pressure.[34] The following steps have proved useful for high yields of membranes.

6. The purified brush border pellet is suspended in the membrane preparation buffer (0.3 M mannitol, 10 mM HEPES, adjusted to pH 7.4 with Tris, 1 mM EGTA) at about 0.5 to 1.0 mg/ml and is exposed to pressures of more than 1600 psi of an inert gas (e.g., N_2) in a Parr cell disruption bomb (Parr Instrument Co., Moline, IL) for 15 min.

7. The suspension is then slowly released from the bomb, resulting in homogenization of the brush border membranes due to shearing and cavitation.

8. The microfilaments are reaggregated by adjusting the $MgCl_2$ concentration to 10 mM. The aggregated material is subsequently removed by centrifugation at 1500 g for 5 min.

9. The brush border membranes remain in the supernatant and can be collected by centrifugation at 45,000 g for 20 min.

The depolymerization of the cytoskeleton in the bomb can be enhanced by inclusion of 0.1 mM ATP and/or micromolar concentrations of free Ca^{2+} in the membrane preparation buffer.

Isolation of Basolateral Plasma Membranes

In contrast to the brush border region, the basolateral plasma membrane of enterocytes lacks unique features that distinguish it from those of other cells. Therefore, the method for isolating this membrane resembles that of the plasma membrane from other, more generic mammalian cells, consisting of a combination of differential and density centrifugations. Actually, to ensure that the isolated plasma membrane originates from enterocytes, it is important to first isolate a relatively pure enterocyte fraction. The procedure by Bjerknes and Cheng[35] works well for small rodents. It entails intraarterial or intracardiac injection of a solution containing 30 mM EDTA to break the attachments of epithelial cells to the basement lamina. The epithelial cells can subsequently be removed from the gut mucosa by either shaking or very light scrapings.

The following protocol is an adaptation of the method by Scalera *et al.*[36] It consists of an initial isolation of entire sheets of epithelial cells and

[34] L. G. Tilney and R. R. Cardell, Jr., *J. Cell Biol.* **47**, 408 (1970).

[35] M. Bjerknes and H. Cheng, *Anta. Rec.* **199**, 565 (1981).

[36] V. Scalera, C. Storelli, C. Storelli-Joss, W. Haase, and H. Murer, *Biochem. J.* **186**, 177 (1980).

subsequent preparation of basolateral plasma membrane from the isolated cells. The actual membrane preparation can, in turn, be subdivided into (1) isolation of a crude, basolateral plasma membrane fraction that is collected as a "fluffy" layer on top of the mitochondrial pellet after centrifugation, and (2) isolation of the final, purified membrane by density centrifugation in a Percoll gradient.

Na$^+$,K$^+$-ATPase or K$^+$-stimulated, neutral phosphatase serves as marker enzyme to assess purification and yield of basolateral plasma membrane. The crude plasma membrane fraction typically is threefold enriched in the marker enzyme at a yield of about 50%. The Percoll gradient provides an additional fourfold purification at a yield of about 25% relative to the homogenate.

Isolation of Enterocytes

1. Rats are anesthetized with sodium pentobarbital ip (about 60 mg/kg body wt).

2. The abdomen is opened by a midline abdominal incision and the small intestine rinsed with 37°, phosphate-buffered 0.9% saline through two small incisions in the gut wall opposite the mesenteries.

3. After the portal vein is severed, about 150 ml of warm (37°), phosphate-buffered 0.9% saline containing additionally 30 mM EDTA, pH 7.4, is injected into the left ventricle. The effectiveness of the intestinal perfusion can be judged by "blanching" of the intestinal wall.

4. The small intestine is removed and inverted over a glass rod. Cells are released by shaking or light scraping of the glass rod, which is suspended in a cold sucrose buffer (250 mM sucrose, 10 mM HEPES adjusted to pH 7.4 with Tris, 0.5 mM Na$_2$EDTA, and 0.1 mM phenylmethylsulfonyl fluoride as serine protease inhibitor).

The isolated cells serve as starting material for the subsequent isolation of plasma membranes.

5. Cells are collected by low-speed centrifugation, resuspended in about 30 ml of the sucrose buffer/animal, and homogenized in a glass–Teflon homogenizer at 600–1200 rpm for 10–25 strokes. The speed and number of strokes must be optimized for each homogenizer with respect to recovery of basolateral plasma membrane marker in the "fluffy layer" (see step 7 below).

6. The heavy particulate material is removed by centrifugation at 2300 g for 15 min and the supernatant retained.

7. A crude basolateral plasma membrane fraction is collected from this

supernatant by centrifugation at 15,000 g for 20 min. The basolateral plasma membrane is enriched in a "fluffy," white layer above the denser, yellow mitochondrial pellet. For greater yield, the heavy, 2300 g particulate material from the previous step can be rehomogenized with a glass–Teflon homogenizer and spun again at 2300 g for 15 min. In this case, the two supernatants are combined before the 15,000 g centrifugation.

8. The white, fluffy layer from step 7 is resuspended in the sucrose buffer, but with 0.1 mM MgCl$_2$ instead of EDTA, and Percoll added to make 12 to 14%.

9. The membrane suspension is centrifuged in a vertical Sorvall SS90 rotor at 27,000 g for 15 min. The basolateral membranes collect in a band about one-fourth down from the top.

10. The band containing most of the marker enzyme is removed and the membranes collected by high-speed centrifugation (>200,000 g for 45 min) on top of a dense, glassy Percoll pellet.

Transport Measurements: General Considerations

The general design principles applicable to transport experiments with membrane vesicles have been discussed previously.[37] Therefore, only specific issues of importance for electrolyte transport are included in this chapter.

One of the problems in electrolyte transport studies is that fixed charges on the inside of vesicles give rise to the so-called Gibbs–Donnan potential which influences the equilibrium distribution of ions across the vesicle membrane. This effect is expressed as apparent binding or exclusion of ions. The magnitude of the potential can be calculated from ion concentration gradients at equilibrium:

$$\Delta\Psi = -(RT/zF)\ln([i_z]_{in}/[i_z]_{out})$$

where R = gas constant, T = absolute temperature, F = Faraday constant, z = charge of ion, $[i_z]$ = concentration of ion i of charge z.

The presence of a Gibbs–Donnan potential can be inferred if several different ions give the same value for $\Delta\Psi$. The apparent ratio of ion concentrations between intra- and extravesicular medium, which converts to $\Delta\Psi$, is equal to "equilibrium uptake of the ion" divided by "the intravesicular space," divided again by "its medium concentration." The space, in turn, can be calculated from the equilibrium uptake of a neutral, nonbinding solute, such as glucose. At equilibrium, the intravesicular concentration of such a neutral solute corresponds to that of the medium so that the

[37] U. Hopfer, this series, Vol. 172, p. 313 (1989).

"intravesicular space" is equal to "uptake" divided by "medium concentration" (e.g., uptake in moles per milligram protein divided by solute concentration in moles per liter to yield liter per mg protein). The intravesicular space of intestinal plasma membrane vesicles typically ranges from 1 to 5 μl/mg protein. For rat small intestinal brush border membranes, an inside-negative Gibbs–Donnan potential of about 10 mV has been inferred from the findings of equilibrium uptakes that are higher for Na^+ and lower for the anions Cl^- and sulfate relative to uncharged D-glucose.[38]

Sodium Ion Transport

Na^+/H^+ Exchange

The existence of Na^+/H^+ exchange as the major form of nutrient-independent Na^+ transport across the brush border membrane of the small intestine is supported by four major types of experimental results.

1. pH gradients can drive concentrative uptake of Na^+, independent from proton diffusion potentials.
2. Sodium ion gradients can drive concentrative proton movements in the absence of diffusion potentials.
3. Na^+/Na^+ exchange rates across the brush border membrane are not dependent on the nature of the anion, excluding Na^+/anion cotransport models.
4. The diuretic amiloride and some of its derivatives, which are inhibitors of Na^+/H^+ exchange in other systems, inhibit Na^+ transport in isolated membrane vesicles.

pH Gradient-Driven Na^+ Uptake

The Na^+/H^+ exchange mechanism implies that a proton gradient can drive uphill Na^+ transport. This prediction can be tested by measuring $^{22}Na^+$ uptake by membrane vesicles in the presence of a pH gradient, whereby the lower pH is on the inside. For such experiments, the following considerations are important.

1. Membrane vesicles must be preloaded with a relatively impermeant buffer such as HEPES or 2-(N-morpholino)ethane sulfonic acid (MES) at a low pH (e.g., 5.5). The preloading can be accomplished by prolonged incubations and homogenizations in such a low-pH buffer. Alternatively,

[38] C. M. Liedtke and U. Hopfer, *Am. J. Physiol.* **242**, G263, G272 (1982).

the membrane vesicles are prepared in a solution containing impermeant buffer components that by themselves have a low pH, but which are adjusted to the higher pH (e.g., 7.4) of the usual membrane preparation buffer with permeant base (e.g., imidazole or ethanolamine). Once the membrane vesicles are isolated, their intravesicular pH can easily be lowered by removing this premeant component of the buffer (e.g., by gel filtration or washing of membranes in a buffer of appropriate low pH and without the permeant base).

2. A pH gradient is established between the intravesicular compartment and the incubation medium at the beginning of the incubation with labeled Na$^+$.

3. The establishment of proton diffusion potentials may have to be prevented by inclusion of high and equal concentrations of K$^+$ in both the intravesicular and extravesicular medium as well as valinomycin to ensure that the K$^+$ conductance is high.

4. Sodium ion transport must be effectively quenched at the termination of the incubation period and vesicles separated from the medium containing the labeled Na$^+$. The quenching can be accomplished by ice-cold solutions containing 0.05 to 0.1 M MgSO$_4$, an impermeant buffer of the same pH as the incubation medium, and sufficient mannitol to bring the osmolarity of the quench solution to that of the incubation medium. Vesicles can be conveniently collected on nitrocellulose filters (0.45 to 0.6 μm).

A Na$^+$/H$^+$ exchanger can utilize the energy of a pH gradient into Na$^+$ uptake into membrane vesicles against a concentration gradient. Since the pH gradient usually dissipates with prolonged incubation periods, the Na$^+$ gradient also collapses with it and Na$^+$ uptake as a function of time presents the picture of an overshoot. The overshooting portion, but not the equilibrium uptake, should be dependent on the pH gradient. Observations of such overshooting uptake are generally considered sufficient to justify the conclusion that the membranes possess functional Na$^+$/H$^+$ exchangers provided the overshoot persists in the absence of a diffusion potential and is not due to time-dependent changes in intravesicular space.

Because of the presence of dissipative proton fluxes (not coupled to Na$^+$ uptake), the buffer capacity of the intravesicular medium must be kept relatively high and the Na$^+$ concentration in the uptake medium relatively low (typical ratio, 50:1).

Figure 3 shows overshooting Na$^+$ uptake by brush border membranes driven by a pH gradient. The initial uptake can be inhibited about 90% by 1 mM amiloride, which is thought to inhibit the transporter. However, this drug also dissipates the pH gradient and this effect may contribute to the

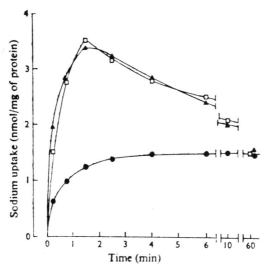

FIG. 3. pH gradient-driven concentrative uptake of Na⁺ by intestinal brush border membrane vesicles. The membranes were preloaded with 50 mM MES/Tris buffer, pH 5.5, followed by incubation in a medium containing 0.5 mM Na$_2$SO$_4$ and (●) 50 mM MES/Tris, pH 5.5; (□) 50 mM HEPES/Tris, pH 7.5; (▲) 50 mM HEPES/Tris, pH 7.5 plus 5 μg of carbonyl cyanide p-trifluoromethoxyphenylhydrazone to increase the proton conductance and set up a proton diffusion potential. Overshooting Na⁺ uptake is seen only in the presence of a pH gradient. In addition, the overshoot is not influenced very much by a diffusion potential. (Reproduced with permission from Ref. 4.)

inhibition of uptake.[39] Other modulators of Na⁺/H⁺ exchange appear to be lipid fluidity in general and free fatty acids. Interestingly, while lipid fluidity correlates positively[40-43] with Na⁺/H⁺ exchange activity, free fatty acids by themselves were shown to inhibit.[44]

Sodium Ion Gradient-Driven Proton Transport

Experimental conditions can also be rigged so that Na⁺ gradients can drive H⁺ transport through the Na⁺/H⁺ exchanger. The proton move-

[39] W. P. Dubinsky, Jr., and R. A. Frizzell, *Am. J. Physiol.* **245,** C157 (1983).
[40] T. A. Brasitus, P. K. Dudeja, and E. S. Foster, *Biochim. Biophys. Acta* **938,** 483 (1988).
[41] P. K. Dudeja, E. S. Foster, R. Dahiya, and T. A. Brasitus, *Biochim. Biophys. Acta* **899,** 222 (1987).
[42] P. K. Dudeja, E. S. Foster, and T. A. Brasitus, *Biochim. Biophys. Acta* **905,** 485 (1987).
[43] P. K. Dudeja, E. S. Foster, and T. A. Brasitus, *Biochim. Biophys. Acta* **859,** 61 (1986).
[44] C. Tiruppathi, Y. Miyamoto, V. Ganapathy, and F. H. Leibach, *Biochem. Pharmacol.* **37,** 1399 (1988).

ments can be conveniently measured by pH changes of the medium with pH electrodes.[4] Experimental design considerations are similar to those for pH gradient-driven Na^+ uptake, except that the roles of Na^+ and protons are reversed. In other words, large Na^+ gradients and Na^+ fluxes as well as low pH buffering capacities are essential to achieve appropriate sensitivities to detect Na^+-driven proton movements.

Na^+/Na^+ Exchange

Na^+/Na^+ exchange experiments are useful under two different types of conditions: (1) tracer $^{22}Na^+$ uptake into vesicles that have been preloaded with high concentrations of nonradioactive Na^+ and (2) $^{22}Na^+$ uptake at chemical equilibrium.

The advantage of preloading with high concentrations of unlabeled Na^+ is that tracer $^{22}Na^+$ is accumulated to high levels, i.e., the transport assay with vesicles is very sensitive. This method was employed with crude membrane vesicles from rat colon to study amiloride-sensitive Na^+ channels.[45,46] The equilibrium condition has the advantage of well-defined concentrations in both intra- and extravesicular compartments, which is particularly well suited for kinetic experiments.[37] At equilibrium, changes in rates of $^{22}Na^+$ movement across the membrane are clearly interpretable in terms of activation or inhibition of transporter(s) without interference from dissipation of gradients across the membrane. The condition was important for studies investigating coupling of transport of Na^+ to that of other solutes.

Early models of electrically neutral NaCl absorption included NaCl cotransport,[47] analogous to Na^+-glucose cotransport. Such cotransport models can be tested because they predict that the obligatory cosubstrate appears as activator of the labeled substrate whose transport is measured. This prediction applies to net flux as well as isotope exchange conditions.[48] NaCl cotransport could be excluded in rat and rabbit brush border membranes on the basis of independence of the rate of Na^+/Na^+ exchange from the magnitude of the Cl^- concentration between 0 and 0.15 M (Fig. 4).[38] This finding provides strong support for parallel Na^+/H^+ and Cl^-/HCO_3^- exchangers, which appear coupled in intact tissue only under steady state conditions of salt transport measurements.

[45] R. J. Bridges, E. J. Cragoe, Jr., R. A. Frizzell, and D. J. Benos, *Am. J. Physiol.* **256,** C67 (1989).
[46] R. J. Bridges, H. Garty, D. J. Benos, and W. Rummel, *Am. J. Physiol.* **254,** C484 (1988).
[47] R. A. Frizzell, M. Field, and S. G. Schultz, *Am. J. Physiol.* **236,** F1 (1979).
[48] U. Hopfer and C. M. Liedtke, *Membr. Biochem.* **4,** 11 (1981) [plus Errata in issue 4 of Vol. 4 (1982)].

Na$^+$/Na$^+$ exchange experiments are carried out by preincubating membrane vesicles under the desired conditions of pH and medium composition, including ^{22}Na$^+$. The preincubation must be sufficiently long so that any pH or ion gradients are abolished. The rate of Na$^+$/Na$^+$ exchange can then be measured during the incubation period by replacing ^{22}Na$^+$ in the medium with unlabeled Na$^+$ and following release of ^{22}Na$^+$ from the vesicles. The effect of an experimental variable, such as the Cl$^-$ concentration in the experiment in Fig. 4, is tested in parallel aliquots of membrane vesicles. The aliquots are treated identically, including measurements of the Na$^+$/Na$^+$ exchange rates, except for preincubation and incubation with different Cl$^-$ concentrations. Although rates of Na$^+$/Na$^+$ exchange can be measured both as ^{22}Na$^+$ uptake or release, ^{22}Na$^+$ release is usually employed because it requires less radiolabel.

pH Gradient-Dissipation Measured with Acridine Orange

One of the methods for measuring pH gradients between medium and the interior of vesicles takes advantage of fluorescence quenching of heterocyclic amines when they are accumulated in vesicles.[49] Suitable dyes are 9-aminoacridine and acridine orange. These amines are accumulated by membrane vesicles preloaded with a pH that is more acid than the medium. The basis of the fluorescent quench of 9-aminoacridine by pH gradients was investigated only recently.[50,51] These studies suggest that the dye method is probably not reliable for quantitative measurements of the kinetics of pH changes since fluorescence decay and pH changes after a pH jump did not coincide. The method has been widely used, however, to probe for existence of intestinal Na$^+$/H$^+$ exchange.[5,8,40,42,43,52] In light of the discovery that the kinetics of fluorescence quench are not identical with intravesicular pH changes, some of the conclusions from the intestinal vesicle studies may have to be reevaluated.

The method consists of preloading the membrane vesicles with a low-pH buffer of an impermeant acid-base pair, subsequent suspension of the vesicles in a medium of higher pH, and equilibration with about $5-10\ \mu M$ acridine orange (excitation, 493 nm; emission. $527-535$ nm). This manipulation results in quenching of acridine orange fluorescence due to dye uptake by the vesicles. With time, the fluorescence recovers. The recovery can be accelerated by Na$^+$ influx from the medium, which is interpreted as Na$^+$/H$^+$ exchange.

[49] H. Rottenberg, this series, Vol. 54, p. 547 (1979).
[50] S. Grzesiek and N. A. Dencher, *Biochim. Biophys. Acta* **938**, 411 (1988).
[51] S. Grzesiek, H. Otto, and N. A. Dencher, *Biophys. J.* **55**, 1101 (1989).
[52] C. B. Cassano, B. Stieger, and H. Murer, *Pflügers Arch.* **400**, 309 (1984).

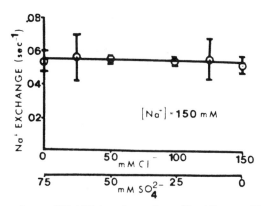

FIG. 4. Independence of Na^+/Na^+ exchange from Cl^-. Aliquots of brush border membranes were preincubated at $25°$ for 75 min containing different concentrations of NaCl or Na_2SO_4 as indicated. The rate of Na^+/Na^+ exchange was subsequently measured from the $^{22}Na^+$ uptake from the medium. The lack of stimulation of Na^+/Na^+ exchange by Cl^- indicates that it is not a cosubstrate for Na^+ transport. (Reproduced with permission from Ref. 38.)

Electrophysiology of Isolated Plasma Membrane Vesicles

The conductive permeabilities of intestinal brush border membranes to ions have been evaluated with potential-sensitive, fluorescent indicators, such as cyanine dyes.[53-55] The studies have been most informative when they were concerned with Na^+–nutrient cotransporters, but the methodology is generally useful to evaluate ion movements associated with electrical currents. However, patch clamping can usually provide more direct information.

Regulation of Na^+/H^+ Exchange

Considerable evidence from intact tissue studies suggests that electroneutral NaCl transport is acutely regulated: Classical secretagogues, such as vasoactive intestinal peptide and cholinergic agonists, decrease this transport, while α-adrenergic agonists increase it. The molecular mechanisms by which the transporter activity is regulated are not known definitively, although some mechanisms appear to persist in isolated brush border membranes. Brasitus' group has demonstrated a positive correlation between V_{max} of Na^+/H^+ exchange and lipid fluidity of the brush border membrane, as measured by anisotropy of fluorescent probes (diphenyl

[53] R. D. Gunther and E. M. Wright, *J. Membr. Biol.* **74,** 85 (1983).
[54] R. D. Gunther, R. E. Schell, and E. M. Wright, *J. Membr. Biol.* **78,** 119 (1984).
[55] E. M. Wright, *Am. J. Physiol.* **246,** F363 (1984).

hexatriene and anthroylstearic acid).[40,42,43] Donowitz and colleagues have provided evidence that the Na^+/H^+ exchanger can be down-regulated by Ca^{2+}/calmodulin-stimulated protein kinase.[56,57]

Chloride Ion Transport

There is considerable evidence from measurements in intact tissues, isolated cells, as well as isolated membrane vesicles that the intestinal brush border membrane contains both Cl^- conductance and electroneutral Cl^-/anion exchange (or HCl cotransport) pathways.[23-26,58,59] The Cl^- conductance, but possibly also Cl^-/anion exchanger(s), are highly regulated in intact cells. The Cl^- conductance is activated by agents that stimulate salt secretion (e.g., cholinergic agents, vasoactive intestinal peptide, histamine, prostaglandins). The simultaneous presence of both electrogenic and electrically silent transport systems in the same membrane can produce difficulties in terms of accurately measuring Cl^- flux through either system or ascribing the observed properties of Cl^- uptake or flux to a particular type of transporter. While there is no evidence for primary active Cl^- transport in vertebrate intestine, Gerencser has obtained data supporting ATP-dependent, concentrative Cl^- uptake by membrane vesicles from *Aplysia* intestinal brush borders.[59-61]

Attempts to distinguish between electrically neutral and electrogenic Cl^- transport pathways have exploited the following features of intestinal electrolyte transport:

1. Cl^-/HCO_3^- exchange activity appears to be more prevalent in the ileum than the jejunum.
2. Stilbene sulfonates, such as 4-acetamido-4′-isothiocyanostilbene-2,2′-disulfonate (SITS) or 4,4′-diisothiocyanostilbene-2,2′-disulfonate (DIDS), appear to inhibit Cl^-/HCO_3^- exchange from the external side, although relatively high concentrations of $1-5$ mM are required.[28,62-64] On

[56] R. P. Rood, E. Emmer, J. Wesolek, J. McCullen, Z. Husain, M. E. Cohen, R. S. Braithwaite, H. Murer, G. W. G. Sharp, and M. Donowitz, *J. Clin. Invest.* **82**, 1091 (1988).
[57] E. Emmer, R. P. Rood, J. H. Wesolek, M. E. Cohen, R. S. Braithwaite, G. W. G. Sharp, H. Murer, and M. Donowitz, *J. Membr. Biol.* **108**, 207 (1989).
[58] M. Montrose, J. Randles, and G. A. Kimmich, *Am. J. Physiol.* **253**, C693 (1987).
[59] G. A. Gerencser, *Am. J. Physiol.* **254**, R127 (1988).
[60] G. A. Gerencser, *Comp. Biochem. Physiol.* **90**, 621 (1988).
[61] G. A. Gerencser, J. F. White, D. Gradmann, and S. L. Bonting, *Am. J. Physiol.* **255**, R677 (1988).
[62] R. G. Knickelbein, P. S. Aronson, and J. W. Dobbins, *J. Membr. Biol.* **88**, 199 (1985).
[63] C. D. Brown, C. R. Dunk, and L. A. Turnberg, *Am. J. Physiol.* **257**, G661 (1989).
[64] M. Vasseur, M. Cauzac, and F. Alvarados, *Biochem. J.* **263**, 775 (1989).

the other hand, micromolar concentrations of SITS inhibit Cl^- conductance from the cytosolic side in vesicles (however, Bridges *et al.* concluded from bilayer studies that colonic Cl^- conductance can be inhibited by SITS and DIDS from the external side[65]).

3. Pretreatment with secretagogues enhances Cl^- conductance present in isolated brush border membranes.

There is evidence for at least two different Cl^-/anion exchangers in the brush border membrane based on substrate specificity and sensitivity to stilbene disulfonate inhibitors (SITS, DIDS): (1) Cl^-/HCO_3^- exchange and (2) Cl^-/oxalate exchange.[38,62,66-68] In addition, a Cl^-/SO_4 exchanger is present in basolateral plasma membranes.[69] The major nature of Cl^-/ HCO_3^- exchange is still debated since it is difficult to experimentally distinguish between Cl^-/HCO_3^- exchange, Cl^-/OH^- exchange, and HCl cotransport. In analogy with the red cell exchanger, Cl^-/HCO_3^- exchange is usually assumed for the intestine. The major experimental evidence for this interpretation is that preloading of vesicles with HCO_3^- stimulates greater overshooting Cl^- uptake in rabbit brush border membranes.[62] However, the interpretation of this result as supporting a Cl^-/HCO_3^- exchange mechanism has been criticized because HCO_3^- preloading could contribute to a pH gradient more favorable for concentrative Cl^- uptake, regardless of the exact mechanism responsible for electrically neutral Cl^- movement.[64]

Chloride ion transport studies in isolated, intestinal membrane vesicles have been carried out in ways analogous to those of Na^+ studies. The same methodologies are useful. Chloride ion uptake by membrane vesicles can be studied with $^{36}Cl^-$. Although the specific activity of radiolabeled $^{36}Cl^-$ is low, it is sufficient for studies down to 1 mM Cl^-. Alternatively, other ions, such as $^{125}I^-$, have been used as alternate substrates for the transporter.[67] Chloride ion transport across brush border membranes is effectively quenched by ice-cold solutions containing 0.05–0.1 M magnesium gluconate, an impermeant buffer such as HEPES or MES, adjusted to the pH of the incubation medium, 1 mM diphenylamine-2-carboxylate, and mannitol to bring the osmolarity to that of the incubation medium. Diphenyl-

[65] R. J. Bridges, R. T. Worrell, R. A. Frizzell, and D. J. Benos, *Am. J. Physiol.* **256**, C902 (1989).
[66] C. M. Liedtke and U. Hopfer, *Biochem. Biophys. Res. Commun.* **76**, 579 (1977).
[67] A. B. Vaandrager and H. R. De Jonge, *Biochim. Biophys. Acta* **939**, 305 (1988).
[68] R. G. Knickelbein, P. S. Aronson, and J. W. Dobbins, *J. Clin. Invest.* **77**, 170 (1986).
[69] C. M. Schron, R. G. Knickelbein, P. S. Aronson, and J. W. Dobbins, *Am. J. Physiol.* **253**, G404 (1987).

amine-2-carboxylate appears to be an inhibitor of both electrogenic and electrically silent Cl^- transport.[70]

Cl^-/HCO_3^- Exchange/pH Gradient-Driven Cl^- Uptake

The activity of Cl^-/OH^- exchange (or HCl cotransport) is measured as rate of $^{36}Cl^-$ uptake in response to a pH gradient which is more alkaline inside the vesicles. Membrane vesicles are preloaded with an impermeant buffer (e.g., HEPES at pH 7.5–7.8) and then exposed to medium with a low pH (e.g., pH 5.0–6.0) in the presence of 1–4 mM $^{36}Cl^-$. If Cl^-/OH^- exchange is sufficiently active, this protocol results in an overshooting $^{36}Cl^-$ uptake, even in the absence of a proton diffusion potential (short circuited by high intra- and extravesicular K^+ and valinomycin).

The involvement of HCO_3^- as one of the physiological substrates for Cl^-/anion exchange is experimentally difficult to evaluate. Because of the conversion of CO_2 into HCO_3^- and vice versa, the presence of CO_2 is required to establish and maintain defined HCO_3^- concentrations. Although the conversion is often slow in bulk solutions, it may be relatively fast in the small volumes (μl) used for membrane incubations. In addition, brush border membranes contain membrane-bound carbonic anhydrase[71] that speeds up the conversion. To test for Cl^-/HCO_3^- exchange, $^{36}Cl^-$ uptake is measured after inclusion of CO_2/HCO_3^- in the preincubation period and establishment of an HCO_3^- gradient (intravesicular > medium).[71] HCO_3^- clearly interacts with one of the Cl^-/anion exchangers since $^{36}Cl^-$ uptake and Cl^-/Cl^- exchange can be inhibited by the presence of HCO_3^-.[67]

Potential-Driven Cl^- Uptake

To probe for the presence of Cl^- channels in vesicles, $^{36}Cl^-$ uptake is measured in response to an inside-positive diffusion potential. Thus, membrane vesicles that are incubated with potassium gluconate (K^+ gradient, outside > inside) and valinomycin produce overshooting $^{36}Cl^-$ uptake. The major arguments for suggesting that this uptake is transporter mediated are inhibition by diphenylamine-2-carboxylic acid and stilbene disulfonates, such as SITS, from the cytosolic side. The ratio of potassium gluconate to Cl^- is usually kept at a ratio of about 50:1, resulting in overshooting Cl^- uptake of three- to fourfold above equilibrium.

[70] L. Reuss, J. L. Constantin, and J. E. Bazile, *Am. J. Physiol.* **253**, C79 (1987).
[71] R. Knickelbein, P. S. Aronson, C. M. Schron, J. Seifter, and J. W. Dobbins, *Am. J. Physiol.* **249**, G236 (1985).

One of the interesting recent observations is that this Cl^- conductance in isolated brush border membrane vesicles depends on the treatment of the membranes[72] as well as exposure of the intact tissue to secretagogues.[26]

Cl^-/Cl^- Exchange

The transporters that catalyze Cl^-/anion exchange and electrogenic Cl^- movement presumably also mediate Cl^-/Cl^- exchange at equilibrium. This experimental condition has been used to demonstrate that Cl^- is not transported by a putative NaCl cotransporter across the brush border membrane. The Cl^-/Cl^- exchange rate was shown to be saturable and independent from the Na^+ concentration.[38] These types of experiments are designed similarly as described above for Na^+/Na^+ exchange at equilibrium: Membrane vesicles are preincubated with about 40 mM labeled $^{36}Cl^-$ and varying Na^+ concentrations (and other conditions as desired) and the rate of $^{36}Cl^-$ efflux is measured during the subsequent incubation period after replacement of $^{36}Cl^-$ in the medium with nonradioactive Cl^-. The Cl^- concentration in the exchange experiments must be higher than in pH and HCO_3^- gradient-driven experiments to provide sufficient radioactive counts that are remaining in the membranes and therewith sufficient experimental accuracy.

Acknowledgements

The work presented from this laboratory was supported by grants from the National Institutes of Health (DK-25170) and the Cystic Fibrosis Foundation.

[72] A. B. Vaandrager, M. C. Ploemacher, and H. R. de Jonge, *Biochim. Biophys. Acta.* **856,** 325 (1986).

[26] Advantages and Limitations of Vesicles for the Characterization and the Kinetic Analysis of Transport Systems

By A. BERTELOOT and G. SEMENZA

Following Kaback's pioneer work,[1] it has become more and more common to use closed and functional membrane vesicles for the study of membrane transport processes.[2] In the particular case of intestinal and renal epithelia to which this chapter will be limited, there is no doubt that the successful development of transporting membrane vesicles[3,4] was a decisive analytical methodological advance which has since been used in many laboratories and has resulted in a vast increase in our knowledge as to the characteristics of various transport systems.

In this chapter we will first present the methodologies involved in both the preparation of functional brush border membrane vesicles and the determination of transport activities. We will next endeavor to discuss the advantages and limitations in the use of vesicles for investigating transport systems, as well as the special problems encountered in kinetic studies. The discussion will be limited to our own experience which is confined almost solely to vesicles from intestinal and renal brush border membranes. It should be emphasized, however, that many of the considerations to follow are likely to hold true for most, if not all, kinds of membrane vesicles. Finally, the interested reader is also referred to previous reviews on similar topics for broader coverage and/or different views of particular aspects.[5-14]

[1] H. R. Kaback, *Fed. Proc., Fed. Am. Soc. Exp. Biol.* **19,** 130 (1960).
[2] J. E. Lever, *Crit. Rev. Biochem.* **7,** 187 (1980).
[3] U. Hopfer, K. Nelson, J. Perrotto, and K. J. Isselbacher, *J. Biol. Chem.* **248,** 25 (1973).
[4] A. G. Booth and A. J. Kenny, *Biochem. J.* **142,** 575 (1974).
[5] U. Hopfer, *Am. J. Physiol.* **233,** E445 (1977).
[6] B. Sacktor, *Curr. Top. Bioenerg.* **6,** 39 (1977).
[7] U. Hopfer, *Am. J. Physiol.* **234,** F89 (1978).
[8] H. Murer and R. Kinne, *J. Membr. Biol.* **55,** 81 (1980).
[9] P. S. Aronson, *Am. J. Physiol.* **240,** F1 (1981).
[10] P. S. Aronson and J. L. Kinsella, *Fed. Proc., Fed. Am. Soc. Exp. Biol.* **40,** 2213 (1981).
[11] R. J. Turner, *J. Membr. Biol.* **76,** 1 (1983).
[12] G. Semenza, M. Kessler, M. Hosang, J. Weber, and U. Schmidt, *Biochim. Biophys. Acta* **779,** 343 (1984).
[13] H. Murer, J. Biber, P. Gmaj, and B. Stieger, *Mol. Physiol.* **6,** 55 (1984).
[14] H. Murer and P. Gmaj, *Kidney Int.* **30,** 171 (1986).

Preparation of Functional Brush Border Membrane Vesicles

The first procedure described was based on the use of EDTA,[3] as originally introduced by Eichholz and Crane,[15] for the isolation of the total brush border. In "second generation" vesicles, advantage is taken of the differential precipitation of intracellular and basolateral membranes with Ca^{2+} or Mg^{2+}.[16] Brush border membrane vesicles prepared via either Ca^{2+} or Mg^{2+} precipitation have been compared by various authors.[17-20] Using rabbit small intestine, Hauser et al.[21] first reported that Mg^{2+} is preferred to Ca^{2+} due to the activation of brush border membrane phospholipase A, which degrades endogenous phosphoglycerides to lysophosphoacylglycerols. However, as discussed by these authors, the possibility that phospholipase activation followed membrane disintegration during lipid extraction could not be ignored.[21] More recent studies suggest that phospholipase activation could occur during freezing and thawing of the intestinal tissue prior to membrane isolation and support the view that Ca^{2+}-prepared membranes are less contaminated by basolateral membranes than are Mg^{2+}-prepared membranes.[20] When assayed on rat intestinal scrapings, phospholipase A activity was found to be higher as compared to isolated cells and higher in the presence of EGTA as compared to Ca^{2+}.[22] The source of enzyme activity was not identified in this study, but activity of both pancreatic and intestinal origin seemed compatible with the data.[22] In this context, it should also be mentioned that guinea pig intestinal phospholipase was not found to be activated by Ca^{2+},[23] and that lysosomal phospholipases do not require Ca^{2+} for activity.[24]

Morphological differences also exist between Ca^{2+}- versus Mg^{2+}-prepared vesicles. Freeze-fracture electron microscopy has shown that the distribution and area density of intramembrane particles on the P face of replicas of brush border vesicles prepared with Ca^{2+} resembled that seen in replicas of intact microvilli.[19] However, when prepared with Mg^{2+} and either Ca^{2+} or Mg^{2+} in the presence of KSCN, the P faces showed striking

[15] A. Eichholz and R. K. Crane, *J. Cell Biol.* **26,** 687 (1965).
[16] J. Schmitz, H. Preiser, D. Maestracci, B. K. Ghosh, J. J. Cerda, and R. K. Crane, *Biochim. Biophys. Acta* **323,** 98 (1973).
[17] K. Verner and A. Bretcher, *Eur. J. Cell Biol.* **29,** 187 (1983).
[18] I. Sabolic and G. Burckhardt, *Biochim. Biophys. Acta* **772,** 140 (1984).
[19] D. J. Bjorkman, C. H. Allan, S. J. Hagen, and J. S. Trier, *Gastroenterology* **91,** 1401 (1986).
[20] H. Aubry, A. R. Merrill, and P. Proulx, *Biochim. Biophys. Acta* **856,** 610 (1986).
[21] H. Hauser, K. Howell, R. M. C. Dawson, and D. E. Bowyer, *Biochim. Biophys. Acta* **602,** 567 (1980).
[22] J. R. F. Walters, P. J. Horvath, and M. M. Weiser, *Gastroenterology* **91,** 34 (1986).
[23] A. Diagne, S. Mitjavila, J. Fauvel, H. Chap, and L. Douste-Blazy, *Lipids* **22,** 33 (1987).
[24] H. Van Den Bosch, *Biochim. Biophys. Acta* **604,** 191 (1980).

aggregation of intramembrane particles and large membrane domains devoid of intramembrane particles.[19] Both clustering of membrane proteins induced by Mg^{2+} precipitation[25] in addition to selective loss of membrane proteins and/or cytoskeletal core proteins[19] in the presence of Mg^{2+} are compatible with these results, but the functional correlates of the morphological differences among various brush border preparations still need to be defined. Whether the reduced leak permeabilities of Mg^{2+}- versus Ca^{2+}-prepared vesicles as reported for kidney brush border membranes[18] is a consequence of intramembrane particle aggregation would be interesting to consider.

For transport studies, it would thus appear difficult at this time to recommend either one or the other of these precipitation methods since both, although still subject to considerable empiricism, have proved valuable in different laboratories. It would, however, seem more advisable, at a time when the biochemical characterization of membrane components is being attempted, to use Mg^{2+}, rather than Ca^{2+}, as the precipitating cation. In fact, Mg^{2+} does not activate (1) brush border Ca^{2+}-dependent proteases, (2) Ca^{2+}-dependent transglutaminase, or (3) Ca^{2+}-dependent phospholipase A. The former two enzyme activities may produce artifactual bands in SDS-PAGE, the last one leads to vesicles containing an artifactually high percentage of lysophospholipids[21] (but see Refs. 20, 22, and 23).

Most of the points to be considered in preparing brush border membrane vesicles have already been discussed in former reviews.[12-14] The procedure which we describe below is based on a version[26] of the Schmitz et al. method.[16] In addition to using Mg^{2+}, rather than Ca^{2+}, a number of minor modifications have been introduced which have proved of value in the laboratory of one of the authors (G.S.) for the 15 or so years that brush border membrane vesicles have been prepared for both research and teaching. Perhaps the most important addition has been the final gel filtration through Sepharose 4B,[27] which doubles the specific activity of brush border marker enzymes and vastly improves the stability of the vesicles.

If vesicles are to be prepared from fresh material, small intestines are removed as soon as possible after sacrifice, rinsed with ice-cold saline, slit lengthwise, and freed of mucus by patting with gross paper towels (NOT with Kleenex!). The mucosa is collected by scraping the lumenal surface firmly with glass slides. The material is either used immediately, or collected and mixed well before freezing by dropping aliquots directly into

[25] M. W. Rigler, G. C. Ferreira, and J. S. Patton, Biochim. Biophys. Acta 816, 131 (1985).
[26] M. Kessler, O. Acuto, C. Storelli, H. Murer, M. Muller, and G. Semenza, Biochim. Biophys. Acta 506, 136 (1978).
[27] J. Carlsen, K. Cristiansen, and B. Bro, Biochim. Biophys. Acta 727, 412 (1983).

liquid nitrogen. The pea-sized bits are stored in tightly closed plastic bottles at $-80°$ for up to 2 months. On the day of use, aliquots are thawed in 300 mM mannitol, 5 mM EGTA, 10 mM Tris-HCl, pH 7.1 buffer (30 ml/ g tissue). The total volume is increased to 1 liter with the same buffer before homogenizing. The steps which follow are identical to those described for the preparation starting from frozen intestine (next paragraph).

In fact, brush border vesicles of nearly equal quality can be prepared from frozen intestine, the only condition being that small intestine be collected, rinsed, and frozen as fast as possible. However, at least for one species, the rat, it has proved impossible to obtain transporting vesicles from frozen tissue. Small intestines are rinsed, slit, and freed of mucus as described in the previous paragraph. Instead of scraping, the entire intestine is frozen in dry ice and stored at $-80°$ for up to a few months. At the day of use, aliquots of the frozen tissue are chopped into small pieces and thawed in 300 mM mannitol, 5 mM EGTA, 10 mM Tris-HCl, pH 7.1 buffer (3 ml/g tissue). During the thawing, the temperature of the mixture should not exceed $4°$. Mucosal cells are removed from the underlying connective tissue by the shearing forces of a vibrator (e.g., Vibromixer; Chemap AG, Mannedorf, Switzerland) and filtered through a Büchner funnel. The filtrate is diluted to a volume of 15 ml/g of original intestine with the same buffer and homogenized for 2 min at full speed in a Waring blender (e.g., Ato-mix; MSE, Crawley, Great Britain).

To the homogenate originating from either frozen or fresh tissue is added concentrated $MgCl_2$ to reach a final concentration of 10 mM which precipitates nonbrush border membranes. After centrifugation at 3000 g for 15 min, the brush border membranes are pelleted from the supernatant at 27,000 g for 30 min. The pellet is resuspended in an appropriate amount of the buffer required for the experiment, thoroughly mixed to homogeneity by five passages with a Teflon Potter-Elvehjem homogenizer (e.g., Dyna-mix; Fisher Scientific Co., Zürich, Switzerland), and can be used at once for transport measurement. Alternatively, the pellet is resuspended and homogenized in 150 mM KCl, 50 mM Tris-HCl pH 7.4 buffer, or any other suitable high ionic strength buffer (approximately 1 ml buffer/4.5 mg membrane protein). The resulting mixture is then chromatographed at $4°$ on Sepharose 4B[27] in the same buffer. The length of the column is always 15 cm but its diameter varies with the sample volume to maintain a constant ratio of 1 ml sample/30 ml settled gel (flow rate, 2.5 ml/hr/cm² of cross-sectional area). The brush border membrane vesicles are recovered in the unretarded volume, are spun down at 30,000 g for 30 min, and washed three times in 300 mM mannitol, 10 mM HEPES-Tris, pH 7.0, or Tris-HCl, pH 7.4, or any other suitable buffer. The Sepharose column is regenerated by removing the retarded soluble proteins with 10 mM Tris-HCl, pH 7.4.

The brush border membrane vesicles are stored in liquid nitrogen (1 ml of the mannitol-containing buffer described above/20–45 mg protein) for months without detectable decrease of transport activity. Shortly before use, the vesicles are carefully thawed, kept on ice, and washed with the buffer required for the planned experiment.

The volumes indicated above have been optimized with the aim of allowing the preparation of as many vesicles as possible in one run, thereby saving both time and material, and providing vesicles from the same batch to be used on different days or even weeks with reproducible results.

For preparing vesicles from renal brush border membranes, we use essentially Booth and Kenny's procedure,[4] with a few minor modifications, as described above. If we wish to differentiate between proximal and distal parts of the renal proximal tubules, we follow Turner and Moran's method.[28]

Transport Measurements

Both optical and radiotracer techniques are available for quantitation of solute transport in brush border membrane vesicles.[2,11–14] When compared, the optical methods would appear to present significant advantages: (1) fast and continuous recording of transport events, (2) high time resolution since slow mixing and separation methods are avoided, and (3) low running costs. Unfortunately, they are also associated with a number of disadvantages which have so far precluded general utilization: (1) available dyes are still rather limited and one must use indirect techniques based on membrane potential or volume determinations, which are thus limited to electrogenic and high flux transport systems; (2) mechanisms of dye responses are not always well understood and can be influenced by numerous factors such as membrane charges, pH, and ionic strength[29–32]; (3) some of the fluorescent dyes must be loaded inside the vesicles and then extracellular probe removed before transport measurement[33]; (4) with intravesicularly loaded dyes, efflux over time will decrease signal-to-noise (S/N) ratios with time[33]; and (5) for absolute activities to be measured, the dye response must be calibrated.[29] These should not discourage the reader from attempting to use such techniques and it is more than likely that

[28] R. J. Turner and A. Moran, *Am. J. Physiol.* **242,** F406 (1982).
[29] A. S. Verkman, *J. Bioenerg. Biomembr.* **19,** 481 (1987).
[30] M. L. Graber, D. C. DiLillo, B. L. Friedman, and E. Pastoriza-Munoz, *Anal. Biochem.* **156,** 202 (1986).
[31] R. Krapf, N. P. Illsley, H. C. Tseng, and A. S. Verkman, *Anal. Biochem.* **169,** 142 (1988).
[32] S. Grzesiek and N. A. Dencher, *Biochim. Biophys. Acta* **938,** 411 (1988).
[33] A. S. Verkman and H. E. Ives, *Biochemistry* **25,** 2876 (1986).

current interest in the development of new fluorescent dyes may soon make optical methods more attractive.

To date, tracer flux measurements according to the so-called "rapid filtration" technique[3] have been most commonly used for transport studies using brush border membrane vesicles. The basic procedure is rather simple and involves three steps[11]: (1) vesicles are combined with an incubation medium containing radioactively labeled ligands or substrates and other constituents as required; (2) the uptake reaction is stopped at established time intervals by either aliquoting the previous mixture in a so-called "stop solution"[3,11] or by directly introducing the stop solution into the uptake medium; (3) the final mixture is collected on a filter which is rapidly washed with stop solution and then counted for radioactivity. Efflux studies can in principle be carried out in the same way using vesicles preloaded with labeled substrate. Obviously, each of the steps involved is associated with a number of problems which have been extensively discussed previously.[2,11-14]

Actually, the most critical limitations are time limitations associated with the mixing of the vesicles and the incubation medium (reliability of the zero time), the sampling over short periods of time (difficulty in measuring true initial rates for fast transport systems), and the duration of the stopping process (leak of accumulated substrates), all of which are related to the manual aspect of the technique. A semiautomatic mixing/diluting apparatus has proved very useful in improving the time resolution to fraction of seconds for both transport and binding measurements.[34] A rapid filtration technique for membrane fragments or immobilized enzymes has also been proposed which allows for time resolutions of the order of 10 to 20 msec at best, but has never been used with brush border membrane vesicles.[35] A further achievement has recently been introduced which overcomes the one-point approach of these two versions of the mechanical rapid filtration technique and allows both fast sampling and rapid filtration to be performed under fully automated conditions.[36] Briefly, vesicles are rapidly injected (5 msec) and mixed (250 msec) with the incubation medium (0.2 – 1 ml) in a chamber under controlled temperature (5 – 45°). At time intervals selected on the keyboard of a computer (any time combination from 0.25 to 9999 sec), a maximum of 18 aliquots (20 – 80 μl) can automatically be sampled (up to four/sec) and injected into the upside chamber of a moving filter holder containing the stop solution.

[34] M. Kessler, V. Tannenbaum, and C. Tannenbaum, *Biochim. Biophys. Acta* **509**, 348 (1978).

[35] Y. Dupont, *Anal. Biochem.* **142**, 504 (1984).

[36] A. Berteloot, C. Malo, S. Breton, and M. Brunette, *J. Membr. Biol.*, submitted (1990).

While the process is going on, the sampled aliquots are automatically filtered and washed under conditions lasting just a few seconds (15–20 sec for one filtration and two washings). The whole sequence is under computer control and is triggered by a signal activated during the injection process. A photocell measures the sampling time which was shown to be directly proportional to the sampling volume (over the range 20–100 μl) and thus allows for standardization of the sampling process. Automatic washing of the incubation chamber also allows runs to be performed every 10 min. Finally, a delay sequence between sampling and triggering of both filtration and washings should permit fast efflux to be measured as well. This version of the "fast sampling, rapid filtration apparatus" (FSRFA) has so far proved very useful for kinetic studies using brush border membrane vesicles from both human[37] and rabbit intestines (see the section (Brush Border) Membrane Vesicles in Kinetic Studies).

Advantages and Limitations in the Use of Vesicles for Investigating Transport Systems

Most of the advantages generally associated with the use of membrane vesicles in the characterization of membrane transport systems have long been recognized and have led to almost universal acceptance of this preparation for such studies. It must be made equally clear, however, that vesicles do also have shortcomings and several limitations which are not always as well appreciated.

As compared to intact whole cells, intestinal rings, or everted sacs, brush border membrane vesicles are much simpler biological preparations and, for this reason, present considerable advantages if trying to localize both transport systems and regulatory events in polarized cells such as those of the intestinal and renal epithelia, or if trying to characterize the properties and kinetics of specific transport systems as well as the possible interactions between them. Some limitations, however, apply, and both the more specific advantages and caveats associated with them can be listed as follows.

1. Vesicles are poor in intracellular enzymes. As such, they cannot metabolize most of the naturally occurring substrates for the different transport systems and the transport of the physiologically relevant substrates can be studied. This contrasts with the situation prevailing in intact tissues where the use of nonmetabolizable substrate analogs is the rule rather than the exception. One can also reasonably assume that the substrates taken up by the vesicles are intact and in a free form, and so vesicles

---

[37] C. Malo and A. Berteloot, *J. Membr. Biol.,* submitted (1990).

are in principle suitable biological preparations for determining energy conversion yields. It should not be forgotten, however, that brush border membranes do possess many membrane-bound enzyme activities, most of which are hydrolases implicated in the final digestion of sugars, proteins, and lipids. Studies on the transport of disaccharides, di- and tri- peptides, and phospholipids may thus prove impossible or inconclusive without further processing of the vesicle preparation. In any case, both the extent of hydrolysis of the free substrate and the amount of hydrolyzed versus intact substrate found in the intravascular compartment should be assessed as a function of incubation time. If either test were to show unacceptable levels of hydrolyzed substrate, alternative strategies including nonmetabolizable analogs, specific enzyme inhibitors, stripping of the membrane enzymes by protease treatment,[38] or use of animal species lacking the corresponding hydrolases could be tried and evaluated as to their success in reducing the extent of hydrolysis.

2. Vesicles are essentially free of cell organelles. Moreover, as low-molecular-weight substrates are washed off during the preparation of the vesicles, the remainders of other cellular contaminants are usually of no consequence in most experimental situations. However, even if brush border membrane vesicles are perhaps among the cleanest organelles that can be prepared, the usual criteria of caution must be exerted when localizing one or another transport system in this membrane. A particularly disturbing contamination in this respect could be represented by vesicles from the basolateral membrane or by "chimeric" vesicles whose membrane derives from both brush border and basolateral domains or components. Also, reorganization of the normally polarized plasma membrane domains following the isolation procedure has been reported to occur when cells are isolated prior to membrane vesicle preparation.[19] In such cases, the Na^+-independent D-glucose transport activity found in brush border membrane vesicles could actually represent the expression of the Na^+-independent D-glucose transport activity normally associated with basolateral membranes. It may thus be necessary to use alternative substrates with specificities known to be selective for either the brush border[39] or the basolateral[40] membranes and/or to use inhibitors specific for the apical[41] or contralumenal[42] glucose transport systems in order to assess this point. More difficult to resolve, however, might be to decide whether the

---

[38] A. Berteloot, R. W. Bennetts, and K. Ramaswamy, *Biochim. Biophys. Acta* **601,** 592 (1980).

[39] G. A. Kimmich and J. Randles, *Am. J. Physiol.* **241,** C227 (1981).

[40] G. A. Kimmich and J. Randles, *J. Membr. Biol.* **27,** 363 (1976).

[41] F. Alvarado and R. K. Crane, *Biochim. Biophys. Acta* **56,** 170 (1962).

[42] G. A. Kimmich and J. Randles, *Membr. Biochem.* **1,** 221 (1978).

Na$^+$-independent D-glucose transport activity found in brush border membrane vesicles is the expression of the Na$^+$-independent operation of the brush border Na$^+$-dependent D-glucose transport system(s) or belongs to a separate entity. This is so because phlorizin inhibition of the Na$^+$/D-glucose cotransporter(s) has an absolute requirement for the presence of Na$^+$.[43] An alternative and complementary strategy to check for this particular point could thus be to study efflux from the vesicles in the presence of Na$^+$ and saturating concentrations of this inhibitor after active loading of the vesicles in the presence of an Na$^+$ gradient.

3. Vesicles are not "alive." As such, they do not impose the restrictions typical of surviving tissues (O$_2$, pH, etc.). It is thus possible to study, for example, a transport system in the absence of O$_2$ and in the presence of thiols, if the stability of the substrate so requires. A corollary limitation is, however, associated with this otherwise interesting property since the functional polarity of the two plasma membrane domains naturally found in intact epithelial cells is lost during the brush border membrane isolation. As a consequence, only transient phenomena (overshoots[3]) can be observed in the case of Na$^+$-dependent, secondary active transport systems due to the absence in the brush border membrane vesicles of the Na$^+$,K$^+$-ATPase necessary for continuous Na$^+$ extrusion in intact cells. Also, since the intravesicular concentrations of Na$^+$ and substrate are always changing with time, integrated Michaelis–Menten equations cannot be used to describe the full time course of the uptake process due to the mathematical limitations in finding analytical expressions of the integrated rate laws.

4. Vesicles have small, negligible unstirred layers. Because the active transport of a substance depends on the concentration of that substance immediately adjacent to the membrane, and because the concentration is affected by both the active transport rate and the presence of unstirred water layers, any measurement of apparent kinetic constants are themselves affected by unstirred water layers.[44] This problem is further complicated *in vivo* by the presence of a membrane surface mucous coat[45] which artificially increases the thickness of the unstirred layer. Since the effect of an unstirred water layer surrounding a spherical membrane decreases as the sphere radius decreases,[46] and since the membrane surface mucous coat is removed during vesicle preparation, unstirred water layer effects can in general be neglected in vesicle studies (the situation might, however,

[43] G. Toggenburger, M. Kessler, A. Rothstein, G. Semenza, and C. Tannenbaum, *J. Membr. Biol.* **40**, 269 (1978).
[44] P. H. Barry and J. M. Diamond, *Physiol. Rev.* **64**, 763 (1984).
[45] K. W. Smithson, D. B. Millar, L. R. Jacobs, and G. M. Gray, *Science* **214**, 1241 (1981).
[46] A. S. Verkman and J. A. Dix, *Anal. Biochem.* **142**, 109 (1984).

be different in highly viscous media). This is shown by the much smaller K_m for D-glucose uptake in rabbit brush border membrane vesicles $(0.1 \text{ m}M)$[43] as compared to those in everted sacs $(5.0 \text{ m}M)$,[47] disks $(14-100 \text{ m}M)$,[48] or biopsies $(20 \text{ m}M)$.[47] Only when subjected to very high shaking rates do these last preparations show a decreased K_m with values of $1.3 \text{ m}M$,[47] $1.9-5.0 \text{ m}M$,[48] and $10 \text{ m}M$[47] for sacs, disks, and biopsies, respectively. However, under comparable conditions, essentially the same K_i values for phlorizin inhibition are found in vesicles and everted rings.

5. Magnesium ion-precipitated brush border membrane vesicles are essentially completely right side out.[13,26] This is probably due to the connections between the membrane and the cytoskeleton and/or to the presence of an external glycocalyx, both of which would decrease the probability of inversion for mechanical as well as steric reasons. More difficult to evaluate, however, is the ratio of intact versus opened or leaky vesicles,[49] the latter would underestimate the true transport capacity. It should not be forgotten either that brush border membrane vesicles are heterogeneous in both form and size, which can be due to the extent of homogenization, but may well reflect a genuine heterogeneity of the original brush borders.[25,50] For this reason, the kinetics of efflux and of tracer exchange are not described by a single exponential,[51,52] and criteria to recognize homogeneity or inhomogeneity of vesicle preparations have been proposed by Hopfer.[52]

6. Brush border membrane vesicles can be stored for months in liquid nitrogen without detectable loss of transport activity while no other preparation can (see the section Preparation of Functional Brush Border Membrane Vesicles). This advantage is highly significant in kinetic work, allowing comparisons in both V_{max} and K_m to be made from a large pool of prepared vesicles. However, it is advisable to check if this also applies to transport systems which have not yet been tested for stability on freezing and thawing.

7. Vesicle studies only need reduced volumes (down to a few microliters) of incubation. Minute amounts of expensive substrates or effectors can thus be used. This advantage may, however, be limited by the specific radioactivity of the available substrates which, in turn, will affect the sensitivity of the assay.

8. Vesicles allow substrate uptake to be measured at far shorter incu-

[47] A. B. R. Thomson and J. M. Dietschy, *Am. J. Physiol.* **239**, G372 (1980).
[48] A. B. R. Thomson and J. M. Dietschy, *J. Membr. Biol.* **54**, 221 (1980).
[49] N. Gains and H. Hauser, *Biochim. Biophys. Acta* **772**, 161 (1984).
[50] G. Perevucnik, P. Schurtenberger, D. D. Lasic, and H. Hauser, *Biochim. Biophys. Acta* **821**, 169 (1985).
[51] U. Hopfer, *J. Supramol. Struct.* **7**, 1 (1977).
[52] U. Hopfer, *Fed. Proc., Fed. Am. Soc. Exp. Biol.* **40**, 2480 (1981).

bation times than with rings. This is especially true when a semi- or fully automated apparatus is used for kinetic work.[34-36] It is thus possible to investigate unstable substrates and effectors under appropriate conditions.

9. Brush border membrane vesicles can in general be considered as "tight." This applies to vesicles prepared as described previously but should be assessed in each case since both the preparation procedure and the nature of the substrate may have to be considered in certain cases. It is thus recommended that one tests each reagent and each vesicle preparation to determine how permeant the vesicles are. With appropriate precautions, it is thus possible to draw conclusions as to the sideness of membrane components.[12]

10. In general, the composition of the media at both sides of the membrane can be fixed by the experimenter much more easily with membrane vesicles than can be done with intact, surviving tissues. This freedom is not unlimited, though, as Mg^{2+}-precipitated vesicles are so stable that it may be difficult to introduce high-molecular-weight compounds into the intravesicular space.[26,53,54] For kinetic work, where the conclusions of the studies rely on precise loading of the vesicles, it is advisable to ascertain whether or not vesicles are fully equilibrated with the appropriate reagents at the start of the experiment. That failure to do so can result in possible artifacts has been reported.[55] In general, most low-molecular-weight substrates, effectors, etc., are assumed to equilibrate completely with the intracellular volume during preincubations of reasonable length (e.g., 1–2 hr at room temperature). This may not be the case actually, even for ions, as recently reported for KCl.[56] Similar results have been observed in the laboratory of one of the authors when comparing vesicle loading with KI and choline iodide. While the former would equilibrate within the time period required for resuspension, the latter required a minimum of 6 hr.[92] Such effects can easily be followed by taking advantage of the membrane potential dependency of Na^+-dependent glucose transport[57] and following transport activity for some time after preparing the vesicles.[92] Failure to load the vesicles at the appropriate concentrations would, in the example given above and when vesicles are incubated with NaI, create an internal negative membrane potential because of the higher external concentration of the very permeant anion I^-. This effect would decrease in time on further incubation of the vesicles in the resuspension medium until reaching stability.

[53] F. S. Van Dommelen, C. M. Hamer, and H. R. De Jonge, *Biochem. J.* **236**, 771 (1986).
[54] M. Donowitz, E. Emmer, J. McCullen, L. Reinlib, M. E. Cohen, R. P. Rood, J. Madara, G. W. Sharp, H. Murer, and K. Malmstrom, *Am. J. Physiol.* **252**, G723 (1987).
[55] F. C. Dorando and R. K. Crane, *Biochim. Biophys. Acta* **772**, 273 (1984).
[56] M. S. Lipkowitz and R. G. Abramson, *Am. J. Physiol.* **252**, F700 (1987).
[57] A. Berteloot, *Biochim. Biophys. Acta* **857**, 180 (1986).

11. Membrane vesicles allow effector studies to be performed with more confidence than with intact tissue preparations. For example, both Na$^+$-free conditions and requirements for extra ions are easier to obtain with brush border membrane vesicles as clearly demonstrated, for example, in the particular case of glutamic acid transport in the rabbit small intestine.[58] However, the quantitative assessment as to the membrane potential dependency of transport systems may be more difficult in vesicle studies since direct recording of potential values by means of microelectrodes is impossible with vesicles because of their small size. Thus, the general principle underlying the generation of membrane potentials consists of the imposition of ion gradients that lead to the formation of diffusive potentials. Both cations in the presence of specific ionophores (K$^+$/valinomycin, H$^+$/FCCP) or anions with different lipophilicities have proved successful for this purpose.[2,10,13,57] It should be noted, however, that such techniques may serve well for qualitative assessments (i.e., electrogenicity of transport mechanisms), but that quantitative studies (i.e., membrane potential dependency of transport processes for kinetic arguments) involve two major problems related to the indirect determination of membrane potential values. The first deals with the generation of membrane potentials of known size and, in the absence of knowledge on the permeability coefficients for all ions present in the incubation medium, has been approached by using the Nernst equation as an approximation to the more general Goldman–Hodgkin–Katz equation, thus assuming that the permeability of one ion greatly exceeds the permeability of all other ions present. Such a situation has been approximated by Kaunitz and Wright,[59] who used variable intra- versus extracellular KCl concentrations in the presence of valinomycin. This same approach was later used by Berteloot[57] and compared to another one involving different intra- versus extravesicular concentrations of a highly permeant anion. It thus appeared that the last procedure offered significant advantages: (a) it is easier to handle than techniques using ionophores which are sparsely soluble in water and must be added as solutions in an organic solvent, thus limiting the concentration of ionophore that can be used; (b) it is insensitive to pH variation, a characteristic not shared by the K$^+$-diffusion potentials induced by valinomycin,[57] (c) it can be used with cotransport systems in which K$^+$ has been shown to participate in the transport mechanism;[60] (d) it may be more accurate in estimating membrane potential differences since K$^+$ permeability in the presence of valinomycin was found to be lower than generally assumed, thus underestimating the true potential as calculated by the

[58] A. Berteloot, *Biochim. Biophys. Acta* **775**, 129 (1984).
[59] J. D. Kaunitz and E. M. Wright, *J. Membr. Biol.* **79**, 41 (1984).
[60] A. Berteloot, *Biochim. Biophys. Acta* **861**, 447 (1986).

Nernst equation;[57] (e) it may thus prove more insensitive to ionic replacement, at least for Na^+, Cl^-, and K^+, which are used very often in vesicle studies. It should be borne in mind, however, that these last two conclusions should be evaluated for each different vesicle preparation since they depend on both the success in finding a highly permeant anion in a particular system and the efficiency of valinomycin in inducing K^+ permeability in that system. The second problem in the quantitative evaluation of membrane potentials deals with the selection of an appropriate probe that would allow their measurements. Radioactive lipophilic cations or anions,[2] membrane potential-sensitive dyes,[13] and high-affinity glucose transport activity[57] have all been suggested and used for this purpose. None of these probes is actually perfect since the first ones bind heavily to brush border membrane vesicles (Berteloot, unpublished), the second ones also suffer from binding problems, are not very sensitive, and respond by unknown mechanisms,[13] while the last one is obviously restricted to vesicle systems having well-characterized, electrogenic, Na^+-dependent glucose transport system(s).[57] It should be stressed, however, that the best of the probes would not solve the main problem, which still consists in the assessment of the validity of using the Nernst equation under particular conditions. For this same reason, the range of membrane potentials that can actually be covered for meaningful quantitative purposes using that approach is rather limited.

12. Membrane vesicles fulfill the three R's (reduce, refine, replace) and thus represent a typical example of positive development at a time of animal protection movements. They reduce the number of experimental animals in inhibitor, drug, and kinetic studies since (a) very little biological material is needed in each experiment, (b) in general fewer and more clear-cut experiments are needed to answer a given question than when using surviving tissues or whole animals, and (c) vesicles prepared from a single animal may allow many tests to be performed and directly compared. Vesicles also lead to a refinement in the experimental design because of their much greater flexibility, as compared to surviving tissues or intact animals. Finally, they often lead to replacement of animals from the animal house with animals from the slaughterhouse. Moreover, membrane vesicles can also be prepared from cultured cells.

13. Membrane vesicles can be used as the starting preparation for a number of biochemical studies. Vesicle preparation is a first and important step toward the identification, purification, and isolation of membrane components, including membrane-bound hydrolases and transporters,[12] lipids,[20,21] or membrane-bound cytoskeletal components.[61] Further negative purification of brush border vesicles can be achieved by removing

[61] M. S. Mooseker, *Annu. Rev. Cell Biol.* **1,** 209 (1985).

(most of) the cytoskeleton[12,62] and membrane-anchored enzymes.[38] The cleaner the membrane preparations, the easier it is to optimize the conditions for solubilizing a membrane component. It should be remembered, however, that the preparation procedure may affect the outcome of the conclusions (see the section, Preparation of Functional Brush Border Membrane Vesicles). Brush border membrane vesicles have also been used for direct determination of carrier molecular weights by radiation inactivation[63] and for the identification of carrier molecules in SDS-PAGE[12] and the determination of molecular mechanisms of different transport systems using chemical modifications by group-specific reagents.[12]

(Brush Border) Membrane Vesicles in Kinetic Studies

Many of the advantages listed in the previous section have indicated the (brush border) membrane vesicles as a "perfect" tool for kinetic studies in trying to understand some of the molecular mechanisms of (secondary active) transport systems. In doing so, most of the theoretical and experimental principles generally applied in classical enzymology have been used. Although justified in general, it should, however, be stressed that the extension of methods devised for enzyme reactions in a homogeneous and liquid phase to transport in membrane vesicles is not straightforward and that this approach also suffers from limitations associated with both the use of the membrane vesicles themselves and the nature of the transport processes as opposed to enzyme reactions. Since several reviews[5-14] have already dealt with different aspects and limitations of kinetic studies as applied to brush border membrane vesicles, it is not our intention to repeat all of these but rather to focus on more specific assumptions inherent in such studies, both theoretical and experimental, while referring the interested reader to previous reviews.

Initial Rate Assumption

Even if obvious, it must first be remembered that kinetic studies are intrinsically model dependent since they aim at comparing the mathematical predictions of different models, as expressed by their rate laws, with a series of experimental results performed under different conditions in order to identify the most probable one. As such, one must be certain that the conditions placed on the mathematical derivations are satisfied in the experiment. So far, the conventional approach to the study of cotransport

---

[62] U. Hopfer, T. D. Crowe, and B. Tandler, *Anal. Biochem.* **131,** 447 (1983).
[63] R. Béliveau, M. Demeule, H. Ibnoul-Khatib, M. Bergeron, G. Beauregard, and M. Potier, *Biochem. J.* **252,** 807 (1988).

kinetics has been to apply the steady state methods in the derivation of kinetic equations for (co)transport models with[64] or without[65] the rapid equilibrium assumption which reduces the number of parameters in the rate equations by assuming that the reactions taking place at the membrane interfaces are faster than those perpendicular to the membrane plane. Also, when brush border membrane vesicles are used, it has been common practice to use the so-called "zero-trans conditions" for influx or efflux measurements, thus assuming that the substrate concentrations are zero on the trans side as compared to the cis side where the radioactive substrates are introduced. It should thus be clearly understood that the initial rate assumption must be satisfied under these specific conditions for both meaningful results and conclusions to be obtained. From a more practical point of view, this means that the initial rate determinations should be performed under conditions where linear uptake is observed.

Although quite simple in appearance, this condition is rather difficult to satisfy when using membrane vesicles, mainly because of their large surface/volume ratio as compared, for example, to isolated cells or intact tissues. This leads very rapidly to significant substrate depletion from the intravesicular space during efflux experiments. Similarly, significant intravesicular accumulation of substrate(s) is rapidly achieved during influx measurements such as to allow trans effects and/or reversal of the reaction over a short incubation period. Hence, in both cases, significant deviations from linearity will be observed very soon after the start of the experiment due to the rapid violation of the initial rate assumption. The situation is still worse under the so-called "gradient conditions," in which gradients for both driving ion and driven substrate are present at the start of the experiment. Early collapses of the chemical gradients for ion and substrate occur in parallel and are accompanied by membrane potential changes. Four principal experimental approaches have been used in trying to deal with this problem.

1. Measure tracer exchange rates under equilibrium exchange conditions to circumvent the problem.[51,52] Although very sound on theoretical grounds, this approach should, however, be used cautiously for the following reasons: (a) The results depend heavily on the successful preloading of the vesicles at the appropriate substrate concentrations[55]; (b) because of the heterogeneity in both the form and size of the vesicles, of the possible scrambling or contamination of the brush border preparation by basolateral membranes, of the presence of multiple transport pathways for the

[64] R. J. Turner, *J. Membr. Biol.* **88**, 77 (1985).
[65] D. Sanders, U.-P. Hansen, D. Gradmann, and C. L. Slayman, *J. Membr. Biol.* **77**, 123 (1984).

same substrate in otherwise homogeneous vesicles, the equilibrium exchange rates will not be described in general by a single exponential[51,52]; (c) the sensitivity of the measurements, in terms of significant differences in measurable radioactivities between different conditions, is fixed by the maximum in uptake differences that one can expect to measure and is limited, under equilibrium exchange conditions, by the size of the intravesicular volume. Thus, the description of multiexponential functions will be very imprecise; (d) in case of multiple transport pathways for the same substrate within the membrane, it may become difficult to analyze the results properly. For example, assuming both Na^+-dependent and Na^+-independent pathways for glucose transport in brush border membranes, the "leak" pathway will not be constant when varying the substrate and/or Na^+ concentrations. One could even select conditions with high Na^+ and low substrate concentrations for which the contribution of the Na^+-dependent pathway to the overall uptake rate would be negligible since trans inhibition by Na^+ of the Na^+-dependent pathway(s) have now been demonstrated.[12,55] A still worse situation may occur if brush border membranes were contaminated by basolateral membrane vesicles since the above conditions would favor expression of the Na^+-independent glucose transport system present in the contaminating membranes; (e) as already discussed by others, the conclusions of such studies are ambiguous from a theoretical point of view since they do not allow distinguishing between random Bi-Bi and ordered Bi-Bi kinetic reaction mechanisms[12,66,67]; (f) experimental conditions that can be used are rather restrictive since they allow only the isotope exchange rates of substrate and Na^+ to be determined. In actual fact, the latter are very imprecise due to the high leak permeability for Na^+ in brush border membrane vesicles.[12]

This is not to say that equilibrium exchange experiments cannot produce valuable information but, rather, that this approach should not be regarded as the sole correct experimental set-up for kinetic studies.

2. Estimate initial transport rates by measuring substrate uptake at one time point taken as soon as possible in order to approximate as closely as possible the true initial rate conditions.[12] Obviously, the validity of this approach is straightforward if two basic hypotheses are validated. First, the chosen time point is on the linear part of the uptake time curve and, next, this condition is fulfilled over the whole range of substrate concentrations and experimental variations within the same experiment. While very easy

[66] J. W. L. Robinson and G. Van Melle, *J. Physiol. (London)* **344,** 177 (1983).
[67] D. A. Harrison, G. W. Rowe, C. J. Lumsden and M. Silverman, *Biochim. Biophys. Acta* **774,** 1 (1984).

to satisfy using a semiautomatic apparatus,[34] these conditions may, however, be more difficult to demonstrate with the manual filtration technique because of its poor time resolution. It should be noted, however, that misleading information as to the contribution of the leak pathway (diffusional component) to the overall transport activity may be obtained with this approach (see "Uptake" versus Transport).

3. Perform multiple uptake measurements over a limited time range and estimate the initial transport rates by polynomial regression of the uptake time course,[55] as seems to be accepted by enzymologists.[68] While having the advantage that no assumption as to the transport mechanism involved must be made,[55,68] it has, however, the disadvantage that the coefficients of the polynomial have no physical meaning.[68] Actually, the only justification for this approach is to consider the polynomial as a limit series of an exponential function, and, as such, may only be valid over a limited time range, close to the origin. Accordingly, it is the experience of one of the authors[57,58,60] that the polynomial fitting approach is very dependent on the number of available data points, on the presence of outliers, and on the degree of curvature in the uptake time courses. In any case, polynomials of order higher than three should not be used.[68] One may also have to decide, and quite arbitrarily, whether the regression should go through the origin. It is the point of view of these authors that this approach should be used (and very cautiously) only when no alternative is available.

4. Estimate the initial transport rates from the slopes of straight lines drawn by linear regression through the uptake time courses determined under the various conditions of an experiment at a very early time period (in the few seconds range).[69] This last condition is actually rather restrictive and may be obtained in most cases only by using special pieces of equipment such as the semi- or fully automated rapid filtration apparatuses described above (see Transport Measurements) or a stopped-flow spectrophoto(fluoro)meter when using optical methods. Obviously, this "dynamic approach" has the advantage of justifying in each experimental condition whether the initial rate conditions are fulfilled. It also allows estimation of the standard errors on the initial rate measurements when using appropriate linear regression software. Finally, it allows the determination of whether the regression line goes through the origin, thus also permitting correction for this uptake component and precise assessment of its mean-

[68] B. A. Orsi and K. F. Tipton, this series, Vol. 63, p. 159.
[69] S. Breton and A. Berteloot, *Physiologist* **31,** A142 (1988).

ing (see further in the next section). Using this approach with the FSRFA, it has recently been shown that the linearity period lies within the first 10% of the time required to reach maximal overshoot values.[69] This conclusion actually applies to both rabbit[69] and human[37] intestinal brush border membrane vesicles when the Na^+-dependent transport of D-glucose is analyzed at a 50 μM concentration over the 5–35° range of temperatures. However, when D-aspartic acid uptake by rabbit vesicles was studied in the presence of an inward Na^+ gradient, the observation of an initial linearity period was dependent on both the temperature and the pH of the incubation medium (A. Berteloot, 1990, unpublished). Actually, upward deviations from linearity were observed at pH 6 and were more important when decreasing the temperature from 40 to 15°. On the other hand, initial linearity over the same range of temperatures was observed at pH 8. Since the linear versus nonlinear behavior did not correlate with the time at which maximum overshoot values were recorded, it is tentatively concluded that these results are compatible with the observation of presteady state kinetics[69] (actually lags) according to the conclusions of a recent theoretical study.[70,71] In any case, this last example clearly shows that initial linearities should not be taken for granted even over very short incubation periods.

Clearly, the authors' preference is for true initial rate determinations under gradient conditions since, as compared to equilibrium exchange, they avoid the interference from trans effects which can be investigated separately. They also allow a broader choice of experimental conditions to be tested with, in most cases, an easy and direct control as to the determination of the leak pathway. Moreover, the steady state assumption should apply under gradient conditions whether the membrane potential and/or the ion and substrate gradients collapse with time provided that linearity in the uptake time courses, which represents in itself a sufficient warranty of the existence of a prevailing steady state, is demonstrated. However, since brush border membrane vesicles have been shown to possess Na^+/H^+ exchange activity, it would be advisable to use amiloride and/or proton ionophores in order to slow down the collapse of the Na^+ gradient and/or build-up of proton gradients through this system.

A few comments also need to be made on efflux measurements. Actually, they still represent a challenge since the amount of substrate that

[70] W. Wierzbicki, A. Berteloot, and G. Roy, Fed., Proc., Fed. Am. Soc. Exp. Biol. 46, 368 (1987).
[71] W. Wierzbicki, A. Berteloot, and G. Roy, J. Membr. Biol. 117, 11 (1990).

can be trapped into the intravesicular space is small, at least in terms of measurable counts per minute, thus making these measurements very imprecise. The so-called "active loading process"[72] may thus look very attractive since it allows this problem to be overcome but suffers from the impossibility of fixing the internal substrate concentrations at values chosen by the researcher. Also, since true zero-trans initial rate conditions would necessitate a complete separation of the vesicles from the loading incubation medium (and simple dilution is obviously imperfect in this respect), it can be expected that significant amounts of substrates may have already leaked out at the start of the efflux experiment, thus making it unlikely that true initial efflux rates can actually be measured. The only possibility for accurate measurements in efflux experiments may be that of using one of the (semi)automated techniques described in a previous section, but these still have to be applied to brush border membrane vesicles and thus cannot be evaluated as yet. Two other complications should also be considered when dealing with efflux experiments. First, internal binding of the substrates, if important and rate limiting for the efflux process, may represent a major obstacle since the kinetics observed for efflux may actually describe the debinding process. In fact, should the debinding and efflux processes proceed at comparable rates, the kinetic parameters for transport would still be considerably distorted. Next, efflux kinetics are again influenced by the heterogeneities in the vesicle preparation and will not in general be described by a single exponential.[12]

These difficulties in determining accurately the kinetics of efflux do represent a very serious drawback for the kineticist since they preclude a full kinetic analysis of the Cleland type to be performed. In fact, they add to other factors which, by themselves, already hamper seriously this type of analysis (multiple transport pathways, vesicle heterogeneity, and Na^+ stoichiometries greater than one, which immensely increase the number of kinetic reaction mechanisms to be considered).

Although all of the above difficulties apply particularly to transport studies using (brush border) membrane vesicles as opposed to enzyme kinetics, one should not forget that all of the experimental precautions that apply to the latter also apply to the former, if one wants to obtain meaningful kinetic studies. Accordingly, all kinetic experiments ideally should be performed under controlled temperature, pH, and ionic strength, and the linearity in initial rates versus protein concentrations should be such as to allow comparisons on a day-to-day basis to be meaningful. Also, since kinetic experiments usually require a few hours to be completed, one

[72] Y. Fukuhara and R. J. Turner, *Am. J. Physiol.* **248,** F869 (1985).

should verify the stability with time and with the incubation temperature of the membrane preparation.

"Uptake" versus Transport

When looking at substrate uptake into (brush border) membrane vesicles, one should realize that the measure so obtained is a composite of different components, both unspecific and specific, that occur simultaneously during the incubation process. Obviously, their contributions to the overall uptake need to be evaluated and separated for meaningful transport kinetic analysis to be obtained. These include the following:

1. The background radioactivity, which is the result of nonspecific, nonsaturable trapping of radioactive substrate in a "dead space" represented by the amounts of radioactivity bound onto the filters and/or trapped in the water space surrounding the vesicles and in leaky vesicles, all of which may not be washed out during the rapid filtration step: "En bloc" correction for these uptake components is usually made by running a zero time point in the presence of vesicles when radioactive substrate and stop solution are added simultaneously. Such a practice cannot be recommended, however, since it assumes (a) "exact" reproducibility from sample to sample, which is essentially not verified, and (b) time independence of the above processes, which may or may not apply in the specific cases under study. The first assumption can be released by using a "quenched" stop solution[3] which contains a differentially labeled space marker or substrate analog (usually an inactive stereoisomer). A better approach, however, which does not make any of the above assumptions, involves a double tracer incubation with these same molecules. For these two corrections to be valid, one should verify the equivalence between ^{3}H- and ^{14}C-labeled spaces. Also, when coincubation with a substrate analog is performed, the simple diffusional component of transport [see point (2) below] may be simultaneously corrected.

2. The simple (passive) diffusion of substrate, which represents the intrinsic leak permeability of the membranes to different substrates, and should demonstrate the following properties: (a) independence, in terms of counts per minute, as to the cold substrate concentrations, or, equivalently, linearity in the v over S plot or absence of saturation; (b) insensitivity to inhibition by known transport inhibitors or substrate analogs. Accordingly, this transport component is usually determined by running a transport experiment at different substrate concentrations in the presence of saturating concentrations of specific inhibitors of the carrier-mediated process(es) and/or under nonenergized conditions in the case of secondary active transport system(s). The slope of the straight line usually obtained by

constructing the corresponding v over S plot is then taken as evidence for the contribution of a diffusive component to total transport and its slope assimilated to the diffusion rate constant k_d. As straightforward as it may seem, we think that such a simple analysis can easily lead to erroneous conclusions since simple diffusion cannot be dissociated from either uncorrected background [point (1) above] or nonspecific binding [point (3) below] based on these criteria alone. This conclusion is best illustrated by recent data using the FSRFA and the dynamic approach for the characterization of glucose transport by human intestinal brush border membrane vesicles.[37] When the v over S plot was constructed from the slopes measured under Na^+ gradient conditions, saturation was obtained directly. It was unequivocally deduced that simple diffusion of glucose does not contribute significantly to total transport. Since this result contrasts with what is usually observed in kinetic studies, the same plot was constructed from initial rates determined according to the one time point approach. In this case, both saturating and diffusional transport components were observed, and it was shown that the latter is actually an artifact due to the noncorrection for the y intercepts. This example clearly demonstrates that the so-called diffusion should only be considered as an operational parameter since it may fail to give any meaningful information as to the real passive permeability of the membrane for a given substrate. It also allows one to seriously question previous negative views as to the leakiness of the membrane vesicles.[49] Moreover, since one is generally interested in analyzing the saturating processes, a detailed delineation of the nonspecific uptake components is usually not required, and the fastest and best way to deal with these components is suggested in a following section (Determination of Kinetic Parameters in Uptake Studies).

3. The unspecific and/or specific substrate binding to intra- and/or extravesicular membrane sites. For kinetic studies, internal binding is usually not a problem during initial rate determinations under true steady state conditions since enough substrate needs to get into the intravesicular space for this component to show a significant contribution. However, the external binding contribution to total uptake should not be assumed to be small *a priori*, and, actually, could be very important, particularly in special circumstances. For example, membrane vesicles have an excess of negative charges,[21] which may make it difficult or even impossible to differentiate between the transmembrane movement and binding of cations with two or more charges. In the case of positively or negatively charged substrates such as amino acids,[73] folic acid,[74] and quaternary

[73] B. Y. L. Hsu, P. D. McNamara, C. T. Rea, S. M. Corcoran, and S. Segal, *Biochim. Biophys. Acta* **863**, 332 (1986).

[74] A. M. Reisenauer, C. J. Chandler, and C. H. Halsted, *Am. J. Physiol.* **251**, G481 (1986).

ammonium bases,[75] binding to membrane components has also been reported. Moreover, since membrane vesicles have a large surface/volume ratio, it may often be impossible to differentiate between transmembrane transport and "uptake" into the lipid bilayer in the case of lipophilic substrates. Several methods are available which allow qualitative evaluation of a contribution of substrate binding to total uptake in (brush border) membrane vesicles; they have been discussed in previous reviews.[12-14] For example, experiments run under accelerated exchange conditions (high intravesicular substrate concentrations) may help in resolving binding from transport since only the latter should be sensitive to these conditions. Absence of binding can also be demonstrated by showing that substrate uptake occurs entirely into an osmotically sensitive intravesicular space,[3] that the amount of substrate associated with the vesicles at equilibrium, i.e., after a long incubation period, is independent of the chemical nature of the substrate and is directly proportional to the extravesicular substrate concentration,[12] and that uptake is abolished in vesicles permeabilized by low concentrations of detergent. Although it is very important to assess whether one deals with transport, binding, or both when performing uptake studies in vesicles, the presence of binding components is not a major obstacle to the kinetic analysis of transport, particularly when using the dynamic approach. For example, a linear relationship between intercepts and substrate concentrations is strong evidence for the presence of a fast, nonspecific binding component and/or of incomplete correction for the background. If a Michaelian relationship is found, however, a fast and specific binding component can be inferred and its kinetic parameters extracted. If a saturable, higher order specific binding component is observed, such a result is compatible with the existence of cooperativity in the fast binding of substrate to the membrane. Alternatively, this binding component may actually represent the amplitude of a presteady state burst,[70,71] the time resolution of which might not have been achieved in the experiment. However, if this last case applies, one may expect that the burst component will disappear in permeabilized vesicles. Finally, it should be emphasized that substrate binding may be difficult to differentiate from transport on the basis of time dependency alone since the former is not necessarily rapid as compared to the latter, contrary to what is generally assumed when analyzing the y intercepts of the uptake time courses. Accordingly, the v over S plots may contain more than one saturable component. In such cases, it might be possible to attribute these sites to binding and/or to transport by further analyzing the kinetics of

[75] K.-I. Inui, H. Saito, and R. Hori, *Biochem. J.* **227**, 199 (1985).

substrate uptake in the absence of transport (permeabilized vesicles, equilibrium uptake, etc.). Obviously, when using the one time point approach, both the intercept and slope components will be included in the measured uptake values, a situation which can lead to rather complex results and may prevent complete evaluation of the data. However, it should work quite well in the simplest cases discussed above.

4. The carrier-mediated transport process(es) can be of either the facilitated or secondary active types and actually represent(s) the system(s) that one wants to study in most cases. The evidence for active transport is best obtained under ionic gradient conditions in which the driving ion (for example Na^+) is replaced by an inactive one (for example K^+). The presence of a transient accumulation of substrate over the equilibrium uptake values, or the so-called "overshoot" phenomenon,[3,76] is usually taken as evidence for secondary active transport. However, for this conclusion to be valid, conditions should be chosen so as to eliminate other possible interpretations such as volume changes or membrane potential effects in the case of charged substrates. It should also be realized that the overshoot may not be observed under the following conditions: transporters with low turnover rates or with special combinations of rate constant values in their molecular mechanisms,[76] high leakiness of the vesicles for the driving ion, too high concentrations of the driven substrate, and transport systems very sensitive to inhibition by trans substrates or requiring additional factors for full expression of activity (ions, cellular components which are removed during the purification, etc.). Still more difficult to resolve and to interpret is the simultaneous presence of both active and facilitated transport pathways for the same substrate in the vesicle preparation, since scrambling of the vesicles and/or contamination of the preparation by other membranes fragments, different expressions of the same transport system under different ionic conditions, and multiple transport systems for a given substrate, either separately or in combination, are all conditions that can lead to rather complex situations.

Determination of Kinetic Parameters in Uptake Studies

As should now be obvious from the above discussion, the interpretation and analysis of uptake kinetics are complicated by the occurrence of, in general, more than one component, usually including both nonspecific and saturable processes. In practical terms, this means that the usual linear transformations of the Michaelis–Menten equation are in fact nonlinear. It is thus current practice to estimate the diffusional component (whether

[76] E. Heinz and A. M. Weinstein, *Biochim. Biophys. Acta* **776**, 83 (1984).

real or just operative) in a separate experiment and to subtract its contribution from the total uptake data so as to extract the saturating component(s) which can now be analyzed by linear regression over a linear transformation of the Michaelis–Menten equation. The Eadie–Hofstee plot is actually preferred for its higher sensitivity to deviations from linearity than the double-reciprocal plot of Lineweaver and Burk and thus serves as a determinant test in assessing the presence of multiple transport pathways.

Although justified on purely mathematical grounds in the case of error-free data, such an analysis is, however, of little value when dealing with experimental determinations for the following reasons: (1) the experimental conditions are not identical when estimating total uptake and diffusion separately and it may be difficult to assess subtle changes in the diffusional component under these two conditions; (2) statistical information on the determination of the diffusional component is lost during the subtraction procedure for which it is implicitly assumed that the k_d value is unique and error free. It is thus lost during the computation of the kinetic parameters for the saturating component(s); (3) even small errors in the estimation of the diffusional component will distort the linearization of a saturating process so as to induce upward (underestimation) or downward (overestimation) deviations from linearity in an Eadie–Hofstee plot. The resulting curves will thus mimic a heterogeneity in binding sites or a negative cooperativity in the first case and a positive cooperativity in the second. Obviously, a more accurate and easier way to deal with such data is to use nonlinear regression analysis and to directly fit the total uptake curve to the (apparently) more complex rate equation which includes both the saturating and the nonsaturating components. All parameters would thus be determined simultaneously from a curve that contains all necessary information.[37,66,77] This point is particularly relevant when considering the high number of second sites (and even third sites) which have been reported for different transport systems by Eadie–Hofstee plots, and one may wonder, in the absence of supporting proof on heterogeneity in transport sites, how many of these would survive a more rigorous analysis by nonlinear regression.

There are also a few other reasons why nonlinear regression analysis of kinetic data should be preferred. These have been discussed in a few recent reviews[78-81] and will only be summarized here: (1) Linear transformations

[77] G. Van Melle and J. W. L. Robinson, *J. Physiol. (Paris)* **77**, 1011 (1981).

[78] F. W. Maes, *J. Theor. Biol.* **111**, 817 (1984).

[79] D. Garfinkel and K. Fegley, *Am. J. Physiol.* **246**, R641 (1984).

[80] G. A. Sagnella, *TIBS* 100 (1985).

[81] J. H. T. Bates, D. A. Bates, and W. Mackillop, *J. Theor. Biol.* **125**, 237 (1987).

may distort (usually magnify) the experimental variability, thus leading to biased estimates for the values of the parameters.[79,80] This is particularly important with the double-reciprocal plot of Lineweaver and Burk, in which the inverse transformation places undue emphasis on the most variable points.[79,80] Similarly, in the Eadie–Hofstee plot the dependent variable v appears on both axes, thus leading to an unavoidable distortion of the variability in the data[80]; (2) nonlinear regression is in general less sensitive to the spacing and number of data points than is linear regression and is a versatile and general curve-fitting procedure which can be used with a variety of functions[80]; (3) the assumptions underlying the linear regression model are in general not satisfied when using either the Lineweaver–Burk or the Eadie–Hofstee plots.[78] For this reason, it has even been claimed that "linear regression analysis should not be applied to linearizing plots in biochemistry and pharmacology"[78]; (4) in the particular case of transport studies, the nonlinear regression approach allows different models to be objectively evaluated without any manipulation of the original data.[66,77] For example, in the case of both saturating and diffusional components, different equations corresponding to diffusion alone, one carrier alone, one carrier plus diffusion, two carriers, two carriers plus diffusion (and more if necessary) could be fitted to the data and the best model chosen according to statistical criteria alone[66,77] (see below). It should be easy for anyone to understand that this procedure is completely different from first trying to correct for a diffusional component before fitting the resulting points to either a one- or a two (or more)-carrier model after linear transformation. Should an underestimation in the diffusion constant be introduced during the subtraction procedure, then any requirement for a two-site fitting could actually give the desired output. This is also true should nonlinear regression be used on the transformed data. This is not to say that linearizing plots should not be used anymore but, rather, that they should be considered only as a quick and useful way to visualize enzyme and transport kinetic data.[82] Although to most individuals linear regression may appear easier to do, this is no longer the case, since very interesting nonlinear regression software is now available which would fit most of the microcomputers currently used for linear regression analysis!

For any serious kinetic study to be carried out, the data must also be subjected to statistical analysis so that the precision of the derived kinetic constants can be evaluated.[83,84] This is particularly important when trying to distinguish between concurrent models or to evaluate the existence of

[82] F. B. Rudolph and H. J. Fromm, this series, Vol. 63, p. 138.
[83] W. W. Cleland, this series, Vol. 63, p. 103.
[84] B. Mannervik, this series, Vol. 87C, p. 370.

more than one transport pathway. In general, the problem of goodness of fit or of which model best describes the observed data should involve analysis of the residuals and the estimated parameters.[78,83,84] Ideally, a satisfactory model should give an error term with constant variance randomly distributed about zero and the residuals should not be correlated with the independent and dependent variables. A satisfactory fit must also provide biologically meaningful values for the parameters. For example, a negative value for any of these would obviously be unacceptable. Finally, in a good fit, the standard errors of the estimated parameters should be small in comparison to the parameter values. It has been suggested that when the standard errors are less than 25% of the values, one can consider the values to be accurately determined, and thus that the term containing this constant is definitely present.[83] Can this type of analysis represent a major obstacle to those who regard statistics as "horrible monsters"? The answer is no, since most of the good software for nonlinear regression analysis also include the statistics mentioned above and allow them to be used in a (very) user friendly way.

A final comment should be made on the presentation of transport kinetic data. In general, kinetic experiments using radiotracer techniques are done at a constant concentration of isotope, while substrate concentrations are adjusted by adding varying amounts of cold substrate. The actual output of such experiments thus consists in decreasing cpm values for increasing substrate concentrations. However, v over S plots are usually constructed after correction for the specific radioactivity calculated at each concentration, so that one shifts from a decreasing curve to an increasing one where the highest v values correspond to the lowest cpm values. Although perfectly acceptable, we think that this approach can easily lead to erroneous interpretations and that a different presentation, closer to the original data, should be chosen. In this approach, which is analogous to the displacement curves used in binding studies, the specific radioactivity is considered as constant and corresponds to that determined at the lowest substrate concentration. The cold substrate is then considered a competitive inhibitor of the saturable process(es) and, accordingly, the v over "cold" S plot is similar to the original cpm data by some scaling factor. The resulting curve can be fitted by nonlinear regression analysis to an equation accounting for the competitive inhibition.[37] It should be emphasized that this approach directly expresses the initial rates in terms of the parameters and so does not involve any transformation of the data as when trying to express the inhibited rates as a function of the rate under pure tracer conditions. These two representations were recently compared in one of the authors' laboratories[37] and the following conclusions were drawn: (1) Identical parameter values are obtained when fitting any of these curves under similar weighting conditions; (2) the second approach is better when trying to visually assess the final fit, particularly when dealing with both high-affinity, low-capacity and low-affinity, high-capacity systems, and in the presence of more than one transport pathway; (3) the overall structure of the data is altered in the classical representation, putting undue emphasis on the high S region while masking the other. This is particularly obvious in the presence of a nonspecific component in which the linear part takes over most of the carrier-mediated transport system(s). However, in the other approach, the v over S plot decreases to a constant value, thus clearly showing the exact contribution of carrier-mediated uptake to the measured values (hence the sensitivity of the assay) and pointing out the impossibility of distinguishing between diffusional and nonspecific binding components from such plots as discussed in a previous section. It can thus be concluded that the second approach could advantageously replace the former without any loss in terms of kinetic parameter estimation, saving time in the processing of the raw data for computational purpose and allowing a better appreciation of the real data.

Multiple Transport Pathways in Kinetic Data

As should now be obvious from the above discussion, good transport kinetics are more difficult to obtain than good enzyme kinetics due to the difficulties associated with both the experimental limitations inherent in the transport measurement in vesicles and the analytical evaluation of the resulting uptake and kinetic data. For these reasons, we cannot conceal our skepticism for "unequivocal" demonstrations as to the existence of two or even three transport systems (plus diffusion!) operating in parallel on a given substrate when these are based solely on curvilinear Eadie–Hofstee plots, and particularly so when the claimed deviation from a Michaelian behavior relies on a single, deviating experimental point! This is not to say that curvilinear Eadie–Hofstee plots can only be the result of poor handling or collection of the data points (see above) but rather that such plots are neither necessary nor sufficient proof for the existence of more than one transporter. First, it is not necessary proof since theoretical studies have shown, in the case of multiple sites, that significant deviations from linearity can only be expected when the V_{max} and K_m values of the individual sites are well separated. Accordingly, linear Eadie–Hofstee plots may actually mask the presence of more than one transport pathway. Nor is it sufficient proof since nonlinear Eadie–Hofstee plots may also be obtained under multiple site situations in which only one carrier is present, including positive cooperativity,[85] hysteresis,[86] and random, steady state addition

[85] K. E. Neet, this series, Vol. 64, p. 139.
[86] K. E. Neet and G. R. Ainslie, this series, Vol. 64, p. 192.

of two substrates on an enzyme[85] or carrier.[12,55,87] This last situation has been very well known among enzymologists since 1966[88] (if not earlier) but does not seem to have attracted the attention of many transport workers. Random Bi-Bi or Bi-Uni kinetic reaction mechanisms may produce apparent cooperativity of the kinetics if the following conditions hold:[85] (1) the two pathways for ternary complex formation have approximately the same individual equilibrium constants so that neither pathway is thermodynamically favored; (2) the rate constant for ternary complex reorientation is of the same order of magnitude or is larger than the other unimolecular rate constants so that rapid equilibrium binding of substrates does not occur; (3) one of the pathways for ternary complex formation is kinetically favored. The apparent cooperativity may thus be conceptually seen as follows. At a fixed concentration of substrate (S) and low concentrations of the other substrate (Na^+ in cotransport systems), the carrier (C) flux will be through the slower $C \rightarrow CS \rightarrow CSNa$ pathway because of the high proportion of CS. As the concentration of Na^+ is increased, the tendency will be for the flux to occur through the faster $C \rightarrow CNa \rightarrow CNaS$ pathway as CNa becomes competitive in concentration. The substrate curve will therefore be nonhyperbolic, and a nonlinear Eadie–Hofstee plot will result. The curvature will depend on the concentration of Na^+ and disappear at saturating Na^+ concentrations. As recently demonstrated in the case of cotransport systems,[87] this simple test should thus allow differentiation between the random steady state and the two-carrier models. Another criterion that should help in differentiating these two models is that increasing internal substrate concentrations should cause uncompetitive inhibition of transport in the former model only.[87] It can be anticipated that similarly misleading kinetic situations may occur for ter-ter or tetra-tetra mechanisms.

In addition to these theoretical considerations, there are other sources of error which may mimic multiplicity of transport systems. For example, even "pure" membrane vesicles may be scrambled (those from basolateral much more so than those from brush border membranes). As there is every good reason to believe that the transport systems are asymmetric,[12] the K_m values of a transport system are likely to be different at the two sides of the membrane. Thus, scrambling will render even otherwise "homogeneous" vesicles kinetically heterogeneous, because right-side-out and inside-out vesicles will expose to the medium faces of the transporter having different kinetic parameters. Curvilinear Eadie–Hofstee plots will be the result.

Since v over S plots alone may fail to give the right answer, they should be complemented by other studies [pH, temperature, stoichiometry, and

[87] D. Sanders, *J. Membr. Biol.* **90,** 67 (1986).
[88] W. Ferdinand, *Biochem. J.* **98,** 278 (1966).

inhibition by either known inhibitors of the putative transporter(s) or substrate analogs, etc.] for a meaningful characterization of the transport pathway(s) to be obtained. Among these, the partial (competitive) inhibition by another substrate is the most often used criterion to assert multiplicity of transport systems acting on a given substrate. The reasoning is deceivingly simple: if the membrane is endowed by, say, two systems transporting amino acid A, and one of the systems transports amino acid B also, saturating concentrations of B will only partially inhibit the transport of A. Straightforward as it may seem, this approach also has limitations. First of all, if a sizeable percentage of vesicles is scrambled, since the substrate specificity of a transport system may be (and usually is) different at the two sides of the membrane,[89] the same transport system may look at the outer surfaces of the two subpopulations of vesicles with substrate binding sites of different specificities. For example, it may interact with both A and B in the right-side-out vesicles but with A alone in the inside-out vesicles (or vice versa). But even if the vesicles are unscrambled, homogeneous, and pure, interactions among Na^+-dependent systems occurring in the same membrane occur through more than fully competitive inhibition alone. In fact, the increase in Na^+ trans and the partial collapse of the Na^+ electrochemical gradient which are brought about by the entry of Na^+ via one Na^+-dependent substrate indirectly inhibits the operation of other, also Na^+-dependent, but otherwise independent cotransporters. The kinetic result is a "partial noncompetitive" inhibition.[90,91] Thus, in the example above, if B, in addition to being another substrate of the Na^+-dependent system for A, is transported by other, also Na^+-dependent system(s), the partial inhibition of A transport by B is the sum of at least a "fully competitive" and of a "partially noncompetitive" inhibition. It thus does not correspond simply to the contributions of the two systems in the transport of A.

This is not to say, of course, that kinetic analysis is useless in testing for multiplicity of transport systems. Rather, we want to remind the reader of pitfalls which are, unfortunately forgotten all too often, that kinetic analysis does not inherently "demonstrate" (it is "compatible with" at best) and that only the use of several, preferably independent, criteria (kinetic, physical separation, biological development, molecular biochemistry, genetics, reconstitution, etc.), can lead to conclusions not liable to fall under Occam's razor.

[89] J. E. G. Barnett, W. T. S. Jarvis, and K. A. Munday, *Biochem. J.* **109,** 61 (1968).

[90] G. Semenza, *Biochim. Biophys. Acta* **241,** 637 (1971).

[91] H. Murer, K. Sigrist-Nelson, and U. Hopfer, *J. Biol. Chem.* **250,** 7392 (1975).

[92] D. D. Maenz, C. Chenu, F. Bellemare, and A. Berteloot, *Biochim. Biophys. Acta,* submitted (1990).

[27] Isolation and Reconstitution of the Sodium-Dependent Glucose Transporter

By PARAMESWARA MALATHI and MARK TAKAHASHI

The lumenal brush border membranes of absorptive epithelial cells in the kidney proximal tubule and the small intestinal mucosa contain sodium-coupled transport systems that are involved in the active absorption of many solutes, including D-glucose.[1,2] The energy for such processes is derived from the electrochemical gradient for Na^+ that is maintained across the lumenal membrane by the Na^+,K^+-ATPase located in the contralumenal membranes of the cell. Highly purified brush border membranes can be easily isolated from small intestinal mucosa and kidney cortex in the form of osmoreactive vesicles which are ideally suited for the study of sodium-dependent glucose transport.[3,4] The electrochemical gradient for sodium normally present across lumenal membranes can be reproduced using membrane vesicles and its magnitude altered by changing the transmembrane concentrations of Na^+ or by manipulating the membrane potential. Several sodium-coupled transport processes have been characterized in brush border membrane vesicles.[5,6] Transport of D-glucose in membrane vesicles is stereospecific, concentrative in the presence of a transmembrane Na^+ gradient, and inhibited by phlorizin and sulfhydryl reagents such as Hg^{2+}, p-hydroxymercuribenzoate, and N-ethylmaleimide.[1,2] While extensive kinetic analysis of glucose transport in both kidney and small intestines has been carried out,[7-10] the molecular characterization of these glucose transporters has yet to be accomplished. Studies utilizing photoaffinity labeling, radiation inactivation, chemical modification, monoclonal antibodies, and purification–reconstruction have been

[1] R. K. Crane, *Rev. Physiol. Biochem. Pharmacol.* **78**, 99 (1977).
[2] G. Semenza, M. Kessler, M. Hosang, T. Weber, and U. Schmidt, *Biochim. Biophys. Acta* **779**, 343 (1984).
[3] J. Schmidt, H. Preisser, D. Maestracci, B. K. Ghosh, J. Cerda, and R.K. Crane, *Biochim. Biophys. Acta* **434**, 98 (1973).
[4] P. Malathi, H. Presier, P. Fairclough, P. Mallett, and R. K. Crane, *Biochim. Biophys. Acta* **534**, 259 (1979).
[5] B. Sacktor, *Curr. Top. Bioenerg.* **6**, 39 (1977).
[6] H. Murr and R. Kinne, *J. Membr. Biol.* **55**, 81 (1980).
[7] U. Hopfer and R. Groseclose, *J. Biol. Chem.* **225**, 4453 (1980).
[8] R. J. Turner, *J. Membr. Biol.* **76**, 1 (1983).
[9] Y. Fukuhara and F. J. Turner, *Biochim. Biophys. Acta* **770**, 73 (1984).
[10] F. C. Dorando and R. K. Crane, *Biochim. Biophys. Acta* **772**, 273.

used in attempts to identify and isolate the transport protein.[2] In a different approach, using the strategy of expression cloning, the Na^+-dependent glucose transporter from rabbit intestine has been cloned.[11] Wright and co-workers have extended these studies to include the human intestine and kidney and conclude that the molecular mass of the transporter in these two tissues ranges between 69,000 and 80,000.[12] This chapter details the methods that have been developed to solubilize, purify, and reconstitute the glucose transporter from kidney cortex; the method is equally well suited for the intestinal brush border membrane with suitable additions of protease inhibitors. Stated briefly, isolated brush border membranes are solubilized with detergent and fractionated on a Mono Q column to yield a highly purified preparation which, when inserted into artificial phospholipid vesicles, exhibits the properties of Na^+-dependent glucose transport observed in native membrane vesicles.

Isolation of Brush Border Membranes

Principle. Cortical cells are lysed in a hypotonic solution and membranes derived from nonbrush border organelles are precipitated with Ca^{2+} and removed by low-speed centrifugation; brush border membranes remain in the supernatant and are sedimented as a pellet by high-speed centrifugation. The procedure utilized[4] is similar to the one described by Schmidt *et al.*[3] for recovering small intestinal brush border membranes.

Procedure. Small experimental animals such as rabbits, guinea pigs, and rats are sacrificed by cervical dislocation and the viscera exposed by ventral resection. The kidneys are perfused with 0.9% NaCl through a catheter introduced into the descending aorta. This exsanguination reduces possible contamination by membranes derived from the formed elements of blood. Blood-free calf kidneys obtained from a local abattoir are also a convenient source of kidney membranes. Isolated kidneys can be frozen at $-70°$ for up to a year without loss of brush border functions. Studies reported in this chapter are based on calf kidneys. The cortex is removed by dissection and processed to prepare brush border membranes. All subsequent operations are carried out at 4°. Each gram of tissue is suspended in 20 ml of lysis buffer consisting of 50 mM mannitol and 2 mM Tris-HCl, pH 7.0 and homogenized using a Sorvall Omnimixer at maximum speed for 4 min. A suitable aliquot of 1 M $CaCl_2$ is added to the homogenate to give a final concentration of 10 mM. The mixture is stirred on a magnetic

[11] M. A. Hediger, M. J. Coady, T. S. Ikeda, and E. M. Wright, *Nature (London)* **330**, 379, 1987.
[12] M. A. Hediger, E. Turk, A. M. Pajor, and E. M. Wright, *Klin. Wochenschrift* **67**, 843, 1989.

stirrer for 10 min and then centrifuged at 3000 g for 10 min. The pellet, consisting of unbroken cells, connective tissue, and Ca^{2+}-aggregated cell membranes, is discarded. The supernatant, containing brush border membranes, is centrifuged at 40,000 g for 30 min. The pellet is resuspended in lysis buffer using a tuberculin syringe and 23-gauge needle and centrifuged again at 40,000 g. This pellet consists of brush border membranes highly enriched in the appropriate enzymatic markers such as alkaline phosphatase, trehalase, leucylnaphthyl amidase, and maltase. The yield of membranes is approximately 5–6 mg membrane-protein/g of cortex.

Extraction of Membranes

Principle. The detergent extraction described for solubilizing the brush border membrane glucose transporter was developed to maximize solubilization while minimizing activity loss; it also permits ready separation of detergent from the membrane proteins for optimal convenience of transport assay and for further purification. Detergents proven to be valuable in membrane solubilization and their properties have been reviewed elsewhere[13-15] and detailed descriptions given regarding approaches to solubilization of membrane proteins.[16]

In our earlier studies, we tested a large number of commercially available detergents and found sodium deoxycholate (DOC) well suited for extraction of membranes. It was also readily removed from the extracts by dialysis or dilution followed by high-speed centrifugation.[17] However, transport activity measured after reconstitution (see below) was only 10–20% that in native membranes. Further manipulation of the extracts did not result in any loss of activity. Some of the data presented in this chapter were derived from DOC extracts of brush border membranes. In recent years *n*-octyl β-D-glucopyranoside (OG) has emerged as a widely used detergent not only because it is easily removed by dialysis or gel filtration but also because biological activity of the extracted components is well preserved. We have found this to be true with the glucose transporter of kidney brush borders.

Procedure. All operations are carried out at 4–5°. Isolated brush border membranes are assayed for protein by a modified Lowry method.[18] Mem-

[13] A. Helenius and K. Simons, *Biochim. Biophys. Acta* **415**, 29 (1975).
[14] C. Tanford and J. A. Reynolds, *Biochim. Biophys. Acta* **475**, 133 (1976).
[15] L. M. Hjedmeland and A. Chrambach, *Electrophoresis* **2**, 1 (1981).
[16] L. M. Hjedmeland and A. Chrambach, this series, Vol. 104, 305.
[17] P. Malathi and H. Preiser, *Biochim. Biophys. Acta* **735**, 314 (1983).
[18] J. H. Waterborg and H. R. Matthews, *in* "Methods in Molecular Biology" (J. M. Walker, ed.), pp. 1–3. The Humana Press, Clifton, New Jersey, 1984.

brane protein (100–200 mg) is suspended in 0.15 M KCl, 20 mM Tris-HCl, pH 7.8 in a capped vial or tube to a final concentration of 5 mg/ml protein. A tuberculin syringe fitted with a 23-gauge needle is very useful for suspending membranes and membrane-derived fractions pelleted by high-speed centrifugation. A stock solution of OG (10% or 0.34 M) in the suspension buffer is added to the membranes to give a final concentration of 1% OG. The container is capped and the contents gently mixed by inverting the container back and forth two or three times. The mixture is transferred to ultracentrifuge tubes and centrifuged at 212,000 g_{av} for 30 min. The supernatant contains 60–70% membrane proteins and little or no detectable glucose transport activity; it can therefore be discarded. The pellet is resuspended using syringe and needle in 20 mM Tris-HCl, pH 7.8 and adjusted to the starting volume. The centrifugation is repeated as above and the supernatant discarded. This step ensures removal of most of the KCl present during the first extraction; subsequent extraction is best accomplished in low ionic strength buffer. The pellet is resuspended in 20 mM Tris-HCl buffer and diluted to half the starting volume to give a protein concentration of 3–5 mg/ml. A stock solution of OG (10% in 20 mM Tris-HCl) is added to a final concentration of 1% OG. The mixture is held in an ice bath and sonicated with a 3/8-in. probe (Biosonic probe; Bronwill Scientific, Rochester, NY) for 1 min at maximum output. The resulting mixture should be nearly clear. It is centrifuged at 212,000 g_{av} for 30 min. The supernatant contains more than 95% of the proteins in the pellet and almost all of the glucose-transporting activity in the starting membranes is recovered in this fraction (see Table I). Further resolution of this extract is achieved in an anion-exchange fast protein and polypeptide liquid chromatography (FPLC) column.

TABLE I
TRANSPORT RECOVERY IN MEMBRANE FRACTIONS

Membrane fraction[a]	Percentage protein recovered	Na+-Dependent glucose transport[b]
Brush border membrane vesicles	100	2.8
ME	96	2.8
ME-1	65	0.3
ME-2	33	10.8
Mono Q fraction	3	81.0

[a] ME, Membranes suspended in 20 mM Tris-HCl and solubilized with 1% OG and sonication; ME-1, first extract of membranes in 0.15 M KCl, 20 mM Tris-HCl with 1% OG; ME-2, second extract of membranes followed ME-1, with 1% OG in 20 mM Tris-HCl; Mono Q fraction, fraction eluted with 0.35 M KCl from a Mono Q column (FPLC).

[b] Nanomoles D-glucose per minute per milligram protein.

FPLC on Mono Q HR 5/5 Anion-Exchange Column

Principle. Mono Q is a strong anion exchanger developed by Pharmacia and designed for FPLC. The charged groups on a beaded hydrophilic matrix consist of quaternary amines. Negatively charged bound proteins are eluted from the positively charged matrix using increasing salt concentration. Details of this and other FPLC matrices are described in a booklet published by Pharmacia (Laboratory Separation Division, Uppsala, Sweden).

Procedure. All operations are carried out in a refrigerated room at 5°. The Mono Q HR 5/5 column is washed and equilibrated in the following sequence:

1. Fifteen milliliters of 20 mM Tris-HCl, pH 7.8 (low ionic strength start buffer)
2. Fifteen milliliters of 1 M KCl in 20 mM Tris-HCl, pH 7.8 (high ionic strength buffer)
3. Fifteen milliliters of high ionic strength buffer containing 0.75% (25.6 mM) OG
4. Thirty milliliters of low ionic strength start buffer containing 0.75% OG

All subsequent buffers run through the column should contain 0.75% OG. Increasing the concentration of OG beyond 0.75% does not increase the efficiency of resolution of the extract on the ion-exchange column. The OG extract is filtered through a 0.22-μm syringe-operated filter unit in order to ensure complete removal of particulate material. The protein concentration in the filtrate is approximately 3–5 mg/ml. The filtrate is injected into the column using a Pharmacia Superloop (10 ml) at the rate of 1 ml/min. Start buffer containing OG is passed through the column at the same rate. A turbid fraction emerges after the void volume (1 ml) and contains no detectable protein. Since this fraction can be dissolved in chloroform, it probably contains membrane lipids. The column is then eluted with 0.2 M KCl. A large fraction containing up to 70% of the applied extract elutes at this salt concentration. When tested for glucose transport activity, none is found in this fraction. The column is washed with the same eluant (5 ml) until no UV absorption at 280 nm is detected. The column is then eluted with 0.35 M KCl. A fraction of protein emerges in a volume of 4–6 ml containing 8–10% of the protein injected into the column. This fraction contains a very high concentration of glucose-transporting activity. Prior to testing for glucose transport, OG must be completely removed from the samples. This is necessary not only because as a detergent it will disrupt the liposome system used in the transport assay (see the next section) but also because it has a high affinity for the glucose-binding site of the glucose transporter (personal observations). The OG can be removed either by gel filtration or dialysis. We dialyze our samples in standard laboratory dialysis tubing overnight at 5° against 0.1 M K_2SO_4, 5 mM HEPES-Tris, pH 7.8. The samples are then centrifuged at 175,000 g_{av} for 60 min. Proteins are pelleted and can be stored for several weeks at $-20°$ without loss of activity.

Reconstitution of Glucose Transport

Principle. Solubilized membrane fractions can be incorporated into liposomes in order to measure the transmembrane flux of solutes.[19,20] Since carrier-mediated glucose transport is dependent on a favorable electrochemical gradient for Na^+, liposomes are formed in media of appropriate ionic composition so that an inside-negative membrane potential can be generated in addition to the Na^+ concentration gradient.

Procedure. Commercially available soybean phospholipids (Asolectin from Associated Concentrates, Woodside, Long Island, NY) are first freed of neutral fats and pigments. Approximately 50 g of crude phospholipids is dissolved in a minimum volume of anhydrous chloroform in 100-ml glass centrifuge bottles. A 10-fold excess of dry methanol is added with mixing. The precipitated phospholipids are centrifuged (200 g for 10 min) and the supernatants discarded. The sediment is rewashed and centrifuged several times with methanol until the supernatant is colorless. The phospholipids are finally washed twice with acetone in which they are resuspended and stored at $-20°$ after flushing with nitrogen. Aliquots of phospholipids used for making liposomes are transferred to preweighted 20-ml glass vials and the acetone removed with a stream of dry nitrogen. The last traces of acetone are removed in a vacuum desiccator. The phospholipid weight is determined and 40 μg valinomycin in ethanol from a stock solution (1 mg/ml) added for every 30 mg of lipid. The ethanol is evaporated off under nitrogen and a buffer containing 5 mM HEPES-Tris, pH 7.5 and 0.1 M K_2SO_4 is added to give 30 mg/ml of phospholipid. The mixture is cooled in a cold water bath and sonicated to clarity at maximum power with a 3/8-in. Biosonik probe. This takes 10–15 min for 10 ml of liposomes. The translucent mixture is filtered through a 0.45-μm filter (Millipore, Bedford, MA) to remove particles of tungsten derived from the probe as well as any lipid aggregates that remain undispersed. The filtrate contains small uni-

---

[19] E. Racker, this series, Vol. 55, p. 699.
[20] E. Racker, *in* "Reconstitution of Transporters, Receptors and Pathological States." Academic Press, New York, 1985.

lamellar liposomes (25–50 nm in diameter) which are stable in the cold (5–10°) for several weeks. Pelleted fractions derived from solubilized membranes are mixed thoroughly with a suitable volume of liposomes (0.5–5 mg membrane protein/ml of liposomes) using a tuberculin syringe and a 23-gauge needle. The mixture, in a sturdy glass tube (Pyrex, 13 × 100 mm), is sonicated in a bath-type sonicator (80 W, 80 kHz, generator model 112SP1G and tank model G112 SPIT from Laboratory Supplies, Inc., Hicksville, NY) to incorporate protein into the membranes. The time required to achieve optimal reconstitution is variable and must be monitored by transport assays for individual batches of liposomes and membrane extracts. A representative test from earlier studies using DOC is depicted in Fig. 1. Increasing the time of sonication results initially in increased incorporation of transport activity followed at later time periods by a loss of transport activity, probably caused by sonic inactivation. In order to establish optimum incorporation, sonication is carried out for 15-sec intervals, D-glucose transport assayed as described below, and the process of sonication and assay continued five or six times; maximum glucose uptake usually occurs after two or three sonication steps. Reconstituted preparations are stable when kept on ice for about 6 hr. Preparations can be stored at −80° or in liquid nitrogen without loss of activity for many months, especially when stabilized with 14% glycerol.

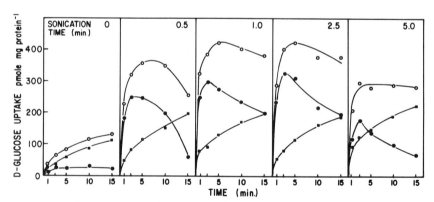

FIG. 1. Influence of sonification upon D-glucose uptake in the reconstituted system. Aliquots of a single batch of solubilized brush border membrane components were sonicated as described (see text) with 0.9-ml aliquots of a single batch of liposomes for the times indicated. Liposomes contained 0.1 M KCl/5 mM HEPES-Tris (ph 7.5) and 1 mM dithiothreitol. Incubation medium contained 5 mM HEPES-Tris (pH 7.5)/0.2 mM D-(UY14HC) glucose, and 0.1 M NaSCN (O) or 0.1 M KSCN (■). Sodium ion-dependent D-glucose uptake (Na⁺-K⁺) (●). (Reproduced with permission from Ref. 21.)

Assay of Na⁺-Dependent D-Glucose Transport

Principle. A suitable Na^+ electrochemical gradient can be obtained by diluting the liposomes containing K_2SO_4 and valinomycin into a medium containing Na^+. Valinomycin-enhanced K^+ efflux establishes an electrical gradient for sodium entry. Alternatively,[21] inside-negative electrical gradients can be augmented using the lipophilic anion SCN^- to enhance the inside-negative membrane potential generated by the passive diffusion of K^+ from K_2SO_4 or KCl-loaded liposomes (valinomycin is not added in these experiments). Data for Fig. 1 were obtained using the latter experimental conditions. Active uptake of glucose is measured as the difference in glucose accumulation from solutions containing radiolabeled glucose containing or lacking Na^+ (Na^+ replaced with K^+ or tetramethylammoniumion). Since chemical and electrical gradients decay with time, linear rates of entry may obtain for only short periods of time.

Procedure. Reaction mixtures are made up so as to yield 50 μM D-glucose labeled with ^{14}C or ^3H (2–5 μCi), 0.1 M Na_2SO_4, and 5 mM HEPES-Tris, pH 7.5 when 175 μl is diluted to 100 μl. Controls lacking sodium contain 0.1 M K_2SO_4 instead of 0.1 M Na_2SO_4, Aliquots (175 μl) are dispensed into tubes (10 × 75 mm) held at 25°. The reaction is initiated by addition of 25 μl of reconstituted liposomes (or membranes preincubated with valinomycin) and arrested by rapid mixing with 2.0 ml of ice-cold 0.15 M KCl. Any soluble compatible with maintaining vesicle integrity may be used to quench the reaction; e.g., isoosmotic solutions of mannitol. A Pasteur pipet is used to transfer the mixture for rapid filtration by suction through a 0.22-μm Millipore filter (0.45 μm if membrane vesicles are assayed). The filter is then washed with suction with 2 × 5 ml of ice-cold 0.15 M KCl to remove nontransported labeled glucose. Transfer, filtration, and washing can be carried out within half a minute. Efflux of accumulated glucose is negligible under the conditions employed. Even repeated washing (4 × 5 ml) of the liposomes trapped on membrane filters does not reduce the uptake measured after a single wash. The washed filters are transferred to vials containing 10 ml of scintillation fluid composed of 25% methylcellosolve, 75% toluene, 10 mg% POPOP, and 300 mg% PPO. The filters dissolve in 30 min in the scintillation fluid. Nonspecific binding of label to the filters and liposomes is exceedingly low ($<10^{-4}$ of total cpm in reaction mixture). In routine testing, uptake is measured in triplicate at 0.05- and 0.1-min time points, in the presence of Na^+ or K^+, the latter serving as a control correcting for diffusional entry and nonspecific binding of D-glucose.

[21] P. Fairclough, P. Malathi, H. Priser, and R. K. Crane, *Biochim. Biophys. Acta* **553**, 295 (1979).

Comments. The uptake of D-glucose into liposomes reconstituted with membrane extracts is shown in Fig. 2, where native membranes were extracted with DOC and the solubilized proteins spun down after dilution and addition of KCl to a final concentration of 0.1 M. Identical results are obtained if OG replaced DOC. Reconstituted liposomes exhibit "overshoot" just as native membrane vesicles do.[22] This is due to a transient accumulation of glucose against its chemical gradient, driven by a transient, large electrochemical Na^+ gradient. Upon dissipation of this gradient, glucose effluxes until equilibrium is reached. The reconstituted membranes show selective stereospecific transport of D-glucose and analogs like D-galactose, 3-O-methylglucose, 6-deoxyglucose, and α- and β-methylglucosides while L-glucose and 2-deoxyglucose are not transported. Phlorizin (200 μm) completely inhibits Na^+-dependent transport in liposomes as it does in native membrane vesicles. The only difference between native membrane vesicles and the reconstituted liposomes lies in the time course of glucose transport. In native membrane vesicles, overshoot is reached within 0.15 min; in reconstituted liposomes glucose uptake is linear for 0.5 min. This probably reflects the greater "leakiness" of native membrane vesicles which dissipates the Na^+ gradient much faster. Table I summarizes the recovery of transport in the fractions derived from the above-described procedures.

The procedures described above are highly reproducible. The specific activity for glucose transport after the Mono Q step is the highest reported heretofore for reconstituted Na^+-coupled transport. Analyses of proteins present at various stages of purification using polyacrylamide gel analyses do not provide evidence for the enrichment of any specific proteins(s) that can be identified as the glucose transporter. Wu and Lever[23] have used similar techniques of solubilization and chromatography and with a specific monoclonal antibody identified a 75-kDa protein as a component of the glucose transporter in pig kidney cortex. Molecular sizes in this range have been reported by other investigators also.[24-26] Our inability to identify a particular protein may reflect the very low copy number of the transporter in the membrane; polyacylamide gel electrophoresis of the highest active preparation viz the 0.35 M KCl eluate from Mono Q columns shows at least seven distinct silver-stained protein bands extending between 25 and 200 kDa with none discernible between 69 and 75 kDA. Further

[22] U. Hopfer, K. Nelson, J. Perrotto, and K. J. Isselbacher, *J. Biol.Chem.* **248**, 25 (1973).
[23] J.-S. R. Wu and J. E. Lever, *Biochemistry* **26**, 5958 (1987).
[24] M. Silverman and P. Speight, *J. Biol. Chem.* **261**, 13820 (1986).
[25] M. Neeb, U. Kunz, and H. Koepsell, *J. Biol. Chem.* **262** (1987).
[26] T. Kitlar, A. I. Morrison, R. Kinne, and J. Deutscher, *FEBS Lett.* **234**, 115 (1988).

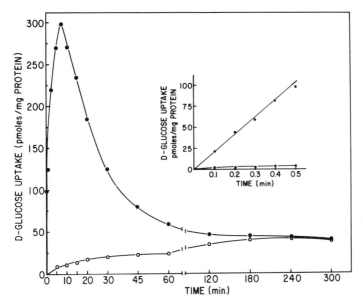

FIG. 2. D-Glucose uptake by membrane extracts reconstituted into liposomes. (●) In the presence of 0.1 M Na$_2$SO$_4$; (○) in the presence of 0.1 M K$_2$SO$_4$. (Reproduced with permission from Ref. 17.)

concentration of the protein(s) is required to establish the identify of the transporter. Radiation-inactivation studies[27] have shown that the functional size of the isolated and reconstituted transporter is 343 kDa, which indicated a multimeric structure for the transporter.

[27] M. Takahashi, P. Malathi, H. Preiser, and C. Y. Jung, *J. Biol. Chem.* **260**, 10551 (1985).

[28] Calcium Transport by Intestinal Epithelial Cell Basolateral Membrane

By JULIAN R. F. WALTERS and MILTON M. WEISER

Introduction

The absorption of calcium ion (Ca^{2+}) by the intestine involves several steps as Ca^{2+} moves through the epithelial cell. The nature of these separate steps, and the role of the vitamin D metabolite 1,25-dihydroxycholecalciferol [1,25(OH)$_2$D$_3$] in their control, have recently been reviewed.[1] To be absorbed, Ca^{2+} must traverse the apical brush border membrane, the intracellular cytoplasm, and eventually the basolateral membrane. Within the cell, certain subcellular organelles have been shown to be capable of sequestering Ca^{2+}; these include the endoplasmic reticulum, the Golgi apparatus, mitochondria, and lysosomes. Additionally, epithelial cells have a specific Ca^{2+}-binding protein (CaBP, calbindin-D$_{9kDa}$ in mammals), which is thought to increase the amount of Ca^{2+} diffusing across the cell by responding either directly to the action of 1,25(OH)$_2$D$_3$ or to increased intracellular Ca^{2+}. As the concentrations of free Ca^{2+} within the cell are submicromolar, whereas those in the lumen of the intestine and in the extracellular fluid are millimolar, the entry of Ca^{2+} is downhill and probably a passive process not requiring energy, though transport systems have been described at the lumenal membrane. However, Ca^{2+} extrusion at the basolateral membrane must be an energy-requiring process since it must move Ca^{2+} against a 10,000-fold concentration gradient. Some of this energy appears to come from the Na^+ gradient created by the Na^+ pump which then drives Na^+/Ca^{2+} exchange, but plasma membranes also possess an ATP-dependent Ca^{2+} pump which extrudes Ca^{2+} from the cell linked to the hydrolysis of ATP. This chapter will describe the methodology for studies of the Ca^{2+} pump in basolateral-enriched membrane vesicles of rat intestinal epithelial cells.

Calcium Ion Uptake by Intestinal Membrane Vesicles

The basic methods adopted in our laboratory for the study of vesicular Ca^{2+} transport will be described first.[2] Factors important in the preparation of membranes and methods which have defined the properties of transport

[1] F. Bronner, D. Pansu, and W. D. Stein, *Am. J. Physiol.* **250**, G561 (1986).
[2] J. R. F. Walters and M. M. Weiser, *Am. J. Physiol.* **252**, G170 (1987).

will be described subsequently. The methodology used in the intestine is similar to that described in other chapters of this series for inside-out vesicles from red blood cells.[3]

Materials. The typical reagents employed in the assay are as follows.

Dilution buffer: 135 mM KCl, 5 mM MgCl$_2$, 10 mM imidazole-acetate, pH 7.5

Calcium buffer: Made up at 2× final concentration in dilution buffer. This 2× buffer contains 1 mM EGTA, ^{45}CaCl$_2$ to give about 100,000 cpm/25 µl (from stock supplied at 10–40 mCi/Mg Ca^{2+}), and additional nonradioactive CaCl$_2$ to give the appropriate total Ca^{2+} which will result in the desired submicromolar free Ca^{2+}. The composition of the Ca^{2+} EGTA buffer is determined by a computer program employing published association constants and methods for the interactions between Ca^{2+}, Mg^{2+}, H$^+$, EGTA, and ATP,[4] and includes corrections for temperature and ionic strength. Three important considerations are (1) the critical dependence on pH of the free Ca$^+$, (2) the purity of the EGTA used (typically 95–97%), and (3) the contaminating levels of Ca^{2+} inevitably found in any laboratory water. These additional amounts of contaminating Ca^{2+}, typically 10 µM, can be measured with a Ca^{2+}-sensitive electrode and included in the calculation of the Ca^{2+} to be added to the buffer to result in the correct total Ca^{2+}. For example, the solution which will give 0.5 µM free Ca^{2+} can be calculated to require approximately 460 µM total Ca^{2+} with 500 µM EGTA, 5 mM Mg^{2+}, 3 mM ATP at pH 7.5. With laboratory water containing 10 µM Ca^{2+}, an additional 450 µM Ca^{2+} needs to be present, so 900 µM Ca^{2+} is added to the 2× concentrated buffer. This Ca^{2+} is obtained from both the ^{45}Ca^{2+} and nonradioactive CaCl$_2$ stock

ATP solution: 30 mM Tris-ATP in dilution buffer adjusted to pH 7.5

Additional compounds, such as inhibitors or stimulatory proteins, are dissolved in dilution buffer and the pH checked

Membrane filters: Nitrocellulose filters, 0.45-µm pore size, 25-mm diameter, soaked in 10 mM CaCl$_2$ for a few minutes before use

Wash solution: Dilution buffer containing 10 mM CaCl$_2$

These reagents may be stored frozen for several months.

Procedure. Transport studies are performed in a final volume of 100 µl. Aliquots of membranes (20 µl) containing 10 to 30 µg of membrane protein in dilution buffer are distributed to disposable glass or plastic tubes

---

[3] T. R. Hinds and F. F. Vincenzi, this series, Vol. 102, p. 47.
[4] O. Scharff, *Anal. Chim. Acta* **109,** 291 (1979).

on ice. Additional compounds may be added in another 20 μl dilution buffer, and then 50 μl of the $2\times$ concentrated calcium buffer is added. These tubes are mixed and kept on ice until the individual uptakes are determined. After a 10-min preincubation at the final temperature (usually $25°$), 10 μl of the ATP solution is added and the tube mixed by a brief vortexing. Following a 1-min incubation, a 90-μl sample is pipetted onto a nitrocellulose disk in a vacuum suction apparatus and washed twice with 2.5 ml of ice-cold wash solution. Filtration is completed in about 15 sec after which the nitrocellulose filter is dissolved in aqueous scintillation fluid and the radioactivity counted in a liquid scintillation counter.

Calculation of Results. In order to determine ATP-independent uptake, ATP is replaced with dilution buffer. Calcium ion uptake in ATP-free solutions is near maximal at 1 min in submicromolar Ca^{2+} buffers and is thought to represent Ca^{2+} binding. ATP-independent Ca^{2+} uptake is subtracted from uptake in the presence of ATP to give ATP-dependent Ca^{2+} transport. Membrane-free blanks are also filtered and give values for binding of Ca^{2+} directly to the filters. An aliquot of the uptake solution is used as an internal standard to convert counts per minute to nanomoles of Ca^{2+} using the total Ca^{2+} present in the Ca^{2+} buffer. Results are usually expressed as 1-min rates, though uptake is linear for at least 2 min. Experiments are performed with at least triplicate determinations.

This procedure is not the only one which will produce satisfactory results. Reagents may be added in different orders and volumes may be altered. For instance, an alternative method has been employed in which the uptake reaction is started by adding the $2\times$ Ca^{2+} buffer. In this case, though, the membranes are not preincubated at the final submicromolar Ca^{2+} concentration, which may be important in determining the actions of stimulatory proteins. Similar values, however, are found for both ATP-dependent and -independent Ca^{2+} uptake.

Effects on Ca^{2+} Uptake of Different Preparations of Basolateral Membrane Vesicles

Membrane Preparation. Basolateral-enriched membrane fractions have been prepared by several similar methods from intestinal cells. The method we use is modified from the one described in 1978.[5] We have found that one must start with isolated cells to demonstrate satisfactorily the ATP-dependent Ca^{2+} pump. This is discussed below. Homogenates of isolated intestinal epithelial cells are prepared using a Polytron (Brinkmann, Westbury, NY), setting 8, for 1 min, and then centrifuged at 1500 g

[5] M. M. Weiser, M. M. Neumeier, A. Quaroni, and K. Kirsch, *J. Cell Biol.* 77, 722 (1978).

for 15 min to remove unbroken cells, nuclei, and brush border membranes. The supernatant is centrifuged at 105,000 g for 15 min to collect a crude membrane fraction. Basolateral membranes are further purified using sorbitol density-gradient centrifugation for 2 hr at 200,000 g and the fractions washed and resuspended in dilution buffer. The fractions with densities similar to, or less than, 40 g/dl sorbitol have been shown to be enriched for the basolateral membrane marker, Na^+,K^+-ATPase activity. Calcium ion transport has been found to be associated with these membranes rather than with the more dense membranes enriched for sucrase activity, a brush border membrane marker (Fig. 1).

Essentially similar membrane preparations have been employed by other groups of workers who have studied basolateral membrane Ca^{2+} transport.[6-8] Enrichment of Na^+,K^+-ATPase activity in excess of 10-fold have been difficult to achieve with intestinal homogenates, and though separation from brush border membranes and mitochondria is relatively easy, contamination with endoplasmic reticulum markers has been more difficult to overcome.

Other properties of the vesicles that may affect Ca^{2+} transport include vesicle size, degree of sealing, and orientation. Size may be estimated from electron micrographs, and quantified more easily from the equilibration volume of molecules such as glucose or mannitol. These give estimates ranging from 1 to 8 μl/mg protein.[2,6] Orientation can be estimated from enzyme latencies, though these determinations may be affected through stimulation of activities by the detergents employed. Most estimates for basolateral membranes give figures of roughly 50% in inside-out and right-side-out orientations.[7] It is assumed that only inside-out vesicles will be in the correct orientation for measuring the function of the basolateral membrane Ca^{2+} pump since substrates are added to the external vesicle surface which, if activated by ATP, would represent the cytoplasmic domain of the membranes in the intact cells.

We have used these vesicles immediately after preparation for Ca^{2+} transport studies. Although Ca^{2+} uptake rates remain stable for several hours when the membranes are kept on ice, substantial loss of activity occurs overnight. Freezing also results in a large and variable reduction of the transport rate, though other workers appear to be able to accept ~50% loss of activity as there was no change in the affinity of the pump for Ca^{2+}.[9]

[6] H. N. Nellans and J. E. Popovitch, *J. Biol. Chem.* **256**, 9932 (1981).
[7] W. E. J. M. Ghijsen, M. D. de Jong, and C. H. van Os, *Biochim. Biophys. Acta* **689**, 327 (1982).
[8] B. Hildmann, A. Schmidt, and H. Murer, *J. Membr. Biol.* **65**, 55 (1982).
[9] W. E. J. M. Ghijsen, C. H. van Os, C. W. Heizmann, and H. Murer, *Am. J. Physiol.* **251**, G223 (1986).

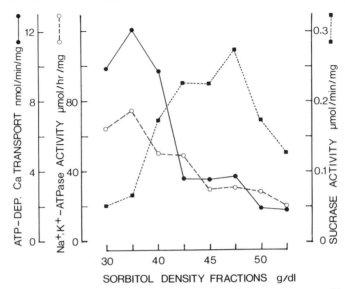

FIG. 1. Distribution of vesicular Ca^{2+} transport and marker enzymes on a sorbitol density gradient. Crude membrane fractions were prepared from rat duodena and separated on a discontinuous sorbitol density gradient.[2] Fractions were collected from the 30 to the 50–60 g/dl interface and washed. ATP-dependent Ca^{2+} was measured at 0.5 μM free Ca^{2+} (see text), and Na^+,K^+-ATPase and sucrase were measured as previously described.[5]

Cell Preparation Methods. The intestine, particularly when compared with other cell types, is sensitive to variations in the methods used to prepare the cells which are homogenized to give subcellular fractions.[10] It is important not to overlook the effects of these factors on the properties of Ca^{2+} uptake in basolateral membrane vesicles. Also important are the differences in epithelial cell function that are found in the various regions of the small intestine and along the crypt–villus axis of differentiation.

We have been able to demonstrate ATP-dependent Ca^{2+} uptake only in vesicles prepared from isolated intestinal cells and not from scrapings of intestinal epithelium.[11] One explanation appears to be the high levels of nonesterified fatty acids found in membranes prepared from scrapings.[12] These fatty acids probably result from lipolysis of other lipids during the homogenization of the scrapings; much higher phospholipase A activity

[10] M. M. Weiser, J. R. F. Walters, and J. R. Wilson, *Int. Rev. Cytol.* **101**, 1 (1986).
[11] J. R. F. Walters, P. J. Horvath, and M. M. Weiser, *in* "Epithelial Calcium and Phosphate Transport: Molecular and Cellular Aspects" (F. Bronner and M. Peterlik, eds.), p. 187. Alan R. Liss, New York, 1984.
[12] J. R. F. Walters, P. J. Horvath, and M. M. Weiser, *Gastroenterology* **91**, 34 (1986).

can be shown in homogenates of intestinal scrapings than in isolated cells. In addition to binding Ca^{2+} and so increasing the ATP-independent Ca^{2+} uptake, there may be direct effects of lipids on the plasma membrane Ca^{2+} pump. Isolated cell preparations may be better in this regard as they are more extensively washed prior to homogenization. This washing may be removing a greater proportion of adherent pancreatic enzymes or inhibiting or removing epithelial cell phospholipases.

Another related problem is the effect of proteolytic enzymatic activity on the Ca^{2+} pump. It has been shown that there are differences in basolateral membrane Ca^{2+} transport from two isolated cell preparations[13] and it was postulated that this was because of differences in amounts of tryptic or other proteolytic activity. Ileal and jejunal membranes seemed more susceptible to this than duodenal; unfortunately this effect could not be prevented by commonly used trypsin inhibitors but only by modifications to the physical methods used for cell isolation.

A third problem in preparing basolateral membranes, particularly related to isolated cell methods, is that redistribution of intracellular and cell surface markers may occur. Consequently, the relative purification factors for basolateral, brush border, and endoplasmic reticulum membranes become lower with prolonged preparation and washing procedures.[10,14] Despite this concern, preparations of "basolateral" membranes show coincident enrichment for Ca^{2+} pump and Na^+,K^+-ATPase activities. Thus, either both activities have redistributed themselves at the same rate and to the same location or the redistribution is selective for the enzyme studied, e.g., alkaline phosphatase. These concerns must be considered in studies of Ca^{2+} transport by intestinal membrane vesicles; unfortunately there is no clearly superior method for cell isolation at present.

Investigation of the Properties of Ca^{2+} Uptake

To confirm that Ca^{2+} uptake by membrane vesicles is transport by an ATP-dependent Ca^{2+} pump, certain basic properties should be investigated to help differentiate ATP-dependent Ca^{2+} transport from Ca^{2+} binding. An absolute dependency on ATP should be shown. The time course of Ca^{2+} uptake by basolateral-enriched membrane vesicles, in the presence and absence of ATP, is shown in Fig. 2. Also shown is that ADP will not substitute for ATP; other nucleotide triphosphates such as GTP or CTP were also ineffective in supporting transport. Estimates of the affinity of the intestinal pump for ATP gave a K_m of under 50 μM.

[13] E. J. J. M. van Corven, M. D. de Jong, and C. H. van Os, Cell Calcium 7, 89 (1986).
[14] C. A. Ziomek, S. Schulman, and M. Edidin, J. Cell Biol. 86, 849 (1980).

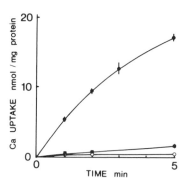

Fig. 2. ATP dependence. Basolateral-enriched membranes were prepared from pooled duodenal villus and crypt cells. Calcium ion uptake was measured at 25° with 0.5 μM free Ca^{2+} in the presence of 3 mM ATP (●) 3 mM ADP (■), and in the absence of ATP or ADP (O). Results are means ± SEM of four determinations. (Reprinted from Ref. 2.)

Calcium ion transport by the Ca^{2+} pump can be most conveniently measured at 25°; uptake is greater at more physiologic temperatures though the rate is linear for a shorter period of time (Fig. 3). However, when experiments are performed on ice, Ca^{2+} uptake is reduced to levels similar to those seen in the absence of ATP. As binding is much less temperature dependent than trans-membrane transport, this result also indicates that Ca^{2+} uptake is by the Ca^{2+} pump.

The affinity of the epithelial cell pump for Ca^{2+} should be determined. This has been studied by several groups. We have found a K_m of 0.3 μM,[2] which is similar to that found by Ghijsen et al.[7] Another group has found a much lower figure[6] but used a different Ca^{2+} buffer system. The free Ca^{2+} and hence the K_m values depend greatly upon the association constants and calculations used in the determination of the Ca^{2+}-buffering solutions. There is the additional problem in that the Ca^{2+}/EGTA complex has been postulated to be recognized by the Ca^{2+} site on the pump, giving an apparent stimulatory effect of EGTA independent of the effect on free Ca^{2+} concentrations.[15] This has not yet been fully resolved for the intestinal basolateral membrane Ca^{2+} pump, but appears to be an explanation for the discrepancy in measured K_m.

Ca^{2+} ionophores, such as A23187, should be used to show that the accumulation of Ca^{2+} by the membranes is in fact intravesicular (Fig. 4). A23187 is dissolved in ethanol or dimethyl sulfoxide (DMSO) and added in micromolar concentrations such that the concentration of the solvent does not exceed 1%. This concentration of ethanol has been shown to not

[15] N. Kotagal, J. R. Colca, and M. L. McDaniel, J. Biol Chem. **258,** 4808 (1983).

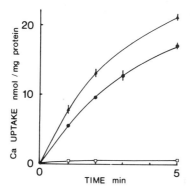

FIG. 3. Temperature dependence. Calcium ion uptake was determined in duodenal baso-lateral-enriched membranes in the presence of 0.5 μM free Ca²⁺ and 3 mM ATP at 35° (▲) 25° (●) and on ice (□). Results are means ± SEM of four determinations. (Reprinted from Ref. 2.)

affect Ca²⁺ uptake. When the ionophore is added to the membranes before ATP, the amount of Ca²⁺ associated with the vesicles is reduced to a level similar to those usually seen in the absence of ATP or when uptake is performed on ice. Further, when A23187 is added after Ca²⁺ has been taken up by the vesicles, the concentrated intravesicular Ca²⁺ will be rapidly released through the membrane pores created by the ionophore.

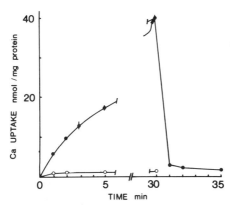

FIG. 4. Effects of A23187 on Ca²⁺ uptake by duodenal basolateral-enriched membrane vesicles. Membranes were incubated at 25° with 3 mM ATP and 0.5 μM free Ca²⁺. Values are the means ± SE of three or four determinations. Uptake was determined for 30 min with A23187 (9 μM) added before ATP and Ca²⁺ (○). Uptake in the absence of A23187 (●); at 30 min (9 μM) A23187 was added and rapidly reduced Ca²⁺ uptake. (Reprinted from Ref. 2.)

Prevention of ATP-dependent Ca^{2+} uptake and release of intravesicular accumulations by A23817 confirms that the Ca^{2+} is concentrated in a soluble form within the vesicles. From determinations of the volume of the vesicles, it can be calculated that the concentration of Ca^{2+} inside the vesicles at equilibrium exceeds 5 mM, so a 10,000-fold gradient can be maintained by the pump.

The subcellular origin of the Ca^{2+}-uptake activity in these membrane fractions may be defined by the actions of established inhibitors of Ca^{2+} pumps. Orthovanadate is a powerful inhibitor of most ATPases and will inhibit membrane Ca^{2+} uptake with a K_i of less than 1 μM, a figure similar to those of other plasma membrane pumps. The mitochondrial Ca^{2+} pump inhibitors such as oligomycin or azide should have no effect. In sodium-free conditions, ouabain or other Na^+ pump inhibitors should have no action on Ca^{2+} accumulation. The effects of Na^+ on Ca^{2+} uptake through the Na^+/Ca^{2+} exchanger found in plasma membranes have been studied[8,16]; the effects of this system on Ca^{2+} uptake can best be avoided by omitting Na^+ from solutions used in the study of the ATP-dependent pump.

Calmodulin, a ubiquitous M_r 16,800 protein, has been shown to stimulate various plasma membrane Ca^{2+} pumps in a direct manner independent of the action of the specific calmodulin-independent protein kinase. As calmodulin binds up to four Ca^{2+} atoms with affinities appropriate for intracellular levels, this is thought to provide a means of control for the Ca^{2+} pump. In the intestine, basolateral membrane Ca^{2+} uptake is also stimulated by calmodulin under appropriate conditions, but the effect is not as great as that described for the erythrocyte pump. The problem appears to be the retained, tightly bound endogenous calmodulin which obscures an effect of added stimulator. Stimulations of 20–50% at 0.5–1 μM free Ca^{2+} can be found with 10 μg/ml calmodulin (about 0.6 μM)[6] though even this relatively small effect is only seen with inclusion of 5 mM EGTA in the homogenization buffer[7] or preincubation with submicromolar Ca^{2+}.[17] The effect of calmodulin appears predominantly to be on the K_m of the pump; in the intestine even in the absence of exogenous calmodulin, the K_m for Ca^{2+} is under 1 μM, similar to the calmodulin-stimulated pump of other tissues. Calmodulin can be detected in basolateral membranes,[9] including those that have been treated with procedures that in other tissues would have removed it. The effects of drugs with calmodulin antagonistic properties have also been investigated; again the results are

[16] W. E. J. M. Ghijsen, M. D. de Jong, and C. H. van Os, *Biochim. Biophys. Acta* **730,** 85 (1983).
[17] J. R. F. Walters, *Am. J. Physiol.* **256,** G124 (1989).

not definitive. It may be that the interaction of calmodulin with the intestinal basolateral Ca^{2+} pump is different from that in the red cell or other tissues, an interaction that may more specifically define this pump.

The intestinal vitamin D-dependent Ca^{2+}-binding protein (CaBP, calbindin-$D_{9\ kDa}$) affects Ca^{2+} absorption and might also be expected to influence the basolateral Ca^{2+} pump. Our studies suggest that this may be the case[17] although others have not found such an action.[9] Concentrations of CaBP are included in the uptake medium at submicromolar Ca^{2+}; with appropriate Ca^{2+} buffering this does not significantly affect the free Ca^{2+} concentrations or specific activity of $^{45}Ca^{2+}$. The effects of CaBP may be dependent on the saturation of the protein with Ca^{2+}; the methodology required to confirm or disprove such an effect of CaBP on the pump remains to be determined.

The rate of Ca^{2+} transport in basolateral membrane vesicles has been shown to be affected by the vitamin D status of the animal.[2,18] Along with CaBP and other actions at the brush border membrane, an increase in basolateral Ca^{2+} transport may represent a method of controlling the overall rate of Ca^{2+} absorption. Rats can be made vitamin D deficient by a combination of a synthetic diet without vitamin D supplementation, a Ca^{2+}-free diet for 2 weeks to deplete stores of vitamin D through conversion to $1,25(OH)_2D_3$, and an environment lit only by incandescent light to prevent synthesis in the skin of vitamin D by the action of UV light.[19] These animals have reduced serum Ca^{2+} concentrations and bone histology compatible with vitamin D deficiency. Intestinal cells and basolateral membrane vesicles have been prepared from these animals by methods detailed above. Purification of marker enzymes, and other studies to show nonspecific effects on vesicles, have been performed with only minor differences found. Vesicular Ca^{2+} transport rates, however, are reduced by almost 50% in the duodenum, an effect which appears to be on the V_{max} and is particularly apparent in membranes of cells at the villus tip. When vitamin D-deficient animals were given repleting doses of $1,25(OH)_2D_3$, 125 ng in ethanol by intravenous injection, the increase after 6 hr in vesicular Ca^{2+} transport was predominantly in the developing cells lower down the villus.[2] Thus, any studies characterizing aspects of Ca^{2+} absorption should investigate the effects of vitamin D status and consider differences related to intestinal regions and to the crypt – villus axis of differentiation.

Ca^{2+}-ATPase enzymatic activity, i.e., the hydrolysis of inorganic phos-

---

[18] W. E. J. M. Ghijsen, and C. H. van Os, *Biochim. Biophys. Acta* **689**, 170 (1982).
[19] J. H. Bloor, A. Dasmahapatra, M. M. Weiser, and W. D. Klohs, *Biochem. J.* **208**, 567 (1982).

phate from ATP, is a function of the Ca^{2+} pump and should be related to the ATP-dependent transport of Ca^{2+}. In the intestine, difficulties have been encountered in making this correlation, partly because of the additional hydrolysis of ATP which, once attributed to alkaline phosphatase activity,[20] appears to be a function of ectonucleotide phosphatase activity.[21] Using similar Mg^{2+} and Ca^{2+} concentrations as those described for uptake measurements, we found that the high level of this Ca^{2+}- and ATP-independent Mg^{2+}-ATPase activity prevented accurate assessment of the small increment of any Ca^{2+}-dependent ATP hydrolysis. Consequently, Ca^{2+}-dependent ATPase activity was best measured at reduced free Mg^{2+} concentrations of 1 μM. This was achieved with a total Mg^+ concentration of 47 μM in the presence of 500 μM EGTA, 3 mM ATP, and 509 μM total Ca^{2+} at pH 7.5. Under these conditions, basolateral Ca^{2+}-ATPase activity was similar to values reported by others, but the properties were not the same as those described for vesicular Ca^{2+} transport.[21] Thus, calcium ion-dependent ATP hydrolysis in these membranes is not solely a function of the Ca^{2+} pump, but represents other enzymatic activities, as has been described in other tissues, including the liver. Conditions for the determination of that portion of ATP hydrolysis due only to the Ca^{2+} pump remain to be determined.

Basolateral Membrane Ca^{2+} Transport in Other Species

Most of the details of the intestinal basolateral membrane Ca^{2+} pump have been determined with experiments in the rat. Some studies have been performed with chick basolateral membranes; these are particularly relevant as this species has been well studied regarding the action of vitamin D. These findings are broadly similar to those with the rat, including a dependence on the vitamin D status of the animal.[22]

Vesicular Transport of Ca^{2+} in Other Intestinal Subcellular Membranes

This chapter has concentrated on the vesicular transport of Ca^{2+} by the basolateral membrane of the intestine. We have not described methods or

[20] W. E. J. M. Ghijsen, M. D. de Jong, and C. H. van Os, *Biochim. Biophys. Acta* **599**, 538 (1980).
[21] T. C. Moy, J. R. F. Walters, and M. M. Weiser, *Biochim. Biophys. Res. Commun.* **141**, 979 (1987).
[22] J. S. Chandler, S. A. Meyer, and R. H. Wasserman, *in* "Vitamin D: Chemical, Biochemical and Clinical Update" (A. W. Norman, K. Schaefer, H.-G. Grigoleit, and D. v. Herrath, eds.), p. 408. de Gruyter, Berlin, 1985.

results of studies conducted by others attempting to measure Ca^{2+} transport at the brush border membrane.[23,24] Only one paper deals with transport by the endoplasmic reticulum in any detail.[25] Though it appears that the Golgi may also have a Ca^{2+} transport system,[11] earlier results were complicated by high binding of Ca^{2+} to nonesterified fatty acids in membranes prepared from intestinal scrapings.[12] In these studies with the intestine, it seems that the major methodological problem remains the preparation of enriched membranes in sufficient quantities, and unaltered so that they reasonably reflect the in vivo condition of the membrane domain. This has been largely true with duodenal basolateral-enriched membranes where vesicular Ca^{2+} transport has most of the properties expected from studies of Ca^{2+} absorption. More recently a transcript for an intestinal plasma membrane Ca^{2+} pump has been detected which responded to vitamin D similarly to that shown for Ca^{2+} uptake by basolateral membranes.[26]

Acknowledgments

This work is supported by NIH Grants AM-32336 and AM-35015, and by the Troup Fund of Buffalo General Hospital, Buffalo, New York.

[23] A. Miller III and F. Bronner, *Biochem. J.* **196**, 391 (1981).
[24] H. Rasmussen, O. Fontaine, E. E. Max, and D. B. P. Goodman, *J. Biol. Chem.* **254**, 2993 (1979).
[25] M. J. Rubinoff and H. N. Nellans, *J. Biol. Chem.* **260**, 7824 (1985).
[26] J. Zelinski, D. E. Sykes, and M. M. Weiser, *Gastroenterology* **98**, A560 (1990).

[29] Electrical Measurements in Large Intestine (Including Caecum, Colon, Rectum)

By ULRICH HEGEL and MICHAEL FROMM

Introduction

The main function of the large intestine is the conservation of water and ions. Electrophysiological methods are therefore of predominant importance in the analysis of physiological as well as of pathological functional states of this organ. Apart from this more applied aspect large intestinal mucosa has increasingly been used as a model to study basic mechanisms of epithelial ion transport which are of relevance also for other tubular organs such as the nephron or excretory ducts of exocrine glands. Several good reviews on this topic have been published in recent years[1-3] (compare also literature quoted in [36], Vol. 171 of this series, by S. A. Lewis and N. K. Wills).

Two problems are involved in electrophysiological transport studies on large intestinal epithelium:

1. The structure of the native mucosa with its many crypts, its mucous surface layer, and its pronounced subepithelium makes it difficult to quantify the electrochemical gradients of transported substances.

2. The large intestinal mucosa exhibits, as do other segments of the GI tract, considerable cellular heterogeneity with respect to cellular function as well as to cellular differentiation. Highly differentiated surface enterocytes probably absorb mainly NaCl, less differentiated juvenile crypt cells probably secrete mainly Cl^-, goblet cells secret mucus, and several interspersed subpopulations of specialized cells are endo- and paracrin active. Until now no definite conclusions have been reached as to which cell population is responsible for which contribution to the observed overall net transport of the large intestinal mucosa and how transport regulation is achieved.

This article will review some electrophysiological techniques suitable for investigating large intestinal epithelial transport. To achieve this we will proceed in a stepwise methodical reduction of the *in vivo* system as follows: *in vivo* measurements in conscious or anesthetized animals or patients, *in vitro* measurements on preparations of intact surviving epithelium, *in vitro* measurements on epithelial monolayers grown from cells of large intestinal origin in cell culture; and methods suitable for evaluating membrane properties of intestinal epithelial cells.

Basic quantities of interest in the electrophysiology of epithelia are as follows (compare also the List of Symbols in the Appendix): transepithelial electrical area resistivity R^e ($\Omega \cdot cm^2$); transepithelial voltage V^e (mV); transepithelial short-circuit current density I_{sc} ($\mu A \cdot cm^{-2}$), which measures the active net ion transport at $V^e = 0$ mV; transmembranal area

---

[1] H. J. Binder and G. I. Sandle, *in* "Physiology of the Gastrointestinal Tract" (L. R. Johnson, ed.), p. 1389, Raven, New York, 1987.

[2] S. G. Schultz, *Annu. Rev. Physiol.* **46**, 435 (1984).

[3] M. Donowitz, *Gastroenterology* **93**, 641 (1987).

[4] In a major part of the electrophysiological literature the quantities R with the dimension $\Omega \cdot cm^2$ are named "resistances." Although this is incorrect because resistance has the dimension Ω, it has nevertheless come into use. In this chapter we shall strictly distinguish between *resistivities*, which are independent of the actual geometry of a measured substrate and which therefore have the character of "material constants," and *resistances,* which are of interest for the actual design of an experimental set-up.

resistivities of apical or basolateral cell membrane R^a or R^b, which together form the transcellular resistivity $R^{tc} = R^a + R^b$ (the cytoplasmic resistivity can normally be neglected since its absolute contribution remains below 1 $\Omega \cdot cm^2$); transmembranal voltages V^a, V^b; and paracellular resistivity R_p.[4]

While these quantities are of primary physiological interest they are not necessarily those which can directly be measured. This will be explained in the following paragraphs.

The measured resistance r (Ω) in a particular experiment on a substrate i is related to its volume resistivity ρ_i (see Appendix) by

$$r_i = \rho_i (\Delta x_i/A) \tag{1a}$$

and to its area resistivity R_i by

$$r_i = R_i/A \tag{1b}$$

with

$$R_i = \rho_i \Delta X_i \tag{1c}$$

Δx signifies the thickness, A the exposed area of the substrate. Therefore, if Δx and A are known, the volume resistivity ρ_i may be evaluated using Eq. (1a). This is not normally done in epithelial electrophysiology since in most applications it would be difficult to evaluate the actual Δx, and it is only of interest how much resistance is presented per unit area by a given structure be it a membrane, an epithelium, a multilayered submucosa, or a total gut "wall." Thus the relation most used is Eq. (1b) in the form $R_i = r_i \cdot A$, in other words, the area resistivity is the measured resistance times the exposed area.

Several authors prefer to give conductance (or conductivities) instead of resistances (or resistivities):

$$g = 1/r \text{ [S]}$$

In this case the international unit S (for "Siemens") should be used instead of $(\Omega)^{-1}$, or "mho."

For area resistivities as quoted in this chapter it is the gross (macroscopic) epithelial area which is meant as reference, unless stated otherwise.

In Vivo Measurements of Transepithelial Voltage (V^e)

In the large intestine of most mammals one must distinguish four functionally different segments: caecum, proximal (ascending, upper) colon, distal (descending) colon, and rectal colon (in humans part of the sigma plus rectum).

The mucosa of the whole large intestine exhibits a lumen-negative transepithelial voltage (V^e). Depending on segment, species, and functional state this voltage is partly generated by one or several of the following active electrogenic transport mechanisms: Na^+ absorption (either inhibitable by amiloride or by phenamil), Cl^- secretion, K^+ absorption, or K^+ secretion. Under *in vivo* conditions these actively generated voltages are modified by diffusion potentials which are generated across the cation-selective passive pathways through the epithelium due to concentration differences of ions between gut lumen and interstitium. Segmental differences have been observed, e.g., in rats,[5–7] rabbits,[8,9] and humans.[7,10–12]

A segment of special interest is the rectal colon, which comprises mainly the retroperitoneal segment and which exhibits *in vivo* V^e values between -10 and -80 mV in correlation to the secretion of aldosterone.[13] In this segment aldosterone stimulates the electrogenic absorption of Na^+.[5,6,14] The distal colon must be considered a transitional region which at low aldosterone levels in the blood behaves like the proximal colon, but which resembles more closely the rectal colon in the presence of high aldosterone levels.[15] The close correlation between rectal V^e and plasma aldosterone concentration led to the proposal of using rectal V^e measurements in humans as a diagnostic parameter for aldosteronism.[12] Because of the wide range of the data its clinical value has, however, been questioned.[16]

Transepithelial voltage measurements *in vivo* require that connections of reversible electrodes be made to both sides of the epithelium. The principle is shown in Fig. 1A. Since the basolateral side (= interstitial = serosal = blood side) of the epithelium is in low resistance contact with the interstitial space it is sufficient to place the electrode of this side (normally the reference electrode RE) somewhere into the interstitium. Among the different techniques used are, e.g.: (1) a saline-filled hypodermic needle placed into the subcutaneous tissue,[10] (2) a saline–agar-filled catheter

[5] C. J. Edmonds and J. C. Marriott, *J. Endocrinol* **39**, 517 (1967).
[6] M. Fromm and U. Hegel, *Pfluegers Arch.* **378**, 71 (1978).
[7] G. I. Sandle and F. McGlone, *Pfluegers Arch.* **410**, 173 (1987).
[8] W. Clauss, H. Schäfer, I. Horch, and H. Hörnicke, *Pfluegers Arch.* **403**, 278 (1985).
[9] J. H. Sellin, H. Oyarzabal, and E. J. Cragoe, *J. Clin. Invest.* **81**, 1275 (1988).
[10] C. J. Edmonds and R. C. Godfrey, *Gut* **11**, 330 (1970).
[11] J. H. Sellin and R. de Soignie, *Gastroenterology* **93**, 441 (1987).
[12] F. Skrabal, *Wien. Klin. Wochenschr.* **89**, Suppl. 78, 3 (1977).
[13] M. Fromm, W. Oelkers, and U. Hegel, *Pfluegers Arch.* **399**, 249 (1983).
[14] C. J. Edmonds and J. Marriott, *J. Physiol.* **210**, 1021 (1970).
[15] M. Fromm, J. D. Schulzke, and U. Hegel, *Pfluegers Arch.* **416**, 573 (1990).
[16] D. G. Beevers, J. J. Morton, M. Tree, and J. Young, *Gut* **16**, 36 (1975).

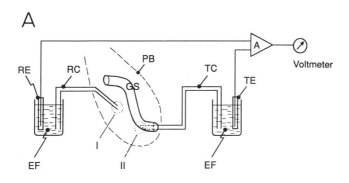

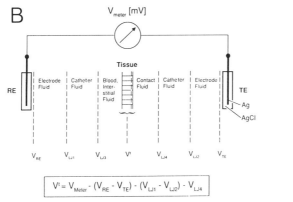

$$V^t = V_{Meter} - (V_{RE} - V_{TE}) - (V_{LJ1} - V_{LJ2}) - V_{LJ4}$$

FIG. 1. Electrochemical chain in voltage measurements across epithelia. (A) Technical set-up. Example of *in vivo* measurement across the gut wall. RE, TE, Reference and test solid state electrodes, respectively (reversible, e.g., Ag/AgCl); EF, electrode fluid (anion reversible with salt of solid state electrode, e.g., KCl); RC, TC, conducting catheters connecting RE and TE to the animal; GS, gut segment; PB, border of peritoneal cavity; A, voltage amplifier (high input impedance); circle I, connection of RC to the interstitium of the animal (see text); circle II, contact of TC to mucosa. (B) Electrochemical chain of set-up shown in (A). V^t, Voltage across tissue preparation; V_{RE}, V_{TE}, electrode potentials conforming to Eq. (1); V_{LJ1-4}, liquid junction potentials conforming to Eq. (2).

placed into the peritoneal cavity,[5] (3) an intravenous saline infusion system,[17] and (4) a small saline container with an open bottom glued to a skin abrasion.[18]

Contact to the apical side of the epithelium would best be made by connecting the electrode to a lumenal perfusion system or at least to a

[17] L. Höjgaard, H. R. Andersen, and E. Krag, *Scand. J. Gastroenterol.* **22**, 847 (1987).
[18] A. I. Stern, D. L. Hagen, and J. I. Issenberg, *Gastroenterology* **86**, 60 (1984).

mucosal irrigation system (see below, R^t measurements). A steady efflux from the lumenal electrode catheter would at the same time clean the mucosa, remove part of the mucus, and keep the surface electrolyte activities constant. Thus, it would establish a "flowing junction" to the mucosa, which is the best method to define the electrolyte concentration at fluid interfaces.

Connection of the apical electrode to a lumenal perfusion system averages values for V^t, R^t, and I_{SC} along a gut segment proportional in length to its "electrical length constant" (see p. 466). In measurements on humans, however, punctual placement of porous contact electrodes under visual control through a colonoscope has also been tried.[10]

Care must be taken that the concentration of the "reversible" ion [the one which is common to the electrode fluid (EF) and the solid phase, as, for example, Cl⁻ with Ag/AgCl electrodes] not change, e.g., by diffusional loss of this ion to the surrounding fluid, because the half-element potential of the electrode–solution interface depends on this concentration:

$$E_{electrode} = E_0 - (RT/F)\ln a_{Cl^-} \qquad (2)$$

where a_{Cl^-} is the chloride activity of the electrode fluid (EF), R is the gas constant, T is temperature, F is the Faraday constant, and E_0 is the standard potential of the electrode. From Eq. (2)—the *Nernst* equation for electrode potentials—it follows practically that a 10-fold activity difference between the electrode fluids of reference and test electrode (EF$_{RE}$ and EF$_{TE}$, respectively, in Fig. 1A) would at $T = 37°$ result in an asymmetry between the electrodes of 61.5 mV [$(RT/F)\ln 10 = 61.5$mV]. It follows also that if the temperatures of RE and TE were 25 and 37°, respectively, an asymmetry of 11.6 mV or roughly 1 mV/°C between them would result merely on the basis of the temperature difference, the warmer electrode being more negative than the colder one. This would be due to the fact that both RT/F and E_0 are temperature dependent, as is shown for the case of Ag/AgCl electrodes in the following tabulation[19]:

Temperature (°C)	E_0 (mV)	$RT \ln 10 \cdot F^{-1}$ (mV)	a_{Cl^-} (at 1 M KCl)
25	220	59.2	0.60
37	214	61.5	0.757

$E_{electrode\ 1}(25°) = 233.0$ mV
$E_{electrode\ 2}(37°) = 221.4$ mV $\Big\}$ $\Delta E_{electrode\ 1}(\Delta T = 12°) = 11.6$mV

[19] "CRC Handbook of Chemistry and Physics." Latest Ed. CRC Press, Boca Raton, Florida, 1989.

A complete scheme of the voltage-measuring chain is given in Fig. 1B. The calculation of all liquid junction potentials (V_{LJ}) follows the Henderson equation:

$$V_{LJ} = -\frac{RT}{F}\left[\frac{(U_1 - V_1) - (U_2 - V_2)}{(U_1' + V_1') - (U_2' + V_2')}\right]\ln\left(\frac{(U_1' - V_1')}{(U_2' - V_2')}\right) \quad (3)$$

where U_1 is the volume conductivity of cation on side 1, V_1 is the volume conductivity of anion on side 1, and U_1', V_1' are the volume conductivities times valencies. For numerical values of U_x and V_x see, for example, Ref. 19.

With respect to Eq. (3), a basic requirement for all voltage measurements is, therefore, the symmetry of the electrode system, at least so long as *absolute* voltages are to be measured.

It may be useful to note that for measurements of the open-circuit voltage across epithelia the sensing electrode need not be positioned close to the epithelium. So long as the input resistance of the voltage amplifier is high compared with the access resistance to the epithelium (electrodes plus bridges plus bath) and so long as there is no low-resistance short circuit between the mucosal and the serosal side, the open-circuit voltage can be measured at any site of the perfusion systems of both sides. This holds for *in vivo* as well as for *in vitro* experiments. As soon as current passes through the epithelium, however, the situation is different. Since current flow produces a voltage drop across the bathing electrolyte, the voltage electrodes should in this case be positioned as close to the epithelium as possible in order to pick up only the voltage drop across the epithelium (refer also to the *in vitro* sections of this chapter).

If connections between the amplifier input and the patient or animal must be made by long leads or salt bridges care should be taken that the loop formed by mucosal connection, animal, and serosal connection does not include too large an area. Otherwise pick-up of interfering electric fields like hum or switching pulses may overload the amplifier input. Keeping the electrode leads closely together, or, much better, shielding of the leads, would eliminate this problem.

The necessary input resistance of the amplifier should be about two to three orders of magnitude higher than the sum of source plus access resistance to achieve an accuracy of 1 to 0.1%. For example, with two saline bridges, each being 1.5 m in length and 2 mm in lumenal diameter, the access resistance would, on the basis of Eq. (1a), amount to $r = 50$ $\Omega \cdot cm \cdot 300$ cm $\cdot [\pi(0.1)^2$ cm$^2]^{-1} = 477$ kΩ. The source resistance of the transepithelial voltage of less than 1 kΩ (see p. 466) could then be neglected. As a result, an electrometer amplifier with input resistance of 10^9 Ω would permit a measurement with less than a 0.1% underestimation. This type of consideration holds for any voltage measurements.

In vivo Measurements of Epithelial Resistivity (R^e)

Epithelial electrical resistivity R^e represents a measure of epithelial lumped ionic "permeability." Its determination *in vivo* is therefore of principal scientific interest as well as of clinical importance in cases of pathological alterations of intestinal mucosa.

Technically, two method have been applied.

1. *Use of Ohm's law:* A known current step Δi is passed through the known area A of a functionally isolated deliminated segment by two "current" electrodes. Simultaneously the current-induced voltage deflection, ΔV, across the gut wall is measured by an additional pair of "voltage" electrodes. The transmural resistivity (i.e., across all layers of the gut wall) is then

$$R^t = \Delta V_T A (\Delta i)^{-1} \tag{4}$$

A prerequisite of this method is a strict proportionality of current and voltage in the range of signal sizes used. This method has been applied to the colon *in vivo* by Edmonds and Marriott[14] who used one end of the lumenal perfusion system as the current electrode. A technique which provided a more homogeneous electrical field across the whole area of exposed gut wall was published by Haag *et al.*[20] They installed an axial Ag/AgCl electrode in the gut lumen. Using current steps of 10- to 50-μA amplitude and 5-sec duration they evaluated a transmural resistivity of $R^t = 121 \pm 5\ \Omega \cdot cm^2$. Linearity of the gut wall had been checked up to 40 μA.[21]

2. *Use of cable theory:* The rationale behind this method is the assumption that an electrolyte solution inside the lumen of a gut segment *in situ* behaves like a cable with poor isolation in a moist environment, the gut wall representing the leaky insulator.

This method was used, in analogy to applications in other tubular physiological structures,[22] by Knauf and co-workers in rat proximal colon.[20,21] They injected current pulses into one end of a cannulated gut segment *in situ* and recorded the generated voltage pulses by use of a movable sensing electrode inside the lumen. In this way they recorded, e.g., voltage signals of about 7 mV at the current injection site and <1 mV at

[20] K. Haag, R. Lübcke, H. Knauf, E. Berger, and W. Gerok, *Pfluegers Arch.* **405**, S67 (1985).

[21] H. Knauf, K. Haag, R. Lübcke, E. Berger, and W. Gerok, *Am. J. Physiol.* **246**, G151 (1984).

[22] U. Hegel, E. Frömter, and T. Wick, *Pfluegers Arch.* **294**, 274 (1967).

the opposite end of the segment. It could be demonstrated that the voltage decay was exponential along the segment and therefore the following formula for the transmural resistivity R^t, derived from cable theory, could be applied.

$$R^t = \alpha^{-1}[(4\pi \lambda^3 r_{eff} \rho^{fl})^{1/2}] \{1 + [(\alpha\lambda\rho^{fl})^{-1}] (\pi r_1^2 r_{eff})\}^{1/2} \qquad (5)$$

where $\alpha = \coth (l/\lambda)$, accounting for the finite length of the cable and practically close to 1; l = length of the segment (cm); λ = electrical length constant, i.e., the distance along which the lumenal voltage signal decays to $1/e = 37\%$ [with two voltage measurements V_0, V^x at $x = 0$ and $x = 1$, respectively, it follows that $\lambda = l \cdot (\ln V^0/V^x)^{-1}$ (cm)]; $r_{eff} = V^0/i$ (Ω); V^0 = voltage at the current injection site (mV); i = total current injected (mA); ρ^{fl} = volume conductivity of the lumenal solution ($\Omega \cdot$ cm^{-1}); r_1 = radius of lumenal pipette.

Introducing all numerical values of theses parameters led Knauf et al.[21] to the following results for rat proximal colon: λ between 0.3 and 0.7 cm, which is roughly two to three times the lumenal diameter; r_{eff} between 200 and 1000 Ω; and $R^t = 128 \pm 16$ $\Omega \cdot$ cm^2.

A prerequisite of both methods for evaluating R^t is that the electrical resistivity be constant along the exposed segment, which may hold for proximal colon but is not true for distal or rectal segments.[23] However, the electrical length constant of the large intestine is short compared with the electrical heterogeneity along the large intestinal axis such that measured values correspond well enough to a local average.

In addition, there are two principal drawbacks involved with R^t measurements in vivo: The first concerns the homogeneity of the current density within the wall of the segment under investigation. The access resistance of the serosal current electrode to the basal side of the mucosa might well be different at different angles of the circumference, depending on the distribution of body fat, blood supply, etc. This factor is not known and may vary between individual animals and in relation to the segment under investigation. As a countermeasure, the serosal electrode has been placed in the form of an Ag/AgCl plate next to the gut wall in the peritoneal cavity with the voltage-sensing electrodes oriented to the same side.[21] The second drawback is the fact that in vivo it is obviously not the transepithelial (R^e) but rather the transmural (R^t) resistance which can be measured, since technically no serosal electrode can be placed in a controlled position directly beneath the basal membrane of the epithelium. A solution of this problem can be approached either by impedance techniques (see p. 472) or by inferences from in vitro stripping experiments (see p. 471).

In Vivo Measurements of Short-Circuit Current (I_{SC})

A quantitative evaluation of active electrogenic net transport by means of I_{SC} measurements requires a transepithelial current measurement under the condition of $V^e = 0$ (cf. the more general remarks on I_{SC} in the section, In Vitro Short-Circuit Current Measurements). Considering the rather short electrical length constant discussed in the preceding section it is evident that clamping of a whole segment of intestine in vivo to 0 mV cannot be achieved with high precision if the clamp is connected only to one end of a lumenal perfusion system.[14] Rather the use of an axial electrode is required, as described above.[20]

An indirect way to evaluate I_{SC} was used by Knauf and co-workers.[21] Since they measured in the same experiments open-circuit transepithelial voltage V^e and, by an independent method R^t, they calculated I_{SC} from the relation $I_{SC} = V^e/R^t$, assuming a linear I/V relation is in the observed range of numerical values. For rat large intestinal gut wall such a linearity has been demonstrated.[14,21]

All in vivo I_{SC} determinations suffer, however, from the same problem as in vivo R^e evaluations, namely, that in both cases the signals involved — zero voltage or ΔV steps — are generated not across the epithelium but rather across the intestinal wall. The size of the error due to this fact and experimental approaches to overcome these problems are discussed in the following section.

In Vitro Electrophysiology of Large Intestinal Epithelium

Large intestinal mucosa of many experimental animals, including mammals, shows surprisingly durable viability when transferred into a physiological bathing fluid. A proper choice of fluid, especially the addition of substrates, seems however to be necessary, e.g., by using the formula given in Table I it was possible in studies on rat rectal mucosa to extend experiments with stimulation of I_{SC} by physiological doses of aldosterone for up to 10 hr.[23,24]

As a rule there are, however, transients of the basic electrical parameters observed during the first hour of in vitro time: e.g., in preparations of rat rectum both V^e and I_{SC} declined from intially high values to a plateau after about 40 min, while R^t remained approximately constant.[25] It turned

[23] J.-D. Schulzke, M. Fromm, and U. Hegel, Pfluegers Arch. 407, 632 (1986).
[24] K. Hierholzer, H. Siebe, and M. Fromm, Kidney Int. 38, in press (1990).
[25] J. D. Schulzke, M. Fromm, U. Hegel, and E. O. Riecken, Pfluegers Arch. 414, 216 (1989).

TABLE I
BATHING FLUID FOR *in Vitro* EXPERIMENTS IN RAT LARGE INTESTINE[a]

Substrate	Concentration (mM)	Substrate	Concentration (mM)	Substrate	Concentration (mM)
Na^+	140	Cl^-	123.8	D-Glucose	10
K^+	5.4	HPO_4^{2-}	2.4	β-Hydroxybutyrate	0.5
Ca^{2+}	1.2	$H_2PO_4^-$	0.6	Glutamine	2.5
Mg^{2+}	1.2	HCO_3^-	21	D-Mannose	10

[a] Gassing: 95% O_2/5% CO_2; pH = 7.4; T = 37°; 290 mOsm kg^{-1}.

out that these initial transients can be suppressed either by the addition of 1 μM tetrodotoxin (TTX) to the serosal bath or by a "total strip," which removes the submucosal neuronal plexus[25] (for technique, see R^e Measurements *in Vitro*, below). The initial transients of V^e and I_{SC} are therefore most likely due to a neuronally mediated stimulation of electrogenic Cl^- secretion as a consequence of a mechanical excitement of the stretch receptors by the preparation procedure and are *not* indicative of an initial deterioration of the *in vitro* preparation.[25]

Ve Measurements *in Vitro*

The general precautions to be taken with V^e measurements have been mentioned above in context with *in vivo* measurements. A special requirement for *in vitro* experiments is to avoid "edge damage." This is the occurrence of low-resistance pathways at the circumference of the preparation either due to mechanical damage of the epithelium at the holding edge or to electrically imperfect sealing. While edge damage leads primarily to erroneous R^t values, it also influences V^e values by providing a partial short circuit. Obviously these effects are all the more serious the higher the R^e value or the smaller the area of the preparation. Techniques to minimize edge damage are a careful control of the mechanical pressure and/or provision of soft silicone rubber fittings. If silicone grease is used to improve electric sealing it should be applied only in minimal amounts since it has the tendency—especially in well-stirred heated chambers—to creep over the exposed area. Edge damage may be detected by voltage scanning along the circumference of the preparation.[26] A more sophisticated proce-

[26] J. T. Higgins, L. Cesaro, B. Gebler, and E. Frömter, *Pfluegers Arch.* **358**, 41 (1975).

dure which not only permits the detection but also the quantification of edge damage has recently been published by Kottra *et al.*[27] Its principle is a comparison of the current density in the center of the chamber at low and high alternating current (AC) frequencies. Further precautions with *in vitro* measurements in general concern control of temperature and pH, effective O_2 gassing, and good circulation of the bath solution. The amplifier input bias current should be adjusted to minimum and the total voltage asymmetry should be measured and corrected.

R^e *Measurements in Vitro.* In vitro R^e measurements have been performed mainly in combination with Ussing-type flux chamber experiments (cf. below, *In Vitro* Short-Circuit Current Measurements). In this context R^e can be evaluated in two possible ways: either by dividing the open-circuit spontaneous voltage V^e by the short-circuit current

$$R^e = V^e/I_{SC} \qquad (6)$$

or by using the voltage clamp to generate a distinct voltage step ΔV^t across the tissue and recording the related Δi. This has already been discussed with the *in vivo* measurements. However, if native gut wall is used without any further preparation all V or ΔV values are not measured exclusively across the epithelium but rather across the following tissue structures: mucous layer, epithelium, and subepithelial tissue layers, including connective tissue of different thickness between the crypts, muscle layers, and serosa (cf. Fig. 2). The respective contributions of these layers to the transmural resistance correspond to their thicknesses and their volume resistivities [cf. Eq. (1)]. The numerical values of all submucosal structures sum up to R^{sub}. The electrical resistivity of the mucus layer can be neglected.

There are at least four different ways to evaluate the contribution of R^{sub} to the resistance in series with the epithelium: (1) selective removal of the epithelial layer, (2) mechanical "stripping" of submucosal tissue layers, (3) microelectrode technique, and (4) AC impedance analysis.

1. An effective method to remove the intestinal epithelial cell layer without use of enzymes which could also attack the submucosal tissue has been published by Weiser.[28] However, to our knowledge no systematic comparison exists between measurements of R^{sub} on gut wall preparations denuded of their mucosa and those made with the use of one of the other methods.

[27] G. Kottra, G. Weber, and E. Frömter, *Pfluegers Arch.* **415,** 235 (1989).
[28] M. M. Weiser, *J. Biol. Chem.* **248,** 2536 (1973).

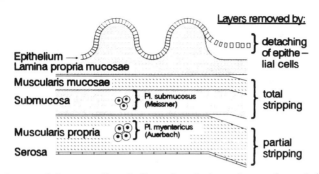

FIG. 2. Layers of the gut wall and effect of methods used to evaluate their fractional contribution to transmural resistance.

2. *Stripping* means removal of submucosal tissue layers by blunt dissection and is performed either as a *partial strip,* which splits the gut wall between submucosa and muscularis propria,[23,29] or as a *total strip,* which splits between the inner and outer layer of muscularis mucosae and leaves only a thin tissue layer of at most 5- to 10-μm thickness directly adjacent to the epithelial basement membrane (cf. Fig. 2). Technically a partial strip is performed best on pieces of gut wall which are kept stretched by clamping one end with the mucosa down onto a plastic plate and pressing a microscope glass slide onto the other free end while pulling it downward. Next a flat incision is made into the muscularis close to the slide edge with a scalpel. Starting from the gaping edge of this cut the muscularis propria is then pulled off by blunt dissection with forceps. The whole procedure can be carried out at room temperature within a few minutes and care should be taken to keep the preparation moist. For performing a total strip the gut segment in our laboratory is first stripped partially, then turned over so that the mucosa is up and — again starting from an incision at the lower end — the mucosa is very gently pushed off from the submucosa, mainly using the finger tips. Experience shows, however, that there are probably as many successful individual variants of the stripping procedures as there are scientists using them.

It is clear that even total stripping does not permit measurement of pure epithelial resistance R^e, e.g., by the current step method, because a small residue of connective tissue remains beneath and mainly between the crypts. It is, however, possible to calculate an average value of the area

[29] R. A. Frizzell, M. J. Koch, and S. G. Schultz, *J. Membr. Biol.* **27**, 297 (1976).

resistivity (and volume resistivity) of the submucosal tissue by correlating, e.g., morphometrically measured layer thickness of stripped tissue with electrically measured alterations in R^t.[23] An extrapolation of such numerical values to the total effective thickness of the submucosal tissue would not, however, give reliable results due to the inhomogeneous composition of submucosal tissue and to its complicated geometry.

3. By use of a microelectrode it is possible to measure experimentally the voltage divider ratio between epithelium and subepithelial tissue and thereby to evaluate epithelial and subepithelial resistivities. For this purpose a voltage-sensing microelectrode is first placed under microscopic observation directly above the epithelial surface in the grounded apical bath. Current steps ΔI are driven through the preparation and the (very small) bath amplitude (ΔV^{fl}) is recorded for later correction. The electrode is then advanced until it has just passed the epithelial cell layer. This is indicated by a sudden drop of the cell potential. In this position the voltage drop across epithelium and bath ($\Delta V^e + \Delta V^{fl}$) is recorded. After further advancement through the preparation the total transmural voltage drop $\Delta V^t + \Delta V^{fl}$ is finally seen. It then holds that

$$R^e = \Delta V^e/\Delta I; \qquad R^{sub} = (\Delta V^t - \Delta V^e)/\Delta I \qquad (7)$$

An experiment of this kind has, e.g., been successfully carried out on rat jejunum using current steps of $\Delta I = \pm 50 \ \mu A \cdot cm^{-2}$, 200-msec duration, and a frequency of 0.1 Hz.[30]

4. The most elegant method to evaluate the epithelial resistivity independent of contributions from submucosal tissue is probably the measurement of the transepithelial electrical impedance. The term "electrical impedance" designates the frequency-dependent current transfer properties of any material and is therefore different in its meaning from "resistance," which refers only to direct current (DC), i.e., to frequency-independent properties. A material without any kind of reactance behaves as a pure resistor, or purely "ohmic."

The special value of impedance analysis for the experimental separation of subepithelial (R^{sub}) from epithelial (R^e) resistance lies in the fact that in the technically easily accessible frequency range between 0 and about 50 kHz only the epithelial cell layer exhibits a noticeable capacitive reactance while the subepithelial tissue—due to the low-resistance shunting of the interstitial fluid all around its cells—behaves practically in a purely ohmic manner.

[30] M. Fromm, J. D. Schulzke, and U. Hegel, *Pfluegers Arch.* **405,** 400 (1985).

Impedance is quantitatively determined by both the DC resistance of the tissue and by its capacitive (and inductive) "reactive" properties. (True inductance does not exist in living matter and is therefore not considered further.) In turn, resistive as well as capacitive reactive components of the tissue can be quantitatively extracted from impedance measurements. Since capacitive reactance is related to the area as well as to structural features of cell membranes, electrical impedance analysis has become a very powerful method in epithelial electrophysiology. For more information on its implications and on experimental details the reader is referred to Refs. 23 and 31–38.

Technically there are two ways to achieve a separation of R^{sub} from R^e by impedance measurements: one a more simple though less precise way, and a second rather sophisticated one which, however, affords at the same time access to much more information on electrical parameters of the system.

The simple way consists of plotting measured impedance data in the complex impedance plane (Nyqvist plot), fitting the data by a semicircle, and evaluating the ohmic resistance in series with the epithelium (R_S) as the intersection of the high-frequency end of the semicircle with the real (ohmic) axis (cf. Fig. 3). This corresponds to an extrapolation of the impedance function to $f = \infty$ by use of an approximation function (see below). The rationale behind this procedure lies in the assumption that at very high frequencies ("very" means high as compared to time constants inherent in the material) the capacitive components act as effective short circuits such that only the ohmic resistances in series with the capacitive reactances are measured. If correction is made for the bath resistance this value can then be taken as the effective series resistance of subepithelial tissue *in vitro*. It should be borne in mind, however, that *in vivo* the effective subepithelial layer ends at the average distance of the capillary lumen, which means practically very close to the basement membrane of the epithelium. At the same time the semicircular fit provides a numerical value for R^e as the difference $R_{DC} - R_\infty$.

[31] E. Schifferdecker and E. Frömter, *Pfluegers Arch.* **377**, 125 (1978).

[32] C. Clausen and N. K. Wills, in "Ion Transport by Epithelia" (S. G. Schultz, ed.), p. 79. Raven, New York, 1981.

[33] H. Goegelein and W. van Driessche, *J. Membr. Biol.* **60**, 199 (1981).

[34] J. J. Lim, G. Kottra, L. Kampmann, and E. Frömter, *Curr. Top. Membr. Transp.* **20**, 27 (1984).

[35] N. K. Wills, *Curr. Top. Membr. Transp.* **20**, 61 (1984).

[36] M. Fromm, C. E. Palant, C. J. Bentzel, and U. Hegel, *J. Membr. Biol.* **87**, 141 (1985).

[37] J. R. Pappenheimer, *J. Membr. Biol.* **100**, 137 (1987).

[38] G. Kottra and E. Frömter, *Pfluegers Arch.* **415**, 718 (1990).

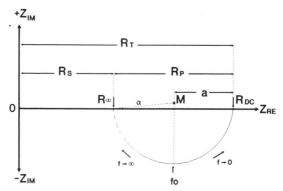

Fig. 3. Impedance locus of a low-resistance epithelium in the complex plane ("Nyqvist plot"). Z_{RE}, Z_{IM}, "Real" (= ohmic) and "Imaginary" (= reactive) components of impedance signals as recorded by two phase-sensitive detectors of 90° phase shift. Since capacitive reactance delays voltage relative to current the function lies in the third quadrant. If the measured structure behaved like the circuit in Fig. 4A a perfect semicircle would result with its center M on the real axis and shifted by $R_S + a$. Mainly due to the presence of a distribution of time constants in biological preparations M lies almost always more or less above the real axis (angle α). R_T, R_S, R_P, as in Fig. 4A; R_{DC}, R_∞, extrapolated intersections of impedance locus for $f \to 0$ and $f \to \infty$ with the real axis; f, frequencies of measurements; f_0, characteristic frequency of substrate (see text); a, radius of the semicircle.

The approximation of the impedance function by a semicircle in the complex plane is based on the supposition that the preparation behaves like the simple network depicted in Fig. 4A because it can be deduced that the impedance locus of a parallel RC element is a semicircle. Empirically this seems to be an acceptable approximation in mammalian large intestine,[23] but also in other epithelia of intermediate tightness such as the gall bladder[31,33,36] and small intestine.[30] Moreover, it has been verified in rat small intestine that semicircle extrapolations of AC measurements and voltage divider experiments with pulsed DC (see above) result within the limits of experimental precision in the same R^{sub} values.[30]

In addition to R^{sub} (= R_∞) and R^e (= $R_{DC} - R_\infty$) the semicircle fit to AC impedance data suggests the evaluation of one more parameter: the frequency f_0 at which the reactive (imaginary) component reaches its maximum and which is called *characteristic frequency*. In case of an underlying simple RC network like the one in Fig. 4A it holds that

$$f_0 = 1/2\pi R_P C_P \tag{8}$$

Since R_P can be extracted from semicircular plots as R_{DC} the effective capacity C_P of the preparation can be evaluated. There has been some

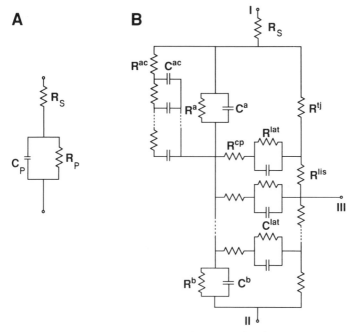

FIG. 4. Networks used to model passive electrical properties of epithelial preparations. (A) Simplest possible equivalent circuit, suitable to model leaky epithelia. R_S, Series sensitivity; R_P, parallel resistivity; C_P, parallel capacitance; $R_S + R_P = R_T$ (total ohmic resistivity). (B) Elaborate "distributed" equivalent circuit as proposed by Kottra and co-workers.[34] Here, infoldings of the apical as well as of the lateral membrane with distributed access resistances (R^{ac}, R^{cp}) to membrane elements (C^{ac}, R^{lat}, C^{lat}) are incorporated in order to model the observed ultrastructure of epithelial cells in more detail. I, II, and III, Trans- and intracellular ports (measuring points).

interest in calculating C_P of different substrates because it promises to provide an electrical measure of the otherwise difficult-to-evaluate exposed membrane area. This approach assumes that cell membranes in general exhibit an area-specific capacitance of about 1 $\mu F \cdot cm^2$. Thus, Lewis and Diamond[39,39a] recommended referring conductances and currents to microfarads (rather than to cm^2). While this is acceptable in the case of tight epithelia it is erroneous in leaky ones such as the large intestine. The shunting of apical and basolateral time constants by the paracellular resistivity R^{sh} ($= R^{tj} + R^{lis}$ in Fig. 4B) leads to the fact that in such epithelia the impedance locus may be found to conform in good approximation to one

[39] S. A. Lewis, and J. M. Diamond, *Nature (London)* 747 (1975).
[39a] S. A. Lewis, D. C. Eaton, C. Clausen, and J. M. Diamond, *J. Gen. Physiol.* **70**, 427 (1977).

semicircle, but that its parameters are complex functions of all apical, basal, and lateral parameters. In these cases, therefore, the apparent C_P by no means represents the geometrical membrane area, and, in addition, C_P could change whenever any of the circuit parameters, e.g., one of the resistances, changed.

The more complex second approach involves a modeling of the microanatomy of the preparation by a detailed electrical equivalent circuit. The most elaborate model so far published is shown in Fig. 4B. By this network it has been attempted to also include the local distribution of reactances along the paracellular pathway with its subdivision into junctional and interspace portions[32,34,40] and some cable-like properties within the apical cell membrane with its infoldings and microvilli.[38] Equations describing such a complex network contain up to 12 parameters. A fit of the related analytical function to measured transepithelial and intracellular impedance data permits then the extraction of most of the parameters, and also a high-quality extrapolation of the impedance locus to R_S at $f \rightarrow \infty$.

In Vitro Short-Circuit Current Measurements

The "electrogenic" active epithelial net transport of either cations or anions corresponds to an electric current which can be measured by the short-circuit technique, which was introduced by Ussing and Zerahn.[41] Its principle is to eliminate all driving forces for passive transport across the preparation by using (1) equal solutions at (2) equal pressure and (3) equal temperature on both sides and (4) to nullify (to "short circuit") any spontaneous "open-circuit" voltage generated by active transport by means of voltage clamp to 0 mV. Under this condition any net flux of charges of one sign which is transported independent of external driving forces can be measured quantitatively as electric current in the external clamp circuit. Because of the precise definition of the electrochemical gradient the Ussing technique is of paramount importance for the analysis of those epithelial and membranal transport processes in which ions are involved — and this is true for nearly all. If net ion transport is present, this may well be the result of a superposition of several different electrogenic transport systems working in parallel. This is, e.g., the case in the lower segment of the large intestine where electrogenic Cl^- secretion, Na^+ absorption, and K^+ secretion or absorption can all be found to be active. An identification of the individual ion flux components is, nevertheless, possible by use of the following techniques: (1) isotopes of the relevant ions are added subse-

[40] C. Clausen, S. A. Lewis, and J. M. Diamond, *Biophys. J.* **26,** 291 (1979).
[41] H. H. Ussing and K. Zerahn, *Acta Physiol. Scand.* **23,** 110 (1951).

quently to both sides and their unidirectional fluxes are quantified by radioactive counting or by mass spectrometry; (2) known specific blockers of the expected transport processes are tested for their efficiency to decrease measured currents as, e.g., amiloride for Na^+ channels, NPPB [5-nitro-2-(3-phenylpropylamino)benzoate] for Cl^- channels, Ba^{2+} for K^+ channels etc.; (3) methods (1) and (2) in combination.

Some technical implications and pitfalls of the short-circuit technique are explained in Fig. 5. If a structure like the gut wall consisting of epithelial and subepithelial tissue of different volume conductivities (ρ^e, ρ^{sub}) is clamped to a voltage V_C, the following trivial relations hold with respect to the definitions of Fig. 5A:

Voltage drop in the bath: $V^{fl} = V_C - V^t$
Transmural voltage: $V^t = V_C - V^{fl} = V^e + V^{sub}$
Transepithelial voltage: $V^e = V_C - V^{fl} - V^{sub}$ (9)
Subepithelial voltage: $V^{sub} = V_C - V^{fl} - V^e$

Thus, by knowing the clamp voltage V_C and measuring the clamp current density I_C the material constants R^t, R^e, and R^{sub} or, if the respective Δx values are known in addition, also ρ^t, ρ^e, and ρ^{sub} can be evaluated on the basis of Ohm's law and Eq. (1).

V^{fl} is also called "bath correction" because this value must always be subtracted since it is due to the unavoidable fluid layer between the voltage electrode tips and the preparation (Δx^{fl} in Fig. 5A). Its size can be estimated by knowing ρ^{fl} and the approximate layer thickness by use of Eq. (1). A numerical example is as follows:

$$\left.\begin{array}{l} I_C = 100 \ \mu A \cdot cm^{-2} \\ \rho^{fl} = \ \ 50 \ \Omega \cdot cm \\ \Delta x^{fl} = \ \ \ 0.2 \ cm \end{array}\right\} V^{fl} = 1 \ mV$$

Thus V^{fl} is rather small with well-positioned electrodes. One should, however, remember that V^{fl} may change if the bath fluid is altered.

While there is no doubt that in a short-circuit experiment the voltage electrodes should be set to V^{fl} in order to achieve 0 mV at the preparation, an intensive debate has taken place on whether the geometrical borders of the preparation or rather the epithelium itself should be clamped to 0 mV. A definite answer has been obtained by Rehm,[42] who made clear that in *in vivo* experiments 0 mV should be established across the epithelium, while in *in vitro* experiments this should be done across the whole preparation. The reason is that the convective compartments, i.e., those for which all

---

[42] W. S. Rehm, *Curr. Top. Membr. Transp.* **7**, 232 (1975).

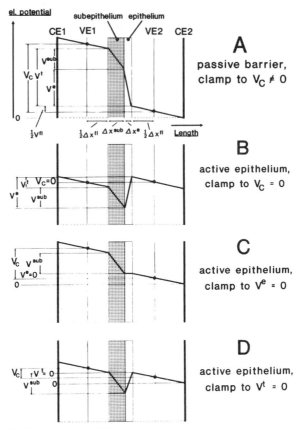

FIG. 5. Profile of electric potential inside an Ussing chamber. Abscissa, Schematic cross-section through measuring chamber; ordinate, absolute electrical potential, zero arbitrarily chosen at current electrode 2. The slope of the potential decay along the cross-section corresponds to the assumption that $\rho^{fl} < \rho^{sub} < \rho^e$. CE1, CE2, Current electrodes; VE1, VE2, voltage electrodes; V_C, clamp voltage, i.e., $V_{VE1} - V_{VE2}$ [for other symbols see Eqs. (9)]. (A) Voltage clamp of a *nontransporting* epithelium to an arbitrary voltage $V_C \neq 0$ mV. (B) Voltage clamp of an *electrogenically transporting* epithelium to $V_C = 0$ mV. (C) Voltage clamp of an epithelium like in (B) such that $V^e = 0$ mV. (D) Voltage clamp of an epithelium like in (B) such that $V^t = 0$ mV.

other physicochemical parameters are defined, *in vivo* border directly on both sides of the epithelium—the basal one being the capillary perfusion beneath the basement membrane—while *in vitro* the "inner" convective compartment starts only at the geometrical border of the preparation. Thus, *in vitro* the whole preparation and not the epithelium only should be clamped to zero.

The reader should, however, be aware of the fact that there is no ideal solution to the short-circuiting problem. The recommended clamping of the whole preparation to 0 mV generates a potential of the size of V^{sub} across the epithelium (Fig. 5D), while clamping the epithelium to 0 mV generates a gradient of the same size across the whole preparation (Fig. 5C). A numerical example using an estimate for ρ^{sub} from Ref. 20 is as follows:

$$\left.\begin{array}{rl} I_C = & 100 \ \mu A \cdot cm^{-2} \\ \rho^{sub} = & 10^3 \ \Omega \cdot cm \\ \Delta x^{sub} = & 0.2 \ cm \end{array}\right\} V^{sub} = 20 \ mV$$

Thus, the deviation of the actual potential across the transporting cell layer from 0 mV may be considerable and depends on—besides the short-circuit current—the thickness and volume resistivity of the submucosal tissue. This is the reason why it is important to remove as much of this tissue as possible, e.g., by stripping.

The effective "under-short circuiting" of the epithelium at $V^t = 0$ leads to an underestimation of the true active net charge transport I_{ion}^{net} by I_{SC} measurements. According to Tai and Tai[43] this error can be corrected by the following formula:

$$I_{ion}^{net} = [(R^{ser} + R^e)/R^e] \ I_{SC} \qquad (10)$$

with $R^{ser} = R^{fl} + R^{sub}$. The accuracy by which I_{ion}^{net} equals I_{SC} is therefore given by the relation R^{ser}/R^e, i.e., if only the epithelial area resistivity R^e is high compared to the sum of those of the bath and subepithelial layers a correction of measured I_{SC} values may be neglected. On the other hand, in preparations of mammalian large intestine where R^{sub} may reach more than 30% of R^e (Ref. 23 and U. Hegel and M. Fromm, unpublished data) this correction must be considered.

Short-circuit experiments are very often combined with isotope flux measurements. For charge balance it is clear that I_{SC} measures net flux of actively transported charge, i.e., that amount of actively transported ions which is not balanced by simultaneously ongoing active transport of ions of opposite sign. Since different ion species i contribute to I_{SC} according to their transference numbers t_i the relation between isotopic net flux of an ion species i, J_i (mol $\cdot$ cm^{-2} $\cdot$ sec^{-1}) and I_{SC} ($\mu A \cdot$ cm^2) is

$$J_i = [t_i/(z_i F)] \ I_{SC} \qquad (11)$$

with z_i being the ionic charge of species i. It is obvious, however, that the isotope flux J_i of Eq. (11) underestimates the true active transport of i also

[43] Y.-H. Tai and C.-Y. Tai, J. Membr. Biol. 59, 173 (1981).

by the inverse of the correction factor of Eq. (10). Thus, also for the correction of isotope flux measurements on large intestine it is essential to know R^{ser}.

A further methodological difficulty with measurements on intestinal mucosa is its nonflat geometry and its local inhomogeneity of active transport and epithelial resistivity.[44] This means that the voltage and flux errors during a short-circuit experiment as treated above differ most likely between the different anatomical and functional substructures of the mucosa. As long as numerical data on such local differences in parameters as J_i, R^e, and R^{sub} are not available it is difficult to estimate the possible size of the related errors.

As a result of an analysis of short-circuiting small intestine *in vitro* Tai and Tai[43] give the following formula for the quotient of the clamping error (V^e deviation from 0 mV during short circuit) between villus (v) and crypt (c) regions:

$$\frac{V^e\,(v)}{V^e\,(c)} = \left[\frac{I_{SC}\,(v)}{I_{SC}\,(c)}\right]\left[\frac{R^e\,(v)\,R^{sub}\,(v)}{R^e\,(c)\,R^{sub}\,(c)}\right]\left[\frac{R^e\,(c) + R^{sub}\,(c)}{R^e\,(v) + R^{sub}\,(v)}\right] \quad (12)$$

One consequence of Eq. (12) is that in a given experiment the relation between clamping errors in crypt regions and villus regions ($\triangleq$ surface epithelium) may vary with the functional state of the mucosa.

Measurements on Epithelial Layers Grown from Cells of Large Intestinal Origin

One possibility of overcoming the problems of cellular heterogeneity as well as the nonflat geometry of native intestinal mucosa could be to grow epithelial monolayers on flat permeable supports from defined homogeneous populations of colonocytes. So far such attempts which have been made with cells isolated and fractionated from native large intestinal mucosa ("primary cultures") have failed[45], in contrast to the considerable success with cells from other systems. Therefore, instead of investigating primary cultures, several groups have started to grow "artificial" intestinal epithelia from established tumorous cell lines which do not present problems with respect to proliferative potency. Some of these lines grow epithelially from the beginning like CaCo-2,[45] T 84[46], and HCA-7[47]; others can be

---

[44] Surface epithelium absorbs NaCl and/or Na^+, crypts secrete Cl^-. In addition, epithelial resistivity seems to be lower in crypt than in noncrypt regions (unpublished observations from own laboratory).

[45] M. Neutra and D. Louvard, in "Functional Epithelial Cells in Culture" (K. S. Matlin and J. D. Valentich, eds.), p. 344. Alan R. Liss, New York, 1989.

[46] J. A. McRoberts and K. E. Barrett, *J. Membr. Biol.* **36**, 325 (1989).

[47] A. W. Cuthbert, S. C. Kirkland, and L. J. MacVinish, *Br. J. Pharmacol.* **86**, 3 (1985).

induced to do so by special culture conditions like HT-29.[48] These cell lines all descend originally from human colonic adenocarcinomas. However, to date it is not clear to what extent they express physiological properties of large intestinal mucosa. A common feature seems a similarity of their transport systems with those of intestinal crypt cells, while other properties such as, e.g., the presence of Na^+-dependent glucose uptake in CaCo-2 cells relate them with the fetal stage of mammalian colonic mucosa. For more information on these cell lines, including electrophysiological experiments, see Ref. 45 and 46.

Measurement techniques on such "artificial epithelia" are not different from what is specified above for *in vitro* experiments on native epithelia: confluent layers grown on permeable membranes are mounted in Ussing-type flux chambers.

A standard technique to obtain such layers is the cup- or porous-bottom dish technique described in detail by Steele *et al.*[49] It consists essentially in cutting flat rings from polycarbonate tubes and cementing the supports to one side of them by use of silicone rubber. Cellulose ester [Millipore filters; Millipore, Bedford, MA] as well as polycarbonate [Nuclepore filters; Nuclepore Corp., Pleasanton, CA] or precipitated and fixed collagen[49] have been used for monolayer support. Such porous bottom cups are also commercially available (as, e.g., Millicell-HA inserts and others). Filters are often coated by collagen, laminin, or special formulas of extracts from native tissues. Many of the tumorous lines grow, however, quite well on uncoated filter membranes.

Before using such filters in an experiment, completeness of confluency must be tested. This can most conveniently be done by monitoring the development of transepithelial resistance R^e after seeding. The time within which a "plateau" of R^e is reached depends on the cell line and varies between 12 hr for the kidney cell line MDCK and about a week for differentiated HT-29 cells[50] if cells are seeded at densities of $5-10 \times 10^5$ cells/cm^2. Repeated controls of R^e and V^e on the same dishes are possible under clean conditions over weeks without too much risk of infection. A simple means for this purpose is a low-frequency ohmmeter with a pair of immersion electrodes, one of which is dipped into the cup while the other contacts the surrounding medium in the culture dish.[51]

[48] A. Zweibaum, M. Pinto, and G. Chevalier, *J. Cell. Physiol.* **122,** 21 (1985).
[49] R. E. Steele, A. S. Presten, J. P. Johnson, and J. S. Handler, *Am. J. Physiol.* **251,** C136 (1986).
[50] K.-M. Kreusel and U. Hegel, unpublished observations from own laboratory.
[51] A commercially available instrument is e.g., the "EVOM" of WPI Inc., New Haven, Connecticut.

Electrophysiology of Cell Membranes of Colonic Cells

Transepithelial transport depends on active and passive characteristics of apical and basolateral cell membranes. With respect to ion transport these are their respective resistivities (R^a, R^b), their voltages (V^a, V^b), and the short-circuit current (I_{SC}). In leaky to moderately tight epithelia such as the large intestine R^a and R^b are shunted by the resistivity of the paracellular pathway, R^{sh}. If there is no *a priori* knowledge on the size of at least one of the resistivity parameters it is impossible to evaluate either R^a, R^b, or R^{sh} in absolute terms from voltage or transepithelial resistivity measurements only. Nevertheless, in order to quantify at least *functional changes* of the membranes, the relative resistivities of apical and basolateral membrane are to be evaluated.

This is technically done by micropuncture. A microelectrode is positioned into an epithelial cell while simultaneously current steps ΔI are passed through the epithelium. The microelectrode then senses the voltage steps which are generated across the cell membranes by the transcellular fraction of ΔI (ΔI^{tc}). Practically, one uses the grounded mucosal bath as reference such that ΔV^a is recorded, and measures at the same time the transepithelial voltage steps ΔV^e (cf. above, the measurement of R^e and R^{sub}). It then holds that $\Delta V^b = \Delta V^e - \Delta V^a$. From Ohm's law it is clear that

$$R^a/R^b = \Delta V^a/(\Delta V^e - \Delta V^a) \qquad (13)$$

independent of th actual sizes of ΔI and ΔI^{tc}.

The quotient R^a/R^b is often referred to as the "voltage divider ratio" because it is nothing but the ratio of the two membrane voltage steps. If one wishes to describe mainly variations of the resistivity of one membrane, mostly R^a, the preference is not to use Eq. (13) but rather to calculate the fractional resistivity of that membrane as given by Eq. (14):

$$f_{R^a} = \Delta V^a/\Delta V^e \qquad (14)$$

Fractional resistivities are recommended for the following reason. In many cases R^a is considerably larger than R^b, which makes R^a/R^b a large number. Small variations of both ΔV^a and ΔV^b transform into a nonnormally distributed scatter of the quotient. In contrast, f_{R_a} values vary maximally between 1 and 0. Here variations in V^a are referred to the more constant number V^e which depends mainly on R^{sh} and which is not measured by microelectrodes but by more stable macroelectrodes. Therefore the scatter is much smaller and more or less linearly related to variations of f_{R_a}.

Criteria for "acceptable" intracellular recordings depend a bit on technical conditions in a given preparation. The following guidelines may be useful: (1) sudden increase of voltage with penetration of the electrode into a cell, (2) voltage variations of less than ±10% over 1 min, (3) less than ±10% variation of f_{R_a} over 1 min, (4) sudden drop to bath value less than ±<2 mV at drawback of the electrode, and (5) no remarkable changes of the electrode resistance by the micropuncture experiment.

With respect to the large intestine the micropuncture technique has been applied not only to native epithelia of proximal and distal colon of rat, rabbit, and man,[52-54] but also to cell culture (see last paragraph). An estimation of the individual resistivities of apical and basolateral cell membranes (R^a, R^b) could be obtained in such experiments by use of the "nystatin method"[39a]: The addition of nystatin to the apical bath reduces R^a within a few minutes to negligible values because this drug forms hydrophil channels in the membranes which are permeable to monovalent ions. By comparing trans- and intracellular recordings before and after the application of nystatin R^a, R^b, and R^{sh} have been extracted.[39a,53]

Two other methods for characterizing enterocyte cell membranes electrically, patch clamp and fluctuation analysis, are described elsewhere in this series (Vol. 171, [14] and [34]). In addition it should be recalled that AC impedance analysis in its latest version[38] as mentioned above in the section, R^e Measurements *in Vitro* (cf. also Fig. 4B), would in principle be able to identify most of the passive electrical parameters of any flat epithelium. So far it has not been applied to intestinal epithelia and it may be questionable whether an adaptation to the additional problems of the nonflat geometry of a gut mucosa could be incorporated.

Finally it remains to point out that a powerful method for the investigation of membrane electrophysiology of cultured cells is also described in this volume in the chapter by Wiederholt and Jentsch ([37], this volume). With the special technique they perfected it is possible to position a Lin-Gerard electrode without damage for hours inside a cell and to record alterations of the membrane potential during repeated changes of the bathing fluid. By then adding successively different specific transport inhibitors or stimulants to the bath it is possible to identify membrane transporters. By varying the concentrations of selected ions in the bath their relative contributions to the membrane conductance can be evaluated on the basis of the Goldman–Hodgkin–Katz equation. This method has been successfully applied, also to monolayers of the intestinal cell line HT-29.[55]

[52] W. Clauss, K. H. Biehler, H. Schäfer, and N. K. Wills, *Pfluegers Arch.* **408,** 592 (1987).

[53] G. I. Sandle and F. McGlone, *Pfluegers Arch.* **410,** 173 (1987).

[54] G. I. Sandle, F. McGlone, and R. J. Davies, *Pfluegers Arch.* **412,** 172 (1988).

[55] W. Ziss, M. Fromm, and U. Hegel, *Pfluegers Arch.* **411,** S1, R81 (1990).

Appendix: List of Symbols

Symbol	Dimension	Quantity, notion
A	cm^2	Area
a	1	Ion activity coefficient
AC		Alternating current
DC		Direct current
E	mV	Electric potential
E_0	mV	Normal potential (as compared to standard H^+ electrode)
F	$A \cdot sec \cdot mol^{-1}$	Faraday constant
G	$S \cdot cm^{-2}$	Area conductivity
g	S	Conductance
I	$\mu A \cdot cm^{-2}$	Current density
i	A	Current
J	$\mu mol \cdot hr^{-1} \cdot cm^{-2}$	Flux
R	$\Omega \cdot cm^2$	Area resistivity
R	$V \cdot A \cdot sec \cdot mol^{-1}$	Gas constant
r	Ω	Resistance
ρ	$\Omega \cdot cm$	Volume resistivity
T	K	Absolute temperature
t	hr, sec	Time
t_i	1	Transference number of ion i
V	mV	Voltage
x	cm	Length
Z	Ω	Electric impedance
z_i	1	Number of charges per ion i

Subscripts (capitals)	Definition
C	Clamp
P	Parallel
S	Series
SC	Short circuit
T	Total

Superscripts (lower case letters: anatomical or directional)	Definition
a	Apical
b	Basolateral
c	Cellular
cp	Cytoplasmic
e	Epithelial
fl	Fluid (bath)
lat	Lateral membrane
lis	Lateral interspace
ser	Series
sh	Shunt path, i.e., paracellular
sub	Supepithelial
t	Tissue (also, transmural)
tj	Tight junction
tc	Transcellular

[30] The Use of Isolated Perfused Liver in Studies of Biological Transport Processes

By Murad Ookhtens and Neil Kaplowitz

The use of isolated perfused liver as a model to study biological phenomena dates back more than a century.[1] However, due to its particular advantages (discussed below) and continuous developments and improvements in methodologies and analytical techniques (tracer techniques, mathematical modeling, etc.), it still remains a valuable tool. Rat liver is the most commonly used model and, after vascular and neural isolation from the carcass, is perfused either *in situ* or *ex situ* in single-pass or recirculating modes. Obviously, in the structural/functional hierarchy of experimental models, perfused liver is the closest to the *in vivo* end of the spectrum, as compared to the isolated cell and vesicle models. Both structural (e.g., imaging) and functional studies are possible with this model. For the latter studies, the kinetics and dynamics of transport, binding and metabolism can be measured with tracer or nontracer techniques.

Criteria for Selection of Perfused Liver Model over Alternative Models

As any experimental model, the perfused liver offers a tradeoff of advantages and disadvantages which need to be weighed carefully in decision making for or against its use in transport studies of interest. Its choice should be based on the following criteria.

Advantages

Structural Integrity. The perfused liver model retains characteristics that are closest to the *in vivo* conditions in terms of its architecture, i.e., hepatocyte polarity (sinusoidal *vs* canalicular domains), zonal inhomogeneities and gradients of certain substrates and enzymes (periportal *vs* pericentral), and vascular integrity.

Functional Integrity. Processes are expected to go on at comparable rates to that *in vivo* in comparison to other models. However, whenever feasible, this point should be investigated and ascertained as directly as possible. Also, in many cases, it might be important to verify the functional

---

[1] L. L. Miller, in "*Isolated Liver Perfusion and Its Applications*" (I. Bartosek, A. Guaitani, and L. L. Miller, eds.), p. 1. Raven, New York, 1972.

integrity of the preparation throughout, or at different times, during an experiment. Such testing could be done easily by using, for example, the uptake of general purpose markers for transport functions, such as glycine or taurocholate, which have high and easily measurable rates of single-pass clearance.

Control and Monitoring. While the model retains properties closest to the *in vivo* conditions, it avoids certain complications that arise in *in vivo* studies, such as effects of humoral (blood-borne) and neural signals that may interfere and produce secondary effects, and measurement complications arising from recirculation of the test substances introduced at the entry to the liver. The isolated perfused liver model, particularly in the single-pass mode, allows near-complete freedom in terms of control and continuous monitoring (electrodes) of input patterns of test compounds (impulse, pulse, step function, etc.), O_2 uptake, temperature, pH, pressure, Ca^{2+} and K^+ levels, and cytosolic marker enzyme release for measurement of lysis.

Ease of Preparation and Measurement. The experimental preparation is a well-established and rather simple one, ordinarily involving the catheterization of the hepatic portal vein for inflow and the thoracic inferior vena cava for outflow (hepatic vein output) routes in the rat. Whenever required, hepatic artery can also be catheterized.[2] Bile duct can be easily catheterized for collection of bile samples or retrograde infusion of substances [now commonly done to infuse acivicin (AT-125) for irreversible inhibition of γ-glutamyltransferase, the enzyme that catalyzes the hydrolysis of glutathione].[3,4] Thus, repeated serial samples can be taken of the following specimens: perfusates (both inflow and outflow) for measurement of uptake or efflux of substances, bile samples, and serial biopsies of the liver tissue, with minimum interference to the perfusion (the authors, using microclamps, routinely remove six or more 50- to 100-mg serial biopsies from 10-g livers during a single perfusion experiment; M. Ookhtens, unpublished method and observations).

Other Factors. There is flexibility for perfusion in single-pass or recirculating modes, and in prograde or retrograde directions. Each method is well suited for certain types of studies; e.g., single-pass perfusion allows the most direct measurement of uptake and efflux of substances which exhibit large one-pass differences across the liver, while recirculating perfusion allows the measurement of rates of uptake or efflux that are not easy to measure in single-pass mode (the cost could be another major determinant of the choice between single-pass vs recirculating modes). Prograde *vs*

[2] J. Reichen, *J. Clin. Invest.* **81**, 1462 (1988).
[3] W. A. Abbott and A. Meister, *Proc. Natl. Acad. Sci. U.S.A.* **83**, 1246 (1986).
[4] N. Ballatori, R. Jacob, and J. L. Boyer, *J. Biol. Chem.* **261**, 7860 (1986).

retrograde perfusions may be used to study processes with distinct zonal distributions and functions.[5] Surface microelectrode measurement techniques can be applied to study various processes in the periportal and pericentral zones.[6] In addition, single-cell micropuncture in the intact organ can be utilized to correlate organ function with electrophysiology.[7]

Disadvantages

Complexity of Interpretation of Data. Although simpler than the *in vivo* models, the perfused liver is still a complex model for transport studies, as compared to isolated cells and, particularly, membrane vesicles. The issues of flow dependence and unstirred layers might have to be considered and taken into account.[8] Moreover, the transport process of interest in most cases may not represent the rate-limiting step in the chain of events in uptake–metabolism–excretion or synthesis–efflux pathways. Then, the rate of the transport process of interest cannot be estimated directly from the steady state differences measured across the liver. Instead, more sophisticated designs, measurements, as well as mathematical modeling representing the different processes, including the transport step of interest, will be required for computation of the rates of the latter from measurements obtained across the liver.

Inhomogeneity of Cell Types. The liver has other cell types in addition to hepatocytes (Kupffer, endothelial, and duct cells). Therefore, in an intact organ it may not be possible to distinguish which cell type is involved and contributes to the process studied.

Other Limitations. Certain aspects of the perfused liver model may pose limitations unacceptable for the intended studies. Main ones are the duration of organ viability, which is normally 2–4 hr depending on the perfusion medium used, and the difficulty of catheterizing the hepatic artery in smaller animals and species.

Costs. Finally, the cost of using the perfused liver model might represent an unattractive tradeoff in terms of the amount of data obtained per liver *vs* the effort invested, as compared to the isolated cell or vesicle models.

Technical Aspects

The technical information presented below highlights the essential factors involved in perfusion of *rat* livers. However, many of the points apply

[5] K. S. Pang and J. A. Terrell, *J. Pharmacol. Exp. Ther.* **216,** 339 (1981).
[6] R. G. Thurman and F. C. Kauffman, *Hepatology* **5,** 144 (1985).
[7] J. G. Fitz and B. F. Scharschmidt, *Am. J. Physiol.* **249,** G56 (1987).
[8] E. L. Forker and B. A. Luxon, *Am. J. Physiol.* **248,** G709 (1985).

equally well to the use of other species. For a comprehensive review of many technical details refer to earlier reviews in this series.[9-13]

Liver Preparation. Following anesthesia and an abdominal incision, the hepatic portal vein (and, if needed, hepatic artery) as well as the bile duct can be catheterized. At this stage, the critical point is to restrict the interruption of hepatic blood flow to no more than a few seconds (preferably ≤ 10 sec) during portal vein catheterization, so that potential hypoxic injury is eliminated, or minimized. Later, after the portal vein (and hepatic artery) cannulation and start of perfusion, an incision in the thoracic area allows the catheterization of the vena cava (through the right atrium of the heart), and ligation of the inferior vena cava above the renal vein, to allow collection of the whole hepatic venous outflow. Tying the portal inflow catheter's tip at ≤ 4 mm from the entrance to the liver (rat) ensures a leak-free perfusion. After catheterizations are complete, the liver might be left *in situ* (with proper draping, to retain moisture, and heating for temperature control), or be excised and placed in a proper container or bath for *ex situ* perfusions.

Perfusion Apparatus. Many earlier publications, including several in this series,[9-13] have presented technical details (and diagrams) of commonly used perfusion apparatuses. We will avoid repetition and discuss only the essential aspects of a well-monitored/controlled perfusion system, an example of which is routinely in operation in our laboratory.[14] The bare essentials of a liver perfusion system are as follow: a temperature-controlled reservoir for the perfusion medium, a main perfusion pump, a gas exchanger (preferably one with no moving parts, if adequate, such as the Hamilton lung[15] or, if necessary for some perfusion media, one with moving parts, such as the rotating cylinder oxygenator in the perfusion apparatus made at Vanderbilt University[10]), a bubble trap, a thermostatically controlled enclosure box to keep the temperatures of the liver (as well as the O_2 exchanger, all the tubing, etc.) under control, plus inflow and outflow perfusate sampling ports. To turn the above system into a well-monitored and controlled one, provisions should be made for continuous or intermittent monitoring/recording and control of pH, perfusion flow

[9] H. Brunengraber, M. Boutry, Y. Daikuhara, L. Kopelovich, and J. M. Lowenstein, this series, Vol. **35**, p. 597.
[10] J. H. Exton, this series, Vol. **39**, p. 25.
[11] H. Sies, this series, Vol. **51**, p. 48.
[12] D. K. F. Meijer, K. Keulemans, and G. J. Mulder, this series, Vol. **77**, p. 81.
[13] W. A. Dunn, D. A. Wall, and A. L. Hubbard, this series, Vol. **98**, p. 225.
[14] M. Ookhtens, K. Hobdy, M. C. Corvasce, T. Y. Aw, and N. Kaplowitz, *J. Clin. Invest.* **75**, 258 (1985).
[15] R. L. Hamilton, M. M. Berry, M. C. Williams, and E. M. Severinghaus, *J. Lipid Res.* **15**, 182 (1974).

rate and pressure, O_2 uptake (preferably by in-line, flow-through oxygen electrodes across the liver), and temperature monitoring/control for both the perfusate (in-line) and the liver (surface electrode). To the above should be added a precision infusion pump for low flow rate delivery of concentrated solutions (e.g., at 20:1 ratio) or test substances into the perfusion line. The site of infusion should be the closest to the liver that does not preclude the complete mixing of test substances with the perfusate prior to entry into the liver. Finally, if necessary, provisions for switching between prograde and retrograde perfusions should be made, by proper arrangement of by-pass tubing and stopcocks, for flow reversal.

Perfusion Media. The following media have been used: whole blood, diluted blood, washed red cells added to buffer, perfluorocarbon emultions, and Krebs buffer (±albumin). The simplest and least costly alternative that is adequate for the purpose at hand should be selected. This is a matter to be settled based on the rationale of the study and preliminary tests to move toward the least cumbersome and costly, but acceptable end of the spectrum of the alternatives. Although perfusion with whole blood might appear to have the advantage of being the closest to the *in vivo* conditions, its handling and potential complications (e.g., hemolysis, or high rate of glycolysis by the erythrocytes) might make it an unattractive and unacceptable choice. Some investigators use a preperfusion step with perfluorocarbons, before switching to the Krebs medium for the actual experiments, to improve viability and function.[16] The necessity for inclusion of energy substrates in the perfusion medium should be verified, especially for long-term perfusions. Among the various media, the Krebs-bicarbonate buffer is the most commonly used one, due to the simplicity of handling and economy. In preparing artificial perfusion media, care should be taken by pregassing and adjustment of pH to near 7.4, before addition of $CaCl_2$, so that it does not precipitate. The inclusion of albumin in these perfusion media is not mandatory and is dependent on the application; e.g., in our studies of hepatic sinusoidal efflux of glutathione, the inclusion of albumin was found to be inconsequential,[14] while in studies involving bilirubin infusions the albumin carrier was necessary.[17] Interpretation of the data from kinetic transport studies involving albumin-bound compounds is complex and has been controversial.[18-20] Recently, it has been reported that under certain circumstances the dissociation of the albumin-

[16] R. A. Weisiger and Wei-Lan Ma, *J. Clin. Invest.* **79,** 1070 (1987).
[17] M. Ookhtens, I. Lyon, J. Fernandez-Checa, and N. Kaplowitz, *J. Clin. Invest.* **82,** 608 (1988).
[18] R. A. Weisiger, *Proc. Natl. Acad. Sci. U.S.A.* **82,** 1563 (1985).
[19] E. L. Forker and B. A. Luxon, *J. Clin. Invest.* **67,** 1517 (1981).
[20] A. W. Wolkoff, *Hepatology* **7,** 779 (1987).

bound ligands may represent the rate-limiting step in the process of uptake.[18]

Flow Rate and Pressure. The choice of flow rate is largely dictated by the adequacy of the rate of oxygenation (preferably ≥ 2 μmol O_2/min/g liver in normal fed rat) and, thus, is different for various media with different O_2-carrying capacity. Media with red cell hematocrit near that of *in vivo*, or high O_2-carrying capacity, such as perfluorocarbon, can be used at ≈ 1.5 ml/min/g liver, which approximates normal hepatic blood flow. Straight buffer-based media, such as Krebs-bicarbonate, require higher rates, i.e., $3-4$ ml/mg/g liver, for adequate oxygenation. It is essential to establish and use a range of flow rates in which the rates of O_2 uptake become independent of the flow and do not influence the measurements. Perfusion pressure increment over the baseline caused by rat livers perfused with buffers is typically ~ 5 cm H_2O. Substantially higher pressures and/or consistently rising pressures during a perfusion indicate problems that need to be addressed and overcome.

Perfusion and Sampling. At the onset of perfusion, a minimum equilibration period of $10-15$ min is allowed, after which collection of specimens may start. Perfusate inflow and outflow samples can be obtained at desired time intervals. To avoid undesirable pressure and flow transients, it is best to sample the inflow perfusate in a manner that avoids causing such transients. This aim may be accomplished by slowing down the rate of intermittent collection of the inflow sample to a safe level or, better yet, by providing a continuously flowing, low-rate sampling port (a continuously dripping port), such that intermittent opening and closing of stopcocks is not called for. Discrete injection of test substances (e.g., bolus injections for indicator dilution studies) should be given in small enough volumes to minimize pressure/flow disturbances. Samplings of the outflowing perfusate and bile do not pose difficulties and can normally be done by gravity (providing a few centimeters of negative pressure by positioning the outflow tips of the catheters below the liver level will prevent the generation of adverse back pressures and facilitate bile flow). Up to six or more serial liver biopsies ($\approx 50-100$ mg) can easily be taken by clamping off small areas at the edges of different liver lobes and excising tissue specimens. This can be done without causing any leaks or injury, and with practically negligible effect on the overall performance of the liver in terms of O_2 uptake, perfusion pressure, lysis and/or necrosis, or any other adverse effects (M. Ookhtens, unpublished observations). In addition, besides collection of samples, parameters such as O_2 uptake, pH, temperature, flow, and pressure can be continuously or intermittently monitored, recorded, and/or controlled in the desired manner.

Methodology for Transport Studies

In choosing a method for studying the desired hepatic transport function(s), as always, one must start with the clear definition of the purpose of the study. Thereafter, the criteria for selection of the perfused liver model presented above should be carefully considered and weighed against other models, such as isolated (fresh or cultured) hepatocytes and isolated sinusoidal or canalicular enriched membrane vesicle preparations. Clearly, the unique advantage of the perfused liver model over the others remains its close relationship to the *in vivo* conditions in terms of architecture and separation of the sinusoidal vs canalicular domains of transport. Thus, if both domains of transport are of interest, the perfused liver becomes a strong contender. After choosing the perfused liver as the experimental model, the design of the type of studies to be conducted will depend on other factors. As an example, the distinction between cellular membrane transport *vs* intracellular transport will influence the choice of the type of studies to be designed. In all cases, any existing knowledge about the transport process of interest should be carefully considered in this decision making. This approach will facilitate the choice between the alternatives and allow the selection of the important aspects of design as follows.

Single-Pass vs Recirculating. The rates of accumulation or removal of substances and the limits of detection and sensitivity of the assays are the determining factors (aside from the cost factor). All other factors aside, single-pass mode remains preferable due to its open-loop characteristics and avoidance of the complications arising from the recirculation of test substances, as well as avoidance of the accumulation of compounds in the perfusate that may have adverse effects on the liver viability and function. However, if single-pass clearance is too low, the precision of the measurement method might preclude its determination from inflow–outflow difference studies.

Time Frame of Study. The choice of the duration of studies should be made based on all the available knowledge, or informed, *a priori* expectation of the fractional transport rates (i.e., half-lives or time constants) of the processes involved. Such estimates will allow one, for example, to predict (simulate) the duration of perfusion required to attain the steady state in constant step infusion studies, or the time frame of a tracer-kinetic study to capture at least one half-life of turnover. In studies using the indicator dilution technique (see below), the time frame of each study is very brief and usually does not exceed 40 sec following the bolus injection (after this period of time, generally the level of tracers in the efferent perfusate becomes undetectable).

Dynamic vs Kinetic Studies. Dynamic (i.e., nonsteady state) designs generate a wealth of critical information that eventually should be integrated into, and be consistent with, one's working hypothesis or model of the system. The transient periods of the on or off phases of the responses in loading experiments represent examples of dynamic testing. However, dynamic designs have limited applicability in transport studies, since the analysis and interpretation of data from such designs are not straightforward and require the use of dynamic and generally nonlinear mathematical models. Dynamic studies may eventually complement the information obtained from steady state, kinetic studies and provide for a more rigorous testing of one's hypothesis (model). However, kinetic approaches and designs (i.e., studies in which pool sizes and mass fluxes are maintained at steady state) are more commonly utilized, at least as the first step, in transport studies since interpretation of the results from such studies, even when requiring mathematical modeling, is much more straightforward and simpler (linear models) as compared to the dynamic ones. Additionally, in many transport studies, the primary intent is to measure the rates of the transport processes of interest. Among the alternatives, the kinetic methods are less complicated and better suited for such studies. These studies can be performed with or without the use of tracers. In either case, conditions should be ensured so that, during the period of kinetic observations, the steady states of pool sizes and transport rates are maintained, or their changes are not substantial (e.g, a near steady state, characterized by ≤ 10% change over the period of study, might well be tolerable).

Commonly Used Kinetic Methods. These types of studies are done while pool sizes and fluxes are maintained at steady state, which are arrived at by using the following: no perturbation, pretreatments with agents that increase or decrease the hepatic pool sizes of interest, or constant infusion with the compound(s) of interest. Two categories of studies, i.e., with or without tracers are common, as discussed below.

Nontracer kinetic methods are used to measure the sinusoidal uptake/efflux or biliary excretion of exogenous or endogenous compounds. Studies to measure net sinusoidal clearance, or biliary excretion, are usually done by perfusion with constant concentration (step infusion) of the compound of interest. Sufficient time is then allowed for the establishment of an adequate near steady state of pool sizes and transport rates, after which the sampling of perfusate, bile, and liver specimens may commence. This type of design yields rather limited information and generally does not allow the separate estimation of the rates of uptake and efflux, or intracellular transport of substances. Analysis of serial liver specimens will yield confirmatory information about whether a steady state was attained by the constant infusion and allow the measurement of the steady state pool sizes.

In exceptionally fortuitous circumstances, when a unidirectional efflux (sinusoidal and/or biliary) occurs from a single, kinetically homogeneous and measurable hepatic pool, the kinetics of the efflux as a function of hepatic pool size can be defined. In fact these aspects, i.e., the asymmetry of the transport process and the kinetic homogeneity of the bulk of the hepatic glutathione pool (~90% cytosolic), were used by the authors to define the saturable and apparently carrier-mediated sinusoidal efflux of hepatic glutathione.[14] In these studies, various near-steady state levels of hepatic glutathione were obtained by treatments with glutathione-depleting or -repleting agents, subsequent to which the kinetics of sinusoidal glutathione efflux was studied.

Kinetic approaches using tracers have gained wide application due to the wealth of additional information they provide as compared to non-tracer kinetic studies. Tracer kinetic studies can be done with several different methods that have been developed and improved over a period of a few decades. The principal challenge that most these methods face is the use of input–output data, obtained by sampling the perfusate and bile specimens, to estimate the "intrinsic" and not directly measurable rates of transport. This approach then almost inevitably calls for the use of mathematical formulations (models), developed based on the investigator's understanding and assumptions about the system and processes involved. Thus, investigators may differ in their understanding of the system and the validity of their assumptions, which may cause differences in their quantitative interpretation of the data. Obviously, any assumption that is directly verifiable by experimentation should be tested for confirmation. In cases in which verification is not possible, at least the effect of alternative assumptions on the quantitative estimates of the transport rates of interest ought to be tested (simulation and sensitivity analyses). A fortuitous outcome is when the quantitative estimates of the rates do not show high sensitivity to the alternate plausible assumptions. When a mathematical model is developed, it is fitted to the data to estimate the desired transport rates. The following is a brief presentation of the highlights and the main characteristics of tracer kinetic methods.

The simplest tracer kinetic studies consist of constant step infusions of tracers into the perfusate at steady state. Then measurements are made on the afferent and efferent perfusates, as well as bile. These types of studies, in addition to the information obtained from nontracer, constant infusion studies, can provide transient tracer kinetic data for the on phase of attaining a steady state for the tracer. Such data may have some usefulness in estimating the initial rates of uptake of compounds from the perfusate tracer profiles (provided that unstirred layers and the extracellular distribution-mixing steps are not rate limiting). The data will allow the calculation of the time constant of the whole system for arriving at steady state and having a mathematical model may allow the fitting of the transient phase of the data to estimate the magnitude of the transport (uptake) rates.

Another approach with tracer kinetic techniques is the pulse-chase method. This method consists of introducing a brief tracer pulse of labeled material over the unlabeled material being perfused at steady state and following its movement through the system, i.e., the rise and fall of the label in different intermediate pools and its metabolism and conversion to other molecular forms. Since both on and off (washout) phases of the dynamics of the tracer movement are studied, more rigorous criteria for fitting one's model to the data must be met, as compared to the constant infusion design. This aspect can contribute to the improved accuracy and precision of the estimates of transport rates. However, this approach heretofore has had very limited application to the perfused liver model, since taking the full advantage of the method will require multiple, serial sampling of the liver tissue to identify the movement of the tracer through the different intrahepatic pools.

In the tracer kinetic studies of the type discussed above, the dynamics of tracer mixing and distribution in the extracellular and Disse space, even though not rate limiting, might still contribute substantially to tracer profiles during the initial transient phases. In fact, if that is the case, precautions should be taken, using marker molecules, to identify and correct for the contributions made to the data by the kinetics of distribution and mixing in these spaces. A method has been developed that deals with these points when measuring the uptake and efflux of compounds in the perfused liver. The method is known as the indicator dilution technique.[21] It involves the introduction of a small bolus of a combination of the labeled compound, whose uptake by the liver is of interest, along with marker molecules in the perfusion line to simultaneously identify the dynamics of tracer mixing-distribution in the extracellular and Disse spaces. Labeled red cells, albumin, sucrose, and sodium have been used as markers of vascular, Disse, and extracellular spaces, as well as labeled water as a freely penetrating compound.[21] This method is obviously suited only for the study of the transport of compounds with large enough single-pass uptake that allows detectable and precise estimation of the fluxes after correction for the distribution-mixing components. The data defining the profiles of the washouts of the different compounds are then fitted with lumped or distributed parameter mathematical models to estimate the rates of sinusoidal uptake (and efflux) of the compound(s) of interest.

Historically, in terms of conceptual approach to modeling the elimina-

---

[21] C. A. Goresky, G. G. Bach, and B. E. Nadeau, *J. Clin. Invest.* **52,** 991 (1973).

tion of substrates by the intact liver, two methods have been proposed and subsequently modified to incorporate later findings (e.g., taking into consideration the binding of substrate to circulating components). These two approaches have been known as the well-stirred (or venous equilibrium) and parallel tube (or sinusoidal) models. Although the latter comes close to describing and incorporating lobular gradients, the former, which views the liver as a single well-mixed compartment, remains useful under certain circumstances. For a detailed discussion of these issues, the reader is referred to Ref. 22.

Finally, an *in vivo* technique has been developed for the measurement of the initial rates of uptake of compounds into tissues directly, using the bolus injection technique, combined with internal marker molecules that readily penetrate the intracellular space (most commonly, ^3H_2O).[23] With this technique, a single, early (10–20 sec postbolus) sample of the tissue allows the measurement of the rates of uptake in relation to the internal marker. This technique has been developed and mainly applied to the studies of uptake of substances across the blood–brain barrier *in vivo*.[23] However, it may have potential usefulness for certain studies of uptake in the isolated perfused liver model. It has been heretofore used in studying the hepatic uptake of amino acids and carbohydrates and receptor-mediated uptake of asialoglycoproteins by rat liver *in vivo*.[24,25]

[22] D. J. Morgan, D. B. Jones, and R. A. Smallwood, *Hepatology* **5**, 1231 (1985).
[23] W. H. Oldendorf, *Brain Res.* **24**, 373 (1970).
[24] W. M. Pardridge and L. S. Jefferson, *Am. J. Physiol.* **228**, 1155 (1975).
[25] W. M. Pardrige, A. J. Van Herle, R. T. Naruse, G. Fierer, and A. Costin, *J. Biol. Chem.* **258**, 990 (1983).

[31] Measurement of Unidirectional Calcium Ion Fluxes in Liver

By Laurent Combettes, Catherine Dargemont, Jean-Pierre Mauger, and Michel Claret

The transfer of extracellular Ca^{2+} through the plasma membrane of mammalian cells is defined as the unidirectional influx. In liver, Ca^{2+} moves into the cells along a decreasing chemical gradient of an amplitude of approximately 2×10^4, Ca^{2+} concentrations in the plasma ([Ca^{2+}]$_o$) and in the cytosol ([Ca^{2+}]$_i$) being around 2 mM and 100 nM, respectively.[1,2]

[1] R. Charest, P. F. Blackmore, B. Berthon, and J. H. Exton, *J. Biol. Chem.* **258**, 8769 (1983).
[2] B. Berthon, A. Binet, J.-P. Mauger, and M. Claret, *FEBS Lett.* **167**, 19 (1984).

Calcium ion influx also is facilitated by a membrane potential of about -35 mV, inside negative.[3] Little information is known about the transport system responsible for the Ca^{2+} influx in liver cells.[4-7] It may result from passive diffusion through specific channels gated by hormones or neurotransmitters as well as from facilitated diffusion mediated by a carrier. There is currently insufficient evidence to support or refute this distinction. Only an electrophysiological method such as patch-clamp[8] of isolated hepatocytes[9] will allows distinguishing between these two systems of Ca^{2+} transport.

The Ca^{2+} entering the cell is taken up by two major organelles which transport Ca^{2+}: the mitochondria and the endoplasmic reticulum. Although their relative importance in controlling $[Ca^{2+}]_i$ has long been debated, the following distinction can be drawn. On one hand, in resting cells mitochondria contain rather little Ca^{2+} but can sequester massive amounts if $[Ca^{2+}]_i$ rises above 1 μM. On the other hand, the endoplasmic reticulum appears to be the physiologically relevant intracellular Ca^{2+} store with a high affinity for Ca^{2+}.[10,11] In addition, this compartment is the target[12,13] (but see also Ref. 14) for the second intracellular messenger inositol 1,4,5-tris-phosphate generated by a variety of hormones or neurotransmitters.[15,16]

At steady state, the permanent passive influx of Ca^{2+} is compensated by an active efflux of Ca^{2+} from the cells. There is evidence from detailed enzymatic and transport studies that the unidirectional Ca^{2+} efflux is driven by a high-affinity Ca^{2+}, Mg^{2+}-ATPase which is active at submicromolar concentrations of cytosol free Ca^{2+}.[17] This transporter is responsible

[3] M. Claret and J.-L. Mazet, *J. Physiol.* **223**, 279 (1972).

[4] J.-P. Mauger, J. Poggioli, F. Guesdon, and M. Claret, *Biochem. J.* **221**, 121 (1984).

[5] G. J. Barritt, *Trends Biochem. Sci.* **12**, 322, (1981).

[6] P. H. Reinhart, W. M. Taylor, and F. L. Bygrave, *Biochem. J.* **223**, 1 (1984).

[7] J.-P. Mauger and M. Claret, *J. Hepatol.* **7**, 278 (1988).

[8] O. P. Hamill, A. Marty, E. Neher, B. Sakmann, and F. J. Sigworth, *Pfluegers Arch.* **391**, 85 (1981).

[9] T. Capiod and D. G. Ogden, *Proc. R. Soc. London, B* **236**, 187 (1989).

[10] G. L. Becker, G. Fiskum, and A. L. Lehninger, *J. Biol. Chem.* **255**, 9009 (1980).

[11] G. M. Burgess, J. S. McKinney, A. Fabiato, B. A. Leslie, and J. W. Putney, Jr., *J. Biol. Chem.* **258**, 5716 (1983).

[12] G. M. Burgess, P. P. Godfrey, J. S. McKinney, M. J. Berridge, R. F. Irvine, and J. W. Putney, Jr., *Nature (London)* **309**, 63 (1984).

[13] S. K. Joseph, A. P. Thomas, R. J. Williams, R. F. Irvine, and J. R. Williamson, *J. Biol. Chem.* **259**, 3077 (1984).

[14] P. Volpe, K. H. Krause, S. Hashimoto, F. Zorzatoa, T. Pozzan, J . Meldolesi, and D. P. Lew, *Proc. Natl. Acad. Sci. U.S.A.* **85**, 1091, (1988).

[15] M. J. Berridge, *Proc. R. Soc. London, B* **234**, 359 (1988).

[16] C. J. Kirk, E. A. Sone, S. Palmer, and R. H. Michell, *J. Recept. Res.* **4**, 489 (1984).

[17] S. Lotersztajn, J. Hanoune, and F. Pecker, *J. Biol. Chem.* **256**, 11209 (1981).

for regulating the resting $[Ca^{2+}]_i$ in a permanent manner, and is the unique process allowing the cell to transfer Ca^{2+} to a vast and unlimited pool.[5] The carrier also plays an essential role in returning $[Ca^{2+}]_i$ to basal level whenever Ca^{2+} influx to the cell and/or Ca^{2+} release from internal stores increase.

Measurement of the Ca^{2+} Unidirectional Influx

Calcium ion fluxes through the plasma membrane have been studied by curve-fitting analysis both in the perfused rat liver or in isolated rat liver cells.[4-6] One of the main limitations in using this method is that it requires a steady state, i.e., constant Ca^{2+} pools and fluxes for the length of the experiment. This is not compatible with the measurement of the action of agents or hormones which promote cell Ca^{2+} redistribution.[18-20]

Another way of estimating the unidirectional Ca^{2+} influx is to determine directly the initial uptake rate of $^{45}Ca^{2+}$ over a period of time which is short compared with the time constant of cell Ca^{2+} exchange. This prevents undue labeling of intracellular Ca^{2+} stores. Under these conditions, Ca^{2+} influx is not altered by a redistribution of Ca^{2+} promoted by the hormones mobilizing Ca^{2+}.[4] The $^{45}Ca^{2+}$ influx is then dependent only on the concentration of the medium Ca^{2+} and on the permeability of the cell plasma membrane.[4,20] Since $[Ca^{2+}]_o$ is experimentally fixed, any change in the initial rate of $^{45}Ca^{2+}$ uptake initiated, for example by hormones, uniquely reflects the change in the translocation rate of Ca^{2+} through the plasma membrane. The procedure routinely used in our laboratory for isolating liver cells from female Wistar rats has been described in detail previously[21] and modified as indicated in Mauger et al.[22] The cells are suspended in Eagle's medium[23] containing 116 mM NaCl, 5.4 mM KCl, 1.8 mM CaCl$_2$, 0.81 mM MgSO$_4$, 0.96 mM NaH$_2$PO$_4$, 25 mM NaHCO$_3$, 292 mg/liter glutamine, 805 mg/liter amino acids, 8.1 mg/liter vitamins, and 1 g/liter glucose. The medium is supplemented with 1.5% gelatin (Difco, Detroit, MI). The suspension is equilibrated at 37° in 250-ml Pyrex Erlenmeyer flasks with continuous gassing with $O_2:CO_2$ (95:5) at pH 7.4 and with constant shaking (60–90 cycles/min).

The $^{45}Ca^{2+}$ uptake is started by adding 50 μl of the cell suspension

[18] J. R. Williamson and J. R. Monck, *Annu. Rev. Physiol.* **51**, 107 (1989).
[19] J. H. Exton, *Hepatology* **8**, 152, (1988).
[20] T. Capiod, J. Poggioli, and M. Claret, in "Hormonal Control of Gluconeogenesis" (N. Kraus-Friedman, ed.), p. 157. CRB Press, Boca Raton, Florida, 1986.
[21] G. M. Burgess, M. Claret, and D. H. Jenkinson, *J. Physiol.* **317**, 67 (1981).
[22] J.-P. Mauger, J. Poggioli, and M. Claret, *J. Biol. Chem.* **260**, 11635 (1985).
[23] H. Eagle, *Science* **130**, 432 (1959).

(3×10^6 cells/ml, i.e., about 6 mg dry wt/ml) to trace amounts of $^{45}CaCl_2$ ($1-3$ $\mu Ci/ml$) and to the hormone.[4] The $^{45}Ca^{2+}$ uptake is determined by taking 100-μl samples at 15, 45, 75, and 105 sec. Each of them is immediately diluted in 4 ml of an ice-cold "washing solution" containing 114 mM NaCl, 5 mM $CaCl_2$, 5 mM Tris/HCl (pH 7.4). The mixture is then filtered through a Whatman GF/C glass filter (25-mm diameter; Whatman, Clifton, NJ) and washed three times with 4 ml of the ice-cold washing solution. Care is taken that the filter does not get into contact with air before the termination of the washing procedure. We have shown[4] that in these conditions, there is no significant loss of intracellular K^+ as determined by flame spectrophotometry. This indicates that the filtration procedure does not alter the ionic permeabilities of rat hepatocytes. The radioactivity associated with the filter is then counted in a scintillation spectrometer (β-matic Kontron) after addition of 3.5 ml scintillation liquid (Hydroluma, Lumac).

The unidirectional influx of Ca^{2+} is determined from the initial rate of $^{45}Ca^{2+}$ uptake by the cells as previously described.[4] The sampling period ($0-105$ sec) is short compared with the time constant ($10-20$ min) of $^{45}Ca^{2+}$ exchange in rat liver cells[4,20] to avoid underestimation of Ca^{2+} influx. It can be calculated from the correction factor $(t/\tau)/[1-\exp(-t/\tau)]$, where τ is the time constant of Ca^{2+} exchange and t the loading time, that after 105 sec the measured influx is underestimated by $6-8\%$, which is in the range of experimental error associated with flux measurements.

Extrapolation of Ca^{2+} uptake to the ordinate gives values other than zero, indicating that a small amount of $^{45}Ca^{2+}$ is resistant to the washing solution containing 5 mM $CaCl_2$. This contaminant Ca^{2+} is also found when the solution contains 0.1 mM La^{3+} or 2 mM EGTA instead of Ca^{2+}. This fraction results in part from $^{45}Ca^{2+}$ bound to the glass fiber filter and in part from $^{45}Ca^{2+}$ bound to damaged cells. In agreement, it has been shown[4] that when the cells are incubated at $1-4°$ to block $^{45}Ca^{2+}$ exchange through the plasma membranes, the radioactivity remaining on the filter is not different from the value obtained by extrapolation to zero time in experiments at $37°$.

Determined in this way, the resting Ca^{2+} influx of rat hepatocytes incubated in 1.8 mM Ca^{2+} ranges from 150 to 300 pmol/min/mg cell dry wt.[4]

Measurement of the Unidirectional Ca^{2+} Efflux

In Intact Cells. The measurement of the unidirectional Ca^{2+} efflux requires preincubation of the cells with $^{45}Ca^{2+}$ for a period long enough to allow the intracellular Ca^{2+} to exchange. The unidirectional Ca^{2+} efflux is determined from the initial rate of the ^{45}Ca^{2+} loss from the cell in a Ca^{2+}-free Eagle's medium as described previously.[24] The isolated rat hepatocytes (3×10^6 cells/ml, i.e., about 6 mg cell dry wt/ml) are suspended in Eagle's medium containing 1 μCi of ^{45}Ca^{2+}/ml until isotopic equilibrium (about 60 min). The ^{45}Ca^{2+} efflux is initiated by adding EGTA (2 mM, pH 7.4) to 1-ml samples to decrease the free external Ca^{2+} from 1.8 mM to 2 μM. Cell samples (200 μl) are removed 30, 60, and 90 sec after adding EGTA, diluted, filtered through Whatman GF/C glass filter (25-mm diameter), washed, and counted as indicated for Ca^{2+} influx experiments.

The rate constant k_{Ca} (min^{-1}) of the unidirectional efflux may be calculated from the relation: $k_{Ca} = 2([Ca^{2+}]_{t_1} - [Ca^{2+}]_{t_2})/[(t_2 - t_1)$ $([Ca^{2+}]_{t_1} + [Ca^{2+}]_{t_2})]$ where $[Ca^{2+}]_{t_1}$ and $[Ca^{2+}]_{t_2}$ are the ^{45}Ca^{2+} content of the cells (nmol/mg cell dry wt) measured at times t_1 and t_2, respectively. The efflux of Ca^{2+}, m_{oCa} (nmol/min/mg cell dry wt), is then equal to

$$m_{oCa} = k[Ca^{2+}]_{t_o}$$

where $[Ca^{2+}]_{t_o}$ is the cell ^{45}Ca^{2+} content at the end of the loading period (isotopic equilibrium). Estimated in this way, the unidirectional efflux from rat hepatocytes is between 130 to 250 pmol/min/mg cell dry wt.[24]

In Inside-Out Vesicles of Plasma Membranes. The efflux of Ca^{2+} measured from intact hepatocytes does not allow estimation of the active transport dependency on the cytosolic factors which control its activity ($[Ca^{2+}]_i$, $[ATP]_i$, $[Mg^{2+}]_i$). An alternative method is to determine the ATP-dependent uptake of Ca^{2+} in inside-out vesicles of plasma membranes.[25,26] Under certain fractionation conditions, plasma membranes undergo spontaneous inside-out vesicularization so that the cytoplasmic surface of the membrane faces the suspending medium and the Ca^{2+} is transported from the external medium into the vesicle.[25] It is then possible to fix the concentrations of Ca^{2+}, ATP, Mg^{2+}, and of all the ligands which control the carrier. Inside-out vesicles of liver plasma membranes are prepared from female Wistar rate according to the method described by Prpic *et al.*[25] After the last centrifugation, the membrane pellet is resuspended in a medium containing 250 mM sucrose, 25 mM HEPES/KOH (pH 7.4) at 1–2 mg protein/ml.

The ^{45}Ca^{2+} uptake is measured according to a modification of the method of Chan and Junger.[26] The membranes (50–100 μg of proteins in 50 μl) are added to 450 μl of a medium containing 100 mM KCl, 20 mM NaCl, 0.96 mM NaH$_2$PO$_4$, 5 mM MgCl$_2$, 25 mM HEPES/KOH (pH 7.15 at 37°), 20 μM CaCl$_2$, 1 μCi ^{45}Ca^{2+}/ml. The medium is supplemented by 5

[24] L. Combettes, B. Berthon, A. Binet, and M. Claret, *Biochem. J.* **237**, 675 (1986).
[25] V. Prpic, K. C. Green, P. F. Blackmore, and J. H. Exton, *J. Biol. Chem.* **259**, 1382 (1984).
[26] K. M. Chan and K. D. Junger, *J. Biol. Chem.* **258**, 4404 (1983).

mM NaN$_3$ to prevent any Ca^{2+} uptake by contaminating mitochondria. Creatine phosphate (5 mM) and creatine phosphokinase (5 U/ml) are added to keep the concentration of ATP constant. The mixture is preincubated 5 min at 37°, then 1.5 mM ATP (disodium salt) is added to start the reaction. At the appropriate times, the reaction mixture is diluted with 4 ml of an ice-cold washing solution containing 250 mM sucrose and 40 mM NaCl, filtered through a Whatman GF/C glass fiber filter (25-mm diameter), and washed three times with 4 ml of ice-cold washing solution. The radioactivity retained on the filter is then counted as described for Ca^{2+} influx experiments. The ATP-dependent ^{45}Ca^{2+} uptake is determined after correction for the radioactivity associated with the filter after incubation of plasma membrane vesicles without ATP. The Ca^{2+} ionophore A23187 (10 μM) solubilized in dimethyl sulfoxide (DMSO)/ethanol solution (1/1, v/v) can be used to release the Ca^{2+} which has been accumulated in the vesicular compartment.

The membrane preparation of inside-out vesicles described here is essentially free of mitochondria.[25] However, it may be contaminated by vesicles of endoplasmic reticulum which likewise accumulate Ca^{2+} in the presence of ATP. Routinely, we use saponin in order to distinguish these two cell compartments. Saponins are steroid glycosides which have been largely used to selectively permeabilize the plasma membrane of a wide variety of cells without affecting the structure and function of the mitochondria and of the endoplasmic reticulum (for reference, see Fiskum[27]) The plasma membrane vesicles (150 μg protein/ml) are incubated as described before. The ATP-dependent ^{45}Ca^{2+} uptake is determined 10 min after addition of increasing concentrations of saponin (Sigma, St. Louis, MO). The results show that the ATP-dependent accumulation of ^{45}Ca^{2+} by vesicles is reduced to 50% by addition of 4 μg/ml of saponin (about 30 μg/mg of membrane proteins) and to less than 10% with 15 μg/ml of saponin (100 μg/mg of proteins). If the experiment is carried out with a preparation of purified microsomes,[28] the ^{45}Ca^{2+} uptake measured in the same experimental conditions as those applied to plasma membrane vesicles is not affected by 20 μg/ml of saponin (140 μg/ml of proteins).[29]

[27] G. Fiskum, *Cell Calcium* **6**, 25 (1985).
[28] A. P. Dawson and R. F. Irvine, *Biochem. Biophys. Res. Commun.* **120**, 858 (1984).
[29] C. Dargemont, M. Hilly, M. Claret, and J.-P. Mauger, *Biochem. J.* **256**, 117, (1988).

[32] Preparation and Specific Applications of Isolated Hepatocyte Couplets

By J. L. BOYER, J. M. PHILLIPS, and J. GRAF

A number of different methods are available for isolating hepatocytes in suspension either for short-term experiments or as preliminary steps in the preparation of hepatocyte monolayer culture systems.[1-7] Although these cell suspensions consist predominantly of single cells, pairs (couplets) of hepatocytes as well as triplets and clusters of more than two or three cells also may be present.[8-10] In contrast to the isolated hepatocyte, which must reestablish its structural polarity only after reaching confluence in monolayer culture systems, a unique feature of the hepatocyte aggregates is their retention of cell polarity during the process of isolation and the ability of selected couplets to express this polarity in short-term tissue culture by forming an enclosed canalicular space or vacuole between the two adjacent cells into which secretion is elaborated.[8-10] Hepatocyte couplets, which are plated at lower cell density then in monolayer preparations, represent a primary secretory unit that enables direct observations of the process of canalicular bile formation at its site of origin. This cell system thus permits the application of a variety of techniques and methodologies available for studying cell function in tissue culture, including phase and Nomarski microscopy and cinephotomicrogaphy as well as fluorescent and electrophysiologic applications.[8-11]

Methods of Isolation

Several different approaches to the isolation of hepatocyte couplets have been used as represented by the authors' laboratories. (Toronto,

[1] S. T. Jacob and P. M. Bhargava, *Exp. Cell Res.* **27**, 453 (1962).
[2] M. N. Berry and D. S. Friend, *J. Cell Biol.* **43**, 506 (1969).
[3] G. Schreiber and M. Schreiber, *Subcell. Biochem.* **2**, 321 (1973).
[4] P. O. Seglen, *Methods Cell Biol.* **8**, 39 (1976).
[5] N. L. Leffert, K. S. Koch, T. Moran, and M. Williams, this series, Vol. 58, p. 536.
[6] C. Guguen-Guillouzo, M. Bourel, and A. Gouillouzo, *Prog. Liver Dis.* **18**, 33 (1986).
[7] B. L. Blitzer, S. L. Ratoosh, C. B. Donovan, and J. L. Boyer, *Am. J. Physiol.* **243**, G48 (1982).
[8] C. Oshio and M. J. Phillips, *Science* **212**, 1041 (1981).
[9] M. J. Phillips, C. Oshio, M. Miyairi, H. Katz, C. R. Smith, *Hepatology* **2**, 763 (1982).
[10] J. Graf, A. Gautam, and J. L. Boyer, *Proc. Natl. Acad. Sci. U.S.A.* **81**, 6516 (1984).
[11] J. Graf, R. M. Henderson, B. Krumpholz, and J. L. Boyer, *J. Membr. Biol.* **95**, 241 (1987).

TABLE I
PREPARATION OF HEPATOCYTE COUPLETS

Steps	Toronto	New Haven	Vienna
Animals	Female Wistar rat, 170–230 g	Male Sprague-Dawley rat, 160–220 g	Male Sprague-Dawley rat, OFA, 150–200 g
Anesthesia	Pentobarbital, 50 mg/kg body wt	Pentobarbital, 50 mg/kg, ip	Thiopental, 80–100 mg/kg, ip
Anticoagulation	—	Heparin, 3.5 IU/g body wt, iv (inferior vena cava)	Heparin, 1000 IU into spleen
Operation			
Cannulation of portal vein	—	16-Gauge cannula	16-Gauge cannula; Venflon
Venous drainage	Subhepatic vena cava severed	Subrenal vena cava severed	Thorax and right atrium opened
Washout perfusion			
Composition	Ca^{2+}- and Mg^{2+}-free Hanks' solution, 0.5 mM EGTA added, prewarmed to 37° and gassed with 100% O_2	0.5 mM EGTA and Ca^{2+}- and Mg^{2+}-free Hanks' solution, 25 mM NaHCO$_3$ added, prewarmed to 37° and equilibrated with 95/5% O_2/CO_2	Ca^{2+}- and Mg^{2+}-free Hanks' solution, 25 mM NaHCO$_3$ added, prewarmed to 39° and equilibrated with 95/5% O_2/CO_2
Flow rate and duration	40 ml/min; 4 min	38 ml/min; 10 min	30 ml/min; 10 min
Collagenase perfusion			
Composition	0.05% collagenase (Sigma type I^a or IV) in L-15 tissue culture medium (GIBCO), gassed with 100% O_2	0.05% collagenase (Boehringer Type A or Sigma type I) in Hanks' solution (+ Ca^{2+}, Mg^{2+}), 25 mM NaHCO$_3$ added, prewarmed and gas equilibrated as above (0.8 units trypsin inhibitor; Sigma) per unit trypsin activity in collagenase batch	Collagenase (Sigma type I) 5° mg in 20 ml Hanks' solution (+ Ca^{2+}, $^{Mg2+}$) + 25 mM NaHCO$_3$ (95% O_2/ 5% CO_2)
Flow rate, duration	25 ml/min; 8 min (60-W heating lamp 10 cm above liver)	38 ml/min; 10 min	10-ml bolus injection, flow then interrupted for 1 min before excision

502

Further collagenase treatment (after liver excision):	None	Place each lobe in 50 ml cold L-15, rinse twice, shred capsule with shallow scissor cuts, use spatula or comb to tease off hepatocytes	2–3 g liver placed in 10 ml collagenase solution, minced with scissors, suspension placed in a shaking water bath at 37°, 60 rpm, under 95/5% O_2/CO_2 for 5 min
Harvesting	Capsule opened, cells suspended into large culture dish by "combing" with forceps and filtering through 60-μM nylon mesh (Tetko). Centrifuge at 80, 45, 20, and 50, in GLC-1, rotor type HL-4 (4 min) in L-15 medium, discarding supernatant. Suspend sediment in 15 ml L-15 medium. Yield is $10-20 \times 10^6$ cells/ml	Filter suspension through 80 and 45 μm Nylon mesh (Tetko). Bring to 100 ml in cold L-15, divide and centrifuge in 50-ml tubes at 500 rpm ($\times 50\ g$) for 2 min; resuspend in 25-ml cold L-15 and centrifuge again. Repeat and resuspend each pellet in 15 ml cold L-15 to obtain a suspension of $5-10 \times 10^6$ cells/ml	10-ml suspension is diluted with 90 ml Hanks' buffer ($+25$ mM NaHCO$_3$, 95/5% O_2/CO_2) and filtered through 60-μm nylon mesh into two tubes with conical bottom; sedimentation proceeds for 10 min, sediment is taken up into 2×50 ml buffer, and sedimented again for 10 min. Sedimented cells are taken up into 2×2 ml L-15 culture medium, giving $\sim 2 \times 10^6$ cells/ml
Viability (Trypan Blue exclusion)	90–98%	81–92%	80–93%
Percentage cells as couplets	5–30%, variable	Up to 30%	10–20%
Incubation media	L-15 (replenished at 2–3 hr) + 10% fetal bovine serum, 10 mM HEPES, Penicillin (100 U/ml), Streptomycin (100 μg/ml)	L-15 or MEM; penicillin (50 U/ml) Streptomycin (50 μg/ml)	L-15
Time of study	Up to 18 hr	4–8 hr	4–8 hr
Characteristics of preparation	Flattened cells, prolonged observation, triplets and clusters also present	High yield of couplets; spherical shape	Low yield, cheap, fast

^{a} Preferred.

Canada, New Haven, Connecticut, and Vienna, Austria). Details of these approaches are provided in Table I for comparative purposes. Each approach uses a modification of the methods of Seglen[4] for the initial isolation procedure.[8-10] Distinctive features of the Toronto group include the use of EGTA in the initial Ca^{2+} and Mg^{2+} free perfusion and in earlier studies the use of type IV rather than type I collagenase mixed with L-15 medium rather than Hank's solution. The L-15 incubation medium also contains 10% fetal calf serum and is replenished at 2–3 hr in culture. With this preparation, cell viability at 18 hr was 90–95%.[9] Both the New Haven and Vienna preparations avoided the use of calf serum in initial studies since its use results in rapid flattening of the cells in culture, making them difficult to micropuncture and to optically section the canalicular space for measurements of canalicular volumes.[10-12] However, it is also possible to maintain the couplets on a biomatrix containing either collagen or Matrigel and in this condition, fetal calf or horse sera and other growth factors may be added without leading to flattening of the cells. An enrichment of couplets up to 30% can be achieved (New Haven method) by filtering the cell suspensions through 45-μm nylon mesh (Tetko, Inc., Elmsford, NY).[12,13] The Vienna method shortens the procedure and diminishes the cost at the expense of a low yield of cells by injecting 10 ml of the collagenase-containing medium into the portal vein and liver. Cell viability by Trypan Blue exclusion is usually >85% with all of these techniques under optimal conditions. Variation in the quality of lots of collagenase continues to be the primary source of difficulty with this procedure, leading to poor viability. The addition of trypsin inhibitors (0.8 U/U trypsin activity) may improve cell function depending on the type and lot of collagenase. Both biomatrix and serum growth factors appear to prolong the viability of hepatocyte couplets in these preparations and may be used depending on the application and the needs for defined rather than complete media.

Optical Systems. After isolation, the couplets are plated onto glass coverslips in 35 × 10 mm plastic dishes at cell concentrations of 0.75 × 10^5/ml to 10^6/ml of L-15 medium and are maintained in the appropriate incubation medium at 37° until placed in a perfusion chamber or modified culture dish for optical measurements. The bottom of the perfusion chamber or culture dish is replaced by a thin plastic or glass coverslip to achieve high optical resolution. Zeiss IM 35 inverted microscopes (Zeiss, Oberkochen, FRG) are used with several different optical systems depending on the experimental strategy. Phase-contrast optics have been used by

[12] A. Gautam, O. C. Ng, and J. L. Boyer, *Hepatology* 7, 216.
[13] J. Boyer, O. C. Ng, and A. Gautam, *Trans Am. Assoc. Physiol.* **98**, 21 (1985).

the Toronto group for time-lapse cinephotomicrography whereas double-differential interference contrast (Nomarski) optics have been used by the New Haven and Vienna investigators since it provides a narrow depth of focus, allowing both precise localization of electrode tips (along the vertical axis) for electrophysiologic studies and sharper focusing of 0.5- to 1-μm optical slices through the canalicular lumen for measurements of canalicular space volume. A Neofluar 63/1.26 oil objective and ×10 eyepieces were also used in the latter studies, as well as a condensor that provides space above the microscope stage for manipulation of microelectrodes. When incident light fluorescence microscopy was needed, a primary filter with 450 to 490-nm window and a secondary filter with a cut-off of 515 nm were used to detect fluorescein. Experimental chambers for direct visualization of the cell couplets were built from Perspex.

Functional Characteristics of Isolated Rat Hepatocyte Couplets

Immediately after isolation, the canalicular spaces between adjacent hepatocytes are collapsed and less than 1 μm in diameter as observed in the intact liver. However, within a few hours the spaces are variably expanded in at least 50% of the couplets examined in preparations with good cell viability. By 4–8 hr, modest expansion (up to 3 μm) in the maximal diameter can be observed in most couplets whereas greater than 3-μm expansion can be observed in ~10%[12,13] (Fig. 1). These canalicular spaces appear to be sealed by tight junction strands that resemble the junctional structure observed by freeze-fracture techniques in the intact liver. Gap junctional particles can also be observed by freeze-fracture techniques at the joining membrane areas.

Electron microscopy, performed in cells adhered to glass coverslips after 4–6 hr reveal hepatocytes with essentially normal ultrastructure although autophagic vacuoles are usually increased in number. An expanded canalicular lumen lined by microvilli is often observed. In some instances (15 of 55 couplets examined), microvilli were limited to the regions adjacent to the tight junction seal, and the pericanalicular ectoplasm was thickened, findings identical to cholestatic morphology in intact liver. Mitochondria, smooth and rough endoplasmic reticulum, and Golgi apparatus usually retain normal ultrastructural appearances.[12,13]

Hepatocyte couplets also retain structural polarity with respect to the organization of their cytoskeletal elements.[16,13a] Fluorescent antibody staining demonstrates aggregates of actin and tubulin filaments in increased amounts in the pericanalicular region in couplets placed in short

[13a] I. Nickola and M. Frimmer, *Cell Tissue Res.* **243**, 437 (1986).

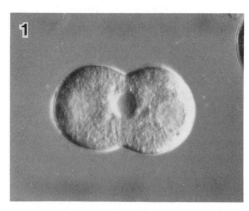

FIG. 1. An isolated rat hepatocyte couplet demonstrating an expanded canalicular vacuole between the two adjacent hepatocytes. Time in culture, 5 hr; Nomarski optics. (×850)

term (4–8 hr) cultures. Similar studies indicate that the ZO-1 tight junction associated protein localizes to the perimeter of the newly developed canalicular space.

The permeability of the canalicular lumen can be assessed by exposing couplets to the electron-dense dye, Ruthenium Red, an ionic 551-Da material that coats surface membranes. Ruthenium Red freely penetrates the canalicular space of 100% of the couplets immediately after their isolation whereas fewer than 10% of canalicular spaces are permeable to Ruthenium Red when expanded and observed 4–8 hr after isolation, suggesting that the spaces reseal and remain relatively impermeable to larger organic solutes as they expand with secretion. The expanded luminal spaces of 10% of couplets are permeable to horseradish peroxidase.[13b] Presumably if couplets do not expand their canalicular lumen they fail to reseal, as reflected by continued penetration of their lumens with Ruthenium Red with time in culture.[12,13]

Couplets that have resealed the canalicular space undergo a cycle of periodic contractions and/or collapse. There is debate as to the precise nature of these reductions in canalicular volume but according to time-lapse cinephotomicrographic studies from the Toronto group, periodic contractions occur at irregular intervals, averaging between 5 and 6 min, during experiments over 16 hr when cells are cultured in L-15 medium and calf serum.[9] In contrast, when observed directly by the New Haven and Vienna group over shorter time intervals in the absence of serum, these spaces appear to collapse only after expanding with secretion.[10,12] The

[13b] M. H. Nathanson, A. Gautam, and J. L. Boyer, *Gastroenterology* **98**, A615 (1990).

former phenomenon appears to be regulated by contractions in the pericanalicular cytoskeleton whereas the latter phenomenon seems related to the development of leaks in the junctional barrier brought about by increases in secretion pressure in the confined canalicular space.[14,14a]

Hepatocyte couplets in culture retain the ability to transport organic anions across both sinusoidal (facing the bathing medium) and canalicular (apical) membranes, as illustrated by their capacity to take up fluorescein, fluorescein-diacetate, or 2'-7'-bis (carboxyethyl)-5,6-carboxyfluorescein (BCECF) from the medium, to hydrolyze their ester bonds, and to excrete the fluorescent product into the bile canalicular lumen.[10,14b] Similarly, bile acids when added to the bathing medium stimulate secretion and expansion of the canalicular space.[14] In addition, both the frequency of contraction as well as the rapidity of collapse of these spaces are stimulated by the addition of bile acids to the medium.[9,14] The canalicular spaces are also sensitive to hypertonic osmotic gradients imposed by additions of sucrose to the medium.[1] Canalicular space volume rapidly diminishes within 1 min after additions of sucrose and to a greater degree than parallel reductions in cell volume, indicating that the junctions sealing the space are more permeant to smaller solutes than the cell membrane, a finding supported by electrophysiologic studies.[11] Thus both the structural properties of the bile canaliculus in the isolated couplets, as well as their capacity to transport bile acids and fluorescent organic anions and to respond to hypertonic media, are remarkably similar to the intact animal or isolated perfused liver.[10,14c,14d]

Canalicular Membrane Rearrangement. While the couplet maintains its apical polarity between the two attached hepatocytes during isolation, the remaining canalicular poles that are now no longer attached to their neighboring cells are reorganized. During the first 4–6 hr in culture, the couplet undergoes a process of apical membrane rearrangement increasing the surface area of the canalicular space by at least 4-fold.[14] This phenomenon can be easily demonstrated histochemically utilizing markers of canalicular membrane enzymes, such as Mg^{2+}-ATPase. Using the Wachstein–Meisel method,[10] reaction product for Mg^{2+}-ATPase can be observed immediately after isolation, encircling each cell in a beltlike configuration characteristic of the location of the canalicular membrane domain in the

---

[14] A. Gautam, O. C. Ng, M. Strazzabosco and J. L. Boyer, *J. Clin. Invest.* **83,** 565 (1989).
[14a] J. L. Boyer, *Hepatology* **7,** 190 (1987).
[14b] S. A. Weinman, H. Thom, U. Schramm, H.-P. Buscher, G. Kurz, and J. L. Boyer, *in* "Trends in Bile Acid Research" (G. Paumgartner, A. Stiehl, and W. Gerok, Eds.), pp. 123–131. Kluwer Academic Publisher, Dordrecht, 1989.
[14c] J. L. Boyer, A. Gautam, and J. Graf, *Semin. Liver Disease* **8,** 308 (1988).
[14d] J. Graf and J. L. Boyer, *J. Hepatol.* **10,** 387 (1990).

intact liver. However, by 4 hr in culture, this reaction product is now localized predominantly to the small portion of the cell surface membrane that represents the remaining apical canalicular membrane between the two adjacent hepatocytes. With time in culture, this remnant staining appears to increase as the canalicular membrane surface area enlarges, a process that is dependent on microfilaments since it can be blocked by cytochalasin D.[12] Histochemical staining for the Golgi enzyme, thiamin pyrophosphatase, also demonstrates reaction product accumulating in the remaining pericanalicular region between the two hepatocytes with time in culture. In contrast, basolateral membrane proteins such as the α subunit of Na^+,K^+-ATPase appear to be localized by immunocytochemical techniques only to the basolateral membrane domain in couplets where the canalicular space is expanded.[14e] Presumably the apical domain is sealed by the tight junctional elements in these cells. Thus this isolated hepatocyte couplet cell culture system maintains polarity of several membrane domain proteins as the apical membrane expands and should be a particularly useful model for studies of mechanisms of membrane polarity such as sorting mechanisms for basolateral and apical membrane proteins.

Other Applications

Quantitative Assessment of Canalicular Bile Secretion. A unique application of this secretory unit is the ability to assess agents that both stimulate and inhibit canalicular bile formation without the confounding influences of blood flow, or the secretory properties of the bile ductules and bile ducts. While this assessment may be made qualitatively by directly visualizing expansions of the space in response to administration of choleretics, quantitative measurements can also be obtained utilizing optical sectioning techniques and reconstructing canalicular volumes over time with the aid of a Dage Series 68 Video Camera (Dage-MTI Inc., Michigan City, IN), video screen, and Zeiss videoplan imaging processing equipment. By transferring 1-μm step images to a video tape, a series of optical sections can be made so that volume measurements of the same bile canaliculus may be obtained every 3–5 min. These images can then be transferred by a video screen at a later date and the area of the perimeter of the canalicular lumen traced with a digitizer and the area of the section computed by image processing equipment, e.g., volume = Σ (areas × 1 μm). Estimates of rate of secretion can then be obtained by calculating the change in volume/unit of time as estimated from the slope of the first two or three

---

[14e] E. S. Sztul, D. Biemesforfer, M. J. Caplan, M. Kashgarian, and J. L. Boyer, *J. Cell Biol.* **104**, 1234 (1987).

sequential measurements.[14] The technique is similar to optical sectioning procedures used by Spring and co-workers in volume measurements of intercellular spaces of epithelia such as the gall bladder.[15]

Application of these techniques have led to the demonstration that ursodeoxycholate and taurocholate have equal choleretic properties at the level of the hepatocyte,[14] that cyclic nucleotides stimulate canalicular secretion,[15a] and that vasopressin and phorbol esters have cholestatic effects at the level of the hepatocyte.[13b]

Video microscopy can also be used to obtain continuous on-line recordings of the maximal cross sectional area of both the canalicular lumen and cells. This procedure is useful for a continuous estimate of cell and canalicular volume and an assessment of their response to stimuli that may affect these parameters.[15b]

Studies of Transcytotic Vesicle Transport. Other studies indicate that the transcytotic vesicle pathway is maintained in the isolated couplet and transverses the hepatocyte from the sinusoidal membrane to the apical bile canalicular membrane by a process dependent on the function of microtubules with a time frame analogous to the intact liver.[16] Thus the isolated hepatocyte couplet system can be utilized to study the process of transcytosis of vesicles in a cell culture system. Couplets adhered to coverslips are exposed for 3 min to HRP and then washed in HRP-free medium and processed at timed intervals for electron microscopy after development of the reaction products as previously described. Quantitative assessment of this tubulovesicle system labeled by HRP can also be obtained by morphometric analysis of the structures containing reaction product in the electron micrographs.[16]

Assessment of Bile Canalicular Motility (Contractions)

Time-lapse cinephotomicrography is performed by the Toronto group with an inverted Zeiss microscope (ICM 405) equipped with phase-contrast optics, a 16-mm movie camera (H 16, RX-5 Bolex), 16-mm reversal films (Plus X 7276, Kodak), and a motor drive system with a time-lapse controller (Hommel Electronics, Toronto, Canada). This equipment is housed in a temperature-controlled room maintained at 38°. Movies are taken at the speed of one frame/15 sec. At every multiple of 100 frames of each movie, a 35-mm still photomicrograph is obtained simultaneously.

[15] B. E. Persson and K. R. Spring, *J. Gen. Physiol.* **79,** 481 (1982).

[15a] M. H. Nathanson and A. Gautam, *Gastroenterology* **94,** A619 (1988).

[15b] J. Graf, L. Schild, J. L. Boyer, and G. Giebisch, *Renal Physiol. Biochem.* **13,** 166 (1990).

[16] S. Sakisaka, O. C. Ng, and J. L. Boyer, *Gastroenterology* **95,** 793 (1988).

The time-lapse movies are started 3–5 hr after cell isolation and 1 hr after changing the medium. Groups of hepatocytes where dilated bile canaliculi are observed are selected for the cine-study. Hepatocytes used for analysis should be viable and show ongoing motility activities at the end of a 16- to 18-hr experimental period.

Thirty-eight hundred movie frames (i.e., 15.8 hr of real time) are examined form a single bile canaliculus to assess the diameter of the canalicular lumen, using an analytic movie projector (photoptical data analyzer 2240A, MK V, L-W International, CA). The length of time required for canalicular contraction is also measured. The minimal diameter of the canaliculus is taken as the point of maximal contraction, at which time the canalicular lumen is closed. The frame numbers between contractions (one/15 sec) obtained from each canaliculi studied are stored and analyzed by a computer (Apple II; Apple Computers, Cupertino, CA). To examine the frequency distribution of contraction intervals, a histogram is then plotted with the number of contractions that occur during 1-min time intervals for each bile canaliculus studied during the 15.8-hr period of cinephotomicrography. Because all bile canaliculi are studied using identical experimental conditions, the graphs of each individual canaliculi can be aggregated with the resultant interval histogram representing the overall contractile activity of normal bile canaliculi. The resultant interval histogram can then be compared to the distribution patterns which are found in time series processes and, in particular, to the Poisson distribution.[9]

Analysis of bile canalicular motility in time-lapse movies reveals several types of cytoplasmic motile activity. Some of these movements are of random Brownian-like nature, while other prominent vacuolar movements are present, especially in the pericanalicular regions. Cell spreading is also noted. Bile canalicular motility interpreted as periods of regular contractile activity are striking,[8,9] and tend to occur following a flurry of pericanalicular vacuolar movements. Canalicular closure appears to be forceful and is frequently accompanied by expulsion of material out of the canalicular lumen. Obviously dilated canaliculi are usually selected at the beginning of the experiments so that alterations in the size of the canalicular lumen can be easily visualized. The mean number of observed canalicular contractions was 34.8 ± 18.7 (mean ± SD) during an experimental period of 15.8 hr. Measured canalicular systole of 239 contractions in 5 bile canaliculi was 60 ± 30 sec (mean ± SD).

The intervals between contractions vary from 1 to 810.25 min. The distribution is skewed and characteristic of a time-series process with a typical Poisson distribution. The most frequent time interval between contractions is 5 min, 30 sec.[9] Both partial and complete contractions are included in this analysis.

Microinjection Techniques

Intracellular Microinjections in Hepatocyte Couplets and Triplets

These injections can be performed according to modified methods of Stockem et al.[17] and Graessmann and Graessmann.[18] Micropipets are prepared from the glass capillaries (outer diameter, 2 mm) (W-P Instruments, New Haven, CT), using a micropipet puller (PG-1, Narishige Scientific Instruments, Tokyo, Japan). The diameter of the tip of the micropipet is 0.5–1.0 μm. After the capillary is pulled, the solution is delivered through its rear open end by means of a 20-gauge Angiocath (Deseret, Sandy, UT). Microinjections are performed using a micromanipulator (MP-1, Narishige) and a microinjector (TN-44, Narishige) on the stage of an inverted microscope (Nikon, Tokyo, Japan). The temperature is maintained at 37° either by superfusion of preheated medium or by using a plastic housing with an incubator (NP-2, Nikon). The dose of injected solution is approximately 10% of the volume of the recipient cell as per the method of Stockem et al.[17] Experiments are rejected if the microinjection technique causes cell damage, such as bleb formation. Successful microinjection can usually be performed 40–100% of the time.[19]

Intracanalicular Micropuncture and Injection

Another unique feature of the isolated couplets is the ability to pass micropipets into the canalicular lumen. Micromanipulators must be mounted so that the microelectrode shaft forms a 30° angle to the plane of the coverslip. The electrode tip is then placed in the center of the field utilizing Nomarski optics, ~50 μm above the coverslip. Dilated canalicular spaces are needed to successfully puncture the space and are placed a few micrometers from the center of the field with each cell oriented to the left and right. A focal plane is chosen that transects the middle of the canalicular lumen. The electrode tip is then moved downward to the same focal plane as the cells but precisely in the center of the field so that the tip is adjacent to the canalicular vacuole, both being sharply in focus. The electrode is then withdrawn, the cells are moved to the exact center of the field so that the canalicular space now occupies the position where the tip of the electrode resided before it was withdrawn. Under direct vision, the electrode is then advanced again by axial movement and inserted into the canalicular space after first traversing the cell cytoplasm as described in

---

[17] W. Stockem, K. Weber, and J. Wehland, *Cytobiologie* **18**, 114 (1978).
[18] M. Graessmann and A. Graessmann, *Proc. Natl. Acad. Sci. U.S.A.* **73**, 366 (1976).
[19] S. Watanabe and M. J. Phillips, *Proc. Natl. Acad. Sci. U.S.A.* **81**, 6164 (1984).

detail elsewhere.[10] If the micropipet is utilized as a microelectrode, then electrical potentials can be recorded from the cell and canalicular lumen during the passage of the electrode.[10] If the micropipet is filled with specific solutions it may be used for microinjections into the canalicular space.

Applications of microinjection techniques in this cell system include the injection of calcium solutions into the cell to assess the effects of this cation on canalicular contractions with and without exposure of the cells to the inhibitors, trifluoperazine and cytochalasin B (Fig. 2). In addition, calcium has been injected in an outer cell of a cell triplet and stimulation of canalicular contractions has been observed in the parent as well as adjacent bile canaliculus. These findings suggest that Ca^{2+} may induce sequential contraction of the two canaliculi, supporting the contention that canalicular motility may be a coordinated purposeful function in liver cells.[20]

To determine that the microinjections are localized in the cell, fluorescein can be loaded into the micropipets and visualized within the cell by fluorescent microscopy after the microinjection. Intracanalicular location of micropipets can also be confirmed by loading the pipet with fluorescein but in this case the fluorescein anion is injected into the canalicular lumen by passing electric current through the pipet (iontophoresis) (Fig. 3).[10]

Fluorescence Microscopy

Cellular uptake and lumenal secretion of fluorescein and other fluorescent anions has been observed by first exposing the cells to medium containing the dyes and then studying secretion in dye-free medium. Fluorescein is readily taken up by the cells through the anion transport system. During washout cellular stain disappears and dye accumulates in the canalicular lumen. Lumenal stain is generally stable until the canaliculus contracts or collapses. Fluorescein diacetate secretion follows the same pattern but uptake may be by diffusion alone. After hydrolysis by an intracellular esterase, the dye becomes fluorescent and is trapped within the cell. Other derivatives have also been used (carboxyfluorescein diacetate, dimethyl-carboxyfluorescein diacetate and BCECF). These compounds reveal pH-dependent spectral shifts and can also be used to measure intracellular pH.

Dye injection has been used to study cell coupling through intercellular communications by microinjection of Lucifer Yellow.[20] Lumenal injection of fluorescein by iontophoresis has been used as described to assess intralumenal microelectrode tip localization.[10]

Fluorescent bile acid derivatives can also be used to study bile acid secretion in this cell system and have been used to demonstrate the role of

---

[20] S. Watanabe, C. R. Smith and M. J. Phillips, *Lab. Invest.* **53**, 275 (1985).

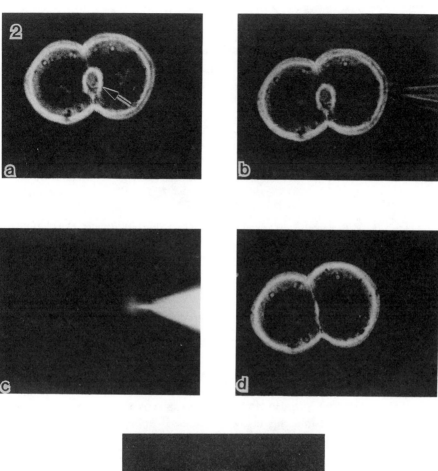

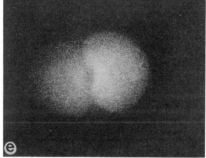

Fig. 2. Normal hepatocytes (no pretreatment). (×440) (a) Open lumen of the bile canali-culus (arrow) can be seen between two (coupled) hepatocytes. Phase-contrast micrograph. (b) The micropipet, which contains the standard calcium solution, is penetrating the hepatocyte on the right. (c) The solution, which contains sodium fluorescein, has been microinjected into the hepatocyte. Note that only the target hepatocyte is visualized. Fluorescence micrograph. (d) Hepatocyte 10 min after the injection of the standard calcium solution. The lumen of the bile canaliculus has closed (i.e., canalicular contraction has occurred). (e) Fluorescent dye has spread into the neighboring cell. Fluorescent intensity of the recipient cell is stronger than intensity of the neighboring cell. [Reproduced with permission from S. Watanabe and M. J. Phillips, *Proc. Natl. Acad. Sci. U.S.A.* **81,** 6164 (1984).]

FIG. 3. Micropuncture of the canalicular space in a liver cell couplet. Fluorescein is injected into the canalicular space by iontophoresis through a micropipet inserted from the upper right (fluorescence microscopy, ×660). [Reproduced with permission from J. Graf, A. Gautam, and J. L. Boyer, *Proc. Natl. Acad. Sci. U.S.A.* **81**, 6516 (1984).]

the intracellular potential in the translocation of bile acids into the canalicular lumen.[14b]

Other fluorescent compounds or fluorescent labelled macromolecules permit additional applications, including intracellular and intracanalicular pH measurements, regulation of internal calcium with Ca^{2+}-sensitive dyes[13b]; pinocytotic-vesicular transport by fluorescent dextrans, and movement of membrane lipids with fluorescent labeled lipids.[20a,20b] The hepatocyte couplet model has also been used to examine the excretion of lipid particles into the canaliculi using a modified quasilight scattering technique.[20c]

Electrophysiology

The formation of canalicular vacuoles which are wide enough to be punctured with microelectrodes has provided the unique opportunity to apply methods of epithelial electrophysiology to the isolated hepatocyte couplet as they have been carried out in transporting epithelia such as the kidney tubule or intestine. Specific applications include (1) determination of transepithelial potential profile,[10] (2) determination of the resistive elements of ionic current flow (basolateral and lumenal membrane, paracellular pathway),[11] (3) determination of partial ionic conductances of the basolateral cell membrane,[11] (4) analysis of conductive properties of inter-

[20a] J. M. Crawford and J. L. Gollan, *Hepatology* **8**, A1260 (1988).
[20b] G. F. Bonner and A. Reuben, *Hepatology* **10**, A599 (1989).
[20c] G. M. Mockel, S. Gorti, T. Tanaka, and M. Carey, *Hepatology* **8**, A1367 (1988).

cellular communications,[11,21] (5) measurements of intracellular[11] and intralumenal ion activities, (6) pH_i regulation by Na^+/H^+ exchange,[22] and (7) analysis of Na^+, K^+-ATPase activity.[11]

For electrophysiological studies aliquots of the final cell suspension $(2-3 \times 10^5$ cells) are placed into 10×35 mm culture dishes containing 3 ml Liebovitz L-15 culture medium and cells are allowed to settle onto coverslips. Couplets are studied after 4–8 hr when large bile canalicular spaces ($>5 \mu m$) have developed. Measurements of potential, resistances, and ion activities are made with standard electrophysiological procedures,[22–24] but fine-tip electrodes should be used in order to minimize artifacts produced by leakage of electrolytes from electrode tips or by membrane damage during impalements, both being particularly critical when puncturing relatively small cells and the smaller canalicular lumen.[10] Patch electrodes may also be used to study cell coupling[21] in this cell system. The transepithelial potential profile was obtained by advancing this microelectrode at an angle of 30° through one cell of the couplet into the lumen. Use of Nomarski optics was essential because the shallow depth of the focal plane permitted precise positioning of the microelectrode tip with reference to the lumenal space in all three dimensions. Intracellular potential readings (-30 to -40 mV) are in agreement with measurements in intact liver and for the first time the canalicular lumenal potential (-5 mV) could be directly determined. The magnitude of this potential is in accordance with the earlier assumption that large impermeant organic anions in bile give rise to a Donnan equilibrium of inorganic electrolytes across the paracellular pathway.[10,11]

Membrane resistances can be measured with two microelectrodes, one being inserted into each cell or into one cell and the lumen. Current pulses are delivered through one electrode and resulting voltage changes are recorded with both electrodes. The resistance of the basolateral cell membrane (0.15 GΩ) is comparable to indirect estimates obtained from intact tissues.[25,26] A lumenal membrane resistance of 0.7 GΩ and a transepithelial resistance of ~20 MΩ were obtained, the latter value indicating that

[21] D. Spray, R. D. Ginzberg, E. A. Morales, Z. Gatmaitin, and I. N. Arias, *J. Cell Biol.* **103**, 135 (1986).

[22] R. M. Henderson, J. Graf, and J. L. Boyer, *Am. J. Physiol.* **252**, G109 (1987).

[23] D. Ammann, "Ion Selective Microelectrodes." Springer-Verlag, Berlin, 1986.

[24] R. C. Thomas, "Ion Sensitive Intracellular Microelectrodes." Academic Press, New York, 1978.

[25] R. D. Purves, "Microelectrode Methods for Intracellular Recording and Ionophoresis." Academic Press, New York, 1981.

[26] J. Graf and O. H. Peterson, *J. Physiol.* **284**, 105 (1978).

the paracellular pathway exhibits a high specific conductance, as observed in other leaky epithelia.[10,11]

Partial ionic conductances (g_i) of the basolateral cell membrane can be obtained by monitoring the intracellular voltage changes during variations of individual ion concentrations in the bath, revealing $g_{K^+} > g_{Cl^-} > g_{Na^+}$.[11]

The conductance of intercellular communications is not voltage sensitive and it is reduced by octanol and by an acidic intracellular pH ($<$ pH 6.6).[21] Intracellular ion activities of Na^+, K^+, and Cl^- are within the range observed in intact liver.[11] Regulation of pH_i by Na^+/H^+ exchange can be demonstrated using ion-selective microelectrodes and the NH_3-pulse technique.[22]

Studies in couplets indicate that the Na^+, K^+-ATPase is electrogenic. The sinusoidal membrane potential is hyperpolarized by activation of the pump through readmission of external K^+ after cells are allowed to gain internal Na^+ in the absence of external K^+. Pump current under this condition is 70 pA/cell and produces a small lumenal hyperpolarization indicating that Na^+, K^+-ATPase is predominantly localized to the basolateral membrane.[11]

Finally, the role of the membrane potential as a driving force for bile acid dependent canalicular fluid secretion has been demonstrated in couplets by impaling rat hepatocytes with microelectrodes, changing the potential by current injection, and assessing changes in canalicular size with optical techniques.[27]

Acknowledgments

Parts of this work have been supported by DK 25636, 36854, and 34989, a Medical Research Council of Canada Grant MT 0785, the Austrian National Bank, and the Hochschule jubilaums Siftung der Gemeinde Wien. The authors also acknowledge the technical assistance of Mrs. Wandee Vetvutanapibul and Oi Cheng Ng.

[27] S. A. Weinman, J. Graf, and J. L. Boyer, *Am. J. Physiol.* **256**, G826 (1989).

[33] Characterizing Mechanisms of Hepatic Bile Acid Transport Utilizing Isolated Membrane Vesicles

By JAMES L. BOYER and PETER J. MEIER

Background

Hepatic clearance of bile acids from blood to bile involves three separate but sequential transport processes: (1) hepatic uptake from portal blood, (2) transcellular transport, and (3) canalicular secretion. Transport across the sinusoidal membrane domain occurs against both concentration and electrical gradients averaging $100-500$ μM and -40 mV, respectively. It is supposedly mediated by two transport systems, one of which is coupled to the chemical gradient for sodium. The mechanisms of transcellular transport are less well understood, but may involve both transport into and transcytotic movement of vesicles from intracellular organelles, such as the endoplasmic reticulum and Golgi apparatus to the bile canaliculus. Excretion of bile acids across the canalicular domain also occurs against significant concentration gradients and is mediated by a carrier that is driven, at least in part, by the intracellular negative electrical potential. Membrane vesicle techniques have been used increasingly to elucidate specific driving forces that move bile acids across these liver cell membrane domains.

Membrane Isolation

Localization of transport systems to specific membrane domains in the hepatocyte is critically dependent on membrane isolation techniques that permit the separation of canalicular (apical) and sinusoidal (basolateral) plasma membrane vesicles in high yield and purity. Several methods have been published for these purposes and are summarized elsewhere in this volume.[1] In order to obtain information on the physiologic direction of ion, pH, and potential gradients it is important to establish the sidedness of the various vesicle preparations. This can best be achieved by freeze-fracture analysis[2-4] or by limited proteolysis or antibody inhibition of ectoenzyme activities (e.g., γ-glutamyltranspeptidase, aminopeptidiase M).[4,5]

[1] P. J. Meier and J. L. Boyer, this volume [34].
[2] K. H. Altendorf and L. A. Staehelin, *J. Bacteriol.* **117**, 888 (1974).
[3] P. J. Meier, A. S. Meier-Abt, C. Barrett, and J. L. Boyer, *J. Biol. Chem.* **259**, 10614 (1984).
[4] W. Haase, A. Schafer, H. Murer, and R. Kinne, *Biochem. J.* **172**, 57 (1978).
[5] M. Inoue, R. Kinne, T. Tran, L. Biempica, and I. M. Arias, *J. Biol. Chem.* **258**, 5183 (1983).

General Techniques for Vesicle Transport Studies

Although a detailed discussion of vesicle transport techniques is beyond the scope of this chapter, the general approaches for studying hepatic bile acid transport mechanisms are comparable to those performed for other ligands in other transporting epithelia.[6-11] In this chapter only problems related to bile acid transport will be summarized.

Physical Properties of Bile Acids

Bile acids are the major product of cholesterol metabolism. They are amphipathic molecules that bind to protein and intercalate into lipid bilayers. Their physical properties depend critically on hydroxylations in the 3-, 7-, and 12-positions of the steroid nucleus, and whether or not the side chain is conjugated with the amino acids taurine or glycine. Cholic acid is a $3\alpha,7\alpha,12\alpha$-hydroxy bile acid, whereas chenodeoxycholic and deoxycholic acid have hydroxyl groups in the $3\alpha,7\alpha$, and $3\alpha,12\alpha$ positions, respectively. Lithocholic acid has only one hydroxyl group in the 3α position. Both the primary bile acids (cholic acid and chenodeoxycholic acid) and secondary bile acids, produced from bacterial 7α-dehydroxylations in the intestine (deoxycholic and lithocholic acid), can be conjugated on the side chain with either taurine or glycine. Ursodeoxycholic acid is similar to chenodeoxycholic acid in structure, except that the 7α-hydroxy group is in the β rather than in the α configuration. Table I summarizes the major species of bile acids and some of their physicochemical properties. The latter are extensively discussed elsewhere[12-16] and are summarized below only to the extent that they present special problems in membrane vesicle transport studies.

Solubility. Various bile acid species differ greatly in their solubility in aequous media (Table I). At low (i.e., monomeric) concentrations (micromolar range) the water solubility is largely dependent on such structural

[6] H. Murer and R. Kinne, *J. Membr. Biol.* **55**, 81 (1980).
[7] G. Sachs, R. J. Jackson, and E. C. Rabon, *Am. J. Physiol.* **238**, G151 (1980).
[8] U. Hopfer, *Am. J. Physiol.* **233**, E445 (1977).
[9] J. E. Lever, *Crit. Rev. Biochem.* **7**, 187 (1980).
[10] R. J. Turner, *J. Membr. Biol.* **76**, 1 (1983).
[11] B. R. Stevens, J. D. Kauritz, and E. M. Wright, *Annu. Rev. Physiol.* **46**, 417 (1984).
[12] A. F. Hofmann and A. Roda, *J. Lipid Res.* **25**, 1477 (1984).
[13] A. Roda and A. Fini, *Hepatology (Baltimore)* **4**, (Suppl.) 725 (1984).
[14] D. M. Small, *in* "The Bile Acids" (P. P. Nair and D. Kritchevsky, eds.), p. 249. Plenum, New York, 1971.
[15] D. J. Cabral, J. A. Hamilton, and D. M. Small, *J. Lipid Res.* **27**, 334 (1986).
[16] A. Roda, A. F. Hofmann, and K. J. Mysels, *J. Biol. Chem.* **258**, 6362 (1983).

TABLE I
PHYSICAL PROPERTIES OF THE MAJOR PHYSIOLOGICAL BILE ACID SPECIES

	Position and orientation of $-OH$ groups	Charge at pH 7.4[a]	$pK_a{}^a$	CMC[b] (0.15 M Na$^+$) (mM)	Water solubility[c] (mM)
Cholic acid	3α, 7α, 12α	—	~5	11	0.273
Taurocholic acid	3α, 7α, 12α	—	<2	6	Very soluble
Glycocholic acid	3α, 7α, 12α	—	~4	11	Soluble
Chenodeoxycholic acid	3α, 7α	—	~6	4	0.027
Taurochenodeoxycholic acid	3α, 7α	—	~2	3	
Glycochenodeoxycholic acid	3α, 7α	—	~4	1.8	
Deoxycholic acid	3α, 12α	—	~6	3	0.028
Taurodeoxycholic acid	3α, 12α	—	~2	2.4	
Glycodeoxycholic acid	3α, 12α	—	~4	2	
Ursodeoxycholic acid	3α, 7β	—	~6	7	0.0009
Tauroursodeoxycholic acid	3α, 7β			2.2	
Glycoursodeoxycholic acid	3α, 7β	—		4	
Lithocholic acid	3α	—	~6	0.5	0.00005
Taurocholic acid	3α	—			
Glycocholic acid	3α	—			

[a] In general, the degree of ionization of an individual bile acid varies according to the Henderson–Hasselbach equation. However, for unconjugated bile acids the actual concentration of the ionized species may be considerably lower than predicted since their pK_a increases above the CMC [H. Igini and M. C. Carey, *J. Lipid Res.* **21,** 72 (1980)] and also varies in the presence of other bile acids, micelles, or phospholipid vesicles.[15] For example, phospholipid vesicles have been shown to increase the pK_a of cholic acid from 4.6 to approximately 7.0, thereby increasing the concentration of the protonated (undissociated) species by a factor of 250 at physiologic pH.[15] In contrast, the pK_a of conjugated bile acids is more stable under various conditions. Consequently, unconjugated bile acids are especially prone to nonionic passive permeation into and across membranes, a process that must be vigorously controlled in vesicle transport studies.
[b] Measured by surface tension.[16]
[c] Data are from Ref. 12.

characteristics as (1) number, position, and orientation of the hydroxyl groups, (2) structure and length of the side chain, (3) degree of ionization, and (4) conjugation with taurine or glycine.[12] In general, the more hydroxyl groups, the more polar the bile acids (order of water solubility: trihydroxy > dihydroxy > monohydroxy bile acids). However, as exemplified by chenodeoxycholic and ursodeoxycholic acid, the configuration of the hydroxyl groups also strongly influences the water solubility (Table I). For the side chain, solubility increases as the side chain is shortened.[13] For practical purposes the most important solubility parameter is ionization. This property is especially relevant for unconjugated bile acids (high pK_a), the solubility of which is greatly increased by alkalinization of the media. In contrast, the solubility of conjugated bile acids is less dependent on bulk pH since their pK_a values are very low (Table I).

Ionization. Dissociation constants (pK_a) range between 5 and 6 for the unconjugated bile acids and are approximately 4 and 2 for the glycine and taurine conjugates, respectively (Table I). Therefore, at physiologic pH all bile acids are supposed to be ionized with a single negative charge on the side chain.[14] However, under *in vitro* conditions decreasing the pH considerably increases the protonated species of unconjugated bile acids. Furthermore, it has been recently shown that the pK_a can increase up to 6.5–7.3 if unconjugated bile acids associate with phospholipid vesicles.[15] Thus, even at physiologic pH high proportions of protonated bile acids may be generated during vesicle transport studies. Since protonated bile acids are barely soluble in water, but more easily permeate into membranes compared to the ionized species, the discrimination between passive phenomena (e.g., binding, diffusion) and active "carrier"-mediated transport processes can become very difficult in vesicle transport studies involving unconjugated bile acids (see below for cholate; Fig. 4). These pitfalls are less important the more polar (i.e., water soluble) and ionized the bile acids are and therefore are virtually absent for the conjugated bile acid taurocholate.

Critical Micellar Concentration (CMC). Bile acids (salts) self-associate to form micelles over a narrow range of concentration, which is usually called the CMC. For natural bile acids the CMC ranges between 0.5 and 11 mM (Table I). At these concentrations bile acids exhibit detergent properties and lead to solubilization of membrane vesicles. Since the CMC is dependent on temperature, pH, and the ionic composition of the media,[16] it is important to routinely control for membrane toxicity in vesicle transport studies even at bile acid concentrations below 200 μM.

Hydrophobicity. All bile acids, whether unconjugated or conjugated with glycine or taurine, absorb strongly to hydrophobic surfaces. These hydrophobic interactions (i.e., binding) are the higher, the less polar the bile acid is (monohydroxy > dihydroxy > trihydroxy bile acids). Furthermore, for unconjugated bile acids the hydrophobic binding is strongly dependent on pH,[12] since, as already mentioned above, protonated bile acids are more lipophilic than ionized bile acids. Consequently, membrane binding of cholic acid has a maximum at pH 6.0 while that of taurocholic acid is independent of pH (Fig. 4).[17,18] Furthermore, bile acids also bind to hydrophobic surfaces of tubes, filters, and/or other amphiphilic solutes, thereby decreasing the concentration of free bile acid in the solution.

[17] L. Accantino and F. R. Simon, *J. Clin. Invest.* **57**, 496 (1976).
[18] G. Hugentobler and P. J. Meier, *Am. J. Physiol.* **251**, G656 (1986).

Therefore, it is mandatory for vesicle transport studies to routinely test for the degree of membrane binding and for the actual free bile acid concentration in solution.

Interactions with Cations. Since ionized bile acids are negatively charged, they easily interact with oppositely charged cations such as Na^+ and Ca^{2+}. Interactions with Ca^{2+} are especially relevant since (1) Ca^{2+} is often included in membrane isolation and/or resuspension media[3,5] (see below), (2) even monomeric taurocholic acid strongly binds Ca^{2+},[19] and (3) certain bile acids have been proposed to function as Ca^{2+} ionophores.[20] Thus, if Ca^{2+} or other physiologic ions are included into assay media, their possible effects on transmembrane transport of bile acids should be controlled by performing parallel experiments in the absence of any physiologic electrolytes.

Consequences for Vesicle Transport Studies

From the above-outlined physicochemical properties of bile acids the following practical conclusions can be drawn for vesicle transport studies:

1. The tracer bile acid to be studied should be freely soluble in water (in monomeric form) and exclusively present as the ionized species in order to minimize nonionic diffusion. Since this can best be achieved with the most polar water-soluble bile acids, it is not surprising that taurocholate has so far been most frequently used in vesicle transport studies.[3,21-33]

2. The bile acid concentration in the assay must be low (usually 1 to 50 μM) in order to increase the "signal" (transport) to "noise" (binding) ratio and to avoid toxic membrane injury.[3,34] In our experience maximal taurocholic and cholic acid concentrations of up to 200 μM can be used

[19] E. W. Moore, L. Celic, and J. D. Ostrow, *Gastroenterology* **83**, 1079 (1982).
[20] D. G. Oelberg and R. Lester, *Clin. Rev. Med.* **372**, 297 (1986).
[21] P. G. Ruifrok and D. K. F. Meijer, *Liver* **2**, 28 (1982).
[22] M. Inoue, R. Kinne, T. Tran, and I. M. Arias, *Hepatology (Baltimore)* **2**, 572 (1982).
[23] M. C. Duffy, B. L. Blitzer, and J. L. Boyer, *J. Clin. Invest.* **72**, 1470 (1983).
[24] M. Inoue, R. Kinne, T. Tran, and I. M. Arias, *J. Clin. Invest.* **73**, 659 (1984).
[25] F. A. Simion, B. Fleischer, and S. Fleischer, *J. Biol. Chem.* **259**, 10814 (1984).
[26] B. L. Blitzer and L. Lyons, *Am. J. Physiol.* **249**, G34 (1985).
[27] H. Lücke, G. Stange, R. Kinne, and H. Murer, *Biochem. J.* **174**, 951 (1978).
[28] R. C. Beesley and R. G. Faust, *Biochem. J.* **178**, 299 (1979).
[29] D. J. Rouse and L. Lack, *Biochim. Biophys. Acta* **599**, 324 (1980).
[30] S. L. Weinberg, G. Burckhardt, and F. A. Wilson, *J. Clin. Invest.* **78**, 44 (1986).
[31] F. J. Suchy, S. M. Courchene, and B. L. Blitzer, *Am. J. Physiol.* **248**, G648 (1985).
[32] J. A. Barnard, F. K. Ghishan, and F. A. Wilson, *J. Clin. Invest.* **75**, 869 (1985).
[33] M. S. Moyer, J. E. Heubi, A. L. Goodrich, W. F. Balistreri, and F. J. Suchy, *Gastroenterology* **90**, 1188 (1986).
[34] P. J. Meier, A. St. Meier-Abt, and J. L. Boyer, *Biochem. J.* **242**, 465 (1987).

with basolateral and of up to 1 mM with canalicular membrane vesicles. The occurrence of membrane toxicity can be routinely controlled by determining tracer bile acid uptake at equilibrium (see below).

3. Because of the required low concentrations only tracer bile acids with high specific radioactivity (>2 Ci/mmol) are suitable for vesicle transport studies.

4. The actual concentration of free bile acid in solution should always be checked by comparing measured counts per minute per milliliter with the theoretically expected concentration of radioactivity.

5. Before transport studies are initiated the degree of overall membrane binding should be roughly estimated by incubating radioactive bile acid (~ 200,000 cpm) with vesicles (50–100 μg protein) at 0°, pelleting the vesicles, and determining the radioactivity remaining in the supernatant. If more than 10% of total radioactivity is bound to the vesicles binding can be assumed to be too high for accurate determination of the transport signal.

6. Nonspecific binding to nitrocellulose filters must be minimized by prewashing them with 1 mM solutions of the respective bile acid.

Practical Procedure

Most of the published studies of the mechanisms of bile acid transport in membrane vesicles have utilized the most polar bile acid [^{3}H]taurocholate and a standard rapid filtration technique. Although there are various modifications of the latter method, we will concentrate on the procedure with which we have the greatest experience.

Solutions

Resuspension of Membranes. Following isolation of membranes the final pellet is best resuspended in a minimum volume of medium that is the desired intravesicular composition at the start of the experiment. An example of a standard membrane suspension medium consists of 0.25 M sucrose, 0.2 mM CaCl$_2$, 5 mM MgSO$_4$, and 10 to 20 mM HEPES/Tris, pH 7.5.[3] The membranes can then be frozen in liquid nitrogen at protein concentrations greater than 3 mg/ml and then utilized for variable periods up to several weeks without loss of transport activity. However, it is essential for each laboratory to determine how long the frozen membrane preparations are stable.

Incubation Media. These should be of similar osmolarity as the membrane resuspension medium and contain the desired extravesicular concentrations of ions and solutes. Thus, a medium corresponding to the above-outlined membrane resuspension solution would be 100 mM sucrose, 100 mM NaCl, 0.2 mM CaCl$_2$, 5 mM MgSO$_4$, and 10 to 20 mM HEPES/Tris, pH 7.5. NaCl can be replaced isoosmotically by any other salts whose effects on bile acid transport one wishes to study.

Cold Stop Solution. This medium is used to stop the transport reaction. It should be isoosmotic to the final incubate, precooled to 4°, and have the similar ionic strength as the incubate. Correspondingly it should contain 100 mM sucrose, 100 mM KCl, 0.2 mM CaCl$_2$, 5 mM MgSO$_4$, and 10 to 20 mM HEPES/Tris, pH 7.5.

Filter Prewash Solution. In order to reduce unspecific binding to the filters it is necessary to prewash the filters with 1 mM bile acid (salt) substrate dissolved in water.

Comments. The above given concentrations of incubation media correspond to the final concentrations in the assay. Thus, appropriate corrections should be made in preparing the various stock solutions for their subsequent dilutions in the assay. Calcium and magnesium are included in the various media to tighten the vesicles. However, it should be realized that calcium inhibits transmembrane transport of taurocholate, especially if it is present on the inside of the vesicles.[35,36] In basolateral rat liver plasma membrane vesicles the inhibitory effect of calcium on Na$^+$ gradient-dependent taurocholate uptake can be counteracted by magnesium or EGTA (Fig. 1). These findings demonstrate that although certain additives may help to preserve the membrane structure their possible effects on the investigated transport process must also be taken into account. Finally, it should be mentioned that the above-given composition of media represents but one possibility and that alternative solutions have also been successfully used to probe for carrier-mediated taurocholate transport.[5,21,22,24–33]

Substrates

Tracer bile acids are supplied by the manufacturers either as the free acids dissolved in methanol/ethanol (NEN Research Products–Du Pont, Boston, MA) or as the sodium salts dissolved in aqueous solutions containing 2% ethanol (Amersham Research Products, Arlington Heights, IL). Since ethanol in concentrations above 0.1–0.3% diminishes the uptake of taurocholate in rat liver plasma membrane vesicles,[36] ethanol must be evaporated (nitrogen stream) before solubilization of the bile acid substrate in the incubation medium. For uptake measurements the required radioac-

[35] J. D. Fondacaro and T. B. Madden, *Life Sci.* **35**, 1431 (1984).
[36] P. R. Mills, P. J. Meier, D. J. Smith, N. Ballatori, J. L. Boyer, and E. R. Gordon, *Hepatology* **7**, 61 (1987).

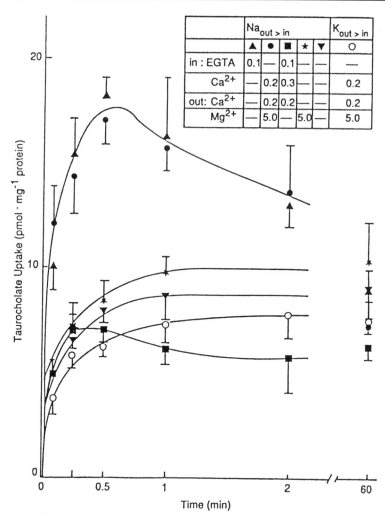

FIG. 1. Effects of Ca²⁺ and Mg²⁺ on sodium-dependent taurocholate uptake in blLPM vesicles. Vesicles were resuspended in 250 mM sucrose and 20 mM HEPES/KOH, pH 7.5 containing additional Ca²⁺ and/or EGTA as indicated. Incubation media were composed of 62.5 mM sucrose, 20 mM HEPES/KOH pH 7.5, 125 mM NaCl (▲,●,■,★,▼), or 125 mM KCl (○) supplemented with various amounts of Ca²⁺ and/or Mg²⁺ as indicated. ³[H]taurocholate was present in 1 μM concentration in the incubation media. Data represent the mean ± S.D. of three experiments in one membrane isolation. Note that low Ca²⁺ concentrations inhibit sodium-dependent taurocholate uptake in basolateral liver plasma membranes but that inclusion of Mg²⁺ in the incubation media completely antagonizes this inhibitory effect (B. Zimmerli and P. J. Meier, unpublished observations).

tivity for a single assay ranges between 200,000 and 500,000 cpm. With the available specific activities between 2 and 5 Ci/mmol for tritiated taurocholic acid and 10–25 Ci/mmol for tritiated cholic acid the resulting tracer concentrations in the incubation media are approximately 1–2.5 and 0.2–0.5 μM, respectively (assuming 50% counting efficiency). If higher substrate concentrations are required unlabeled bile acid should be dissolved in methanol or ethanol, added to the tracer solution, and the methanol/ethanol mixture evaporated under a nitrogen stream. Alternatively, bile salts (i.e., sodium salt) could be directly dissolved in the required concentration in the incubation medium. Less polar bile acids can be solubilized in dimethyl sulfoxide or propylene glycol, both of which are considerably less toxic to membranes than ethanol.

Uptake Experiments

Thawing of Membranes. On the day of study, the membranes are rapidly thawed by immersing tubes into a 37° water bath. The membranes are revesiculated with a tight Dounce homogenizer (type B; Dounce, Wheaton, NJ) utilizing 30 up-and-down strokes and/or passing the membranes 20 times through a 25- to 27-gauge needle. The suspension is then diluted to a protein concentration of 2 to 4 mg/ml in membrane resuspension medium and the vesicles kept on ice until the start of the assay.

Assay. All transport studies are carried out in a temperature-controlled water bath. Twenty microliters of membrane suspension (40–80 μg protein) is transferred into 5-ml glass or plastic tubes and preincubated at the desired reaction temperature for at least 5 min. Uptake is initiated by adding 80 μl of incubation medium containing the labeled bile acid ligand. After the desired time period the reaction is terminated by adding 3.5 ml of ice-cold (4°) stop solution. The membrane vesicle-associated ligand is then separated from the free ligand by rapid filtration (1 ml/sec) through a 0.45-μm nitrocellulose filter presoaked in deionized water and prefiltered with 3 ml of 1 mM unlabeled bile acid. The filter is additionally washed twice with 3.5 ml of cold stop solution. The filters containing the trapped vesicles are then dissolved in an appropriate scintillation fluid and the filter-associated radioactivity counted in a liquid scintillation counter. All measurements are usually carried out in triplicate and confirmed in several membrane preparations.

Because of the high membrane-binding properties of bile acids (see above) it is critical to always prepare appropriate membrane/filter-binding blanks for all uptake series. Our preferred method is to determine "zero time" blanks by keeping 20 μl of membrane suspension on ice and adding 80 μl of precooled (4°) radioactive incubation medium to the side of the tubes followed by mixing at the time of addition of the cold stop solution.

All other steps are exactly the same as described above. Alternative techniques for determining the blank value have utilized membrane vesicles that have been collapsed by adding high concentrations of raffinose or cellobiose.[22]

Calculation of Bile Acid Uptake. The absolute amount of substrate taken up into the vesicles is calculated according to the following formula:

$$\frac{(X - B) \times \text{pmol substrate in assay}}{\text{cpm}_{\text{total}} \times \text{protein in assay (mg)}} = \text{pmol mg}^{-1} \times \text{time interval}$$

where X is the vesicle-associated radioactivity after the various timed incubations, B is the "zero time" blank (membrane/filter binding), and $\text{cpm}_{\text{total}}$ is the total amount of radioactivity in assay.

Assessment of Binding versus Transport

The extent of membrane binding can be assessed by determining the effect of media osmolality on bile acid uptake at equilibrium (usually 60 min or longer). Increasing the concentration of an impermeant nontransportable and nonmetabolized solute such as raffinose or cellobiose progressively decreases the intravesicular volume and correspondingly also the membrane-enclosed amount of free substrate. Figure 2 illustrates a representative study for taurocholate and compares these values with alanine and glucose, two substrates with low membrane-binding affinity. Extrapolation of the values to theoretical zero intravesicular volume provides an assessment of the magnitude of binding, while the upward slope of the line denotes uptake into a closed vesicular space. If the line were horizontal with the y axis, no uptake would have occurred. Differentiating between intravesicular and extravesicular binding can also be assessed by using cold stop solutions and washing the vesicles with media containing a 100-fold excess of unlabeled bile acid (Fig. 2) or by washing the vesicles in distilled water and determining the amount of radioactivity lost from the membrane.[23] Often, most of the binding occurs on the inside of the vesicles. Binding can be reduced by raising the ionic strength of the membrane suspension and incubation media.

Assessment of Driving Forces

In general active transmembrane solute movement can be mediated by primary (ATP dependent), secondary (Na^+ gradient and/or electrical potential dependent), or tertiary (coupled dual-exchange mechanisms) active transport processes.[37] For bile acids only secondary active transport sys-

[37] P. S. Aronson, *Am. J. Physiol.* **240**, F1-F11 (1981).

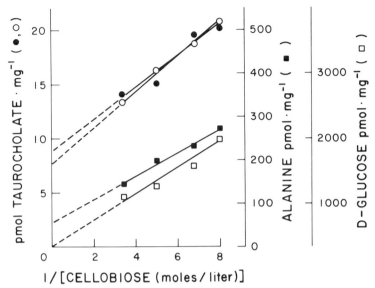

FIG. 2. Effect of medium osmolality on taurocholate, alanine, and D-glucose uptake into canalicular liver plasma membrane vesicles at equilibrium. Membrane vesicles were incubated in 100 mM sucrose, 100 mM NaCl, 0.2 mM CaCl$_2$, 5 mM MgSO$_4$, 10 mM HEPES/Tris, pH 7.5, and varying concentrations of cellobiose. Uptakes of [^{3}H]taurocholate (1 μM), [^{3}H]alanine (0.2 mM), and [^{14}C]glucose (0.9 mM) were determined at 25° after 120 min. For taurocholate, uptake was also quenched by inclusion of 100 μM unlabeled taurocholate in the cold stop solution (O). The regression lines were calculated by least-squares analysis and extrapolated (dotted line) to theoretical zero intravesicular volume. (Reproduced with permission from Ref. 3.)

tems involving Na$^+$, pH, and electrical potential gradients have been described so far.

Ion Gradients. The effect of out-to-in Na$^+$ gradients on the uptake of bile acids can be determined by substituting 100 mM NaCl with 100 mM KCl, LiCl, CsCl, tetramethylammonium chloride, or choline chloride. Stimulation of uptake by an out-to-in Na$^+$ gradient (Fig. 3) suggests the presence of a secondary active Na$^+$-coupled transport system. This possibility can be more definitely assessed by comparing the initial bile acid uptake rates under tracer exchange conditions where the effects of Na$^+$ are compared with K$^+$ or Li$^+$ at equal cation concentrations inside and outside the vesicles.[3] Tracer exchange conditions avoid problems created by cation diffusion potentials that may stimulate the initial rate of transport of the bile acid anion.[6-11] By this means Na$^+$/taurocholate cotransport can be established in basolateral, but not in canalicular, rat liver plasma membrane vesicles.[3]

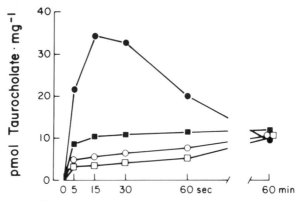

FIG. 3. Uptake of taurocholate into basolateral (sinusoidal) rat liver plasma membrane vesicles. Freshly isolated membrane vesicles were resuspended in 0.25 M sucrose, 0.2 mM CaCl$_2$, 5 mM MgSO$_4$, 10 mM HEPES/Tris, pH 7.5 ($\bullet$,$\bigcirc$) or, in order to preload the vesicles with the various salts, in 100 mM sucrose, 0.2 mM CaCl$_2$, 5 mM MgSO$_4$, 10 mM HEPES/Tris, pH 7.5, supplemented with either 100 mM NaNO$_3$ ($\blacksquare$) or 100 mM KNO$_3$ ($\square$), respectively. Frozen and quickly thawed vesicles were incubated in 100 mM sucrose, 0.2 mM CaCl$_2$, 5 mM MgSO$_4$, 10 mM HEPES/Tris, pH 7.5, and 100 mM NaCl ($\bullet$), KCl ($\bigcirc$), NaNO$_3$ ($\blacksquare$), or KNO$_3$ ($\square$). (Reproduced with permission from Ref. 3.)

Electrical Membrane Potential. The effects of membrane potential on transmembrane transport of bile acids can be assessed in several ways. Utilizing counteranions with varying membrane permeability characteristics (SCN$^-$, NO$_3^-$, Cl$^-$, SO$_4^{2-}$, gluconate), initial uptake rates of bile acids can be compared.[3,23] If uptake is greater with more permeant anion, transport may be facilitated by the inside-negative diffusion potential. Conversely, if uptake is inhibited, the potential may serve as a driving force to promote efflux of the bile acids from the inside of the vesicle. Valinomycin-induced K$^+$-diffusion potentials provide a more definitive approach.[3] Vesicles must first be preloaded with potassium gluconate (100 mM) and uptake rates of the bile acid compared in the presence and absence of valinomycin (10 μg/mg protein in ethanol or dimethyl sulfoxide) and in the presence and absence of 100 mM potassium gluconate in the incubation medium. Valinomycin is a K$^+$ ionophore and, in the absence of extravesicular K$^+$, permits the rapid diffusion of K$^+$ out of the vesicles, creating a transient negative potential inside the vesicle. When potassium gluconate (100 mM) is present on both sides of the vesicles, then the membrane is voltage clamped by valinomycin and an electrical potential cannot develop. In general, voltage clamping is a useful technique to avoid confounding effects of membrane potential changes in the assessment of ion and pH gradients as driving forces.

pH Gradients. By loading the vesicles with an alkaline medium of high buffer strength (e.g., 180 mM sucrose, 40 mM tetramethylammonium, 60 mM K$^+$, 0.2 mM calcium, 5 mM magnesium, 110.4 mM gluconate, 70 mM Tris, and 70 mM HEPES, pH 8.0) and incubating them in an acidic buffer system (e.g., 184 mM sucrose, 40 mM tetramethylammonium, 60 mM K$^+$, 0.2 mM calcium, 5 mM magnesium, 110.4 mM gluconate, 30 mM Tris, 14 mM HEPES, and 90 mMMES, pH 6.0) the effects of inside alkaline pH gradients on bile acid uptake can be investigated.[18,24,38] In these experiments voltage-clamped conditions should be used in order to avoid interference from membrane potential changes. While transmembrane pH gradients have no effect on uptake of taurocholic acid,[23] a Ph$_{in > out}$ gradient stimulates cholic acid uptake into basolateral rat liver plasma membrane vesicles (Fig. 4).[18,38] Although this pH gradient-driven cholate uptake has been interpreted as representing mediated hydroxyl/cholate exchange, recent, more detailed investigations demonstrated that it is in fact due to nonionic diffusion of the protonated species followed by intravesicular trapping of the cholate anion.[15,39] These findings emphasize the difficulties in applying pH gradients to transport studies with unconjugating bile acids.

Anion Exchange. In order to test for anion/bile acid exchange mechanisms the vesicles can be resuspended in media containing millimolar concentrations of physiologic anions such as bicarbonate, chloride, or sulfate and tracer bile acid uptake studied in the presence of various in-to-out anion gradients. These experiments must be performed in the presence of the protonophore FCCP [carbonylcyanide p-(trifluoromethoxy)phenylhydrazone], valinomycin, and equimolar extra- and intravesicular K$^+$ concentrations in order to prevent interfering effects of pH and/or electrical potential differences. Although these methods have led to the detection of an anion/bile acid exchange transport system in intestinal basolateral membrane vesicles,[30] to date a similar system has not been definitely described in liver plasma membrane vesicles.

Efflux Experiments

Depending on the sidedness of membrane vesicles, efflux rather than uptake may represent the physiologic direction of transport (e.g., transport in canalicular membrane vesicles).[3] In this situation the vesicles should be preloaded with labeled bile acid in order to carry out efflux experiments. Preloading is accomplished by adding cold and labeled substrate (10 μM) to the thawed membranes prior to vesiculation. Following tight Dounce

[38] B. L. Blitzer, Ch. Terzakis, and K. A. Scott, *J. Biol. Chem.* **261**, 12042 (1986).
[39] C. Caflisch, B. Zimmerli, J. Reichen, and P. J. Meier, *Biochim. Biophys. Acta* **1021**, 70 (1990).

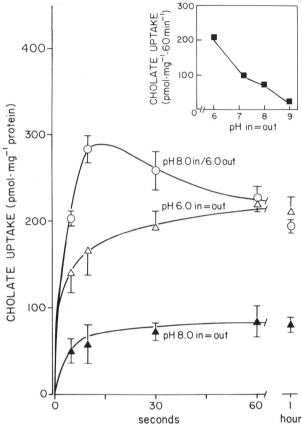

FIG. 4. pH gradient-driven cholate uptake into basolateral (sinusoidal) rat liver plasma membrane vesicles. Vesicles were resuspended in 180 mM sucrose, 40 mM tetramethylammonium, 60 mM K$^+$, 0.2 mM Ca^{2+}, 5 mM Mg^{2+}, 110.4 mM gluconate, 70 mM Tris, 70 mM HEPES, pH 8.0 (O,▼) or 184 mM sucrose, 40 mM tetramethylammonium, 60 mM K$^+$, 0.2 mM Ca^{2+}, 5 mM Mg^{2+}, 110.4 mM gluconate, 30 mM Tris, 14 mM HEPES, and 90 mM MES, pH 6.0 (△). The same buffer systems were also used to investigate [^{3}H]cholate (10 μM, 25°) uptake under either pH-equilibrated (△,▲) or pH gradient (O) conditions. The inset figure demonstrates the association of [^{3}H]cholic acid with the vesicles after 60 min of incubation at various equilibrated pH values (pH-dependent binding). All vesicles were voltage clamped with valinomycin. (Reproduced with permission from Ref. 18.)

homogenization, the membranes are preincubated for 20–30 min at 25° before being placed on ice to ensure maximal intravesicular concentrations of the bile acid. Twenty microliters of membranes is then diluted 10-fold into 180 μl of substrate-free incubation medium. Efflux studies are difficult to perform as the label can diffuse rapidly from the membranes and signal-to-noise ratios are smaller than with uptake studies. Nevertheless, this is a useful technique to examine the effects of countertransporters and inhibitors on carrier-mediated transport systems (Fig. 5).[3,34]

Characteristics of Carrier-Mediated Bile Acid Transport

In general, the conclusion that an identified bile acid transport pathway is mediated by a membrane-associated carrier must rely on several rather than just one characteristic parameter.

Temperature Dependency. Usually temperature changes affect active, carrier-mediated transport mechanisms to a greater extent than passive diffusional processes. A sensitive method to characterize the temperature

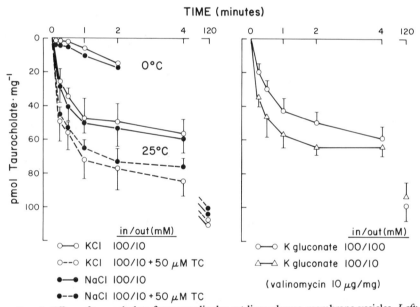

FIG. 5. Efflux of taurocholate from canalicular rat liver plasma membrane vesicles. *Left*: The membranes were resuspended in 100 mM sucrose, 0.2 mM CaCl₂, 5 mM MgSO₄, 10 mM HEPES/Tris, pH 7.5, and either 100 mM NaCl (●) or 100 mM KCl (○). The vesicles were preloaded with 10 μM [³H]taurocholate and tracer efflux determined at 0 and 25°. The incubation medium contained 100 mM tetramethylammonium chloride instead of NaCl and KCl and plus (– –) or minus (—) 50 μM unlabeled taurocholate (TC). *Right*: Vesicles were resuspended in 100 mM potassium gluconate, 100 mM sucrose, 0.2 mM CaCl₂, 5 mM MgSO₄, 10 mM HEPES/Tris, pH 7.5. The vesicles were preloaded with 10 μM [³H]taurocholate, treated with valinomycin, and incubated in the presence of 100 mM potassium (○) or tetramethylammonium gluconate (△) in the extravesicular medium. (Reproduced with permission from Ref. 3.)

dependency of a given transport function is the construction of so-called Arrhenius plots.[40]

Saturation Kinetics. Saturability of initial transport rates with increasing substrate concentrations is a *sine qua non* condition for a carrier-mediated process. However, in these experiments it is very crucial to test for true (i.e., linear) uptake rates at all measured substrate concentrations. This is especially relevant for bile acids since initial uptake rates may increase by increasing their concentration. For example, while pH gradient-driven uptake of cholic acid is apparently saturable at 5 sec,[38] 1-sec measurements increase linearly with increasing cholate concentrations.[39] In order to correctly detect such variability in initial uptake rates it is best to perform three rather than one early uptake measurement and to plot the slopes of the initial uptake measurements against the various substrate concentrations.[41] The intrinsic advantage of this approach is that the slopes of the individually determined regression lines are independent of the degree of binding and consequently no "zero time" blanks must be subtracted.[41] Furthermore, since bile acids have detergent properties, great care must be taken to avoid concentrations that damage the membrane. Thus, equilibrium uptake measurements should always be included and bile acid concentrations excluded from the analysis if they lower equilibrium uptake values.

Cis Inhibition. Substrates that compete for a transport carrier will inhibit uptake (or efflux) when placed on the same side of the membrane as the labeled bile acid. If cis inhibition occurs the type of inhibition (competitive or noncompetitive) should be analyzed in order to decide whether or not a given inhibitor is a cosubstrate for the carrier.

Trans Stimulation. Countertransport or trans stimulation is examined by placing 5- to 100-fold higher concentrations of cold ligand on the opposite side of the membrane. If the transport is carrier mediated, the rate of uptake or efflux should occur more rapidly in the presence of cotransported substrate on the trans side of the membrane. This approach has been successfully used to definitely establish that transcanalicular transport of taurocholate is a carrier-mediated process (Fig. 5)[3] as well as to delineate its specificity for various bile acid substrates.[34] However, the latter studies also demonstrated that two hydrophobic bile acids (i.e., dihydroxylated deoxycholic and chenodeoxycholic acids) exhibit both cis and trans inhibition, suggesting that their binding properties are too high to accurately determine transport signals.

[40] M. D. Honslay and K. K. Stanley, "Dynamics of Biological Membranes." Wiley, New York, 1982
[41] K. D. Kröncke, G. Fricker, P. J. Meier, W. Gerok, T. Wieland, and G. Kurz, *J. Biol. Chem.* **261,** 12562 (1986).

Summary and Conclusions

Utilizing the above-outlined approaches, mechanisms of hepatic bile acid transport have been characterized in membrane vesicles of rat liver, particularly for the conjugated trihydroxy bile acid, taurocholic acid. Uptake across the sinusoidal membrane is carrier mediated and coupled to the transmembrane sodium gradient.[22,23,26] This carrier has an apparent K_m between 30 to 50 μM and a V_{max} between 4 to 6 nmol mg^{-1} protein min^{-1}.[23,26] Furthermore, Na$^+$ gradient-dependent sinusoidal uptake of taurocholate can be stimulated by low concentrations of albumin.[26] There is controversy as to whether the process is electrogenic.[21-23] Although photoaffinity labeling studies indicate that an additional carrier for Na$^+$-dependent bile acid uptake is also present at the sinusoidal membrane,[42] this carrier has so far not been characterized in membrane vesicles. The proposition that pH gradient-driven furosemide-sensitive cholic acid uptake into sinusoidal membrane vesicles may represent carrier-mediated hydroxyl/cholate exchange[38] must be revised on the basis of the recent findings that (1) true initial uptake rates are not saturable; (2) pH gradient-driven cholate uptake is also found in liposomes; and (3) furosemide also inhibits pH gradient-driven cholate uptake in liposomes.[39] The mechanisms of transcellular transport of bile acids have been studied less extensively, but Na$^+$-independent carrier-mediated taurocholic acid transport has been demonstrated in purified subcellular fractions such as rat liver microsomes and Golgi membranes.[43,44] Finally, transport studies in canalicular rat liver plasma membrane vesicles indicate that canalicular excretion of bile acids is also a carrier-mediated process that may be driven, at least in part, by the physiologic electrical potential gradient,[3,24] and that preferentially transports trihydroxy and conjugated dihydroxy bile acids.[34]

Acknowledgment

These studies were supported by the Swiss National Science Foundation (Grants 3.983.0.84 and 3.992.0.87) and United States Public Health Service Grants DK-25636 and DK-36854.

[42] Th. Wieland, M. Nassal, W. Kramer, G. Fricker, U. Bickel, and G. Kurz, *Proc. Natl. Acad. Sci. U.S.A.* **81,** 5232 (1984).
[43] F. A. Simion, B. Fleischer, and S. Fleischer, *J. Biol. Chem.* **259,** 10814 (1984).
[44] A. Reuben and R. Allen, *Gastroenterology* **98,** 19 (1990).

[34] Preparation of Basolateral (Sinusoidal) and Canalicular Plasma Membrane Vesicles for the Study of Hepatic Transport Processes

By Peter J. Meier and James L. Boyer

Hepatocytes represent highly polarized secretory cells that exhibit efficient transport of a wide variety of endogenous and exogenous compounds from blood into bile. In order to maintain these vectorial transport processes hepatocytes localize distinct membrane transport systems on their various surface domains. While sinusoidal and lateral surfaces (i.e., the "basolateral" pole of hepatocytes) provide for efficient exchange of various ions, organic solutes, and proteins with blood plasma, the bile canalicular or apical pole of the cell is separated from the plasma space by tight junctions and is highly specialized for the primary secretion of bile. In order to be able to separately study basolateral and canalicular membrane transport processes without interference of intracellular metabolic events, it is necessary to selectively isolate basolateral (blLPM) and/or canalicular (cLPM) liver plasma membrane vesicles.

Since the original description of isolation of plasma membranes from rat liver by Neville,[1] numerous major technical modifications have been introduced in order to improve yield and purification of membranes as well as reproducibility of the procedure.[2-4] Most of these techniques involve mild homogenization of the liver to prevent bile canaliculi from fragmentation followed by isolation of a "mixed" plasma membrane fraction from an initial low-speed (i.e., "crude nuclear") pellet. This fraction is enriched in intact bile canaliculi and lateral membrane sheets with some attached sinusoidal membrane fragments.[5-8] Extensive studies by Evans and co-workers have shown that up to 30 mg "bile canalicular enriched" plasma membrane protein can be isolated from a total of 100–120 g liver

---

[1] D. M. Neville, Jr., *J. Biophys. Biochem. Cytol.* **8**, 413 (1960).

[2] P. Emmelot, C. J. Bos, R. P. van Hoeven, and W. J. van Blitterswijk, this series, Vol. 31, p.76.

[3] N. N. Aronson, Jr., and O. Touster, this series, Vol. 31, p. 90.

[4] W. H. Evans, *in* "Laboratory Techniques in Biochemistry and Molecular Biology" (T. S. Work and E. Work, eds.), Vol. 7, Part I. Elsevier/North-Holland Biomedical, Amsterdam, 1978.

[5] W. H. Evans, *FEBS Lett.* **3**, 237 (1969).

[6] M. H. Wisher and W. H. Evans, *Biochem. J.* **146**, 375 (1975).

[7] C. S. Song, W. Rubin, A. B. Rifkind, and A. Kappas, *J. Cell Biol.* **41**, 124 (1969).

[8] A. L. Hubbard, D. A. Wall, and A. Ma, *J. Cell Biol.* **96**, 217 (1983).

by a rate zonal centrifugation technique.[4-6] Furthermore, subsequent tight homogenization followed by density-gradient centrifugation in sucrose resulted in "light" and "heavy" plasma membrane subfractions of preferential canalicular and basolateral origin, respectively.[6,9] However, these studies required specialized zonal rotors and lengthy ultracentrifugation steps (up to 16 hr). In addition, modest enrichments of characteristic marker enzyme activities (i.e., 5'-nucleotidase, leucinaminopeptidase) indicate that the presumptive canalicular membrane subfraction, as isolated in these studies, was still contaminated with basolateral membranes to a large extent and/or the respective enzyme activities were partially inactivated during the prolonged centrifugation steps. Although recent alternative isolation methods have been proposed, these procedures are associated either with insufficient purification,[3,10-16] low yields,[12,14,15,17,18] or they require antibodies against specific canalicular membrane components.[8,18] Hence, for *in vitro* transport studies isolation procedures are still required that result in both high yields and high purification of basolateral and canalicular membrane subfractions within reasonable working hours. The procedure described here represents an extensive modification of the Evan's procedure using a combination of rate zonal flotation and high-speed discontinuous sucrose gradient centrifugation techniques.[19] The method permits the simultaneous isolation of blLPM and cLPM vesicles from the same homogenate in a yield sufficient for a large number of individual transport studies to be carried out in both plasma membrane subfractions.

Isolation Procedure

The method is essentially based on two sequential centrifugation steps (Fig. 1). First, rate zonal flotation is used to prepare a "mixed" plasma membrane fraction that exhibits the same composition and morphology as the bile canalicular enriched fraction described by others.[5-8] Second, after

[9] W. H. Evans, *Biochim. Biophys. Acta* **604**, 27 (1980).
[10] G. Toda, H. Oka, T. Oda, and Y. Ikeda, *Biochim. Biophys. Acta* **413**, 52 (1975).
[11] M. M. Fisher, D. L. Bloxam, M. Oda, M. J. Phillips, and I. M. Yousef, *Proc. Soc. Exp. Biol. Med.* **150**, 177 (1975).
[12] B. F. Scharschmidt and E. B. Keeffe, *Biochim. Biophys. Acta* **646**, 369 (1981).
[13] E. G. Loten and J. C. Redshaw-Loten, *Anal. Biochem.* **154**, 183 (1986).
[14] P. Gierow, M. Sommarin, Ch. L. Larsson, and B. Jergil, *Biochem. J.* **235**, 685 (1986).
[15] R. E. Poupon and W. H. Evans, *FEBS Lett.* **38**, 134 (1979).
[16] R. J. Epping and F. L. Bygrave, *Biochem. J.* **223**, 733 (1984).
[17] J. L. Boyer, R. M. Allen, and Oi Cheng Ng, *Hepatology (Baltimore)* **3**, 18 (1983).
[18] L. M. Roman and A. L. Hubbard, *J. Cell Biol.* **98**, 1497 (1984).
[19] P. J. Meier, E. S. Sztul, A. Reiben, and J. L. Boyer, *J. Cell Biol.* **98**, 991 (1984).

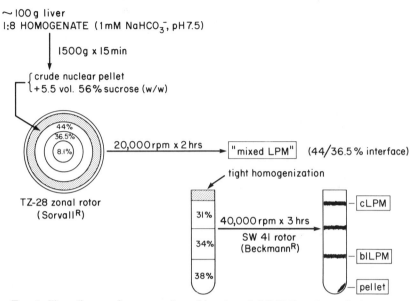

FIG. 1. Flow diagram for preparation of basolateral (blLPM) and canalicular (cLPM) plasma membrane vesicles from rat liver.

tight homogenization of mixed LPM, basolateral and canalicular vesicles are separated by high-speed centrifugation through discontinuous sucrose gradients.

Animals. Male Sprague-Dawley rats (200–250 g) are housed in a constant temperature–humidity environment with alternating 12-hr light and dark cycles for at least 4 days prior to use. During this time all animals are allowed access to food and water *ad libitum.* On the day of the experiment, the rats are killed by decapitation between 7:30 and 8:30 AM.

Solutions. NaHCO₃ (1 m*M*), pH 7.4; 0.25 *M* sucrose (8.1%, w/w); 31% sucrose (w/w; $d = 1.1318$); 34% sucrose (w/w; $d = 1.1463$); 36.5% sucrose (w/w; $d = 1.1587$); 38% sucrose (w/w; $d = 1.1663$); 44% sucrose (w/w; $d = 1.1972$); 56% sucrose (w/w; $d = 1.2623$).

All solutions are prepared 24 to 48 hr before use and stored at 4°. Densities of the sucrose solutions are adjusted at room temperature by refractometry (Abbe-3L refractometer; Bausch and Lomb, Rochester, NY).

Step 1: Isolation of Mixed LPM. Routinely, 10–12 fed animals are killed and their lives (110–115 g) rapidly removed and chilled on ice. Ten-gram portions of liver are cut into small pieces, washed three times in 80 ml cold 1 m*M* NaHCO₃ with a loose (type A) Dounce homogenizer (seven up-and-down strokes). The homogenate is further diluted to 1800–

2000 ml with cold NaHCO$_3$ and filtered twice through two layers of 60-grade cheesecloth. Centrifugation at 1500 g for 15 min (GSA rotor, E. I. Du Pont de Nemours and Co., Inc., Sorvall Instruments Division, Newton, CT) gives a "crude nuclear pellet" that is resuspended in 5.5 vol of 56% sucrose and stirred for 15 min to disrupt membrane aggregates (Fig. 1). The sample is then loaded onto a 100-ml cushion of 56% sucrose with a variable speed pump into the zonal rotor TZ-28 (Sorvall Instruments) and overlayed with 400 ml of 44% sucrose and 200 ml of 36.5% sucrose, respectively. Finally, the rotor is filled to its total volume capacity (1350 ml) with 0.25 M (8.1%) sucrose. During the entire loading procedure the rotor is run at 3000 rpm. The completed discontinuous sucrose gradient system is centrifuged at 20,000 rpm for 120 min (Fig. 1). After slow deceleration to a complete stop, 70 15-ml fractions are collected from the bottom of the rotor and routinely analyzed for turbidity (absorbance at 700 nm). The bulk of plasma membrane fragments is normally contained in fractions of the third turbidity peak (44/36.5% sucrose interface).[19] These fractions are combined and diluted with NaHCO$_3$ to 1000 ml. The suspension is then centrifuged at 7500 g for 15 min. The resulting pellet is gently resuspended (by vortex mixing) in 250 ml NaHCO$_3$, and the material representing the mixed LPM resedimented at 2700 g for 15 min.

Step 2: Separation of Basolateral (blLPM) and Canalicular (cLPM) Subfractions. Mixed LPM are diluted with 0.25 M sucrose to a total volume of 25 ml and homogenized with a tight type B glass–glass Dounce homogenizer by 50 strokes (1 stroke equals up and down). Mixed LPM (3.5 ml) are layered on top of a three-step sucrose gradient consisting of 4 ml 38%, 2.5 ml 34%, and 2.5 ml 31% sucrose (Fig. 1). The tubes are centrifuged at 40,000 rpm (195, 700 g) for 3 hr in a Beckman SW41 rotor Beckman Instruments, Inc., Palo Alto, CA). This results in three distinct bands and a pellet (Fig. 1). The material on top of the 31% sucrose layer represents cLPM whereas the membranes at the 34/38% interface correspond to blLPM (Fig. 1). These two LPM subfractions are carefully collected with a plastic Pasteur pipet, diluted in 0.25 M sucrose, and sedimented at 105,000 g for 60 min. The pellets are resuspended in the appropriate buffer medium by repeated (20 times in and out) suctioning through a 25-gauge needle. Alternatively, vesiculation of the membrane fragments can be performed by 30 strokes in a tight-fitting glass–glass or Teflon–glass homogenizer. The vesiculated membranes are then stored in liquid nitrogen for up to 4 weeks without loss of transport functions.

Properties of Isolated blLPM and cLPM Subfractions

Morphology. Transmission electron microscopy revealed that both LPM subfractions are composed of membrane vesicles, although blLPM still contained some unbroken lateral membrane sheets (Fig. 2a). In addition, mitochondrial membrane fragments were occasionally found in the blLPM subfraction, a finding consistent with the slight enrichment over homogenate of succinate cytochrome c reductase activity (Table I). Correspondingly, intravesicular volumes are approximately twofold lower in blLPM ($\sim$1 μl mg^{-1} protein) than in cLPM ($\sim$2 μl mg^{-1} protein) vesicles as estimated from equilibrium uptakes of [^{14}C]glucose.[20] Freeze-fracture analysis indicated that 72% of blLPM[21] and 77% of cLPM vesicles[20] are oriented right side out, i.e., the outer vesicle surface corresponds to the extracellular membrane area *in vivo*. However, since the basolateral subfraction is regularly contaminated with canalicular vesicles to approximately 10%[19] the true right-side-out orientation of basolateral vesicles might range between 60 and 65%.

Recoveries. Following the above-outlined procedure it is possible to simultaneously isolate approximately 0.19 mg blLPM and 0.10 mg cLPM protein/g liver from the same homogenate (Table I). Since the capacity of the method extends to a total of 115 g liver (wet wt), the maximum yields of blLPM and cLPM protein range around 22 and 12 mg, respectively. These are the highest recoveries of simultaneously isolated basolateral and canalicular LPM reported so far.

Degree of Purification. Table I summarizes the protein recovery and enzymatic characteristics of the isolated LPM subfractions. Compared to the homogenate blLPM are slightly enriched with mitochondria, but three- to fivefold deenriched with respect to microsomes, lysosomes and Golgi complex. In contrast, cLPM are virtually free of mitochondrial fragments, but contain slightly more microsomal, lysosomal, and Golgi membranes than the blLPM subfraction. Since endoplasmic reticulum represents 24%, Golgi complex 1%, lysosomes 2%, and mitochondria 16% of total homogenate protein,[22] it can be calculated that the total contamination of blLPM and cLPM with these intracellular organelles accounts for approximately 38% ($24 \times 0.3 + 1 \times 0.2 + 2 \times 0.3 + 16 \times 1.9$) and 23% ($24 \times 0.7 + 1 \times 0.4 + 2 \times 1.1 + 16 \times 0.2$) of total protein, respectively.

In regard to domain-specific plasma membrane markers the blLPM subfraction is 24- to 45-fold enriched over homogenate in Na$^+$,K$^+$-ATPase activity, contains this enzyme's α-subunit,[23] and localizes glucagon-stimu-

[20] P. J. Mier, A. St. Meier-Abt, C. Barrett, and J. L. Boyer, *J. Biol. Chem.* **259**, 10614 (1984).
[21] R. H. Moseley, P. J. Meier, P. S. Aronson, and J. L. Boyer, *Am. J. Physiol.* **250** G35 (1986).
[22] D. M. Neville, *in* "Biochemical Analyis of Membranes" (A. H. Maddy, ed.), p. 27. Chapman & Hall, London, 1976.
[23] E. S. Sztul, D. Biemesderfer M. J. Caplan, M. Kashgarian, and J. L. Boyer, *J. Cell Biol.* **104**, 1239 (1987).

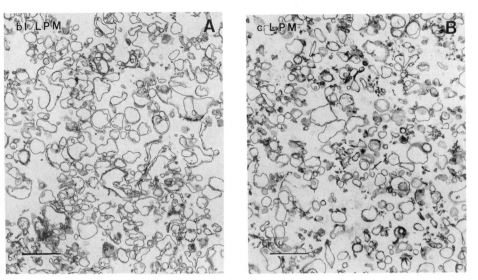

FIG. 2. Electron micrographs of (A) isolated basolateral (blLPM) and canalicular (cLPM) rat liver plasma membrane subfractions. (Bars, 1 μm; ×14,000)

latable adenylate cyclase activity (Table I). In contrast, these marker enzymes could not be detected in highly purified cLPM that are additionally characterized by 50- to 90-fold enrichments of classical canalicular marker enzyme activities such as alkaline phosphatase, aminopeptidase (leucylnaphthylamidase), γ-glutamyltranspeptidase, and 5'-nucleotidase (Table I).[19] These results indicate that the isolated cLPM are virtually devoid of basolateral (sinusoidal) contaminants. In contrast, based on the fourfold enrichment of the canalicular specific marker enzyme aminopeptidase (leucylnaphthylamidase), the blLPM subfraction is contaminated with canalicular membranes to approximately 10%.

Comments

Suitability of Marker Enzyme Activities for Assessment of the Separation between blLPM and cLMP. In order to correctly delineate the functional polarity of the hepatocellular surface domains with respect to ion transport processes, it is necessary to routinely test the purification of isolated blLPM and cLPM vesicles. For this purpose marker enzymes must be used that have unequivocally been demonstrated to be selectively localized at either the canalicular or the basolateral pole of hepatocytes. Aminopeptidase (leucylnaphthylamidase) has been shown to be specific for the canalicular membrane by both enzyme activity measurements as well as

TABLE I

PROTEIN RECOVERY AND ENZYMATIC CHARACTERIZATION OF BASOLATERAL (blLPM) AND CANALICULAR (cLPM) SUBFRACTIONS[a]

	blLPM	cLPM
Protein recovery (mg/g liver) (193)	0.19 ± 0.06	0.10 ± 0.03
Marker enzymes for intracellular organelles[b]	Relative specific activities[c]	
Succinate cytochrome c reductase (mitochondria) (55)	1.9 ± 0.9	0.2 ± 0.4
NADPH cytochrome c reductase (microsomes) (32)	0.2 ± 0.2	0.7 ± 0.4
Acid phosphatase (lysosomes) (8)	0.3 ± 0.2	1.1 ± 0.3
UDPgalactosyltransferase (Golgi) (8)	0.2 ± 0.3	0.4 ± 0.4
Plasma membrane enzyme markers[b]		
Na^+,K^+-ATPase activity (53)	34 ± 11	Not detectable[d]
Alkaline phosphatase (12)	12 ± 4	71 ± 21
Leucylnaphthylamidase (33)	4 ± 3	42 ± 9
γ-Glutamyltranspeptidase (7)	15 ± 5	60 ± 12
5′-Nucleotidase (5)	11 ± 5	64 ± 5
Glucagon-stimulatable adenylate cyclase (3)[e]	Present	Not detectable
α Subunit of Na^+,K^+-ATPase[f]	Present	Not detectable

[a] Results are given as the means $\pm$ SD with the number of experiments in parentheses.

[b] Marker enzyme activities were determined according to the following procedures: Succinate and NADPH cytochrome c reductases [G. L. Sottocasa, B. Kuylenstierna, L. Ernster, and A. Bergstrand, *J. Cell Biol.* **32**, 415 (1967)], acid phosphatase [T. L. Rothstein and J. J. Blum, *J. Cell Biol.* **57**, 630 (1973)], UDPgalactosyltransferase [B. Fleischer and M. Smigel, *J. Biol. Chem.* **253**, 1632 (1978)], Na^+,K^+-ATPase [B. F. Scharschmidt, E. B. Keeffe, N. M. Blankenship, and R. K. Ockner, *J. Lab. Clin. Med.* **93**, 790 (1979)], alkaline phosphatase [E. B. Keeffe, B. F. Scharschmidt, N. M. Blankenship, and R. K. Ockner, *J. Clin. Invest.* **64**, 1590 (1979)], leucylnaphthylamidase [J. A. Goldbarg and A. M. Rutenburg, *Cancer (Philadelphia)* **2**, 283 (1958)], γ-glutamyltranspeptidase [M. Orlowski and A. Meister, *Biochim. Biophys. Acta* **73**, 676 (1963)], 5′-nucleotidase [J. Avruch and D. F. H. Wallach, *Biochim. Biophys. Acta* **233**, 334 (1971)].

[c] Relative specific activity = specific activity ratios, LPM subfractions/homogenate.

[d] Absence of any detectable Na^+,K^+-ATPase activity in cLPM has been achieved in 2 out of 3 in a total of over 500 membrane preparations.

[e] A. F. Stewart, K. L. Insogna, D. Goltzman, and A. E. Broadus, *Proc. Natl. Acad. Sci. U.S.A.* **80**, 1454 (1983).

[f] Determined by immunoblotting using monoclonal antibodies [E. Sztul, M. Caplan, D. Biemesderfer, L. Barrett, M. Kashgarian, and J. L. Boyer, *J. Cell Biol.* **104**, 1239 (1987).

immunolocalization studies.[24] Therefore, this enzyme is well suited for estimating the contamination of blLPM with cLPM. In contrast, according to histochemical and various subcellular fractionation studies, the Na^+,K^+-ATPase is selectively localized at the basolateral mem-

[24] L. M. Roman and A. L. Hubbard, *J. Cell Biol.* **98**, 1488 (1984).

brane.[15,17,19,25-27] In addition, recent monoclonal antibody studies have also found the α subunit of Na$^+$,K$^+$-ATPase to be exclusively localized at the basolateral membrane domain.[23] Although two other immunolocalization studies found the α subunit at the canalicular membrane as well,[28,29] further purification of the antibody preparation resulted in a complete loss of canalicular immunoreactivity in one study.[28,29a] Since in isolated plasma membrane subfractions enzyme activity as well as the α subunit of Na$^+$,K$^+$-ATPase are associated with blLPM, but not with cLPM,[23] the absence of any detectable Na$^+$,K$^+$-ATPase activity is an adequate criterion to exclude significant basolateral contamination of cLPM.[29b] Thus, measurements of aminopeptidase and Na$^+$,K$^+$-ATPase activities are useful tests for routine screening of the degree of cross-contamination of liver plasma membrane subfractions.

Reproducibility of the Isolation Method. Based on the determination of leucine aminopeptidase and Na$^+$,K$^+$-ATPase activities, blLPM with approximately 10% canalicular contamination and cLPM devoid of measurable Na$^+$,K$^+$-ATPase activity can be prepared in two out of three experiments. Successful separation appears to most critically depend on (1) the use of fed animals, (2) rat weights of 200 g, and (3) tight homogenization of the mixed LPM. The latter step is especially important since insufficient dissociation of canalicular membranes from lateral membrane sheets considerably decreases the final yield of cLPM. Although we recommend 50 strokes (1 stroke equals up and down) with a tight type B glass–glass Dounce homogenizer, alternative methods (e.g., sonication, motor-driven Teflon–glass homogenizer) can also be used. It should be realized, however, that too-harsh homogenization conditions can damage the membrane integrity which can best be controlled by determining the recovery of marker enzyme activities in the isolated LPM subfractions.

Practicability of the Procedure. Although the described procedure involves zonal centrifugation, the handling of the TZ-28 reorienting density gradient zonal rotor (Sorvall) is considerably easier than the use of classical high-speed zonal rotors. For example, the TZ-28 rotor does not require an

---

[25] B. L. Blitzer and J. L. Boyer, *J. Clin. Invest.* **62,** 1101 (1978).

[26] P. S. Latham and M. Kashgarian, *Gastroenterology* **76,** 988 (1979).

[27] M. Inoue, R. Kinne, T. Tran. L. Biempica, and I. M. Arias, *J. Biol. Chem.* **258,** 5183 (1983).

[28] S. Takemura, K. Omori, K. Tanaka, K. Omori, S. Matsuura, and Y. Tashiro, *J. Cell Biol.* **99,** 1502 (1984).

[29] D. B. Schenk and H. L. Leffert, *Proc. Natl. Acad. Sci. U.S.A.* **80,** 5281 (1983).

[29a] Y. Fashiro, K. Omori, A. Yamamoto, *J. Histochem. Cytochem.* **36,** 221 (1988).

[29b] M. Sellinger, C. Barrett, Ph. Malle, E. R. Gordon, and J. L. Boyer, *Hepatology (Baltimore)* **11,** 223 (1990).

ultracentrifuge and can be unloaded statically (nonturning position). The total centrifugation time of the procedure is $6\frac{1}{2}$ hr permitting high-yield isolation of both LPM subfractions in 1 day (total preparation time ~ 11 hr).

Furthermore, the procedure can easily be scaled down to one rat liver (~10 g net wt). For this purpose the initial crude nuclear pellet is resuspended in 2.2 vol of 70% (w/w) sucrose. After stirring for 15 min, 12 ml of the suspension is filled into SW 28 rotor (Beckman Instruments, Inc., Palo Alto, CA) tubes and overlayered with 10 ml 44% and 10 ml 36.5% (w/w) sucrose. The tubes are filled to the top with 0.25 M sucrose and the gradient system centrifuged at 26,000 rpm for 90 min. Mixed LPM are recovered from the 44/36.5% sucrose interface and subjected to the same washing, tight homogenization, and high-density sucrose centrifugation steps are described above. By this means approximately 1 mg cLPM and 3 mg blLPM protein can be isolated from one liver with virtually the same degree of purification as demonstrated for the high-capacity zonal centrifugation method (Table I).

Suitability of Isolated blLPM and cLPM Vesicles for in Vitro Transport Studies. Table II summarizes several receptors and transport systems identified and partially characterized in the two isolated LPM subfractions. The findings demonstrate that both vesicle preparations are well suited for investigating hepatocellular membrane transport functions. Furthermore, the results illustrate the polar nature of hepatocyte plasma membrane transport processes. This supports the concept that the mechanisms and driving forces involved in overall transport of solutes from blood into bile are similar to vectorial transport mechanisms in other epithelial cells.

Comparison with Other Isolation Procedures

In addition to the described technique, other methods have recently been developed for isolation of highly purified basolateral and canalicular LPM vesicles. Although we have no direct experience with these alternative procedures, a short description of the various principles and techniques involved might help in choosing the most adequate method for a given purpose.

Procedures of Inoue and co-workers[27,30] describe the isolation of sinusoidal or canalicular membrane vesicles. *Sinusoidal* vesicles are prepared from a postnuclear supernatant by sucrose-Ficoll density gradient centrifugation. Seventy percent of the vesicles are oriented right side out and

---

[30] M. Inoue, R. Kinne, T. Tran, and I. M. Arias, *Hepatology (Baltimore)* **2,** 572 (1982).

TABLE II
Receptors and Transport Systems Identified in
Basolateral (bILPM) and Canalicular
(cLPM) Vesicles[a]

	bILPM	cLPM
Intact IgA$_2$ receptor[b] ("secretory component")	+	−
Bile acid-binding polypeptides[c]	M_r 54,000 M_r 48,000	M_r 100,000
ATP-dependent Ca^{2+} uptake [d]	+	−
Na$^+$/alanine cotransport[20]	+	+
Na$^+$/taurocholate cotransport[20]	+	−
Na$^+$,K$^+$-dependent glutamate uptake[e]	−	+
Na$^+$/H$^+$ antiport[21]	+	−
Cl$^-$/HCO$_3^-$ exchange[f]	−	+
OH$^-$/SO$_4^{2-}$ exchange[g]	+	−
HCO$_3^-$/SO$_4^{2-}$ exchange[h]	−	+

[a] +, Present; −, not detectable.

[b] Determined by immunoblotting using a monospecific antibody.[19]

[c] Identified by photoaffinity labeling of bILPM and cLPM with the sodium salt of the photolabile derivative (7,7-azo-3,12-dihydroxy-5-[^{3}H]cholan-24-oyl)-2-aminoethanesulfonic acid [W. Kramer, U. Bickel, H. P. Buscher, W. Gerok, and G. Kurz, *Eur. J. Biochem.* **129,** 13 (1982); St. Ruelz, G. Fricker, G. Hugentobler, K. Winterhalter, G. Kurz, and P. J. Meier, *J. Biol. Chem.* **262,** 11324 (1987)].

[d] C. Evers, G. Hugentobler, R. Lester, P. Gmaj, P. J. Meier, and H. Murer, *Biochim. Biophys. Acta* **939,** 542 (1988).

[e] N. Ballatori, R. H. Moseley, and J. L. Boyer, *J. Biol. Chem.* **261,** 6216 (1986).

[f] P. J. Meier, R. Knickelbein, R. H. Moseley, J. W. Dobbins, and J. L. Boyer, *J. Clin. Invest.* **75,** 1256 (1985).

[g] G. Hugentobler and P. J. Meier, *Am. J. Physiol.* **251,** 6656 (1986).

[h] P. J. Meier, J. Valantinas, G. Hugentobler, and I. Rahm, *Am. J. Physiol.* **253,** 6461 (1987).

exhibit an internal volume of 1 μl mg^{-1} protein. The recovery is 0.3 mg protein/g liver with a maximum yield of 7.5 mg protein. Na$^+$,K$^+$-ATPase activity is 20-fold enriched over homogenate. Total contamination with intracellular organelles approximates 34%. The preparation time is less than 5 hr. *Canalicular* vesicles are isolated from the crude nuclear pellet using nitrogen cavitation and Ca^{2+} precipitation techniques. Virtually 100% of the vesicles are reported to be oriented right side out (intravesicular volume 1 μl mg^{-1} protein). Recovery and maximal yields are 0.12 mg/g liver and 3.0 mg, respectively. Canalicular marker enzyme activities are 50- to 60-fold enriched over homogenate, whereas the relative specific activity for Na$^+$,K$^+$-ATPase is low (2.5 ×). Contamination with intracellular organelles is around 15%. Isolation time is 4 to 5 hr.

Comments. Since sinusoidal vesicles are prepared from a postnuclear supernatant while canalicular vesicles are isolated from the crude nuclear pellet, this method should theoretically also permit isolation of both LPM subfractions from the same homogenate. However, comparative studies from such simultaneously isolated sinusoidal and canalicular vesicles have not been reported so far and, therefore, the practicability and total preparation time required for isolation of both LPM subfractions are not known (rough estimation: 7 to 10 hr). Furthermore, the contamination of the LPM subfractions with Golgi-derived membranes has not been assessed. Nevertheless, these vesicle preparations have been successfully used to assess the polar distribution of a variety of plasma membrane functions, including insulin and asialoglycoprotein receptors (sinusoidal),[31] Na$^+$/H$^+$ antiport (sinusoidal),[32] and transport of taurocholate[30,33] and glutathione[34-37] and the multidrug resistance cation pump (Gp 170; canalicular).[37a]

The method of Blitzer and Donovan[38] involves Percoll gradient centrifugation and permits the preparation of 0.64 mg/g liver basolateral vesicle protein within 4 hr. The maximal yield is between 12 and 16 mg. Seventy-

[31] I. M. Arias, *in* "Progress in Liver Diseases" (H. Popoer and F. Schaffner, eds.), Vol. 8, p. 145. Grune & Stratton, Orlando, Florida, 1986.

[32] I. M. Arias and M. Forgac, *J. Biol. Chem.* **259**, 5406 (1984).

[33] M. Inoue, R. Kinne, T. Tran, and I. M. Arias, *J. Clin. Invest.* **73**, 659 (1984).

[34] M. Inoue, R. Kinne, T. Tran, and I. M. Arias, *Eur. J. Biochem.* **138**, 491 (1984).

[35] M. Inoue, R. Kinne, T. Tran, and I. M. Arias, *Eur. J. Biochem.* **134**, 467 (1983).

[36] T. Akerboom, M. Inoue, H. Sies, R. Kinne, and I. M. Arias, *Eur. J. Biochem.* **141**, 211 (1984).

[37] M. Inoue, T. M. Akerboom, H. Sies, R. Kinne, T. Trans, and I. M. Arias, *J. Biol. Chem.* **259**, 4998 (1984).

[37a] Y. Kamimoto, Z. Gaitmaitan, J. Hsu, and I. M. Arias, *J. Biol. Chem.* **264**, 11693 (1989).

[38] B. L. Blitzer and C. B. Donovan, *J. Biol. Chem.* **259**, 9295 (1984).

five percent of the vesicles are oriented right side out. Na⁺,K⁺-ATPase activity is 28-fold enriched over homogenate. Total contamination with intracellular organelles approximates 46%.

Comments. This method is fast and does not require ultracentrifugation. However, it requires removal of the Percoll before transport studies can be performed and results in an approximately 30% contamination of basolateral vesicles with endoplasmic reticulum. The vesicles have so far mainly been used for characterization of the basolateral bile acid uptake systems[38-40] as well as the ATP-dependent Ca⁺ pumps.[41]

The highest purification of canalicular membrane vesicles (153-fold enrichment of aminopeptidase) has been reported by Hubbard *et al.,*[8,18] whose isolation procedure involves immunoadsorption on anti-aminopeptidase antibody-coated *Staphylococcus aureus* cells. Although the procedure has not been used on a preparative scale (protein yield ~45 μg), it could theoretically be scaled up 10- to 100-fold provided high enough amounts of antibodies are available.

Acknowledgment

Theses studies were supported by the Swiss National Science Foundation (Grants 3.983.0.84 and 3.992.0.87) and United States Public Health Service Grants DK-25636 and DK-36854.

[39] B. L. Blitzer and L. Lyons, *Am. J. Physiol.* **249** G34 (1985).
[40] B. L. Blitzr, Ch. Terzakis, and K. A. Scott, *J. Biol. Chem.* **261**, 12042 (1986).
[41] B. L. Bitzer, B. R. Hosteller, and K. A. Scott, *J. Clin. Invest.* **83**, 1319 (1989).

Section II

Other Epithelia

[35] Electrogenic and Electroneutral Ion Transporters and Their Regulation in Tracheal Epithelium

By Carole M. Liedtke

Introduction

Tracheal epithelium has been intensely studied in recent years because it expresses the electrolyte transport defect of cystic fibrosis, a genetic disease characterized by abnormal epithelial Na^+ and Cl^- transport. Electrolyte transport in the airway epithelium is critical in hydrating mucus, maintaining the osmolarity of respiratory tract fluid, and humidifying inspired air. To accomplish this, tracheal epithelium can absorb or secrete fluid by regulating transport of the electrolytes Na^+ and Cl^-. During fluid absorption, Na^+ is actively transported down its electrochemical gradient from the lumen to the submucosa through an apical Na^+ channel while Cl^- "passively" moves through the tight junctions down the paracellular pathway. Fluid secretion, on the other hand, is characterized by the "active" net transport of Cl^- from submucosa to lumen through a basolateral NaCl cotransporter and apical Cl^- channel. Charge transfer across the apical membrane is offset by K^+ efflux through a basolateral channel. Chloride secretion can be rapidly elicited by hormones and local mediators.

Particular attention has focused on electrogenic pathways that operate during Cl^- secretion. Electrogenic pathways, or channels, mediate the net movement of an ion and its charge across the plasma membrane. Table I lists methods utilized to investigate Cl^- and K^+ channels and their regulation in airway epithelial. Because of the electrical nature of ion channels, they are readily detected using techniques of electrophysiology, such as Ussing chamber-type experiments, intracellular microelectrode techniques, and patch-clamp technology. Researchers using these techniques have identified apical Cl^- channels in airway epithelium as a critical site of expression of the cystic fibrosis defect.[1-3] Investigators also use radioactive tracers thought to be specific for apical Cl^- channels, such as ^{125}I, as a quick, less technically oriented method to screen Cl^- channel activity and to study Cl^- channels in a population of cells.[4] Recent measurements of the rate of ^{125}I efflux from cultured canine tracheal epithelial cells demonstrate

[1] R. A. Frizzell, G. Rechkemmer, and R. L. Shoemaker, *Science* 233, 558 (1986).

[2] M. J. Welsh and C. M. Liedtke, *Nature (London)* 322, 467 (1986).

[3] M. Li, J. D. McCann, C. M. Liedtke, A. C. Nairn, P. Greengard, and M. J. Welsh, *Nature (London)* 331, 358 (1988).

[4] J. P. Clancy, J. D. McCann, M. Li, and M. J. Welsh, *Am. J. Physiol.* 258, L25 (1990).

TABLE I
DETECTION OF CL⁻ AND K⁺ ELECTROGENIC PATHWAYS IN
AIRWAY EPITHELIUM

Channel	Cell source	Method	Reference
Cl⁻	Intact tissue	Ussing chamber	$a-d$
		Microelectrode	e,f
	Cell culture	Ussing chamber	g,h
		Microelectrode	i
		Patch-clamp technology	j,k
		^{125}I efflux	l
		^{36}Cl⁻ uptake	m
K⁺	Intact tissue	Ussing chamber	n
		Micropuncture	e
	Cell culture	Patch-clamp technology	k

[a] F. J. Al-bazzaz and Q. Al-Awqati, *J. Appl. Physiol.* **46**, 111 (1979).
[b] R. C. Boucher, M. J. Stutts, and J. T. Gatzy, *J. Appl. Physiol.* **51**, 706 (1981).
[c] S. R. Shorofsky, M. Field, and H. A. Fozzard, *Am. J. Physiol.* **72**, 105 (1983).
[d] J. Gliemann, K. Osterlind, J. Vinten, and S. Gammeltoft, *Biochim. Biophys. Acta* **286**, 1 (1972).
[e] P. L. Smith and R. A. Frizzell, *J. Membr. Biol.* **77**, 187 (1984).
[f] M. J. Welsh, *J. Clin. Invest.* **71**, 1392 (1983).
[g] D. L. Coleman, I. K. Tuft, and J. H. Widdicombe, *Am. J. Physiol.* **246** C355 (1984).
[h] J. R. Yankaskas, C. U. Cotton, M. R. Knowles, J. T. Gatzy, and R. C. Boucher, *Am. Rev. Respir. Dis.* **132**, 1281 (1985).
[i] M. J. Welsh, *J. Membr. Biol.* **91**, 121 (1986).
[j] R. A. Frizzell, G. Rechkemmer, and R. L. Shoemaker, *Science* **233**, 558 (1986).
[k] M. J. Welsh and C. M. Liedtke, *Nature (London)* **322**, 467 (1986).
[l] J. P. Clancy, J. D. McCann, M. Li, and M. J. Welsh, *Am. J. Physiol.* **248**, L25 (1990).
[m] M. J. Stutts, C. U. Cotton, J. R. Yankaskas, E. Cheng, M. R. Knowles, J. T. Gatzy, and R. C. Boucher, *Proc. Natl. Acad. Sci. U.S.A.* **82**, 6677 (1985).
[n] M. J. Welsh, *Am. J. Physiol.* **245**, C248 (1983).

Ca²⁺-dependent activation of apical Cl⁻ channels that requires intracellular but not extracellular Ca²⁺.[4] These findings support and extend previous research in which Ca²⁺-dependent Cl⁻ channel activity in membrane

[5] M. Li, J. D. McCann, M. P. Anderson, J. P. Clancy, C. M. Liedtke, A. C. Nairn, P. Greengard, and M. J. Welsh, *Science* **244**, 1353 (1989).

patches was linked to diacylglycerol- and Ca^{2+}-dependent protein kinase C.[5]

In contrast to electrogenic pathways for electrolyte transport, electroneutral transporters mediate the net translocation of an ionic species without a net charge transfer. Examples of electroneutral transport pathways that are thought to operate in the tracheal epithelium include an NaCl(K) cotransporter[6-8] and Cl⁻/anion exchange.[9] Because of their electroneutrality, these transport pathways are not readily detected with electrophysiological techniques. Rather, they are studied using radiolabeled electrolyte fluxes. We use two methods for assessing an NaCl(K) cotransporter in airway epithelium: radioisotopic Na^+,K^+, or Cl⁻ flux in isolated airway epithelial cells and bioelectric properties of a monolayer of cells grown on a filter support. In the latter method, the role of a cotransporter in vectorial ion transport and consequences of cotransporter activation on net ion movements across the monolayer are assessed. Both methods allow kinetic analysis of electrolyte transport, an advantage useful in modeling cotransport mechanisms.

Radioisotopic Fluxes

Transport of Na^+ and Cl⁻ in dispersed cells can be readily measured using radiotracers, either as uptake or efflux of $^{22}Na^+$ or $^{36}Cl^-$. Our approach in studying tracheal epithelial transport pathways is to use tracer efflux because it potentially allows detection of all electrolyte transport pathways. Radiotracer efflux is accomplished by preloading cells with either $^{36}Cl^-$ or $^{22}Na^+$ to achieve radioisotopic equilibrium, then initiating movement of radiolabel into transport medium lacking radiolabel. By controlling the composition of the transport medium, one can observe electrolyte transport or tracer exchange for unlabeled electrolyte. Net electrolyte transport, as seen in Cl⁻ secretion, is achieved by utilizing transport medium of a lower Cl⁻ concentration than the preincubation medium, that is, generating a Cl⁻ electrochemical gradient similar to that in the *in vivo* state. Tracer exchange occurs when the prelabeling and transport medium have the same physiological salt composition. With this experimental approach, radiotracer will move from the cell by any transport pathway operative under the selected conditions. Depending on the driving force for electrolyte transport, a transporter can operate in reverse directions.

[6] C. M. Liedtke, *Am. J. Physiol.* **257**, L125 (1989).
[7] C. M. Liedtke, *Am. J. Physiol.* in press.
[8] C. M. Liedtke, M. Romero, and U. Hopfer, in "Cellular and Molecular Basis of Cystic Fibrosis" (G. Mastella and P. M. Quinton, eds.), p. 307. San Francisco Press, San Francisco, California, 1988.
[9] A. Elgavish and E. Meezan, *Biochem. Biophys. Res. Commun.* **152**, 99 (1988).

To identify and characterize transport processes operative during radiotracer efflux, probes for specific transport pathways are utilized. For example, ouabain, an inhibitor of Na^+,K^+-ATPase, will reduce tracer exchange because the driving force for electrolyte movements, the Na^+ electrochemical gradient, is perturbed. More specific blockers allow detection of transport pathways that directly contribute to tracer efflux (exchange). Loop diuretics, such as furosemide and bumetanide, block Cl^--dependent cotransport systems, such $NaCl$, KCl, or $NaKCl_2$ cotransport, and the disulfonic stilbene (SITS) inhibits $Cl^-/OH^-(Cl^-/HCO_3^-)$ exchange pathways.

Intrinsic to the effectiveness of these agents is their specificity for a particular transport pathway. Some blockers, such as amiloride and its analogs, inhibit mechanistically diverse Na^+ transport pathways at different effective concentrations. Thus, low (10^{-7} M) concentrations of amiloride block Na^+ channels but at higher concentrations of drug (10^{-4} M), Na^+/H^+ exchange is also blocked. Other transport blockers display a sensitivity in isolated membrane, such as membrane vesicles or membrane patches, or in specific cell types or cell lines that may not be realized in intact cells. These types of inhibitors include Ba^{2+} for K^+ channels and diphenylamine carboxylic acid (DPC) and 5-nitro-2-(3-phenylpropylamino)benzoate (NPPB) for Cl^- channels.

We measure time-dependent tracer exchange, to characterize NaCl(K) cotransport and its regulation by α-adrenergic agonists in dispersed cells. Tracer exchange is particularly useful because the rate of exchange, as the dependent variable, becomes a function of only one experimental variable. The experimental variable may be temperature, pH, hormone, Ca^{2+} concentration, ionophore, or other variable. Tracer transport is assessed by determining the initial linear transport rates of tracer efflux.

Cell Isolation

Airway epithelial cells are isolated from adult male New Zealand White rabbits, weighing 3–4 kg, from human trachea obtained at autopsy, or from nasal polyps obtained at surgery. Methods for cell isolation are those devised by this laboratory[10,11] with modifications for the different types of tissue.

1. Rabbit trachea is transported in 50 ml Hanks' balanced salt solution buffered with 10 mM N-2-hydroxyethylpiperazine-N'-2-ethanesulfonic acid (HEPES)-OH, pH 7.4 (HPSS). The trachea is placed in a Petri dish

[10] C. M. Liedtke, *Am. J. Physiol.* **251,** C209 (1986).
[11] C. M. Liedtke and B. Tandler, *Am. J. Physiol.* **247,** C441 (1984).

containing HPSS and is dissected free of extratracheal tissue using a fine forceps and scissors. The tissue is cut into four to eight pieces and transferred to a 50-ml polypropylene test tube containing the first digestion enzyme solution.

2. Human trachea and nasal polyps are retrieved in sterile medium consisting of a 1 : 1 (v/v) mixture of Ham's F-12 and Dulbecco's modified Eagle's medium (DMEM) supplemented with (per milliter) 50 μg tobramycin, 100 μg streptomycin, and 100 U penicillin. Trachea is treated as described above for rabbit tissue. Nasal polyp specimens often include extraepithelial material and vary considerably in size. Polyps are placed in a Petri dish flooded with HPSS, dissected free of extraepithelial tissue, and cut into small pieces by a mincing action with sharp scissors. The pieces of tissue are rinsed thoroughly with HPSS, then transferred to a 50-ml polypropylene conical tube containing the first digestion enzyme solution.

3. The first digestion enzyme solution is prepared just before use according to the formulation in Table II. Solutions for use with human tissue is supplemented with 2 mg dispase/ml, standard antibiotics (see above), and 200 mg piperacillin/ml.

DNase and RNase are effective in minimizing loss of cells that adhere to nucleic acids released from damaged or dying cells. The sulfhydryl agent dithiothreitol reduces the viscosity of secreted mucus which interferes with the recovery of dispersed cells. Ibuprofen diminishes the synthesis of cyclooxygenase products that may stimulate mucus secretion. Likewise, atropine is used to prevent muscarinic stimulated mucus secretion by endogenous cholinergics. EDTA, at this low concentration, apparently loosens the epithelium from the underlying submucosa by chelating divalent cations required as attachment cofactors.[12]

4. Rabbit tissue is incubated for 60 min at 32° and human tissue overnight at room temperature, then, after addition of EDTA, at 37° for 15 min. Collagenase is markedly sensitive to temperatures between 18 and 37° hence the longer incubation time at room temperature. Below 18°, collagenase is practically inactive. Cells recovered from the overnight treatment are comparable to those obtained at 32° incubation as assessed by viability, proportion of ciliated cells, and β-adrenergic mediated elevations in cAMP levels.

5. Tracheal tissue is transferred to a Petri dish containing HPSS and the epithelium is dissected from the lumenal surface using a stainless steel spatula or angled probe. HPSS serves to continuously hydrate the tissue. To evaluate the efficiency of the dissection technique, the surface epithe-

---

[12] J. M. Sherman, P.-W. Cheng, B. Tandler, and T. F. Boat, *Am. Rev. Respir. Dis.* **124,** 476 (1981).

TABLE II
DIGESTION ENZYME SOLUTIONS FOR AIRWAY EPITHELIAL CELL ISOLATION

Stock solution	Stock concentration	First incubation		Second incubation	
		Final volume	Concentration	Final volume	Concentration
Ca^{2+},Mg^{2+}-HPSS	—	10 ml	—	20 ml	—
Collagenase	—	675 U	135 U/ml	2700 U	135 U/ml
DNase	1 mg/ml	100 μl	10 μg/ml	200 μl	10 μg/ml
RNase	10 mg/ml	50 μl	50 μg/ml	100 μl	50 μg/ml
Dithiothreitol	—	1 mg	50 μg/ml	1 mg	25 μg/ml
EDTA	0.66 M	100 μl[a]	3.3 mM	—	—
EGTA	0.40 M	—	—	100 μl	2 mM
Atropine	1 mM	100 μl	1 M	100 μl	5 μM
Ibuprofen	10 mM	10 μl	1 mg	10 μl	2.5 μM

[a] Omit for incubations with human trachea.

lium can be visualized under a dissecting microscope during the dissection procedure or afterward. The surface epithelium is removed as sheets of cells that can easily be viewed by phase microscopy. The epithelial sheets are immediately transferred to the second digestion enzyme solution, prewarmed to room temperature.

The solution with nasal polyps is diluted with 20 ml Ca^{2+},Mg^{2+}-free HPSS and repeatedly pipetted with a 10-ml sterile, plastic pipet from which the tip is removed. After tissue pieces settle to the bottom of the test tube, the supernatant is simultaneously decanted and filtered through a 70-μm Nitex nylon filter. Filtered cells are recovered by centrifugation in an IEC (Needham Heights, MA) Centra 7R centrifuge at 850 g for 10 min at 18° and resuspended in the second digestion solution.

6. The second digestion enzyme solution is prepared just before use according to the formulation in Table II. EGTA is used to chelate Ca^{2+} ions that are required for the integrity of the tight junction. Opening of the tight junction in the presence of proteolytic enzymes is thought to accelerate digestion of protein essential structurally for tight junctions.

Incubation time may vary from 40 to 60 min at 32°. One can easily monitor this incubation step by periodically observing an aliquot of cell suspension by phase microscopy. Our preference is to terminate the enzyme incubation when we observe single cells or clumps of two to six cells. Further incubation to achieve a uniform incubation time may affect cell viability.

7. Dispersed airway epithelial cells are recovered by centrifugation at

850 g for 5 min at 18°, then resuspended by gently mixing in 10 ml Hanks' balanced salt solution buffered with 10 mM HEPES, pH 7.4.

To remove erythrocytes, cell debris, and large clumps of cells, the suspended cells are centrifuged over a cushion of 60% Percoll. Percoll (6 ml) is added to the bottom of the tube with a syringe and needle. Cells are centrifuged at 850 g for 10 min at 18°, recovered as a band at the buffer-Percoll interface, and diluted to 40 ml with HPSS supplemented with 1% fatty acid-free bovine serum albumin. Cells are pelleted by centrifugation and washed twice with HPSS plus 1% fatty acid-free bovine serum albumin to remove traces of enzymes and Percoll. The final cell pellet is resuspended in 1 ml HPSS plus 1% fatty acid-free bovine serum albumin and incubated at 25° to allow reequilibration with divalent cations.

8. Cells are counted with a hemocytometer and viability is routinely assessed by staining with the fluorescent dyes acridine orange and ethidium bromide.[13] Living cells admit only acridine orange and fluoresce green while dead cells or cells with a compromised plasma membrane allow entry of ethidium bromide. Ethidium bromide binds to DNA more strongly than acridine orange and hence causes a red/orange fluorescence.

Electrolyte Transport Measurements

General. Airway epithelial cells are used the same day for transport experiments. Cells are concentrated to at least 15×10^6 cells in 1 ml HPSS plus 1% fatty acid-free bovine serum albumin. With this approach, smaller quantities of radioisotopes are needed to achieve a high concentration of radioisotope. Experiments are conducted at 25° to reduce Na$^+$,K$^+$-ATPase activity. At physiological temperatures of 37°, radioisotopic electrolyte fluxes are too rapid for kinetic analysis.

Cells are pretreated with transport blockers, enzyme inhibitors, or hormone receptor antagonists by adding 1/1000 (v/v) of concentrated stock solutions of agent to the cell suspension. Addition of this low volume of agent prevents dilution of extracellular fluid and correspondingly extracellular radioisotope. Thus, steady state electrolyte distribution is maintained.

Initiation of Radioisotopic Electrolyte Transport Uptake: An aliquot of cell suspension (1.6 to 3.2×10^6 cells) is transferred to a 15-ml polypropylene conical tube containing prewarmed transport medium of the same composition as the suspension medium plus 1.5 to 2.0 μCi radioisotope (^{22}Na$^+$, ^{36}Cl$^-$, or ^{86}Rb$^+$) and drugs, as indicated. Cells preincubated with transport blockers, enzyme inhibitors, or hormone receptors antagonists

---

[13] S. K. Lee, J. Singh, and R. B. Taylor, *Eur. J. Immunol.* **5**, 259 (1975).

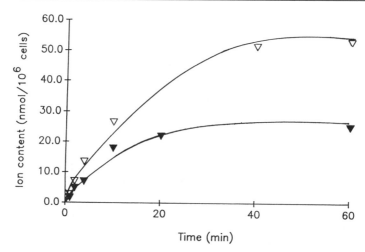

FIG. 1. Time course of $^{22}Na^+$ and $^{36}Cl^-$ uptake in rabbit tracheal epithelial cells at 25°.[6] To initiate radioisotope uptake, aliquots of cells were transferred to Hanks' balanced salt solution buffered with 10 mM HEPES-OH, pH 7.4, and containing either 3.0 μCi $^{22}Na^+$ or 1.5 μCi $^{36}Cl^-$. At the indicated times, an aliquot of cells was recovered for radioactive counts. Each data point represents results from at least two separate experiments. Symbols: ▼, $^{22}Na^+$; ▽, $^{36}Cl^-$.

are added to transport medium that contains the same concentration of agent as the concentrated cells.

Efflux: Cells are preequilibrated with radioisotopes by adding 1.5 μCi $^{36}Cl^-$ or $^{22}Na^+$ and incubating for 60 min at 25°. Preliminary experiments on the uptake of radiolabeled ions indicate that the selected preincubation time was sufficient to achieve radioisotopic equilibrium (Fig. 1). Cells to be pretreated with agents are firsts preequilibrated with radioisotopes.

To initiate efflux of radiolabel, an aliquot of preequilibrated cells (1.6 to 3.2 × 10⁶ cell) is transferred to a 15-ml polypropylene conical tube containing ninefold excess prewarmed transport medium of the same composition as the preincubation medium but lacking radiolabeled and containing drugs of interest.

Termination of Transport. Radiolabeled electrolyte transport is stopped by centrifugation of an aliquot of cell suspension (2.5 to 5.0 × 10⁵ cells) through 200 μl of cooled phthalate oil (1 : 1.75, dioctyl phthalate : dibutyl phthalate, v/v). Phthalate oils allow rapid, quantitative separation of airway epithelial cells from suspension.[14,15] Composition of the phthalate oil

[14] P. A. Andreasen, B. P. Schaumburg, K. Osterlind, J. Vinten, S. Gammeltoft, and J. Gliemann, *Anal. Biochem.* **59**, 610 (1974).

[15] J. Gliemann, K. Osterlind, J. Vinten, and S. Gammeltoft, *Biochim. Biophys. Acta* **286**, 1 (1972).

is determined empirically by testing mixtures of component phthalate oils. Dioctyl ($d = 0.981$) and dibutyl ($d = 1.043$) phthalate oils are commonly used in varying proportions to achieve oil mixtures that are tested to support the aqueous phase while allowing cells to sediment through the oil phase.

To assess the amount of extracellular fluid trapped in the pelleted cells, the phthalate oil is overlayerd with 25 μl HPSS containing 0.05 μCi[^{14}C]mannitol.

Phthalate oils and extracellular marker are added to polypropylene microcentrifuge tubes. The tubes are precooled in an 8° water bath and placed in a microfuge just before addition of cell suspension. Cells are centrifuged in a microfuge (5415; Beckman, Fullerton, CA) for 9 sec. A 10 μl aliquot of the upper aqueous layer is sampled for radioisotopic counts and the tube transferred to a dry ice–alcohol bath.

Radioactive Counting and Data Analysis. After the aqueous and oil phases are frozen, the tip of the tube, containing the cell pellet, is cut off directly into a 5-ml scintillation vial. Five milliliters of scintillation fluid (Ecolume; ICN, Irvine, CA) is added to each vial and the vial capped and incubated at room temp for 18 hr. After this time period, vials are counted in a beta scintillation counter (Beckman LS5801). Counts are collected over a 5-min counting interval and used to calculate the radiolabeled ion content of the cell pellets. Tracer content is corrected for extracellular space and expressed as nanomoles ion/10^6 cells. Data are graphed as tracer content vs time, shown for Na$^+$ and Cl$^-$ uptake in Fig. 1 and efflux in Fig. 2A and B.

The rate of tracer uptake and $t_{1/2}$ of uptake is derived from the relationship in Eq. (1):

$$C_t = C_\infty e^{kt} \tag{1}$$

where C refers to the radioisotope content of the cells at equilibrium (∞) or a time point (t) after initiation of transport and k is a rate constant.

The initial rate of radioisotopic electrolyte efflux is determined by least-squares analysis of initial linear portion of the time curves (see Fig. 2).

Early time points are critical for sophisticated kinetic analysis of data. Implicit in this is the detection of sufficient counts from experiment to experiment and cell preparation to cell preparation to obtain reproducible results. Adjustments in the amount of radioisotope and in the number of cells per time point may be necessary to achieve this goal.

Electrophysiology

Electrophysiology is a method used to evaluate the net electrolyte transport between two aqueous compartments separated by an epithelium.

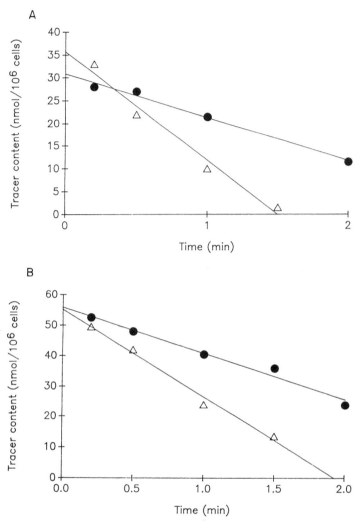

FIG. 2. Efflux of radioisotopic label from rabbit tracheal epithelial cells at 25°.[6] Cells were isolated, resuspended in Hanks' balanced salt solution buffered with 10 mM HEPES-OH, pH 7.4, and preincubated with either ^{22}Na$^+$ or ^{36}Cl$^-$ for 60 min at 25° to achieve radioisotopic equilibrium. Efflux of radiolabel was initiated and aliquots of cells recovered for radioactive counts as described under Termination of Transport. Data are calculated as nanomoles radioisotope/10^6 cells and plotted against time. (A) ^{22}Na$^+$ efflux: ●, Basal; △, l-epinephrine, 10 μM. (B) ^{36}Cl$^-$ efflux: ●, Basal; △, l-epinephrine, 10 μM.

In the airway epithelium, access to the tissue or to the basolateral plasma membrane is limited, making some electrophysiological measurements difficult to obtain and to interpret. However, we can now reconstitute the epithelium by growing isolated airway epithelial cells as a monolayer in *in vitro* culture on a permeable filter support. Primary cultures[5,16] or retroviral immortalized cell cultures[17] retain the properties of native epithelium. These cell cultures develop tight junctions, form a monolayer that polarizes with the basolateral surface toward the filter and apical surface facing the culture media, retain hormone receptors, and express hormonal stimulation of intracellular second messengers and transepithelial ion transport.

Applying electrophysiological techniques to cell cultures allows a more detailed study of the role of apical and basolateral channels in transepithelial electrolyte movements. The same techniques are valuable in assessing the role of electrically neutral $NaCl(K)$ cotransport to vectorial ion transport. This is accomplished by combining simultaneous bioelectric measurements with transepithelial radioisotopic electrolyte flux measurements to distinguish and quantitate electrically neutral electrolyte movements.

Cell Isolation

Airway epithelial cells are isolated aseptically as described above but without a Percoll purification step. Tissue is retrieved in sterile medium and dissected under a sterile bench top or laminar flow hood with gloved hands. Retrieval and digestion enzyme solutions are filtered sterilized by passage through a 0.2-μm filter. Solutions are supplemented with (per milliliter) 50 μg tobramycin, 100 μg streptomycin, and 100 U penicillin.

An alternative cell isolation technique that yields a comparable number of viable cells is dissociation in protease enzyme. Tissue pieces are incubated at $4°$ for $18-36$ hr in 20 ml Ca^{2+},Mg^{2+}-free HPSS containing 20 mg protease and 2 mg DNA. The enzyme solution is gassed to saturation with 5% CO_2, 95% O_2 before filter sterilization. After $18-36$ hr, tissue pieces are transferred to a sterile Petri dish and the surface epithelium gently scraped from the underlying submucosa with a stainless steel spatula. Scrapings are transferred to sterile HPSS and collected by centrifugation. Epithelial cells either singly or in small clumps are suspended in the enzyme solution. To recover these cells, the enzyme solution is diluted to 40 ml final volume with HPSS and centrifuged at 850 g for 5 min to collect epithelial cells.

Scrapings and cells are resuspended in culture medium, collected by centrifugation for 10 min at 125 g, and washed twice with culture medium.

[16] M. J. Welsh, *Physiol. Rev.* **67**, 1143 (1987).
[17] J. W. Jacobberger, U. Hopfer, C. M. Liedtke, M. F. Romero, and T. L. Sladek, *Ped. Pulmon., Suppl.* **4**, 126 (1989).

The final cell pellet is resuspended in culture medium and counted using a hemocytometer.

Cell Culture

Biomatrix. Epithelial cells appear to require a biological matrix for *in vitro* growth. There are a number of biological matrices that support the growth of airway epithelial cells, including the following:

Human placental collagen (type VI; Sigma St. Louis, MO): Prepare by dissolving 50 mg collagen in 10.2% glacial acetic acid. Filter sterilize the collagen solution before using to coat filters

Rat tail collagen: Prepare from tendon shafts of rat tails. Collect, weigh, and soak tissue overnight in 70% alcohol, then rinse three times with sterile deionized water. Extract collagen from tissue by overnight cold incubation in sterile 0.1% glacial acetic acid at a ratio of 1 g tissue/300 ml acid. Decant the extract into 50-ml sterile conical tubes and centrifuge at 500 g for 30 min to pellet large fragments of undissolved tissue. Decant the supernatant into 15-ml sterile conical tubes and store at $-70°$ until use

Other commercially available products, such as vitrogen

Filter Supports. Culture plastic inserts of 0.4-μm pore size and 12-mm diameter are purchased from Millipore (Bedford, MA) and are coated twice with collagen. Filters are air dried between coatings. To achieve a more uniform coating, the collagen solution is diluted with ethanol to a final alcohol concentration of 20%.

The collagen is glutaraldehyde fixed according to the method of Whitcutt.[18] Working glutaraldehyde solution is prepared by diluting 25% aqueous stock solution (Sigma, grade I) to 2.5% in deionized water just before use. Sufficient working solution is added to cover filters. After a 1-hr incubation at room temperature, the filters are washed twice with deionized water. The filters are covered with sodium borohydride prepared just before use to a final 0.1% concentration. After incubation for 18 hr at 4°, sodium borohydride is discarded and the filters rinsed at least four times with cold deionized water.

Filters are stored refrigerated in 70% alcohol in a 24-well tissue culture cluster dish. On the day of use, filters are transferred to a sterile 12-well tissue culture cluster dish and allowed to dry under a laminar flow hood.

Culture Protocol. Isolated cells are suspended in prewarmed Ham's F-12:DMEM (1:1) then preplated at 37° in a 5% CO_2 incubator for 1–2 hr in T75 tissue culture plastic flasks to remove fibroblasts which quickly

---

[18] M. J. Whitcutt, K. B. Adler, and R. Wu, *In Vitro Cell. Dev. Biol.* **24,** 420 (1988).

adhere to the plastic surface. Nonadherent cells remain suspended in the medium and are recovered with the medium.

Cells are seeded onto dry filters at a density of 1 to 1.5×10^6 cells/filter in 100 μl culture medium. After a 10-min incubation at room temperature, 100 μl culture medium is added to the filter well and 600 μl to the outside well. Cultures are incubated at 37° in a humidified atmosphere of 95% air–5% CO_2. Culture medium is changed every 3 days.

The growth of the cells as a monolayer is assessed by microscopy and fluorescent staining. Cell cultures are screened by phase microscopy for the presence of cell–cell boundaries that are indicative of a monolayer. Fluorescent dye staining with propidium iodide is particularly useful for establishing the extent of monolayer growth. Propidium iodide is sequestered in the cell cytosol but does not kill cells. After staining and observing the characteristic green fluorescence of the dye, the medium is changed and cell cultures returned to the CO_2 incubator.

Culture Medium. Cell culture medium consists of a 1:1 mixture of Ham's F-12 and Dulbecco's modified Eagle medium (GIBCO, Grace Island, NY) supplemented with the same antibiotics as the HPSS and insulin (5 μg/ml), hydrocortisone (50 μM), epidermal growth factor (4 ng/ml), fibronectin (4 μg/ml), selenium (0.03 μM), and 5% newborn calf serum.[19] The culture formulation supports the retention of differentiated properties of tracheal epithelial cells as assessed by morphology, expression of keratins, increased cAMP levels caused by β-adrenergic agonists, and release of PGE_2 mediated by α-adrenergic receptors.[19]

Electrophysiological Measurements

Probe Recording. Electrical resistance (R) and open circuit voltage (V) of the epithelial cell monolayer is easily assessed using a Millicell (Bedford, MA) electrical resistance system (ERS). We routinely select cultures for experimental use on the basis of resistance or its inverse, conductance (G). The ERS provides a good screening technique for resistance without the need for an Ussing chamber and related electronic equipment. The major problem with the Millicell ERS system is drift in electrode basal readings with time.

Short-circuited ERS electrodes are stored and equilibrated in phosphate-buffered saline (PBS). If cultures are to be returned to the incubator, the electrodes are sterilized by immersion in 70% alcohol for 15 min, then the meter and electrode are exposed to UV light for 20 min. The electrodes are transferred to the same medium the cells are grown in and allowed to equilibrate for 15 min at ambient temperature. The ERS is adjusted for the culture medium resistance by blanking the voltage potentiometer to zero.

---

[19] C. M. Liedtke, *Am. J. Physiol.* **255**, C760 (1988).

On the day of use, blank filters (not seeded with cells) and filters with cell monolayers are transferred to a fresh, sterile 12-well tissue culture cluster dish and preequilibrated in culture medium or other selected phsyiological salt buffers for 10 min at ambient temperature. Resistance and voltage readings are first done on blank filters, then on cell cultures. Care must be taken to avoid touching the electrode to the sides of the wells or the bottom of the filter ring. Abnormally high resistance readings usually are associated with air bubbles under the filter or debris inside the wells.

Transepithelial potential difference is calculated from voltage readings (Eq. 2) as

$$V_T = (V_M - V_b) \qquad\qquad (2)$$

where V refers to potential difference and the subscripts T, M, and b denote transepithelial, monolayers, and blank values. V_T is expressed in millivolts.

Resistance readings are converted to conductance according to Eq. (3):

$$G_T = [1000/(R_M - R_b)]/A \qquad\qquad (3)$$

where R_M and R_b represent the resistance readings of the monolayer and blank, respectively. A is the area of the filter, or a value of 0.6 cm^2 for the filters used in this laboratory. Conductance, or G_T, is expressed as mS/cm^2. Short-circuit current (I_{SC}) is calculated using Ohm's Law, $I = VG$, and is expressed as $\mu A/cm^2$.

The development of a resistance constitutes a critical step in the functioning of a cell monolayer capable of vectorial ion transport. The kinetics of resistance development can be established from daily ERS measurements on cells seeded onto Millipore filters. Typical results from one set of experiments is shown in Fig. 3. I_{SC} and G_T developed slowly until reaching peak values after 8 to 10 days in culture, then I_{SC} declined, causing V_T to decrease. These findings suggest that, for the conditions of cell cultures used in this experiment, the cultures are ready for continuous bioelectrical recordings or transepithelial flux measurements between 7 and 12 days in culture.

Continuous Recording. Detailed measurement of I_{SC} over a time course is achieved using a modified Ussing chamber which is custom built at our institution to accommodate Millipore CM filter cups. One advantage of designing and constructing chambers in our own facility is the ability to modify the basic chamber to allow use of other types of filter supports or culture well inserts which are now commercially available from a number of companies.

The chamber system is connected to two current and two potential

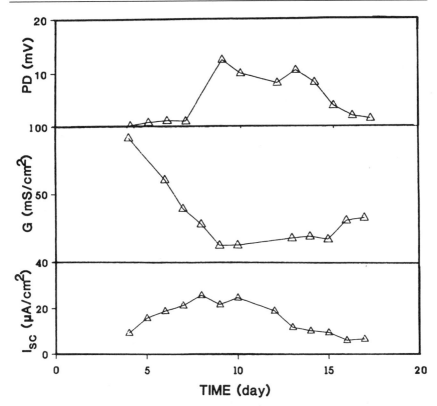

FIG. 3. Bioelectric characteristics in rabbit tracheal epithelial cells grown on Millipore filters. Cells were isolated and seeded in culture as described in the text. After 4 days in culture, electrochemical resistance system (ERS) readings on resistance and voltage were performed daily for 16 days. Short-circuit current (I_{sc}) was calculated from Ohm's law and conductance (G) as the inverse of resistance. PD, Potential difference.

electrodes and to ports for continuous perfusion of the apical and basal compartments. The electrodes are also built at this institution. Voltage electrodes are Ag/AgCl wires bathed in a concentrationed NaCl solution which is encased in a plastic jacket with a ceramic tip. Current electrodes are platinum wires which have been treated with chloroplatinic acid to increase their surface area. The resulting "blackened" platinum wires reduce the accumulation of gases released during high currents in the basolateral bathing solution. The bubbles of gas cause net spurious readings. Pairs of electrodes (voltage and current) are exposed to apical and basolateral bathing solutions and connected through a ground wire to a preamplifier. The preamplifier serves to amplify signals to the calibrated range of the automatic voltage clamp apparatus (University of Iowa, model 710C-1).

The output of the voltage clamp apparatus is connected (1) in parallel to an oscilliscope and chart recorder and (2) to an AT-compatible computer with data acquisition hard- and software. The oscilliscope is used to monitor AC noise during experiments. Because data are recorded to both the chart recorder and computer, we can obtain copies of hard data during the course of an experiment and later printout and analyze stored data in the computer.

Because cell cultures are used repeatedly, a sterile environment during experimental periods is required. The Ussing chamber is routinely used in a sterilized "premie"incubator, which allows control of temperature and CO_2 levels in the environment.

The electrophysiological parameters (V and I_{SC}) are measured with an automatic voltage clamp. The voltage clamp is used to pass sufficient current across the monolayer to clamp V_t to zero. G is calculated from the change in current in response to a voltage pulse of 0.1 to 1 mV of 1–2 sec. It is important to emphasize the I_{SC} is monitored continuously during the voltage pulses, hence the small voltage pulse compared to that used in intact tissue.

Ringer solutions for perfusion of apical and basolateral surfaces of the cell monolayer are gassed continuously with 5% CO_2. Before recording from monolayers, baseline readings and adjustments are done on a blank, unseeded, collagen-coated, glutaraldehyde-fixed Millicell filter. Test cultures replace the blank filter and, after baseline electrophysiological readings are made, drugs, transport blockers, or other additives are added. Either concentrated stock solutions are added directly into the apical chamber or the perfusate is changed to a Ringer solution of the desired final concentration. The latter method is preferred for exposing the basolateral surface to additives.

Transepithelial radioisotopic fluxes are accomplished by switching either the apical or basolateral bathing solutions to Ringer solutions containing the radioisotope of interest. Aliquots of perfusate from the opposite chamber are collected in a automated fraction collector and sampled for radioactivity. Electrolyte flux through the paracellular pathway is assessed qualitatively from G and by pyranine dye permeability. Pyranine is a trisuslfonic acid with a hydrated radius approximately an order of magnitude greater than that of Na^+. The perfusion rate is carefully controlled to prevent application of sufficient hydrostatic pressure to allow spurious movement of dye through the paracellular pathway.

Future Directions

Recent advances in cell isolation and culture and in electrophysiology as applied to airway epithelial cell physiology have enabled us to ascertain, in more detail, how this epithelium transports electrolytes and, to some extent, how this epithelium controls electrolyte movements. We can now approach questions on intracellular mechanisms of electrolyte transporter regulation. This can be accomplished by combining tools from molecular biology, biochemistry, and pharmacology with electrolyte transport methodology to investigate the molecular nature of transporters and signal transduction at the plasma membrane and intracellular levels.

Acknowledgments

This research was supported in part by grants from the National Cystic Fibrosis Foundation and from the National Institutes of Health (DK-27651).

[36] Transformation of Airway Epithelial Cells with Persistence of Cystic Fibrosis or Normal Ion Transport Phenotypes

By James R. Yankaskas and Richard C. Boucher

Introduction

Until 5 years ago, the pathophysiologic mechanisms unique to cystic fibrosis (CF) were difficult to study because of lack of *in vitro* specimens or animal models. The development of primary epithelial cell cultures from affected organs provided an important research tool. Cultured CF respiratory[1-6] and sweat duct[7-9] epithelial cells demonstrate the abnormal regulation of an apical membrane chloride permeability that characterizes this disease. These tissue culture studies indicate that the abnormalities described *in vivo* are primary effects of the abnormal gene rather than

[1] J. R. Yankaskas, M. R. Knowles, J. T. Gatzy, and R. C. Boucher, *Lancet* **1**, 954 (1985).

[2] J. R. Yankaskas, C. U. Cotton, M. R. Knowles, J. T. Gatzy, and R. C. Boucher, *Am. Rev. Respir. Dis.* **132**, 1281 (1985).

[3] R. A. Frizzell, G. Rechkemmer, and R. L. Shoemaker, *Science* **233**, 558 (1986).

[4] M. J. Welsh and C. M. Liedtke, *Nature (London)* **322**, 467 (1986).

[5] J. H. Widdicombe, M. J. Welsh, and W. E. Finkbeiner, *Proc. Natl. Acad. Sci. U.S.A.* **82**, 6167 (1985).

[6] N. J. Willumsen and R. C. Boucher, *Am. J. Physiol.* **256**, C226 (1989).

[7] G. Collie, M. Buchwald, P. Harper, and J. R. Riordan, *In Vitro* **21**, 597 (1985).

[8] C. M. Lee, F. Carpenter, T. Coaker, and T. Kealey, *J. Cell Sci.* **83**, 103 (1986).

[9] P. S. Pedersen and E. H. Larsen, *ICRS Med. Sci.: Libr. Compend.* **14**, 1159 (1986).

secondary effects of infection or inflammation. Comprehensive studies of CF pathogenesis are still limited by the low availability of cystic fibrosis tissues and by the limited life of epithelial cells in primary culture. Cell lines derived from cystic fibrosis and normal epithelial cells have the potential to provide ample research material for genetic, physiologic, and biochemical studies. Such cell lines may be used to test the ability of normal counterparts of candidate genes to complement or reverse the physiologic abnormalities of the recessive CF genes and may be used to develop and test new therapeutic modalities.

To be useful, a disease-specific cell line must have sufficient proliferation capability to substantially increase cell availability, express the abnormal gene, and exhibit phenotypic properties of interest. Developing such a cell line requires a supply of the affected tissues, a transforming agent that can increase cell proliferation without markedly changing differentiated cell properties, a means to introduce the transforming agent into the cells, selection criteria for cell lines that incorporate the transforming gene, and a means to assay the phenotypic characteristics. The following sections describe specific techniques for tissue procurement and cell culture, inserting a selected transforming gene with a retroviral shuttle vector, selecting for transforming gene expression and formation of intercellular tight junctions, and characterization of phenotypic properties. These procedures have been used to develop cystic fibrosis (CF/T43) and control (NL/T4) cell lines, which exhibit abnormal and normal regulation of an apical membrane chloride permeability, respectively.[10]

Tissue Procurement, Epithelial Cell Isolation, and Primary Culture

Progressive airway obstruction from thick secretions, inflammation, and bacterial infection is currently the major cause of morbidity and death in cystic fibrosis.[11] *In vivo* measurements[12,13] and *in vitro*[14,15] studies of intact excised CF and normal airway tissues have identified abnormalities in regulation of a chloride permeability and sodium absorption in CF epithelia. These differences are manifest in nasal, tracheal, and bronchial tissue. Because of the high incidence of nasal polyposis in cystic fi-

[10] A. M. Jetten, J. R. Yankaskas, M. J. Stutts, N. J. Willumsen, and R. C. Boucher, *Science* **244**, 1472 (1989).

[11] L. M. Taussig, L. I. Landau, and M. I. Marks, and L. M. Taussig, "Cystic Fibrosis," p. 115. Thieme-Stratton, New York, 1984.

[12] M. Knowles, J. Gatzy, and R. Boucher, *N. Engl. J. Med.* **305**, 1489 (1981).

[13] M. Knowles, J. Gatzy, and R. Boucher, *J. Clin. Invest.* **71**, 1410 (1983).

[14] M. R. Knowles, M. J. Stutts, A. Spock, N. Fischer, J. T. Gatzy, and R. C. Boucher, *Science* **221**, 1067 (1983).

[15] R. C. Boucher, M. J. Stutts, M. R. Knowles, L. Cantley, and J. T. Gatzy, *J. Clin. Invest.* **78**, 1245 (1986).

brosis,[16,17] this tissue is more frequently available than lower respiratory tract tissues and provides more epithelial cells than can be isolated from the sweat glands in a typical skin biopsy. Nasal tissues are relatively free of bacterial and fungal infection compared to tracheal tissues obtained during autopsy. Consequently, surgically excised nasal and bronchial tissues were used to develop airway epithelial cell lines. Detailed studies of ion transport properties of cultured cystic fibrosis and normal airway epithelial cells[2-4,18-20] derived from excised specimens have defined the phenotypic ion transport properties of CF.

Cystic fibrosis center directors and otolaryngologists identify patients undergoing elective nasal surgery. Excised cystic fibrosis or normal nasal polyps and turbinates[2,21] are rinsed in sterile saline solution in the operating room to remove blood and debris and transported to the laboratory at 4° in Joklik's minimum essential medium (JMEM) supplemented with antibiotics (penicillin, 50 U/ml; streptomycin, 50 μg/ml; and gentamicin, 40 μg/ml). Epithelial cells are prepared for culture by the following protocol: (1) place specimens in chilled (4°) modified Eagle's medium (MEM containing 0.1% protease XIV (Sigma, St. Louis, MO) and 100 μg/ml deoxyribonuclease (Sigma); (2) after 24–48 hr, add 10% (v/v) fetal bovine serum (FBS) to neutralize the protease; (3) concentrate the detached cells by centrifugation (800 g, 5 min) and wash in MEM plus 10% FBS; (4) determine cell number with a hemocytometer and cell viability by Trypan Blue exclusion; (5) concentrate cells by a final centrifugation (800 g, 5 min) and resuspend in a plating solution of supplemented F-12 medium[21,22] (F12-7X) that contains insulin (5 μg/ml), epidermal growth factor (20 ng/ml), triiodothyronine (3×10^{-8} M), endothelial cell growth supplement (7.5 μg/ml), transferrin (5 μg/ml), hydrocortisone (10^{-6} M, all from Collaborative Research, Lexington, MA) and cholera toxin (10 ng/ml; Sigma); (6) plate cells on uncoated plastic tissue culture dishes or on culture substrates suitable to planned experiments; (7) after overnight incubation (37°, 5% CO_2, 95% air, 98% humidity), remove nonadherent cells by gentle washing. Refeed with supplemented medium three times per week.

[16] H. Schwachman, L. L. Kulczycki, H. L. Mueller, and C. G. Flake, *Pediatrics* 30, 389 (1962).

[17] B. Taylor, J. N. G. Evans, and G. A. Hope, *Arch. Dis. Child.* 49, 133 (1974).

[18] R. C. Boucher, C. U. Cotton, J. T. Gatzy, M. R. Knowles, and J. R. Yankaskas, *J. Physiol. (London)* 405, 77 (1988).

[19] N. J. Willumsen, C. W. Davis, and R. C. Boucher, *Am. J. Physiol.* 256, C1045 (1989).

[20] M. Li, J. D. McCann, C. M . Liedke, A. C. Nairn, P. Greengard, and M. J. Welsh, *Nature (London)* 331, 358 (1988).

[21] R. Wu, J. Yankaskas, E. Cheng, M. R. Knowles, and R. Boucher, *Am. Rev. Respir. Dis.* 132, 311 (1985).

[22] N. J. Willumsen, C. W. Davis, and R. C. Boucher, *Am. J. Physiol.* 256, C1033 (1989).

Transforming Agent Selection and Retroviral Infection

Viral oncogenes,[23] hybrid viruses,[24,25] and viral genes[10,25] have been used to transform human airway epithelial cells. Chemical carcinogens[26] have been effective transforming agents for rodent airway epithelial cells but have been less effective with human cells. The gene coding the simian virus 40 large T (SV40T) protein was selected for these studies because it can increase cell proliferation and has been shown to immortalize human keratinocytes while causing minimal changes in phenotypic characteristics.[27] An origin of replication-deficient (ori$^-$) SV40T mutant[28] was used to minimize autologous viral replication.

Different means are available to insert the selected gene into cells. Hybrid adeno/SV40 viruses can infect airway epithelial cells,[25] but the adenoviral genes may complicate the transforming effects of SV40T or result in virus-producing cell lines. Genes incorporated into plasmids may be introduced into cells by a variety of transfection techniques, including DEAE dextran,[29] calcium phosphate[30] or strontium phosphate[31] coprecipitation, lipofection,[32] and electroporation,[33] but these techniques have relatively low efficiency. DNA can be microinjected[34] directly into cell nuclei, but relatively few cells can be injected in a reasonable time period. To increase transformation efficiency, an amphotrophic retroviral shuttle vector was used to transfer an ori$^-$ SV40T gene into primary cultures of human airway epithelial cells. The ΨAM packaging cell line harboring the recombinant retrovirus pZipneoSV(X)1/SV40T was obtained from P. S. Jat (MIT, Cambridge, MA).[35,36] Retroviral infections of 10^5 cells/60-mm

[23] G. H. Yoakum, J. F. Lechner, E. W. Gabrielson, B. E. Korba, L. Malan-Shibley, J. C. Willey, M. G. Valerio, A. M. Shamsuddin, B. F. Trump, and C. C. Harris, *Science* **227**, 1174 (1985).

[24] R. R. Reddel, Y. Ke, B. I. Gerwin, M. G. McMenamin, J. F. Lechner, R. T. Su, D. E. Brash, J-B. Park, J. S. Rhim, and C. C. Harris, *Cancer Res.* **48**, 1904 (1988).

[25] B. J. Scholte, M. Kansen, A. T. Hoogeveen, R. Willemse, J. S. Rhim, A. W. M. Van Der Kamp, and J. Bijman, *Exp. Cell Res.* **182**, 559 (1989).

[26] D. G. Thomassen, T. E. Gray, M. J. Mass, and J. C. Barrett, *Cancer Res.* **43**, 5956 (1983).

[27] J. S. Rhim, G. Jay, P. Arnstein, F. M. Price, K. K. Sanford, and S. A. Aaronson, *Science* **227**, 1250 (1985).

[28] M. B. Small, Y. Gluzman, and H. L. Ozer, *Nature (London)* **296**, 671 (1982).

[29] J. H. McCutchan and J. S. Pagano. *J. Natl. Cancer Inst.* **41**, 351 (1968).

[30] F. L. Graham and A. J. Van der Eb, *Virology* **52**, 456 (1973).

[31] D. E. Brash, R. R. Reddel, M. Quanrud, K. Yang, M. P. Farrell, and C. C. Harris, *Mol. Cell. Biol.* **7**, 2031 (1987).

[32] P. L. Felgner, T. R. Gadek, M. Holm, R. Roman, H. W. Chan, M. Wenz, J. P. Northrop, G. M. Ringold, and M. Danielsen, *Proc. Natl. Acad. Sci. U.S.A.* **84**, 7413 (1987).

[33] M. C. Iannuzzi, J. L. Weber, J. Yankaskas, R. Boucher, and F. S. Collins, *Am. Rev. Respir. Dis.* **138**, 965 (1988).

[34] A. Graessmann, M. Graessmann, and C. Mueller, this series, Vol. 65, p. 816.

[35] C. L. Cepko, B. E. Roberts, and R. C. Mulligan, *Cell* **37**, 1053 (1984).

[36] P. S. Jat, C. L. Cepko, R. C. Mulligan, and P. A. Sharp, *Mol. Cell. Biol.* **6**, 1204 (1986).

dish were carried out at 37° for 2 hr in 2 ml of a 1 : 1 mixture of keratino-cyte growth medium (KGM, Clonetics Corp., San Diego, CA) and retrovi-rus-containing (2×10^3 cfu/ml) RPMI 1640 with 5% FBS. Virus was then diluted by addition of 3 ml KGM. Two days later, the cells were exposed to the neomycin analog G418 (50 μg/ml) for 10 to 14 days. After this time, G418-resistant colonies were isolated with cloning cylinders and passaged to individual tissue culture flasks for further culture in KGM medium. One to six G418-resistant colonies/60-mm dish were obtained. Colonies were fed twice weekly and passaged during exponential growth. Colonies with continued proliferation at passage 3 were cyropreserved and aliquots screened for development of secondary selection features.

Secondary Selection: Formation of Tight Junctions and
   Transepithelial Resistance

   Specialized epithelial functions such as barrier formation and vectorial transport of solutes depend on formation of tight junctions, which separate the plasma membrane into apical and basolateral regions.[37] Since the ion transport properties that characterize cystic fibrosis are dependent on such epithelial cell polarization, cell lines that form functional tight junctions (and hence develop a transepithelial resistance) are most likely to demon-strate vectorial transport of sodium and/or chloride ions. Such differentia-tion is partially dependent on the culture substrate. Permeable collagen matrix support (CMS) culture dishes[2] support the best differentiated ion transport properties in primary cell cultures. These matrices are cast in a beveled orifice (4.5-mm diameter) in the bottom of a 2.5-cm-diameter polycarbonate cup by the following protocol. (1) Dissolve type III calfskin collagen (Sigma Cat. No. C-3511, 15 mg/ml) in 0.2% acetic acid at 37°. (2) Add 0.5 vol of 2.5% glutaraldehyde and cool in an ice bath until jelling begins. (3) Apply a bead of collagen/glutaraldehyde around the orifice of the inverted polycarbonate cup, and drag mixture across the orifice with a Pasteur pipet, forming a thin sheet. (4) Dry in air (3–4 hr). (5) Apply a second coat of collagen (without glutaraldehyde) over the bottom surface of the matrix. Dry overnight. (6) Apply a third coat of collagen (without glutaraldehyde) to the inside surface of the matrix. Dry overnight. (7) Expose bottom side of CMS to two drips of 2.5% glutaraldehyde for 5 min. (8) Rinse four times with bicarbonate containing saline solution. Dry overnight. (9) Sterilize unit in 70% ethanol for 30 min. Rinse in sterile culture medium. Incubate overnight at 37° in culture medium to test sterility. (10) CMS dishes can be stored in physiologic saline at 4° for 3–4 weeks.

---

[37] M. Cereijido, A. Ponce, and L. Gonzalez-Mariscal, *J. Membr. Biol.* **110,** 1 (1989).

Prior to seeding, CMS dishes are incubated (37°, overnight) in 35-mm-diameter culture dishes with culture medium in both the inner cup and the outer dish. Medium in the inner cup is removed by aspiration and epithelial cells (removed from plastic tissue culture dishes with trypsin/EDTA) in KGM medium are seeded at 2×10^6 cells/cm^2 in 10 to 30 μl of medium. Additional medium is added to the inner cup after 24 hr, and the cells are washed vigorously at 48 hr. Cell attachment and confluency are monitored by phase-contrast microscopy. Transepithelial electric potential and resistance are measured by established techniques.[22]

Characterization

Expression of the transforming SV40T antigen and epithelial cell-specific keratins by standard Western blot and hybridization techniques[10] tests transforming gene expression and the epithelial nature of the cells. The persistence of the cystic fibrosis gene[38] in the CF/T43 cell line was confirmed by amplifying the exon 10 region of the cystic fibrosis transmembrane regulator (CFTR) gene by polymerase chain reaction (PCR)[39] and hybridizing with oligonucleotide probes specific to presence of deletion of the Phe^{508} codon. CF/T43 cells are homozygous for Phe^{508} deletion. Similarly, expression of the CFTR messenger RNA was confirmed by Northern hybridization of messenger RNA extracted from CF/T43 cells by Chirgwin's cesium chloride method[40] and an oligonucleotide probe.

Phenotypic characterization may be performed with any technique that differentiates CF and normal airway epithelial cells. For the CF/T43 and NL/T4 cell lines, chloride conductance of the apical membrane (G_{Cl^-}) was assessed with transepithelial and chloride-selective intracellular microelectrodes.[22] Compared to NL/T4 cells, CF/T43 cells have a reduced G_{Cl^-} in the basal state. This conductance is not activated by cAMP-dependent agonists, but is increased by Ca^{2+}-mediated agonists. These findings were confirmed with single-channel patch-clamp techniques.[10]

Summary

These studies demonstrate the feasibility of transforming human airway epithelial cells while inducing only modest changes in function. Fea-

[38] J. R. Riordan, J. M. Rommens, B-S. Kerem, N. Alon, R. Rozmahel, Z. Grzelczak, J. Zielenski, S. Lok, N. Plavsic, J-L. Chou, M. L. Drumm, M. C. Iannuzzi, F. S. Collins, and L-C. Tsui, *Science* **245**, 1066 (1989).
[39] R. K. Saiki, S. Scharf, F. Faloona, K. B. Mullis, G. T. Horn, H. A. Erlich, and N. Arnheim, *Science* **230**, 1350 (1985).
[40] J. M. Chirgwin, A. E. Przybyla, R. J. MacDonald, and W. J. Rutter, *Biochemistry* **18**, 5294 (1979).

tures central to the pathophysiology of cystic fibrosis, i.e., abnormal regulation of a chloride permeability in the apical cell membrane, appear to be preserved in the CF/T43 transformed cell line. This work indicates that additional cystic fibrosis and normal cell lines may be developed, as well as epithelial cell lines from other diseases of interest. In addition to SV40T gene, temperature-sensitive viral genes,[41] or genes driven by inducible promoters (e.g., glucocorticoids, heavy metals[42]) may produce cell lines in which proliferation or differentiation can be controlled. For example, the temperature-selective SV40A gene is expressed in cells cultured at the permissive (33°) temperature but is degraded at the nonpermissive (40°) temperature.[41] Thus, the transfected gene may induce proliferation to increase cell number, and then be suppressed to permit expression of a differentiated phenotype.

Out strategy of initially selecting clones by G418 resistance and then selecting clones that develop functional tight junctions (and a transepithelial resistance) was useful in identifying a cell line with highly differentiated phenotypic properties. Cell lines that do not form transepithelial resistances may be valuable for studies that do not depend on cell polarization. Initial evidence suggests that some of the differentiated properties of CF/T43, i.e., formation of functional tight junctions and a transepithelial resistance, are lost at late passages. Although these properties may be a function of the culture medium constituents and the nature of the culture substrate, it is possible that the transforming gene caused further mutations that resulted in subclones that have increased growth rate but different phenotypic properties. The continuing expression of the CFTR messenger RNA in these late passage cells indicates that they will be an important tool for future CF research.

[41] J. Y. Chou, *Proc. Natl. Acad. Sci. U.S.A.* **75**, 1409 (1978).
[42] O. Hurko, L. McKee, and J. G. E. M. Zuurveld, *Ann. Neurol.* **20**, 573 (1986).

[37] Cell Culture of Bovine Corneal Endothelial Cells and Its Application to Transport Studies

By Michael Wiederholt and Thomas J. Jentsch

Introduction

Ion transport studies using cultured epithelial cells have become increasingly popular in the past few years. Advantages of studies of cultured cells as opposed to investigations of the tissue *in situ* include the availability of a large number of uniform cells and the excellent possibilities of controlling and changing their environment. In addition, growing the cells on a suitable support allows for the application of techniques which would be difficult or impossible with intact tissue. In this chapter, after describing the establishment of corneal endothelial cell cultures, we will describe two such techniques we have used with these cells, namely, intracellular microelectrode technique[1-3] and ion flux measurements across the plasma membrane of intact cells.[4] Recently, we have also determined intracellular pH in these cells using a pH-sensitive, intracellularly trapped dye.[5] These methods can, of course, be applied successfully to other culture systems, as we and others have demonstrated.

Culture of Bovine Corneal Endothelium

Bovine corneal endothelium is easily established in culture, since these cells can be isolated without problems of contamination by other cell types and since they grow well in general purpose cell culture media. Furthermore, since these cells secrete an extensive extracellular matrix onto the bottom of the culture dish,[6,7] which has also been used as a physiological support for other cells,[6,8] they grow on their natural support and therefore very closely resemble their *in vivo* counterparts. Thus, extrapolation of data obtained with the culture system to the *in vivo* situation should be comparatively unproblematic.

In establishing primary cultures of bovine corneal endothelium, we have closely followed the method of MacCallum *et al.*[7] Other methods, such as explant techniques,[9,10] have also been described in the literature. Though Gospodarowicz *et al.* stress the importance of fibroblast growth factor (FGF) for the growth of these cells,[6,11] we chose to work with cells at

---

[1] T. J. Jentsch, M. Koch, H. Bleckmann, and M. Wiederholt, *J. Membr. Biol.* **78,** 103 (1984).

[2] T. J. Jentsch, S. K. Keller, M. Koch, and M. Wiederholt, *J. Membr. Biol.* **81,** 189 (1984).

[3] T. J. Jentsch, H. Matthes, S. K. Keller, and M. Wiederholt, *Pfluegers Arch.* **403,** 175 (1985).

[4] T. J. Jentsch, T. R. Stahlknecht, H. Hollwede, D. G. Fisher, S. K. Keller, and M. Wiederholt, *J. Biol. Chem.* **260,** 795 (1985).

[5] T. J. Jentsch, C. Korbmacher, I. Janicke, D. G. Fischer, F. Stahl, H. Helbig, H. Hollwede, E. J. Cragoe, Jr., S. K. Keller, and M. Wiederholt, *J. Membr. Biol.* **703.** 291 (1988).

[6] D. Gospodarowicz and C. Ill, *Exp. Eye Res.* **31,** 181 (1980).

[7] D. K. MacCallum, J. H. Lillie, L. J. Scaletta, J. C. Occhino, W. G. Frederick, and S. R. Ledbetter, *Exp. Cell Res.* **139,** 1 (1982).

[8] D. Gospodarowicz, D. Delgado, and I. Vlodavsky, *Proc. Natl. Acad. Sci. U.S.A.* **77,** 4094 (1980)

[9] D. Gospodarowicz, A. L. Mescher, and C. R. Birdwell, *Exp. Eye Res.* **25,** 75 (1977).

[10] C. Arruti and Y. Courtois, *Exp. Eye Res.* **34,** 735 (1982).

[11] L. Giguère, J. Cheng, and D. Gospodarowicz, *J. Cell Physiol.* **110,** 72 (1982).

a low passage number in the absence of high concentrations of FGF, because this might yield cells closer to the *in vivo* situation. Eyes from 1- to 4-year-old steers are obtained from a local slaughterhouse up to 1 hr after the animal's death and are transported to the laboratory in ice-cold Ringer's solution. In the laboratory, they are fixed with three needles on a support (cork or styrofoam) and rinsed with sterile phosphate-buffered saline (PBS). An incision is made just at the rim of the cornea until aqueous humour leaves the eye. Using a fine scissor and tweezers, the cornea is cut out. Care should be taken not to scratch the endothelial surface and to avoid a rim of sclera remaining attached to the cornea (danger of contamination with other cell types). The cornea is then put into a supporting plastic cup with its endothelial surface facing upward, and sterile PBS (containing calcium) is added to avoid drying of the endothelial surface. After the desired number of corneas has been prepared, they are held with tweezers and gently rinsed with sterile Ca^{2+}-free PBS using a syringe. They are put back onto the plastic support, which has been dried in the meantime to increase the adhesion to the cornea. A few drops of trypsin/EDTA solution (0.05%/0.02%) is added in the cup formed by the cornea, taking care not to allow the solution to make contact with the edge (however, even being less careful we have only very rarely seen fibroblasts growing in our cultures). After incubation for about 20 min at room temperature, corneal endothelial cells are harvested using an Eppendorf pipet (100 μl) with a sterile tip. To increase the yield of cells, it is important to gently rub the surface of the endothelium with the pipet tip. Sometimes this leads even to a detachment of fragments of Descemet's membrane with endothelial cells attached to it, thus coming close to the explant technique. The cell suspension is pipetted into a 25-cm² tissue culture flask (Nunc, Roskilde, Denmark) which had been previously filled with cell culture medium and was preequilibrated with 5% CO_2. The culture medium is DMEM (Dulbecco's modified Eagler's medium) containing 2 mM glutamine, 10% fetal calf serum, 100 U/ml penicillin, and 100 μg/ml streptomycin. To avoid fungal contamination, we have routinely included 2.5 μg/ml amphotericin B during the first few days, which was later discontinued since it is likely to interfere with ion transport processes. Source of all tissue culture components was Biochrom KG (Berlin, FRG). The cultures are incubated at 37° in a humidified 5% CO_2 atmosphere. The medium was changed twice a week. After about 5 to 10 days the primary cultures form confluent monolayers of hexagonal epithelial cells which cover the entire bottom of the 25-cm² flask. The cells are then detached by trypsinization (0.05% trypsin, 0.02% EDTA) and are propagated using a split ratio of 1 : 3. The cells may be passaged more than 10 times, but in order to avoid alterations in culture we have used only cells between the first and the fifth passage. For measurements of intracellular potentials, the cells were seeded on 60-mm tissue culture dishes. For determination of isotopic uptake, we mixed the cells trypsinized from 12 80-cm² flasks (second passage, obtained from 4 eyes) and seeded them into about 100 25-cm² flasks to obtain a uniform, large population of cells. For determinations of intracellular pH (pH_i), the cells were seeded on plastic Leighton tubes (Costar, Cambridge, MA).

Measurements of Intracellular Potentials

The use of the culture system represented a significant advance for the study of the plasma membrane voltage of corneal endothelial cells,[1-3] since it has been previously virtually impossible to measure stable potentials while changing extracellular fluid composition in the intact tissue.[12,13] This was also true for a study using cultured rabbit corneal endothelial cells.[14] With the present system, however, it has become possible to change the extracellular fluid bathing the cells several dozens of times while recording the potential of a single cell (stable up to several hours), and we have applied the same method with success to a variety of other cell types (e.g., lens epithelial cells,[15] retinal pigment epithelial cells,[16] and BSC-1,[17] an established kidney epithelial cell line). Recently, others have used a technique similar to ours to measure intracellular potentials of MDCK cells.[18] The key to this success is probably the mechanical stability, which is achieved by growing the cells on a rigid plastic support and by advancing the electrode on the plastic surface, and thereby fixing it relative to the cell.

The micropuncture set-up is installed on a vibration-damped table inside a Faraday cage. The main parts are the fluid delivery system (pump, valves, and electronics for their control), a temperature-controlled chamber (including a flow chamber pressed on the tissue culture dish), and the recording system (microelectrode amplifier, stepper, control electronics, and pen recorder). The following sections will describe these items in detail.

---

[12] M. Wiederholt and M Koch, *Exp. Eye Res.* **27**, 511 (1978).

[13] J. J. Lim and J. Fischberg, *Exp. Eye Res.* **28**, 619 (1979).

[14] M. M. Jumblatt, *Vision Res.* **21**, 45 (1981).

[15] T. J. Jentsch, B. von der Haar, S. K. Keller, and M. Wiederholt, *Exp. Eye Res.* **41**, 131 (1985).

[16] S. K. Keller, T. J. Jentsch, M. Koch, and M. Wiederholt, *Am. J. Physiol.* **250**, C124 (1986).

[17] T. J. Jentsch, H. Matthes, S. K. Keller, and M. Wiederholt, *Am. J. Physiol.* **251**, F954 (1986).

[18] M. Paulmichl, G. Gstraunthaler, and F. Lang, *Pfluegers Arch.* **405**, 102 (1985).

Flow Chamber

For the measurements, the tissue culture dish is inserted into a plexiglas chamber (Fig. 1A). The lower part of the chamber (d) is heated by circulating water from a water bath (Haake type FE, Berlin, FRG). This lower chamber is electrically isolated from the upper chamber by a thin Plexiglas plate. It is essential to connect it to ground via an inserted copper wire (not shown in the figure) in order to avoid capacitive coupling of the line frequency into the recording system. The upper chamber (c) is also filled with water (thermally equilibrated with the lower chamber), which reaches up to the bottom of the tissue culture dish and bathes the tubings (f) delivering the saline solutions for superfusing the cells. A Plexiglas flow chamber (Fig. 1, e, and Fig. 1B for more detail) is tightly pressed onto the bottom of the tissue culture dish (l) by springs (r) pressing against the closed lid of the chamber (not shown). In order to avoid scratching of the cell monolayer when inserting the flow chamber and closing the lid, it is held in place by a frame (not shown) fitted into the chamber. The flow chamber isolates a small channel (s) (width, 1.5 mm; length, 30 mm) from the rest of the culture dish. For better sealing, a thin rubber layer (q) is glued to the bottom of the chamber. The cells growing on the dish inside this area can be rapidly superfused by up to eight different test solutions, which are let in at one end of the chamber (b). At the other end, the solution is removed solely by hydrostatic pressure, which ensures a smooth and vibration-free removal of the liquid. This is achieved by connecting the end of the channel by a rigid tubing (k) filled with saline to a second container (g), whose height relative to the culture dish can be adjusted. The fluid level inside this container is kept constant by the position of the outflow tubing. An essential feature is the design of the outlet of the chamber. At the end of the chamber, just before the inserted tubing for fluid removal, the top of the channel is covered by a piece of Plexiglas (p) which reaches down to the bottom of the channel within approximately 1 mm. By capillary forces, this ensures that no air is sucked into the tubing, which would clog it. This, however, will only work when the suction force, which can be regulated by the height of the second container, is not too strong.

Fluid Delivery

The solutions are delivered by a syringe pump (Braun Melsungen type Unita I, Braunschweig, FRG), which, in contrast to peristaltic pumps, guarantees a very smooth operation. Due to the tiny dimensions of the flow chamber, a moderate pump rate (60 ml/hr) is sufficient to ensure a rapid fluid exchange at the cell (90% exchange in less than 2 sec). Six glass

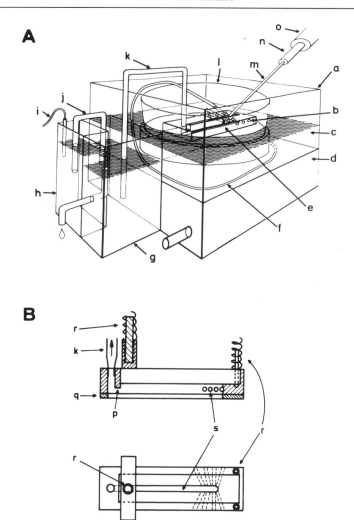

Fig. 1. (A) Experimental set-up for micropuncture of cultured monolayer cells (modified from Ref. 1); (B) details of flow chamber. a, Temperature-controlled chamber; b, inlets into flow chamber; c, upper compartment filled with water; d, lower, electrically insulated chamber with inlets for circulating temperature-controlled water; e, flow chamber [see (B) for details]; f, tubings; g, overflow chamber; h, compartment filled with 0.5 M KCl; i, counterelectrode, connected to ground; j, agar bridge; k, tubing for removal of fluid; l, 60-mm-diameter tissue culture dish with attached cell monolayer; m, microelectrode; n, microelectrode holder with integrated Ag/AgCl element; o, preamplifier, connected to stepper; p, protruding piece of plastic which prevents air bubbles from entering the tubing (k); q, rubber sealing; r, springs; s, channel of flow chamber.

syringes (50 ml each) are propulsed in parallel. Shortly after the syringe pump, fluid flow is switched by corrosion-resistant three-way electromagnetic valves (Lucifer type 133A, Geneva, Switzerland). Fluid flow is either directed into the chamber (one solution at a time), or is directed into waste containers (rest of solutions). Since separate tubings lead from each valve to the flow chamber (inlets b), solution mixing occurs next to the cell and solution exchange is fast. Care should be taken to adjust the outlets of the tubings leading to the waste containers to the height of the flow chamber. Otherwise, due to differences in hydrostatic pressure and due to the elasticity of the plastic tubings, fluid will be either sucked out of, or injected into, the chamber during a switching process. The main advantage of the electromagnetic valves is the automation of the set-up. To this end we have installed a personal computer connected to an interface which operates the valves. A simple BASIC program specifies the planned time course of the solution exchanges for the experiment. In addition to the ease in operation, this ensures an exact timing of the solution exchanges. A disadvantage of the electromagnetic valves is that they generate a small pressure peak (toward the flow chamber) when switching fluid flow to the waste containers. This will lead to an ejection of fluid while the second valve is already switched on, thus leading to a slow-down in solution exchange. To avoid this problem, we have programmed the computer to open the second valve only about 0.8 sec after the preceding valve has been shut off.

For our studies, which were largely concerned with effects of HCO_3^- and pH, it was essential to closely control the partial pressure of CO_2 in the superfusing solution. This was achieved by a system of concentric tubings. The inner tubing, which is gas permeable (Silastic, type 602-135, Dow Corning, Midland, MI), was constantly perfused with the appropriate gas mixture, while the outer tubing (Tygon, type R-3603, Norton Plastics, Columbus, Ohio, which is poorly permeable to gas) served to conduct the saline to the chamber. Since the inlet of the inner tubing into the outer tubing was located only 5 mm from the entrance (b) to the flow chamber, a precise control of pCO_2 (and pH) in the superfusing solution is obtained. The temperature is controlled by immersing several turns of the tubing in the upper chamber of the temperature-controlled flow chamber. Since the fluid flow rate is low, this is sufficient for thermal equilibration.

Electrical Measuring System

For the measurements of plasma membrane voltages we used conventional microelectrodes (m) made of borosilicate glass with internal filament (o.d. 1 mm, i.d. 0.58 mm) (Hilgenberg, Malsfeld, FRG), which were pulled on a horizontal microelectrode puller (Narashige type PD-5, Tokyo,

Japan) and were filled with 0.5 M KCl. When measured in Ringer's solution, their resistances ranged between 60 and 140 MΩ. They were connected via a KCl-filled microelectrode holder (n) equipped with a reversible Ag/AgCl element (WPI, Hamden, CT) to the headstage (preamplifier) (o) of a high impedance amplifier (input resistance $R_i > 10^{10}$ Ω) (M701; WPI, Hamden, CT). The output signal of the amplifier is recorded on a pen writer. The preamplifier is fixed to a mechanical stepping device (Nanostepper "Heidelberg," Science Trading, Frankfurt, FRG) mounted on a micromanipulator (Leitz, Wetzlar, FRG). The stepper quickly advances the microelectrode in micrometer steps. Since it uses a step motor coupled to a gear, it is faster than hydraulic steppers and thus gives higher yields of successful impalements.[19] In contrast to piezosteppers,[20] which provide even faster movements, the Nanostepper is capable of advancing the electrode repeatedly over a range of 20 mm. Our method of puncture depends very much on this capability. We carefully lower the microelectrode onto the tissue culture dish inside the channel of the flow chamber (at an angle of about 30 to 40°), until a negative potential is recorded on the pen writer attached to the microelectrode amplifier. Sometimes this already yields a stable potential. In most cases, however, we must advance the electrode further on the tissue culture dish by use of the stepper. In this way, it is possible to advance the electrode, which glides (probably on its shank close to the tip) on the surface of the plastic dish, by several millimeters, and to impale dozens of cells in a row. Therefore, even if the probability of getting a stable impalement of one single cell is low, repeated puncture using the same electrode is a relatively convenient way to obtain a good recording. To do this, we have built an electronic device ("cell finder") which does the search for a stable impalement automatically. It is similar to a previously published device,[21] but has some additional features. A voltage comparator compares the output voltage of the microelectrode amplifier with an adjustable threshold voltage. When the measured voltage is more positive than the threshold, it triggers the stepper with a preset frequency (about 10 Hz) to advance the microelectrode in micrometer steps. The number of steps is counted and—converted by a digital-to-analog converter— recorded in parallel to the measured potential on the pen recorder. When the measured voltage exceeds the preset threshold (we usually use about −20 mV), the stepper stops. In a restart mode, the cell finder again begins to advance the electrode when the potential has fallen below the threshold for more than 20 sec (this delay is provided to allow for a recovery of the

[19] U. Sonnhof, R. Forderer, W. Schneider, and H. Kettermann, *Pfluegers Arch.* **392,** 295 (1982).
[20] M. Fromm, P. Weskamp, and U. Hegel, *Pfluegers Arch.* **384,** 69 (1980).
[21] W. S. Marshall and S. D. Klyce, *Pfluegers Arch.* **391,** 258 (1981).

plasma membrane voltage after the initial damage inflicted to the membrane). As an option, a buzzer provides an acoustical signal when the voltage is below the threshold. This is especially useful in signaling the operator a loss of the cell during a computerized experiment using the electromagnetic valves. The counterelectrode (i) (an Ag/AgCl half-element; WPI, Hamden, CT) is put into a 0.5 M KCl solution (h), which is connected via a 3 M KCl-saturated agar bridge (j) to the overflow chamber (g) and is thus in electrical contact with the saline superfusing the cells.

To provide electrical shielding, the flow chamber, preamplifier, and stepper are located inside a Faraday cage (on a vibration-damped table). It is good practice to keep the tubings leading from the syringe pump into the Faraday cage as short as possible to avoid electrical coupling into the recording system. As mentioned before, the lower part of the temperature-controlled chamber (d) should be grounded to avoid electrical coupling by the tubings leading to the water bath. Care should be taken not to allow any contact of the superfusion saline to metal parts connected to ground, since this will result in offset voltages (formation of half-elements).

Comments on the Method of Cell Puncture

The above-described method allows us to obtain impalements which are stable for up to several hours, enabling us to perform a large number of solution exchanges while recording the potential of a single cell. A representative experiment, showing evidence for an electrogenic sodium-bicarbonate symport in cultured bovine corneal endothelial cells, is shown in Fig. 2. This success depends very much on the mechanical stability of the system. In addition to the vibration-damped table (a standard feature), we believe that this is due to the fact that the electrode is held in place (fixed relative to the cell) by friction on the plastic surface of the dish. Though the electrode is repeatedly advanced on the dish, we almost never have problems with broken tips [though they sometimes clogged (increase of resistance)]. This might be due to the geometry of the electrode: the very tip will probably not be in contact with plastic, but will be held on top of it due to the conical form of the tip.

A potential disadvantage is that we destroy the cell monolayer by repeatedly advancing the electrode. Thus, by destroying the previously impaled cell, we create an additional access of saline to the basolateral surface of the cell, making a distinction between apically and basolaterally located ion transport processes virtually impossible. However, with cultured cells this problem might only be tackled when growing the cells on a permeable support (such as a filter or a dialysis membrane) which could be superfused independently on both sides. Such a system, including a solu-

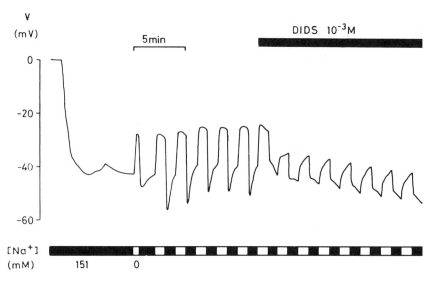

FIG. 2. Representative experiment showing the response of the plasma membrane voltage V of a cultured bovine corneal endothelial cell to changes in sodium concentration and the effect of DIDS (modified according to Ref. 2). In the presence of 46 mM HCO$_3^-$ and 5% CO$_2$, sodium concentration was repeatedly changed from 151 to nominally 0 mM (replaced by 46 mM choline and 105 mM N-methyl-D-glucamine). Sodium removal led to a fast depolarization, and readdition to a hyperpolarization. This effect could be largely blocked by 1 mM DIDS (2,2'-diisothiocyano-4,4'-disulfonic acid stilbene). The effect of DIDS was essentially irreversible after 10 min. These are typical characteristics of an electrogenic Na$^+$/HCO$_3^-$ symport.

tion exchange as fast as with the present one, remains to be devised, and difficulties with mechanical stability might be expected.

Isotopic Flux Measurements

Measurement of isotopic uptake into monolayers of cultured cells is an established technique which has been used by a number of laboratories.[22-28] Basically, the cell monolayer is incubated for the desired time

[22] M. J. Rindler, M. Taub, and M. H. Saier, Jr., *J. Biol. Chem.* **254,** 11431 (1979).
[23] P. Rothenburg, L. Glaser, P. Schlesinger, and D. Cassel, *J. Biol. Chem.* **258,** 4883 (1983).
[24] C. Frelin, P. Vigne, and M. Lazdunski, *J. Biol. Chem.* **258,** 6272 (1983).
[25] N. E. Owen and M. L. Prastein, *J. Biol. Chem.* **260,** 1445 (1985).
[26] G. L'Allemain, S. Paris, and J. Pouysségur, *J. Biol. Chem.* **260,** 4877 (1985).
[27] T. J. Jentsch, B. S. Schill, P. Schwartz, H. Matthes, S. K. Keller, and M. Wiederholt, *J. Biol. Chem.* **260,** 15554 (1985).
[28] T. J. Jentsch, P. Schwartz, B. S. Schill, B. Langner, A. P. Lepple, S. K. Keller, and M. Wiederholt, *J. Biol. Chem.* **261,** 10673 (1986).

with the uptake solution containing the radioisotope of interest. The cells are then rapidly rinsed with ice-cold stop solution (e.g., 100 mM MgCl$_2$), dissolved (e.g., by 0.1 M NaOH), and the radioactivity is measured. Since these are published standard techniques, we will focus on special problems encountered when measuring HCO$_3^-$-dependent isotopic uptake with corneal endothelial cells,[4] and ways to solve them.

While with corneal endothelium *in situ* transepithelial flux measurements have been performed in several laboratories[29-31] and have yielded important data, information on specific plasma membrane transport processes is easier to obtain by determination of intracellular uptake of label. This, however, is virtually impossible with *in situ* preparations due to the limited amount of material, and, most importantly, due to the attached stroma which will take up the majority of label unspecifically. Cells cultured on plastic, on the other hand, both provide a large amount of uniform cells and very much restrict the amount of extracellular space. However, since corneal endothelial cells actively secrete an extracellular matrix (ECM) in culture,[6,7] extracellular entrapment of label is still a problem. Indeed, in our experience, the background level of uptake is much less with cells which do not secrete such an extensive matrix (e.g., BSC-1, an established kidney epithelial cell line,[27,28] and primary cultures of bovine retinal pigment epithelium[32]). To minimize this problem, we use the cells only 5 days after seeding. Another problem, especially with primary cultures, is the quantitative difference in uptake between different experiments, even when expressed as uptake per cell (in our experience variation by a factor of two or three). Thus, to obtain quantitative results (e.g., determinations of apparent K_m values) it is essential to perform the experiment on a single-cell preparation. We have therefore mixed a large number of cells (12 80-cm^2 flasks, second passage, obtained from 4 different donor eyes) and seeded them evenly into approximately 100 25-cm^2 flasks for single flux experiment.

Since our studies were mostly concerned with effects of bicarbonate on ^{22}Na$^+$ uptake,[4,27,28] we had to control carefully the pCO$_2$ and (pH) of the incubation solutions. This problem is even more formidable in experiments designed to determine apparent K_m values for bicarbonate. In this case, to keep extracellular pH constant, we equilibrate the cells with up to six different CO$_2$ concentrations in a single experiment. This is done by using 25-cm^2 tissue culture flasks (instead of dishes or cluster plates, which

[29] S. Hodson and F. Miller, *J. Physiol. (London)* **263**, 563 (1976).

[30] J. J. Lim and H. H. Ussing, *J. Membr. Biol.* **65**, 197 (1982).

[31] D. S. Hull, K. Green, Z. M. Boyd, and H. R. Wynn, *Invest. Ophthalm. Visual Sci.* **16**, 883 (1977).

[32] S. K. Keller, T. J. Jentsch, I. Janicke, and M. Wiederholt, *Pfluegers Arch.* **411**, 47 (1988).

are used by most groups when determining uptake in the absence of HCO_3^-) which are filled individually with the appropriate premixed CO_2/ air mixture. This is repeated at every change of incubation solution, including the addition of the radioactive uptake solution. This "hot" solution is kept in a small Erlenmeyer flask in a circulating water bath to keep the temperature at 37°. The appropriate gas mixture (water-saturated and preheated to 37°) is continuously blown at a low rate (controlled by a flow meter) on the surface of the (preequilibrated) solution in the flask.

Since the corneal endothelium produces an extensive ECM, it is useless to refer uptake to milligrams protein as done in other studies with other cells.[22,24,25] Therefore, we refer uptake to the surface area and to the number of cells. The latter is determined by photographing representative (stained) cell monolayers and counting the cells, since determination by a Coulter counter is difficult (even after prolonged incubation with trypsin, the cells often form clumps).

Recent Developments and Conclusion

Recently, we have constructed a photometer[5] which allows the continuous determination of intracellular pH in monolayers of cells exploiting the pH-sensitive absorbance of 5 (and 6)-carboxydimethylfluorescein. Using this technique, we have investigated the regulation of intracellular pH in BSC-1 kidney cells,[33] retinal pigment epithelial cells,[32] and more recently in cultured bovine corneal endothelium.[5] This technique is described in these papers and an excellent article on the properties of this dye and its suitability as a probe for pH_i is available.[34]

In conclusion, the use of a cell culture system has proved to be extremely useful for the investigation of ion transport processes in the corneal endothelium. It has opened the way for methods which are not, or not easily, applicable to the intact tissue and has provided important information on the properties of electrogenic sodium bicarbonate symport.

Acknowledgment

Supported by DFG (Wi 328/11).

[33] T. J. Jentsch, I. Janicke, D. Sorgenfrei, S. K. Keller, and M. Wiederholt, *J. Biol. Chem.* **261**, 12120 (1986).
[34] J. R. Chaillet and W. F. Boron, *J. Gen. Physiol.* **86**, 765 (1985).

[38] Methods for Studying Eccrine Sweat Gland Function in *Vivo* and in *Vitro*

By KENZO SATO and FUSAKO SATO

Introduction

In man the eccrine sweat gland is one of the major cutaneous appendages and is distributed over the entire skin surface. Although the principal function of eccrine sweating is thermoregulation during exercises or in high ambient temperatures, the diverse roles of eccrine sweating in regulation of cutaneous homeostasis are still poorly understood. In contrast to the sweat glands in the hairy skin, those in the palms and soles respond mainly to emotional, rather than thermal, stimuli. It remains to be fully elucidated whether those latter sweat glands are pharmacologically identical to those in the hairy skin. The eccrine sweat gland is one of the target organs in cystic fibrosis.

A recent flurry of research into the pathogenesis of cystic fibrosis facilitated the need to further study this cutaneous gland using various *in vivo* and *in vitro* methodologies. The eccrine sweat gland is made up of the duct and the secretory coil. The duct consists of two layers of cells; lumenal and basolateral cells. The secretory coil consists of three cell types; clear (or secretory), dark, and myoepithelial.[1] Thus in selecting methodologies, these anatomical complexities must also be borne in mind.

In *Vivo* Methods to Study Sweat Gland Function

Visualization of Sweat. Visualization of sweating is often required for diagnosis and evaluation of abnormal sweating such as hyperhidrosis and anhidrosis or dysfunction of the autonomic nervous system.[2] Sweat water is readily visualized by iodine–starch reaction (Minor's method,[3] Wada's method,[4] one-step visualization method[5]), or by reaction with Bromphenol Blue powder,[6] pyrogallol,[7] ferric hydroxide,[7] and Quinazarin.[8] We find that

---

[1] K. Sato, W. H. Kang, K. Saga, and K. T. Sato, *J. Am. Acad. Dermatol.* **20**, 537 (1989).
[2] K. Sato, W. H. Kang, K. Saga, and K. T. Sato, *J. Am. Acad. Dermatol.* **20**, 713 (1989).
[3] V. Minor, *Dtsch. Z. Nervenheilkgd.* **101**, 302 (1927).
[4] M. Wada and T. A. Takagaki, *Tohoku J. Exp. Med.* **49**, 282 (1948).
[5] K. T. Sato, A. Richardson, D. E. Timm, and K. Sato, *Am. J. Med. Sci.* **31**, 528 (1988).
[6] I. Sarakany and P. A. Gaylarde, *Br. J. Dermatol.* **80**, 601 (1968).
[7] J. Zahejsky and J. Rovensky, *J. Soc. Cosmet. Chem.* **23**, 775 (1972).
[8] L. Guttman, *Proc. R. Soc. Med.* **35**, 77 (1941).

the one-step method,[5] which only involves spraying iodinated starch powder (prepared by adding 0.5–1 g of iodine crystals to 500 g soluble starch in a tightly capped bottle) with large cotton balls (or a body powder applicator) or an atomizer, is especially useful in clinical settings. The iodine–starch reaction is also used for obtaining sweat imprint papers,[9] but the imprint (which is the mirror image of the pore pattern on the skin) can be obtained only from flat body surfaces (e.g., trunk, palms, and soles) or small area of skins. We routinely prepare the iodine-impregnated imprint papers by exposing about 100 sheets of ordinary Xerox paper to a gram of iodine crystals in an airtight jar for a week or longer. The imprints are either immediately photocopied or tightly sealed in a plastic bag (with or without coating the paper with white petrolatum) for prolonged storage. The plastic impression method[10] allows us to obtain a three-dimensional negative impression of the skin with emerging sweat droplets visualized as small holes corresponding to the sweat pores. Thus the method is instrumental in relating the sweat pores to the surface topography of the skin.[1]

Determination of the Sweat Rate. It should be noted that the only informative measure of the sweat gland function that can be determined *in vivo* is the maximal sweat rate, i.e., the higher the sweat rate, the larger the sweat gland and the higher the pharmacological sensitivity.[11] Some of the methods still useful for determining the maximum sweat rate include the water vapor analyzer[12,13] (perhaps the most sensitive method but is applicable only to local pharmacological sweating), the filter paper method,[14] collection of sweat droplets under mineral oil,[15,16] and the anaerobic bag method[17] (which is also instrumental in collecting the least contaminated sweat samples for biochemical analysis of sweat ingredients).

Ductal Function in Vivo. Sweat is formed in two steps: formation of nearly isotonic primary fluid by the secretory coil and partial reabsorption of NaCl (or modification of other sweat ingredients) by the duct.[1] Furthermore, since ductal absorptive and excretory functions are dependent on the sweat flow rate, it is essential that concentrations of sweat ingredients be expressed relative to the sweat rate. Attempts have been made by several investigators to indirectly determine ductal Na^+ absorptive capacity from

[9] J. H. Thaysen and I. L. Schwarz, *Fed. Proc., Fed. Am. Soc. Exp. Biol.* **11**, 161 (1952).
[10] I. Sarkany and P. A. Gaylarde, *Br. J. Dermatol.* **80**, 601 (1968).
[11] K. Sato and F. Sato, *Am. J. Physiol.* **245**, R203 (1983).
[12] H. R. M. Van Gasselt and R. R. Vierhout, *Dermatologica* **127**, 255 (1963).
[13] K. Sato and F. Sato, *J. Lab. Clin. Med.* **111**, 511 (1988).
[14] I. L. Schwartz and J. H. Thaysen, *J. Clin. Invest.* **35**, 115 (1956).
[15] S. W. Brusilow, *J. Lab. Clin. Med.* **65**, 513 (1965).
[16] K. Sato and F. Sato, *J. Clin. Invest.* **73**, 1763 (1984).
[17] T. C. Boysen, S. Yanagawa, F. Sato, and K. Sato, *Am. J. Physiol.* **56**, 1302 (1984).

the curves of Na^+ excretion rates plotted on the y axis and sweat rates on the x axis, where the slope of the line at the highest sweat rate range is the Na^+ concentration of the primary fluid and the x intercept, designated as free water clearance, is the measure of the ductal Na^+ absorption rate.[18-20] The underlying assumption for this plotting method is that ductal Na^+ absorption saturates at the high sweat rate range. Unfortunately, such an assumption has not yet been proved.

Determination of the electrical potential in sweat droplets relative to the dermal potential by Quinton and Bijman[21] has led them to discover that the cystic fibrosis sweat duct has a higher lumen-negative potential which is most likely due to the decreased permeability of the ductal membrane to Cl^-. They placed an electrode filled with 3 M KCl agar in a sweat droplet or in the superfusate placed on the skin which was conducted to an electrometer (610 B from Keithley, Cleveland, OH, or any other amplifier with high input impedance) via an Ag-AgCl wire. The reference side is the abraded skin on the forearm/saline agar bridge/saturated KCl solution/an Ag-AgCl wire connected to the ground. The asymmetry potential was corrected by determining zero potential over an additional small abraded skin spot created near the sweat droplet inside the oil-filled test site.[22]

In Vitro Methods to Study Ion Transport of the Sweat Gland. Skin Biopsy: Skin specimens are usually obtained from volunteers under local anesthesia. Since no data are available on the effect of local anesthetics such as 1% lidocaine on subsequent *in vitro* glandular functions, we avoid direct infusion of lidocaine into the surgical site. We therefore use local block anesthesia by infusing 1% lidocaine about 5 cm proximal to the surgical site. After excisional biopsy, the skin is gently stretched with forceps as it is sliced, freehand, with a sharp blade. The slices are rinsed in several changes of well-oxygenated Ringer's solution kept at 10°. In rhesus or patas monkeys, palmar skin is excised after tranquilizing the animals with intramuscular injection of 30 mg/kg Ketamine (Parke-Davis) and 0.03 ml/kg of Inovar-vet (a mixture of fentanyl, 0.4 mg/ml, and droperidol, 20 mg/ml) (Pitman-Moore, Inc.) and the surgical wound closed with subcuticular embedding sutures.[11,16]

Isolation of Single Sweat Glands and Separation into the Duct and the Secretory Coil. Under a stereomicroscope at ×30 magnification, a tissue slice is transilluminated in a dissection chamber[23] containing Krebs-Ringer bicarbonate solution (KRB: 125 mM NaCl, 5 mM KCl, 1.2 mM MgCl$_2$,

[18] I. L. Schwartz and J. H. Thaysen, *J. Clin. Invest.* **34**, 114 (1955).
[19] G. W. Cage and R. L. Dobson, *J. Clin. Invest.* **44**, 1270 (1965).
[20] K. Sato, *Rev. Physiol. Biochem. Pharmacol.* **79**, 51 (1977).
[21] P. M. Quinton and J. Bijman, *N. Eng. J. Med.* **308**, 1185 (1983).
[22] J. Bijman and P. M. Quinton, *Am. J. Physiol.* **247**, C3 (1984).

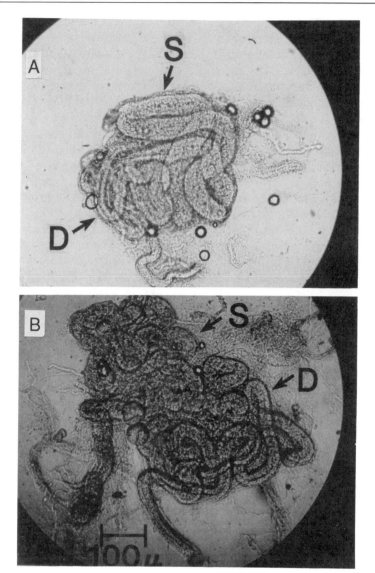

FIG. 1. Isolated eccrine sweat glands. 1, Isolated from rhesus monkey palm; 2, from human forearm. The magnification is the same for both glands. Note that the duct (D) is easily discernible from the secretory portion (S) by its smaller diameter and the presence of lumenal cuticular border.

1.0 mM CaCl$_2$, 25 mM NaHCO$_3$, 1.2 mM NaH$_2$PO$_3$, and 5.5 mM glucose, 10 mg/100 ml bovine or human albumin: pH is 7.48 at 95% O$_2$/5% CO$_2$) constantly kept at 10°. The sweat gland is readily identified in subcutaneous connective tissues (in human skin) or in subcutaneous fat lobuli (in monkey palm skin) as a small translucent mass about 200 to 500 μm in diameter. Sweat glands are teased out of the subcutaneous tissues using two pairs of sharp forceps (e.g., No. 5 Swiss jeweller's forceps whose tips are polished to a fine point on a grinding paper). Care is taken not to directly hold the glands with forceps during isolation. The isolated gland at this point consists of the secretory coil, the coiled (or proximal) duct, and a short segment of the distal (or straight) duct (Fig. 1). Thus it is often necessary to further separate the gland into the duct and the secretory coil. This is a tedious procedure which requires some practice and dexterity, especially when the separated segments must be absolutely free of physical damage.

Figure 2 is a schematic illustration for such a technique of uncoiling the duct and the secretory portion. Glass hooks are easily constructed on a microforge commercially available from various manufacturers such as Narishige (Greenvale, NY, or Tokyo, Japan) or Stoelting (Wood Dale, IL). It takes an hour to separate the duct or the secretory portion without damage (Fig. 3). However, the separation is quicker if the tubules are to be used for metabolic or biochemical studies, or for preparation of dissociated cells where minor damage to the tubules is still acceptable.

Sweat Induction in Vitro. Construction of pipets and a pipet holder[23] is modified from those of Burg *et al.*[24] originally developed for microperfusion of isolated nephrons. The sweat induction pipet (Figs. 4 and 5) we routinely use is constructed from a glass capillary (2.5-mm o.d., wall thickness 0.25 mm) which is pulled from a commercial 1.0-cm Pyrex glass tubing (VWR Scientific) using a Narishige PY-6 pipet puller. However, any commercial thin-walled glass capillary tubing with a diameter larger than 2 mm is equally satisfactory. The glass capillary is first pulled concentrically with a Narishige PV-4 micropipet puller and is further pulled on a microforge to reach the desired diameter at the constriction site (about 40 μm for sweat induction from the duct and about 70 μm for the secretory coil) (see Fig. 5). The sweat induction pipet thus constructed is about 8 cm long (pipet C in Fig. 4). The pipet tip is siliconized by dipping it in hexamethyldisilazane (Sigma, St. Louis, MO) for a few seconds and baking it at 200° for 5 min. Uncured Sylgard 184 (Dow Corning, Midland, MI) is then suctioned to fill about 200 μm long from the tip. The distal half of the

[23] K. Sato, *Am. J. Physiol.* **225**, 1147 (1973).
[24] M. B. Burg, J. Grantham, M. Abramov, and J. Orloff, *Am. J. Physiol.* **210**, 1293 (1966).

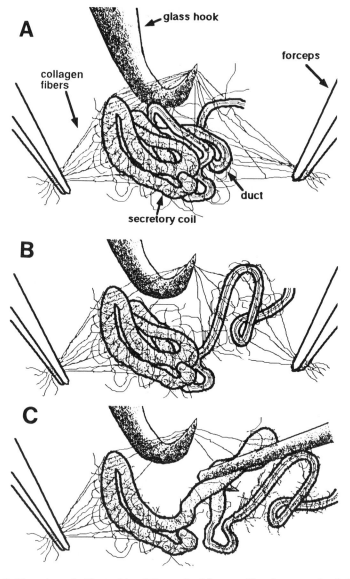

FIG. 2. The schematic illustration of the method for uncoiling the sweat gland. Under a stereomicroscope at ×40–×70 magnification, periglandular collagen fibers are held with forceps and small bundles of stretched fibers are teased away one at a time with a sharp glass hook immobilized on a micromanipulator (A). This procedure is repeated until most of the fibers connecting each tubular loop are removed (B). A second, smaller glass hook is instrumental in holding the tubules when the collagen fibers are no longer available for holding with forceps (C). The tubule should not be stretched or pinched with forceps.

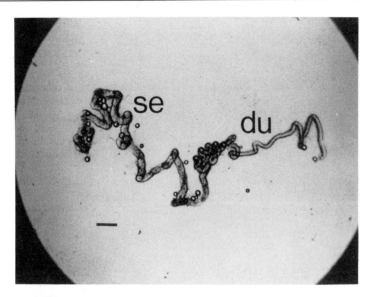

FIG. 3. An uncoiled human eccrine sweat gland. se, Secretory coil; du, duct. (Bar = 100 μm.)

pipet above Sylgard is then back filled with paraffin oil (mineral oil) that has been equilibrated with water vapor. The open end of the duct or the secretory coil from which sweat samples are to be collected is suctioned into the tip of the pipet C (induction pipet, see Fig. 4) mounted on a pipet holder. A small amount of Ringer's solution coincidentally suctioned into the pipet during this procedure is removed by aspiration with an inner

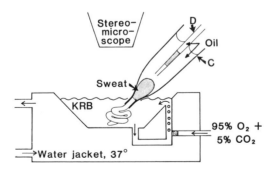

FIG. 4. Schematic illustration of a sweat induction pipet (C), a sampling pipet (D), and an incubation chamber thermostatted by a water jacket. Pipets C and D are mounted on a pipet holder equipped with a rack and pinion.

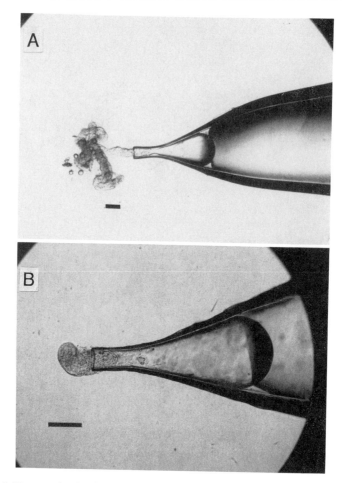

Fig. 5. Photographs showing induction of sweat secretion from isolated secretory coils. (A) Human secretory coil; (B) monkey secretory coil. Note sweating is induced from a short secretory segment removed from the tip (blind end). The sweat sample is seen as a ballooning droplet in the oil-filled sweat induction pipet. (Bars = 100 μm.)

pipet. The sweat sampling pipet (D in Fig. 4), which is also filled with mineral oil, is then inserted into pipet C for subsequent sweat collection. All of the above preparations are done at about 10° during which time no sweat or Ringer's solution enters the pipet. When the temperature of the bath is increased to 37°, some sweat glands show a transient sweat secretion but such sweating subsides spontaneously or after washing the bath a few times. Addition of cholinergic agents such as methacholine (MCh) induces persistent sweat secretion (see Fig. 5) which is reversibly inhibited by atropine[25] or removal of extracellular Ca^{2+}.[26] In our experience, possible artifacts such as leakage of incubation medium into the pipet have never been a concern so long as the sweat gland is held snugly by the pipet. In fact, deliberately damaging the gland at this point, say by pinching with a forceps, stops inflow of sweat into the pipet partially because the pipet interior has a larger hydrostatic pressure than the periglandular medium (which prevents inflow of incubation medium into the pipet). Further support for the lack of contamination of incubation medium into the pipet includes a mild dilatation of the ductal or secretory tubular lumen which is noted with movement of small debris during stimulation of sweating and cessation of inflow of fluid into the pipet with atropine, cyanide, ouabain, removal of Ca^{2+} from the bath, quinidine, or low medium temperature.

Perfusion of the Sweat Duct in Vitro. The technique for perfusion of the sweat duct is almost identical to that used for isolating nephrons,[24] which has now become the favorite technique for studying membrane transport by a number of nephrologists throughout the world. Thus we will deal with some of the technical notes pertinent only to the sweat gland. Perfusion of the sweat duct was first attempted by Mangos[27] and was later used more extensively by Quinton and Bijman, which led to their discovery of abnormal Cl⁻ transport in the cystic fibrosis sweat duct.[21] In our laboratory, isolation of the duct is performed as illustrated in Fig. 2, where complete removal of collagen fibers, avoidance of any physical damage to the tissue, and selection of relatively straight segments are the keys to the success in ductal perfusion. Furthermore, selection of well-acclimatized subjects for skin biopsy is also important because the sweat glands from poorly acclimatized individuals tend to be much smaller[11] and more fragile. Likewise, the rhesus monkey palm sweat duct which has a small lumenal diameter (≤ 5 μm) due to its narrow rigid cuticular border was also found to be unsuitable for cannulation. For our purposes, the outer diameter of the lumenal pipet tip (pipet B in Fig. 6) is $8 - 10$ μm and that of pipet A, $30 - 40$ μm. We usually use a double-channeled coaxial capillary for construction of pipet B where one of the channels, filled with Ringer's solution, is used only for perfusion of Ringer's solution and for determination of lumenal potential via an Ag/AgCl wire. For both channels, the perfusates were infused from the tip (about 100-μm diameter) of the inserted thin polyethylene tubings (F in the upper panel of Fig. 6) (pulled manually from the Eppendorf disposable 200-μl yellow pipets over a small flame) placed near

---

[25] K. Sato and F. Sato, *Am. J. Physiol.* **240**, R44 (1981).
[26] K. Sato and F. Sato, *Am. J. Physiol.* **241**, C113 (1981).
[27] J. M. Mangos, *Tex. Rep. Biol. Med.* **31**, 651 (1973).

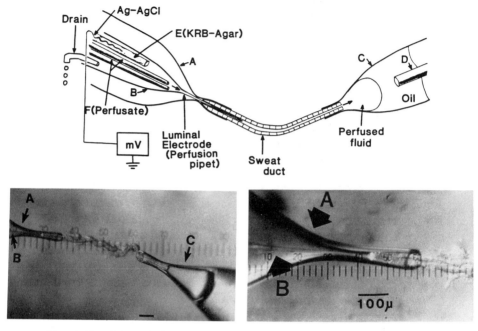

FIG. 6. *Top:* Schematic illustration of the method of ductal perfusion using the Burg method.[24] *Bottom:* Left, perfusion of a human proximal duct; right, its close-up view (bars = 100 μm). A, Outer perfusion pipet; B, lumenal pipet; C, collection pipet; D, sampling pipet. Note accumulation of perfusate in the oil-filled C pipet. The lumenal potential can be determined by the Ringer/agar tubing inserted into a single channel pipet via an Ag/AgCl wire connected to an amplifier. This configuration is used when the liquid junction potential is not a problem. However, a coaxial pipet (or the θ glass pipet) is used as pipet B for perfusion of gluconate or other perfusates where the liquid junction potential is significant. In such a case, one of the channels is used for perfusion of Ringer solution and for determination of lumenal potential via an Ag/AgCl wire (without interposing Ringer/agar) and the other channel for perfusion of substitution solutions through polyethylene tubings connected to pressurized oil-filled syringes.

the shoulder of the B pipet. These thin tubings were connected to oil-filled syringes pressurized by metal blocks placed on the pestles. The flow rate was adjusted by the pressure on the pestles and the amount of the drain so that the flow rate ranged from 50 to 100 nl/min. These perfusion and drain tubings, as well as the Ag/AgCl wire, are sealed at the posterior end of pipet B using a fast-curing epoxy glue whereas the junction between pipets A and B is sealed with a silicone or Teflon gasket in the pipet holder.[24] The criteria for successful ductal perfusion include the dependence of lumenal electrical potential difference (PD) on the bath temperature, hyperpolari-

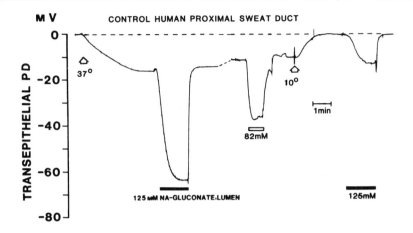

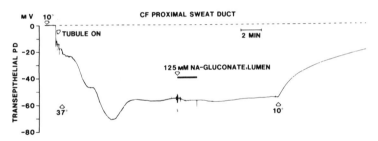

FIG. 7. Lumenal electrical potential (PD) of continuously perfused control human (upper) and cystic fibrosis (CF) sweat ducts (lower). In the control, the resting lumenal PD is less than −20 mV, which hyperpolarizes with reduction of lumenal Cl⁻ concentration (i.e., during gluconate perfusion). In contrast, the resting lumenal PD is much higher in the CF duct and does not further hyperpolarize during perfusion with gluconate. The observation is in agreement with that of Quinton and Bijman.[21] Also note the dependence of lumenal PD on the temperature in the bath.

zation of lumenal PD by lumenal perfusion of low Cl⁻ solution (e.g., by replacing with gluconate), and persistence of lumenal potential (see Fig. 7). As in nephrons, it is possible to combine glass microelectrode techniques and cable analysis with lumenal perfusion in order to determine the voltage divider ratio, the membrane potential, and the transepithelial and membrane resistance.[28]

Electrophysiology of the Secretory Coil. Since the secretory coil is densely surrounded by collagen fibers, blood vessels, and nerve fibers, and

[28] K. Sato, *Clin. Res.* **34**, 778A, (1986).

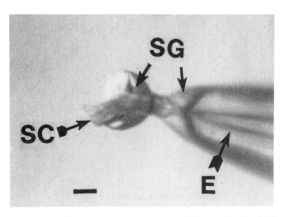

Fig. 8. Determination of lumenal electrical potential in an isolated distal blind end of human secretory coil (SC). Note the cuplike flare of the outer pipet containing Sylgard 184 (SG) used for insulation of the ductal lumen from the bath. E, Lumenal electrode. (Bar = 100 μm.)

manual removal of these fibrous structures is only partial at best, electrical insulation of the lumen from the bath is the most critical step in determining the lumenal potential. For example, a simple cannulation of the gland as shown in Fig. 5 will not yield a lumen-negative electrical potential during cholinergic stimulation unless Sylgard 184 is used for insulating the lumen from the bath.[29,30] We also use a special holding pipet with a cuplike flare at the tip (see Fig. 8). The lumenal electrode is filled with Ringer-agar and connected via an Ag/AgCl wire to a Keithley 604 electrometer. If the electrical insulation is satisfactory, the lumenal potential is unaffected by the position of the electrode tip although we usually place the tip only a few hundred micrometers from the cut end of the coil. When the lumenal electrode is inserted too deep into the lumen, the strenuous contraction of the secretory coil by cholinergic stimulation causes the secretory coil to be dislocated from the insulation cup, thus abolishing the lumenal PD.

Unlike the nephrons or the sweat ducts, we found it impossible to simultaneously determine membrane potential using a glass micorelectrode and the lumenal potential using the above cannulation experiment mainly because impalement of a secretory cell through the well-developed outer myoepithelial layer is possible (although very infrequently) only when a short secretory coil segment is immobilized under tension[31] which

[29] K. Sato and F. Sato, *Am. J. Physiol.* **242,** C360, (1982).
[30] M. D. Burg and J. Orloff, *Am. J. Physiol.* **219,** 1714, (1970).
[31] K. Sato, *Pfluegers Arch.* **407**(S2), S100, (1986).

precludes the luminal cannulation. Holding the secretory coil under tension is also necessary to keep the electrode tip from dislocating by myoepithelial contraction during cholinergic stimulation.

Superfusion of Isolated Secretory Coils. The lumen of the isolated secretory coil is narrow and almost collapsed, and any remaining lumenal space is filled with nearly isotonic fluid. In addition, the basolateral membrane of the secretory cell where most of the ionic movement occurs during pharmacological stimulation is exposed to the incubation bath. Although the secretory coil is made up of three types of cells, the ionic movements in different cell types appear to occur in parallel.[32] These characteristics of the secretory coil provide an excellent model system for studying the ionic mechanism of stimulus–secretion coupling by simply superfusing the isolated secretory coils and determining the change in ionic composition of the effluent superfusate as shown in the upper panel of Fig. 9. The superfusion chamber, which is constructed from 0.8-mm o.d. glass capillary on a microforge, is about two mm long and has an internal diameter of 0.4 mm with a constriction at its orifice (to achieve a tight seal with a superfusion tubing) and a flared distal portion (to accommodate ion-sensitive electrodes). The entire unit is secured on a small glass or epoxy block (table) with a slow-curing epoxy glue. The secretory coils are placed inside the chamber on a polyethylene rod (about 0.15-mm o.d.) with thorns which are created on the rod by shaving with a No. 11 surgical blade. This simple device effectively prevents clogging of the distal end of the chamber during superfusion. As many as eight superfusion polyethylene tubings (for eight superfusates) can be installed on a holder secured on a micromanipulator. Each tubing is connected to a perfusion syringe on a Harvard syringe pump. Perfusion speed is usually 1 μl/min. The superfusion capillary is submerged in the bath (Ringer's solution) kept at 37°. The ion-sensitive electrode is constructed by fusing a small porous alumina granule (Sigma), about 50 μm in diameter, to the tip of the electrode with a microforge.[33] After siliconizing the electrode tip as described earlier, a section about 500 μm from the tip is filled by suction with various ion-sensitive resins (purchased from WP-I, New Haven, CT). The electrode is filled with appropriate solutions as recommended by WP-I (e.g., 100 mM KCl for K$^+$ electrode and 100 mM NaCl for Na$^+$ or Cl$^-$ electrode) and connected to a Keithley amplifier (No. 602 or 604) via an Ag/AgCl wire and calibrated. The number of secretory coils inserted into the chamber in each experiment varies with the size of the sweat gland but is usually from 5 to 10 secretory coils.

[32] K. Saga and K. Sato, *J. Membr. Biol.* **107**, 13 (1989).
[33] K. Saga, F. Sato, and K. Sato, *J. Physiol. (London)* **405**, 205 (1988).

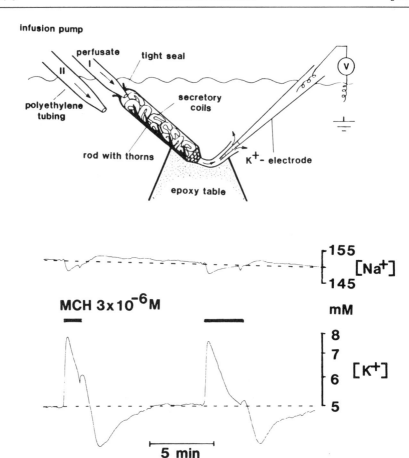

FIG. 9. Superfusion of isolated secretory coils for studying ionic movement into and out of the secretory coil cells during methacholine (MCH) stimulation. *Top:* Superfusion chamber containing isolated secretory coils. *Bottom:* Na⁺ and K⁺ concentration in the efflux superfusate as simultaneously determined by K⁺ and Na⁺ electrodes. Rhesus secretory coils were used in the experiment shown in the lower panel. Perfusates are changed simply by inserting a new polyethylene tubing until it forms a tight seal with the constriction ring at the entrance of the superfusion chamber. The result (in the lower panel) shows that methacholine stimulation of the secretory coil is associated with K⁺ efflux from and Na⁺ influx into the secretory coil.

Determination of Tissue $[Na^+]$ and $[K^+]$ by Microtitration. Although the above superfusion method provides the most accurate time course of ionic movement into and out of the sweat gland during pharmacological stimulation, the result is only qualitative (not quantitative unless total glandular volume is determined). Thus the method is not suitable for studying the

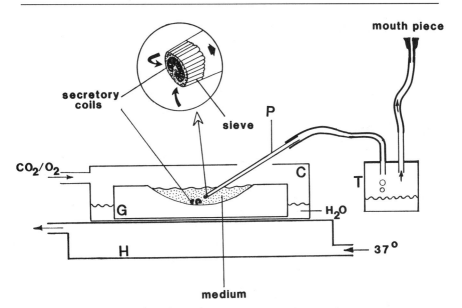

FIG. 10. Schematic illustration of the method for incubating a single segment of the secretory coil and extracting cellular ions for microtitration. In the circle, the tip of the pipet (P) is shown holding a secretory coil by mouth suction. The suctioned liquid is trapped in a small fluid trap (T). C, A transparent polycarbonate cover to partially trap CO_2/O_2 and moisture; G, a glass slide with a concavity containing isolated secretory coils; H, thermostatted water jacket.

effect of some lipophilic agents that impair ion-sensitive resins. In such cases, the direct microtitration of extracted cellular ions is in order. The method is modified from that of Sudo and Morel,[34] which was used for determination of tissue electrolytes in isolated nephrons (Fig. 10). Again we take advantage of the fact that the secretory coil lumen is very narrow and collapsed so that the contribution of the lumenal (extracellular) electrolytes to the total tissue electrolytes is minimal and constant before and during MCh stimulation. The volume of a single secretory coil segment is estimated by measuring its diameter and the tubular length from the Polaroid pictures of the gland taken before preincubation or from the image captured by a video camera. The segment of the secretory coil is first preincubated in KRB for 5 min at 37° in the concavity of a glass slide placed in a moist chamber constantly equilibrated by 5% CO_5/95% O_2, then transferred to a new incubation medium (containing appropriate drugs to be studied) in an adjacent concavity. After incubation, the coil is

[34] J. Sudo and F. Morel, *Am. J. Physiol.* **246**, C407 (1984).

picked up by suction on a small sieve (400-mesh copper grid secured by epoxy glue, about 1 mm in diameter) secured at the open end of a pipet tip. The periglandular ions are easily removed by immersing the sweat gland held at the tip of the pipet in a cold isotonic Tris buffer (pH 7.4, 150 mM Tris-HCl, 1.2 mM MgCl$_2$, and 1.0 mM CaCl$_2$) or 300 mM mannitol solution under constant suction for 1 sec, during which the loss of intracellular ions is negligible.[32] The secretory tubule is picked up with sharp forceps and placed in a 100-nl droplet of 10% trichloroacetic acid (TCA) or 1 N HNO$_3$ (if Cl$^-$ is also determined) in hydrated mineral oil in a moist chamber and Na$^+$, K$^+$, and Cl$^-$ extracted overnight at 4°. [Na$^+$] and [K$^+$] in the TCA extract are determined by helium glow spectrophotometry using standard solutions prepared in 10% TCA and placed adjacent to the TCA for sweat gland extraction and taking other precautions recently discussed by us.[35] Chloride ion is titrated using the WP-I model ET-1 chloride titrator according to the manufacturer's instruction manual.

Preparation of Dissociated Sweat Gland Cells. Dissociated sweat gland cells can be used for cell culture, determination of intracellular [Ca^{2+}] and other ions, studies of cell volume regulation, patch clamping, and others. A minimum of 100 to 200 isolated whole sweat glands are first incubated with 0.75 mg/ml type I collagenase in Krebs' bicarbonate buffer also containing 10 mM HEPES buffer (HEPES-KRB, pH 7.5) for 15 min and then thoroughly washed in fresh (collagenase-free) HEPES-KRB. Lowering [Ca^{2+}] in HEPES-KRB to 0.1 mM is often helpful at this stage if digestion of periglandular connective tissue by collagenase alone is not sufficient. This brief collagenase digestion makes the subsequent manual separation of the secretory coils and the ducts considerably easier. Nevertheless, the tedious manual separation of the secretory coils from the ducts is necessitated because all attempts at separating the mixed cells into the secretory coil cells and the ductal cells by Percoll gradient centrifugation have failed (presumably because the secretory cells and the duct cells may have similar density). After manual removal of the ducts and the secretory coils, cross-contamination of the ductal and secretory cells is practically nonexistent.

The subsequent method of cell isolation is a modification of our previous method.[36] Briefly, isolated secretory coils or the ducts are incubated in KRB containing 0.1 mM Ca^{2+}, 0.75 mg/ml type I collagenase, and 100 μg/ml DNase I for 5 min at 37°. The glands are then incubated for 10 min in two changes of 500 μl Ca^{2+}- and Mg^{2+}-free KRB containing 2 mM EGTA. A second enzyme digestion is performed for 30 to 60 min in KRB

[35] K. Sato and F. Sato, *Am. J. Physiol.* **252,** R1099 (1987).
[36] K. Sato and F. Sato, *Am. J. Physiol.* **254,** C310 (1988).

containing 0.1 Ca^{2+}, 1.25 mg/ml collagenase (or 0.125 mg/ml type XIII trypsin from Sigma), penicillin and streptomycin mixture (30 U/ml), and 100 μg/ml DNase. The cells are dispersed from the digested glands by repeatedly passing them through a sieve, which is a tungsten grid for electron microscopy (3-mm o.d., 100 mesh) epoxy glued to the tip of a disposable plastic pipet. Dispersed cells are subsequently passed through a finer sieve (200–300 mesh) and centrifuged at 900 g for 2 min. The cells are then suspended in KRB containing 4% bovine albumin. The electron microscopy of the dispersed cells obtained in this way shows more than 60% secretory cells, less than 30% dark cells, and about 10% myoepithelial cells. The mixed dispersed cells thus prepared can be used for determination of intracellular $[Ca^{2+}]$ using microscopic fluorometry with Fura-2,[37] for patch clamping,[38] or for the study of cell volume regulation[39] because the different cell types can be identified by fluorescent microscopy (the presence of autofluorescent lipofuscin granule in the clear cell) or under differential interference contrast (dark cell granules for the dark cell; characteristic appearance of the myoepithelial cell). When necessary, the clear cells are partially purified by discontinuous Percoll gradient (80–12.5%) centrifugation of the dispersed cells at 1000 g for 30 min, which removes most of the myoepithelial cells and more than half of the dark cells in addition to cellular debris. The secretory cell-rich fraction is found at the interface of 60 and 65% Percoll. More than 95% of these cells exclude Trypan Blue.[35]

[37] K. Sato and F. Sato, *Clin. Res.* **37**(2), 354A (1989).
[38] K. Sato and F. Sato, *Clin. Res.* **37**(2), 639A (1989).
[39] Y. Suzuki, K. Sato and F. Sato, *Clin. Res.* **37**(4), 970A (1989).

[39] Isolation, Voltage Clamping, and Flux Measurements in Lepidopteran Midgut

By William R. Harvey, Dwight N. Crawford, and Daniel D. Spaeth

Introduction

A unique K^+ pump in certain insect epithelia actively transports potassium out of the cells across the apical plasma membrane to the lumen. In the lepidopteran midgut the electrogenic transport of K^+ produces a transepithelial potential difference (PD) as high as 180 mV (lumen positive).

Both transport and potential are oxygen dependent and ouabain insensitive.[1] The K^+ pump is essential for nutrient uptake[2] as well as for pH and ion regulation by the lepidopteran midgut.[3] It drives ion and water movements in salivary glands, labial glands, Malpighian tubules, and recta of certain insects.[4] It contributes to the resting potential in insect sensory sensilla.[5] An unusual vacuolar-type ATPase has been identified and characterized in lepidopteran midgut.[6,7] This enzyme resides in the same plasma membrane segment as the K^+ pump and shares many characteristics with the pump. In this chapter we describe biophysical techniques for studying the K^+ pump in the isolated *Manduca sexta* midgut. In the following chapter Wieczorek *et al.*[8] describe biochemical techniques for studying the K^+-ATPase in this same tissue.

The short-circuit current (I_{sc}) measures the active transport of potassium from the blood side (B) to the lumen side (L) of the isolated lepidopteran midgut under the conditions specified below. The relationship between I_{sc} and transport is obtained by applying the Ussing[9] flux ratio equation to K^+ fluxes across the midgut.

$$\phi K^+_{BL}/\phi K^+_{LB} = ([K^+]_B/[K^+]_L)^{(zF/RT)(\psi_B - \psi_L)}$$

ϕK^+_{BL} and ϕK^+_{LB} are the unidirectional K^+ fluxes from blood side to lumen side and vice versa, $[K^+]_B$ and $[K^+]_L$ are the concentrations of K^+ in blood side and lumen side solutions, $\psi_B - \psi_L$ is the potential difference (PD) in volts of the blood side with respect to the lumen side and $zF/RT = 38.9$ V^{-1} at $25°$ (where F is the Faraday constant, R is the gas constant, T is the temperature, and z is the ionic charge number).

To obtain short-circuit condition, $[K^+]_B$ is set equal to $[K^+]_L$ and $\psi_B - \psi_L$ is set to zero. With no electrochemical gradient across the midgut, the flux ratio, $\phi K^+_{BL}/\phi K^+_{LB}$, is expected to be 1. However, the ratio is not 1 because K^+ is actively transported from blood side to lumen side. Under open-circuit conditions in the standard solution described below this net

[1] W. R. Harvey, M. Cioffi, J. A. T. Dow, and M. G. Wolfersberger, *J. Exp. Biol.* **106**, 91 (1983).
[2] B. Giordana, F. V. Sacchi, and G. M. Hanozet, *Biochim. Biophys. Acta* **692**, 81 (1982).
[3] J. A. T. Dow and W. R. Harvey, *J. Exp. Biol.* **140**, 455 (1988).
[4] J. A. T. Dow, *Adv. Insect Physiol.* **19**, 187 (1986).
[5] U. Thurm and J. Kuppers, *in* "Insect Biology in the Future" (M. Locke and D. S. Smith, eds.), p. 735. Academic Press, New York, 1980.
[6] H. Wieczorek, M. G. Wolfersberger, M. Cioffi, and W. R. Harvey, *Biochim. Biophys. Acta* **857**, 271 (1986).
[7] H. Schweikl, U. Klein, M. Schindlbeck, and H. Wieczorek, *J. Biol. Chem.* **264**, 11136 (1989).
[8] H. Wieczorek, M. Cioffi, U. Klein, W. R. Harvey, H. Schweikl, and M. G. Wolfersberger, this volume [40].
[9] H. H. Ussing, *Acta Physiol. Scand.* **19**, 43 (1950).

TABLE I
AGREEMENT BETWEEN THE I_{sc} AND THE NET K⁺ FLUX
IN THE ISOLATED *Manduca sexta* MIDGUT[a]

Time	ϕK^+_{BL}	ϕK^+_{LB}	Net K⁺ flux	I_{sc}
20	484 ± 56	9 ± 1	475 ± 56	615 ± 25
30	452 ± 36	14 ± 3	438 ± 36	499 ± 18
40	443 ± 22	16 ± 1	427 ± 22	433 ± 18
50	383 ± 13	17 ± 3	366 ± 13	390 ± 19
60	373 ± 27	17 ± 3	356 ± 27	356 ± 20
70	342 ± 14	21 ± 2	321 ± 14	331 ± 22
80	295 ± 14	23 ± 3	272 ± 14	308 ± 24
90	254 ± 48	23 ± 2	231 ± 48	287 ± 26

[a] All experimental values, mean ± SEM, expressed as
$\mu A/Cm^2$; $n = 6$.

K⁺ flux is accompanied by an equivalent flux of Cl⁻ in the same direction, thereby preserving electroneutrality. The Cl⁻ flows across the membrane in response to the electrical driving force created by the charge separation as K⁺ is pumped. However, under short-circuit conditions $[Cl^-]_B = [Cl^-]_L$ and $\psi_B - \psi_L = 0$ so there can be no passive net flux of Cl⁻ from blood side to lumen side. Instead Cl⁻ moves from the solution to the Ag/AgCl (current) electrode on the blood side, electrons are propelled by a battery through an external circuit to the current electrode on the lumen side, and Cl⁻ moves from this electrode into the solution on the lumen side $[AgCl + e^- \rightarrow Ag° + Cl^- (aq)]$. The current measured on a microammeter as the electrons flow through the external circuit is equivalent to the summed net flux of all actively transported ions.

The lepidopteran midgut can actively transport all of the alkali metal ions (i.e., Li⁺, Na⁺, K⁺, Rb⁺, and Cs⁺) from blood side to lumen side and also can transport Ca^{2+} and Mg^{2+} from lumen side to blood side. However, in the standard solution all alkali metal ions except K⁺ are left out and the Ca^{2+} and Mg^{2+} concentrations are so low that their transport accounts for less than 5% of the I_{sc}. Therefore the I_{sc} is equivalent to the rate of active K⁺ transport by the gut (Table I) provided that the K⁺ concentration of the bathing solution is below 64 mM. Other solutions, designed to resemble larval blood, minimize the characteristic I_{sc} decay found with isolated midguts.[10-12] However, the equivalence between the I_{sc} and the net K⁺ flux has not been investigated in any of these other solutions.

[10] M. G. Wolfersberger and K. M. Giangiacomo, *J. Exp. Biol.* **102**, 199 (1983).
[11] M. V. Thomas and T. E. May, *J. Exp. Biol.* **108**, 273 (1984).
[12] P. Parenti, B. Giordana, V. F. Sacchi, G. M. Hanozet, and A. Guerritore, *J. Exp. Biol.* **116**, 69 (1985).

The original Ussing and Zerahn[13] chamber was adapted for the insect midgut as a cylinder by Harvey and Nedergaard,[14] as a flat sheet by Wood and Moreton,[15] and as a flat sheet mounted in a removable holder by Dow *et al.*[16] The Dow chamber has recently been modified by Crawford to accommodate smaller tissue samples and for use in flux studies. Numerous special chambers have been developed, e.g., for measuring oxygen uptake and I_{sc} concomitantly.[17] The techniques were originally worked out for *Hyalophora cecropia* but have been applied to many different lepidopteran insects such as *Manduca sexta,*[18,19] *Spodoptera eridania,*[11] and more recently to *Lymantria dispar.* However, the technical problems in short circuiting all lepidopteran midguts are the same; the gut is fragile, it has low electrical resistance, and it produces a high PD and I_{sc}.

Materials

The chamber (Fig. 1F) was constructed by Mr. Francis J. Mansell, College of Arts and Sciences Machine Shop, Temple University, Philadelphia, Pennsylvania. Silver for Ag/AgCl electrodes, 0.5 mm thick, is from T. B. Hagstoz and Son, Philadelphia, Pennsylvania. Gas regulators are from Matheson, Inc., (Secaucus, NJ) model 8-540 for O_2, model 8-580 for N_2. Gas lines to the chamber are amber latex tubing, $\frac{1}{4}$-in. bore and $\frac{1}{16}$-in. wall. Pressure regulating columns are Pyrex tubing, 5-cm o.d., 120 cm high, rubber stoppered at the bottom, water filled to 20 cm of the top, with a Pyrex T-tube and 8-mm Pyrex tubing to within 5 cm of the bottom. Gas washing bottles are Pyrex, 250-ml capacity, with a fritted glass dispersion element. The voltmeter, with an input resistance of approximately 10^{10} Ω, is model 602 from Keithley Instruments, Cleveland, Ohio. Calomel electrodes, K401, are from Radiometer America, Inc., Westlake, Ohio. Bridges from the calomel electrodes to the chamber are number 7446 Intramedic polyethylene tubing, 0.082-in. o.d., 0.062-in. i.d., from Clay Adams Division of Becton, Dickinson, and Company, Parsippany, New Jersey. Glass capillary tubes are Blue-Tip-Plain capillary tubes number Z629-B63, 1.5-mm o.d., 1.1-mm i.d. from A. H. Thomas Company, Philadelphia, Pennsylvania. These capillary tubes are inserted into one end of a 30-cm length

[13] H. H. Ussing and K. Zerahn, *Acta Physiol. Scand.* **23,** 110 (1951).
[14] W. R. Harvey and S. Nedergaard, *Proc. Natl. Acad. Sci. U.S.A.* **51,** 757 (1964).
[15] J. L. Wood and R. B. Moreton, *J. Exp. Biol.* **77,** 123 (1978).
[16] J. A. T. Dow, W. R. Harvey, M. G. Wolfersberger, and B. Boyes, *J. Exp. Biol.* **114,** 685 (1985).
[17] W. R. Harvey, J. A. Haskell, and K. Zerahn, *J. Exp. Biol.* **46,** 235 (1967).
[18] D. F. Moffett, *J. Exp. Biol.* **78,** 213 (1979).
[19] J. T. Blankemeyer and W. R. Harvey, *J. Exp. Biol.* **77,** 1 (1977).

of the polyethylene tubing. The assembled bridges are filled with standard solution containing 3% agar using a 10-ml syringe equipped with a 16-gauge hypodermic needle. Rubber seals for the bridges are National 002 0-rings from Federal Mogul Corporation, Detroit, Michigan. The microammeter, model 630-NA, is a battery-powered VOM with 5% accuracy from Triplett Corporation, Bluffton, Ohio. See Figure 1 for circuit details of the voltage clamp. A circuit diagram for our automatic voltage clamping device is available upon request. Additionally, a low noise clamping device suitable for fluctuation analysis is described by Van Driessche and Lindemann.[20] Hemostats are Halstead, straight, mosquito forceps with fully serrated jaws, 5 in. long, number 18-2300; *dissecting scissors* are Stille, straight, sharp point, 5½ in. long, number 37122; and the forceps are Pilling, tissue type with serrated tips 4¾ in. long, number 18-1270; all from Pilling Company, Fort Washington, Pennsylvania.

Chemicals are reagent grade. Radioisotopes are ^{42}KCl in water and ^{86}RbCl in HCl with a specific activity > 5 μCi/μl. The HCl is evaporated off and the ^{86}RbCl is resuspended in distilled water at 0.1 μCi/μl. Compressed gases, O_2 and N_2, are laboratory grade. *Manduca sexta* larvae are purchased as eggs or third instar larvae from Carolina Biological Supply Company, Burlington, North Carolina.

The standard solution is 32 mM KCl, 1 mM CaCl$_2$, 1 mM MgCl$_2$, 5 mM Tris-Cl, and 300 mM sucrose at pH 8.3. KCl (32 mM) approximates the larval blood concentration and yields intermediate values of I_{sc} and net flux. Tris-Cl (5 mM) controls pH while 1 mM CaCl$_2$ and 1 mM MgCl$_2$ maintain tissue integrity. Sucrose (300 mM) is slightly hypersomotic to larval blood. The sucrose concentration is held constant when ionic concentrations are altered because substances with $M_r < 340$ do not contribute to maintaining the osmotic pressure of the tissue.[18] The standard solution is oxygenated for 20 min prior to use.

Short-Circuiting *Manduca sexta* Midgut

Apparatus for Measuring PD and I_{sc}. The midgut chamber, PD-measuring circuit, I_{sc}-measuring circuit, and accessory services are shown in Fig. 1F. The Dow chamber is shown with the holder inserted, ready for short circuiting. To assemble the apparatus, the chamber is mounted on a ring stand, gas supply lines are connected, the calomel electrodes are connected via the voltage clamp to an electrometer, the Ag-AgCl electrodes are connected via the voltage clamp to a microammeter, and a vacuum line is provided.

[20] W. Van Driessche and B. Lindemann, *Rev. Sci. Instrum.* **49**, 52 (1978).

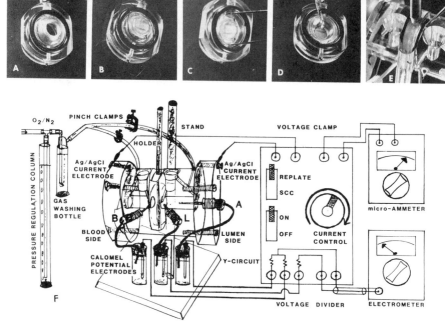

FIG. 1. Apparatus for short-circuit and flux measurements in isolated lepidopteran midgut. (A–D) Details of the midgut being mounted on the aperture within its holder; (E) filling of the aperture void volumes; (F) details of the midgut chamber along with electrical connections to the voltage clamp, microammeter, and electrometer. Diagram (F) also shows provisions for the gassing and stirring of solutions within the chamber.

Gas Flow Rate. The holder, with aperture in place, is inserted into the chamber which is then filled to the base of the funnel tapers with oxygenated standard solution. The reduction valve on the oxygen gauge is set to 10 psi and the needle valve is cautiously opened until gas bubbles just exit the glass tubing in the pressure regulation column. The gas flow to each chamber half is adjusted with pinch clamps until gas bubbles enter each lift pump so fast that they are barely visible as distinct entities. The O_2 lines are then clamped off with hemostats. The nitrogen supply is adjusted in the same way and then clamped off. These gas adjustments are not touched during the measurement of the I_{sc}.

Asymmetry PD. The asymmetry PD and bridge alignment are measured and adjusted with the aperture in the chamber and the O_2 lines opened. The PD is read from the electrometer and double the value is recorded too compensate for its halving by a "Y" circuit in the voltage clamp. A PD >10 mV usually indicates an air bubble in one of the agar bridges. A PD remaining after air bubbles are removed is usually due to asymmetry of the three calomel electrodes. If this so-called asymmetry PD exceeds 2 mV, then one calomel electrode at a time is replaced until a matched set is obtained.

Bridge Alignment. With the aperture in place and the O_2 lines open, the alignment electrode, A (Fig. 1F, on right), is adjusted to yield the asymmetry PD as a 500-μA current is applied in either direction.

Preparation of Chamber and Holder to Receive Midgut. The holder is removed, the O_2 lines are clamped off, and the chamber is refilled to the base of the funnel tapers (funnel details, Fig. 1E). The set screw in the holder is loosened and the aperture is removed to receive the gut (Fig. 1A). The holder is lightly coated with petroleum jelly on the surface where the tissue-laden aperture is inserted and sealed latter. The holder and aperture are then chilled in crushed ice.

Preparation of Larva and Dissecting Materials. Two or three feeding fifth instar larvae, weighing 4.5 to 6.0 g, are immobilized in crushed ice for 20 to 40 min. Fifty milliliters of standard solution and a 1-ml glass pipet, 10 cm long, with a fire-polished tip are chilled along with the larvae. A Lucite dissecting block ($10 \times 10 \times 1$ cm) is chilled in a freezer ($-20°$), then held on a small styrofoam box during the dissection. Dissecting scissors and forceps are placed nearby. Several loops of 3-0 surgical cotton are prepared, each with a double overhand knot.

Midgut Isolation. Careful preparation is important so the midgut can be isolated quickly. No more than 3 min should elapse from the first incision to the final wash on the chamber. Approximately 10 ml of chilled standard solution is pooled on the cold Lucite block. A chilled larva is placed in the pool. The midsection of the larva is isolated by cutting just behind the most posterior set of walking appendages and just ahead of the most anterior set. To manipulate the larva, the 1 ml fire-polished pipet is inserted through the exposed lumen cavity from front to back. The integument is opened by cutting along the dorsal midline, avoiding the underlying gut tissue. The integument is folded back, exposing the tracheae which attach the integument to the gut. The gut tissue is rinsed well with chilled solution. The tracheae are cut close to the midgut, first on one side then the other. The isolated gut is rinsed and opened to a flat sheet by cutting along its long axis. The peritrophic membrane and enclosed gut contents are removed, followed by a final rinse. The gut tissue must not freeze to the plate during the dissection.

Placing Gut in Holder. The ends of the gut are grasped and placed, blood side down, so as to cover the entire aperture of the holder with either the anterior, middle, or posterior region.[21] The tissue is secured to the aperture with a loop of thread (Fig. 1B). The aperture is pressed into the holder (Fig. 1C), the set screw is tightened, and the gut is rinsed with chilled solution. To test for leaks, three to four drops of solution are first pipetted gently on the lumen side of the horizontally held aperture then on the blood side of the tissue (Fig. 1D). The gut is rejected if a drop forms on the underside of the tissue.

Inserting Holder into Chamber. The holder, oriented with blood side to the left, is inserted until only one-eighth of the aperture is visible above the chamber. Air is then forced out of the aperture by filling both sides alternately with standard solution using a Pasteur pipet (Fig. 1E). Finally, the holder is seated in the chamber and the O_2 lines are unclamped. To wash the tissue, approximately 60 ml of standard solution (20 × half-chamber volume) is introduced into the blood side funnel and removed by suction slowly enough (ca. 1 min) so that the tissue is thoroughly washed with fresh solution. The procedure is repeated on the lumen side. These washes with cool solution are thought to remove substrate and digestive enzymes from the mucosal surface and hemolymph from the opposite surface of the midgut. After the washes, the solution level is set to the bottom of the funnel tapers.

Monitoring I_{sc}. Immediately after the final wash the transepithelial PD is recorded and brought to the asymmetry PD using the voltage clamp. The clamp is adjusted manually to keep the PD at the asymmetry PD and the I_{sc} is recorded from the microammeter. As an alternative, an autoclamp can be used to maintain the transepithelial PD at the asymmetry PD (see the section, Asymmetry PD, above). The gut is rejected if the I_{sc} is below 200 μA/cm^2 after a 30-min equilibration period.

Preparation for Subsequent Experiments. After experiments are concluded, AgCl equivalent to the Cl⁻ displaced during calibration and experimentation is replated slowly. Between experiments the chamber is filled with the standard 32 mM KCl solution (sucrose omitted) and all gas lines are removed from the chamber openings.

Test Experiments

Three simple experiments confirm that the isolated midgut and voltage-clamping apparatus are functioning correctly.

Interrupted Stirring. The I_{sc} is recorded for 65 min, then the stirring gas supply to both sides is clamped off. After 20 min of recording without stirring, the hemostats are removed and recording is continued for another

[21] M. Cioffi and W. R. Harvey, *J. Exp. Biol.* **91**, 103 (1981).

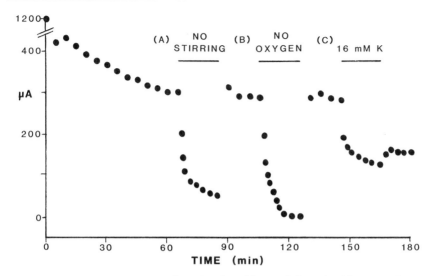

F<small>IG</small>. 2. Test experiments to confirm that the midgut and short-circuiting apparatus are functioning properly: (A) interrupted stirring, (B) oxygen replaced by nitrogen, (C) 32 mM KCl standard solution replaced by a 16 mM KCl variant.

20 min. Typically the I_{sc} drops abruptly, but reversibly, with interrupted stirring (Fig. 2A).

Oxygen Lack. The stirring gas is changed from O_2 to N_2 by moving the hemostat from the N_2 to the O_2 line. Typically after a 2- to 3-min lag time, the I_{sc} drops and eventually becomes zero in nitrogen (Fig. 2B). After 20 min in nitrogen, oxygen stirring is restored and the I_{sc} returns to expected levels following a minor overshoot.

Decreased [K⁺]. The K⁺ concentration is decreased to 16 mM by washing with a 16 mM KCl variant of the standard solution. Typically the I_{sc} is depressed sharply in 16 mM KCl and the depression is partially reversed when standard 32 mM KCl solution is restored (Fig. 2C).

Flux Measurements: Agreement of I_{sc} and Net K⁺ Flux

Agreement between the I_{sc} and net K⁺ flux across the isolated *Manduca sexta* midgut in the standard solution confirms that the I_{sc} is equivalent to the net K⁺ transport. After the midgut has been short circuited for 10 min, 5 μCi of ⁴²KCl is added to one side of the chamber and a 10-μl standard is taken. Then 100-μl samples are taken at regular intervals from the opposite side and replaced with fresh standard solution. A second 10-μl standard is taken at the end of the experiment from the original side of the chamber.

The samples are counted without cocktail, by the Cerenkov effect, in a liquid scintillation counter. At the end of each experiment the holder is removed from the chamber and the tissue covering the 0.5-cm^2 aperture is cut out, washed for 10 sec, blotted dry, and weighed. Values of ϕK_{BL}^+ and ϕK_{LB}^+ are calculated by assuming that the specific activity of the radioisotope flowing across the tissue into the sampling volume is the same as that in the solution to which it is introduced[22] and are corrected both for tracer removed during sampling and for isotope decay.[21] The data from six K$^+$ flux experiments are shown in Table I. The tracer steady state is achieved between 30 and 40 min. It is clear that between 40 and 70 min the I_{sc} agrees with the net K$^+$ flux within experimental error.

Acknowledgments

We thank Dr. Michael G. Wolfersberger for his critical reading of the manuscript, Mr. Robert M. Simons for the experiments reported in Fig. 2, and Ms. Qian-Wei Tan for help in preparing illustrations. This work is supported in part by National Institutes of Health Research Grant AI 22444, NIH Biomedical Research Support Grant 2 S07 RR07115-17, and Temple University's Research Incentive Fund.

[22] G. deHevesey, *Enzymologia* **5**, 138 (1938).

[40] Isolation of Goblet Cell Apical Membrane from Tobacco Hornworm Midgut and Purification of Its Vacuolar-Type ATPase

By Helmut Wieczorek, Moira Cioffi, Ulla Klein, William R. Harvey, Helmut Schweikl, and Michael G. Wolfersberger

Introduction

The larval lepidopteran midgut is a one-cell-thick epithelium composed of two major cell types, columnar cells and goblet cells.[1] An electrogenic alkali metal ion pump is located in the apical domain of the plasma membrane of the less abundant goblet cells.[2] ATP-dependent and electrogenic proton transport across this membrane,[3] mediated by a vacuolar-type

[1] E. Anderson and W. R. Harvey, *J. Cell Biol.* **31**, 107 (1966).
[2] J. A. T. Dow, B. L. Gupta, T. A. Hall, and W. R. Harvey, *J. Membr. Biol.* **77**, 223 (1984).
[3] H. Wieczorek, S. Weerth, M. Schindlbeck, and U. Klein, *J. Biol. Chem.* **264**, 11143 (1989).

ATPase,[4,5] is thought to energize secondary active proton/alkali metal ion antiport, resulting in net electrogenic transport of alkali metal ions.

In this chapter the only known method for isolation of the goblet cell apical membrane from larval lepidopteran midgut is described. Also described is the purification of the ATPase associated with this membrane as well as procedures for measuring both membrane-bound and purified solubilized ATPase activity. A detailed description of reliable and sensitive colorimetric methods for measurement of inorganic phosphate and protein is included.

Isolation of the Goblet Cell Apical Membrane[6]

Reagents

Buffer A: 0.25 M sucrose, 5 mM Na-EDTA, 5 mM Tris-HCl (pH 8.1)
Sucrose (37%, w/w, 5 mM Tris-HCl (pH 8.1)
Sucrose (45%, w/w), 5 mM Tris-HCl (pH 8.1)
Na-EDTA (5 mM), 5 mM Tris-HCl (pH 8.1)
EGTA (0.32 mM), 5 mM Tris-MOPS (pH 7.6)
EGTA (0.32 mM), 5 mM Tris-MOPS, 3.2 mM 2-mercaptoethanol, 0.1% bovine serum albumin (pH 7.6)

Procedure[7]

Midguts are isolated from fifth instar *Manduca sexta* larvae as described in the preceding chapter[8] and rinsed with ice-cold buffer A. The sheet of tissue appears as a thick anterior region, a thin central region, and a very thick posterior region.[9] The anterior and central regions are removed and discarded. The posterior region is composed of six strips of highly folded epithelial tissue separated by narrow unfolded strips, along each of which runs a longitudinal muscle. The isolated posterior midgut segment is placed lumen side down and the Malpighian tubules are re-

[4] H. Wieczorek, M. G. Wolfersberger, M. Cioffi, and W. R. Harvey, *Biochim. Biophys. Acta* **857**, 271 (1986).
[5] H. Schweikl, U. Klein, M. Schindlbeck, and H. Wieczorek, *J. Biol. Chem.* **264**, 11136 (1989).
[6] M. Cioffi and M. G. Wolfersberger, *Tissue Cell* **15**, 781 (1983).
[7] Whereas each larva will yield approximately 5 μg of goblet cell apical membrane protein, it is futile to attempt this preparation with fewer than nine larvae.
[8] W. R. Harvey, D. N. Crawford, and D. D. Spaeth, this volume [39].
[9] M. Cioffi, *Tissue Cell* **11**, 467 (1979).

moved. Using microscissors, the tissue is then cut along each side of each longitudinal muscle and the strips of epithelial tissue are unfolded. The unfolded strips are cut into small, approximately square pieces of tissue. As they are obtained, the tissue pieces are placed into a 1.5 × 10 cm round-bottom plastic centrifuge tube which contains approximately 5 ml of buffer A and is stored on ice. The tissue squares from three larvae are collected into one tube. Tissue and tissue extracts are maintained between 0 and 4° at all times.

The tip of the sonicator[10] probe is placed in the middle of a suspension of tissue squares from three larvae. The sonicator is set at level 6 and the suspension is sonicated for about 8 sec. The tissue pieces are allowed to settle and the cloudy solution is removed by aspiration. The tissue pieces are washed three times by resuspension in fresh cold buffer A. The washed tissue squares are resuspended in 1 to 2 ml of cold buffer A. The suspension of yellowish brown tissue pieces is vigorously drawn into and expelled from a Pasteur pipet 10 times. The solution becomes cloudy and yellowish in color. The disrupted tissue pieces should have a pale translucent yellow color. If they have a definite brown tint, pipetting should be continued until the brown tint is no longer evident but not to the point of complete color loss or tissue disintegration. The suspension is filtered through four layers of surgical gauze that have been saturated with buffer A. The residue is discarded and, without delay, the filtrate is centrifuged at 250 g for 2 min (RC-5B centrifuge with SS-34 rotor or equivalent; Sorvall, Newtown, CT). The supernatant is discarded and the pellet is stored on ice under buffer A while the tissue squares in the other tubes are disrupted. The total elapsed time from the beginning of the dissection of larvae to the collection of 250 g pellets should not exceed 1.5 hr.

When all dissected midguts have been processed to 250 g pellets, the pellets are resuspended in buffer A, combined, and centrifuged at 5000 g for 5 min. The supernatant is discarded and the pellet resulting from 12 to 15 larvae is resuspended in 1 ml of buffer A. Up to 0.5 ml of this suspension is layered onto one linear 37 to 45% sucrose gradient. The gradients are centrifuged at 25,000 rpm for 17 hr in a swinging bucket rotor (type SW-41 or equivalent; Beckman, Fullerton, CA). The second band from the top of each gradient (band 2, Fig. 1) is collected. The bands are pooled and diluted stepwise with four times their volume of 5 mM Na-EDTA, 5 mM Tris-HCl (pH 8.1).

---

[10] Typical large powerful laboratory sonicators are not suitable for use in this preparation. A sonicator with a finely controllable low-end output, such as the MS-50 Microson (Heat Systems-Ultrasonics, Inc., Farmingdale, NY) is required. The times and instrument settings for sonication given here apply only to the MS-50 Microson in our laboratory. They must be determined empirically for each instrument.

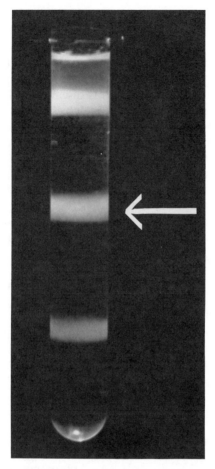

FIG. 1. Result of typical gradient centrifugation step in the isolation of goblet cell apical membrane vesicles. The arrow points to the band which is collected.

For solubilization and purification of the ATPase, the suspension is centrifuged at 10,000 g for 30 min. The supernatant is discarded, the pellet is resuspended in buffer A, and again centrifuged at 10,000 g for 30 min. The pellet may be solubilized immediately or it may be frozen in liquid nitrogen and stored for several weeks at $-30°$ before use.

For further purification of the goblet cell apical membrane, the diluted band 2 suspension is divided into aliquots of approximately 4 ml and each aliquot is sonicated for 4 sec using setting 6. The sonicated suspension is centrifuged at 10,000 g for 30 min. The supernatant is decanted into a clean centrifuge tube and stored on ice. The pellet is resuspended in 3–4 ml of buffer A, sonicated as described above, and centrifuged for 30 min at 10,000 g. The supernatant is combined with the previous 10,000 g supernatant. The pellet is resuspended, sonicated, and centrifuged as before. The three 10,000 g supernatants are combined and centrifuged at 30,000 g for 1 hr. The supernatant is discarded. The pellet consists of highly purified goblet cell apical membrane vesicles.

Freshly isolated goblet cell apical membrane vesicles exhibit only a limited ATPase activity.[11] Their ATPase activity is increased greatly by subjecting the vesicles to an osmotic shock.[4] The 30,000 g pellet resulting from 12 larvae is resuspended uniformly in 6 ml of 0.32 mM EGTA, 5 mM Tris-MOPS (pH 7.6), by vigorously drawing the suspension into and expelling it from a Pasteur pipet. The suspension is then centrifuged for 1 hr at 100,000 g. The pellet is resuspended in 0.32 mM EGTA, 5 mM Tris-MOPS, 3.2 mM 2-mercaptoethanol, 0.1% bovine serum albumin (pH 7.6) and used for the assay of membrane-bound ATPase activity.

Purification of the Goblet Cell Apical Membrane ATPase[5]

Reagents

Tris-HCl (16 mM), 0.32 mM EDTA, 9.6 mM 2-mercaptoethanol, 0.05–0.25% $C_{12}E_{10}$ (pH 8.1)
Sucrose (10, 20, 30, 40, and 50%, w/v), 16 mM Tris-HCl, 0.32 mM EDTA, 9.6 mM 2-mercaptoethanol, 0.01% $C_{12}E_{10}$ (pH 8.1)

Procedure

Fresh or frozen pellets obtained from band 2 (ca. 50 μg/larva) are solubilized at 4° in 16 mM Tris-HCl, 0.32 mM EDTA, 9.6 mM 2-mercaptoethanol, 0.05–0.25% $C_{12}E_{10}$ (pH 8.1). Approximately 2 mg of $C_{12}E_{10}$ is required for 1 mg of membrane protein. The suspension is mixed gently for several seconds until it becomes uniformly opaque, then centrifuged at 100,000 g for 60 min at 4°. The brownish pellet, which contains more than 90% of the contaminating mitochondrial ATPase, is discarded. The supernatant, which contains more than 90% of the goblet cell apical membrane ATPase and approximately 70% of the band 2 protein, is layered onto a discontinuous 10 to 50% sucrose density gradient (10% steps, 1 ml each) and centrifuged at 220,000 g for 1 hr at 4° in a vertical rotor (Sorvall type TV 865 or equivalent). The 30% sucrose fraction contains the purified goblet cell apical membrane ATPase (Fig. 2). Yield is approximately 9 μg/larva. Centrifugation of the 30% sucrose fraction on a second gradient or

---

[11] W. R. Harvey, M. Cioffi, J. A. T. Dow, and M. G. Wolfersberger, *J. Exp. Biol.* **106**, 91 (1983).

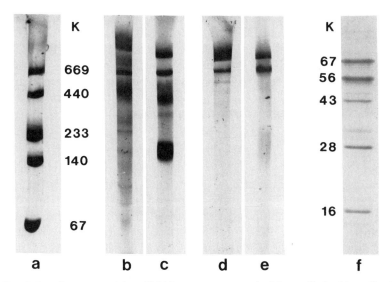

FIG. 2. Protein pattern of the 100,000 g supernatant and of the purified goblet cell apical membrane ATPase. (a–e) Native microgradient PAGE. a, Standard proteins; b and c, supernatant of 100,000 g centrifugation (0.6 μg); d and e, purified ATPase from 30% sucrose fraction (0.2 μg); f, SDS-PAGE slab gel, subunits of purified ATPase (4 μg). Lanes a, b, d, and f are stained for total protein;[5] lanes c and e are stained for ATPase activity.[5]

anion-exchange chromatography (Mono Q column; Pharmacia, Piscataway, NJ) does not lead to further purification.

Substrate and inhibitor specificities of the purified ATPase[5] are similar to those of the membrane-bound ATPase.[4] The enzyme consists of at least five subunits with relative molecular masses of 67K, 56K, 43K, 28K, and 16K (Fig. 2f). The 67K and 56K subunits are major bands in SDS-PAGE of the purified goblet cell apical membrane.[11] The purified ATPase crossreacts with antibodies against vacuolar-type ATPases.[12] Due to its inhibitor specificity, its polypeptide composition, and its immunological properties this enzyme belongs to the class of vacuolar-type ion-transporting ATPases.

Assay of ATPase Activity[4,5]

Composition of Incubation Mixtures

The incubation mixture for assay of the activity of membrane-bound ATPase contains approximately 1 μg membrane protein, 1 mM Tris-ATP,

[12] U. Klein, S. Weerth, B. Egerer, and H. Wieczorek, *Verh. Dtsch. Zool. Ges.* **82,** 207 (1989).

1 mM MgCl$_2$, 20 mM KCl, 50 mM Tris-MOPS, 0.1 mM EGTA, 1 mM 2-mercaptoethanol, and 0.3% bovine serum albumin (pH 8.0).

The incubation mixture for assay of the activity of purified ATPase contains approximately 0.5 μg of enzyme protein, 1 mM Tris-ATP, 1 mM MgCl$_2$, 20 mM KCl, 50 mM Tris-MOPS, 0.1 mM EDTA, 3 mM 2-mercaptoethanol, and 0.003% C$_{12}$E$_{10}$ (pH 8.1). Inclusion of 1 mM phosphoenolpyruvate and 3 U/ml of pyruvate kinase is optional.

Procedure

Incubation mixtures have a total volume of 0.16 ml and are prepared in 1.5-ml disposable microcentrifuge tubes. Each mixture is prepared in duplicate or triplicate. The mixtures are preincubated without substrate for 5–20 min at 30°. Reactions are started by addition of ATP and incubated for 2 to 16 min at 30°. Reactions are stopped by placing the assay tubes in liquid nitrogen. Zero-time controls are dropped into liquid nitrogen immediately after addition of substrate. Substrate hydrolysis is a linear function of extract concentration and, for membrane-bound enzyme, of incubation time. However, the solubilized ATPase is extremely sensitive to product inhibition[13] and linear dependence of substrate hydrolysis on time for incubations longer than 4 min can be achieved only by use of an ATP-regenerating system.[5]

The specific activity of the membrane-bound ATPase is about 1.3 μmol min^{-1} mg protein^{-1}, indicating an approximately 15-fold purification with respect to the pellet of a crude midgut extract.[4] The specific activity of the purified enzyme is approximately 3.5 μmol min^{-1} mg protein^{-1}. The low apparent enzyme activity purification factor is likely due not only to partial inactivation during the purification but also to the high relative molecular mass of the native ATPase (oligomers of about 900K and 600K, Fig. 2a and d).

Inorganic Phosphate Determination

Inorganic phosphate is measured as a complex of phosphomolybdate with Malachite Green,[14] modified as follows.[15]

Reagents

Trichloroacetic acid (TCA), 20%
Sulfuric acid, 7.8%

[13] M. G. Wolfersberger, D. D. Spaeth, and W. R. Harvey, *Physiologist* **30**, 157 (1987).
[14] L. Muszbeck, T. Szabo, and L. Fesus, *Anal. Biochem.* **77**, 286 (1977).
[15] H. Wieczorek, *J. Comp. Physiol.* **148**, 303 (1982).

Malachite Green solution: 18.5 mg Malachite Green in 25 ml 1% polyvinylalcohol (28/20)

Acid molybdate solution: On day of use add one part 24% sulfuric acid and three parts 200 mM sodium molybdate to six parts of distilled water

Procedure

Pipet 0.1 ml of acid molybdate solution into clean 1.5-ml disposable microcentrifuge tubes, one tube for each sample to be assayed. Timing of the following steps, given below for four samples,[16] is critical. Numbers in parentheses refer to the timing of samples 2, 3, and 4. Samples are stored in a freezer until use.

$t = 0$ (0.50, 1.00, 1.50) min:	Add 50 μl of TCA to sample, place in 70° water bath for 15 sec, mix, place in microcentrifuge
$t = 2.00$ min:	Centrifuge samples for 1 min
$t = 3.50$ (4.00, 4.50, 5.00) min:	Transfer 60 μl of supernatant from sample into the corresponding tube of acid molybdate solution and mix
$t = 3.75$ (4.25, 4.75, 5.25) min:	Add 30 μl of Malachite Green solution and mix
$t = 5.75$ (6.25, 6.75, 7.25) min:	Add 200 μl of 7.8% sulfuric acid and mix

Store samples at 22–25° and repeat procedure on the next four samples. The absorbance of each sample at 625 nm should be read between 70 and 110 min after addition of TCA to the sample. The inorganic phosphate content of each sample is calculated using a standard curve. Ten nanomoles of inorganic phosphate in a 0.16-ml ATPase assay mixture should produce an absorbance at 625 nm of approximately 0.7. The zero-time controls typically have absorbances between 0.2 and 0.3. Internal standards should be run if the influence on the phosphate determination of compounds added to ATPase assay mixtures is questionable.

---

[16] If two people are available to perform this assay the samples can be processed at a rate of 1/min. At approximately 3.25 min the first person removes the first four samples from the centrifuge and gives them to the second person. At 4.00 min the first person begins treatment of the next four samples while the second person completes treatment of the first four samples.

Protein Determination

Protein concentrations are determined using Amido Black.[17] This assay is highly sensitive and free from interference by any substance, including detergents, we have used so far.[5]

Reagents

Acetic acid/methanol (10%/90%)
Amido Black solution: 26 mg Amido Black dissolved in 100 ml acetic acid/methanol (10%/90%)
NaOH (0.1 N)

Procedure

Amido Black solution (300 μl) is added to 50 μl of protein sample in a 1.5-ml disposable microcentrifuge tube. After 5 min at room temperature the mixture is centrifuged for 4 min in a microcentrifuge. The supernatant is discarded, the pellet is resuspended in 500 μl 10% acetic acid/90% methanol and centrifuged as before. The washing procedure is repeated twice. The washed pellet is dissolved in 350 μl of 0.1 N NaOH and the absorbance of the solution at 615 nm is determined. Four micrograms of bovine serum albumin produce an absorbance of approximately 0.3.

Acknowledgments

We thank Mr. Daniel D. Spaeth for his valuable comments on this manuscript and for preparing Fig. 1. This work was supported in part by National Institutes of Health Research Grants AI 09503 and 22444, Deutsche Forschungsgemeinschaft Grant Wi 698/2, USDA Competitive Research Grant 87-CRCR-1-2487, NIH Research Support Grant 2 S07 RR07115-17, and Temple University's Research Incentive Fund.

[17] N. Popov, M. Schmitt, S. Schulzeck, and H. Matthies, *Acta Biol. Med. Ger.* **34,** 1441 (1975).

[41] Methods for the Study of Fluid and Solute Transport and Their Control in Insect Malpighian Tubules

By S. H. P. MADDRELL and J. A. OVERTON

Insect Malpighian tubules form a very important part of the insect excretory system, the successful functioning of which is vital to the ability of insects to survive in the terrestrial environment and to feed on a great variety of foodstuffs.[1,2] They are therefore worth detailed study if we are to understand insect physiology and survival. Insect Malpighian tubules also deserve attention in that they are almost ideal for studying many aspects of epithelial fluid and solute transport. This derives in part from the fact that in the insect they carry out a great variety of transport and excretory functions,[2] at rates which can be controlled and which may be extraordinarily fast (Table I). For example, the cells of some tubules can secrete fluid at a rate equivalent to their own volume in as little as 10 sec. However, their usefulness also stems from their simple anatomy. Each tubule is in the form of a blind-ending hollow epithelial cylinder, the walls of which are one cell thick. The extracellular basement membrane which encloses them is usually thin and although some tubules have external muscles they do not usually form a coat, but only a discrete longitudinal spiral that does not interfere with access to most of the epithelial cells. Since, in addition, isolated tubules survive and function for long periods (up to 24 hr) under suitable *in vitro* conditions, they make very convenient experimental material. Just as important is that most insects can be maintained in culture easily, cheaply and in large numbers. Virtually all insects have between 2 and 200 tubules though most have 4, 5, or 6 tubules.

The small size of Malpighian tubules (the much studied *Rhodnius* tubules are 50–100 μm in diameter and 45 mm in length) means that only tiny quantities of bathing fluids are required for experiments and much smaller quantities of drugs, hormones, or radioactive tracer substances can be used than is the case of comparable studies on other epithelia.

Because insects have an open circulatory system and the hemolymph is circulated relatively slowly, their Malpighian tubules are not tightly packed into a compact organ as are, for example, mammalian kidney tubules, but they lie relatively freely in the hemocoel, maintained in position only

[1] J. A. Ramsay, *J. Exp. Biol.* **35,** 871 (1958).
[2] S. H. P. Maddrell, *J. Exp. Biol.* **90,** 1 (1981).

TABLE I
FUNCTIONS AND THEIR CONTROL IN MALPIGHIAN TUBULES

Function	Control	Degree of stimulation
Fluid secretion	Diuretic hormone	Up to 1000×
Transport of monovalent inorganic ions (Na^+, K^+, Cl^-)	Diuretic hormone, natriuretic hormone	Up to 1000×
Transport of SO_4^{2-}	Induced by SO_4^{2-} in diet	Up to 100×
Transport of Mg^{2+}	Induced by Mg^{2+} in diet	Up to 100×
Accumulation of Ca^{2+} in mineralized concentrations	Regulated by Ca^{2+} level in diet	Up to 10×
Transport of acylamides (such as p-aminohippurate)	Regulated by protein level in diet	Up to 20×
Transport of sulfonic acids	Presumed regulated by effect of diet	?
Transport of uric acid	Regulated by changes in diet	Up to 30×
Allantoin	None known	—
Alkaloids (such as nicotine and atropine)	None known	—
Cardiac glycosides (such as ouabain)	Regulated by glycoside level in diet	—
Reabsorption of KCl	Diuretic hormone	Up to 1000×
Reabsorption of sugars	None known	—

loosely by their tracheal supply. Their dissection and removal is therefore comparatively simple. With practice, a set of four Malpighian tubules can usually be removed from an insect in less than 5 min.

A particular advantage of Malpighian tubules over other insect epithelia is that they have a very high surface area/volume ratio because of their long narrow shape. They function in unstirred saline at rates not greatly inferior to the *in vivo* rates; preparations of lepidopteran midguts, for example, require both sides of the epithelia to be swept by continuously circulated saline equilibrated with air containing 70% O_2 (see [39], by Harvey *et al.*, this volume).

Tubules function by secreting fluid into the lumen into which other solutes diffuse or are transported.[2] An advantage of this is that merely by collecting the secreted fluid, one can continuously monitor transepithelial transport of any desired solute. This is the basis of the classical method for studying Malpighian tubules devised by Ramsay.[3] In this method single tubules are put in a small drop of suitable saline under liquid paraffin (mineral oil) and the cut end of the tubule pulled out and held out of the drop. Secreted fluid emerges from the cut end or from a cut in the wall of

[3] J. A. Ramsay, *J. Exp. Biol.* **31,** 104 (1954).

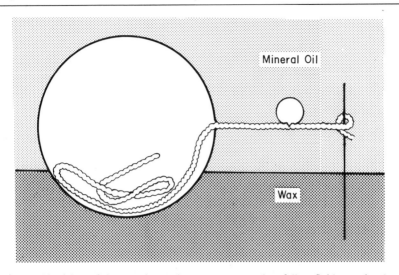

Fig. 1. Side view of the experimental arrangement used to follow fluid secretion by an isolated Malpighian tubule from *Rhodnius*. The bathing drop of saline is put in an indentation in the wax base. The cut end of the tubule is hitched round a fine glass or steel pin pushed into the wax. Any fluid secreted by the tubule emerges through a cut made in the wall of the tubule.

the tubule. A convenient arrangement, scarcely changed from Ramsay's original one, is shown in Fig. 1.

Saline for Use with Insect Malpighian Tubules

Most insect tissues are very tolerant of relatively large changes in the extracellular fluid bathing them. Perhaps this results from the evolution of insects as small animals.[2,4] This seems particularly the case for Malpighian tubules. Those of *Rhodnius*, for example, will secrete fluid at high rates in bathing fluids varying from 60 to 1000 mOsm/liter[5] in fluid of pH ranging from 6 to 8.2, in fluid lacking calcium ions and even in a saline where all the Na^+, K^+, and Cl^- has been replaced by NH_4^+ and NO_3^-. However, their survival *in vitro* is, not surprisingly, improved in salines whose concentration reflects more closely that of the hemolymph of the insect. In particular, function can usually be maintained for extended periods only if the saline contains an energy supply, such as glucose. Finally, some tubules

[4] S. H. P. Maddrell, *in* "A Companion to Animal Physiology" (C. R. Taylor, K. Johansen, and L. Bolis, eds.). Cambridge Univ. Press, Cambridge, 1982.
[5] S. H. P. Maddrell, *J. Exp. Biol.* **52,** 71 (1969).

for largely unknown reasons require the presence of amino acids in the saline bathing them. The composition of salines used for *in vitro* preparations of Malpighian tubules from different insect species may be found in the relevant papers cited below.

Dissection of Malpighian Tubules

In general terms, the technique for dissecting and removing Malpighian tubules is straightforward. In nearly all insects, the tubules are held in the animal only by their tracheal connections so that once the abdomen has been opened under dissecting saline, the tubules can be freed by cutting, or breaking, the tethering trachea.

Tubules from Rhodnius

To illustrate in more detail how dissection proceeds, let us follow the technique that has been evolved for removing Malpighian tubules from *Rhodnius.* The account given here is based on a full description given in Maddrell.[6]

Adult *Rhodnius* are large but the abdomen contains ovaries or testes and accessory glands in addition to the gut and Malpighian tubules. It is more convenient to use fifth stage insects taken 1–3 weeks after the moult from the previous instar. Each instar yields four tubules, equivalent in all known aspects. An insect is killed by decapitation or by crushing the head and pinned down in a dissecting dish 70–100 mm in diameter. If pins bent at right angles near the point are used all three legs on one side can be held with one pin. The insect is next covered with saline to a depth of 5–7 mm to allow easy dissection. The roof of the abdomen is cut anteriorly, peeled back with fine watchmaker forceps, and pinned open. The gut anterior to the rectum is pulled out straight and cut or torn away where it joins the anterior part of the rectum. Now the coiled mass of tubules is unraveled with fine glass rods or fine pins held in pin holders in a manner difficult to describe and best learned by practice. Essentially, loops of tubule are enlarged by moving apart two pins or rods put into each loop. In this way tracheal connections are peeled away from the tubules. Proceed in this manner until that pair of tubules is completely unraveled. Repeat the procedure on the other side of the insect to yield two more tubules. The upper fluid-secreting parts of the tubules are more opaque than the more proximal parts. If they are cut through just below the obvious junction between the two, the short length of transparent lower tubule can be used

[6] S. H. P. Maddrell, *in* "Experimental Techniques in Entomology" (T. A. Miller, ed). Springer-Verlag, Heidelberg, 1980.

to identify the cut end. The freed tubules can be transferred on a pin or rod to a drop of saline under liquid paraffin and arranged as shown in Fig. 1. Provided the saline includes a stimulant, 5-hydroxytryptamine (5-HT), or the diuretic hormone, the tubule starts to secrete within 3–5 min. As it swells, it is easy to hole the tubule at a suitable point on the length between the saline drop and the anchoring glass rod, so that the secreted fluid accumulates and is readily accessible for both analysis and volume measurements.

The tubules from *Rhodnius* are comparatively easy to dissect in that the tubules themselves are tougher and less brittle than the tracheae that anchor them. The tracheae can thus be safely stretched and broken without harm to the tubules. In many other species of insect the tracheae are stronger and more resilient and the tubules more easily damaged, so that dissection is more difficult. However, tubules from a wide variety of insects have been successfully removed and made to function *in vitro*, using techniques more or less similar to that described above for *Rhodnius*. If care is taken to cut tracheal connections and to minimize any stretching of the tubules, it is likely that tubules from any desired insect can be successfully removed. The details of the dissection of tubules from a particular insect are best consulted in the original papers. A selection of relevant papers is as follows: *Carausius morosus*,[3] *Dysdercus fasciatus*,[7] *Calliphora erythrocephala*,[8] *Calliphora vomitoria*,[9] *Glossina morsitans*,[10] *Periplaneta americana*,[11] *Pieris brassicae*,[12] *Schistocerca gregaria*,[13] and *Aedes campestris*.[14] Most more recent accounts use procedures avowedly similar to the ones described in these papers.

In Vitro Techniques for Malpighian Tubules

Fluid Secretion

Once removed from the insect, Malpighian tubules can be persuaded to secrete fluid by arranging them in a drop of suitable saline under liquid paraffin in a fashion similar to that shown in Fig. 1. The arrangement shown in Fig. 1 depends on the cut end of the tubule adhering to the pin,

[7] M. J. Berridge, *J. Exp. Biol.* **44**, 553 (1966).
[8] M. J. Berridge, *J. Insect Physiol.* **12**, 1523 (1966).
[9] G. Knowles, *J. Exp. Biol.* **62**, 327 (1975).
[10] J. D. Gee, *J. Exp. Biol.* **64**, 357 (1976).
[11] R. R. Mills, *J. Exp. Biol.* **46**, 35 (1967).
[12] S. W. Nicholson, *J. Exp. Bol.* **46**, 35 (1976).
[13] S. H. P. Maddrell and S. Klunsuwan, *J. Insect. Physiol.* **19**, 1369 (1973).
[14] J. E. Phillips and S. H. P. Maddrell, *J. Exp. Biol.* **61**, 761 (1974).

glass or steel, stuck in the wax near the drop of bathing saline. Looping a length of tubule once or twice round the pin is usually sufficient to ensure this. With tubules from some insects, though, this method often fails and the cut end of the tubule slides off the pin and is pulled back into the bathing drop by surface tension. To combat this it is best to tie a very fine length of silk around the end of the tubule and use this to anchor the tubule in the required position. Alternatively, and especially useful where the exact position of the tubule relative to its bathing drop is important, the cut end is held in fine forceps clamped shut and carried on a micromanipulator.

Occasionally it is found that no fluid emerges from the cut made in the wall of the tubule outside the bathing drop. One possible explanation is the rate of fluid secretion is so slow that the lumen of the tubule, which may be deflated after dissection, has not yet filled. Alternatively, there may be local damage to the tubule at some point, or points, along its length and secreted fluid escapes through a hole in the wall.

If fluid secretion is slow, it is of course possible that this is characteristic of the tubules from the particular insect in which case more time must be allowed before secreted fluid will emerge. In other cases it may be that the saline is inappropriate in some respect. Many tubules such as those of the blowfly, *Calliphora*, secrete K^+-rich fluid and they require at least 20 mmol liter^{-1} K^+ in order to secrete at easily measurable rates.[15] Other tubules, such as those of *Rhodnius*, secrete very slowly indeed in the absence of a stimulant. Such tubules can be induced to secrete much faster in the presence of the naturally occurring stimulant hormone.[12,16-18] Mimics or agonists may be found; 5-hydroxytryptamine elicits rapid fluid secretion by tubules in several insects.[19] Alternatively, cyclic AMP can be included in the saline at $0.1 – 1$ mmol liter^{-1} (care needs to be taken that the pH of the saline is adjusted back to near neutrality) and this will often greatly accelerate tubule function. Finally, it is worth checking that the oxygen supply to the tubule is not unnecessarily limiting. The oxygen content of salines is usually adequate, but they may need bubbling with air or even with oxygen. If this is done, there is another danger to be guarded against. If the saline is buffered with $HCO_3^-/H_2PO_4^-$, aeration will remove CO_2 and the saline will rapidly become too alkaline; inclusion of a little Phenol Red will allow the pH level to be monitored. Oxygenation may also be insufficient if

[15] M. J. Berridge, *J. Exp. Biol.* **48,** 159 (1968).
[16] S. H. P. Maddrell, *Nature (London)* **194,** 605 (1962).
[17] L. M. Schwartz and S. E. Reynolds, *J. Insect Physiol.* **25,** 847 (1979).
[18] S. H. P. Maddrell, *in* "Insect Biology in the Future" (D. S. Smith and M. Locke, eds.). Academic Press, New York, 1980.
[19] S. H. P. Maddrell, D. E. M. Pilcher, and B. O. C. Gardiner, *J. Exp. Biol.* **54,** 779 (1971).

the bathing saline drop is too small—we use drops in excess of 100 μl for a single *Rhodnius* tubule 30 mm long and 80 μm in diameter. Finally, if different regions of a tubule are clumped together, oxygen supply by diffusion may be too slow to satisfy the needs of so much tissue in close proximity. Unraveling the tubule so that no part lies too close to any other part will alleviate this problem.

That Malpighian tubules may suffer problems in oxygen supply is not surprising. *In vivo*, they receive oxygen through the tracheal system, which is able to supply oxygen at very high rates.[20] *In vitro* they are deprived of this and need to rely instead on diffusion from the bathing saline aided by the fact that they expose a very high surface area relative to their volume. The problem is, presumably, less for smaller tubules and for tubules whose metabolism is slow.

If the tubule is not starved of oxygen nor is in an inappropriate saline and yet fails to secrete fluid, then it is likely that it has been inadvertently damaged in dissection and removal. A convenient way to check this is slowly to pull the tubule by its cut end further and further out of its bathing drop of saline. A hole in the wall of the tubule will then often become apparent from fluid emerging through it, visible as a small droplet on the length of the tubule. The tubule can then be allowed to be taken back into its bathing drop by surface tension until the holed area is just outside the drop. Remarkably enough, tubules will often secrete fluid satisfactorily even when one or more cells are badly damaged and detach from the acellular basement membrane. Evidently the membrane is a sufficient barrier to bulk flow of fluid through it and secreted fluid instead escapes through the downstream cut. The danger here is that the tubule wall is much more permeable to solutes and so its use would give a false picture of wall permeability. This point is taken up below.

Studies of the Permeability of the Tubule Wall

The function of Malpighian tubules in excretion embraces that of allowing solutes present in the hemolymph to pass passively by diffusion into the primary excretory fluid in the lumen. Obviously the properties of the pathway through which diffusion occurs are of critical importance. To investigate these properties *in vitro* is potentially very simple. The composition of the secreted fluid is compared with that of the bathing fluid and ratio of the concentration of a solute in the secreted fluid, S, to that in the bathing medium, M, depends on the wall permeability. Ramsay[1] analyzed the situation and showed that this ratio, S/M, depended not only on the

---

[20] T. Weis-Fogh, *J. Exp. Biol.* **41**, 229 (1964).

permeability but also the area of the tubule wall and the rate of fluid secretion in a fashion given by the following formula.

$$S/M = b/(a + b)$$

where b is the permeability of the wall (in nl min^{-1} mm^{-2}) and a is the rate of fluid secretion per unit area of wall (also in nl min^{-1} mm^{-2}). From examination of the formula it is clear that to obtain accurate estimates of the permeability of the wall to highly permeant substances, one needs to use tubules secreting fluid as fast as possible in order that S/M should not be too close to 1. Similarly, for the least permeant substances, a slower rate of fluid secretion is desirable as then the concentration of the solute in the secreted fluid is more easily measured; at high rates of fluid secretion it may be near zero. The easiest way of determining the concentrations of solute in secreted fluid and medium is to use radioactive tracers. Obviously one must take the precaution of checking that the radioactivity in the secreted fluid is still that of unchanged solute. This is best done by chromatography, in which one compares how tracer solute as supplied behaves in comparison with radioactivity in the secreted fluid.[21] Substances such as amino acids or sugars are among those most likely to be affected by metabolism in the cells.

Possible Active Transport of Solutes

Values for permeability significantly higher than for substances of similar size, shape, and charge will suggest that the particular solute studied is actively transported into the lumen. If the rate of fluid secretion is slowed then active transport may be tentatively confirmed by finding that the lumenal concentration is now higher than in the medium. Final confirmation depends on showing that such elevation of concentration is not achieved by the electrical gradient; in other words one must look at the electrochemical potential gradient which combines both the electrical potential difference and the concentration gradient (strictly, the activity gradient). Active transport of substances related to p-aminohippuric acid (PAH) or alkaloids can often be shown by using known inhibitors of the process.[21]

Occasionally it may be found that a substance seems to have an unusually low permeability through the tubule wall. This may indicate active reabsorption of solute from the tubule lumen. Knowles[9] showed just such an effect in studies on sugar permeation. Glucose and, to a lesser extent, trehalose were underrepresented in the fluid secreted by Malpighian tu-

[21] S. H. P. Maddrell and B. O. C. Gardiner, *J. Exp. Biol.* **64**, 267 (1976).

bules of *Calliphora*. More glucose penetrated if its concentration were raised in the medium or if the tubules were treated with phloridizin, an inhibitor of sugar transport.

Validation of *in Vitro* Permeability Studies

If *in vitro* studies of permeability of Malpighian tubules are to be useful, they must provide accurate estimates of the actual permeabilities of the tubules *in vivo*. The obvious danger, already alluded to, is that isolated tubules may be damaged to some extent and so be made more permeable. If one repeatedly measures the permeability to a solute known neither to be actively transported nor to cross the wall through the cells, such as sucrose,[22] it is found that the permeability varies but has a distinct minimum at which most determinations cluster. This is particularly the case if great care is taken in dissection not to damage the tubule, as by cutting all the tracheal connections rather than tearing them away from the tubule. One may reasonably conclude then that the minimum value reflects the *in vivo* situation. In *Rhodnius*, the truth of this assertion can be checked more readily. Radioactive sucrose is injected into the hemolymph of a freshly fed instar, and left to circulate to reach a uniform concentration in the hemolymph. Samples of urine are collected as they are produced at regular intervals and from their volume and times at which they were collected the rates of fluid secretion by each tubule can be calculated. There are four similar tubules. If samples of hemolymph are then taken through a cut leg and these and the urine collections assayed for their radioactive content, the S/M ratio can be determined. Ramsay's formula then allows the permeability to be calculated (each upper tubule has an area of 9 mm^2).

Of course, this assumes the upper fluid-secreting parts are the only areas in the excretory system permeable to sucrose, and that the primary secretory fluid is not concentrated in the downstream parts of the excretory system. These are reasonable assumptions, as the lower tubules are shorter than the upper and have lower permeabilities[23] and the rectum has a low surface area-to-volume ratio and fluid spends little time in it. The values determined *in vivo* are closely similar to the minimal values determined *in vitro*.[22]

To repeat this exercise in insects with tubules that secrete more slowly would be difficult or impossible. What needs to be done in such other insects is to inject radioactive tracer and measure its concentration in the

---

[22] M. J. O'Donnell, S. H. P. Maddrell, and B. O. C. Gardiner, *Am. J. Physiol.* **246**, R759 (1984).

[23] S. H. P. Maddrell and J. E. Phillips, *J. Exp. Biol.* **62**, 671 (1975).

hemolymph when uniform. Rapid dissection and the collection of samples of fluid from the tubules' lumena by micropuncture[24] allows the S/M ratio to be determined. The dimensions of the tubule need to be measured and the *in vivo* rate of fluid secretion determined, when the permeability to the injected tracer can be calculated. Obviously, this procedure is difficult and one may have to rely on finding the minimum *in vitro* permeability of a particular solute.

Once, however, one knows the permeability of an undamaged tubule to a particular solute, this can be used to check that an individual tubule has an unaltered permeability before it is used to measure the rate of permeation of other solutes.

It is worth reminding ourselves here that arguably the most crucial of the functions of a Malpighian tubule is to remove potential toxins from the hemolymph automatically by passive diffusion.[1,2] Knowledge of the permeability, area, and rates of fluid secretion is therefore central to understanding the insect excretory system.

Determining Whether Solutes Cross the Walls of Malpighian Tubules by Paracellular or Transcellular Routes

Perhaps because of their pronounced capability for rapid transport of fluid and of a variety of substances, the cell membranes of Malpighian tubules are enormously amplified.[25,26] The relative areas of the intercellular spaces and of the cell membrane facing the bathing medium in *Rhodnius* are in the ratio of 1 : 120,000.[27] As a result, any solute at all able to cross the cell membranes is potentially able to cross the wall of the tubule through the cells at rates comparable to the rate it transverses the wall via the intercellular spaces, the paracellular route. Thus, mannitol, used in many vertebrate studies as an extracellular space marker (see, for example, Pitts[28]), crosses Malpighian tubules of *Rhodnius* up to 50 times faster transcellularly than paracellularly.[22,29] The evidence that a solute may cross the tubule wall passively by a transcellular route rests to a large extent on finding significant concentrations of that solute in the cell cytoplasm. In turn, to show this, it is convenient to use radioactive test substances, and to maximize the uptake of the substance into the tubule cells, the tubule

---

[24] J. A. Ramsay, *J. Exp. Biol.* **29**, 110 (1952).
[25] S. H. P. Maddrell, *Curr. Top. Membr. Transp.* **14**, 427 (1980).
[26] M. J. O'Donnell, S. H. P. Maddrell, H. leB. Skaer, and J. B. Harrison, *Tissue Cell* **17**, 865 (1985).
[27] H le B. Skaer, S. H. P. Maddrell, and J. B. Harrison, *J. Cell Sci.* **88**, 251 (1987).
[28] R. F. Pitts, "Physiology of the Kidney and Body Fluids." Year Book, Chicago, 1952.
[29] M. J. O'Donnell and S. H. P. Maddrell, *J. Exp. Biol.* **110**, 275 (1984).

lumen is perfused with the same tracer-labeled saline as that bathing its external surface. The technique for this is described below in the next section. After a period of perhaps 15–20 min to allow equilibration, nonradioactive saline made visible with dye (Amaranth or Lissamine Green) is rapidly driven through the perfusing cannula so as to flush out the radioactive saline in the tubule lumen. This need only take 1–2 sec. The tubule is then cut through close to the cannula and rapidly pulled through two or three drops of tracer-free saline so as to wash the external surface of the tubule free from the radioactive bathing medium. The tubule is then put in a drop of distilled water of about 20 μl in volume to disrupt the tubule osmotically and allow the cellular contents to escape into the water. The radioactive content of the tubule and drop is then measured by conventional scintillation counting techniques to give an estimate of the number of radioactive counts in the tubule cells at the time when they were harvested. Any one experiment should be done in two parts, each using one of a pair of radioactive substances, one thought to cross the tubule wall only paracellularly (i.e., through the intercellular clefts) and the other one being tested to see if it may cross transcellularly as well as paracellularly. If the experimenter is made unaware of the identity of the solutions, unconscious bias of the results can be avoided. Substances able to penetrate into the cells give total counts of the order of 10 times higher than do substances not so capable.[22] The small number of counts found with substances not able to permeate into cells results from incomplete removal of surface contamination during the necessarily rapid washing of the two tubule surfaces. More lengthy washing would allow the loss of counts from within the cells.

The significance of these types of studies is that since Malphigian tubules expose a large surface area to the hemolymph, they are ideally suited to carry out biochemical processing of substances from the hemolymph. To be able to do this they must either have specific transport processes for uptake of metabolites or have a significant permeability to them.

Technique for the Perfusion of the Lumen of Malpighian Tubules

For a variety of reasons, it is often useful to be able to perfuse fluid along the lumen of a Malpighian tubule. This could be in order to gain swifter access to substances entering the lumen, to wash away the existing fluid there, or to remove the gradients that normally exist between lumenal and bathing fluids.

The simplest technique for perfusing a length of tubule is to hold it firmly in some way while a glass pipet or cannula of suitable dimensions is advanced with a micromanipulator so that it cuts through the tubule wall and passes on down the lumen. If the cannula is tapered it can be advanced until it fills the lumen, so that fluid passed in through the cannula moves along the lumen and does not leak back out through the cut in the tubule wall.

To produce a suitable cannula, one can first pull a microelectrode from a glass pipet 1–1.2 mm in outside diameter. If the glass wall is thin, the resulting cannula is less likely to become blocked by any small particles in the perfusing fluid, but naturally the cannula is less robust. The microelectrode, which of course has an unusually fine tip, can be converted to a cannula merely by breaking off the tip with a fine pair of forceps. This is best done under a dissecting microscope fitted with an ocular micrometer so that a suitable tip diameter is achieved—perhaps one/third of the diameter of the tubule to be perfused. Better still is to chamfer the tip until it is at the right diameter using a suitable grinding wheel. The resulting cannula is sharper and the angled orifice less likely to become blocked.

A system which allows the cannula to be connected to a motor-driven syringe is relatively easily constructed. In it a cannula is held with an 0-ring into a small plastic chamber which has fluid driven into it from a syringe emptied by a motor driving a micrometer advance. The chamber can have extra parts so that the fluid in it can be changed very rapidly or so that (an) electrode(s) can be put in contact with the perfusing fluid.

Such a perfusing system is particularly useful with tubules whose rate of fluid secretion is very low. We had, for example, to perfuse fluid through the lumena of Malpighian tubules from larvae of *Manduca sexta*.[21] These tubules, partly perhaps because of their lobed shape, produced insufficient fluid even in 8 hr for any to emerge from the cut ends. By perfusing fluid through them we were able to show that they had rapidly transported radioactive nicotine into the lumen.

A Less Artificial Way of Lumenal Perfusion

Underlying the use of the cannulation and perfusion technique is the belief that the normal functioning of the perfused tubule is not disturbed. We have recently found evidence that perfused tubules of *Rhodnius* are very sensitive to the composition of the fluid passed into their lumena; rapid sodium transport is often much reduced in perfused tubules.[30]

An alternative way of achieving the same end—that is, to cause a flow of fluid through the lumen of a length of tubule under study—is shown in Fig. 2. Essentially fluid secreted by one part of a tubule is used to wash out

[30] S. H. P. Maddrell and J. A. Overton, *J. Exp. Biol.* **137**, 265 (1988).

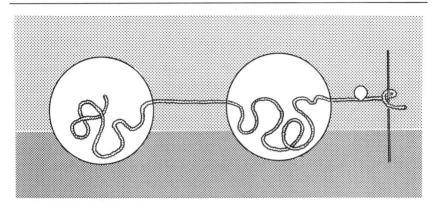

FIG. 2. The arrangement of an isolated Malpighian tubule in two bathing drops so that fluid secreted by the upper part of the tubule (in the left-hand drop) flushes the lumen of the lower part of the tubule (in the right-hand drop).

the lumen of another, usually more proximal (downstream), part of the tubule. Since in this arrangement fluid flow is virtually normal, tubule function is much less likely to be disturbed. We have used the technique, for example, to study sodium transport and PAH transport by single cells of *Rhodnius* Malpighian tubules.[30] It has also been useful to carry fluid through downstream parts of a tubule which do not themselves secrete fluid[23] and whose function may well depend on the composition of the lumenal fluid.

Measurement of Transepithelial Potential

Examination of Fig. 1 suggests a simple way of measuring the transepithelial electrical potential difference (TEP) of Malpighian tubules. It consists simply of siting two electrodes, one in the bathing drop and one in the drop of secreted fluid. This technique was used in early response to various treatments.[31] A convenient modification of this approach was to isolate a tubule into a specially designed chamber whose different compartments could be perfused with saline with a central section filled with oil.[32]

While the technique described in these papers is undoubtedly convenient, two recent publications have shown that there are potential flaws in its use. Basically, the technique depends on the resistance of the fluid

[31] S. H. P. Maddrell, *Adv. Insect Physiol.* **8,** 199 (1971).
[32] M. J. Berridge and W. T. Prince, *J. Exp. Biol.* **56,** 139 (1972).

column in the lumen being so low that one can ignore the effects of potentials developed by the tubule in the oil gap or in the bath containing the cut or any electrical leakage along the outside of the length of tubule in the oil. Both in *Aedes egypti*[33] and in *Onymacris plana*,[34] the lumenal resistances are sufficiently high as to introduce large errors. The only safe ways to measure TEP are (1) by pushing a conventional microelectrode through the wall into the lumen of the tubule, or (2) by having an electrode in the fluid filling a cannula that is placed in the lumen of the tubule. It turns out that with *Rhodnius*, the tubule diameter is sufficiently large that the errors in using that simple technique are negligible,[29] but with any new tubule it would be wise to use one of the two techniques that involve an electrode in the lumen. There are advantages and disadvantages peculiar to each of these techniques. If the lumenal electrode is in a perfusing cannula, the lumen can be irrigated with the same saline as in the outside bath, which eliminates transepithelial concentration differences as voltage sources, and the passage of the cannula a significant distance down the lumen means that the TEP is measured across an untouched part of the wall. However, as mentioned earlier, tubule function may be affected by the presence in the lumen of fluid different from that occurring there naturally. Insertion of a microelectrode directly through the wall into the lumen avoids this difficulty but one is then measuring the TEP across a part of the wall which is damaged to some extent. For a full quantitative discussion of the problems, the reader should refer to the extensive analysis by Aneshansley *et al.*[33]

Measurements of Osmotic Permeability in Malpighian Tubules

Fluid transport by Malpighian tubules may be very fast. They may secrete fluid equivalent to the volume of the cells every 10 sec, and it has recently been shown that tubule cells in *Aedes egypti* can secrete its own cell content of Na^+ every 6 sec.[33] Just how movements of ions and water are coupled together is controversial (see Hill[35] and Diamond and Bossert[36]), but one plausible mechanism is by osmosis.[37,38] One of the main difficulties has been ignorance of the osmotic permeability of the epithelial cells. It is possible to measure this parameter using a method[37] in which

[33] D. J. Aneshansley, C. E. Marler, and K. W. Beyenbach, *J. Insect Physiol.* **35**, 41 (1989).
[34] L. Isaacson and S. W. Nicolson, *J. Exp. Biol.* **141**, (1989).
[35] A. E. Hill, *Proc. R. Soc. London, Ser. B* **190**, (1975).
[36] J. L. Diamond and W. H. Bossert, *J. Gen. Physiol.* **50**, 2061 (1967).
[37] M. J. O'Donnell, G. K. Aldis, and S. H. P. Maddrell, *Proc. R. Soc. London, Ser. B* **267** (1982).
[38] D. L. S. McElwain, *Proc. R. Soc. London, Ser. B* **222**, 363 (1984).

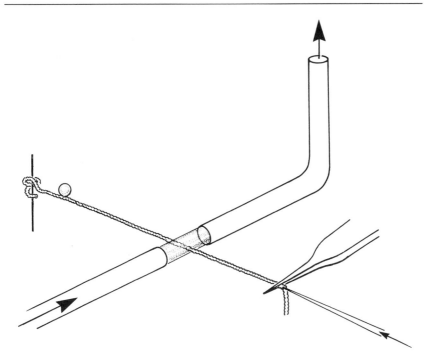

FIG. 3. Arrangements used to perfuse fluids of different osmotic concentrations through the lumen and over the external face of an isolated Malpighian tubule.

streams of saline solution of known rate and known different osmotic concentrations, are passed rapidly across both lumenal and basal (outside) surfaces, so that unstirred layers are kept small. We discussed earlier how fluid may be perfused along the lumen. Perfusion across the external surface is, on the face of it, more difficult. However, this can be achieved by a method which depends for its success on the small size of Malpighian tubules. In it, saline flows down one stainless steel tube (i.d. 0.9 mm), emerges, crosses a 1-mm gap, and is collected in a second steel tube (Fig. 3). This is done under a layer of liquid paraffin and surface tension is so strong at this size range that the column of fluid passing from one tube to the other is kept as a constant cylinder. To study the osmotic permeability of *Rhodnius* tubules we arranged a perfused length of tubule in the column at right angles to its length. We perfused the tubule with standard saline, (osmotic concentration 340 mOsm liter^{-1}), at about 240 nl min^{-1}, while the external fluid flowed at 20 μl min^{-1} and had an osmotic concentration of 170 mOsm liter^{-1}. Fluid entered the tubule at a rate determined by measuring the rate at which fluid emerged from the tubule downstream and subtracting from it the rate at which fluid was passed into it upstream. The osmotic permeability is then easily calculated by dividing the inward osmotic flux per unit area (in cm^3 cm^{-2} s^{-1}) by the mean difference in osmotic concentration between the two flows of fluid; one can assume that this is the mean of the osmotic gradients at the upstream and downstream ends of the external perfused segment.

In principle, this method should be easily applied to any Malpighian tubule. For fuller consideration of the theoretical background to the method see O'Donnell *et al.*[36] So far its use supports the idea that osmotic coupling of solute and water fluxes explains how Malpighian tubules secrete fluid, though other coupling cannot be ruled out.

[42] Physiological Approaches for Studying Mammalian Urinary Bladder Epithelium

By Simon A. Lewis and John W. Hanrahan

Introduction

Epithelia are an integral component of a diverse group of organ systems. Their prime function is to maintain plasma homeostasis by selectively absorbing, secreting, or excluding solutes and water. While many epithelia modify the plasma composition by active transport, a few function as an impermeable wall between blood and external environment. The best example of such a storage organ, i.e, a relatively impermeable barrier, is the mammalian urinary bladder. Although it absorbs sodium at a low rate, the bladder's main physiological function is to store urine and maintain large ionic and osmotic gradients which are generated by the kidney. This barrier function is easily lost when the tissue is handled, therefore care must be taken when studying it either *in vivo* or *in vitro*.

The purpose of this chapter is to describe the methods used to measure the physiological properties of the mammalian urinary bladder epithelium *in vitro*.[1-3] This involves the selection of a bathing solution similar in composition to plasma, development of a method to remove the muscle layers from the epithelium with minimal traumatization of the tissue, design of an *in vitro* chamber into which the epithelium is mounted with

[1] S. A. Lewis and J. M. Diamond, *J. Membr. Biol.* **28,** 1 (1976).
[2] S. A. Lewis and J. L. C. deMoura, *Nature (London)* **297,** 685 (1982).
[3] S. A. Lewis, M. S. Ifshin, D. D. F Loo, and J. M. Diamond, *J. Membr. Biol.* **80,** 135 (1984).

minimal damage, and choice of an electrical recording system to assess the transepithelial properties of the tissues in a nonintrusive manner. In addition to these transepithelial methods, our laboratory has recently used the patch-clamp technique to study ion channels in the basolateral membrane of rabbit urinary bladder epithelium.[4-6] In this chapter we discuss methods that were used during this work, emphasizing modifications that are unique to this particular tissue. General patch-clamp methodology has been described in detail elsewhere.[7,8]

We have studied the rabbit urinary bladder because it has a structure characteristic of other mammalian bladders. It is composed of three distinct cell layers, a lower germinal layer of small cells 15 μm in diameter, an intermediate layer of fused lower cells (30-μm diameter) and a superficial cell layer of fused intermediate cells ranging from 60 to 120 μm in length and 5 to 10 μm in height. The apical membranes of the superficial cells have an unusual structure, being composed of polygonal plaques approximately 1 μm^2 and 12 nm in thickness. These plaques occupy approximately 73% of the apical surface area with the remaining 27% being normal lipid bilayer.[9] The cytoplasm of the superficial cells contains many disk-shaped vesicles attached to each other and to the apical membrane by microfilaments. The disk-shaped vesicles are composed of two opposing plaques joined at the perimeter by a lipid bilayer. It has been proposed that during bladder expansion these vesicles fuse into the apical membrane to increase the surface area and storage capacity of the bladder.[10]

Solutions

Transepithelial Experiments

Control Ringer's (in mM): NaCl, 111.2; KCl, 5.8; NaHCO$_3$, 25; KH$_2$PO$_4$, 1.2; CaCl$_2$, 2; MgSO$_4$, 1.2; D-glucose, 11.1, pH 7.4, at 37° aerated with 95% O$_2$/5% CO$_2$

Zero Na$^+$ Ringer's: The same as above with all Na$^+$ replaced with K$^+$

[4] J. W. Hanrahan, W. P. Alles, and S. A. Lewis, *J. Gen. Physiol.* **84**, 30a (1984).
[5] J. W. Hanrahan, W. P. Alles, and S. A. Lewis, *Proc. Natl. Acad. Sci. U.S.A.* **82**, 7791 (1985).
[6] J. W. Hanrahan, W. P. Alles, and S. A. Lewis, *Biophys. J.* **47**, 11a (1985).
[7] O. P. Hamill, A. Marty, E. Neher, B. Sakmann, and F. J. Sigworth, *Pfluegers Arch.* **391**, 85 (1981).
[8] D. P. Corey and C. F. Stevens, in "Single Channel Recording" (B. Sakmann and E. Neher, eds.), p. 53. Plenum, New York, 1983.
[9] L. A. Staehelin, F. J. Chlapowski, and M. A. Bonneville, *J. Cell Biol.* **53**, 73 (1972).
[10] B. D. Minsky and F. J. Chlapowski, *J. Cell Biol.* **77**, 685 (1978).

Zero HCO_3^- Ringer's: The same as control except all $NaHCO_3$ is replaced with NaCl

Zero K^+ Ringer's: The same as control except all K^+ is replaced with Na^+

Zero Cl^- Ringer's: The same as control except all Cl^- is replaced with gluconate and methanesulfonate

Amiloride (a gift from Merck-Sharpe and Dohme, Rahway, NJ) and silicone high-vacuum grease (Dow Corning, Midland, MI) are also used

Single-Channel Experiments

Normal bathing saline (in mM): NaCl, 140; KCl, 6; $MgCl_2$, 2; $CaCl_2$, 2; HEPES, 10; pH 7.2

Pipet solutions: K"X_v^-", 150 (where "X_v^-" is Cl^-, Br^-, I^-, F^-, SCN^-, NO_3^-, acetate$^-$, or gluconate$^-$); HEPES, 10; EGTA, 0.08; pH 7.2

Cell Dissociation and Culture. Ham's F-12 and M199 media are mixed in the ratio 1:1 and supplemented with penicillin (100,000 U/liter), streptomycin (100 mg/liter), hydrocortisone (1 mg/liter), and insulin (1 mg/liter). The medium is usually made hypertonic for dissociating and storing cells by adding 30 mM sucrose. For culturing cells, the medium is supplemented with 5% fetal calf serum (GIBCO, Grand Island, NY).

Apparatus

To study the physiological properties of the mammalian bladder in a controlled fashion it must be studied *in vitro*.

Transepithelial Experiments

The bladder is excised after sacrificing a rabbit, the three layers of adhering muscle are removed, and the epithelium is placed between two compartments (hemichambers) that permit solution stirring and replacement, temperature control, and bubbling with an O_2/CO_2 mixture. A tight seal is made between the hemichambers without damaging cells where the edges of the hemichamber openings contact the epithelial sheet.

Figure 1a is a diagram of the chambers. These are now available from E. W. Wright (Guilford, CT). The vise, illumination, solution changing, aeration, and stirring system are also shown. Figure 1b is a diagram of the rack used to stretch the bladder epithelium when removing the muscle layers.

The most important features of this chamber design are as follows:

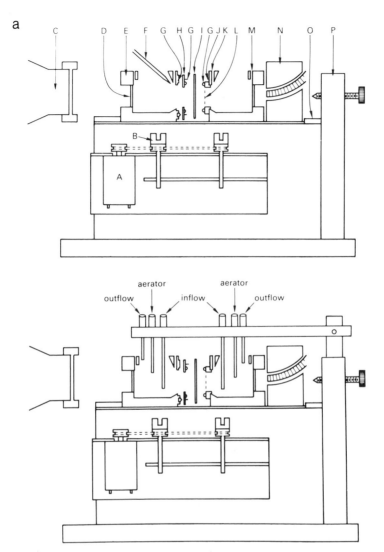

FIG. 1. (a) *Top*: Schematic representation of the chambers used to study the rabbit urinary bladder. A, DC motor (6 V) to drive the external magnets (B) which in turn drive magnetic spin bars in the chambers (not shown); C, horizontally positioned stereo zoom dissecting microscope; D, optical window; E, hemichamber with external water jacket (not shown) for temperature control; F, microelectrode entering chamber through the open top; G, silicone grease rings; H, plastic ring with 20 pins around circumference; I, the epithelium; J, plastic ring with nylon support mesh (L); K, access port for voltage-measuring electrode; M, access port for current-passing electrode; N, light pipe holder (light source not shown); O, platform with V-shaped tracks for holding and aligning chamber; P, vise for pushing rear hemichamber into contact with front hemichamber. *Bottom*: The solution changing and aeration systems. The solution inflow line is connected to a 60-ml syringe placed about 50 cm above the chambers, and the outflow line is connected via a reservoir bottle to a vacuum line. The aerators are connected to a gas cylinder (5% CO_2/95% O_2) via a wash bottle, which saturates the gas with water vapor. (b) Schematic of the dissecting rack. 1, Fixed plastic block with six pins; 2, top templates; and 3, movable plastic block with six pins.

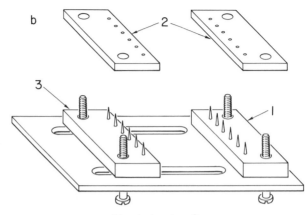

Fig. 1. *(continued)*

1. Temperature control is achieved by circulating water through external water jackets which surround each hemichamber. When the water bath and circulator are grounded using a small amount of salt solution in the water bath, each chamber has its own built-in electrical shield. During microelectrode experiments this has the advantage of reducing 60-Hz pickup by high-impedance electrodes.

2. The bottom of each chamber is flat so that small Teflon-coated magnetic bars can be used to stir the solution bathing each side of the tissue. These spin bars are driven by external magnets coupled to a 6-V DC motor.

3. Optical glass windows are placed at the end of each hemichamber so that the epithelium can be observed using a horizontally mounted stereo zoom dissecting microscope and transilluminated using a fiber optic light source.

4. The tops of the chambers are open to permit easy access for microelectrodes, aeration lines, and solution inflow pipes (gravity feed) and outflow pipes (vacuum assisted).

5. The lower edges of each chamber have triangular grooves running the full length, which fit into corresponding side tracks (one is fixed, the other is laterally adjustable) on the chamber platform. This system maintains the parallel alignment and relative distance between each half-chamber without requiring constant pressure to hold the chambers together. Once it is placed in the tracks, the rear chamber is pushed using a thumb screw toward the front chamber, which abuts against a stop.

6. Voltage-measuring Ag/AgCl wires (or agar bridges connected elec-

trically to Ag/AgCl wires) have an access port about 2 mm away from the opening of each hemichamber. Current-passing Ag/AgCl wires (or agar bridges connected to Ag/AgCl wires) have an access port at the rear of each hemichamber.

7. The epithelium is mounted on a ring having an internal opening of 2 cm² (i.e., its inside diameter is 1.6 cm), an outside diameter of 3 cm, and a circle of 20 pins located 1.3 cm from center. A circle of silicone grease is applied between the pins and internal circumference. This layer of grease acts as a cushion for the mucosal surface of the epithelium and thus reduces cell damage, while the pins (on which the tissue is impaled) maintain a constant degree of stretch. The ring fits into a groove and covers the opening of the front hemichamber. The ring is held in place using silicone grease.

8. The opening of the rear chamber is covered by a second ring (made of Delrin) having an external diameter of 2.1 cm. A layer of nylon mesh is stretched over the ring and held in place by a second, larger ring of 2.5-cm o.d. The nylon acts as a firm support for the serosal (blood) side of the epithelium. A layer of silicone grease is applied between the inner and outer circumference of the ring to cushion the epithelium from damage, and the assembly is held in place over the opening of the hemichamber by silicone grease.

Single-Channel Experiments

The chamber is constructed from a rectangular block of Plexiglas which fastens to the microscope stage. A 1-cm diameter hole is drilled through the center of the block. The side of the hole nearest the pipet holder is beveled at an angle of 40°. The bottom of the Plexiglass block is recessed 0.5 mm and a glass coverslip is fastened into the depression using silicone sealant (RTV, Dow Corning). A small Plexiglas block is also used to hold the agar bridge electrode and suction line in the chamber at a sufficient angle to be clear of the microscope condenser. Agar bridge electrodes are prepared by filling polyethylene tubing (i.d. 0.38 mm, o.d. 109 mm) with 4% agar dissolved in 150 mM KCl. Pipets are fabricated from borosilicate glass capillaries (0.95-mm i.d., 1.2-mm o.d.) using a two-stage puller (PP-83, Narishige Scientific Instruments Laboratory, Tokyo, Japan). Both ends of each capillary tube are briefly fire polished in a Bunsen burner flame before pulling so the pipet does not scrape AgCl from the wire electrode which is contained within the holder.

A three-dimensional hydraulic drive (MO-103, Narishige) is used to position the headstage. Mechanical micromanipulators are generally preferred for cell-attached recording; however, in our experiments involving excised patches, the drift caused by this hydraulic manipulator has not been a problem. The headstage with manipulator is fastened by a rotatable collar to a small post near the chamber to facilitate changing pipets. Cells are viewed using an inverted microscope (BioStar 1820; Bausch and Lomb) mounted on a passively damped antivibration platform (Vibraplane). The microscope and table are enclosed in a Faraday cage made from aluminum screen.

Procedures

Transepithelial Experiments

After sacrificing the rabbit, the abdominal cavity is exposed and the bladder removed by cutting at the junction of the ureters. The bladder is then slit open and rinsed three times in normal Ringer's solution. One cut edge is placed (muscle side up) on a row of six pins attached to a fixed plastic block (Fig. 1b) and a top template is overlaid and held tightly in place by two wing nuts. The other cut edge is hooked on a row of six pins that project from a second, movable, plastic block. After positioning the second top template over the pins, the movable block is pulled until the bladder cannot be stretched further without tearing and then secured using wing nuts. These steps are performed with the rack immersed in a Ringer's solution-filled Petri dish. The Petri dish is then positioned under a dissecting microscope and muscle is carefully dissected away from the underlying connective tissue using rat-toothed forceps and shallow, transverse cuts with a razor blade (Gillette blue blades if possible). In this manner the three muscle layers are removed as long strips. When an adequate area has been cleared of muscle a glass slide and ring that has pins pointing upward are placed beneath the tissue and lifted, impaling the epithelium. The ring is freed by cutting the tissue around its outer edge and then mounted in the front hemichamber. This hemichamber is placed in the vise (already containing the rear hemichamber) and the rear hemichamber is gently brought into contact by tightening the back screw. Then tubes from an external circulating water bath are connected to the chamber's water jacket, the chambers are filled with Ringer's solution, magnetic spin bars are placed in the bottom of each chamber, voltage-measuring and current-passing electrodes are connected to the voltage clamp, and aeration tubes (syringe needles) are placed in each hemichamber.

It is important to periodically determine whether an adequate seal has been achieved between the two silicone grease layers and the epithelium. If the seal is inadequate and there is a leak of solution, the chambers can be brought closer together by tightening the back screw. This usually stops the leak by smoothing out the silicone grease gasket, but care must be taken not to overtighten the chamber, as this damages the epithelium.

In our studies we have found that the method of changing or refreshing the bathing solutions is critical. Two approaches can be used. The first is simply to remove all the bathing solution and replace it with new solution. The second is to perform an isovolumic solution change by adding (by gravity feed) and removing solution (by using vacuum) so as to maintain chamber solution volume and solution levels constant. In this regard, the mucosal solution (front hemichamber) is maintained approximately 2 mm higher than the serosal solution. This small hydrostatic pressure pushes the epithelium against the supporting mesh which covers the opening of the rear hemichamber. Immobilizing the epithelium in this manner is essential if one wants to impale the cells with microelectrodes.

Of the two solution changing protocols listed above, the first is totally unacceptable as it causes large changes in epithelial resistance, transport characteristics, and cannot be used in the presence of microelectrodes. The second method (isovolumic changes) does not produce the above undesirable artifacts, and if care is taken a cell impalement can be maintained throughout a solution change.

Single-Channel Experiments

Our single-channel studies have been carried out using excised membrane patches. Except for a few preliminary experiments, no attempt was made to maintain cells at 37° immediately prior to and during experiments.

The bladder is dissected as described above; however, instead of mounting it on the rack shown in Fig. 1b it is placed in a Petri dish (100-mm diameter) containing 15 ml of serum-free culture medium so that the epithelial surface faces upward. One microscope slide is used to hold the tissue while the other is used to gently scrape small sheets of cells into the medium. After removing the bladder the dish is tilted so that excess medium can be aspirated off and the remaining 2–3 ml containing cells is placed in culture dishes (see below) or into a 20-ml glass vial containing 1 ml (~400 U) high-purity collagenase solution [e.g., type CLSPA (Worthington, Freehold, NJ), or type VII (Sigma, St. Louis, MO)]. A 5-min exposure at room temperature (22°) usually causes partial dissociation of the cells such that a few cells from lower cell layers remain attached to the surface cells (Fig. 2a). This is useful because adhering intermediate cells can then be used to identify the basolateral membrane of surface cells during experiments. The cells are washed by filling the vial with medium, allowed to settle for 5 min, and the excess is siphoned off and replaced with

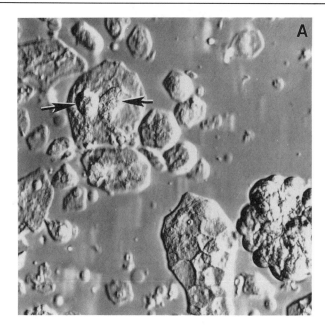

FIG. 2. (A) Partially dissociated cells from rabbit urinary bladder epithelium. Arrows show intermediate cells adhering to a surface cell. (B) Urinary bladder epithelial cells after 15 days in primary culture.

fresh medium that has been equilibrated with 95% O_2/5% CO_2. Cells are usually placed on ice (exposed to the same gas mixture *without* bubbling) and transferred to the experimental chamber in small aliquots as required. It is important to wash cells in the recording chamber several times with normal bathing saline and to maintain positive pressure inside the pipet before attempting to form seals.

We have studied *subconfluent* primary cultures (i.e., just before reaching the stage shown in Fig. 2b) to see if the apical membrane might still contain basolateral ion channels. Glass coverslips are washed in concentrated detergent, rinsed in a series of three ethanol baths (70, 70, and 95%), flamed, and then placed in the bottom of 35-mm tissue culture dishes. A thin layer of bovine dermal collagen (Vitrogen 100; Collagen Corp., Palo Alto, CA) which has been diluted sixfold with sterile 0.013 N HCl solution is applied to each coverslip and allowed to dry in a laminar flow hood. The collagen is rehydrated by filling dishes with complete medium and the dishes are allowed to stand in the incubator overnight. The medium is replaced next day and two drops of cell suspension, obtained as described above, are plated in each dish. The cultures are fed every 2 or 3 days and cells usually cover the glass coverslip completely in 15–19 days.

FIG. 2. (*Continued*)

Electrical Recording

Transepithelial Experiments

The electrical properties of the intact epithelium are measured using a commercially available four-electrode current/voltage clamp (Warner Instruments, Hamden, CT). Unlike patch clamps or excitable membrane clamps, which use two electrodes, the typical epithelial clamp has a pair of electrodes for measuring voltage (differential measurement) and a second pair for passing current (one is attached to a virtual ground current-to voltage converter). Four electrodes are needed because the large currents required in transepithelial voltage clamping would rapidly polarize the

Ag/AgCl electrodes of a two-electrode clamp and this would result in erroneous measurements of transepithelial potential.

While Ag/AgCl electrodes can be placed directly into the chambers when both sides of the epithelium are bathed with symmetrical chloride-containing solutions, this is not possible when the Cl⁻ activity on one side of the tissue differs from the opposing side, because they sense Cl⁻ activity in addition to transepithelial voltage. The remedy to this situation is to use electrodes that are reasonably insensitive to bathing solution composition; e.g., 3 M KCl agar bridges connected to the Ag/AgCl electrodes via 3 M KCl solution. Should the use of such bridges be required, it is advisable to use wide-diameter tubing and keep them as short as possible to minimize their impedance and thus 60-Hz pickup.

The current and voltage outputs of the current/voltage clamp are connected to a strip chart recorder, storage oscilloscope, and microcomputer. Transepithelial voltage (or current) pulses are generated by microcomputer and fed to the current/voltage clamp. Using a computer for pulse generation and data collection is preferred but not essential as the pulses can be generated using a stimulator and the resulting deflections measured using the storage oscilloscope.

There are four important parameters that must be routinely measured. These are transepithelial resistance (R_T), spontaneous transepithelial potential (V_T), short-circuit current (I_{sc}), and epithelial capacitance, (C_T). We will consider each of these parameters in turn and where appropriate point out possible pitfalls or artifacts.

Transepithelial Resistance. The total resistance between voltage-sensing electrodes is calculated using the voltage deflection measured during a current pulse or current deflection measured during a voltage step. To obtain the transepithelial resistance from this value it is necessary to subtract the resistance between the epithelium and the agar bridges (i.e., series resistance).

Two approaches can be used to measure series resistance: The first is to set up chambers without mounting an epithelium, fill the chambers with each solution to be used during subsequent experiments, and measure the resistance between the voltage-sensing electrodes. This resistance can then be subtracted from that measured after mounting the epithelium. This approach is time consuming and neglects any resistance that may be contributed by subepithelial tissue, notably the serosa. Also, positioning of the voltage-measuring electrodes must be precise and reproducible. We prefer a second approach which can be performed while the tissue is in the chamber and does not require electrode repositioning; it involves measuring the time-dependent response to a current or voltage step, and interpreting the response using an equivalent circuit model for epithelial and solution impedances (Fig. 3).

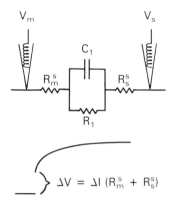

FIG. 3. Equivalent circuit of the solution resistance (R_s) and the epithelial impedance. R_m^s is the mucosal series resistance and R_s^s is the serosal series resistance. For simplicity the epithelium has been represented by a single resistor and parallel capacitor. An idealized voltage response to a square current pulse is shown below. The instantaneous voltage increase to a square current pulse (ΔI) is equal to $\Delta I(R_s^s + R_m^s)$ and the subsequent hyperbolic increase in voltage is due to the time-dependent charging of the membrane capacitor.

The equation that describes the time-dependent voltage response $[V(t)]$ to a square current pulse (ΔI) is

$$V(t) = \Delta I(R_s) + \Delta I R_1 (1 - e^{-t/R_1 C_1})$$

where R_s is the series resistance, R_1 is the transepithelial resistance, and C_1 is the transepithelial capacitance. Immediately after a pulse is applied $(t = +0)$, $V(t = +0)$ is equal to $\Delta I R_s$. After this step change the voltage increases hyperbolically and reaches a steady state $V(t = \infty) = \Delta I(R_s + R_1)$. Using a high-speed, storage oscilloscope one can measure the instantaneous voltage change and calculate R_s directly from Ohm's law. Similarly, the equation that describes the time-dependent current response $[I(t)]$ to a voltage clamp step is

$$I(t) = \frac{v}{R_1 + R_s} = \left(\frac{V}{R_s} - \frac{V}{R_1 = R_s} \right) e^{-t(R_1 + R_s/R_1 R_s C_1)}$$

Immediately after the voltage pulse $(t = +0)$, $I(t = +0)$ equals V/R_s, knowing the voltage clamp step and measuring the current we can calculate the series resistance. In practice, series resistance measurement is more difficult and prone to instrumentation problems when voltage pulses are used than when current pulses are used. With both pulse methods it is important to determine the frequency response characteristics of the current-passing and voltage-measuring systems because the pulse must be square and have a rise time less than 5 μsec.

Transepithelial Potential (V_T). The most straightforward parameter to measure in bladder epithelium (any epithelium for that matter) is the transepithelial potential. Under steady state conditions, the presence of such a spontaneous potential when the epithelium is bathed in symmetrical solutions implies there is active (energy-utilizing) net transport of charge across the epithelium. The measurement of this charge movement will be discussed below. Before measuring transepithelial potential one must measure the asymmetry in the voltage-sensing electrodes. Typically the asymmetry should not exceed 2 mV, otherwise the Ag/AgCl wires must be recoated. The magnitude of the spontaneous voltage is the difference between the measured voltage and the small asymmetry potential. The asymmetry should be checked at the end of the experiment. Commercially available voltage clamps have a built-in offset voltage source which allows one to zero out these asymmetries.

Short-Circuit Current (I_{sc}). The concept of the short-circuit current revolutionized the study of epithelial transport and still represents the most basic tool in the arsenal of methods available to epithelial physiologists. Ussing and Zehran[11] showed that if one applies sufficient current across the epithelium to reduce the spontaneous potential to zero when it is bathed in symmetrical solutions, the amount of applied current (I_{sc}) equals the net flow of actively transported charge across the epithelium. Tight junctions, which bind the cells together to form a sheet, do not possess active transport systems, therefore I_{sc} must also equal net ion transport through the cells although short-circuit current measurements alone do not reveal which ionic species is being transported. By Kirchoff's current law, the net current flowing across the apical membrane must equal that flowing across the basolateral membrane.

Series resistance is not normally a significant fraction of the total resistance measured in the bladder (80 $\Omega \cdot cm^2$ series compared to 20,000 $\Omega \cdot cm^2$ tissue) but it is always a possible source of error that might lead to underestimates of transport rate.

Transepithelial Capacitance. The capacitance of the epithelium provides an estimate of epithelial surface area. Whereas the bathing solution acts as a pure resistor, the apical and basolateral membranes of an epithelium can each be modeled as a resistor and capacitor in parallel. The resistive pathway reflects charge flowing through membrane-spanning ionic channels while the capacitive component reflects charging the capacitance of the lipid bilayer. The resistive component can be altered by channel blockers and by modulators of their (time-averaged) conductances, but capacitance can be altered only by varying membrane surface

[11] H. H. Ussing and K. Zehran, *Acta Physiol. Scand.* **23,** 110 (1951).

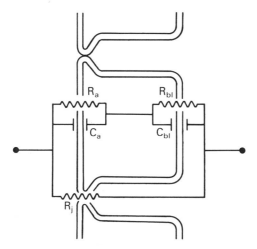

Fig. 4. The simplest equivalent circuit which can describe the passive electrical properties of the rabbit urinary bladder epithelium. R_a, R_{bl}, and R_j are the apical, basolateral, and junctional resistances, respectively; C_a and C_{bl} are the apical and basolateral membrane capacitance values. Note that R_a may be up to 50 times greater than R_{bl} and $R_j > R_a$. Consequently, the voltage response to a square current pulse is determined mainly by the capacitance and resistance of the apical membrane and parallel junctional resistance.

area. In biological tissues, 1 cm² of flat tissue area is approximately equal to 1 μF of capacitance.[12] Measuring capacitance allows one to determine membrane surface area and test whether variations in membrane area are synchronized with changes in transport rate or mechanical stress.

Membrane capacitance is measured by applying a current wave form and recording the voltage response. Capacitor values can be extracted by fitting the observed voltage response to the response expected for a morphologically based equivalent circuit model composed of resistors and capacitors. For example, Fig. 4 is the simplest equivalent circuit that could describe the response obtained with mammalian urinary bladder epithelium. To date three different current wave forms have been used for these measurements: square current pulses,[1] current sinusoids,[13] and a pseudorandom current signal.[14]

For rapid measurements of capacitance, we apply a square current pulse and store the resulting voltage response on an oscilloscope screen. We determine the length of time it takes the voltage to reach 63% of its steady state value because this time constant is equal to the product of the

[12] R. Fettiplace, D. M. Andrews, and D. A. Haydon, *J. Membr. Biol.* **5,** 227 (1971).

[13] C. Clausen, S. A. Lewis, and J. M. Diamond, *Biophys. J.* **26,** 291 (1979).

[14] C. Clausen and T. E. Dixon, *Curr. Top. Membr. Transp.* **20,** 47 (1984).

epithelial resistance and capacitance. Dividing the time constant (units of seconds) by the resistance (units of ohms) yields the capacitance of the epithelium in farads.

This approach is valid only if the product of the apical membrane resistance and capacitance (i.e., time constant) is much greater than the basolateral membrane time constant. This condition is fulfilled by the bladder, and the capacitance value obtained is an estimate of the apical membrane surface area.[2] The validity of this method has been verified using more sophisticated techniques for measuring capacitance.[13]

We routinely use capacitance measurements to normalize tissue resistance and short-circuit current to actual membrane area. Normalizing allows comparison of these parameters among tissues. Capacitance measurements have demonstrated that surface area of the urinary bladder is increased by the insertion of cytoplasmic vesicles into the apical membrane during mechanical stretch.[15]

Single-Channel Experiments

A patch clamp (Yale Mark V) equipped with a 10-GΩ feedback resistor is used for single-channel recording. Previously, data were collected using an FM adapter (Vetter, DC-1000T, Rebersburg, PA) and stored on tape using a dual-capstan stereo cassette deck (Aiwa, AD-3800U, Japan); however, a digitizing unit (Neurodata, DR-384, New York, NY) and video cassette recorder (Hitachi, VT1100A, Japan) are now used because they provide better high-frequency response and larger storage capacity. In addition to recording the patch-clamp output directly on tape, the data are also filtered using an eight-pole Bessel filter, observed on a storage oscilloscope, and sometimes digitize (Tekmar, TM-DA100, Cleveland, OH) at 2 or 4 kHz and stored in a microcomputer (North Star, Horizon, Berkeley, CA) equipped with a RAM disk (Semi Disk Systems, Beaverton, OR). Pulses (~0.2 mV) for monitoring the development of a seal are supplied by a conventional stimulator. Records containing 16,284 data values are displayed in segments of 2048 points and the baseline is calculated between two locations set using cursors. The baseline is subtracted from the data, occasionally edited to remove noise, and then analyzed to give a histogram of current amplitudes. A threshold is set at one-half the open channel current (calculated from the amplitude histogram) and the data are reanalyzed for open and closed durations.

The solutions bathing either side of the membrane are kept bionic; i.e., during studies of the anion channel the pipet contains 150 mM KCl while the bath contains potassium salts of other anions (e.g., bromide, iodide,

[15] S. A. Lewis and J. L. C. deMoura, *J. Membr. Biol.* **82,** 123 (1984).

thiocyanate) at the same concentration. Potentials arising at the junction between the agar bridge (150 mM KCl) and other bath solutions are estimated relative to a flowing KCl junction as follows: A large patch pipet filled with 3 M KCl is placed in the bath, the current flowing between pipet and KCl agar/reference electrode is zeroed, and the voltage required to clamp the current at zero is monitored while potassium chloride in the bathing solution is replaced by other potassium salts. The voltage deflections caused by substituting other anions for chloride agree with predictions based on their relative mobilities in those instances where ionic mobilities are available in the literature.

During patch-clamp experiments, the probability of forming a seal is approximately 0.5 although this varies from day to day despite efforts to keep all the experimental conditions constant. Once a seal is obtained, the probability of observing an anion channel is about 0.3; however, there is some tendency for anion channels to be clustered, with patches containing no channels or two to three channels occurring more often than those with only a single channel. It is occasionally possible to record anion- and K$^+$-selective channels in the same patch. The two channel types can be distinguished easily by their very different kinetics, conductances, and also by their different sensitivities to drugs. For example, the anion channel can be inhibited by DIDS (4,4'-diisothiocyanostilbene-2,2'-disulfonic acid) without affecting the behavior of the K$^+$ channel (Fig. 5A and B), and the K$^+$ channel can be blocked by barium without affecting the anion-selective channel (data not shown). Basolateral potassium channels are seen in patches of surface membrane from subconfluent bladder cultures but not from confluent monolayers, consistent with *in vitro* development of normal epithelial polarity. Rabbit urinary bladder does not have K$^+$-selective or anion-selective conductances in the apical membrane.

Assessing Epithelial Viability during Transepithelial Experiments

Several criteria are used to assess the viability of the urinary bladder *in vitro*. One might expect it to have high electrical resistance because the bladder functions as a barrier to ionic diffusion. Also, the bladder actively absorbs sodium and this should generate a spontaneous lumen-negative electrical potential. Unlike most other preparations, the bladder has a settling-in period of 30–90 min, during which the transepithelial potential and resistance *increase* from 300 $\Omega \cdot \text{cm}^2$ and −2 mV, respectively, to stable values of 20,000 $\Omega \cdot \text{cm}^2$ and −40 mV. The length of this settling-in period is inversely related to the time taken to dissect the tissue and mount it in the chambers, suggesting that the tissue is traumatized by dissection. To reduce traumatization the dissection should be performed as quickly as

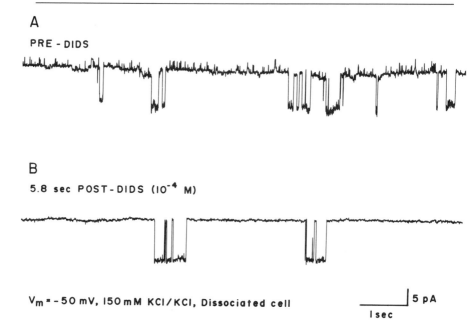

FIG. 5. (A) Single-channel recording from an excised, inside-out patch of membrane from a freshly dissociated surface cell. This patch contains one low conductance anion channel (64 pS; which is mostly open) and one high-conductance K^+ channel (200 pS). (B) Adding DIDS inhibits the 64-pS channel without affecting the K^+ channel.

possible. Small rhythmic fluctuations of the potential indicate incomplete dissection of the muscle layers. Epithelial capacitance should remain constant over the course of the experiment. We have found that swelling the bladder cells increases apical capacitance because it causes fusion of cytoplasmic vesicles into the apical membrane.[9]

Transport Properties of the Bladder

In this section some of the basic transport properties of the mammalian urinary bladder epithelium are described.

1. The bladder has amiloride-sensitive Na^+ channels in the apical membrane and an apical leak pathway which is not blocked by amiloride. The leak pathway appears to be 58% more permeable to K^+ than Na^+.[10] Mucosal addition of 10^{-4} M amiloride (final concentration) increases the transepthelial resistance and decreases the transepithelial potential and I_{sc}. These effects of amiloride are rapid and reversible. However, the I_{sc} is slightly less ($\sim 3\%$) than the rate of active Na^+ reabsorption from lumen to

blood; there is probably a small K^+ secretion from blood to lumen through the apical leak pathway.[16]

2. The ability of the bladder to absorb Na^+ depends on the composition of the serosal bathing solution. Complete removal of HCO_3^- (by replacement with Cl^-; "zero HCO_3 Ringer's") or Cl^- (by replacement with gluconate; "zero Cl^- Ringer's") decreases V_T and I_{sc}, and increases transepithelial resistance.[1] The decrease in I_{sc} results from reduced apical Na^+ permeability as well as depolarization of the basolateral membrane. Preliminary evidence suggests that depolarization of the basolateral membrane during HCO_3^- or Cl^- removal is caused by decreased K^+ permeability of the basolateral membrane.

In summary, it is clear that Cl^- and HCO_3^- are required in the serosal bathing solution for optimal rates of transport.

Advantages and Potential Sources of Error

The mammalian urinary bladder offers many advantages over other Na^+-transporting epithelia that are widely used as model preparations. The apical cell layer seems not to be coupled to the lower two germinal cell layers, and this apical layer is composed of only one cell type. The cells are very large (> 60-μm diameter and 10-μm height) and therefore are convenient for microelectrode experiments and patch clamping of either membrane. Using a homogeneous population of large cells simplifies the interpretation of transepithelial transport measurements as well as intracellular microelectrode data. If care is taken to dissect away the three muscle layers, stable microelectrode impalements can be maintained for up to 1 hr. The large cells of the apical layer in conjunction with high tight junction resistance makes the mammalian bladder one of the least permeable structures to water and ion movements and thus ideal for electrophysiological studies. On the other hand, its low permeability means that even small leaks can have dramatic effects on the transport properties of the tissue, so that great care must be exercised when mounting it in the chambers.

Another potential source of problems occur when changing the bathing solutions. At no time should solution levels be allowed to vary by more than 2 mm, and the mucosal solution must be maintained at a higher level than the serosal solution at all times. The reasons are twofold: First, silicone grease seals are sensitive to perturbations in fluid level and will leak if hydrostatic pressures are maintained. Second, as mentioned above, the epithelium itself is sensitive to mechanical distension; stretching causes

[16] S. A. Lewis and N. K. Wills, *J. Physiol. (London)* **341**, 169 (1983).

them to be removed. This is important because the permeability properties of "old" channels, which have been exposed to proteolytic enzymes in the urine, differ from those in newly inserted vesicular membrane.[17] Incorporation of new vesicles and removal of old channels will cause marked changes in transepithelial transport rate.[2,3] Vesicle translocation is itself an interesting phenomenon which is being examined electrophysiologically. These studies should yield information on the mechanisms by which epithelial cells selectively shuttle a particular transporter only to the apical or basolateral membrane.

[17] S. A. Lewis and W. Alles, *Proc. Natl. Acad. Sci. U.S.A.* **83**, 5345, (1986).

[43] Electrophysiological Methods for Studying Ion and Water Transport in *Necturus* Gall Bladder Epithelium

By G. ALTENBERG, J. COPELLO, C. COTTON, K. DAWSON, Y. SEGAL, F. WEHNER, and L. REUSS

Introduction

In this chapter techniques used for the study of ion and water transport mechanisms in isolated *Necturus* gall bladder epithelium are summarized. With appropriate modifications, these methods can be applied to other epithelia as well as to nonepithelial preparations. The emphasis is on techniques, but we also have attempted to provide orientation for the nonspecialist on the rationale of the experimental approaches utilizing these methods. Finally, we briefly illustrate the application of each technique in identifying and characterizing mechanisms of ion and water transport across individual cell membranes, and their regulation. Compared to other methods, electrophysiological techniques have major advantages in both accuracy and time resolution, and permit measurements in a nondestructive fashion. Several excellent reviews on electrophysiological techniques have been published and should be consulted for topics not covered in this chapter.[1-5]

[1] F. J. Alvarez-Leefmans, F. Giraldez, and J. M. Russell, *in* "Chloride Channels and Carriers in Nerve, Muscle and Glial Cells" (F. J. Alvarez-Leefmans and J. M. Russell, eds.), p. 3. Plenum, New York, 1990.
[2] R. D. Purves, "Microelectrode Methods for Intracellular Recording and Ionophoresis." Academic Press, London, 1981.
[3] B. Sakmann and E. Neher, "Single Channel Recording." Plenum, New York, 1983.

Conventional Microelectrode Techniques

Conventional intracellular microelectrodes are used in epithelial transport studies to measure cell membrane voltages, electrical resistances, and ion selectivities. Since their introduction by Ling and Gerard,[6] glass microelectrodes filled with concentrated salt solutions have been widely used in epithelial and nonepithelial cells. The most frequently used electrodes are micropipets made of borosilicate glass and filled with 3 M KCl.[2,4,5]

Preparation of Conventional Microelectrodes

For ease of filling, we prefer borosilicate glass with an inner fiber. Glass tubing of 1-mm o.d. and 0.5-mm i.d. is convenient, but other dimensions are also adequate. As a preliminary step, the glass is cleaned in boiling ethanol, rinsed with distilled water, dried in an oven, and stored packed in aluminum foil.

Pipets are pulled from the prepared tubing using a horizontal microelectrode puller. For most applications we prefer the Narishige PD-5 puller (Narishige, Tokyo, Japan). However, for impalements through the basolateral membrane, very fine tips are necessary and can be obtained with a Brown-Flaming puller (model P-87; Sutter Instrument Co., San Rafael, CA). Each capillary yields two pipets. Immediately after pulling, pipets are positioned tip up in a jar, with their back ends immersed in the chosen filling solution (e.g., 3 M KCl). A good device for this purpose is the jar manufactured by WPI (World Precision Instruments, New Haven, CT). It is important to keep the filling solution clean. We routinely boil the 3 M KCl solution, complete the volume with boiled distilled water, keep the stock in a refrigerator, and filter the solution in the microelectrode-filling jars about once a week. The microelectrode tips fill rapidly by the capillary action of the inner fiber; the shank is easily filled with a 2-in. 30-gauge needle. The pipet is connected to an Ag-AgCl electrode, i.e., either a commercial pellet in a holder, which permits tight insertion of the filled glass microelectrode (WPI), or a chloridized silver wire that can be inserted in the shank and sealed in place with dental wax (Kerr, Emeryville, CA).

Effectual microelectrodes for impaling *Necturus* gall bladder epithelium from the apical surface have resistances of 20 to 50 MΩ when filled with 3 M KCl and immersed in NaCl Ringer; finer tips, with resistances of

[4] N. B. Standen, P. T. A. Gray, and M. J. Whitaker, "Microelectrode Techniques." The Company of Biologists Limited, Cambridge, 1987.

[5] R. C. Thomas, "Ion-sensitive Intracellular Microelectrodes: How to Make and Use Them." Academic Press, London, 1978.

[6] G. N. Ling and R. W. Gerard, *J. Cell. Comp. Physiol.* **34**, 383 (1949).

40 to 100 MΩ, are necessary for basolateral impalements. In our hands, leakage of electrolyte (from the microelectrode into the cell) does not result in artifactual measurements of membrane voltages, probably because the *Necturus* gall bladder epithelial cells are electrically coupled,[7,8] and therefore the volume of distribution of the putative leak is large. The intracellular membrane voltage and resistances do not differ appreciably among microelectrodes filled with 0.5 to 3 M KCl, 4 M potassium acetate, 1 M sodium formate, and a variety of other filling solutions. In addition, use of the organic phase microelectrode described by Thomas and Cohen[9] yields identical results. This microelectrode, constructed as an ion-selective microelectrode (see below), is filled with a cation-selective exchanger [potassium tetrakis (*p*-chlorophenylborate) dissolved in *n*-octanol] and does not distinguish between K^+ and Na^+. The organic phase microelectrode measures membrane voltage provided that $[Na^+]$ plus $[K^+]$ are similar in the intra- and extracellular compartments, and that the sum remains constant in the cell during the measurement.

Tissue Mounting Techniques

For apical (mucosal) impalements, the gall bladder is excised, opened, pinned onto a cork ring, and mounted horizontally, apical side up, in the assembly shown in Fig. 1. For basolateral (serosal) impalements, the tissue is mounted serosal side up, and part of the subepithelial tissue is dissected away with sharp steel needles under stereoscopic observation at about ×50. Finally, the gall bladder is placed in the chamber shown in Fig. 2, which allows for separate superfusion of apical and basolateral surfaces. To avoid electrical shunting derived from damage at the edge of the preparation, the edge is covered with a ring of Sylgard (kit 184, Dow Corning Co., Midland, MI) coated on its lower surface with a thin layer of silicone grease. This ring provides a convenient way of restricting the volume of the upper bathing solution. The chamber is held firmly on the stage of an inverted microscope situated on a vibration-isolation table. A Faraday cage is convenient to reduce electric noise, but not indispensable if shielding and grounding are carefully done.

To facilitate interpretation of the results, it is convenient to change the solution composition rapidly at the membrane surface (Table I). Superfusion of the upper compartment is effected by gravity inflow, while outflow occurs via a glass pipet connected to vacuum. Solution reservoirs are connected via plastic tubes to two six-way valves (Rheodyne, Inc., Cotati,

[7] E. Frömter, *J. Membr. Biol.* **8**, 259 (1972).
[8] L. Reuss and A. L. Finn, *J. Membr. Biol.* **25**, 115 (1975).
[9] R. C. Thomas and C. J. Cohen, *Pfluegers Arch.* **390**, 96 (1981).

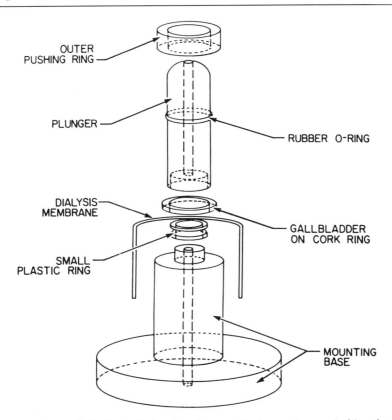

FIG. 1. The small plastic ring is placed on the mounting base and covered with a piece of dialysis membrane secured with rubber bands. At least 50 holes are punched in the membrane with a small pin. The gall bladder is stretched, pinned on a cork ring, centered over the small plastic ring, and pressed down with a plunger. The pins holding the gall bladder to the cork ring are removed, the cork ring is lifted, and a rubber O-ring is pushed down around the outside of the plunger and held in place by an outer pushing ring so that when the plunger is removed, the O-ring seals the tissue onto the small plastic ring. Excess tissue and dialysis membrane are cut away, and the inner ring is fitted into a flat plastic annulus (2.5-cm o.d., 1.0-cm i.d.) which is then positioned in the chamber.

CA). The outlets from these valves are in turn connected to a four-wing Teflon rotary valve (Rheodyne, Inc.) equipped with a pneumatic actuator to exchange the two streams, one of which flows to the preparation while the other flows to waste. The actuator is controlled by a solenoid valve (Rheodyne, Inc.). Because the volume of the upper compartment is small (0.1–1 ml), a complete solution change can be achieved in a few seconds. The lower compartment of the chamber is closed, has a volume of ≈ 0.8

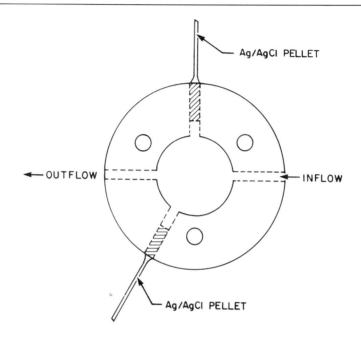

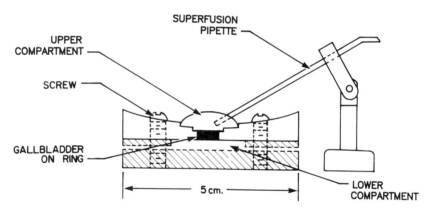

FIG. 2. Gall bladder chamber. The upper drawing represents a top view of the bottom half of the chamber. The lower drawing represents a cross-sectional view of the assembled chamber.

ml, and is perfused at a rate of up to 40 ml/min by a siphon system. To reduce noise due to antenna effects of the fluid columns, air gaps (e.g., drop counters from intravenous infusion sets) are positioned near the chamber in the inflow and outflow lines.

TABLE I
BASIC BATHING SOLUTIONS USED IN ELECTROPHYSIOLOGIC STUDIES OF *Necturus*
GALL BLADDER

	NaCl– Ringer	1 mM HEPES- Ringer[a]	Na$^+$-free Ringer	Cl$^-$-free Ringer
[Na$^+$][b]	100.5	110.0	0[c]	100.5
[K$^+$]	2.5	2.5	2.5	2.5
[Ca^{2+}]	1.8	1.0	1.8	1.8
[Mg^{2+}]	1.0	0	1.0	1.0
[TMA$^+$]	0	0	100.5	0
[HCO$_3^-$][d]	10.0	0	10.0	10.0
[Cl$^-$]	98.1	113.5	98.1	0
[Cyclamate]	0	0	0	98.1
[HEPES]	0	1.0	0	0
[H$_2$PO$_4^-$]	0.5	0	0.5	0.5

[a] The solution is equilibrated with air overnight and the pH is generally adjusted to 7.6 by titration.
[b] Concentrations are given in millimolar units.
[c] 0, Solutions nominally free of the indicated ion.
[d] Solutions with 10 mM [HCO$_3^-$] are equilibrated with 1% CO$_2$; the final pH is about 7.66.

For acceptable impalements, adequate micromanipulators are essential. We have been successful with Narishige hydraulic microdrives (MO-103, Narishige), motorized manipulators (MM33, Stoelting, Chicago IL), and piezoelectric drives (Inchworm, Burleigh Instruments, Inc., Fishers, NY). We currently use the Narishige models mounted on manual micromanipulators (M-152, Narishige) to facilitate coarse positioning.

Electrophysiologic Apparatus

The microelectrodes are connected to the input stage of a high-input impedance (10^{12} Ω) and low leakage current ($<5 \times 10^{-12}$ A) electrometer. Negative capacity compensation, provisions for current application via a microelectrode, and the existence of circuitry for monitoring microelectrode resistance are useful.

The transepithelial voltage (V_{ms}) is measured via macroelectrodes positioned in the mucosal and serosal bathing solutions. We routinely use an Ag-AgCl pellet as reference electrode, positioned in the lower half of the chamber and separated from the bathing solution by a short 3% agar-Ringer bridge. In the upper half of the chamber, the electrode is a calomel half-cell connected to the solution via a flowing, saturated KCl bridge ending in a glass fiber plug (tip diameter ~1 mm; Ultrawick, WPI) to reduce KCl leakage into the solution. If superfusion is slow or must be stopped during the experiment, a static agar-Ringer bridge can be used instead of the KCl bridge.

To measure the transepithelial electrical resistance and the ratio of cell membrane resistances (Fig. 3), transepithelial current pulses are passed between the two fluid compartments. Constant-current pulses of known intensity are provided by stimulus isolation units, and the frequency and duration are controlled by a stimulator. The current source is connected to an Ag-AgCl pellet in the lower half-chamber, and an Ag-AgCl wire in the upper compartment. If superfusion is slow or must be stopped during the experiment, the wire must be connected to the solution via an agar-Ringer bridge to reduce contamination of the solution with Ag⁺.

The output of the electrometers can be low-pass filtered, digitized with an analog–digital converter, and stored and displayed on an IBM-compatible computer. The voltage records are displayed in parallel, if desired, on a conventional strip chart recorder, or an oscilloscope. Data acquisition is carried out with a commercial system (Asyst, Macmillan Software Co.,

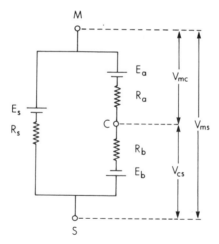

Fig. 3. Steady state electrical equivalent circuit for *Necturus* gall bladder epithelium. M, C, and S denote mucosal bathing medium, cell, and serosal bathing medium, respectively. Each element of the circuit is represented by a Thévenin electrical equivalent, i.e., an electromotive force (E) in series with a resistance (R). Subscripts a, b, and s denote apical cell membrane, basolateral cell membrane, and paracellular (shunt) pathway, respectively. To estimate cell membrane resistances, only passive elements of the circuit are considered. [From L. Reuss, E. Bello-Reuss, and T. P. Grady, *J. Gen. Physiol.* **73**, 385 (1979), with permission.]

New York, NY). Each channel is usually sampled at 10 Hz. The data can be analyzed with customized programs that permit calculations of membrane voltages, transepithelial resistance, ratio of cell membranes resistances, and other parameters.

Transepithelial and Cell Membrane Voltages and Resistances

Validation of impalements is a crucial step, particularly in the case of small epithelial cells, in which the risk of artifacts is serious. We use the following impalement validation criteria: (1) The voltage change recorded by the electrode upon impalement is abrupt and monotonic; (2) after withdrawal the microelectrode potential is within 2 mV of the value before impalement; (3) the voltage deflection measured by the microelectrode during transepithelial current application decreases on impalement, indicating that the electrode tip has passed a resistive barrier; and (4) the potential measured on impalement is stable within 2 mV for at least 1 min. Once the electrophysiologic properties of the tissue are known, they can be exploited to validate the impalements further and to assess the condition of the cells. For instance, we frequently assess the ion selectively of the membrane (e.g., by measuring the magnitude of the depolarization produced by raising extracellular $[K^+]$), and the value of the ratio of cell membrane resistances (apical/basolateral, $R_a/R_b > 5$ in gall bladders bathed in HCO_3^-/CO_2-Ringer).

The apical membrane voltage (V_{mc}) is the difference in electrical potential between the cell interior and the mucosal solution; the basolateral membrane voltage (V_{cs}) is the voltage between the cell and the serosal bathing solution (conventionally, the adjacent external solution is the reference).

The transepithelial resistance (R_t) is determined by passing current pulses across the epithelium (see the section, Electrophysiologic Apparatus) and measuring the resulting change in V_{ms}:

$$R_t = \Delta V_{ms}/I_t \tag{1}$$

where I_t is the applied current.

The current–voltage relationship of the epithelium is linear within the range of $\pm 200 \ \mu A \cdot cm^{-2}$. Steady state voltage deflections of sufficient amplitude are obtained with pulses of $50 \ \mu A \cdot cm^{-2}$ and 1-sec duration. Because of the geometry of the preparation–chamber assembly, it is necessary to correct carefully for series resistances. This is conveniently done by subtracting the voltage deflection produced by current pulses of the same amplitude, keeping all electrodes in the same positions, after the epithelium is removed by scraping or by exposure to detergent. This procedure also corrects for inhomogeneities of current flow.

The "fractional resistance" of the apical membrane $[f_{R_a} = R_a/(R_a + R_b)]$ and/or the resistance ratio R_a/R_b are determined by comparing the voltage deflections elicited across individual cell membranes (measured with an intracellular microelectrode) and across the whole epithelium. Since all cells have the same conductive properties, the currents across apical and basolateral membranes are equal. Hence,

$$R_a/R_b = \Delta V_{mc}/\Delta V_{cs} \tag{2}$$

where ΔV_{mc} and ΔV_{cs} are the voltage deflections elicited by the current across the apical and the basolateral membrane, respectively.

Assessment of the absolute resistances of the epithelial barriers (apical membrane, basolateral membrane, and paracellular pathway, see Fig. 4) requires intracellular cable analysis. To this end, an intracellular microelectrode is used to pass current pulses of intensity i_0, generally 10 to 40 nA, 1-sec duration. Another intracellular microelectrode measures the elicited intracellular voltage deflection (V_x) at varying distances (x) from the current injection site. The cells are well coupled and, under control conditions, the space constant for spread of current within the epithelial sheet is

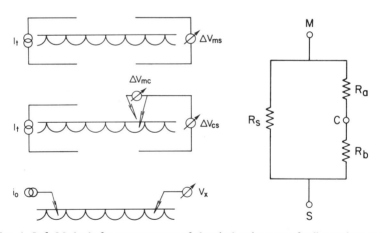

FIG. 4. *Left:* Methods for measurement of electrical resistances of cell membranes and paracellular pathway. Top, transepithelial resistance: $R_t = \Delta V_{ms}/I_t$, where I_t is applied current (per cm^2 of tissue). Middle, ratio of cell membrane resistances: $R_a/R_b = \Delta V_{mc}/\Delta V_{cs}$. Bottom, cable analysis that allows calculation of R_z, from $R_z = R_a \cdot R_b/(R_a + R_b)$; i_0 is the applied intracellular current, and V_x is the intracellular voltage change induced by i_0 at distance x. From these equations, R_a, R_b, and R_s can be calculated. *Right:* Passive equivalent electrical circuit. M, C, and S are mucosal, cell, and serosal compartments, and R_s, R_a, and R_b are paracellular, apical membrane, and basolateral membrane resistances. [From L. Reuss, *in* "Membrane Transport in Biology," Vol. IVB (G. Giebisch, D. C. Tosteson, and H. H. Ussing, eds.), p. 853. Springer-Verlag, Berlin and New York, 1979, with permission.]

ca. 400 μm.[7,8,10] The epithelium is modeled as a flat, thin, infinite sheet with low internal resistance and high resistance of the cell membranes. Both bathing solutions are grounded, so that current flow proceeds from the current injection site to either the neighboring cells or through either cell membrane to ground. Under these conditions, V_x is described by Eq. (3):

$$V_x = AK_0(x/\lambda) \tag{3}$$

where A (mV) is a parameter dependent on the current and the input resistance of the epithelium, K_0 is a zero-order Bessel function of the second kind,[11] λ is the space constant, and x is the interelectrode distance.

The best fit of the data to the function K_0 yields estimates of A and λ, from which the electrical resistance to current flow out of the epithelial sheet (R_z, equivalent resistance to the two cell membranes in parallel) and the internal resistance of the sheet (R_i) can be calculated:

$$R_z = 2\pi A\lambda^2/i_0 \tag{4}$$
$$R_i = 2\pi A/i_0 \tag{5}$$

Combining determinations of R_t [Eq. (1)], R_a/R_b [Eq. (2)], and R_z [Eq. (4)], the equivalent circuit shown in Fig. 4 can be solved for the three unknowns, namely R_a, R_b, and R_s.[7,8]

A complete cable analysis experiment requires V_x measurements in at least six cells to permit reasonable fit of the data to the Bessel function. This procedure takes several minutes and therefore prevents rapid assessment of changes in cell membrane resistances. To circumvent this limitation, we have developed a two-point cable analysis method that permits rapid determination of the direction of change of R_z during an experimental perturbation.[12]

Determinations of Ion Selectivities of the Cell Membranes

The passive circuit analysis methods illustrated above can be extended further to assess the ionic selectivities of the cell membranes. A simplified equivalent circuit of the epithelium is necessary to accomplish this end (Fig. 4). When the solutions have identical composition on both sides of the tight junctions, the value of the equivalent electromotive force (emf) of the paracellular pathway (E_s) is zero, and the steady state voltages are described by Eqs. (6) and (7).

[10] J. Stoddard and L. Reuss, *J. Membr. Biol.* **102**, 163 (1988).
[11] M. Abramowitz and I. A. Stegun, eds., "Handbook of Mathematical Functions." Dover, New York, 1965.
[12] K.-U. Petersen and L. Reuss, *Am. J. Physiol.* **248**, C58 (1985).

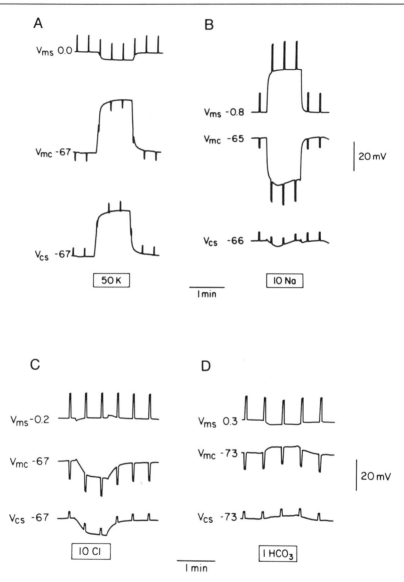

FIG. 5. Electrodiffusive permeability of the apical membrane of *Necturus* gall bladder epithelium. The four panels depict isomolar ion substitutions in the mucosal bathing solution. In all cases, the control bathing medium was NaCl-Ringer. During 1-min periods (bars), one of the ion concentrations was changed to that indicated: (A) K$^+$ was elevated (Na$^+$ lowered); (B) Na$^+$ was lowered (replaced with tetramethylammonium, TMA$^+$); (C) Cl$^-$ was lowered (replaced with cyclamate); (D) HCO$_3^-$ was lowered (replaced with cyclamate). For abbreviations and polarity conventions, see text. Control voltages are given (in millivolts) at the beginning of each trace. Liquid junction potential changes are sizable only in experiment

$$V_{mc} = \frac{E_a(R_b + R_s) + E_b R_a}{R_a + R_b + R_s} \tag{6}$$

$$V_{cs} = \frac{E_a R_b + E_b(R_a + R_s)}{R_a + R_b + R_s} \tag{7}$$

Therefore, the emf values of the cell membranes (E_a and E_b) can be computed if the membrane voltages and resistances are measured.

Similar equations can be derived for the case of rapid ionic substitutions in one of the bathing solutions, such as those illustrated in Fig. 5, assuming that all effects on voltages and resistances are restricted to the ipsilateral cell membrane and the junctions.[8,13]

In the case of a substitution on the apical side only, the membrane voltages are given by[8]

$$V'_{mc} = \frac{E'_a(R_b + R'_s) + R'_a(E_b - E'_s)}{R'_a + R_b + R'_s} \tag{8}$$

$$V'_{cs} = \frac{E_b(R'_a + R'_s) + R_b(E'_a - E'_s)}{R'_a + R_b + R'_s} \tag{9}$$

and in the case of basolateral substitutions, the expressions are[13]

$$V''_{mc} = \frac{E_a(R'_b + R'_s) + R_a(E'_b - E'_s)}{R_a + R'_b + R'_s)} \tag{10}$$

$$V'_{cs} = \frac{E'_b(R_a + R'_s) + R'_b(E_a - E'_s)}{R_a + R'_b + R'_s} \tag{11}$$

where prime (') denotes values after the ion substitution.

This kind of analysis permits assessment of the dependence of the membrane emf on external ion concentration and yields, in principle, a measure of the ionic selectivity of the membrane. However, the recent demonstration of voltage-sensitive membrane conductances in *Necturus*

---

[13] L. Reuss, *J. Membr. Biol.* **47**, 239 (1979).

(C). The brief voltage deflections in the traces are caused by transepithelial current pulses. Raising [K$^+$] causes a large depolarization of both membranes and a mucosa-negative change in V_{ms}. These effects are consistent with a dominant apical membrane P_{K^+} and a junctional $P_{K^+} > P_{Na^+}$. Lowering [Na$^+$] causes a large mucosa-positive change in V_{ms}, a large hyperpolarization of V_{mc}, and virtually no change of V_{cs}. These results indicate that at both the junctions and the apical membrane $P_{Na^+} > P_{TMA^+}$. Lowering [Cl$^-$] causes cell membrane hyperpolarization, a result which is contrary to the expectation for an electrodiffusive Cl$^-$ pathway. Lowering [HCO$_3^-$] at constant PCO$_2$ causes a slight cell membrane depolarization. [From L. Reuss, *Physiol. Rev.* **69**, 503 (1989), with permission.]

gall bladder[14-16] imposes restrictions in its use, particularly when the ionic substitution causes cell membrane depolarization.

In special cases the study of cell membrane ion selectivity is simplified considerably. When one of the cell membrane resistances is very high compared to the other one, the circuit equations become more manageable. For instance if $R_a/R_b > 10$, the contribution of E_a to the basolateral membrane voltage is small, and V_{sc} is approximately

$$V_{cs} \simeq \frac{E_b(R_a + R_s)}{R_a + R_b + R_s} \tag{12}$$

and since $R_a \gg R_b$, the equation reduces further to

$$V_{cs} \simeq E_b \tag{13}$$

Under these conditions, the ionic selectivity of the basolateral membrane can be assessed directly from the changes in V_{cs} elicited by changes in basolateral solution ion concentrations, which can be analyzed by the Goldman–Hodgkin–Katz equation.[17,18]

$$V_{cs} \simeq \frac{RT}{zF} \ln \left[\frac{[K^+]_o + (P_{Na^+}/P_{K^+})[Na^+]_o + (P_{Cl^-}/P_{K^+})[Cl^-]_i}{[K^+]_i + (P_{Na^+}/P_{K^+})[Na^+]_i + (P_{Cl^-}/P_{K^+})[Cl^-]_o} \right] \tag{14}$$

We have used this approach to study the selectivity of the basolateral membrane of tissues incubated in 10 mM HCO$_3^-$/1% CO$_2$-Ringer solution, in which $R_a \gg R_b$,[15] and the selectivity of the apical membrane in tissues exposed to elevated levels of cAMP, in which the apical membrane resistance falls by induction of a large Cl$^-$ conductance (and $V_{mc} \simeq E_a$).[19]

Measurement of Net Fluid Transport

A particular application of conventional microelectrodes is the measurement of the rate of net fluid transport.[20] The gall bladder is mounted apical side up, covered with a thin layer of Ringer solution, and overlaid with mineral oil. A conventional microelectrode is advanced through the oil layer until it makes contact with the aqueous solution, closing the circuit. The change in height of the fluid layer (Δh) can be calculated by recording the position of the microelectrode. If the aqueous fluid compartment is a cylinder, the rate of fluid absorption (for instance in

[14] J. F. Garcia-Díaz, W. Nagel, and A. Essig, *Biophys. J.* **43**, 269 (1983).
[15] J. S. Stoddard and L. Reuss, *J. Membr. Biol.* **103**, 191 (1988).
[16] Y. Segal and L. Reuss, *J. Gen. Physiol.* **95**, 791 (1990).
[17] D. E. Goldman, *J. Gen. Physiol.* **27**, 37 (1943).
[18] A. L. Hodgkin and B. Katz, *J. Physiol. (London)* **108**, 37 (1949).
[19] K.-U. Petersen and L. Reuss, *J. Gen. Physiol.* **81**, 705 (1983).
[20] L. Reuss, *J. Gen. Physiol.* **84**, 423 (1984).

$\mu l \cdot cm^{-2} \cdot hr^{-1}$, which is equivalent to $\mu m \cdot hr^{-1}$) can be calculated according to Eq. (15):

$$J_v = \Delta h / \Delta t \qquad (15)$$

This method has the advantages that the measurements are performed under experimental conditions closer to those used in intracellular microelectrode experiments, and that the time resolution is better than that of other techniques such as gravimetric and volumetric.[21]

Ion-Selective Microelectrode Techniques

Ion-selective electrodes are used to measure intra- and extracellular ionic activities.[1,2,4,5] Subtracting the voltage output of the reference microelectrode from that of the ion-selective microelectrode yields a voltage proportional to the chemical activity of the ion of interest. We have used ion-selective microelectrodes to measure intracellular activities of Na^+, K^+, H^+, Cl^-, and tetramethylammonium (TMA^+). Among other applications, measurements of intracellular Na^+, Cl^-, and H^+ activities, combined with determinations of the changes in mucosal solution pH, have allowed us to establish the mechanism of salt uptake by the apical membrane of *Necturus* gall bladder epithelium and to specify aspects of its regulation. Microelectrodes sensitive to quaternary ammonium compounds have been used to measure cell volume changes and to study the mechanism of transepithelial water transport.

To estimate intracellular ion activities, we have used simultaneous impalements with two single-barrel microelectrodes, or impalement with one double-barrel microelectrode. Since the epithelial cells in *Necturus* gall bladder are tightly coupled electrically, the steady state membrane voltage and its changes are uniform in neighboring cells. Therefore, it is possible to obtain reliable measurements of intracellular ionic activities in this preparation by impaling cells with two single-barrel microelectrodes, conventional and ion selective, separated from each other by less than ca. 600 μm. However, recently we have preferred the use of double-barrel microelectrodes, which reduces experimental error. If technical difficulties make simultaneous impalements with single-barrel microelectrodes necessary, these impalements must be validated. Basically, we use two validation criteria: (1) the steady state voltage deflections produced across the cell membranes upon passage of a transepithelial current pulse must be equal in the two impaled cells, and (2) brief mucosal solution substitutions of Na^+ with K^+ must result in rapid membrane voltage changes of the same amplitude in both cells. The first criterion is used in every impalement,

---

[21] J. M. Diamond, *in* "Handbook of Physiology" (W. Heidel and C. F. Cope, eds.), p. 2451. American Physiological Society, Washington, D.C. 1968.

and the second one occasionally; the latter assumes that the intracellular activity of the measured ion does not change appreciably during brief exposure of the cells to the high-K^+ solution.

For ion-selective microelectrodes, the input impedance of the electrometer must be $>10^{14}$ Ω and the leakage current $<10^{-14}$ A, because the resistance of these microelectrodes is $>10^9$ Ω, usually $80-200 \times 10^9$ Ω. Conventional microelectrodes or reference barrels of double-barrel microelectrodes are connected to an electrometer appropriate for conventional microelectrodes (see Electrophysiologic Apparatus). Data acquisition is done as described for conventional microelectrodes.

Preparation of Ion-Selective Microelectrodes

Intracellular Microelectrodes. Pipets for single-barrel microelectrodes are pulled as described for conventional microelectrodes and have similar tip resistances when filled with 3 M KCl.[22] Once pulled, the pipets are positioned tip up on a perforated metal base sitting in a petri dish. The assembly is covered by an inverted glass jar and placed in an oven at 200° for $1-2$ hr. This procedure dehydrates the glass surface, thereby facilitating its silanization, necessary both to render the inner glass surface hydrophobic prior to filling with ion-selective sensors, which are also hydrophobic, and to reduce electrical shunting along the inner surface of the glass. For silanization, hexamethyldisilazane ($0.1-0.2$ ml; Sigma Chemical Co., St. Louis, MO), is ejected from a syringe onto the hot metal plate, thus permitting exposure of the pipets to silane vapors for at least 60 min at 200°. The assembly containing silanized pipets is cooled to room temperature prior to beginning the filling procedure.

Using a microliter syringe with a 2-in. 30-gauge needle, pipet tips are filled from the back with approximately $0.5-1.0$ μl of the appropriate sensor (see Table II). If the pipets are left covered in an horizontal position for at least 15 min, completion of tip filling will generally occur spontaneously. Final length of the sensor columns is usually about $500-1000$ μm. Any air bubbles present can be removed by gentle heating with a microforge and/or by introducing thin fibers (e.g., a hand-pulled glass fiber) through the shanks. At this point, pipets are backfilled with the appropriate reference solution (see Table II), and Ag-AgCl wires are inserted into the electrodes (see the section, Preparation of Conventional Microelectrodes).

Pipets for double-barrel ion-selective microelectrodes are pulled from double-barrel (laterally fused) glass with inner filament (each barrel 1.0-mm o.d., 0.43-mm i.d.; Hilgenberg, Malsfeld, Federal Republic of Germany) on a horizontal puller (PD-5, Narishige). The double-barrel glass is

[22] L. Reuss, P. Reinach, S. A. Weinman, and T. P. Grady, *Am. J. Physiol.* **244**, C336 (1983).

TABLE II
SENSORS AND FILLING SOLUTIONS FOR ION-SELECTIVE MICROELECTRODES

ISM[a]	Sensor[b]	Filling solution	
		ISM	Ref. barrel[c]
K^+	5 mg potassium tetrakis (p-chlorophenylborate) + 0.1 ml 3-nitro-O-xylene	NaCl-Ringer	1 M sodium formate/10 mM KCl[d]
TMA^+/TBA^+	5 mg potassium tetrakis (p-chlorophenylborate) + 0.1 ml 3-nitro-O-xylene[e]	NaCl-Ringer	1 M sodium formate/10 mM KCl
Cl^-	Corning Cl^- exchanger 477913	NaCl-Ringer	1 M sodium formate/10 mM KCl
H^+	Hydrogen ionophore I-cocktail A	40 mM KH_2PO_4, 23 mM NaOH, 15 mM NaCl	3 M KCl
Na^+	Sodium ionophore I-cocktail A	NaCl-Ringer	3 M KCl

[a] ISM, Ion-selective microelectrodes.

[b] Sensors for the K^+- and Cl^--selective microelectrodes are from Corning Medical (Medfield, MA). Sensors for H^+- and Na^+ selective microelectrodes are from Fluka (Ronkonkoma, NY). The commercial cocktails are ready to use, but appropriate mixtures can be prepared from the ion sensors and other compounds (see Reuss et al.[22]) as indicated for the K^+- and TMA^+/TBA^+-selective microelectrodes.

[c] Ref. barrel, Reference barrel of the double-barrel microelctrode.

[d] One molar sodium formate/10 mM KCl is used for double-barrel microelectrodes; 3 M KCl can be used in the single-barrel reference microelectrode.

[e] Ten percent (w/v) polyvinylchloride and tetrahydrofuran (20%, v/v) are used only in extracellular electrodes.

heated, twisted 360°, allowed to cool for 20 sec, and then pulled.[23] In order to adjust the puller settings, it is convenient to fill some micropipets with 3 M KCl and check the tip resistances, which should be 50–100 MΩ. After pulling, the back of one of the barrels is broken by inserting a sharp blade between both barrels. The back end of the short barrel is touched with a droplet of distilled water, which rapidly fills the tip by capillarity, blocking it and thus preventing its silanization. The back end of the long barrel is then touched with a droplet of hexamethyldisilazane, which fills the tip by capillarity. Immediately, the microelectrodes are placed tip up on a preheated metal base sitting on a hot plate. The silane reacts with the glass of the long barrel while the water column, which takes longer to evaporate,

[23] L. Reuss, J. L. Constantin, and J. E. Bazile, Am. J. Physiol. 253, C79 (1987).

prevents silanization of the other barrel. During this procedure, a stream of hot air from a hair dryer is directed toward the micropipets to flush away the silane vapor that escapes from the tips. After 1–2 hr the micropipets are allowed to cool; then the silanized barrels are injected with the appropriate ion-selective sensor and the reference barrels with appropriate filling solutions (see below). Once the tips are filled and bubbles are removed with a microforge, the reference and the silanized barrels are backfilled. The appropriate filling solutions for each barrel vary according to the type of electrode and are listed in Table II. Finally, the double-barrel ion-selective microelectrodes are connected to electrometer probes with Ag-AgCl wires (see the section, Preparation of Conventional Microelectrodes).

Double-barrel TMA$^+$-selective microelectrodes[24] are constructed using the same procedure as for other ion-selective microelectrodes, but in most instances these microelectrodes do not sense low [TMA$^+$] (<20 mM) unless their tips are beveled or broken by impaling the tissue several times. Useful electrodes have resistances of 80–150 MΩ (reference barrel) and 5–10 GΩ (TMA$^+$-selective barrel) when measured in NaCl-Ringer solution. Some electrodes lose sensitivity to TMA$^+$ and/or have severe voltage drift during an impalement and have to be discarded. The measurements of TMA$^+$ activity can also be made with two single-barrel microelectrodes (see the section, Ion-Selective Microelectrode Techniques); however, due to the small signal (e.g., 20% change in cell volume results in ≃ 5-mV change in signal[25]) it is imperative that the differential voltage be stable and the noise level low.

Extracellular Ion-Selective Microelectrodes. Single-barrel ion-selective microelectrodes for measuring TMA$^+$ and tetrabutylammonium (TBA$^+$), are prepared as described above, except that the tips are broken to a diameter of ≃ 2–3 μm before silanization. To this end, the pipet is advanced, under microscopic observation, against a polished stainless steel surface. Afterward, the pipets are silanized and filled with the sensor (see Table II) as described above (see the section, Preparation of Intracellular Ion-Selective Microelectrodes). Polyvinylchloride (PVC) is added to the sensor to polymerize the resin and prevent its loss from the large tip of the microelectrode. The pipets are placed in a jar with desiccant for 36–48 hr to allow the PVC to polymerize. Electrodes are backfilled with NaCl-Ringer's solution (see Table II). Resistances range from 1 to 5 GΩ when immersed in NaCl-Ringer solution.

Extracellular pH electrodes are constructed by fusing Corning pH-sensitive glass (Microelectrodes, Inc., Londonderry, NH) to the tips of lead–

---

[24] E. Neher and H. D. Lux, *J. Gen. Physiol.* **61,** 385 (1973).
[25] C. U. Cotton, A. M. Weinstein, and L. Reuss, *J. Gen. Physiol.* **93,** 649 (1989).

glass pipets. The lead glass (2.0-mm o.d., 1.3-mm i.d.; Corning 0120) can be pulled by hand or with a vertical puller (model 51511, Stoelting Co.) to produce a pipet having a tip diameter of <0.5 mm. A small amount of pH-sensitive glass is melted onto the wire of a microforge and is subsequently fused to the pipet tip. The heat is rapidly increased, and a thin bulb of pH-sensitive glass is formed by forcing air from a syringe which is attached to the back of the lead-glass pipet with Silastic tubing (Dow Corning). This procedure is carried out under microscopic observation to obtain a bulb diameter of 0.5–1.0 mm. The electrodes are backfilled with a solution containing 5 mM NaH$_2$PO$_4$/Na$_2$HPO$_4$ plus 100 mM NaCl, pH 7.0. Bubbles trapped in the neck and bulb can be removed by inserting a thin glass filament into the bulb, by heating the air bubble with the microforge, or by gently tapping the barrel. An Ag-AgCl wire is inserted and sealed with wax. When stored in a low-pH buffer these electrodes have a lifetime of several months. Acceptable slopes and response times are obtained when electrodes have resistances of 5 GΩ or less.

Calibration of Ion-Selective Microelectrodes

Intracellular ion activities, in the case of K$^+$, H$^+$, and Cl$^-$, can be calculated using the Nicolsky–Eisenman equation[26]:

$$V_i^* = E_0 + S \log(a_i + k_{ij} a_j) \qquad (16)$$

where V_i^* is the measured voltage difference between the ion-selective microelectrode and the reference microelectrode, S is the Nernstian slope (ideally 2.303 RT/zF), a_i is the activity of the ion i, a_j is the activity of the interfering ion, and k_{ij} is the selectivity coefficient (for j over i). E_0 is a constant that depends on the temperature and comprises the junction potential and the reference electrode potentials. Activity coefficients are calculated by the Debye–Hückel formalism.[27]

The method of calibration depends on the characteristics of the sensor and the conditions of use of the electrodes. If S and k_{ij} are constant, i.e., independent of ion substitutions, k_{ij} can be determined by single-ion substitutions, and calculation of a_i is done according to Eq. (16) (see details below, for K$^+$ and Cl$^-$-selective microelectrodes). Further, if k_{ij} is near 0 (see details below for pH-sensitive electrodes) the term $k_{ij} a_j$ is negligible, and Eq. (16) reduces to the Nernst equation. On the other hand, if S or k_{ij} are not constant (which is the case with Na$^+$-selective microelectrodes) the

[26] G. Eisenman, "Glass Electrodes for Hydrogen and Other Cations, Principles and Practice." Dekker, New York, 1967.

[27] D. Ammann, "Ion Selective Microelectrodes." Springer-Verlag, Berlin, 1986.

behavior of the electrodes is not described adequately by Eq. (12), and empirical calibrations are recommended. These calibrations are best carried out with solutions mimicking those in which the measurement will be performed. Specific details of calibration procedures are given below.

Potassium Ion Microelectrodes. Since the selectivity coefficient of the liquid-exchanger-based K^+-selective microelectrodes for K^+ over Na^+ is high ($k_{K^+,Na^+} \simeq 0.02$) and $a_{K^+}^i$ (intracellular K^+ activity) is high relative to $a_{Na^+}^i$ the error in the calculation of $a_{K^+}^i$ resulting from neglecting $a_{Na^+}^i$ is very small, <1 mM at $a_{Na^+}^i = 10$ mM, to about 1.5 mM when $a_{Na^+}^i = 75$ mM. However, the contribution of $a_{Na^+}^o$ (extracellular Na^+ activity) to $V_{K^+}^*$ must be taken into account. Note that for $a_{Na^+}^o = 75$ mM and $a_{K^+}^o = 2$ mM, the electrode "senses" an activity equivalent to 3.5 mM K^+ ($2 + 0.02 \cdot 75$). The slope of the K^+-selective microelectrodes can be determined in pure KCl solutions by varying $[K^+]$ from 0 to 150 mM, taking into account the changes in activity coefficient with variations in ionic strength.[27] The selectivity coefficient k_{K^+,Na^+} can then be obtained by mixing KCl and NaCl ($[KCl] + [NaCl] = 100$ mM), and determining the best fit to the Nicolsky–Eisenman equation.

Chloride Ion Microelectrodes. Chloride ion-selective microelectrodes can be calibrated in pure NaCl solutions, with appropriate correction for ionic strength (see above). Since k_{Cl^-,HCO_3^-} is about 0.05, the HCO_3^- interference is negligible at $a_{Cl^-}^i$ greater than 15 mM but becomes significant when $a_{Cl^-}^i$ is low, $a_{HCO_3^-}^i$ is high, or both. Changing $[Cl^-]$ in the calibration solutions in the presence of a fixed HCO_3^- background is one possible way to calculate k_{Cl^-,HCO_3^-}. Usually, an $a_{HCO_3^-}$ close to the intracellular value is chosen and a_{Cl^-} is modified changing $[KCl]$ or $[NaCl]$ in the solutions. k_{Cl^-,HCO_3^-} can be calculated from the best fit to the Nicolsky–Eisenman equation. From k_{Cl^-,HCO_3^-} and $a_{HCO_3^-}^i$, $a_{Cl^-}^i$ can be calculated according to Eq. (16). The $a_{HCO_3^-}^i$ can be estimated from the intracellular pH assuming that intra- and extracellular pCO_2 values are the same.

pH Electrodes. Calibration of the pH-sensitive electrodes is simple because of the absence of significant interfering ions. They can be calibrated with external solutions of pH between 6 and 8 buffered with *N*-2-hydroxyethylpiperazine-*N'*-2-ethanesulfonic acid (HEPES).

Sodium Ion Microelectrodes. Sodium ion-selective microelectrodes based on neutral carriers exhibit significant K^+ and Ca^{2+} interferences. We calibrate these electrodes in nominally Ca^{2+}-free solutions, with a fixed KCl concentration (K^+ activity $\simeq a_{K^+}^i$, i.e., 90 mM), varying NaCl from 0 to 50 mM. The $a_{Na^+}^i$ is determined empirically by interpolation onto the calibration curve obtained for each particular electrode. This curve represents the difference between $V_{Na^+}^*$ in every calibration solution and $V_{Na^+}^*$ in Ringer (usually NaCl-Ringer or low-buffer NaCl-Ringer, see Table I).

The Ca^{2+} interference is significant in the extracellular fluid, because $k_{Na^+,Ca^{2+}} > 1$, but the interference upon impalement is negligible ($a^i_{Ca^{2+}} \simeq 100-200$ nM).

TMA$^+$/TBA$^+$ Electrodes. Quaternary ammonium-selective electrodes are made with the same ion exchanger as the K$^+$-selective microelectrodes but, given the high selectivity of this sensor for quarternary ammonium compounds, interferences are negligible. Selectivity ratios are: TMA$^+$/K$^+$ = 10^2-10^3, TMA$^+$/Na$^+$ = 10^3-10^4, TBA$^+$/Na$^+$ or TBA$^+$/K$^+$ > 10^6. We calibrate intracellular TMA$^+$-selective microelectrodes with solutions of 120 mM KCl and TMACl ranging from 0.1 to 20 mM. Extracellular electrodes are calibrated with NaCl-Ringer plus 0.5 to 20 mM TMACl or TBACl.

Examples of Experimental Uses of Ion-Selective Microelectrodes

Ion-selective microelectrodes have been used successfully in *Necturus* gall bladder epithelium to investigate ion and water transport mechanisms and their regulation. Specific applications include (1) measurement of steady state intracellular ionic activities and definition of electrochemical driving forces across individual cell membranes,[28] (2) measurement of changes in ionic activities on experimentally modifying the electrochemical driving force,[20,29] (3) estimation of net ion fluxes across the apical or the basolateral membrane, using the "initial rates" of change in intracellular ionic activities upon rapidly changing the electrochemical driving force; this approach has been used to estimate the contribution of specific transport mechanisms at the apical and basolateral membranes to transepithelial salt absorption[20,23,30] and to carry out kinetic studies of the apical membrane Na$^+$/H$^+$ and Cl$^-$/HCO$_3^-$ exchangers,[31-33] (4) determination of the effects of regulatory (e.g., çAMP) and pharmacological agents (e.g., ouabain)[19,32-34] on intracellular ionic activities and ionic fluxes, (5) measurement of changes in mucosal solution pH under control conditions, on ionic substitutions, and after exposure to regulatory or pharmacological agents,[12,32,33,35] (6) estimation of the thickness of the unstirred fluid layer on the apical surface of the tissue,[36] (7) measurements of changes in cell

[28] L. Reuss and S. A. Weinman, *J. Membr. Biol.* **49**, 345 (1979).

[29] L. Reuss and J. L. Costantin, *J. Gen. Physiol.* **83**, 801 (1984).

[30] L. Reuss and J. S. Stoddard, *Annu. Rev. Physiol.* **49**, 35 (1987).

[31] G. A. Altenberg and L. Reuss, *J. Gen. Physiol.* **95**, 369 (1990).

[32] L. Reuss, *J. Gen. Physiol.* **90**, 172 (1987).

[33] L. Reuss and K.-U. Petersen, *J. Gen. Physiol.* **85**, 409 (1985).

[34] L. Reuss, *Physiol. Rev.* **69**, 503 (1989).

[35] S. A. Weinman and L. Reuss, *J. Gen. Physiol.* **80**, 299 (1982).

[36] C. U. Cotton and L. Reuss, *J. Gen. Physiol.* **93**, 631 (1989).

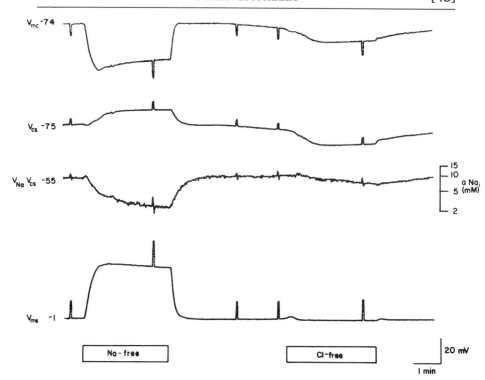

FIG. 6. Effects of mucosal Na⁺ removal (left) and mucosal Cl⁻ removal (right) on membrane voltages and intracellular Na⁺ activity. Note the large difference in the rates of fall of cell Na⁺ in response to these perturbations. [From L. Reuss, *J. Gen. Physiol.* **84,** 423 (1984), with permission.]

volume on changing osmolality of bathing solutions,[25,37] and estimation of hydraulic water permeability of apical and basolateral membranes.[25]

In the next section, we briefly illustrate a few applications of ion-selective electrodes to address problems pertaining to transepithelial salt and water transport. Further details can be found in the references listed above.

Identification of Ion Transport Mechanisms. We have used ion-selective microelectrodes to identify ion transport mechanisms at the apical and the basolateral cell membranes. Typical examples are the studies of the apical membrane transport mechanisms involved in salt reabsorption. To ascertain whether NaCl uptake across the apical membrane is via cotransport (Na^+/Cl^- or $K^+/Na^+/Cl^-$), or via ion exchange (e.g., parallel Na^+/H^+ and Cl^-/HCO_3^- exchangers), we carried out experiments with

[37] L. Reuss, *Proc. Natl. Acad. Sci. U.S.A.* **82,** 6014 (1985).

intracellular Na^+, Cl^-, and pH-sensitive microelectrodes, and extracellular pH-sensitive macroelectrodes. As shown in Fig. 6, removing mucosal solution Na^+ reduces $a^i_{Na^+}$ rapidly and reversibly; the initial rate of decrease in $a^i_{Na^+}$ upon mucosal solution Cl^- removal is much slower than upon Na^+ removal. Similarly, the decrease in $a^i_{Cl^-}$ is much faster when Cl^- is removed than when Na^+ is replaced with tetramethylammonium.[20] These results indicate that NaCl entry is not by direct coupling between Na^+ and Cl^- fluxes (Na^+/Cl^- or $Na^+/K^+/Cl^-$ cotransport), but via parallel Na^+ and Cl^- pathways. Other experiments indicate that these parallel pathways are Na^+/H^+ and Cl^-/HCO_3^- exchangers.[20,29,34,35,38]

Studies of Water Transport. Changes in cell water volume can be measured with good temporal resolution and sensitivity by introducing an impermeant ion into the cells and measuring its activity with an ion-selective microelectrode. We have used TMA^+ as the probe, and TMA^+-selective microelectrodes to monitor its intracellular activity during experimental perturbations.[25,37]

The normally impermeant cation TMA^+ is loaded into the cytosol by transient exposure of the apical cell membrane to nystatin in the presence of mucosal $(TMA)_2SO_4$ Ringer solution. Sulfate does not penetrate nystatin pores, and is therefore used in order to prevent cell swelling. The loading procedure can be monitored by observing the membrane depolarization and the fall in the cell membrane resistance ratio or by measuring intracellular $[TMA^+]$ with a microelectrode.[25] After nystatin removal, mucosal perfusion with $(TMA)_2SO_4$-Ringer solution is continued until membrane voltages and R_a/R_b recover. Under these conditions, the intracellular $[TMA^+]$ is usually between $5-15$ mM, which can be accurately sensed with insignificant interference. The application of this technique to measure changes in cell volume is illustrated in the bottom panels of Figs. 7 and 8.

To calculate cell membrane osmotic water permeability, the time course of the change in osmolality at the cell surface must be known. Since the diffusion coefficients for TBA^+ and sucrose are similar, TBA^+ can be used as a tracer to estimate changes in sucrose concentration. For instance, if the bulk solution [sucrose] and $[TBA^+]$ are simultaneously increased, the time courses for arrival of each solute at the apical cell membrane should be similar. Therefore, by measuring the TBA^+ activity adjacent to the cell membrane we can calculate the change in [sucrose] and thus the osmolality as a function of time after the solution change. In principle, a variety of osmoticant/probe pairs could be used if their diffusion coefficients are similar and the extracellular ion-selective microelectrode has sufficient

[38] S. A. Weinman and L. Reuss, *J. Gen. Physiol.* **83,** 57 (1984).

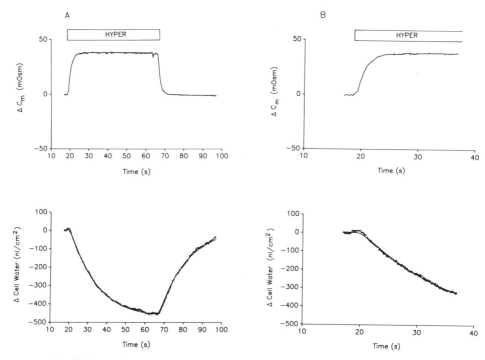

FIG. 7. Changes in mucosal solution osmolality, cell water volume, and the fitted change in cell water volume in response to exposure to a hyperosmotic mucosal solution. (A) The mucosal solution osmolality was increased from 206 to 244 mOsm/kg water. The initial cell water volume was assumed to be 2870 nl/cm^2. The data for the complete experiment ($\simeq 70$ sec) were fit to the model described by C. U. Cotton, A. M. Weinstein, and L. Reuss, *J. Gen. Physiol.* **93**, 649 (1989). The estimates of apical and basolateral hydraulic permeabilities are 0.87×10^{-3} and 0.09×10^{-3} cm sec^{-1} (Osm/kg)$^{-1}$, respectively. (B) Same experiment as above except that the fit spans only 20 sec. The estimates of apical and basolateral hydraulic permeabilities are 0.90×10^{-3} and 0.10×10^{-3} cm sec^{-1} (Osm/kg)$^{-1}$, respectively. [From C. U. Cotton, A. M. Weinstein, and L. Reuss, *J. Gen. Physiol.* **93**, 649 (1989), with permission.]

selectivity and adequate time response for the probe molecule. Figures 7 and 8 show simultaneous records of experimentally induced changes in osmolality at the apical cell surface and the elicited changes in cell water volume. Apical and basolateral membrane osmotic water permeabilities can be estimated from data such as those in Figs. 7 and 8.[25]

Patch-Clamp Techniques

To identify and characterize ion channels which underlie macroscopic conductive pathways in *Necturus* gall bladder epithelium, we utilize the patch-clamp technique. The method is based in the formation of a high-

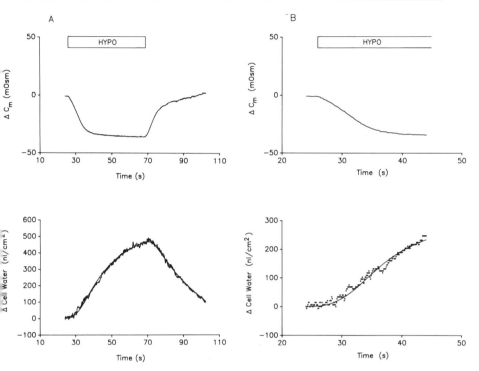

FIG. 8. An example of the change in mucosal solution osmolality, cell water volume, and the fitted change in cell water volume in response to exposure to a hyposmotic mucosal solution. (A) The mucosal solution osmolality was decreased from 206 to 163 mOsm/kg water. The initial cell water volume was assumed to be 2870 nl/cm². The data for the complete experiment ($\simeq 70$ sec) were fit to the model described by C. U. Cotton, A. M. Weinstein, and L. Reuss, *J. Gen. Physiol.* **93**, 649 (1989). The estimates of apical and basolateral hydraulic permeabilities are 0.72×10^{-3} and 0.08×10^{-3} cm sec⁻¹ (Osm/kg)⁻¹, respectively. (B) Same experiment as above except that the fit spans only 20 sec. The estimates of apical and basolateral hydraulic permeabilities are 0.69×10^{-3} and 0.17×10^{-3} cm sec⁻¹ (Osm/kg)⁻¹, respectively. [From C. U. Cotton, A. M. Weinstein, and L. Reuss, *J. Gen. Physiol.* **93**, 649 (1989), with permission.]

resistance seal between the tip of a micropipet and a patch of membrane, providing an electrical and a chemical barrier between the inside and the outside of the pipet. When the seal resistance exceeds ca. 5 GΩ ("giga-seal"), currents through single-ion channels can be recorded without major interference from background noise. For channel types amenable to study by patch-clamp techniques, information on channel selectivity, conductive properties, and regulation of intracellular and/or extracellular mediators can be obtained. For reviews of the technique and its application to epithelial tissues, we recommend Sakmann and Neher[3] and Rae and Levis.[39]

Preparation of Micropipets

Micropipets are made from Corning 7052 capillary glass (1.5-mm o.d., 1.0-mm i.d.; Friedrick and Dimmock, Millville, NJ), borosilicate glass (1.5-mm o.d., 1.1-mm i.d.; TW150F-6, WPI), or from blue coded tip microhematocrit glass (1.4-mm o.d., 1.2-mm i.d.; Fisher Scientific, Pittsburgh, PA). All capillary glass is cleaned before use. Micropipets are pulled in two stages on a vertical puller (PP-83, Narishige). The first pull lengthens the pipet by 9 mm, and leaves a central, narrowed region 0.2–0.3 mm in outer diameter. The second pull separates two pipets with tip diameters of 1–2 μm. To reduce noise associated with the capacitance of the pipet wall, a thin coat of Sylgard 184 (Dow Corning) is applied with a needle, extending 1 to 10 mm from the pipet tip. Sylgard is cured by placing the coated part of the pipet in the center of a heated coil, until it hardens without cracking. Soft glasses have the advantage that the pipet tips do not have to be fire polished. However, soft glass has a lower bulk resistivity and releases di- or polyvalent cations that might alter channel properties.[40] We currently prefer borosilicate glass. These pipets are fire polished under microscopic observation at ×400, passing current pulses through a glass-coated platinum/iridium wire positioned about 100 μm from the pipet tip. Pipets made from Corning 7052 or TW150F-6 borosilicate glass are Sylgard coated and fire polished; pipets made from soft glass are only Sylgard coated. Useful pipets have resistances of 7–10 MΩ with 10 mM HEPES-Ringer in the bath and KCl-Ringer in the pipet.

General Experimental Procedure

Patch pipets are mounted on an ad hoc holder with a port for applying suction; the holder is connected to the headstage of the LIST EPC-7 patch clamp. The headstage is mounted on a three-axis hydraulic micromanipulator (MO-103, Narishige). The chamber with the preparation (Figs. 2 and 9) is positioned on the stage of an inverted microscope which sits on a vibration-isolation table. The microscope and the table are enclosed in a Faraday cage.

To obtain a seal, the pipet is placed onto the cell under microscopic observation until the cell membrane and the glass tip are in contact and then suction is applied to the pipet interior. During seal formation, resist-

[39] J. L. Rae and R. A. Levis, *Mol. Physiol.* **6,** 115 (1984).
[40] G. Cota and C. M. Armstrong, *Biophys. J.* **53,** 107 (1988).

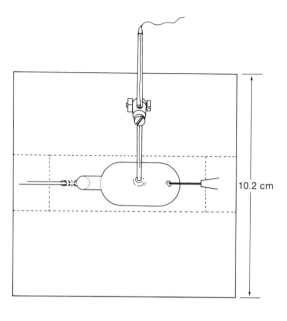

10.2 cm

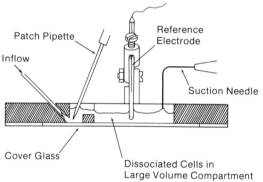

Patch Pipette

Inflow

Reference
Electrode

Suction Needle

Cover Glass

Dissociated Cells in
Large Volume Compartment

Fig. 9. Experimental chamber for patch clamping of basolateral membranes or dissociated epithelial cells. Two compartments (large = 5 ml and small = 0.5 ml) separated by a 0.5-cm partition are milled into a block of Lucite. The chamber is also equipped with a flexible holder for the reference electrode, and a tunnel for fluid inflow. A glass microscope slide glued to the bottom permits microscopic observation. *Top:* Top view. *Bottom:* Cross-sectional view taken from the side of the chamber. Scaling is approximate. The chamber is depicted during superfusion of an excised membrane patch. Fluid enters into the small compartment via the inflow port, and flows over the partition into the large compartment, where it is removed by continuous aspiration through a stainless steel needle connected to vacuum. Seals are formed on basolateral preparations or dissociated cells in the large compartment, during a brief cessation of superfusion. After excision, the flow into the small compartment is resumed. Once overflow establishes a fluid bridge between the two compartments, the membrane patch is transferred to the small compartment, where it is superfused with the test solutions at a rate of 3–5 ml/min. The chamber was designed by Daniel Lang (Department of Physiology and Biophysics, University of Texas Medical Branch, Galveston, TX).

ance at the pipet tip is assessed by monitoring the currents elicited by voltage pulses of 1–50 mV, applied to the stimulator input of the patch clamp with a Pulsar 7+ stimulator (Frederick Haer and Co., Brunswick, ME). Success in the formation of gigaseals varies with the preparation. Gigaohm seals are rarely obtained on the apical membrane of intact *Necturus* gall bladder epithelium but occur in ca. 30% of the attempts when patching dissociated cells. Dissociated cells have the disadvantage of the uncertainty of the channel origin (apical or basolateral) but they are an excellent preparation to characterize the channels. A preparation that allows for a high yield of gigaseal formation in basolateral membrane has been recently developed (see below).

Single-channel currents can be recorded in three configurations. In the cell-attached configuration, currents are recorded *in situ* with the membrane patch remaining continuous with the tissue preparation. The cell-free inside-out configuration is achieved by rapidly withdrawing the patch pipet. Here, the intracellular aspect of the membrane patch is exposed to the experimental bath, whose composition can be changed in order to study channel conduction and/or regulation. The cell-free, outside-out mode is achieved by withdrawing a cell-attached patch from the cell, awaiting membrane vesicle formation at the pipet tip, and selectively disrupting the portion of the vesicle facing the pipet solution. Here, the extracellular face of the membrane is exposed to the bath. See Sakmann and Neher for further details.[3]

The composition of the solutions varies according to the design of the experiment. The solutions used most frequently are depicted in Table III. In general, for studies in the cell attached configuration the preparation is bathed in 10 mM HEPES-Ringer. KCl-Ringer and potassium gluconate-Ringer solutions are used to determine anion conduction in excised patches (pipets filled with KCl solution). Current–voltage (I-V) relations in inside-out patches are measured with "physiological", low-K^+, high-Na^+ solution (pipets contain 10 mM HEPES-Ringer), and the Ca^{2+} sensitivity or pH effects are examined using Ca^{2+}-buffered and pH-buffered solutions, as appropriate.

Patch-clamp experiments in *Necturus* gall bladder epithelial cells have been carried out on apical and basolateral membranes of the epithelium as well as on dissociated cells.

Patch-Clamp Experiments in Apical Membrane

The gall bladder is mounted apical side up in the modified Ussing chamber used for microelectrode studies (Fig. 2). Successful experiments require well-stretched tissues and enzymatic treatment. The apical surface

TABLE III
SOLUTIONS USED IN PATCH-CLAMP EXPERIMENTS

	10 mM HEPES-Ringer	KCl-Ringer[a]	Potassium gluconate-Ringer	Low Na+, high K+ Ringer	Ca2+-buffered Ringer
[Na+]	104.0	0	0	10.0	10.0
[K+]	2.5	106.5	106.5	96.5	96.5
[Mg2+]	1.0	1.0	1.0	1.0	1.0
[Ca2+	1.0	1.0	1.0	1.0	Variable[b,c]
[Cl−]	106.0	106.0	4.0	106.0	Variable[d]
[Gluconate]	—	—	102.0	—	—
[HEPES][e]	10.0	10.0	10.0	10.0	10.0
[Sucrose][f]	10.0	10.0	10.0	Variable	Variable
EGTA or BAPTA	—	—	—	—	3.0

[a] This solution, omitting CaCl$_2$, is the pipet-filling solution used in most experiments. The osmolarity of the pipet solution is generally 5 – 10% lower than that of the bathing solutions.

[b] At a fixed EGTA (ethylene glycol-O-O'-bis(2-aminoethyl)-N,N,N',N'-tetracetic acid) or BAPTA (1,2-bis(2-aminophenoxy) ethane-N,N,N,$'N'$-tetraacetic acid) concentration, Ca^{2+} is added to obtain the desired free Ca^{2+} concentration. The total [Ca^{2+}] can be calculated from the stability constants, but it is advisable to measure free Ca^{2+} because of variations in the purity and water content of the chemicals.

[c] S. M. Harrison and D. M. Bers, *Biochim. Biophys. Acta* **925**, 133 (1987).

[d] Solution Cl$^-$ concentration varies between 92 and 98 mM, depending on Ca^{2+} buffer used and total Ca^{2+} concentration.

[e] Solution pH is generally adjusted to 7.4 with NaOH and KOH; if pH of Ca^{2+}-buffered solutions has to be adjusted to different values, we recommend BAPTA over EGTA, because of the relative pH insensitivity of its Ca^{2+}-binding constants.

[f] Sucrose is added in order to adjust the osmolality of the solutions to 230 mOsm/kg.

of the tissue is exposed to 10 mM HEPES-Ringer containing 1 mg/ml hyaluronidas (type IV or V, Sigma Chemical Co.) for 1 hr, and then rinsed for at least 15 min. This treatment has no effect on the electrical properties of the tissue.[16] The mucosal bath is grounded through an Ag-AgCl wire immersed in a 10 mM HEPES-Ringer agar bridge. The liquid junction potentials developed during ion substitutions are calculated or measured by conventional methods[12] and corrected if necessary.

Patch pipets are gently lowered onto the apical surface; application of suction to the pipet interior generally results in the formation of a 50- to 100- MΩ seal. Gigaohm seals are formed infrequently.

Patch-Clamp Experiments in Basolateral Membrane

Gall bladders are pinned apical side up on a Sylgard-coated Petri dish. The epithelium is scraped with a polished stainless steel blade (edge ca. 7 mm in length). A continuous flap of epithelial cells is obtained by a single, smooth, horizontal movement of the blade, which is held vertically, applying gentle pressure on the epithelium. The flap is gently pushed onto a plastic coverslip coated with Cell-Tak (Collaborative Research, Inc., Bedford, MA; see below), apical side facing the coverslip, by applying a stream of Ringer solution to the basal side through a Pasteur pipet. The flap is then cut at its base with the coverslip, and transferred to the experimental chamber (Fig. 9). Complete epithelial sheets of up to 1 cm^2 can be obtained. Folds of the epithelial layer and portions that do not adhere to the coverslip are removed by applying suction through a 100-μm o.d. glass pipet mounted on a micromanipulator. This maneuver also removes adherent basement membrane, increasing markedly the success rate of seal formation. Seal resistances of 5 to 10 GΩ were obtained in ca. 10% of the attempts.[41]

The adherent surface is prepared by cutting a plastic coverslip (25 × 25 mm) into two identical right-angled trapezoids. About 8 μl of Cell-Tak is applied near the smallest edge (7 mm in length). The coverslips are air dried, rinsed in ethanol and water, air dried again, and stored at 4°. Best adhesion is obtained if they are used within a week

Patch-Clamp Experiments in Dissociated Cells

The gall bladders are pinned apical side up on a Sylgard-coated Petri dish as described above. One or several sheets of epithelial cells are scraped from the tissue with a glass coverslip and collected in a 15-ml centrifuge tube. The sheets are treated for 7 min with hyaluronidase (type IV or V, Sigma), 1 mg/ml in 10 mM HEPES-Ringer. Dispersed cells are centrifuged for 30 sec at 50 g, and the pellet is resuspended in 10 mM HEPES-Ringer and transferred to the large compartment of the experimental chamber shown in Fig. 9.

Four categories of cellular and extracellular material can be visualized: (1) isolated spherical cells, which settle rapidly to the bottom of the chamber, exclude the dye Trypan Blue, and are easily patchable, (2) broken cells, which settle to the bottom of the chamber, but do not exclude Trypan Blue, (3) epithelial sheets, made up of undissociated material, and (4) extracellular debris, visible as a gelatinous network, which floats over

[41] F. Wehner, L. Garretson, K. Dawson, Y. Segal, and L. Reuss, *Am. J. Physiol.* **258**, C1159 (1990).

the cellular material. In order to wash away the extracellular debris we perfuse the chamber with 10 mM HEPES-Ringer for 5–10 min at a rate of 10–20 ml/min.

Patch pipets are gently lowered onto the surface of the cells under microscopic observation. In most cases, the angle of approach is 45°, and seals are formed on the top hemisphere of the cells. Gentle suction is generally sufficient to form seals, but "hard" suction is sometimes necessary. Seal resistances in the range 5–20 GΩ are routinely obtained but more than 50% of the attempted excisions result in vesiculation of the patch at the pipet tip, evident as distortion or disappearance of single-channel currents visible in the cell-attached mode.[42] Vesicles can be disrupted by applying extreme holding voltages, briefly moving the pipet tip through the air–solution interface, or applying slight positive or negative pressure to the pipet tip. In most cases, channel activity cannot be recovered.

Inside-out patches are usually stable for 15–60 min and, in some cases, for over 2 hr. Based on the Ca^{2+} and voltage sensitivities and/or the effect of blockers we estimate that outside-out patches are obtained in about 10% of the excised patches.

Recording Techniques

Single-channel currents are amplified with a gain of 50 or 100 mV/pA. The output from the patch clamp is stored on video tape at wide bandwidth (0 to 20 kHz). For real-time observation, the output of the patch clamp is low-pass filtered at 1–2 kHz with an eight-pole Bessel filter, connected to an oscilloscope. The filter output is also connected to a strip chart recorder, which provides a convenient experimental log for subsequent analysis. The holding voltage of the pipet is continuously recorded on the digital instrument recorder and on the strip chart recorder.

Data Analysis

Single-channel records allow for analysis of basic properties of ion channels, such as single-channel conductance, ion selectivity, and gating. In addition, the method yields information about the rates of opening and closing, the presence of one or more conductance levels, and the effects of blockers on these properties. We summarize here only the methodology used to analyze the basic channel properties.

The ion selectivity of the channel can be determined from the reversal

[42] O. P. Hamill, A. Marty, E. Neher, B. Sakmann, and F. J. Sigworth, *Pfluegers Arch.* **391,** 185 (1981).

potential (V_r, voltage at which $i = 0$) and the ionic composition on both sides of the patch.[43] Information on single-channel conduction properties can be obtained from unitary current amplitudes, determined at different membranes voltages, under voltage-clamp conditions. Current–voltage (I-V) relations are usually determined manually. For each holding voltage, short segments of data containing well-resolved openings and closures are displayed onto a strip chart recorder (corner frequency 100 Hz) running at high speed (25 mm/sec), and single-channel currents are measured with or without computer assistance (see below). The unitary conductance of the channel at any membrane voltage is given by the slope of the I-V relation.

For patches containing one or more channels of the same type, single-channel open probability is estimated using segments of digitized data, 10–60 sec in duration, according to Eq. (17):

$$P_o = (1/N_c) \sum_{j=1}^{N_c} jP_j \qquad (17)$$

where j is a summation index, P_j is the fraction of the record during which j channels are open, and N_c is the total number of channels in the patch. This expression assumes that channels gate identically and independently, i.e., in a binomial fashion. Comparison between P_j values determined experimentally and those predicted from the binomial distribution reveals qualitative agreement for more than 80% of the multiple-channel patches. Channels in the remaining patches clearly deviate from binomial predictions.

Estimates of P_o, and some I-V relationships, are determined with computer assistance. Segments of data stored on video tape are low-pass filtered at 1–2 kHz and digitized using an analog-to-digital (A/D) conversion system. Operation of the A/D board is controlled by the ADCIN software routine (written by J. Pumplin). Data are transferred to the hard disk of an IBM-compatible microcomputer. When the parameter of interest does not depend strongly on filtering, e.g., estimation of single-channel P_o, the rate of digitization is equal to or twice the analog corner frequency. In those experiments in which it is essential to replicate accurately the filtered signal, e.g., study of drugs that induce flickering in the currents through the channels,[44] digitization rates are 5- to 10-fold higher than the corner frequency of the analog filter.

Analysis is carried out using SEGMENTS, FORTRAN package written at the Department of Physiology and Biophysics of the University of Texas

[43] B. Hille, "Ionic Channels of Excitable Membranes." Sinauer, Sunderland, Massachusetts, 1984.
[44] D. Colquhoun and F. J. Sigworth, in "Single Channel Recording" (B. Sakmann and E. Neher, eds.), p. 191. Plenum, New York, 1983.

Medical Branch (D. C. Eaton, D. Busath, and W. C. Law), for a VAX 11/750 computer (Digital Equipment Corporation, Maynard, MA). Data files are transferred from the microcomputer to the VAX computer using KERMIT, a data transfer and terminal emulation package (Columbia University Center for Computer Activities, New York, NY). Data files and SEGMENTS output files are transferred to magnetic tape for permanent storage. Numerous other software packages for single-channel analysis are available commercially.

Ion Channels in Gall Bladder Epithelium

A large-conductance (maxi) K^+ channel has been identified in the apical membrane using patch-clamp techniques. The channel is activated by depolarization and elevations in free Ca^{2+} concentrations on the cytosolic surface, and is blocked by Ba^{2+}. The permeability ratio (K^+/Na^+) is about 20, and the conductance in symmetrical KCl solutions is ca. 160 pS, and independent of voltage.

Since gigaseal formation is rarely obtained on the apical membrane of the intact epithelium, we have used dissociated cells to access apical membrane maxi-K^+ channels. The ion selectivity, unitary conductance, and Ba^{2+} sensitivity of maxi-K^+ channels in dissociated cells indicate that these channels are similar to those observed in the intact apical membrane, suggesting their apical origin.[16] The voltage dependence and Ca^{2+} activation of these channels, as well as the effects of blockers (TEA^+ and Ba^{2+}) from either side of the membrane, are easier to study in dissociated cells because of the higher yield of high-resistance seals. Figure 10 shows the results obtained in an excised patch from a dissociated epithelial cell. In symmetrical KCl solutions, the *I-V* relation was linear with a slope conductance of 199 pS and a reversal potential of 0 mV. There were no changes in either parameter when Cl^- was replaced with gluconate (after correction for liquid junction potential), demonstrating that the channel is anion impermeable.

From the single-ion conductance, the P_o at the appropriate membrane voltage, and the estimated channel density, we calculated that the maxi-K^+ channels account for only 17% of the apical membrane conductance under basal conditions. These channels are responsible for the increase in apical membrane conductance produced by membrane depolarization or elevation of intracellular Ca^{2+} levels.[16]

Preliminary studies on dissociated gall bladder epithelial cells have revealed a small-conductance (10 pS) Cl^- channel, evident after elevating intracellular cAMP levels.[45] These channels may underlie the cAMP-induced apical membrane conductance.[12]

[45] Y. Segal and L. Reuss, *FASEB J.* **3**, A862 (1989).

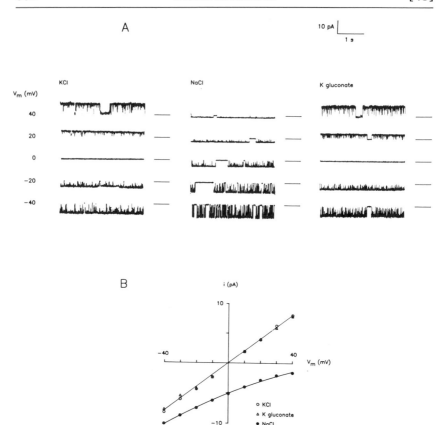

FIG. 10. Ion selectivity of a maxi-K$^+$ channel in an excised (inside out) patch from a dissociated cell. *I-V* relations were obtained with 100 m*M* KCl solution in the pipet. Bath solutions contained 1 m*M* Ca^{2+}. Voltages are referred to the pipet solution. (A) Channel activity in the presence of 100 m*M* KCl, 97.5 m*M* NaCl/2.5 m*M* KCl, or 100 m*M* potassium gluconate in the bath. Baselines are indicated to the right of the current records. (B). *I-V* relations obtained under the three sets of ionic conditions. Reducing internal K$^+$ from 100 to 2.5 m*M* shifted the reversal potential from 0 to >60 mV, while replacing KCl with potassium gluconate had no effect on the *I-V* relation. The lines represent fits to the Goldman–Hodgkin–Katz equation. [From Y. Segal and L. Reuss, *J. Gen. Physiol.* **95,** 791 (1990), with permission.]

A different high-conductance K$^+$ channel, activated by internal Ca^{2+} and blocked by internal Ba^{2+}, has recently been identified in the basolateral membrane.[41] This channel coexists with another K$^+$ channel of lower conductance, and with an anion-selective (Cl$^-$ conductive) channel. Characterization of these channels is under way. It is likely that they account for the baseline K$^+$ and Cl$^-$ conductances of the basolateral membrane.

Acknowledgments

Supported in part by National Institutes of Health Grants DK 38588 and DK 38734. G.A. Altenberg was supported by Consejo Nacional de Investigaciones Científicas y Técnicas de la República Argentina, and F. Wehner by the Deutsche Forschungsgemeinschaft (Grant We 1302/1-1).

[44] Toad Urinary Bladder as a Model for Studying Transepithelial Sodium Transport

By MORTIMER M. CIVAN and HAIM GARTY

Introduction

Much of the information currently available concerning Na^+ transport across tight epithelia has been obtained by study of isolated anuran preparations: urinary bladders and skins. These tissues can be mounted in chambers as flat sheets, enormously facilitating efforts both to clamp the transepithelial parameters (such as voltage and external ionic composition) and to measure transepithelial fluxes of charge, solutes, and water. These epithelial sheets are easily prepared, relatively inexpensive, and share many of the functional characteristics of distal mammalian nephrons, which are far less accessible to experimental manipulation. As first recognized by Ussing and Zerahn,[1] anuran skin and urinary bladder also present the investigator with an enormous technical advantage in the measurement of transepithelial Na^+ transport ($J_{Na^+}^{ms}$, from mucosa to serosa). When the transepithelial potential ($\Delta\psi$, serosa positive with respect to mucosa) is clamped to 0 mV, and the tissue is bathed with identical solutions on its two surfaces, the tissue is considered to be in the short-circuited state. Under these conditions, the transepithelial current is termed the short-circuit current (I_{sc}) and, under many conditions, is equal to the net flux of Na^+ across these epithelial models. This equality of current and Na^+ flux permits the investigator to monitor $J_{Na^+}^{ms}$ instantaneously, without the need to resort to time-consuming, expensive, and relatively imprecise measurements of bidirectional $^{22}Na^+$ and $^{24}Na^+$ radioisotope fluxes across the preparation.

Much of the earlier study of Na^+ transport across isolated anuran epithelia was actually performed with frog skin.[2] However, the presence of

[1] H. H. Ussing and K. Zerahan, *Acta Physiol. Scand.* **23**, 110 (1951).
[2] H. H. Ussing, *Handb. Exp. Pharmakol.* **13**, 1 (1960).

five to seven layers of epithelial cells within the epidermis was initially a source of concern to investigators. The multiple layers were thought possibly to obscure the physical significance of transepithelial measurements. For this reason, the development of the urinary bladder of the toad as a model both for Na^{+} [3] and for water transport [4] was welcomed as providing a considerably simpler model system. This preparation consists of one to two layers of epithelial cells. In recent years, this anatomic advantage has appeared to be less substantive than first perceived. Specifically, measurements of intracellular electrolyte contents by electron probe X-ray microanalysis [5] and of the electrical potential and resistance profile across frog skin [6,7] have indicated that to a large extent frog skin functions as a syncytium (approximating a single extended cell) under many conditions. Furthermore, despite the presence of multiple cell layers, the relative chronology of intracellular and transepithelial events can be measured, at least within uncertainties of tens of seconds. [8] On the other hand, the rich cellular heterogeneity and underlying smooth muscle of the urinary bladder present complexities largely absent from frog skin.

This chapter consists of three parts. First, we note the considerations involved in measuring short-circuit current, since the ease of performing this measurement is the principal *raison d'être* for using toad bladder as a model of transepithelial Na^{+} transport. Second, we compare the advantages and disadvantages displayed by frog skin and toad bladder in bringing a number of biophysical techniques to bear on studying $J_{Na^{+}}^{ms}$. Finally, we consider in some detail the use of vesicles in studying Na^{+} transport across urinary bladder because of the considerable success achieved with this approach in this tissue.

Measurement of Short-Circuit Current

Figure 1 presents a cartoon of a short-circuited preparation. A urinary bladder is considered to be mounted (in a plane normal to the page) between the two halves of an Ussing chamber. Fluid is aerated in separate mucosal and serosal reservoirs (not pictured), and circulated past the two surfaces of the tissue. The electrical potential is sensed by salt bridges placed close to each of the surfaces; each bridge consists of polyethylene tubing filled with a saline solution containing 2–4% agar. The presence of

[3] A. Leaf, J. Anderson, and L. B. Page, *J. Gen. Physiol.* **41,** 657 (1958).

[4] P. J. Bentley, *J. Endocrinol.* **17,** 201 (1958).

[5] R. Rick, A. Dörge, E. von Arnim, and K. Thurau, *J. Membr. Biol.* **39,** 313 (1978).

[6] W. Nagel, *Nature (London)* **264,** 469 (1976).

[7] J. DeLong and M. M. Civan, *J. Membr. Biol.* **72,** 183 (1983).

[8] E. Kelepouris, Z. S. Agus, and M. M. Civan, *J. Membr. Biol.* **88,** 113 (1985).

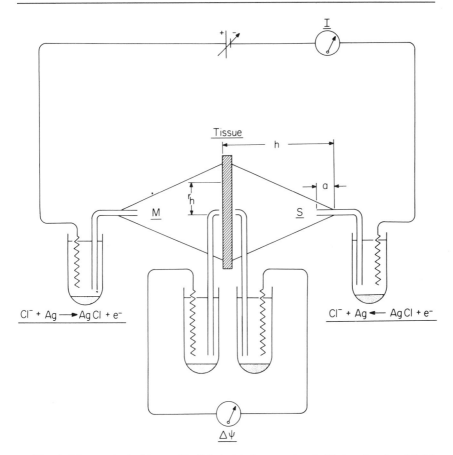

FIG. 1. Measurement of transepithelial electrical parameters in Ussing chamber (M. M. Civan, "Epithelial Ions and Transport: Application of Biophysical Techniques." Wiley, New York, 1983). The tissue is bathed with mucosal (M) and serosal (S) solutions, and is displaced by a distance (h) from the apex of each conical half-chamber. The difference in electrical potential ($\Delta\psi$) between the two ends of agar salt bridges is conveyed to a voltmeter through Ag/AgCl or calomel half-cells. Current (I) is passed from a voltage source through chlorided silver wires and agar salt bridges placed at the ends of the chamber at a distance (a) from the apices of the conical half-chambers. The reactions are indicated for each half-cell during the course of passing positive current from M to S. The parameter (r_h) is the radial distance of an annulus of tissue from the geometric center of the planar sheet. (Copyright ©1983 by John Wiley & Sons; reprinted by permission of John Wiley & Sons, Inc.)

the agar prevents emptying of the saline solution into the bath; the ends of the tubing are frequently pulled out manually to reduce the diameter of the tips of the bridge and further increase the resistance to flow of the saline solution out of the bridge. Similar current-passing agar bridges are placed at the far ends of the chamber (displaced by a distance h from the plane of the tissue presented in the figure). Simple circuits appropriate for voltage clamping the preparation are easily built[9] and are also widely available commercially.

The other ends of the voltage-sensing agar bridges are coupled to a well-defined half-cell. In the figure, the ends are immersed in a saturated solution of KCl, which also bathes the surface of a silver wire coated with AgCl; alternatively, an $Hg/HgCl_2$ half-cell would serve just as well. The half-cell reaction is indicated in the figure. Since the Cl^- activity of the half-cell is fixed, the activity of the Ag^+ is also fixed by the solubility product of AgCl, and the reference potential of the half-cell is well defined. The ends of the silver wires not coated with AgCl are then coupled electrically to the inputs of a voltmeter; the instrument should have a high input impedance to reduce the magnitude of the current flowing through the half-cell, which might otherwise shift the half-cell reference potential.

The measured difference in potential $\Delta\psi$ actually reflects several terms in addition to the true transepithelial potential (ψ^{ms}):

$$\Delta\psi = \psi^{ms} + \Delta\psi_R + \Delta\psi_{jcn}^{m,s} + \Delta\psi_{jcn}^{cells} \tag{1}$$

where $\Delta\psi_R$ is the voltage drop produced by current flow across the external Ringer's solution included between the tissue surfaces and the voltage-sensing agar leads, $\Delta\psi_{jcn}^{m,s}$ is the difference in junction potentials between the voltage-sensing bridges and the mucosal and serosal bathing solutions, and $\Delta\psi_{jcn}^{cells}$ is the difference in the junction potentials between the agar bridges and half-cell solutions.

We can minimize $\Delta\psi_R$ by placing the tips of the voltage-sensing bridges as close as possible to the tissue surfaces. Just this precaution alone usually suffices for tissues carefully mounted without edge damage since the intrinsic resistivity of the tissue is of the order of $k\Omega\cdot cm^2$. However, when the transepithelial resistance R_T is experimentally reduced to values comparable to the series resistance R_R of the solution between tissue and voltage-sensing leads, the contribution of $\Delta\psi_R$ to the measured voltage across the total preparation can be substantial. This point can be appreciated from simple electrical circuit analysis.[10] When sufficient current is passed to short circuit R_T and R_R in series, the transepithelial potential will be reduced from its open-circuited value (ψ_{OC}^{ms}) to a lower but non-zero value (ψ_{SC}^{ms}) given by Eq. (2):

$$\psi_{SC}^{ms}/\psi_{OC}^{ms} = [1 + (R_T/R_R)]^{-1} \tag{2}$$

[9] D. R. DiBona and M. M. Civan, *J. Membr. Biol.* **12**, 101 (1973).
[10] M. M. Civan, "Epithelial Ions & Transport: Application of Biophysical Techniques." Wiley, New York, 1983.

Thus, if the resistance across the epithelial cells is an order of magnitude greater than that across the Ringer's solution, $(\psi_{SC}^{ms}/\psi_{OC}^{ms})$ is (1/11) or 0.09, and the epithelial layer itself is more than 90% short circuited. On the other hand, if R_T and R_R are approximately equal, short circuiting the series arrangement of tissue and solution will depolarize the epithelium by only 50%. Under these latter circumstances, it is necessary to make corrections in I_{SC}; circuits for this purpose are readily available (e.g., Ref. 9).

The best approach for minimizing $\Delta\psi_{jcn}^{m,s}$ is to fill the voltage-sensing bridge with the identical saline solution comprising the Ringer's medium. Under these conditions, there can be no junction potential since the ionic composition in the bridge and in the bathing medium are identical. This is clearly the preferred technique for studying a tissue (like the iris-ciliary body) whose transepithelial potential is very low (≈ 1 mV). However, for toad urinary bladder, this approach is often unsatisfactory when the need arises to perfuse the tissue with different solutions during a single experiment.

A common alternative technique for minimizing $\Delta\psi_{jcn}^{m,s}$ is to fill the agar bridges with 3 M KCl solution. The rationale for this approach can be appreciated from a consideration of the Henderson equation, which has been found applicable for estimating the liquid junction potential not only in simple aqueous solutions, but at the interface between micropipets and the intracellular fluids, as well[11]:

$$\psi_{jcn} = -\left[\sum_i u_i(a_i'' - a_i')/\sum_i z_i u_i(a_i'' - a_i')\right](RT/F)\left[\ln(\sum_i z_i u_i a_i''/\sum_i z_i u_i a_i')\right] \quad (3)$$

ψ_{jcn} is the potential within the agar bridge relative to the Ringer's solution, u_i, z_i, and a_i are the mobility, valence, and chemical activity of the ith ion, and double prime ($''$) and prime ($'$) refer to the phases within the bridge and external medium, respectively. By making a_{K^+}'' and a_{Cl^-}'' much larger than the activities of the ions in the Ringer's solutions, the first term in brackets on the right-hand side of Eq. (3) reduces to $[(u_{K^+} + u_{Cl^-})/(u_{K^+} - u_{Cl^-})]$, so that

$$\psi_{jcn} \propto -[(u_{K^+} + u_{Cl^-})/(u_{K^+} - u_{Cl^-})] \quad (4)$$

The value of u_{K^+} is very nearly equal (and opposite in sign) to that of u_{Cl^-}, so the term in brackets in Eq. (4) is approximately 0.02, leading to suppression of the junction potential. However, depending upon the precise value of the second term in brackets on the right-hand side of Eq. (3), ψ_{jcn}

---

[11] L. G. Palmer and M. M. Civan, *J. Membr. Biol.* **33**, 41 (1977).

can still be several millivolts different from zero. The final contribution of $\Delta\psi_{jcn}^{m,s}$ in Eq. (1) to the measured value of $\Delta\psi$ will depend upon the symmetry of the bathing media and agar bridges on the mucosal and serosal surfaces. To the extent that the compositions are identical on the two sides, the difference in junction potential $\Delta\psi_{jcn}^{m,s}$ will be zero.

The final term ($\Delta\psi_{jcn}^{cells}$) contributing in Eq. (1) to the measured transepithelial potential ($\Delta\psi$) is the difference in liquid junction potential established between the agar bridges and the KCl solution of the Ag/AgCl half-cells (Fig. 1). Each of the two junction potentials established at the two half-cells can be separately reduced to zero by using 3 M KCl both in the agar bridge and in the half-cell solution. However, with evaporation of water over time, the KCl concentration of the half-cell solution will gradually increase. Rather than replace the reference fluid each day, it is more convenient to establish a well-defined potential difference at each interface by using saturated KCl in the half-cell and 3 M KCl solution in the agar leads. The algebraic difference between these two similar junction potentials ($\Delta\psi_{jcn}^{cells}$) is then negligible.

Thus far, we have been concerned with factors which can distort estimations of the transepithelial potential ψ^{mc} derived from measurements of the transmural potential $\Delta\psi$ conducted across the central point of the exposed tissue area. However, unless the current density is uniform throughout the tissue, the passage of current will produce a smaller change in potential ($\Delta\phi$) at a radial displacement r_h (Fig. 1) than the polarization ($\Delta\phi$)$_0$ induced at the geometric center. $[(\Delta\phi)/(\Delta\phi)_0]$ is a function of the chamber geometry and the resistances of the tissue and bathing media. Figure 2 presents $[(\Delta\phi)/(\Delta\phi)_0]$ as a function of r_h for different geometries and values of resistances. Under certain experimental conditions (trajectory d, Fig. 2), heterogeneity of current density can introduce a significant variation in transepithelial voltage within the tissue. At least two solutions can address this problem. One approach is to restrict the maximum diameter of exposed tissue surface to $\leq [(1/2)(h - a)]$, where the parameters h and a are defined in Fig. 1. Alternatively, the current density can be made more uniform by, for example, mounting the bladder between cylindrical rather than conical half-chambers. Instead of short circuiting the tissue from point sources at the apices of cones (Fig. 1), current can then be passed with disks of platinized platinum, whose diameter equals that of the exposed area, producing current density of high uniformity.[12]

[12] F. C. Weinstein, J. J. Rosowski, K. Peterson, Z. Delalic, and M. M. Civan, *J. Membr. Biol.* **52**, 25 (1980).

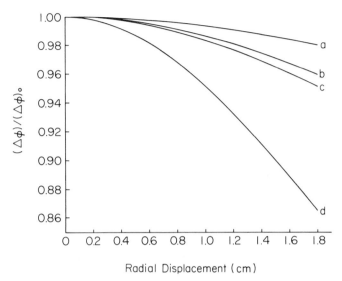

FIG. 2. Effect of increasing radial displacement on transepithelial potential (Civan[10]). Passage of current across the tissue of Fig. 1 produces a change $(\Delta\phi)_o$ in transepithelial potential measured at the center of the sheet. Annuli of tissue at increasing radial distances (r_h) from the center are subjected to decreasing polarizations $(\Delta\phi)$ because of the nonuniform current density applied to the tissue. All four graphs have been calculated from Eq. (3.5) of Ref. 10. Curves (a) and (b) are calculated for a chamber in which $h = 6.1$ cm and $a = 0.9$ cm; (c) and (d) are calculated for a chamber foreshortened by a factor of two. In all cases, the resistivity of the serosal solution is taken to be 81.5 $\Omega \cdot$ cm, that of a 115 mM NaCl solution. In graphs (a) and (c), the resistivities of tissue and mucosal medium have been taken to be 6250 $\Omega \cdot$ cm^2 and 81.5 $\Omega \cdot$ cm, respectively, while the graphs of (b) and (d) have been calculated for values of 450 $\Omega \cdot$ cm^2 and 602.6 $\Omega \cdot$ cm, respectively. These two sets of values are commonly observed, and do not constitute upper and lower bounds for all permissible trajectories. (Copyright ©1983 by John Wiley & Sons; reprinted by permission of John Wiley & Sons, Inc.)

Relative Advantages of Toad Bladder and Frog Skin

Over the past three decades of intensive study, a number of differences have been described concerning the transport characteristics of frog skin and toad urinary bladder. For example, mucosal acidification produces different effects on $J_{Na^+}^{ms}$ across toad bladder and frog skin (from at least certain species of frog) (summarized in Ref. 13) and net acid secretion has thus far been documented only for toad bladder.[14,15] Despite these and

[13] H. Garty, E. D. Civan, and M. M. Civan, J. Membr. Biol. 87, 67 (1985).
[14] L. W. Frazier and J. C. Vanatta, Biochim. Biophys. Acta 241, 20 (1971).
[15] J. H. Ludens and D. D. Fanestil, J. Biol. Chem. 223, 1338 (1972).

other differences, the urinary bladder of the toad[16] and the skin of the frog[2] are functionally very similar, indeed. Both are tight epithelia which largely regulate net Na^+ transport by modifying apical Na^+ entry into the cells from the mucosal media through electrodiffusive channels. Both tissues also respond to many of the same hormones and inhibitors. In addition, the net transport of Na^+ across both tissues is generally far greater (in the short-circuited state) than the net movement of other ions. Because of these close functional similarities, use of one or the other model system for studying transepithelial Na^+ transport is guided primarily by the technical advantages and disadvantages of the two preparations.

The major strength of both epithelial preparations is the ease with which transepithelial parameters can be measured. In this respect, there is little to choose between the two tissues. One exception is the need to evaluate the contribution by the skin glands to the transepithelial fluxes of solutes. It is now clear that the glands can produce net transport of both Ca^{2+}[17] and Cl^-.[18] Fortunately, the transport contributions of the glands can be abolished by stripping the epidermis from the underlying dermis, following preincubation with collagenase.[19] Therefore, study of the minor ionic contributions to the total net transport of ions across frog skin is best conducted with such split skin preparations. With this exception, transepithelial measurements of short-circuit current and total epithelial resistance, and transepithelial fluctuation analysis in the presence of amiloride (to characterize the apical Na^+ channels),[20,21] have been conducted successfully with both preparations. However, it has long been clear that in order to understand the basis and regulation of Na^+ transport across the entire epithelium, it is essential to measure the electrical potential and resistance profiles across the cell and the ionic composition of the intracellular fluid. The ease with which such intracellular measurements can be performed is very different in toad bladder and frog skin.

Sodium transport across tight epithelia proceeds in two steps[22]: Na^+ entry from the mucosal medium into the cell in response to the electrochemical gradient for Na^+ across the apical membrane, and Na^+ extrusion from the cell into the serosal medium by the basolateral Na^+/K^+-exchange pump. It is clearly essential to measure the intracellular Na^+ activity and

[16] A. D. C. Macknight, D. R. DiBona, and A. Leaf, *Physiol. Rev.* **60**, 615 (1980).
[17] F. N. Ziyadeh, E. Kelepouris, M. M. Civan, and Z. S. Agus, *Am. J. Physiol.* **249**, F713 (1985).
[18] I. G. Thompson and J. W. Mills, *Am. J. Physiol.* **244**, C221 (1983).
[19] R. S. Fisher, D. Erlij, and S. I. Helman, *J. Gen. Physiol.* **76**, 447 (1980).
[20] B. Lindemann and W. Van Driessche, *Science* **195**, 292 (1977).
[21] B. Lindemann, *Annu. Rev. Physiol.* **46**, 497 (1984).
[22] V. Koefoed-Johnsen and H. H. Ussing, *Acta Physiol. Scand.* **42**, 298 (1958).

electrical potential in order to quantify the electrochemical driving forces acting on Na^+ movement across each of the plasma membranes. It is also necessary to measure intracellular Ca^{2+} activity ($a^c_{Ca^{2+}}$) and intracellular pH (pH^c), given the putative regulatory roles of both Ca^{2+} and H^+ within the cell. In order to measure these parameters, at least three major biophysical techniques have been applied to isolated frog skin and toad bladder: intracellular recording with reference micropipets and ion-selective microelectrodes, nuclear magnetic resonance (NMR) spectroscopy, and electron probe X-ray microanalysis (EPMA).

In principle, intracellular recording with micropipets and microelectrodes is an exceedingly powerful tool, particularly when coupled with transepithelial electrical measurements. Such electrometric measurements can define the electrochemical driving forces and electrical resistances to conductive ionic flow across each of the two series plasma membranes. The small size of these and most other epithelial cells ($= 10 \ \mu$m in diameter) presents a significant technical challenge. Nevertheless, intracellular recordings can be obtained with frog skin which can be stable for hours following a single impalement.[23,24] Such successful impalements of frog skin have been reproduced in a number of other laboratories. Furthermore, intracellular electrometric measurements have permitted estimation of the intracellular activities of Na^+, K^+, Cl^-, H^+, and Ca^{2+}.[7,8,25,26] In contrast, stable, reproducible intracellular impalements have been much more difficult to perform in the urinary bladder of the toad because of the shallow depth of the epithelial cells and the difficulty of completely abolishing smooth muscle activity of the underlying submucosa without also altering the transport properties of the epithelium.[7,27] Thus, toad bladder is a far less favorable preparation for intracellular electrophysiologic studies than is frog skin.

Toad bladder is also an unfavorable preparation for studies based on NMR spectroscopy. Because of the substantial presence of submucosal smooth muscle, the epithelial cells contribute only 10–20% of the total intracellular Na^+, K^+, Cl^-, and water measured in this tissue.[28] Thus, while some information can be obtained by NMR analysis of whole toad bladder, it is necessary to isolate the epithelial cells in order to conduct rigorous

[23] S. I. Helman and R. S. Fisher, *J. Gen. Physiol.* **69**, 571 (1977).

[24] W. Nagel, *Pfluegers Arch.* **365**, 135 (1976).

[25] W. Nagel, J. F. Garcia-Diaz, and W. McD. Armstrong, *J. Membr. Physiol.* **61**, 127 (1981).

[26] M. E. Duffey, E. Kelepouris, K. Peterson-Yantorno, and M. M. Civan, *Am. J. Physiol.* **251**, F468 (1986).

[27] J. T. Higgins, Jr., B. Gebler, and E. Frömter, *Pfluegers Arch.* **371**, 87 (1977).

[28] A. D. C. Macknight, M. M. Civan, and A. Leaf, *J. Membr. Physiol.* **20**, 365 (1975).

NMR analyses of their intracellular fluids.[29] The extent to which the isolation procedures disrupt epithelial function is unclear. In contrast, frog skin is a far more favorable preparation because of the relatively acellular nature of the underlying dermis, and the ease with which the intact functioning epithelium can be isolated. In this tissue, the intracellular contents of Na^+ have been measured by $^{23}Na^+$ NMR,[30] the intracellular signals of inorganic phosphate, ATP and phosphocreatine and the intracellular pH have been studied by ^{31}P NMR,[31-35] and the intracellular pH has also been measured by ^{19}F NMR analysis.[33]

Toad bladder also presents a technical challenge to attempts to measure the intracellular electrolyte and water contents by electron probe X-ray microanalysis. However, in this case, the technical difficulty is less formidable. The problem arises from the need to section the tissue at low temperatures ($-60°$ or below). Cooling the sample below the eutectic temperature halts diffusion during the course of the sectioning, but complicates thin sectioning of the tissue. Sectioning can be facilitated by freeze substitution, a technique widely used by electron microscopists. However, the freeze-substitution procedure may possibly lead to significant translocations of water and solute. Cryosectioning of toad bladder proves to be considerably more difficult than with frog skin, because of the thin diaphanous structure of the former tissue. The much thinner epithelium of toad bladder, together with its thinner layer of supporting connective tissue, substantially increase the risk of shattering during the cryosectioning. This problem has been addressed by Rick et al.[36] by lining the chuck holding the sample with a pocket of indium; this approach has also been successfully applied by others.[37] Thus, although cryosectioning of the toad bladder remains more tedious than for frog skin,[5] EPMA can be successfully applied to both model systems.

The considerations of the preceding paragraphs of this section indicate that measurements of the intracellular fluids of the transporting cells are greatly facilitated by studying frog skin rather than toad urinary bladder.

[29] M. Bond, M. Shporer, K. Peterson, and M. M. Civan, *Mol. Physiol.* **1**, 243 (1981).

[30] M. M. Civan, H. Degani, Y. Margalit, and M. Shporer. *Am. J. Physiol.* **245**, C213 (1983).

[31] L.-E. Lin, M. Shporer, and M. M. Civan, *Am. J. Physiol.* **243**, C74 (1982).

[32] R. L. Nunnally, J. S. Stoddard, S. I. Helman, and J. P. Kokko, *Am. J. Physiol.* **245**, F792 (1983).

[33] M. M. Civan, L.-E. Lin, K. Peterson-Yantorno, J. Taylor, and C. Deutsch, *Am. J. Physiol.* **247**, C560 (1984).

[34] L.-E. Lin, M. Shporer, and M. M. Civan, *Am. J. Physiol.* **248**, C177 (1985).

[35] M. M. Civan and K. Peterson-Yantorno. *Am. J. Physiol.* **251**, F831 (1986).

[36] R. Rick, A Dörge, A. D. C. Macknight, A. Leaf, and K. Thurau, *J. Membr. Biol.* **39**, 257 (1978).

[37] M. M. Civan, T. A. Hall, and B. J. Gupta, *J. Membr. Biol.* **55**, 187 (1980).

However, our understanding of Na^+ transport also depends upon our ability to study the properties of the ionic channels within the apical and basolateral plasma membranes. In this aim, both toad bladder and frog skin possess certain advantages for experimental study. Measuring membrane currents by patch clamping plasma membranes is currently one of the very most widely utilized biophysical techniques.[38] The great power of the approach is the ability to monitor single-channel events, either when the whole epithelium is intact (cell-attached mode) or when the patch is studied in isolation from the rest of the cell and tissue (excised patch mode). The practicability of applying the technique depends upon the ability to bring the surface of the polished patching pipet into extremely close contact with the membrane surface. Patch clamping has recently been found to be feasible for studying the basolateral membranes of split frog skin.[39] The basolateral surface is of interest, in part because of the presence of the great bulk of the K^+ channels, which are important in establishing the baseline intracellular potential, and in thus regulating the electrical force driving apical Na^+ entry. In contrast, patch clamping of the toad urinary bladder is more difficult, albeit feasible,[39a] presumably because of the relative inaccessibility of the basolateral surface of the mucosal epithelium and because of the extensive glycocalyx on the apical surface.

Another widely used technique for studying the transport characteristics of isolated plasma membranes is the preparation of membrane vesicles. General considerations concerning, and the preparation of, such vesicles are presented elsewhere in this series.[40] Here, we are specifically concerned with the use of vesicles in studying Na^+ transport across the two model epithelia under consideration. In this respect, the urinary bladder of the toad constitutes a far more favorable preparation. The major problem posed by frog skin in this regard is the presence of the *stratum corneum*, the keratinized outer cell layer. It is likely this layer which complicates efforts to separate the epithelial cells early in the preparative procedure, markedly reducing the yield of satisfactory vesicles. Given the particular suitability of applying the technique to toad urinary bladder, and the considerable information which has already been obtained with this approach, the use of vesicles from this tissue is considered in detail in the concluding section of the chapter.

---

[38] O. P. Hammill, A. Marty, E. Neher, B. Sakmann, and F. J. Sigworth, *Pfluegers Arch.* **391**, 85 (1981).

[39] R. Yantorno and M. M. Civan, *Biophys. J.* **49**, 160a (1985).

[39a] S. Frings, R. D. Purves, and A. D. C. Macknight, *J. Membr. Biol.* **106**, 157 (1988).

[40] H. Garty and S. J. D. Karlish, this series, Vol. 172, p. 155.

Sodium Ion Transport across Vesicles from Toad Bladder

As noted in the previous section, the advantage of studying vesicles lies in isolating the Na^+ channels from the many ionic and biochemical factors likely to regulate channel permeability.

At least two different strategies can be followed in preparing vesicles from toad bladder. Preparative steps can be introduced in order to partially purify the heterogeneous population of vesicles initially obtained.[41] Alternatively, one can deal with a highly heterogeneous population by relying more heavily on the specific Na^+ channel inhibitor amiloride to define the flux of interest.[42] This latter approach simplifies and speeds preparation of vesicles, increases the overall yield of functional channels, and reduces the number of animals required for each experiment.

Once the vesicles have been prepared, two different strategies may also be applied to measure Na^+ transfer between the intra- and extravesicular spaces. The most direct approach is to measure Na^+ fluxes with the compositions of the fluids inside and outside the vesicles approximating those of the intra- and extracellular fluids, respectively (for vesicles oriented right side out). This, however, poses a significant technical difficulty which can be readily appreciated from Eq. (5):

$$k = (A/v)\,(P) \simeq (3/r)(P) \qquad (5)$$

which presents the rate coefficient (k) for Na^+ flux as a function of the surface area (A), volume (v), and membrane permeability (P). Approximating both whole cells and vesicles as spheres, (A/v) reduces to ($3/r$), where r is the radius. Since r is roughly 20-fold smaller for the vesicles than for the whole cells of toad bladder, the corresponding rate coefficient should be one to two orders of magnitude faster. One solution to this problem is to improve the time resolution of the detecting equipment, by using a fast-reaction apparatus.[43]

A simpler alternative is essentially to increase the amount of Na^+ loaded into the vesicles. The time courses of the $^{22}Na^+$ uptake and subsequent efflux are thereby prolonged, permitting measurements to be conducted simply and precisely with inexpensive equipment.[42] The Na^+ loading can be enhanced in several ways.[40] One approach is to establish a K^+ gradient across the vesicular walls. With the subsequent addition of the potassium ionophore valinomycin, the chemical gradient produces a strong electrical gradient in the physiologic direction (cell negative to mucosal medium), favoring Na^+ uptake. This approach mimicks the phys-

[41] H. Chase and Q. Al-Awqati, *J. Gen. Physiol.* **77**, 693 (1981).
[42] H. Garty, B. Rudy, and S. J. D. Karlish, *J. Biol. Chem.* **258**, 13094 (1983).
[43] H. S. Chase and Q. Al-Awqati, *J. Gen. Physiol.* **81**, 643 (1983).

iologic state, not only in the orientation of the electrical gradient, but also in establishing an internal space initially rich in K$^+$ and poor in Na$^+$.

Study of vesicles has already provided considerable information concerning not only conductive apical entry of Na$^+$, but also Ca^{2+}/Na$^+$ exchange[41] and conductive K$^+$ uptake[44] across the basolateral membrane. However, like all other experimental techniques, it can be misapplied or the results misinterpreted. Here, we shall consider six caveats in studying vesicles. The primary concern in using broken cell preparations is that the isolation procedures may alter the properties of the channels or delete critical gating sites. This does not appear to be a problem with current preparations of apical vesicles, in view of their high sensitivity to external amiloride ($K_I \simeq 20$ nM).[45]

A second consideration in studying vesicles is simply another aspect of the fundamental strength of the technique. Vesicles can be used to assess the direct effects of agents applied *in vitro*, in isolation from cytoplasmic factors. However, even when experimental perturbations exert significant effects on the Na$^+$ channels, the expression of these effects may be either undetectable or altered in the vesicle suspension. An example of this phenomenon is the effect of intravesicular pH (pH^i). In the absence of Ca^{2+} within the vesicles, pH^i has little effect on the Na$^+$ flux over the physiologic range.[13] However, when the intravesicular Ca^{2+} activity is increased to approximate that of the intracellular fluids, pH^i is found to regulate Na$^+$ movement through the amiloride-sensitive channels, presumably by altering the charge of Ca^{2+}-sensitive gating sites.[46]

A third caveat in studying vesicles is that experimental perturbations can exert direct effects on Na$^+$ transfer which have no physiologic significance, whatsoever. For example, by increasing the conductance of parallel channels, an agent can alter the electrical driving force for Na$^+$ uptake. These changes might have little relevance to the intact cell, which could buffer such effects by the operation of multiple other conductive channels (principally K$^+$-selective channels). It is prudent to assess this possibility by measuring the changes in the driving forces, as well as in the fluxes of Na$^+$.

A fourth concern in interpreting the flux data must be the awareness that the Na$^+$ influx is generally measured under highly unphysiologic conditions. For example, when a K$^+$ gradient and valinomycin are used to establish the electrical driving force, the external Na$^+$ concentration ($c^o_{Na^+}$) must be kept very low ($= 0.2 \mu M$). Increasing $c^o_{Na^+}$ by as little as 50 μM significantly depolarizes the apical vesicles and reduces Na$^+$ uptake.[13] For-

[44] H. Garty and M. M. Civan, *J. Membr. Biol.* **99**, 93 (1987).
[45] H. Garty and C. Asher, *J. Biol. Chem.* **260**, 8330 (1985).
[46] H. Garty, C. Asher, and O. Yeger, *J. Membr. Biol.* **95**, 151 (1987).

tunately, the effect of the perturbation on Na^+ permeability can also be assessed by studying Na^+ efflux under more physiologic conditions; in such efflux studies, the external Na^+ concentration is elevated only after first preloading the vesicles.

A fifth serious caveat in studying fluxes, either with vesicles or with the patch-clamp technique, is the problem of biological sampling. Data obtained with the vesicles alone can never fully exclude the possibility that the channels under study are unrepresentative of the population of Na^+ channels present in the apical membranes. For this reason, correlation of the data with results obtained with the whole tissue is essential.

Finally, it will be recognized that vesicles can also be used to detect covalent changes in the Na^+ channels produced in the whole tissue, before disrupting the epithelial cells.[45] However, in interpreting such effects, it is important to consider the possibility that the preincubation has altered the mechanical fragility and/or elasticity of the apical membranes, leading to an altered volume distribution of the heterogenous vesicle population. As noted above, increasing the internal volume of vesicles having Na^+ channels will also increase the Na^+ uptake even in the absence of changes in the Na^+ permeability or electrochemical driving force. It should be emphasized that, when care is taken to address this and the above caveats, study of vesicles in suspension can provide considerable information clarifying ion transport across the apical and basolateral membranes of toad bladder.

Summary

Sodium ion transport across tight epithelia has been investigated particularly extensively by studying two model systems: the urinary bladder of the toad and the frog skin. The greatest advantage presented by these models is the capability of monitoring net transepithelial Na^+ flux simply, precisely, and instantaneously by measurement of the short circuit current (I_{SC}). Many of the caveats involved in the measurement are discussed in detail.

In order to fully characterize the forces driving Na^+ movement across the series apical and basolateral membranes, it is necessary to measure intracellular potential and ionic composition. Such measurements are far more easily conducted with frog skin than with toad bladder, using the major biophysical techniques currently available.

Regulation of transepithelial Na^+ movement across tight epithelia is largely conducted at the apical membranes. This regulation can be clarified by study of the isolated Na^+ channels in membrane vesicles. Such vesicles are far more easily prepared from toad urinary bladder than from frog skin.

The strengths and potential misappropriations of this technique are considered in detail.

Acknowledgments

The work presented was supported in part by research grants from the U.S. National Institutes of Health (AM 20632) and the U.S.–Israel Binational Science Foundation (84-00066).

[45] Methods to Detect, Quantify, and Minimize Edge Leaks in Ussing Chambers

By L. G. M. GORDON, G. KOTTRA, and E. FRÖMTER

Introduction

The overall transport properties of flat sheet epithelia such as frog skin, urinary bladder, or intestinal epithelia are usually (conveniently) investigated by mounting the tissue between two fluid-filled half-chambers and measuring net fluxes, isotope fluxes, transepithelial potentials or electrical currents between both fluid compartments. In honor of the pioneering work of Hans Ussing this double-compartment chamber is usually referred to as an Ussing chamber. A problem with such experiments is to know whether or not the chamber is tight, i.e., whether the tissue is tightly sealed to the chamber wall or whether a shunt path persists between both fluid compartments around the tissue edge. Such shunts would allow unphysiologically high fluxes to be observed and would decrease the recorded values of open circuit potential (V_t) and transepithelial resistance (R_t) in comparison with the true electrical properties of the tissue, leaving only the determination of the short-circuit current (I_{SC}) unaffected.

This problem was first mentioned by Ussing,[1] who described that in order to tighten the chamber, uniform pressure was required, which he achieved by placing the tissue between the ground edges of two cylindrical celluloid cups and applying pressure via parallel plates. In a later publication he noted that "gentle" pressure sufficed to make the apparatus water tight. However, whether macroscopic tightness to water flow also guaranteed electrical tightness remained unclear and how much pressure was necessary remained unclear, too. In reality we are dealing with two prob-

[1] H. H. Ussing, *Acta Physiol. Scand.* **17**, 1 (1948).

lems: (1) too little pressure leaves the preparation unsealed (edge leak) and (2) too much pressure may injure the tissue under the edge of the chamber (edge damage), which again imposes a leak pathway between the fluid compartments. While Ussing seems to have overcome the problem by using a chamber which exposes a large skin surface area, this is not possible with all preparations. But it was not until 1968 that the first systematic attempts were made to investigate the problem of edge leak (or damage) in Ussing chambers in some greater detail.

Direct Demonstration of Edge Leaks

Working with chambers of different size Dobson and Kidder[2] noticed that the measured V_t and the calculated R_t (in $\Omega \cdot cm^2$ of exposed skin area) increased with the ratio of surface area to circumference of the exposed tissue while the measured I_{SC} was unaffected. Moreover, using a phase-contrast microscope these authors noticed that the appearance of the tissue in the last 0.7 mm from the compressed rim was altered, such as to exhibit enlarged intracellular spaces. This was taken as evidence for edge damage although otherwise no direct signs of cellular damage were detectable.

A similar observation was subsequently made by Walser,[3] who found lower V_t and R_t values of toad urinary bladders mounted in Ussing chambers as compared to measurements on a sac preparation in which the excised bladder sac was only partially immersed into Ringer solution, while the rim was being held up in the air by means of a small Lucite ring (see also Civan et al.[4]). To further demonstrate the occurrence of edge leaks in Ussing chambers, Walser[3] replaced the mucosal bath solution with a gel, added $^{22}Na^+$ or $^{24}Na^+$ to the serosal medium, and found, after some time had elapsed, that the radioactivity was higher in pieces of gel that had been excised from the edge region as compared to pieces from the central region.

These observations were later confirmed by Helman and Miller,[5,6] who also noticed that V_t and R_t depended on the size of the exposed surface area inside the chamber and who developed a special mounting technique to reduce or avoid edge leaks which will be discussed further below. In 1975, eventually Higgins et al.[7] directly demonstrated the occurrence of leaks in conventionally mounted tissue preparations by scanning the rim of the chamber with a voltage-recording microelectrode (see Fig. 1).

[2] J. G. Dobson and G. W. Kidder, Am. J. Physiol. 214, 719 (1968).
[3] M. Walser, Am. J. Physiol. 219, 252 (1970).
[4] M. M. Civan, O. Kedem, and A. Leaf, Am. J. Physiol. 211, 569 (1966).
[5] S. I. Helman and D. A. Miller, Am. J. Physiol. 225, 972 (1973).
[6] S. I. Helman and D. A. Miller, Am. J. Physiol. 226, 1198 (1974).
[7] J. T. Higgins, L. Cesaro, B. Gebler, and E. Frömter, Pfluegers Arch. 358, 41 (1975).

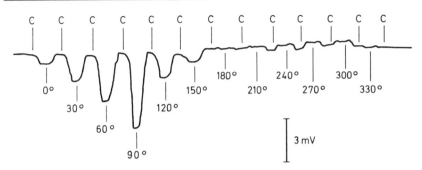

FIG. 1. Voltage scan of mucosal fluid compartment in a leaky preparation with toad bladder. A trace record of an experiment in which a microelectrode was moved around in the mucosal fluid compartment from center to edge and back to center and the potential difference against the serosal fluid compartment was continuously monitored. Locations are indicated at the center of the chamber by C and the periphery by angles of polar coordinates. The bladder was mounted without O-ring and without sealing agent. Potential difference and resistance were 55 mV and 8.4 $k\Omega \cdot cm^2$ and a leak was detected at the edge between 0 and 150°. [J. T. Higgins, L. Cesaro, B. Gebler, and E. Frömter, *Pfluegers Arch.* **358**, 41 (1975).]

Although these different experiments had clearly proved the occurrence of edge leaks in conventional Ussing chamber preparations none of the techniques was readily applicable for screening purposes, i.e., to identify whether a given preparation was leaky and if so, how important from a quantitative point of view the leak conductance was. A first step in this direction was made by comparing individual measurements with a set of control data as initially proposed by Higgens *et al.*[7] Plotting R_t versus V_t of a great number of control measurements on urinary bladder of *Necturus* the authors found a hyperbolic relationship with the more leaky preparations lying toward the left inner margin of the distribution and the more tight preparation toward the right outer margin (see Fig. 2). Comparison of individual data with this plot, which apparently does not only hold for urinary bladders of *Necturus,* served as a first help in judging whether a given preparation was more or less leaky or tight. However, the method did not allow any definite statement to be made on the leakiness of a tissue and more importantly was incapable of quantifying an edge leak conductance when present.

In low-resistance epithelia, such as gall bladder, small intestine, or choroid plexus, which have specific resistances approximately 100 times lower than those of frog skin or urinary bladder, the problem of edge leaks may also be expected to be 100 times less disturbing. On the other hand, these tissues may be more easily injured so that edge leaks from disruption of cell layers may become more important. Another problem is that low-

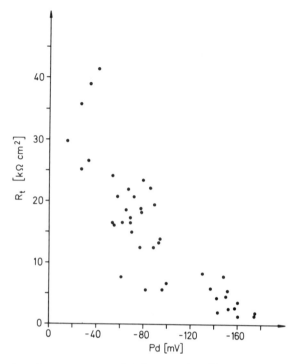

F<small>IG</small>. 2. Relationship between transepithelial potential difference (Pd) and specific transe-pithelial resistance in *Necturus* bladders. Plotted are the first stable readings obtained after mounting of 43 urinary bladders of *Necturus*. [J. T. Higgins, L. Cesaro, B. Gebler, and E. Frömter, *Pfluegers Arch.* **358,** 41 (1975).]

resistance epithelia do not generate high transepithelial voltages so that the approach of Fig. 2 cannot be applied to judge whether a low resistance of a given tissue is likely to be artifactual or real. Nevertheless, some evidence for the presence of edge leaks in some preparations of *Necturus* gall bladder has been obtained.[8]

Quantification of Edge Leaks

In view of 40 years of history of Ussing chambers and of their edge leak problems it is amazing that thus far no method has been published that would readily allow edge leaks to be recognized and quantified. This is particularly astonishing since the determination of paracellular shunt con-ductances (g_{sh}) is of great importance for all epithelial research and since all

[8] G. Kottra, G. Weber, and E. Frömter, *Pfluegers Arch.* **415,** 235 (1989).

attempts to quantify g_{sh} will be in error if leaks are present ($g_{leak} \neq 0$). Studying the effect of various leak artifacts on impedance measurements in gall bladder epithelium, Gordon et al.[9] concluded that leaks produced a change in the distribution of current density across epithelia. They found that the distribution of current was dependent on the size and position of the leak and, in the presence of leaks, on the frequency of the alternating current (AC). The next step was to use this information to quantify edge leaks for the proper correction of measured transepithelial resistances.

In a preparation with an edge leak, a flow of direct current (DC) through the leak will be greater than through an equal area of undamaged tissue; i.e., for DC an edge leak reduces the current density at the center of the preparation. On the other hand the cellular membranes which offer a resistive pathway to low-frequency currents also provide a capacitive shunt to high-frequency AC so that the electrical properties of the leak become less distinguishable from those of the remaining tissue as the frequency increases; i.e. as the frequency of the AC increases, the current density across the tissue becomes more uniform. If the central current density increases with increasing frequency then a leak must be present at the chamber edge.

These principles set the stage for the quantification of leak conductances. Kottra et al.[8] have shown that for the measurement of leak conductances in an Ussing chamber, a knowledge of the distribution of the electrical field inside the chamber is required as a function of the absolute transepithelial resistance. The calculated field distribution also depends on the geometry of the chamber as well as on the size, shape, and position of the electrodes. Here we shall discuss in outline the experimental and theoretical aspects of the quantification of edge leaks and refer the reader to Kottra et al.[8] for details.

The technique is applicable to preparations in which the epithelium lies horizontally between the Ringer's solutions of two cylindrical half-chambers[7] (Fig. 3). This configuration with an open upper chamber allows the insertion along the chamber axis (a) of two electrodes: (1) a shielded, broken tipped electrode[10] (b) placed just above the epithelium (f), and (2) an Ag/AgCl pellet reference electrode placed near the fluid surface at the top of the chamber (c). The first electrode was connected to an electrometer amplifier and the latter electrode or the Ag/AgCl grid electrode (e) in the lower chamber was alternatingly used to ground the preparation. In addition each chamber contained platinized Pt-ring electrodes (d) for pass-

[9] L. G. M. Gordon, G. Kottra, and E. Frömter, this series, Vol. 171, p. 642.
[10] G. Kottra and E. Frömter, *Pfluegers Arch.* **396**, 156 (1982).

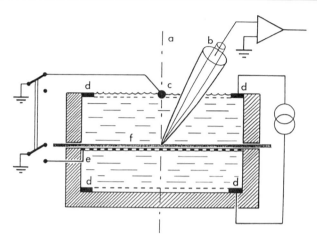

FIG. 3. Schematic view of experimental set-up. For detail see text. [G. Kottra, G. Weber, and E. Frömter, *Pfluegers Arch.* **415**, 235 (1989).]

ing current across the tissue from a ground-free regulated constant current source.

In the method described by Kottra *et al.*[8] a composite sine wave current of 20 frequencies and of equal amplitude were passed across the preparation and two sets of voltages were recorded by the microelectrode. These sets were produced by the same signals but measurements were made with respect to either the Ag/AgCl grid electrode or the Ag/AgCl pellet in the upper bath. After fast Fourier transform of the measured voltages and applied currents, the transepithelial impedances and the frequency-dependent drop in voltage along the axis of the upper fluid bath were calculated, respectively. Values of the frequency-dependent parameters, when extrapolated to zero and infinite frequencies, were used in the analysis. This procedure was convenient with the apparatus at hand, although parameter values at only two frequencies were required. (Kottra *et al.*[8] mentioned that the accuracy of the method could be improved with a tuned amplifier and consecutive measurements made while passing one low- and one high-frequency current across the tissue.)

Since the amplitude of the applied alternating current was independent of frequency, the ratio of the apparent bath resistances $(R_{bath})_{f \to 0}/(R_{bath})_{f \to \infty}$ equals the ratio of the induced bath potentials $(\Delta V_{bath})_{f \to 0}/(\Delta V_{bath})_{f \to \infty}$. However, the ratio of the true bath resistances is unity so that the ratio of the apparent bath resistances is due to a variation of the current density with frequency at the center of the chamber. Indeed, the above voltage ratio is equal to the inverse of the corresponding current density ratio and where values of the apparent bath resistance ratio are less than unity an edge leak is present.

To evaluate the magnitude of the edge leak it is necessary to compare the experimental data generated using the Laplace equation

$$\partial^2 V/\partial X^2 + (1/X)\partial V/\partial X + \partial^2 V/\partial Y^2 = 0 \tag{1}$$

in conjunction with other equations representing the boundary conditions imposed by the experimental set-up. $V = f(X, Y)$ is the voltage within the fluid cylinder of the upper chamber in which X and Y are, respectively, the radial and vertical coordinates. A number of boundary conditions relate specifically to the chamber, e.g., (1) for all values of Y, $\partial V/\partial X = 0$ at the wall and at the axis of the upper chamber and (2) since a ring electrode is used for carrying current at the upper surface, the vertical component of the field ($\partial V/\partial Y$) at the free aqueous surface is zero. Other boundary conditions relate to the resistivity, areas, and location of the intact and damaged cells. The equipotential lines produced in the chamber fluid by a particular electrode arrangement in the absence (a) and presence (b) of an edge leak are illustrated in Fig. 4a and b.

The practicability of this approach was tested in experiments performed on frog skins where the transepithelial and apparent bath impedances were measured before and after deliberate induction of edge leaks.[8] The results are shown in the form of Nyquist plots in Fig. 5. Comparison of panels a and b show that upon leak induction the recorded transepithelial resistance decreased as evidenced by contraction of the epithelial impedance semicircle (measurements E) while simultaneously the data points of ΔV_{bath} (represented in measurements F as apparent equivalent fluid resistances) dispersed from a single dot (in Fig. 5a) into a semicircle (in Fig. 5b). This disposition indicates that the ΔV_{bath} changes with the frequency of the applied current, and a closer look at the data reveals that the bath potential response to high-frequency current remains essentially constant while the response to low-frequency currents decreases when the preparation becomes leaky.

The coordinates of the points of the loci, E and F, vary with the magnitude of the edge leak. The basis of the method of leak measurement and correction is the intimate relationship between the two loci. If from the same tissue E_{leak}, F_{leak} and E, F are two corresponding pairs of loci of which three are established, then the fourth can be determined, i.e., by obtaining E_{leak} and F_{leak} experimentally and by choosing F as the corresponding locus of the hypothetically leak-free state (a dot with coordinates identical to that found at the high-frequency end of F_{leak}) E can be determined theoretically.

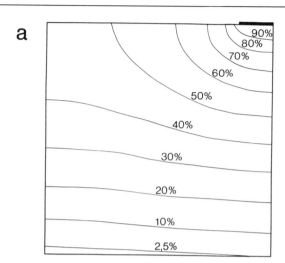

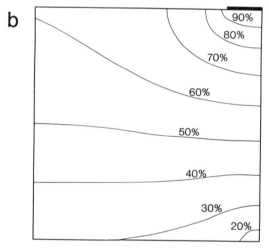

FIG. 4. Calculated electrical field in Ussing chambers without (a) and with (b) circumferential edge leak (DC case). Depicted are equipotential lines representing the step-wise drop (in percentage units) of the electrical potential in the chamber as a function of radial distance (abscissa, full length = 0.5 cm) and chamber height (ordinate, full height = 0.5 cm). The left border of each panel represents the chamber axis and right border the chamber wall. R_t was taken as 10 kΩ·cm^2 in (a) and (b) and in (b) the leak resistance of 25.5 kΩ was confined to a ring of width 25 μm at the circumference. The resulting absolute voltages at the lower left corner of (a) and the lower right corner of (b) were, respectively, 0.993 and 0.987, which may be compared to the voltages of 1.0 at the upper current electrode and 0 below the epithelium. The upper electrode was a ring of 0.08-cm width. [G. Kottra, G. Weber, and E. Frömter, *Pfluegers Arch.* **415**, 235 (1989).]

a

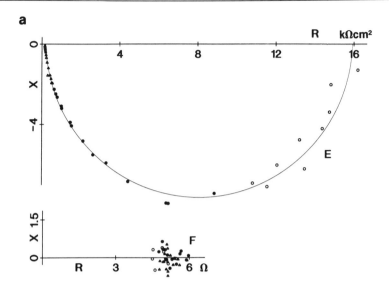

b

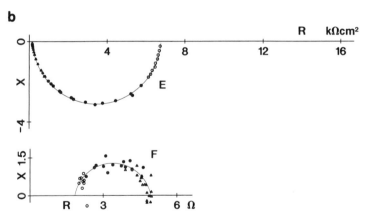

FIG. 5. Electrical impedance and frequency-dependent bath potential measurements on a frog skin in the Ussing chamber before (a) and after (b) deliberate induction of an edge leak. The impedance measurements (E) are depicted as real (R) versus imaginary (X) components (both in $k\Omega \cdot cm^2$) and the same convention has been used for the bath potential measurements (F, after division by the transepithelial current following fast Fourier transform), except that the data are presented in ohms both on the real and imaginary axes. Different symbols indicate frequency ranges as follows: (O), <10 Hz; (●), 10–100 Hz; (△), 100–1000 Hz; (▲), >1 kHz. Semicircles with center suppression were fitted by inspection. Note that the bath potential measurements fall onto a single dot in (a) but after induction of edge leak become frequency dependent and disperse into a semicircle in (b). [G. Kottra, G. Weber, and E. Frömter, *Pfluegers Arch.* **415**, 235 (1989).]

The locus E provides information concerning the parameters of the epithelium in the leak-free condition and in particular the transepithelial resistance, R_t.

With reference to Fig. 5, the measured transepithelial resistivity is given by the difference in the "real" coordinates of the end points of the extrapolated locus on the real axis. Kottra et al.[8] found that the theoretical or true transepithelial resistivity (R_t) could be calculated from measured transepithelial resistivity (R_t') by the transformation

$$R_t' (1 - X_0)/((\Delta V_{\text{bath}})_{f \to 0}/(\Delta V_{\text{bath}})_{f \to \infty}) - X_0) \to R_t \qquad (2)$$

where $(\Delta V_{\text{bath}})_{f \to 0}/(\Delta V_{\text{bath}})_{f \to \infty}$ is the correction factor based on the transformation of locus F and where X_0 is the correction factor relating to the characteristics of the Ussing chamber and to R_t itself (Fig. 6). Although the dependency of X_0 on R_t introduces a problem in the correction procedure,

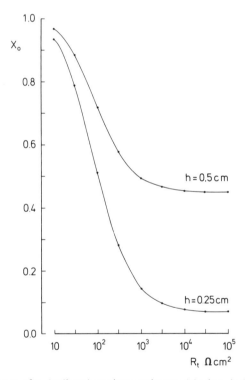

FIG. 6. Dependence of x_0 (ordinate) on tissue resistance (abscissa, in logarithmic scale in $\Omega \cdot cm^2$). Calculations are made for chambers of 1-cm diameter and two different heights, filled with Ringer's solution. Note that x_0 becomes independent of tissue resistance above $\approx 1\ k\Omega \cdot cm^2$. [G. Kottra, G. Weber, and E. Frömter, Pfluegers Arch. 415, 235 (1989).]

it appears to be confined to tissues of resistivity lower than $1 \text{ k}\Omega \cdot \text{cm}^2$. (For further details on the correction procedure refer to Kottra et al.[8]).

Methods to Reduce or Avoid Edge Leak and Edge Damage

As mentioned above the first attempt to develop a mounting technique that could regularly yield perfect seals was made by Helman and Miller.[5] Based on experience with sealing isolated renal tubules in a glass pipet by means of Sylgard[11] they glued two solid Lucite gaskets to either side of the frog skin, then sealed the gaskets in recesses of a Lucite chamber with liquid Sylgard. The advantage of this approach is that virtually no pressure is applied to the tissue at the place where the seal is formed since the liquid Sylgard is held in place essentially by coherence and adherence to the tissue surface. Helman and Miller[5] tested the validity of their mounting technique by measuring V_t, R_t, and I_{SC} in chambers which allowed increasing tissue surface areas to be exposed during the measurement. With the use of Sylgard the previously observed dependence of V_t and R_t on tissue area was no longer seen. What appears to be the best design of the chamber edge for this type of seal formation is shown in Fig. 7A.

Since liquid Sylgard may sometimes float away to cover the water surface of the chamber, particularly if experiments are performed at 37° rather than at room temperature, alternative approaches have been developed by others. Either gaskets of semicured Sylgard were prepared which sealed the tissue by slight pressure against the apical surface or, following the same principle, Sylgard was directly positioned onto the edge of a half-chamber and semicured under warming after addition of some hardener (50% of quantity normally recommended). Alternatively, instead of Sylgard, ordinary silicone grease has been used in virtually the same way. It has the advantage of being readily available in graded viscosities which can be either mixed or sandwiched and it is directly applicable without precuring. A simple test for the choice of the grade of silicone grease is in its ability to be drawn into threads when examined for stickiness. With the latter three techniques, pressure is required to form the seal as depicted in Fig. 7B. In some experiments this pressure has been directly measured.[12] While toad urinary bladder needed and tolerated several kilopascals of clamping pressure *Necturus* urinary bladder tolerated only 20–50 kPa and cultured collecting duct epithelium of rabbit needed about 6 kPa.

A problem with all techniques reported above is that the results are not always fully reproducible. Moreover, it is by no means clear that leaks are

---

[11] S. I. Helman, J. J. Grantham, and M. B. Burg, *Am. J. Physiol.* **220**, 1825 (1971).
[12] P. Gross, W. W. Minuth, W. Kriz, and E. Frömter, *Pfluegers Arch.* **406**, 380 (1986).

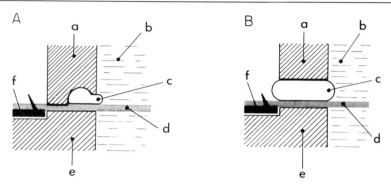

FIG. 7. Designs of Ussing chamber. Details of construction at chamber edge: a, Upper chamber wall; b, upper chamber fluid; c, Sylgard (liquid or semi-cured) or sticky silicone grease; d, epithelium; e, lower chamber wall; f, mounting frame. In (A) variability of mounting pressure is reduced. Liquid Sylgard is kept in position by housing. In (B) mounting pressure is variable.

definitely excluded. Indeed none of the techniques has been carefully tested for the persistence of small residual edge leaks because no handy and highly sensitive test was hitherto available. Indeed, that the mounting technique with Sylgard or silicone grease is not foolproof but may be associated with the persistence of small leaks, can be seen in Table I.

TABLE I
MEASURED (R'_t) AND CORRECTED (R_t) TRANSEPITHELIAL RESISTANCE OF FROG SKINS IN USSING CHAMBER

Skin/measurement	R'_t (kΩ·cm²)	$(\Delta V_{bath})_{f \to 0}/(\Delta V_{bath})_{f \to \infty}$	r_{leak}[a] (kΩ)	R_t (kΩ·cm²)
1/1	15.77	1.00	∞	15.77
1/2	6.63	0.40	3.23	19.19
1/3	2.92	0.26	1.16	14.68
2/1	6.32	0.54	4.05	12.56
2/2	2.60	0.27	1.04	12.66
2/3	1.95	0.20	0.71	15.60
3/1	6.20	0.89	16.97	7.02
3/2	4.45	0.68	4.02	6.87
3/3	4.00	0.53	2.49	8.20

[a] r_{leak} is the absolute value of the leak resistance calculated from Eq. (2) and the relation, $R'_t = R_t A r_{leak}/(R_t + A r_{leak})$, where A is the area of the epithelium. [G. Kottra, G. Weber, and E. Frömter, *Pfluegers Arch.* **514**, 235 (1989).]

Although all skins had been mounted under great care using silicone grease to seal the chambers and although the transepithelial resistances were so high that the preparations would have been readily accepted as leak free in virtually every laboratory, anywhere in the world, the determination of the frequency-dependent bath voltage clearly revealed the presence of edge leaks in two out of three preparations.

Concluding Remarks

In this chapter we have reviewed the techniques to detect and to "prevent" edge leaks when mounting epithelia for biophysical studies in Ussing chambers and we have described a technique which can readily be applied to quantify residual leaks. With the help of this technique it will be possible in the future to properly correct measured transepithelial resistances and potential differences for the effect of edge leaks or edge damage. This aspect is particularly important for all studies which attempt to quantify the paracellular shunt conductances in high resistance epithelia because hitherto it was not possible to decide whether any measurable noncellular conductance originated from tight junctional pathways or was caused by edge leak artifacts.

The question of how leaks or edge damage arise and what they consist of has not been considered in detail yet, except in the case of edge damage where inflated lateral spaces have been observed in the phase microscope. They may arise from the inability of absorbed fluid to proceed along the lateral spaces into the compressed subepithelial compartment. The idea that cells are damaged and pinched off, to leave open gaps across which the leak current can pass, seems less likely because no supporting microscopic evidence has been obtained and because only a relatively small number of destroyed cells would suffice to generate the commonly observed leaks. Let us, for example, consider a tight epithelium of $R_t = 10$ k$\Omega \cdot$cm^2 which has been mounted in a cylindrical chamber of 1-cm^2 area with an edge leak of 10 kΩ so that we measure a value of $r_t = 5$ kΩ. If a single cell has the dimension of $20 \times 20 \times 20$ μm and is filled with Ringer's solution of 100 $\Omega \cdot$cm resistivity, we calculate that the entire leak resistance could be generated by the loss of 20 cells, while the outermost cell ring of the preparation contains $\sim$2000 cells. It appears unlikely that only 20 cells out of 2000 can be destroyed. Alternatively it might be that many more cells are only partially damaged or that the pressure that builds up within the lateral spaces and distends them, eventually disrupts the tight junctions, similarly the formation of seals or the persistence of leak pathways between the apical surface of the epithelial cells and the sealing material are only poorly understood. In theory extremely thin fluid layers would suffice to generate the commonly observed leak conductances. It is possible that sealing is not so much achieved by squeezing the fluid out from the slit that extends between apical cell membrane and sealing material but rather by transporting the ions out of this fluid. This is at least suggested by the observation that sealing of toad urinary bladder whose surface is usually covered with mucus, takes time, and may reach a final level only after 1–2 hr.

Acknowledgments

The work was supported by the Deutsche Forschungsgemeinschaft and the Medical Research Council (New Zealand).

[46] Amphibian Nephron: Isolated Kidney and Cell Fusion

By H. Oberleithner, B. Gassner, P. Dietl, and W. Wang

Introduction

It has long been recognized[1] that the amphibian kidney can serve as a unique model for investigating renal transport mechanisms. As in the mammalian kidney there is morphological[2] and functional[3] segmentation along the nephron which allows the experimenter to study individual transport properties in detail. A major breakthrough occurred when intracellular microelectrode techniques were applied to individual nephron portions.[4-6] With the advent of ion-sensitive microelectrodes the amphibian kidney preparation was pushed further into the center of experimental activities.[4-9] There are indeed several crucial advantages over the mamma-

[1] A. N. Richards, "Methods and Results of Direct Investigations of the Function of the Kidney." Williams & Wilkins, Baltimore, Maryland, 1929.

[2] R. Taugner, A. Schiller, and S. Ntokalou-Knittel, *Cell Tissue Res.* **226,** 589 (1982).

[3] S. Long and G. Giebisch, *Yale J. Biol. Med.* **52,** 525 (1979).

[4] E. E. Windhager, E. L. Boulpaep, and G. Giebisch, *Proc. 3rd Int. Congr. Nephrol., Washington, D.C., 1966* **1,** 35 (Karger, Basel/New York, 1967).

[5] T. Hoshi and F. Sakai, *Jpn. J. Physiol.* **17,** 627 (1967).

[6] G. Giebisch, D. Cemerikic, H. Oberleithner, W. Guggino, and B. Biagi, *in* "Ion Transport by Epithelia," p. 163. (S. G. Schultz, ed.). Raven, New York, 1981.

[7] D. Ammann, "Ion-Selective Microelectrode." Springer-Verlag, Heidelberg, 1986.

[8] M. Fujimoto and T. Kubota, *Jpn. J. Physiol.* **26,** 631 (1976).

[9] W. B. Guggino, H. Oberleithner, and G. Giebisch, *J. Gen. Physiol.* **86,** 31 (1985).

lian preparation: (1) the kidney can be easily prepared for study; (2) various portions of the nephron (proximal tubule, diluting segment, collecting duct) are accessible to microelectrode techniques; (3) cells are large enough to allow intracellular impalements in minutes; (4) the amphibian kidney has a dual blood supply, i.e., the peritubular space and the lumenal compartment can be perfused separately; and (5) preparation is sturdy enough to allow experiments of more than 12 hr without compromising cell function. At the same time, basic transport properties such as glucose and amino acid transport,[10,11] H^+ and HCO_3^- transport,[12–15] and steroid hormone-regulated K^+ transport[16] are comparable with those in the mammalian kidney.

If the experimenter's major interest is the electrophysiological evaluation of the intracellular compartment and its regulatory function in controlling membrane transport processes, stable intracellular measurements with microelectrodes are often required over long periods while the extracellular fluid is well controlled. This requirement led us to develop a cell about 10 times larger in diameter than the single epithelial cell of the intact epithelium. These "giant" cells can survive for many hours in well-defined media while micromanipulations are performed inside the cells.

In this chapter we will give a detailed description of the isolation and perfusion of the frog kidney, which has served for some years as a reliable experimental model in various studies.[17–20] Then we will focus on a novel technique for fusing single epithelial cells to form giant cells. This experimental model could be a valuable tool for future research in various scientific disciplines.

[10] F. Lang, G. Messner, W. Wang, M. Paulmichl, H. Oberleithner, and P. Deetjen, *Pfluegers Arch.* **401**, 14 (1984).

[11] G. Messner, H. Oberleithner, and F. Lang, *Pfluegers Arch.* **404**, 138 (1985).

[12] W. F. Boron and E. L. Boulpaep, *J. Gen. Physiol.* **81**, 29 (1983).

[13] W. Wang, G. Messner, H. Oberleithner, F. Lang, and P. Deetjen, *Pfluegers Arch.* **401**, 6 (1984).

[14] Y. Matsumura, S. Aoki, and M. Fujimoto, *Jpn. J. Physiol.* **35**, 741 (1985).

[15] G. Planelles, A. Kurkdjian, and T. Anagnostopoulos, *Am. J. Physiol.* **247**, F932 (1984).

[16] H. Oberleithner, M. Weigt, H.-J. Westphale, and W. Wang, *Proc. Natl. Acad. Sci. U.S.A.* **84**, 1464 (1987).

[17] H. Oberleithner, F. Lang, G. Messner, and W. Wang, *Pfluegers Arch.* **402**, 272 (1984).

[18] H. Oberleithner, P. Dietl, G. Münich, M. Weigt, and A. Schwab, *Pfluegers Arch.* **405**, S110 (1985).

[19] H. Oberleithner, *Pfluegers Arch.* **404**, 244 (1985).

[20] H. Oberleithner, G. Münich, A. Schwab, and P. Dietl, *Am. J. Physiol.* **251**, F66 (1986).

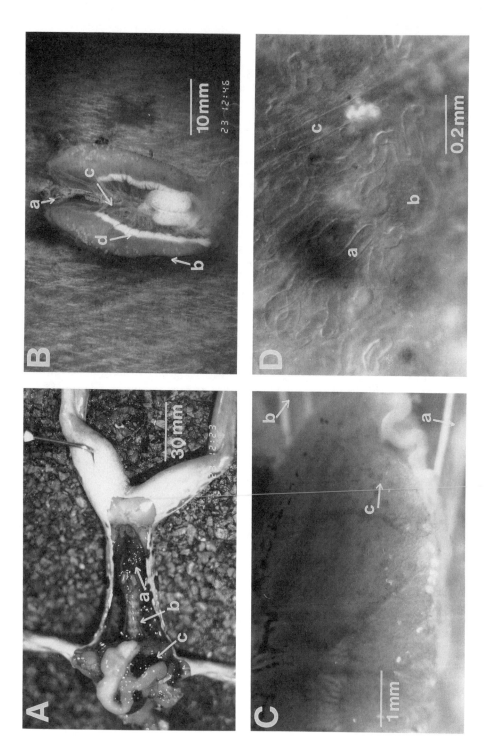

The Isolated Perfused Frog Kidney

How to Isolate the Kidneys

To perfuse the lumenal and the peritubular compartments adequately it is necessary to isolate the kidney from the rest of the organism without significant damage to the epithelial tissue and the blood vessel supply. This can be done best in *Rana esculenta* or *Rana pipiens*. Amphibia such as *Xenopus laevis*, *Amphiuma*, or *Necturus* are less suitable because of large amounts of connective tissue in and close to the kidney. After the frog is decapitated and pithed, the abdomen is opened. The intestines are carefully removed until both kidneys are visible at the ventral surface of the spinal cord. With a fine forceps the aorta and the superior poles of the kidneys are lifted up. Then, with the aorta as a bridle, the kidneys are pulled off the spinal cord and removed with fine scissors. They are then transferred onto a flat Plexiglas dish and cannulated. Figure 1 shows some photographs of the kidneys of *Rana pipiens* before and after cannulation.

How to Perfuse the Kidneys

For electrophysiological experiments, it is of paramount importance to maintain well-defined conditions at the two sides of the tubule wall, the lumenal (apical) and the peritubular (basolateral) cell membranes. If level flow conditions on both sides of the epithelium are required, then both lumenal and peritubular compartments must be perfused as fast as possible to avoid the problems of an unstirred layer. If static head conditions are to be achieved, the lumen flow rate must be zero while the peritubular compartment is perfused rapidly. These requirements can be met only if there are no major leaks in the kidney preparation. Figure 2 summarizes the individual steps for an adequate perfusion. The portal veins (one or both, depending on whether experiments are done only in one or in both kidneys) are cannulated with hand-pulled polyvinyl chloride (PVC) tubes. These cannulas deliver the peritubular perfusate into the capillary network of the kidney. Next, the caudal ending of the aorta is cannulated. This tubing delivers the lumen perfusate. In order to force the fluid through the glomeruli and the tubule compartment, the cranial endings of the aorta

---

FIG. 1. Isolation of the kidney of *Rana pipiens*. (A) *In situ* view: a, Kidneys; b, aorta; c, intestines. (B) The isolated kidneys (ventral surface): a, Aorta; b, portal vein of the right kidney; c, caudal vein; d, interrenal aldosterone-producing tissue. (C) Ventral kidney surface (caudal portion): a, Tubing into the portal vein; b, tubing into the aorta; c, glomerulus. (D) Ventral kidney surface, higher magnification: a, Diluting segment; b, glomerulus; c, vein. (The photographs were kindly provided by A. Schwab.)

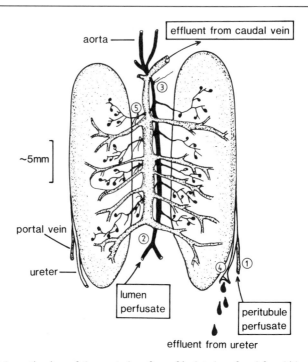

FIG. 2. Schematic view of the ventral surface of isolated perfused frog kidney. (1) Insert tubing into the portal vein; (2) insert tubing into the aorta; (3) close the cranial endings of the aorta; (4) open the ureter wide to release the aortic (lumenal) perfusate; (5) close the arteries branching off into the *right* kidney when performing the experiments only in the *left* kidney. (Reproduced from Ref. 17.)

must be ligated. If this step is successfully performed, the kidneys expand when fluid is allowed to enter. Finally, the outlets (ureters) are opened wide to allow the perfusate to escape. If studies are limited to one kidney, the arterial branches of the other one can be ligated to direct the aortic perfusate into the kidney under study.

How to Make a Perfusion System

First, we prepare a piece of 10-cm PVC tubing, making one end suitable to fit into the vein or the aorta, and the other suitable to take up at least four polyethylene (PE-50) tubes (5–16 cm). These tubes are glued into the "mouth" of the PVC tubing at one end and at the other end they are connected to appropriate hypodermic needles built into a tiny rack in front of the kidney preparation. CO_2-tight tubes (2 m) connect the needles with glass bottles containing the various perfusion solutions. The perfusion pressure can be varied from 2 m H_2O to zero by lifting the reservoirs up or down. In addition, three-way valves are incorporated into the CO_2-tight tubes to start and stop the individual perfusions. Lissamine Green dye (5 g/liter) is used in a perfusion solution to test for adequate perfusion (rate of perfusion, homogeneous distribution of the perfusate in the kidney, leaks, etc.).

How to Find Different Tubule Segments

If studies are performed in the intact kidney, only those tubules can be selected for electrophysiological measurements that are close to the surface. The dorsal surface of the kidney is almost entirely composed of proximal tubule segments. The length of an individual nephron portion, located at the surface and visible under low microscopic magnification ($\times$100), is about 500 μm. This is long enough to perform cable analysis.[10,11] One of the unique and exciting features of frog kidney is the fact that the ventral surface is composed mainly of diluting segments.[1,3,7-20] These segments have the functional characteristics of the thick ascending limb of Henle's loop in the mammalian kidney[21,22] but, in contrast to the mammalian preparation, are located at the ventral surface (Fig. 1). They are in close contact with glomeruli, which are frequently found at the surface of the lower pole of the kidney and which can be impaled easily with pipets. Visible at the lateral edges of the kidneys are collecting ducts, which later join to form the ureter. The aldosterone-producing interrenal tissue[23] is clearly visible as yellow insulae lined up along the ventral surface of the kidneys. Electrophysiological studies can be performed for at least 12 hr when the kidneys are adequately perfused with amphibian Ringer's solution containing glucose as a substrate.

Fusion of Renal Epithelial Cells

Reasons for Isolating Renal Epithelial Cells

Intracellular studies in the intact tubule are beset with various problems: (1) cells are small (10 μm) and impalements quite often leaky; (2) intracellular impalements for hours are not feasible; (3) paracellular current loops lead to misinterpretations of cell membrane properties; and (4) the visibility of intracellular events (e.g., vesicle formation) is rather limited.

[21] M. B. Burg and N. Green, *Am. J. Physiol.* **224** (3), 659 (1973).
[22] R. Greger, *Physiol. Rev.* **65** (3), 760 (1985).
[23] C. Maser, P. A. Janssens, and W. Hanke, *Gen. Comp. Endocrinol.* **47**, 458 (1982).

Reasons for Fusing Single Cells to make Large Cells

1. We observed that isolated single epithelial cells die seconds after impalement with microelectrodes.[24] This is probably due to the unfavorable volume-to-surface ratio of a small isolated cell (small volume, large surface).

2. Large cells allow micromanipulations (e.g., dye injection and insertion of more than one microelectrode) not feasible in the single cell.

3. Eventually different cell types can be fused. Then, transport mechanisms operating in parallel in one individual cell can be studied.

Methods Available for Cell Fusion

The polyethylene glycol (PEG) method has become an important tool to fuse cells of various origin. Fusion of myeloma cells with spleen cells has led to the derivation of permanent cell lines producing homogeneous antibodies.[25] Plant protoplasts,[26] mouse blastocysts,[27,28] human erythrocytes,[29] and many more (the list is far from complete) have been successfully fused by the PEG method. There are excellent articles dealing with the more theoretical aspects of the fusion process itself.[29-34] Another promising approach is the so-called electric pulse-induced fusion.[35,36] This elegant technique has been used successfully in cultured cell lines.[37-40] We have applied the PEG method to fuse epithelial cells, isolated from individ-

---

[24] H. Oberleithner, B. Schmidt, and P. Dietl, *Proc. Natl. Acad. Sci. U.S.A.* **83** (10), 3547 (1986).

[25] G. Galfre, S. C. Howe, C. Milstein, G. W. Butcher, and J. C. Howard, *Nature (London)* **226**, 550 (1977).

[26] K. N. Kao and M. R. Michayluk, *Planta* **115**, 355 (1974).

[27] M. A. Eglitis, *J. Exp. Zool.* **213**, 309 (1980).

[28] A. Spindle, *Exp. Cell Res.* **131**, 465 (1981).

[29] S. Knutton, *J. Cell Sci.* **36**, 61 (1979).

[30] F. S. Cohen, *Science* **217**, 458 (1982).

[31] J. W. Wojcieszyn, R. A. Schlegel, K. Lumley-Sapanski, and K. A. Jacobson, *J. Cell Biol.* **26**, 151 (1983).

[32] R. J. Westerwoudt, *J. Immunol. Methods* **77**, 181 (1985).

[33] G. Poste and A. C. Allison, *Biochim. Biophys. Acta* **300**, 421 (1973).

[34] J. A. Lucy, *Cell Surf. Rev.* **5**, 267 (1978).

[35] U. Zimmermann and J. Vienken, *J. Membr. Biol.* **67**, 165 (1982).

[36] J. Vienken, U. Zimmermann, H. P. Zenner, W. T. Coakley, and R. K. Gould, *Biochim. Biophys. Acta* **820**, 259 (1985).

[37] C. Finaz, A. Leferee, and J. Teissie, *Exp. Cell Res.* **150**, 477 (1984).

[38] J. Teissie, V. P. Knutson, T. Y. Tsong, and M. D. Lane, *Science* **216**, 537 (1982).

[39] T. Ohno-Shosaku and Y. Okada, *J. Membr. Biol.* **85**, 269 (1985).

[40] M. M. S. Lo, T. Y. Tsong, M. K. Conrad, S. M. Strittmatter, L. D. Hester, and S. H. Snyder, *Nature (London)* **310**, 792 (1984).

ual tubule segments, to form large cells. Our goal was to obtain cells as large as possible with cell membrane properties still intact.

How to Isolate and Fuse Cells

The methods for isolation and fusion of cells of the frog diluting segment have been published previously.[24] Here we will summarize the crucial steps and include some illustrations that give further insight into the fusion process itself. Figure 3 displays the single steps from kidney isolation to cell fusion.

Step A. Kidneys are removed using the methods described above.
Step B. Both kidneys are perfused via the aorta (see also Fig. 2) with

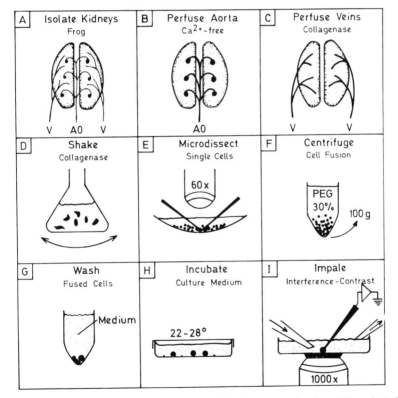

Fig. 3. Isolation and fusion. Illustration of the single steps starting from kidney isolation to the intracellular impalement of a fused cell with microelectrodes (Reproduced from Ref. 24.)

TABLE I

PARAMETERS THAT DETERMINE THE YIELD OF ISOLATION OF CELLS FROM RENAL
TUBULES OF THE FROG KIDNEY

Protease[a]	Concentration (%, w/v)	Treatment time (min)	Ca^{2+} (mM)	Temperature (°C)	Isolation yield
Trypsin	0.5	30	0	28	−
	0.5	120	0	4	+
	0.025	120	0	4	−
	0.025	10	0	4	−
	0.025	10	0 + 2 mM EDTA	28	−
	0.025	30	1.0	28	−
Collagenase I	0.1	10	1.5	28	−
	0.025	15	1.5	28	+
	0.05	10	1.5	28	+
	0.05	45	1.5	28	+
	0.05	60	0.1	28	+
Pronase	0.1	10	1.5	28	−
Pronase/dispase	0.1/0.1	30	1.5	28	−
Collagenase II	0.2	60	1.5	28	−

[a] Characterization of the proteases: trypsin (40 U/mg; Serva, Heidelberg, FRG), Pronase (5 U/mg; Serva), dispase (6 U/mg; Boehringer, Mannheim, FRG), collagenase I (180–300 U/mg; Sigma, Deisenhofen, FRG), collagenase II (140–150 U/mg; Seromed, Berlin, FRG).

Ca^{2+}/Mg^{2+}-free amphibian Ringer's solution (50–100 ml for 10 min; hydrostatic pressure: 100 cm H_2O). The Ringer's solution is composed of (mM) 87 NaCl, 3 KCl, and 10 HEPES titrated to a pH of 7.8 with 0.1 M NaOH. According to previous reports[41,42] this treatment should break up the intercellular junctions. Furthermore, 5×10^{-5} M furosemide is present in this perfusate to prevent cell swelling.

Step C. The kidneys are perfused with amphibian Ringer's solution containing Ca^{2+} (1.5 mM), Mg^{2+} (1.0 mM), and collagenase (for details see Table I) to dissolve the intercellular matrix. By means of fine scissors, superficial portions of 2–3 mm² (thickness, 1 mm) are removed from the ventral surface of the kidney. These kidney fragments contain almost exclusively diluting segments (see also Fig. 1).

Step D. The kidney fragments are transferred to a glass vial with 1 ml of the collagenase-containing kidney perfusate and are shaken gently (70 strokes/min) for up to 60 min at standardized temperature (28°).

Step E. The kidney fragments are microdissected with needles in the

[41] P. O. Seglen, *Exp. Cell Res.* **76**, 25 (1973).
[42] J. Graf, A. Gautam, and J. L. Boyer, *Proc. Natl. Acad. Sci. U.S.A.* **81**, 6516 (1984).

presence of some 100 μl of the collagenase-containing solution. Thus, the epithelial structure is dissolved and the individual cells can be collected with a Pasteur pipet.

Step F. Cells are transferred into a conical tube containing the fusion medium. This consists of 30% (w/v) polyethylene glycol (PEG, M_r 4000) dissolved in Leibovitz-15 medium. The fusion medium is diluted to about 170 mOsm, buffered with 10 mM HEPES, and titrated to an apparent pH of 8.1–8.6. Since the pH measurements in the fusion medium by means of pH-sensitive glass microelectrodes are unreliable (probably because of the interaction of PEG with the glass electrode), the fusion medium is titrated with 1 M NaOH until the indicator Phenol Red (present in the medium) changes completely from yellow to red. The cells are centrifuged (100 to 200 g) while exposed to PEG. Then the PEG is aspirated off the cell pellet. Total PEG exposure time is 3 min.

Step G. Serum-free Leibovitz-15 medium (diluted to 200 mOsm) is gently added (1 ml) on top of the cell pellet. The pellet is split into small fragments by tapping the tube. The cells are then washed by gently shaking the tube for a minute or two.

Step H. The cells are transferred onto a thin microscope coverslip coated with polylysine (poly-L-lysine, 0.1 g/liter; Serva, Heidelberg, FRG). They are placed inside a polyvinyl chloride ring (diameter 25 mm; 1 mm in height) that has previously been glued on the coverslip. More culture medium is then added (1 ml), and the cells are allowed to rest for at least 4 hr.

Step I. Giant cells are impaled with microelectrodes while circumfused rapidly. Recently we have added 10^{-4} M ATP-MgCl$_2$ to the medium to provide the cells with enough energy during the process of cell fusion.[43]

How Many Cells Form a Giant Cell

It can be 2 and it can be 100 cells fused together. Unfortunately, we are still unable to control this parameter satisfactorily. Figure 4 gives some information on the number of cells fused to form a giant cell. Staining the nuclei with the fluorescent dye acridine orange (1 g/liter) makes the nuclei visible.[44] A completely fused and spherical cell is shown in Fig. 5. Such cells can be impaled with microelectrodes several times without being significantly damaged. In our hands a giant cell must have a diameter of at least 50 μm to survive an intracellular impalement. Please note that cells in the intact epithelium can be as small as 2 to 5 μm (e.g., cells of the cortical

[43] M. E. Stromsky, K. Cooper, G. Thulin, M. J. Avison, K. M. Gaudio, R. G. Shulman, and N. J. Siegel, *Am. J. Physiol.* **250**, F834 (1986).
[44] R. G. O'Neil and R. A. Hayhurst, *Am. J. Physiol.* **248**, 1449 (1985).

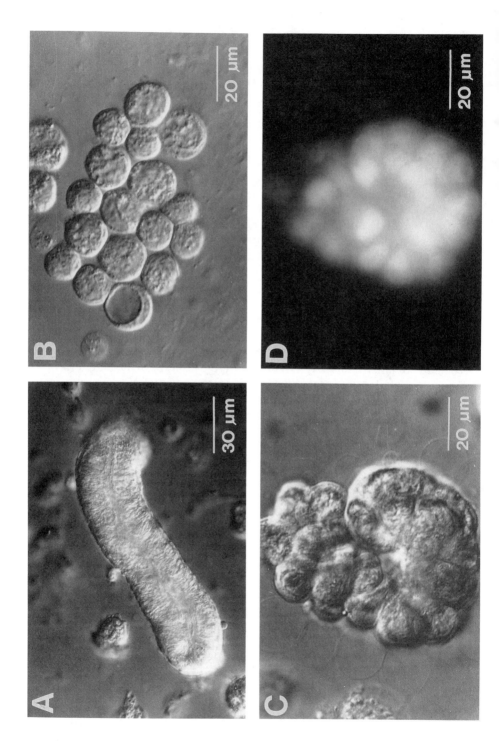

FIG. 5. Giant cell fused from single cells of the frog diluting segment. The photograph was taken about 6 hr after PEG treatment. Cell fusion seems complete. Differential interference contrast microscopy does not disclose any cytoplasmic compartmentation.

thick ascending limb of Henle's loop from the mammalian kidney) and still be impaled successfully.[22] This is probably due to low-resistance pathways from one cell to another which allow large ionic fluxes at the moment of the impalement. Then, the impalement damage is compensated.

Visualization of the Fusion Process

We tried to observe the process of cell fusion under the microscope. This can be nicely done by applying differential interference-contrast microscopy with high optical resolution (Zeiss IM 35, inverted microscope, objective lens 100/1.25 oil). Figure 6 shows how three cells of the frog diluting segment fuse. Fusion is induced by PEG and becomes visible *after*

FIG. 4. Tubule–single cell–fused cell. (A) Isolated amphibian tubule. (B) Isolated tubule cells. (C) Incomplete cell fusion, 10 min after PEG treatment. (D) Fused cell as in (C) but applying fluorescence microscopy. Nuclei are stained with acridine orange. (Excitation wavelength: 495 nm.)

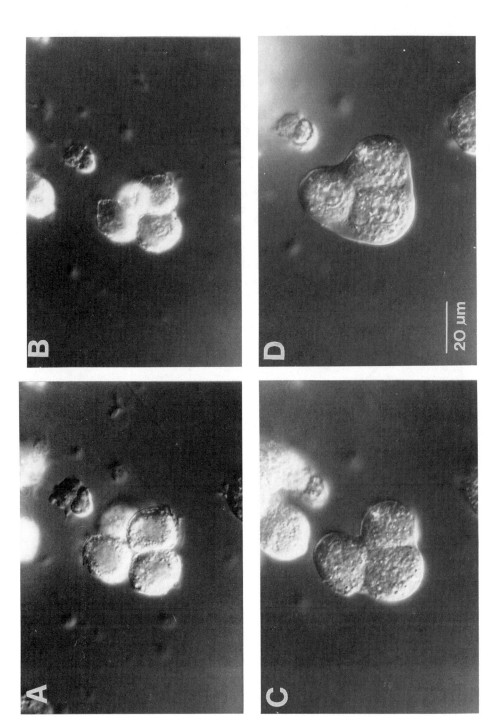

20 μm

culture medium is substituted for PEG. Figure 7 demonstrates that PEG-induced cell shrinkage is a dramatic event. Here, a giant cell fused from a cultured dog kidney cell line (MDCK cells) is exposed several times to PEG. The cell shrinks within seconds but regains its previous shape within seconds when PEG is replaced by culture medium.

Complete vs Incomplete Cell Fusion

Cells aggregate when exposed to PEG but quite often do not fuse. Two examples are given in Fig. 8. MDCK cells were treated with PEG. Iontophoretic injection of the fluorescent dye Lucifer Yellow stains the cytoplasm of single cells.[45] The dye is not distributed equally in the cell cytosol. This indicates incomplete fusion (i.e., cell aggregation). Two examples of complete fusion are given in Fig. 9. Again, MDCK cells were fused with PEG. Injection of Lucifer Yellow stains the cytoplasm immediately and completely.

We are still trying to increase the yield of completely fused cells. Obviously, there are several variables, from the biological preparation to the quality of the enzymes used to separate tubules and cells, that cannot be readily controlled. Tables I and II show various conditions applied during isolation and fusion. The + and − symbols give some indication of whether the results were positive or negative. Cell isolation is labeled positive (+) if (1) a *relatively* large number of cells can be isolated (the *absolute* number depends on the size of renal tissue removed for dissection); (2) cells are healthy (Trypan Blue uptake negative, no cell swelling, no "blebs," cells stick firmly to the coverslip); and (3) single cells survive for several hours in culture medium. Cell fusion is labeled positive (+) if (1) at least one giant cell is formed with a diameter of >50 μm; (2) the cytoplasm of the fused cell is homogeneous (as judged by Lucifer Yellow injection, differential interference contrast microscopy; stable cell membrane potentials when impaled with microelectrodes); and (3) giant cells survive for at least 4 to 6 hr (postfusion).

There are several crucial parameters that determine the yield of fusion.[32] We have found that the fusion procedure must be well adapted to the biological preparation. Thus, the above-described methods can be used

---

[45] W. W. Stewart, *Nature (London)* **292**, 17 (1981).

Fig. 6. Three epithelial cells from the frog diluting segment are fused under microscopic observation. (A) Three cells located close to each other before PEG treatment. (B) Superfusion of PEG. The cells shrink. (C) Three minutes later PEG is replaced by culture medium. The cells swell and fuse. (D) Ten minutes later the three cells have formed a common cell membrane.

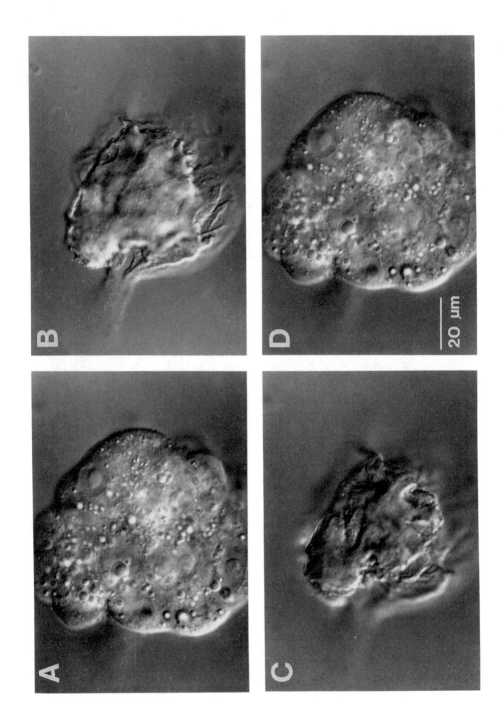

20 μm

TABLE II
THE FUSION MEDIUM AND ITS RELATION TO THE FUSION YIELD[a]

PEG (M_r)	PEG concentration (%, w/v)	Ca^{2+} concentration (mM)	pH	Osmolality (mOsm)	Exposure time (min)	Fusion yield
1500	30	1.0	8.1	260	3	+
1500	50	1.0	8.1	340	3	−
4000	30	1.0	8.1	170	3	+
4000	50	1.0	8.1	204	3	−
4000	30	1.0	8.1	170	10	−
4000	30	3.0	8.1	170	3	−
4000	30	1.0	7.4	170	3	−

[a] PEG is dissolved in Leibovitz-15 medium. pH is titrated with 1 M NaOH. The osmolality is adjusted with distilled water.

as a basic guideline for cell fusion but "fine tuning" is absolutely required for success. In *Table III* we give some directions and suggestions on how to overcome problems that will definitely occur every now and then in isolating and fusing renal epithelial cells.

Fusion of Renal Tubules

Recently, we modified the technique of single-cell fusion and adapted the fusion procedure to the specific geometry of the renal tubule. Our goal was to fuse neighboring cells in the *intact tubule* so that we should end up with a giant cell consisting of several hundred single cells. The method is quite simple and is now worked out sufficiently well that it can be described. The details are as follow:

Step A. Kidneys are removed using the methods described above.
Step B. Both kidneys are perfused via the aorta (see Fig. 2) with amphibian Ringer's solution (50–100 ml for 10 min; hydrostatic pressure: 100 cm H_2O). In contrast to the method for the isolation of single renal cells this solution contains Ca^{2+} and Mg^{2+}. Therefore, the intercellular junctions remain intact.

FIG. 7. Visualization of PEG treatment. (A) About a dozen cultured dog kidney cells (MDCK cells) have been fused by PEG. (B) A second time PEG is added. This photograph was taken immediately after exposure to PEG (50% PEG dissolved in Leibovitz-15 medium). (C) Cell shrinkage has approached a steady state about 30–60 sec later. (D) Replacement of PEG by culture medium allows the cell to regain its previous size and shape (the cultured MDCK cells were kindly provided by Dr. G. Gstraunthaler, Univ. Innsbruck, Austria).

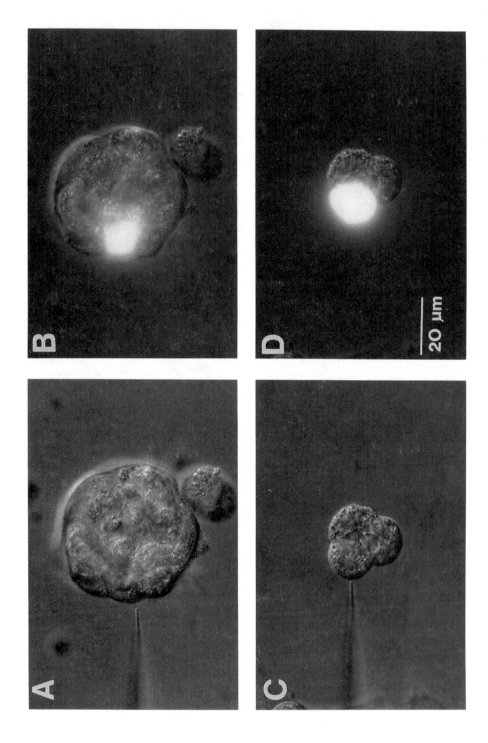

Step C. The kidneys are perfused with collagenase-containing Ringer's solutions via the peritubular veins to denudate the basal and basolateral cell surfaces enzymatically. The enzyme concentrations are similar to those used in the single-cell fusion technique. Superficial tissue is removed from the ventral surface of the kidney and incubated for about 30 min in the collagenase-containing Ringer's solution.

Step D. The kidney fragments are microdissected by needles and single tubules are cautiously transferred onto a thin microscope coverslip coated with poly-L-lysine (*vide supra*).

Step E. A few minutes later, after the tubule segments have settled on the glass surface that serves as the bottom of the perfusion chamber, PEG is superfused until dramatic cell shrinkage is visible (Fig. 10). Three minutes later the PEG is replaced by Leibovitz-15 medium, allowing the tubules to regain their original size.

Step F. The tubules are allowed to rest for *at least* 24 to 48 hr in culture medium at 4°. Over this time period fusion is completed. Injection of the fluorescent dye Lucifer Yellow allows one to distinguish between complete and incomplete fusion (Fig. 11). The giant cells are about 150 μm in length and 60 μm in diameter. Some of them become spherical (diameter of about 100 μm).

This specific fusion technique has the advantage that mixing of different cell types is excluded if, of course, the fused nephron segment itself is homogeneous. The transport proteins of the lumenal cell membrane (K^+ channels, Na^+/H^+ exchanger, $Na^+/K^+/Cl^-$ cotransporter) are found in the cell membranes of these giant cells since they remain sensitive to drugs and inhibitors known to interfere with those transport systems. Although we have not yet characterized the function of these cells completely, they could become a useful experimental model.

Future Perspectives

The giant cell fused either from single cells of renal epithelium or from intact tubules could serve as a valuable model in future research. At the

---

FIG. 8. Demonstration of an incomplete fusion (MDCK cells). (A) Fused cell about 4 hr after PEG treatment. Microelectrode filled with the fluorescent dye Lucifer Yellow (0.1 M) just before impalement at left. (B) Lucifer Yellow has been injected iontophoretically (10 pulses of 5×10^{-8} A, duration 1 sec). The dye is visible only in a sharply limited space of the cell. (C) Three cells have obviously fused. (D) The injection of Lucifer Yellow discloses that fusion has been incomplete. (A and C) Interference contrast; (B and D) Combination of interference contrast with fluorescence microscopy (excitation wavelength: 495 nm).

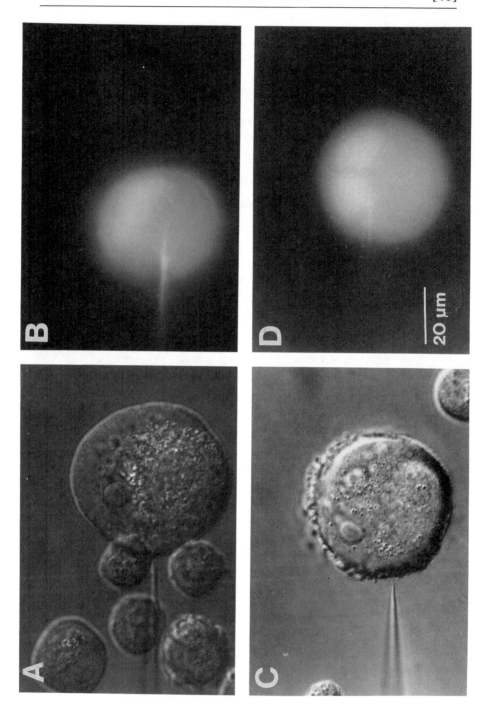

TABLE III
SUGGESTIONS FOR OVERCOMING ISOLATION AND FUSION PROBLEMS

Problem	Why	How to solve
Tubules isolated instead of cells	Intercellular junctions still intact	Perfuse the preparation with Ca^{2+}-free solutions properly
Small number of cells that die immediately after microdissection.	Enzyme treatment insufficient (concentration too low, exposure too short, enzyme activity reduced); cells are mechanically damaged	Use fresh enzymes, increase concentration and/or exposure time. Dissect gently
Large number of cells that die immediately after microdissection.	Enzyme treatment overdone (concentration too high and/or exposure time too long, enzyme contaminated), cell membrane structures are affected	Reduce enzyme treatment, use a fresh batch of enzymes
Cells die immediately after PEG treatment	pH of the fusion medium too low	Adjust pH to values above 8.0, use Phenol Red as indicator
Cells die immediately after PEG treatment	Medium added too rigorously after PEG treatment (cells swell too rapidly)	After removal of PEG add medium "slowly" on top of the cell pellet and tap "gently"
After PEG treatment, single cells alive but not fused	Enzymatic denudation of the cell membranes insufficient (poor cell-to-cell contact)	Increase enzyme treatment
	Cell number too small (physical contact missing)	Increase cell number
Cell fusion incomplete	Cell membranes not "clean" enough	Increase enzymatic treatment
Cells do not stick to the glass coverslip	Glass coverslip not coated	Coat with poly-L-lysine; use the L, not the D form
Cells stick but get loose after hours while resting in culture medium	Cell material from dead cells covers the coverslip and "attacks" the intact cells	Rinse the glass coverslip with medium 30 min after the cells have attached to it. Dead material will go, healthy cells will stay

FIG. 9. Two examples for complete fusion (MDCK cells). (A) Fused cells among single cells. The microelectrode at left is filled with the fluorescent dye Lucifer Yellow. (B) The cell is impaled and Lucifer Yellow is injected. The cell cytosol is stained homogeneously. (C and D) Another example of complete fusion. (A and C) Interference contrast. (B and D) Fluorescence microscopy.

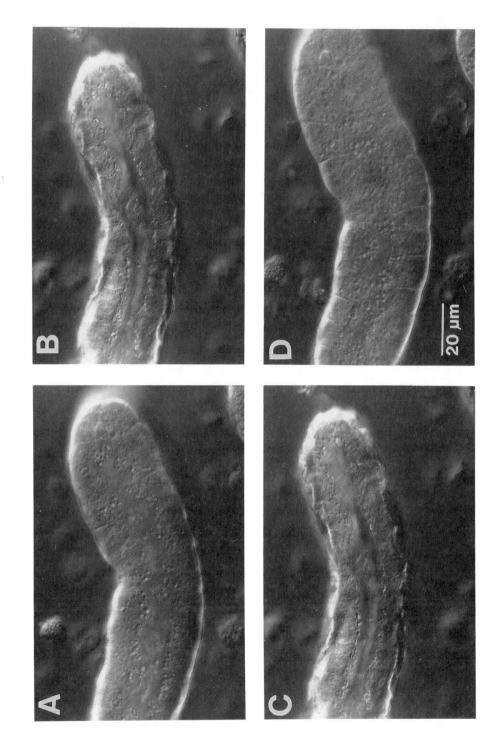

moment we see at least four major areas in which the giant cell could advance to the experimental model of choice:

1. Research which needs intracellular micromanipulations (as micro-injection of dyes and substances that cannot penetrate the cell membrane)

2. Research in which long-term intracellular measurements by micro-electrodes are required while the extracellular milieu of the cell under study is well controlled. We have recently used this model successfully to monitor intracellular pH with pH-sensitive microelectrodes for more than 60 min while the cell was exposed to the corticosteroid hormone aldosterone[16]

3. Research in which the cell membrane properties of one and the same cell need to be evaluated over days while the cell is exposed to specific stimuli (e.g., hormones). The cells can be impaled repeatedly without compromising cell function.

4. Research in which cells of different origins are fused to study transport mechanisms that usually do not exist in parallel in one individual cell. We recently have fused proximal and distal tubule cells from the amphibian kidney and found the rheogenic Na^+/glucose cotransporter and the electroneutral $Na^+/K^+/2Cl^-$ cotransporter operating in parallel in one and the same cell (H. Oberleithner, unpublished observation).

Still, we must keep in mind that after cell fusion the major characteristic of an epithelial cell, namely the cell polarity, has obviously vanished. Thus, investigating the *direction* of transepithelial transport can be adequately performed only in the intact tubule.

FIG. 10. Fusion of renal tubules. (A) Diluting segment before PEG treatment. (B) One minute after addition of PEG. (C) Three minutes after addition of PEG. (D) The same segment after replacement of PEG by culture medium. Note that the individual cell borders are still visible after PEG treatment. They disappear within the following 24 to 48 hr.

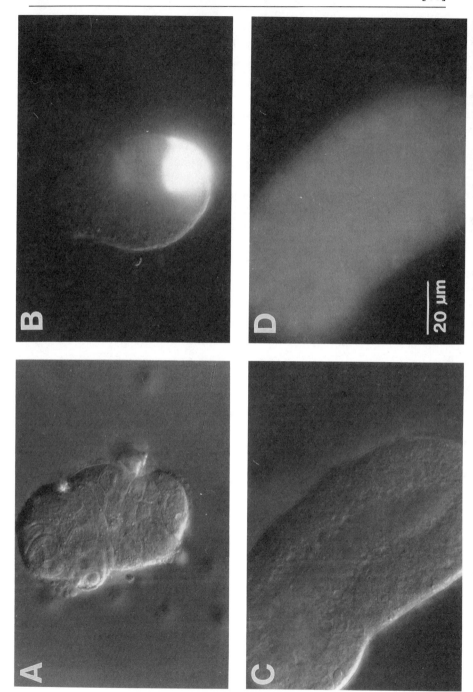

Acknowledgments

The authors thank Dr. S. Silbernagl and Dr. G. Giebisch for their encouragement and enthusiastic support while establishing the fusion technique, and Dr. W. Dantzler for constructive criticism of the paper. We gratefully acknowledge the collaboration with Dr. G. Gstraunthaler and Dr. W. Pfaller (Dept. of Physiology, University of Innsbruck, Austria), who provided the MDCK cells. The fusion experiments were designed in collaboration with Dr. U. Zimmermann and J. Schmidt (Dept. of Biotechnology, University of Würzburg, FRG). The photographs were processed by Mrs. M. Schulze, and the manuscript was typed by Mrs. I. Ramoz and I. Schönberger.

The study was supported by Deutsche Forschungsgemeinschaft, SFB 176-A6.

FIG. 11. Incomplete and complete fusion of renal tubules. (A) Fused tubule 24 hr after PEG treatment. Please note the individual cell borders inside the cytoplasm. (B) Lucifer Yellow has been injected iontophoretically (10 pulses of 5×10^{-8}A, duration 1 sec). The dye is captured in a sharply limited space, indicating *incomplete* fusion. (C) Another fused tubule 24 hr after PEG treatment. No individual cell borders are visible in the cytoplasm. (D) The same fused tubule after Lucifer Yellow injection (same amount as above). There is instantaneous and complete staining of the cytoplasm, indicating *complete* fusion. Note that the nuclei are lined up along the cell membrane.

[47] Turtle Colon: Keeping Track of Transporters in the Apical and Basolateral Membranes

By David C. Dawson and Dean Chang

The Study of Parts

Our nearly 20-year romance with the turtle colon began, as many things do in science, by accident. In 1973 one of us (D.C.D.) had just arrived at the University of Iowa as an Assistant Professor of Physiology and Biophysics having completed a 2-year postdoctoral stint in the laboratory of Peter F. Curran at Yale University. In Peter Curran's lab we studied active Na^+ absorption using the colon of an amphibian, the toad, *Bufo marinus*.[1] We chose the toad colon because it appeared as though it might be a good model system in which we could study the subcellular "parts" involved in active Na^+ absorption. We were looking, in a sense, for a better frog skin.

The initial attempt to continue studies of the toad colon at Iowa was thwarted by the Iowa winter: our first shipment of toads from Colombia, South America arrived frozen solid! At this point it seemed important to study these parts, regardless of their origin, so we called Phillip Steinmetz, a long-time student of turtles who had recently arrived as Iowa's new Chief of Nephrology. He assured us that turtles possessed a colon and within a few days we had determined that the turtle colon transported Na^+ better than its amphibian counterpart and that, as an additional bonus, there was much more of it. Furthermore, sharing the animals made good ecological sense so it was resolved to continue the pursuit of active Na^+ absorption using the turtle colon.

Seventeen years later it appeared that the turtle colon was a good choice for a model for Na^+ absorbing epithelium. It is a prodigious Na^+ transport machine with rates of absorption far exceeding those in the frog skin or toad bladder. The epithelium is a single layer of cells, lacking the crypts and folds characteristic of mammalian colon. The isolated tissue is quite hardy and can be maintained *in vitro* for several days and subjected to a variety of experimental maneuvers (such as apical permeabilization) while maintaining its functional properties. In addition salt absorption is subject to both stimulatory (aldosterone) and inhibitory (cholinergic) control. In this chapter we will describe our attempts to identify and characterize the

---

[1] D. C. Dawson and P. F. Curran, *J. Membr. Biol.* **28**, 295 (1976).

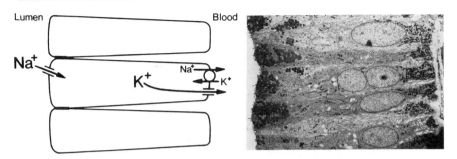

FIG. 1. *Right:* Electron micrograph showing the columnar cells which compose the epithelial cell layer of the turtle colon. *Left:* The Koefoed–Johnsen–Ussing model for active Na⁺ absorption. Sodium ion is presumed to enter the cell across the terminal membrane via amiloride-sensitive Na⁺ channels and exit the cell across the basolateral membrane via the Na⁺,K⁺-ATPase.

individual transporters which comprise the molecular basis for the ion transport properties of the turtle colon.

One of the vexing problems associated with epithelial cells is their tendency to lose their highly differentiated properties when they are isolated from one another. In order to obtain information about individual apical and basolateral transporters it has been necessary, therefore, to design experimental strategies which permit some degree of functional isolation of the transporters of interest while preserving the structural integrity of the cell layer. In several instances the turtle colon has proved to be uniquely suited to this type of experimental approach.

Transepithelial Transport

Cellular and Noncellular Paths

Figure 1 shows an electron micrograph of the colonic epithelial cell layer juxtaposed with the model used by Koefoed-Johnsen and Ussing[2] to describe active Na⁺ absorption by frog skin. The long columnar cells of the turtle colon form a single, flat cell layer and, as might be expected, ions can traverse the cell layer by way of cellular and paracellular paths. In our initial description of the properties of the colon[3] we used ³H-labeled mannitol as a marker for the paracellular path and concluded that solute diffusion between the cells behaved much as it would in free solution. This

[2] V. Koefoed-Johnsen and H. H. Ussing, *Acta Physiol. Scand.* **42**, 298 (1958).
[3] D. C. Dawson, *J. Membr. Biol.* **37**, 213 (1977).

and later studies[4] were all consistent with the notion that the paracellular path behaved as a watery leak with no special selectivity properties. With identical solutions bathing both sides of the tissue the transepithelial electrical potential difference would constitute a driving force for net transepithelial movement of Na^+, K^+, and Cl^- via this extracellular leak path.

The transepithelial potential difference across isolated colonic sheets, stripped of musculature, varies from about 30 to 130 mV, lumen negative, and the transepithelial resistance is in the range of 500 to 200 $\Omega \cdot cm^2$, placing this epithelium in the class of "moderately tight" epithelia. Our earliest estimates of cellular and paracellular conductances, 0.6 and 1.4 mSc/cm^2, respectively, were probably biased by the tissue damage which is inevitable in conventional Ussing chambers. More recent experiments suggest that values of 1.48 for the cellular conductance and 0.65 for the shunt conductance are more typical.

Active Transport of Na^+ and K^+

Electrogenic Na^+ absorption is the dominant transport process in the turtle colon, comprising greater than 90% of the short-circuit current (I_{sc}) which is virtually abolished by mucosal amiloride. Sodium ion absorption is also abolished by serosal ouabain. Absorption is stimulated by aldosterone[5] but is inhibited by muscarinic cholinergic agonists[6] as well by adenosine and histamine.[7] The basis for the small residual I_{sc} observed in the presence of mucosal amiloride is, at present, unknown.

Potassium ion is both actively absorbed and actively secreted by turtle colon[4] but at rates which are of the order of 100-fold less than that of Na^+. In some tissues the *net* flow was absorptive under short-circuit conditions, but little is known about this process except that it was blocked by mucosal (but not serosal) ouabain and orthovandate (M + S). The secretory transport, when present, was blocked by amiloride under short-circuit conditions, a result which was taken to reflect an effect of the reduced turnover of the basolateral Na^+, K^+-ATPase and hyperpolarization of the apical membrane potential. Potassium ion secretion was enhanced by serosal barium and blocked by mucosal barium, suggesting that the net rate of secretion depended on the ratio of the apical to basolateral K^+ conductances. As expected for a conductive transport process the cellular secretory rate was highly dependent on transmural potential.[8] The normal, lumen-negative open cicuit potential enhanced K^+ secretion so that *net* trans-

[4] D. R. Halm and D. C. Dawson, *Am. J. Physiol.* **246**, C315 (1984).
[5] D. R. Halm and D. C. Dawson, *Pfluegers Arch.* **403**, 236 (1985).
[6] C. J. Venglarik and D. C. Dawson, *Am. J. Physiol.* **251**, C563 (1986).
[7] C. J. Venglarik. Cholinergic Regulation of Salt Absorption by Turtle Colon: Dual Control of Sodium and Chloride Transport. Ph.D. Thesis, University of Michigan, 1988.
[8] D. R. Halm and D. C. Dawson, *Am. J. Physiol.* **247**, C26 (1984).

mural K^+ flow *in vivo* would be expected to be secretory. The relatively small rate of K^+ secretion (compared to Na^+ absorption) may indicate that not all of the Na^+-absorbing cells contain the apical K^+ channels necessary for the exit of K^+ toward the lumen, or simply that apical K^+ conductance is quite small in all of the cells.

A Cellular Cl⁻ Leak?

The initial description[3] of the transport of NaCl by the colon produced no evidence for any *net* movement of Cl^- under short-circuit conditions, but in some experiments there was the suggestion of a cellular component of the transmural flux of $^{36}Cl^-$. More recent experiments of Venglarik and Dawson[9] suggested, in fact, that a relatively small proportion of turtles ($\sim 5\%$) exhibited a substantial, transcellular Cl^- leak pathway which had several interesting properties. The appearance of the Cl^- leak was accompanied by a dramatic increase in tissue conductance which could not be attributed to the paracellular path. The conductive leak was highly Cl^- selective as judged by Cl^- currents (induced by imposing transmural Cl^- gradients) and transmural $^{36}Cl^-$ fluxes. The Cl^- leak path was inactivated by carbachol (as was the net active transport of Na^+) but was specifically inhibited by experimental maneuvers which were expected to elevate cellular cyclic AMP (phosphodiesterase inhibitors, forskolin, cAMP derivatives). These maneuvers did not affect Na^+ absorption. Net Cl^- transport occurred only in the presence of an applied driving force, suggesting that the presence of this highly anion-selective leak pathway could facilitate NaCl absorption under open circuit conditions. The salt absorptive process could be envisioned as being driven by active Na^+ absorption (the engine of salt absorption) but also dependent on the conductance of the anion-selective leak path. In addition, the inhibitory regulation of salt absorption by cholinergic agonists was seen as having a dual mechanism: inhibition of active Na^+ absorption and attenuation of the Cl^- leak path. Although these were intriguing thoughts, further study of this transmural path has been frustrated by the fact that it appears in such a small percentage of the turtles examined.

Apical Ion Channels

As expected from the large component of electrogenic Na^+ absorption the dominant components of the apical membrane of the turtle colon appears to be amiloride-sensitive, Na^+-selective channels. We examined these initially by measuring initial rates of $^{22}Na^+$ uptake from the mucosal

---

[9] C. J. Venglarik and D. C. Dawson, *Fed. Proc., Fed. Am. Soc. Exp. Biol.* **46**, 496 (1987).

bath[10] and later using blocker-induced Na^+ channel noise.[11] Thompson and Dawson[10] showed that a 30-sec tracer uptake from the mucosal path provided an estimate of the rate coefficient for $^{22}Na^+$ entry across the apical membrane of the transporting cells. At low mucosal Na^+ concentrations the entry rate coefficient was highly correlated with the amiloride-sensitive, transmural conductance as expected if the apical membrane was the major resistance barrier in the transporting cells. Other studies provided evidence that these apical channels were highly permeable to Li^+ but that the presence of Li^+ did not markedly affect Na^+ entry, as if the two ions did not interact in the conduction process.[12] This result would be consistent with the notion that the apical amiloride-sensitive channels can accommodate either Na^+ or Li^+ but contain only one ion at any instant.

Kirk and Dawson[13] obtained evidence that the apical Na^+ conductance can be modulated in response to changes in intracellular Na^+ concentration. Experimental maneuvers which would be expected to lead to an increase in cytosolic Na^+ concentration produced an inhibition of the Na^+ conductance of the apical membrane. In ouabain-treated tissues this inhibition could be reversed by simply depleting cellular Na^+.

Wilkinson and Dawson[11] applied the techniques of Van Driessche and Lindemann[14] to study blocker-induced fluctuations in apical Na^+ currents. Using the weak Na^+ channel blocker, CDPC,[15] they observed blocker-induced Lorentzian components in the power density spectra which were indicative of reversible blockade of open Na^+ channels. Under control conditions single-channel Na^+ currents averaged 0.43 pA and the density of Na^+ channels was 250×10^6 channels/cm^2. Carbachol, which inhibited Na^+ absorption, caused a decrease in the number of open Na^+ channels but did not alter the single-channel current. The result was consistent with the notion that inhibitory cholinergic control of Na^+ absorption was associated with decrease in the conductance of both the apical and the basolateral membranes such that the fractional resistance was not greatly altered.

Noise analysis also produced evidence for apical K^+ channels.[16] In some instances it was possible to detect a "spontaneous" Lorentzian component in the power density spectrum with a corner frequency of about 15 Hz. This Lorentzian appeared to be unrelated to Na^+ absorption. It was abolished by mucosal barium, however, and the plateau values were aug-

---

[10] S. M. Thompson and D. C. Dawson, *J. Membr. Biol.* **42,** 357 (1978).

[11] D. J. Wilkinson and D. C. Dawson, *FASEB J.* in press (1989).

[12] S. M. Thompson and D. C. Dawson, *J. Gen. Physiol.* **72,** 269 (1978).

[13] K. L. Kirk and D. C. Dawson, *Pfluegers Arch.* **403,** 82 (1985).

[14] W. Van Driessche and B. Lindemann, *Nature (London)* **282,** 519 (1979).

[15] S. I. Helman and L. M. Baxendale, **95,** 647 (1990).

[16] D. J. Wilkinson and D. C. Dawson, *Am. J. Physiol.* **259,** 668 (1990).

mented by applying a serosa-positive transmural potential. This behavior suggested that these fluctuations emanated from the apical channels which mediated the potential-sensitive K^+ secretion documented by Halm and Dawson.[8]

Basolateral Transporters

Permeabilized Cell Layers

In most studies of isolated sheets of turtle colon the tissue is stripped of circular and longitudinal musculature, but a layer of connective tissue and a muscularis mucosa remains on the serosal side. The thickness of the layer is roughly 5 to 10 times that of the single layer of mucosal cells so that the shortest route to the basolateral membrane is through the apical membrane.[17,18] For this reason we have explored a number of approaches to chemically modifying the apical membrane so that it no longer constitutes a significant barrier to transcellular ion flow. The object of any such strategy is to reach a compromise between two goals: that of eliminating the apical membrane as a permeability barrier and that of restricting or controlling the changes in intracellular composition which occur as a result of the permeabilization process. To this end we used the pore-forming, polyene antibiotic, amphotericin B, to permeabilize the apical membrane to monovalent cations. The principal advantage of amphotericin as a permeabilizing tool is its relative selectivity. The resulting pores are relatively nonselective for monovalent cations so that the membrane becomes freely permeable to Na^+ and K^+. The pores appear to be impermeable to divalent cations, however, and six-carbon sugars. Monovalent anions are only about one-seventh as permeant as cations but the Cl^- permeability is sufficient to increase salt entry into the cells. One consequence of this is that cells exposed to amphotericin will gain salt if bathed by mucosal NaCl or KCl, a fact exploited by Germann et al.[19] to induce swelling in turtle colon cells. In our earliest experiments we investigated a number of different anion replacements in order to find one which would attenuate cell swelling but would also preserve cellular transport. We settled on benzene sulfonate and used this in a number of experiments although we later learned that this anion may, itself, promote some degree of cell swelling.[19] It is interesting that the effects of amphotericin appear to be completely reversible even though permeabilization can induce substantial changes in cell volume and cellular ionic composition.

[17] D. C. Dawson, *Curr. Top. Membr. Transp.* **28**, 41 (1987).
[18] D. C. Dawson, D. J. Wilkinson, and N. W. Richards, *Curr. Top. Membr. Transp.* **37**, 191 (1990).
[19] W. J. Germann, S. A. Ernst, and D. C. Dawson, *J. Gen. Physiol.* **88**, 253 (1986).

Na$^+$/K$^+$ Pump and Potassium Conductance

Our first experiments with amphotericin-permeabilized colon were designed to determine if the basolateral membranes contained the two elements basic to the Koefoed–Johnsen–Ussing model, a basolateral K$^+$ conductance and an Na$^+$/K$^+$ exchange pump.[20] A K$^+$ conductance was identified by imposing a transmural K$^+$ gradient and noting that the resulting K$^+$ current was blocked by serosal barium. Using tissues treated with serosal barium (but not ouabain) it was then possible to show that in the absence of transmural gradients a current appeared which exhibited an obligatory dependence on mucosal Na$^+$ and serosal K$^+$ and was inhibited by ouabain. Transmural flux determinations using these permeabilized tissues confirmed that the current was due to opposing net Na$^+$ and K$^+$ fluxes in the ratio 3Na$^+$:2K$^+$. Halm and Dawson[21] arrived at the same estimate of the pump stoichiometry by studying the kinetics of activation by mucosal (cellular) Na$^+$ and serosal K$^+$. Kirk and Dawson[22] investigated the properties of the basolateral K$^+$ conductance and found that under the conditions of these experiments (benzene sulfonate as the anion) it was possible to observe basolateral currents carried by K$^+$, Rb$^+$, or Th$^+$, all of which were blocked by serosal barium. Furthermore, ^{42}K$^+$ fluxes revealed coupling between the flows of tracer and abundant species which was indicative of single filing.

Subsequent experiments by Germann *et al.*[23] revealed that the basolateral K$^+$ conductance (g_K) observed in tissues bathed by benzene sulfonate ringers was more complex than we had initially guessed. Germann *et al.*[23] showed that it was possible to identify two distinct components to the basolateral K$^+$ conductance in permeabilized cells. One did not discriminate between K$^+$ and Rb$^+$, was blocked by serosal barium (but not by quinidine or lidocaine), was inactivated by cholinergic agonists, and was present under conditions which prevented the cells from swelling. Because normal active transport was supported equally well by K$^+$ or Rb$^+$ and was inactivated by cholinergic agonists[6] this conductance was dubbed the "resting" K$^+$ conductance. If the cells were swollen by the entry of salt or urea, a second conductance was activated which could be distinguished by a marked preference for K$^+$ over Rb$^+$, sensitivity to quinidine and lidocaine, and marked single-filing behavior. In addition this conductance could be activated by swelling in cell layers which had previously been exposed to carbachol to inactivate the resting g_K. Germann *et al.*[19] pro-

[20] K. L. Kirk, D. R. Halm, and D. C. Dawson, *Nature (London)* **287**, 237 (1980).

[21] D. R. Halm and D. C. Dawson, *J. Gen. Physiol.* **82**, 315 (1983).

[22] K. L. Kirk and D. C. Dawson, *J. Gen. Physiol.* **82**, 297 (1983).

[23] W. J. Germann, M. E. Lowy, S. A. Ernst, and D. C. Dawson, *J. Gen. Physiol.* **88**, 237 (1986).

posed that this second conductance was due to a second population of channels which was not active under resting conditions but was activated by cell swelling. In retrospect, it seems clear that the earlier results of Kirk and Dawson[22] represented the properties of a mixed channel population. The Rb^+ currents are likely to have been due to the resting conductance whereas the strong single-filing probably reflected a component of the swelling-induced g_K.

Dawson, Van Driessche, and Helman[24] used noise analysis to investigate the possible induction of basolateral K^+ channels by cell swelling. They showed that under conditions of cell swelling it was possible to detect a lidocaine-induced, Lorentzian component in the power density spectrum of the basolateral K^+ current. Under nonswelling conditions this lidocaine-induced Lorentzian was absent. This result was consistent with the notion that cell swelling activated a population of lidocaine-blockable channels in the basolateral membrane which was not active under normal osmotic conditions. Analysis of the blocker-induced noise provided an estimate of 20 pS for the single-channel conductance. Richards and Dawson[25] used single-channel recording techniques to investigate K^+ channels in isolated colonic cells. In cells bathed in KCl-Ringer's solution they identified a 20-pS channel which was blocked by lidocaine and quinidine and appears to be the molecular basis for the swelling activated g_K.

Venglarik and Dawson[6] investigated the possible role of the resting g_k in the inhibitory cholinergic control of Na^+ absorption. They showed that cholinergic agonists, either exogenous or released from submucosal nerves, inactivated both active Na^+ absorption and basolateral g_K. The cholinergic response of g_K could be duplicated by exposing the tissue to a calcium ionophore, ionomycin, suggesting that perturbing intracellular calcium activity was sufficient to trigger the chain of events which led to the inactivation of basolateral g_K. Richards and Dawson,[26,27] however, had recorded in isolated cells several types of channels which were *activated* by cytoplasmic calcium, suggesting that calcium might be involved in both activation and inactivation of components of the basolateral g_K.

Cl⁻ Channels

Venglarik et al.[28] identified a basolateral Cl⁻ conductance in amphotericin-permeabilized colonic cells. This conductance had several properties

[24] D. C. Dawson, W. VanDriessche, and S. I. Helman, *Am. J. Physiol.* **254**, C165 (1988).
[25] N. W. Richards and D. C. Dawson, *Am. J. Physiol.* **251**, C85 (1986).
[26] N. W. Richards and D. C. Dawson, *Biophys. J.* **51**, 344a (1987).
[27] N. W. Richards and D. C. Dawson, *FASEB J.* **3**, A1149 (1989).
[28] C. J. Venglarik, J. L. Keller, and D. C. Dawson, *Fed. Proc., Fed. Am. Soc. Exp. Biol.* **45**, 2082 (1986).

which suggested that it is related to the transcellular Cl⁻ leak which is present in a small percentage of colons studied. The conductance, which could be identified by ion substitution, was inactivated by muscarinic cholinergic agonists, as was the resting basolateral K^+ conductance. Basolateral g_{Cl^-}, however, was specifically inhibited by experimental maneuvers which are expected to raise the cellular levels of cyclic AMP. Exposure to forskolin, phosphodiesterase inhibitors [3-isobutyl-1-methylxanthine (IBMX), for example] or cAMP derivatives (8 cAMP) all produced inactivation of g_{Cl^-} but had little or no effect on g_K. As opposed to the transcellular Cl⁻ leak this basolateral conductance was present in virtually all cell layers examined. At present, however, the physiological role of this conductance and the significance of the possible regulation by cAMP have yet to be determined.

Role of Intracellular Ca²⁺: Digitonin-Permeabilized Cells

Questions concerning a possible role for cytoplasmic calcium in the control of basolateral g_K led Chang and Dawson[29] to develop a preparation of the isolated turtle colon which was apically permeabilized by the detergent, digitonin. Exposure of the apical membranes to digitonin (20 μM) led to the release of the cytoplasmic enzyme, lactate dehydrogenase (LDH), from the cells and rendered the apical membrane highly permeable to monovalent and divalent cations as well as to organic buffers for pH (PIPES) and calcium (EGTA). Somewhat surprisingly the basolateral membrane retained its functional integrity despite this chemical onslaught. Chang and Dawson[29] speculated that the long columnar shapes of the epithelial cells may have contributed to this result.

In an attempt to at least partially mimic the intracellular milieu Chang and Dawson employed a mucosal solution which consisted primarily of potassium aspartate along with an EGTA (5 mM) buffer system designed so that the free calcium concentration could be readily controlled in the range of 10^{-9} to 10^{-6} M.[30] The pH was buffered to 6.6 to exploit the optimal Ca²⁺-buffering range of EGTA. In the presence of about 10^{-9} M mucosal calcium digitonin permeabilization did not lead to a transcellular K^+ current. Raising the free calcium to 10^{-7} M, however, led to the prompt development of a K^+ current which could be rapidly inactivated by adding additional EDTA to reduce the free calcium. This type of result led to the conclusion that the basolateral membranes contained a K^+ conductance which could be activated by increases in cytosolic calcium. Similar experiments conducted using K^+-free solutions in the presence of a Cl⁻ gradient

[29] D. Chang and D. C. Dawson, J. Gen. Physiol. **82**, 281 (1988).
[30] D. Chang, P. S. Hsieh and D. C. Dawson, Comput. Biol. Med. **18**(5), 351 (1988).

were used to identify a calcium-activated basolateral Cl⁻ conductance. Both the K^+ and Cl⁻ conductances exhibited spontaneous inactivation in the continued presence of calcium. The nature of this inactivation is not understood at present.

The activation of basolateral conductances in digitonin-permeabilized cells was also sensitive to cytoplasmic pH.[31] Raising the cytosolic pH from 6.6 to 7.4 shifted the calcium activation dose response to the left so that there was greater activation at any given concentration of free calcium. These results suggested that the calcium activity and intracellular pH could interact to determine the degree of activation of basolateral conductances, and that the mechanism of this interaction is a proton modification of the calcium activation site. Regulation of basolateral conductance by pH is particularly interesting in light of the demonstration of a basolateral Na^+/H^+ exchanger[32] which could couple intracellular pH to transmembrane Na^+ gradients.

The observation of a basolateral K^+ conductance which could be activated by calcium in digitonin-permeabilized cells raised the question of the relation of such conductances to the previously defined "resting" and "swelling activated" basolateral g_K. Interestingly, the calcium-activated K^+ conductance did not match up perfectly with either of these. Calcium-activated K^+ currents could be carried by Rb^+, but were blocked by quinidine. The Ca^{2+}-activated g_K was blocked by barium, but *only* from the cytoplasmic side, whereas both of the K^+ currents identified in amphotericin-treated cells were blocked by serosally applied barium. In an effort to identify the channels that were the basis for the calcium-activated conductance Richards and Dawson[27] exploited the observation that in digitonin-treated cells Ca^{2+}-activated g_K was blocked by n-phenyl anthranilic acid (DPC). Somewhat surprisingly this effect turned out to be relatively specific. The compound did not block the 20-pS swelling activated channel nor did it block a "big K^+" channel recorded from isolated colonic cells. The compound did, however, block a very flickery, inwardly rectifying channel which could be recorded from isolated colonic cells. The channel was activated by cytoplasmic calcium, and could play some role in the resting K^+ conductance of the basolateral membranes.

Na^+/H^+ Exchange and Na^+ Channels

Kirk and Dawson[33] showed that it was possible to measure transcellular Na^+ currents and fluxes using colons which had been treated with ouabain

[31] D. Chang, D. C. Dawson, *FASEB J.* **2**, A1284 (1988).

[32] M. A. Post and D. C. Dawson, *FASEB J.* **4**, A549 (1990).

[33] K. L. Kirk and D. C. Dawson, *J. Gen. Physiol.* **82**, 497 (1983).

to abolish active transport and exposed to a steep, serosa to mucosa Na^+ gradient (112 mM : 2 mM). In this condition a mucosal amiloride-sensitive, cellular component of the S to M Na^+ current and transmural $^{22}Na^+$ flux was evident. Analysis of transcellular Na^+ flow in the absence of active transport suggested that the cells normally responsible for active Na^+ absorption contained, in the basolateral membranes, a cation exchanger which could carry out $Na^+ : Na^+$ or $Na^+ : Li^+$ exchange and a cation channel conductive to Na^+ and Li^+.

Kirk and Dawson[33] proposed that basolateral Na^+/Li^+ exchange was the basis for the ouabain-sensitive active Li^+ absorption described by Sarracino and Dawson.[34] The free energy for Li^+ transport was thought to reside in the basolateral Na^+ gradient maintained by the Na^+,K^+-ATPase which, as shown by Halm and Dawson,[21] was not activated by cytoplasmic Li^+. In more recent experiments Post and Dawson[32] explored the properties of basolateral Na^+ transport elements using amphotericin-permeabilized cell layers. They showed that the basolateral membrane contains two sodium-selective, amiloride-inhibitable transport activities which may or may not reside in the same membrane protein. One is an Na^+-selective conductance and the other is an Na^+/H^+ exchanger. Interestingly, the activity of both of these transporters was modulated by cell volume. Exchange flow and conductive flow were largest in shrunken cells and were markedly attenuated by cell swelling. The presence of both a *swelling* activated K^+ conductance and *shrinkage*-activated Na^+/H^+ exchange (and Na^+ conductance) may suggest that these elements represent the two poles of a push–pull system for volume regulation.

A Model for Salt Absorption

Figure 2 shows a working model which incorporates most of the transport "parts" which have been identified in the turtle colon epithelial cells. The basic elements of the Koefoed–Johnsen–Ussing model remain the foundation for the ability of the cell to carry out active, transcellular transport of Na^+. In our current model we envision the resting basolateral K^+ conductance as serving to recycle much of the K^+ which enters the cell via the Na^+,K^+-ATPase, although a relatively small fraction leaks out across the apical membrane to produce K^+ secretion. A negative feedback, "self-regulation" loop is shown to suggest that increases in cytosolic Na^+ concentration can act, via some as yet unidentified mechanism, to shut down the apical Na^+ conductance. Possible pathways for inhibitory cholinergic regulation are shown in keeping with the evidence that the agonist-induced decline in active absorption is accompanied by decreases in both

[34] S. M. Sarracino and D. C. Dawson, *J. Membr. Biol.* **46**, 295 (1979).

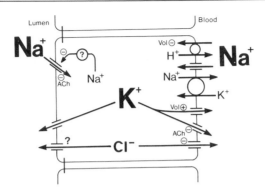

Fig. 2. Current working hypothesis for the disposition of cellular transporters in the Na^+-transporting cells of the turtle colon. See text for details.

basolateral g_K and apical g_{Na^+}. The hypothetical model cell is indicated as possessing Cl^- channels in the apical and basolateral membranes, although the apical channel is shown as being tentative because the *transcellular* conductance is observed so infrequently. If present, the series arrangement of Cl^- channels would provide a conductive, transcellular Cl^- leak path which could augment NaCl absorption. In keeping with the observations of Venglarik et al.[28] the basolateral Cl^- conductance is also shown as being attenuated as a result of muscarinic agonist binding to membrane receptors.

Depending on the prevailing osmotic conditions at least three additional transport elements might be present in the basolateral membranes. Cell swelling is associated with the activation of a specific basolateral conductance which is blocked by quinidine or lidocaine. This same condition inactivates two Na^+-selective elements, an Na^+/H^+ exchanger and an Na^+-selective conductance, whereas the latter are fully activated in shrunken cells. It is possible that these transport elements play a role in the maintenance of cellular homeostasis as well as the regulation of transcellular transport. These mechanisms could play a role in the regulation of cell volume but it might also be that cell volume is one component of the intracellular coupling network which permits "crosstalk" between the apical and basolateral membranes.

Acknowledgments

The research described herein was supported by the National Institute for Diabetes, Digestive and Kidney Disease, The Cystic Fibrosis Foundation, and the American Heart Association of Iowa and Michigan. Steve Thompson, Dave Luneman, Kevin Kirk, Dan Halm, Jane Nelson, Bill Germann, Jeff Keller, Chuck Solc, Dan Le Gault, Charles Venglarik, Neil Richards, Marc Post, Dan Wilkinson, Melinda Lowy, and Nancy Kushman all played an important role in the studies described herein.

[48] Ion Transport of Marine Teleost Intestine

By MARK W. MUSCH, SCOTT M. O'GRADY, and MICHAEL FIELD

The intestine of the winter flounder, *Pseudopleuronectes americanus*, has a homogeneous epithelium of absorptive cells without the crypt–villus distribution seen in the intestines of warm-blooded animals. Occasional mucus-bearing cells, basal cells, and endocrine cells can be found, but these are far less numerous than goblet cells or enterchromaffin cells in mammalian or avian intestine. Absorptive cells possessing a well-defined brush border comprise greater than 90% of the epithelial cells.[1,2] This chapter deals exclusively with studies of winter flounder intestine. Other teleosts may express different transport properties in their intestines and no claim is made that the properties shown here are universal or even necessarily typical.

When stripped of its external muscle layers, flounder intestine has a relatively flat geometry.[1] The lack of cryptoid elements in flounder intestine may be responsible for its inability to secrete water and electrolytes. Evidence has been gathered from other intestinal preparations that secretion of water and electrolytes originates from the crypt region.[3] Since flounder intestine only absorbs salt,[4] it is particularly useful for dissecting the cellular processes responsible for electrolyte absorption. Its ion transport properties are strikingly similar to those of the thick ascending limb of Henle's loop in mammalian kidney and it therefore serves as an excellent model system for studying the latter.

Morphologically and functionally, flounder intestine differs only minimally along its length, which contrasts sharply with mammalian intestine and mammalian renal tubules. Unless a section of tissue is taken very near the diverticula of the stomach, the entire length of the intestine exhibits a transepithelial electric potential difference (PD) between 3 and 6 mV, serosa negative. As discussed below, the current needed to nullify this PD (I_{sc}) is an accurate measure of the salt-transporting activity of the tissue.

[1] M. Field, K. J. Karnaky, Jr., P. L. Smith, J. E. Bolton, and W. B. Kinter, *J. Membr. Biol.* **41,** 265 (1978).

[2] R. L. Curtis, J. S. Trier, R. A. Frizzell, N. M. Lindem, and J. L. Madara, *Am. J. Physiol.* **246,** C77 (1984).

[3] M. J. Welch, P. L. Smith, M. Fromm, and R. A. Frizzell, *Science* **218,** 1219 (1982).

[4] M. Field, P. L. Smith, and J. E. Bolton, *J. Membr. Biol.* **55,** 157 (1980).

Transepithelial Ion Fluxes and Dilution Potentials

Flounder intestine stripped of its external muscle layers can be mounted in an Ussing-type chamber and bathed on both sides in a Ringer's solution which differs only slightly from that characteristically used with mammalian tissues. The composition generally used is (in mmol/liter): 160, NaCl; 5, KCl; 1.25, $CaCl_2$; 1.1, $MgCl_2$; 0.3, NaH_2PO_4; 1.65, Na_2HPO_4; 10, glucose; 5, EPPS (N-2-hydroxyethylpiperazine propanesulfonic acid), pH 8.0. The solution is bubbled with room air. A bicarbonate buffer can also be used, but one should adjust the pH to about 8.0. We did this with a 20 mM HCO_3 solution gassed with 99% O_2, 1% CO_2. The PD and I_{sc} of flounder intestine are greater at pH 8.0 than at 7.4, indicating a greater rate of ion transport at the more alkaline pH.[4] The tissue should also be maintained below 15°. We now routinely run experiments at 7°. Transport rates at this temperature are not lower than at higher temperatures. If flounder intestine is provided with glucose (10 mM) as a substrate, it will maintain its initial PD and I_{sc} for at least 24 hr.

The PD across founder intestine is generally between −3 and −6 mV (serosa negative), but appears to be modulated by the feeding behavior of the fish. Freshly caught or fed flounder will often exhibit only a small PD and I_{sc}. When starved for 3 days, however, the PD increases (becomes more negative) and therefore so does the absorption of salt and water.

Omission of either Na^+, K^+, or Cl^- from the mucosal bathing solution will abolish the PD and I_{sc} (see Table I),[5-8] The mucosal side addition of a loop diuretic such as furosemide or bumetanide,[9] the serosal side addition of a permeable analog of cGMP such as 8-Br-cGMP,[10] or of a Ca^{2+}-related stimulus such as carbamylcholine or substance P, or of ouabain[1] will also abolish the PD. The mucosal side addition of the K^+ channel blocker Ba^{2+},[5,11] or the serosal side addition of an analog of cAMP,[4,10] decreases the PD and I_{sc} but does not completely abolish them. Each of these maneuvers is helpful in understanding the mechanism of salt absorption in flounder intestine.

[5] M. W. Musch, S. A. Orellana, L. S. Kimberg, M. Field, D. R. Halm, E. J. Krasny, Jr., and R. A. Frizzell, *Nature (London)* **300**, 351 (1982).

[6] S. M. O'Grady, M. W. Musch, and M. Field, *J. Membr. Biol.* **91**, 33, (1986).

[7] S. M. O'Grady, M. Field, N. T. Nash, and M. C. Rao, *Am. J. Physiol.* **249**, C531 (1985).

[8] S. M. O. Grady, H. R. De Jonge, A. B. Vaandrager, and M. Field, *Am. J. Physiol.* **254**, C115 (1988).

[9] R. A. Frizzell, P. L. Smith, E. Vosburgh, and M. Field, *J. Membr. Biol.* **46**, 27 (1979).

[10] M. C. Rao, N. T. Nash, and M. Field, *Am. J. Physiol.* **246**, C167 (1984).

[11] R. A. Frizzell, D. R. Halm, M. W. Musch, C. P. Stewart, and M. Field, *Am. J. Physiol.* **246**, F946 (1984).

TABLE I
TRANSMURAL Na$^+$ AND Cl$^-$ FLUXES[a]

Condition	$J_{ms}^{Na^+}$	$J_{sm}^{Na^+}$	$J_{net}^{Na^+}$	$J_{ms}^{Cl^-}$	$J_{sm}^{Cl^-}$	$J_{net}^{Cl^-}$	I_{sc}	Ref.
A. 170 mM Na$^+$, 150 or 170 mM Cl$^-$								
Control	13.6	11.6	1.96	8.43	2.97	5.46	−3.40	1
Na$^+$ free(m and s)	—	—	—	6.32[b]	6.15[b]	0.17[b]	−0.16[b]	1
Cl$^-$ free(m and s)	13.8	13.9	−0.11[b]	—	—	—	0.04[b]	1
Control	—	—	—	6.97	2.63	3.54	−1.78	5
K$^+$ free (m only)	—	—	—	3.93[b]	2.49	1.44[b]	−0.93[b]	5
B. 50 mM Na$^+$, 180 mM Cl$^-$								
Control	3.96	1.82	2.15	8.11	3.79	4.45	−2.70	6
Bumetanide (m)	2.15[b]	1.67	0.46[b]	3.21[b]	2.99	0.39[b]	−0.21[b]	6
Control	4.08	2.59	1.49	6.95	3.58	3.33	−3.20	7
8-Br-cGMP (s)	2.56[b]	2.33	0.23[b]	2.94[b]	2.48[b]	0.47[b]	−0.30[b]	7
Control	3.50	2.41	1.10	9.74	5.50	4.24	−3.30	8
8-Br-cAMP (s)	3.36	3.16	0.21[b]	10.75	10.24[b]	0.78[b]	−0.90[b]	8

[a] Results shown are mean transmural mucosa (m)-to-serosa (s) and s-to-m Na$^+$ and Cl$^-$ fluxes (in μEq/hr/cm^2) from cited references. $J_{net} = J_{ms} - J_{sm}$. Fluxes were measured under short-circuit conditions.

[b] Different from control, $p < 0.05$.

Transepithelial unidirectional fluxes of radioisotopes can be measured by short circuiting the tissues (usually 60–100 μA/cm^2 is required), adding radioisotope to one side (generally 1 μCi), and measuring the steady state appearance of radioisotope on the opposite side.[1] The mucosa-to-serosa (J_{ms}) and serosa-to-mucosa (J_{sm}) fluxes of Na$^+$, Cl$^-$, and K$^+$ or Rb$^+$ have been measured and are shown for various conditions in Tables I and II. Rubidium ion was found to quantitatively substitute for K$^+$.[11] Tables I and II contain a number of critical observations: (1) The I_{sc} is approximately equal to the difference between $J_{net}^{Cl^-}$ and $J_{net}^{Na^+}$; (2) $J_{net}^{Cl^-}$ is consistently greater than $J_{net}^{Na^+}$; (3) $J_{net}^{Cl^-}$ is entirely dependent on the presence of Na$^+$ and vice versa; (4) omission of K$^+$ from the mucosal side inhibits $J_{net}^{Cl^-}$ and I_{sc}; therefore it presumably also inhibits $J_{net}^{Na^+}$; (5) K$^+$ (i.e., Rb$^+$) is actively secreted under normal conditions. This process is blocked by ouabain and furosemide (both of which also inhibit Na$^+$ and Cl$^-$ transport, although by different mechanisms) and is dependent on the presence of Cl$^-$. Addition of the K$^+$ channel blocker Ba^{2+} to the mucosal side reverses $J_{net}^{K^+}$ from secretion to absorption; thus a mechanism for active K$^+$ absorption also

TABLE II
TRANSMURAL Rb$^+$ (K$^+$) FLUXES^a

Condition	$J_{ms}^{Rb^+}$	$J_{sm}^{Rb^+}$	$J_{net}^{Rb^+}$	I_{sc}	Ref.
Control	0.40	1.18	−0.78	−3.4	11
Ouabain (s)	0.66^b	0.66^b	0.00^b	0.0^b	11
Cl$^-$ free (m and s)	0.40	0.52^b	−0.11^b	0.0^b	11
Furosemide (m)	0.45	0.59^b	−0.14^b	0.1^b	11
Barium (m)	1.12^b	0.53^b	0.59^b	−0.8^b	11
Barium (m) plus furosemide (m)	0.50	0.48^b	0.02^b	−0.2^b	11
Control	0.30	1.09	−0.79	−3.2	10
8-Br-cGMP (s)	0.56^b	0.79^b	−0.23^b	0.2^b	10
8-Br-cAMP (s)	0.51	0.85^b	−0.34^b	−1.6^b	10

^{a} Results shown are mean transmural m-to-s and s-to-m Rb$^+$ fluxes (in μEq/hr/cm^2) from cited references. Transmural fluxes of ^{86}Rb$^+$ are equal to those of ^{42}K$^+$ as shown by Frizzell et al.[11] $J_{net} = J_{ms} - J_{sm}$. Fluxes were measured under short-circuit conditions.
^{b} Different from control, $p < 0.05$.

exists; (6) judging from the magnitudes of the serosa to mucosa unidirectional fluxes, passive permeability to Na$^+$ is substantially greater than passive permeability to Cl$^-$. This is reinforced by dilution potentials (see below). Rubidium ion appears to have about the same passive permeability as does Na$^+$; (7) "loop" diuretics, when added to the mucosal side, abolish $J_{netinng}^{Cl^-}$, $J_{net}^{Na^+}$, and $J_{net}^{K^+}$; the last is true whether $J_{net}^{K^+}$ is absorptive or secretory; (8) cyclic GMP also abolishes $J_{net}^{Cl^-}$ and $J_{net}^{Na^+}$ and greatly reduces $J_{net}^{K^+}$. The effect of cGMP is perfectly mimicked by atrial natiuretic factor,[7] a known stimulator of guanylate cyclase.[12] Marine teleosts are able to maintain body fluid osmolalities less than half that of sea water by absorbing salt and water isotonically in the gut and then selectively excreting the salt through their gills. When euryhaline fish enter waters of low salinity, they greatly diminish their intestinal fluid absorption and they cease to excrete salt through their gills. Atrial natiuretic factor may play a major regulatory role in this adaptation, although this is at present speculative; (9) finally, cGMP and cAMP have opposite effects on $J_{net}^{Cl^-}$, and therefore presumably passive Cl$^-$ permeability, the former reducing it and the latter increasing it.

Another way to check the cation selectivity of the epithelium is to measure dilution potentials.[13] If the mucosal medium is diluted with isoos-

[12] S. A. Waldman, R. M. Rapoport, and F. Murad, *J. Biol. Chem.* **259**, 14332 (1984).
[13] E. J. Krasny, Jr., J. Madara, D. DiBona, and R. A. Frizzell, *Fed. Proc., Fed. Am. Soc. Exp. Biol.* **42**, 1100, (1983).

motic mannitol in H_2O, an increase in the negativity of the potential difference (PD) is observed. For an untreated tissue the change in PD is generally from about -5 to -25 mV with a $1:6$ dilution. This change in PD is also observed when the tissue has first been treated with cGMP, indicating that active transport does not play a role. If, however, the tissue is first treated with cAMP, no change in PD occurs following dilution of the mucosal medium with isoosmotic mannitol. This observation is consistent with the change in serosa to mucosa Cl^- flux caused by cAMP and indicates that cAMP somehow selectively increases passive Cl^- permeability. Whether this is due to a change in the permselectivity of the intercellular junctions (zonae occludens) or to a change in cellular Cl^- permeability or both is not entirely clear.

Initial Rates of Uptake (Influxes) of Na^+, Cl^-, and Rb^+

To measure unidirectional fluxes from the mucosal medium into the epithelium, rates of uptake must be determined over short time intervals (45 sec for Na^+ and Cl^-, 2 min for K^+ and Rb^+) so that they represent initial rates. Flounder intestinal mucosa is well suited for such experiments because it is a relatively flat epithelium, rendering diffusional distances from the bulk medium to the epithelial surface small. Sections of mucosa are mounted in Ussing-type chambers modified to provide open, low-volume mucosal reservoirs and gas-lift reservoirs for perfusion of the serosal surface with oxygenated Ringer's solution. The chambers we now employ are modifications of those published previously.[9,14] [^{3}H]Polyethylene glycol is added as a marker for contaminating adherent medium. A short time after radioisotopes are added, the tissue is quickly punched out, rinsed, blotted, and extracted. Radioisotopes of monovalent ions are both absorbed into the cells and into the paracellular spaces. Since at least some of the cellular uptake occurs through the transport system of interest, it can be blocked with a specific inhibitor and the magnitude of the carrier or channel-mediated influx is the difference in total influx in the presence and absence of the inhibitor. Table III shows influxes of Na^+ and Cl^- in the presence and absence of bumetanide and Rb^+ influxes in the presence and absence of furosemide, 8-Br-cGMP, Na^+, and Cl^-. It can be seen that the bumetanide-inhibitable Cl^- influx has twice the magnitude of the bumetanide-inhibitable Na^+ influx and that furosemide, 8-Br-cGMP, Na^+ omission, and Cl^- omission all inhibit Rb^+ influx to the same extent. Furthermore, Rb^+ influx is far greater than $J_{ms}^{Rb^+}$ (see Table II), indicating that most of the entering Rb^+ is recycled to the mucosal medium. Electrophysiologic

[14] H. N. Nellans, R. A. Frizzell, and S. G. Schultz, *Am. J. Physiol.* **225,** 1131 (1074).

TABLE III
Na^+, Cl^-, AND Rb^+ INFLUXES[a]

Condition	J_{me}	J_{me}	Ref.
A. Na^+ and Cl^- influxes	Na^+	Cl^-	
Control	4.0	8.3	6
Bumetanide (m)	2.0^b	3.9^b	6
B. Rb^+ influxes	Rb^+		
Control	3.05		5
Furosemide	0.70^b		5
Na^+ free	0.69^b		5
Cl^- free	0.64^b		5
Control	2.88		10
8-Br-cGMP (s)	0.66^b		10

[a] Results shown are mean influxes (in $\mu Eq/hr/cm^2$) from the mucosal solution into the epithelium (J_{me}) of Na^+ and Cl^- or Rb^+ from cited references. All fluxes were measured under short-circuit conditions.
[b] Different from control, $p < 0.05$

experiments (see below) indicate that this occurs through Ba^{2+}-inhibitable K^+ channels. In experiments shown in Table 3 simultaneous measurements were made on tissues from the same animals of NaCl influx and RbCl influx.[6] The ratio of bumetanide-inhibitable influx for Cl^-/Na^+ was 2.2 and for Cl^-/Rb^+, 1.8, indicating an overall stoichiometry of $1Na^+:Rb^+:2Cl^-$. Measurements of Rb^+ influxes at varying Na^+, Rb^+, and Cl^- concentrations indicated $K_{0.5}$ values of 5, 4.5, and 20 mM, respectively and Hill coefficients of 0.9, 1.2, and 2.0, respectively. There thus appears to be a high degree of cooporativity between the two Cl^--binding sites.

Electrophysiological Studies

Flounder intestine is well suited for measurements of intracellular electric potentials and ion activities. Microelectrode impalements yield stable potentials for hours. Frizzell and co-workers have recently published a detailed analysis of the conductance properties and electric potential profile of this epithelium.[15,16] The salient features of this analysis are the following: (1) 96% of the transepithelial conductance appears to be paracellular; (2) apical membrane conductance is dominated by a Ba^{2+}-sensi-

[15] D. R. Halm, E. J. Krasny, Jr., and R. A. Frizzell, *J. Gen. Physiol.* **85**, 843 (1985).
[16] D. R. Halm, E. J. Krasny, Jr., and R. A. Frizzell, *J. Gen. Physiol.* **85**, 865 (1985).

tive K$^+$ channel; (3) basolateral membrane conductance is dominated by a
Cl$^-$ channel. There is no evidence for a K$^+$ channel in this membrane; (4)
the PDs across the apical and basolateral membranes, which differ by less
than 5 mV and are of about 60-mV magnitude, lie between the equilibrium potentials for K$^+$ (76 mV) and Cl$^-$ (40 mV). This indicates the
existence of driving forces for K$^+$ secretion across the apical membrane and
Cl$^-$ absorption across the basolateral membrane; (5) equivalent circuit
analysis suggests that the apical conductance is sufficient to account for the
observed rate of K$^+$ secretion but that the basolateral membrane conductance can account for only about one-half of the observed rate of Cl$^-$
absorption. This suggests that the remaining Cl$^-$ absorption occurs through
a KCl cotransport process; (6) inhibition of Na$^+$/K$^+$/Cl$^-$ cotransport,
whether due to a loop diuretic, cGMP, or Na$^+$ or Cl$^-$ replacement, leads to
hyperpolarization of the cell potential. Since the cotransport is electroneutral, the hyperpolarization is most likely due to the accompanying decrease
in channel-mediated Cl$^-$ exit across the basolateral membrane. The hyperpolarization brings cell K$^+$ into equilibrium, explaining the loss of K$^+$
secretion (see Table II). Cell Cl$^-$ also falls, approaching its equilibrium
value; (7) inhibition of Na$^+$/K$^+$/Cl$^-$ cotransport also leads to a marked
decrease in the Cl$^-$ conductance of the basolateral membrane, the explanation for which is currently obscure. Whatever the regulatory explanation,
however, the decrease in Cl$^-$ conductance limits the loss of intracellular
ions and water, thereby helping to maintain cell homeostasis.

A Model for Salt Absorption across Flounder Intestine

The above flux and electrophysiologic data suggest the general model
depicted in Fig. 1. It has five major elements: (1) an electrically neutral
Na$^+$/K$^+$/Cl$^-$ cotransport system located in the brush border membrane
that exhibits a transport stoichiometry of $1:1:2$, (2) a Ba^{2+}-sensitive K$^+$
channel in the brush border through which the K$^+$ entering by cotransport
can recycle, (3) Na$^+$,K$^+$-activated ATPase in the basolateral membrane, (4)
Cl$^-$ permeability in the basolateral membrane that involves a conductive
element (i.e., a Cl$^-$ channel) and also KCl cotransport, (5) a cation-selective, leaky paracellular pathway through which Na$^+$ diffuses to partially
balance the Cl$^-$ absorbed through Na$^+$/K$^+$/Cl$^-$ cotransport. The Na$^+$,K$^+$-
ATPase maintains a low intracellular Na$^+$ concentration, thereby creating
a driving force for Na$^+$ entry into the cell. The presence of an Na$^+$/K$^+$/Cl$^-$
cotransporter in the apical membrane couples Cl$^-$ entry to that of Na$^+$ and
permits Cl$^-$ to accumulate in the cell above electrochemical equilibrium.
This provides an energetically efficient way of transporting large quantities
of NaCl since the electroneutral cotransport system effectively uncouples

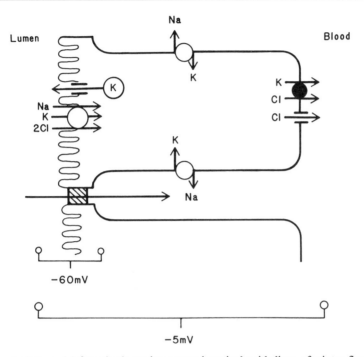

FIG. 1. Cell model for salt absorption across intestinal epithelium of winter flounder. When bathed with identical Ringer's solution on both sides, flounder intestine generates a serosa-negative electrical potential difference of 3 to 6 mV and the potential across the cell plasma membrane averages −60 mV. For detailed discussion of the cellular transport properties, see text.

apical Na⁺ and Cl⁻ entries from the electric potential of the cell. Although this uncoupling diminishes the large driving force for Na⁺ entry, it generates a driving force for Cl⁻ entry since the medium Cl⁻ concentration is much greater than cell Cl⁻. Furthermore, since 2Cl⁻ are transported for each Na⁺, the driving force for Cl⁻ entry is doubled. Finally, since 2Cl⁻ are transported with each Na⁺, less ATP is utilized by the Na⁺ pump to transport a given amount of salt than would be the case if the Na⁺:Cl⁻ stoichiometry were 1:1.

[49] Shark Rectal Gland

By Patricio Silva, Richard J. Solomon, and Franklin H. Epstein

Introduction

The rectal gland of the spiny dogfish *Squalus acanthias* plays a major role in salt homeostasis, secreting a concentrated salt solution in response to volume stimuli.[1-3] The composition of the secretion in comparison to that of the shark plasma and that of the sea is shown in Table I. The concentration of the secretion is approximately twice that of plasma but its osmolality is the same by virtue of the markedly reduced urea content. The potential difference across the gland is −6 to −15 mV. The gland, therefore, secretes chloride against an electrical and chemical gradient.

The mechanism by which chloride is secreted is termed "secondary active" transport because it is indirectly powered by Na^+,K^+-ATPase. The currently accepted mechanism for chloride transport is shown in Fig. 1 and can be described as follows: two chloride ions enter the cell together with a sodium and a potassium ion through an electrically neutral chloride, sodium, and potassium cotransporter.[4-7] The energy for this entry step is provided by the downhill gradient for sodium directed into the cell (280 mM outside, 20 mM inside). This downhill gradient for sodium is maintained by the sodium pump. Chloride leaves the cell passively through a chloride conductance present on the lumenal side of the cell following a favorable electrical gradient.[8] Potassium is recycled across the basolateral surface of the cell through a potassium channel.[9] Sodium cannot leave the cell across the lumenal surface because of the steep opposing electrochemical gradient. Sodium enters the lumen by traversing a paracellular pathway.[4]

[1] J. W. Burger, *Physiol. Zool.* **35**, 205 (1962).
[2] J. N. Forrest, P. Silva, A. Epstein, and F. H. Epstein, *Bull. Mount Desert Island Biol Lab.* **13**, 41 (1973).
[3] R. Solomon, M. Taylor, S. Sheth, P. Silva, and F. H. Epstein, *Am. J. Physiol.* **248**, R638 (1985).
[4] P. Silva, J. Stoff, M. Field, L. Fine, J. N. Forrest, and F. H. Epstein, *Am. J. Physiol.* **233**, F198 (1977).
[5] P. Silva and M. A. Myers, *Am. J. Physiol.* **250**, F516 (1986).
[6] J. Eveloff, R. Kinne, E. Kinne-Saffran, H. Murer, P. Silva, F. H. Epstein, J. Epstein, J. S. Stoff, and W. B. Kinter, *Pfluegers Arch* **378**, 87 (1978).
[7] J. Hannafin, E. Kinne-Saffran, D. Friedman, and R. Kinne, *J. Membr. Biol.* **75**, 73 (1983).
[8] R. Greger and E. Schlatter, *Pfluegers Arch.* **402**, 63 (1984).
[9] P. Silva, J. A. Epstein, M. A. Myers, A. Stevens, P. Silva, Jr., and F. H. Epstein, *Life Sci.* **38**, 547 (1986).

TABLE I
COMPOSITION OF PLASMA AND RECTAL GLAND
SECRETION OF THE DOGFISH AND SEA WATER[a]

	Plasma	Rectal gland	Sea water
Sodium	286	500	440
Potassium	7	10	10
Chloride	296	510	490
Calcium	2.6	1	10
Magnesium	3.7	1	51
Urea	350	10–20	0
Osmolality	1018	1018	975

[a] Values are millimolar except osmolality (in mOsm/liter). In our experience, the osmolality of the plasma of the dogfish is the same as that of the rectal gland secretion and 20–50 mOsm/liter greater than sea water.

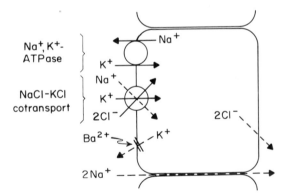

FIG. 1. Model for the transepithelial transport of chloride across the rectal gland of the shark. Chloride moves into the cell against an electrochemical gradient. The entry step for chloride proceeds via a postulated coupled neutral sodium, potassium and chloride carrier. The stoichiometry of this carrier is $2Cl^-:1Na^+:1K^+$. The energy for this entry step is provided by the gradient for sodium directed into the cell. This gradient, favoring the movement of sodium into the cell, is maintained by the activity of the Na^+,K^+-ATPase that pumps sodium out of the cell into the basolateral spaces. Chloride leaves the cell following an electrochemical gradient that favors its efflux across the lumenal border. Sodium enters the lumen via a paracellular pathway. Potassium recirculates back to the basolateral spaces through a potassium conductance.

The secretory response of the gland to volume expansion is under humoral control,[10] which is demonstrable in the gland *in situ* and also in an isolated gland connected only by a catheter conveying blood supply from a volume-expanded fish. Chloride secretion is stimulated by cyclic AMP, vasoactive intestinal peptide (VIP), and adenosine.[11–15] Atriopeptin, secreted by the heart in response to volume stimuli, induces the release of VIP stored in nerve fibers within the gland that in turn stimulates the gland to secrete salt.[16–17]

The rectal gland is rich in Na^+,K^+-ATPase and was the source used by Hokin and collaborators to purify the enzyme and analyze its subunits.[18]

Gross Anatomy of the Rectal Gland

The rectal gland of elasmobranchs is located at the caudal end of the peritoneal cavity. The gland is a digitiform organ attached by a mesentery to the vertebral column and by its caudal end to the dorsal surface of the cloacal end of the rectum (Fig. 2). The gland has a single artery, vein, and duct. The rectal gland artery arises from the dorsal aorta, travels down the leading (cephalic) edge of the mesentery and, upon reaching the rectal gland approximately one-third from the cephalic end of the gland, divides into anterior and posterior branches. Both branches give rise to secondary branches perpendicular to the main ones that travel the surface of the gland immediately beneath the capsule. The secondary branches give off smaller short arteries that give rise to sinusoids. The sinusoids travel parallel to the radially arranged tubules and converge into a venous sinus that runs along the center of the gland within the central duct. This venous sinus exits as the rectal gland vein at the caudal end of the gland and

[10] R. Solomon, M. Taylor, J. Stoff, P. Silva, and F. H. Epstein, *Am. J. Physiol.* **246**, R63 (1984).

[11] J. S. Stoff, P. Silva, M. Field, J. Forrest, A. Stevens, and F. H. Epstein, *J. Exp. Zool.* **199**, 443 (1977).

[12] J. S. Stoff, R. Rosa, R. Hallac, P. Silva, and F. H. Epstein, *Am. J. Physiol.* **237**, F138 (1979).

[13] D. Erlij, P. Silva, and P. Reinach, *Bull. Mount Desert Island Biol. Lab.* **18**, 92 (1979).

[14] J. N. Forrest, D. Rieck, and A. Murdaugh, *Bull. Mount Desert Island Biol. Lab.* **20**, 152 (1980).

[15] J. N. Forrest, Jr., F. Wang, and K. W. Beyenbach, *J. Clin. Invest.* **72**, 1163 (1983).

[16] R. Solomon, M. Taylor, D. Dorsey, P. Silva, and F. H. Epstein, *Am. J. Physiol.* **249**, R348 (1985).

[17] P. Silva, J. S. Stoff, R. J. Solomon, S. Lear, D. Kniaz, R. Greger, and F. H. Epstein, *Am. J. Physiol.* **251**, F99 (1987).

[18] L. E. Hokin, in "Membrane Transport Processes" (D. C. Tosteson, Y. A. Ovchinnikov, and R. Latore, eds.), Vol. 2, p. 399. Raven, New York, 1978.

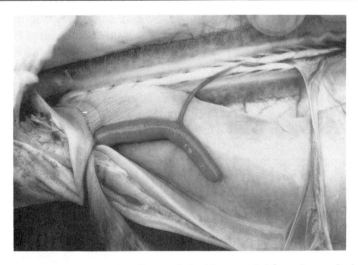

Fig. 2. The rectal gland *in situ*. The rectal gland is suspended from the vertebral column by a short mesentery. The gland is attached to the rectum by its distal or caudal end, at the left. The artery can be seen on the leading end of the mesentery. The posterior intestinal vein arises as the rectal gland vein.

continues in a cephalad direction along the dorsum of the rectum as the posterior intestinal vein. Glandular tubules open into a central duct that extends the length of the gland along its center and also exits at its caudal end. The duct of the gland transverses the wall of the rectum, reversing direction as it does so, and opens into the lumen of the rectum.

Experimental Preparations and Uses

The rectal gland duct can be catheterized *in vivo* and the secretion collected in the free-swimming fish.

The rectal gland can be removed and perfused *in vitro* with an artificial solution or cross-perfused with blood from another fish. Perfusion of the rectal gland is facilitated by its single artery, vein, and duct.

Rectal gland tubules can be teased apart and perfused *in vitro*.

Rectal gland cells can be separated and studied as isolated cells.

Rectal Gland Perfusion

To remove the gland for perfusion, the artery, vein, and duct must be preserved. The gland with its mesentery and the portion of the rectum extending from the cloaca to approximately 2 cm proximal to the attachment of the gland to the rectum are removed *en bloc.* The gland does not need to be perfused immediately, and several minutes can elapse between removal and the start of the perfusion, particularly if the gland is kept on ice.

In male or female specimens of *Squalus acanthias* weighing 1 to 3 kg, the artery is generally 2 to 3 cm long, depending on the size of the animal, and large enough to permit catheterization with PE90 polyethylene tubing. We use catheters that are 15 cm long. Occasionally, double arteries are found that can still be catheterized with a single catheter if the bifurcation occurs some distance from the origin at the dorsal aorta. It is useful, when removing the gland, to cut the mesentery at its insertion along the vertebral column; this small precaution secures the maximal length of the artery and, should the artery bifurcate before reaching the gland, permits sufficient length to allow for its catheterization. We usually tie the catheter in place with a 0 silk ligature. After taking special care that the catheter is free of bubbles, perfusion is started immediately after the catheter is tied in place. The vein is also catheterized with PE90 polyethylene tubing. The tubing is heat flared and inserted, flared end first, into the cut end of the vein at the open end of the rectum. We usually color the flared end of the catheter with an indelible black marker because it allows us to position the end of the catheter in the vein just before it enters into the gland. The vein and the duct exit the gland together. From the site of their exit and for 1 to 1.5 cm cephalad from the gland the duct is immediately under the vein, separated from it only by their respective walls. Therefore care must be exerted to avoid perforation of the wall of the vein and duct when catheterizing either the vein or the duct. A very large duct flow is usually a sign that a communication has been established between the duct and the vein, and the duct fluid should be tested for a chloride concentration approaching that of the perfusate. The duct opens into the lumen of the rectum in a cephalad direction. To catheterize the duct, the rectum is opened longitudinally along its ventral edge (away from the gland). The duct opening will be found along a line in the center of the open rectum. The duct opening can be identified by two parallel mucosal folds that end in a hoodlike opening approximately 1 to 1.5 cm from the insertion of the gland. The duct is catheterized with a heat-flared PE90 polyethylene catheter blackened at the tip. The tip of the catheter is positioned immediately outside the gland. Both the duct catheter and vein catheter are tied in place tightly with a single 0 silk ligature placed around the entire opened segment of the rectum. The flared ends of the catheters hold them in place.

The glands are then placed in an all-glass perfusion chamber like the one shown in Fig. 3, through which fresh sea water is circulated, as a simple constant-temperature device. The perfusion chambers were built for us by

FIG. 3. The rectal gland in the perfusion chamber. The gland is sitting in a groove in the top of the chamber. The arterial catheter can be seen in the background. The vein and duct catheter can be seen coming out of the gland toward the left.

Wilbur Scientific Company, Inc. (Boston, MA) but their design is simple enough so that any glassblower can make them. In order to supply sufficient oxygen for maximal rates of secretion when using a perfusion medium devoid of red cells, it is usually necessary to perfuse at a pressure of 40 mmHg, approximately twice the normal pressure in the dorsal aorta of *Squalus acanthias in vivo.* As illustrated in Fig. 4, O_2 utilization of a stimulated gland may approach 3 μmol O_2/g/min, requiring a perfusion rate of 2 ml/g/min of a solution nearly saturated with oxygen at 1 atm of pressure at 15°. We perfuse the gland by gravity. Rectal gland secretion is related to the perfusion pressure but this effect is of small magnitude and does not modify the effect of secretagogues.

To quantify the rate of perfusate flow, the venous outflow is collected directly in a graduated cylinder. The rate of arterial flow is usually slightly greater than the rate of venous flow, because of mesenteric anastomoses of the rectal gland artery that leak perfusate through the cut edge of the mesentery.

Rectal gland secretion contains mucus that can sometimes block the catheter that is in the duct. It is important, therefore, to monitor the collection of rectal gland secretion continuously. If fluid secretion stops completely it is usually due to a mucus plug in the catheter. Generally this can be displaced and flow reestablished by gentle suction at the open end of the catheter. If rectal gland secretion is less than 20 μl/min, it is collected in

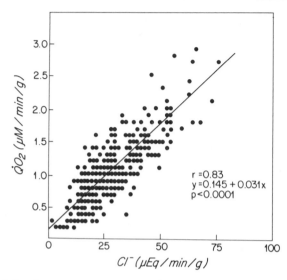

FIG. 4. Relation between oxygen consumption and chloride secretion in the isolated perfused rectal gland. At maximal rates of chloride secretion the consumption of oxygen approximates the maximal oxygen carrying capacity of the perfusate. [Reproduced by permission from P. Silva, J. S. Stoff, R. J. Solomon, R. Rosa, A. Stevens, and J. Epstein, *J. Membr. Biol.* **53**, 215 (1980).]

calibrated 100-μl pipets. These offer the convenience that PE90 tubing fits tightly inside the pipet and therefore collections and measurements can be made easily. When the rate of secretion is greater, it is collected in graduated 1-ml pipets or, if still greater, in graduated test tubes. Collections are timed over periods of arbitrary length until the secretion rate is stable. After changes in the perfusate composition it is important to continue the collections until secretion rate is again stable.

Transglandular potential difference can be measured across the gland by 1 M KCl agar bridges, one placed in continuity with the rectal gland secretion in the collection tube and another in contact with the perfusate or the surface of the gland, using a high-impedance voltmeter and appropriate reference electrodes. Care must be exercised to ensure that there is no electrical continuity between the bridges.

Preparation of the Perfusate

The solution used for perfusion of the rectal gland is a solution of the same ionic composition and osmolarity as the shark plasma containing (in g/liter): NaCl, 16.36; KCl, 0.298; NaHCO$_3$, 0.672; KH$_2$PO$_4$, 0.027; CaCl$_2$, 0.368; Na$_2$SO$_4$, 0.036; MgCl$_2$, 0.61; urea, 75.5. Before the solution is gassed, its pH is approximately 9, therefore the calcium is not added until the pH has been lowered to the normal pH of the shark plasma (7.6). The pH of the perfusate is adjusted to 7.6 by gassing with a mixture of 99% O_2/1% CO_2; this maneuver also saturates the solution with oxygen. Glucose (5 mM) is added as a substrate. Once the solution has been gassed and the calcium and glucose added it is placed in glass aspirator bottles that serve as reservoirs during the perfusion. The aspirator bottles are connected to the arterial catheter of the gland using sterile intravenous fluid administrations sets equipped with self-sealing ports and a three-way stopcock. The self-sealing port is useful for injections directly into the perfusate, and the three-way stopcocks are used to change perfusate composition rapidly.

Ex Vivo Perfusion

Ex vivo perfusion is the technique by which an isolated rectal gland is perfused with blood from a live fish (Fig. 5). This technique has a number of advantages compared to the *in vitro* perfused rectal gland. First, because blood rather than shark's Ringer's solution is the perfusate, gland perfusion occurs at normal rates of blood flow and may therefore be more physiologic. Metabolic processes may be more reflective of *in vivo* conditions; for example, arteriovenous oxygen differences across the *ex vivo* rectal gland are similar to *in vivo* values. Secondly, the effects of an experimental variable can be tested separately in the whole animal and in the rectal gland. For example, a substance can be infused into the donor fish and the effects on the *ex vivo* rectal gland observed. The same substance can then be infused directly into the *ex vivo* rectal gland and the result compared. This model thus permits the investigator to separate direct effects on the gland from systemic effects.

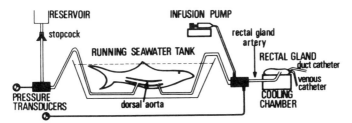

FIG. 5. Schematic representation of the *ex vivo* rectal gland preparation. The *ex vivo* gland is perfused with the blood of a donor fish. The preparation permits the comparison between the responses of the *ex vivo* gland and that of the donor fish after physiological and pharmacological maneuvers on the donor fish.

The disadvantages of this technique are the fact that two animals are needed for each experiment compared to one animal using either the *in vitro* perfused gland or the *in situ* catheterized gland. Additionally, the *ex vivo* gland, like the *in vitro* gland, is denervated, a factor which may be of importance in certain types of experiments.

Pithing

Ex vivo perfusion can be performed using either a physically restrained or pithed shark as donor. The advantages of the pithed preparation is the prevention of the unpredictable and often technically catastrophic gross muscular activity of the donor which can upset the experimental model, resulting in the need to restart collection periods, reinsert catheters, and even sacrifice additional animals for replacement of the rectal gland to be perfused (see also, Modifications of the *ex vivo* Perfusion Preparation, below). The disadvantage of the pithing procedure is that some animals will not spontaneously resume respiratory movements of the gills following pithing. Small animals seem to be more sensitive to this phenomenon. In addition, careful attention to maintaining a midline approach in the pithing procedure seems to minimize this adverse effect. While we have tried to obviate hypoxia by running sea water through the mouth and out the gill slits of such animals, this does not appear to be entirely satisfactory. Hypoxia may still occur and is probably a result of shunting of blood away from the gill lamellae in response to the spinal shock of pithing.

To perform the pithing, one needs a scalpel (#11 blade) and a stiff, straight wire (~1-mm diameter), somewhat longer than the length of the fish and sharpened to a tip at one end. The fish is quickly removed from the water and restrained by hand, ventral side down, on a firm, flat surface. A small midline incision is made in the tip of the snout approximately 5 mm in depth. The sharpened tip of the wire is then inserted through the incision just dorsal of the midplane of the snout. The wire is slowly passed caudally through the brain into the spinal canal. This is the most difficult and important part of the procedure. The more the wire can be directed without excessive probing into the spinal column, the less is the risk that the animal will fail to gill adequately following the procedure. Once in the spinal canal, the wire can be quickly pushed the full length of the fish. The animal will immediately become flaccid. The wire can be then quickly removed and the animal returned to sea water. If gilling movements are not immediately present, stimulation of gilling by forcing sea water into the mouth or through the opercular openings will sometimes remedy the situation. If no gilling is obvious after a few minutes, it is unlikely that the animal will recover. Either the animal should be sacrificed and the rectal gland used as the *ex vivo* gland (see below) or an attempt can be made to prevent hypoxia by perfusing the gill arches with running sea water. Hypoxia can be determined directly with arterial sampling and an oxygen electrode but can also be observed as a bluish mottling of the ventral surface of the animal which is normally a uniform white color.

Establishing Donor Conditions

The successfully pithed animal or restrained animal is next prepared for the *ex vivo* perfusion technique by inserting the necessary catheters for blood perfusion of the *ex vivo* rectal gland and intravascular sampling/infusion of the donor. The donor fish is placed with its ventral side facing the surface of the water. The caudal one-half of the fish is then raised above the surface of the water by elevation of the tail. It is important to make sure that the gills and opercula remain in the sea water. A Touhy needle, #16, bevel facing cephalad, is then inserted in the midline of the tail section and advanced at a slight angle until the cartilaginous structures of the skeleton are met. This is the location of the dorsal aorta. Removal of the trocar of the needle should be followed by the appearance of blood from the barrel of the needle. If no blood is present, reinsertion of the trocar and careful probing will usually be successful. Once blood flows freely from the needle, PE 90 tubing can be inserted through the needle barrel and advanced slowly cephalad. Once the catheter is in place, the Touhy needle is removed over the free end of the catheter. The catheter is then flushed and 1000 U of sodium heparin is given through the catheter to anticoagulate the donor systemically. For the perfusion catheter, we usually advance the tubing as far as the upper abdomen (lateral fins externally). A second catheter can then be inserted into the same vessel caudad to the first catheter using the identical procedure. This catheter must be inserted caudal to the first, otherwise the Touhy needle will sever the first catheter. Blood may not flow as freely from this second catheter because the first catheter may partially obstruct such flow. Careful suction with a syringe will then tell you when you are in the vessel. The PE 90 tubing is advanced only to the midabdomen so that any solution infused into the donor fish will pass through the entire circulation before entering the more proximally placed perfusion catheter. This second catheter may also be used to monitor dorsal aortic pressure. An appropriate-gauge needle and three-way stopcock can be placed at the free end of each of the catheters to control the flow of blood.

Neither catheter needs to be secured in the pithed fish and the shortest length can be used for the *ex vivo* rectal gland. In a restrained fish preparation, we leave a longer length of catheter between the donor fish and the perfused gland in case the donor suddenly swishes its tail or breaks through the restraints.

The free end of the perfusion catheter is next attached either directly to the *ex vivo* rectal gland or, as we prefer, to one port of a three-way flow system. We have used autoanalyzer stream dividers for this purpose but any arrangement is acceptable. One port is the inflow from the donor fish; one port is the outflow to the *ex vivo* rectal gland. The third port can be used for the infusion of substances which the investigator wants the *ex vivo* gland alone to receive. This port may also be used for the monitoring of rectal gland perfusion pressure.

Establishing the ex Vivo Perfused Gland

Removal of the rectal gland from a fish is accomplished as described under Rectal Gland Perfusion and all catheterizations are essentially identical with those for the *in vitro* perfused gland. The arterial catheterization is accomplished first to establish perfusion of the *ex vivo* gland as quickly as possible.

Free flow from the catheters in the duct and the vein is established, sometimes with suction using a needle and syringe. The gland is then placed in a glass perfusion chamber as described above. The perfusion chamber and rectal gland must be placed at the level of the heart of the donor fish to ensure physiologic perfusion pressure. Raising or lowering the chamber will permit the investigator to manipulate perfusion pressure according to the experimental design.

Collections

Timed collections of duct fluid and venous blood are made according to the experimental protocol. Periods of 20–30 min are usually necessary to collect sufficient venous blood and duct fluid under basal conditions. Blood can be returned to the donor fish periodically via the second, more caudal catheter so as to maintain normal systemic hemodynamics in the donor fish. When venous blood contains substances which the investigator does not want the donor fish to receive, an equal volume of shark Ringer's solution can be used if the experimental design is not influenced by the systemic hematocrit. Alternatively, prior to the experiment blood can be removed from the fish which is sacrificed for the *ex vivo* rectal gland, and later used as replacement for the blood lost by the donor fish.

Modifications of the ex Vivo Perfusion Preparation

The basic preparation can be modified to permit experimental protocols with various types of controls. For example, using a large donor fish, we have established two *ex vivo* rectal glands perfused simultaneously. One gland can then act as a control for the other which is subjected to the experimental variable. This model requires more frequent return of blood to the donor fish.

It is also possible to catheterize the *in situ* gland of the donor fish and collect duct fluid simultaneously with collections from the *ex vivo* gland. This model allows the investigator to study the effects of denervation per se as this is the major difference between the *in situ* and *ex vivo* rectal glands.

Isolated Rectal Gland Cells

Isolated cells offer many advantages for the study of many different aspects of cellular metabolism. Cellular respiration, substrate utilization, product formation, ATP levels, and enzyme activity can be directly and easily measured. Changes in cell potential, pH, or calcium concentration can be monitored in a cuvette using fluorescent dyes. Ion transport can be evaluated by measuring intracellular electrolytes or transport-dependent oxygen consumption. The effect of changes in transport activity on cell morphology can be measured with minimal fixation artifacts and the functional and morphological effects of pharmacological and toxic agents can be followed. Finally, binding of different ligands can be assessed in the absence of the constraints imposed by tissue organization and a variably labeled interstitial space. A disadvantage is that cell function and membrane surface characteristics may be altered by the separation procedures.

Isolated epithelial cells have been obtained from a wide variety of tissues, including heart, liver, and the cortex and medulla of the kidney. Isolated rectal gland cells respond to agents that increase chloride transport by the intact rectal gland by increasing their rate of transport dependent oxygen consumption. Oxygen consumption in these cells is sensitive to ouabain, which inhibits Na^+,K^+-ATPase, and to furosemide and related diuretics that block chloride transport. Oxygen consumption is also sensitive to the removal of sodium or chloride from the incubation solution. These characteristics mirror those of the intact gland.

Cell Separation

Rectal glands of two sharks are removed and perfused *in vitro* as described above, for about 10 min, in order to remove red blood cells from the gland. The glands are then perfused with 10 ml of perfusate solution containing in addition 0.2% (w/v) collagenase, 0.25% (w/v) hyaluronidase, and 10% (v/v) fetal calf serum. After the perfusion is stopped the glands are sectioned longitudinally and minced into 0.5- to 1-mm cubes with either a

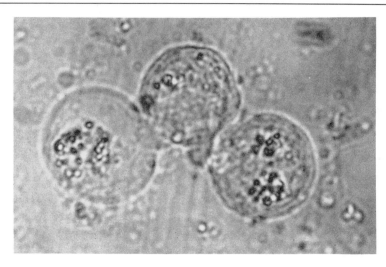

FIG. 6. Isolated rectal gland cells. Light microscopy of the cells in the digestion solution. The cells are now spherical, having lost the columnar shape they have in intact tubular epithelial. The granular structures are mitochondria. The cells reaggregate in solution and can attach to glass.

hand-held razor blade or, if available, a McIlwain tissue slicer (Brinkmann Instruments, New York). The minced tissue is placed in a 250-ml Erlenmeyer flask in 100 ml of a solution of the same composition as that of the perfusate solution but containing 40 mM HEPES buffer, pH 7.6 rather than bicarbonate and also containing 0.2% (w/v) collagenase, 0.25% (w/v) hyaluronidase, and 10% (v/v) fetal calf serum. The tissue is incubated at room temperature for 45 min. The tissue suspension is constantly stirred with a magnetic stirrer at a low speed sufficient to keep the tissue from settling at the bottom of the flask. The digestion process is constantly monitored by examining the digest under phase-contrast microscopy to note the presence of separated cells (Fig. 6). The tissue digest is centrifuged at 50 g for 1 min in a 50-ml round-bottom plastic tube in a refrigerated centrifuge to remove the undigested tubules; the supernatant is then centrifuged at 500 g for 3 min to harvest the cells. The cells are washed twice with the same solution but without collagenase or hyaluronidase and kept on ice until used.

Author Index

Numbers in parentheses are footnote reference numbers and indicate that an author's work is referred to although the name is not cited in the text.

Gratzl, M., 281, 283(8), 286, 297(8), 298, 299(7)
Gray, D. W. R., 195
Gray, G. M., 343, 344(10), 345(10), 348, 417
Gray, M. A., 266, 267(66), 269
Gray, P. T. A., 651, 663(4)
Gray, T. E., 568
Green, K. C., 499, 500(25)
Green, K., 581
Green, N., 715
Green, R., 266, 267(73), 268(73), 269(73), 270
Greenblatt, D., 72
Greengard, P., 281, 549, 550, 567
Greenwell, J. R., 266, 267(66), 269
Greep, J. M., 263
Greger, R., 266, 267(78, 79), 269(78, 79), 715, 721(22), 754, 756
Gregory, R. A., 257
Grey, D. W. R., 194
Grills, B. L., 228
Grinsberg, M., 224
Grinstein, S., 45, 51(16), 290, 356
Grodsky, G. M., 228
Groen, A. K., 108
Groseclose, R., 438
Gross, D. J., 227, 228, 233(25, 27)
Gross, D., 221, 233
Gross, P., 707
Grossman, M. I., 143, 258
Gruber, W. D., 16, 18(7)
Grynkiewicz, G., 38, 60(2), 63(2), 318, 320(13)
Grzesiek, S., 403, 413
Gstraunthaler, G., 574, 725
Guerritore, A., 601
Guesdon, F., 496, 497(4), 498(4)
Guggino, W. B., 710, 715(9)
Guggino, W., 710
Guguen-Guillouzo, C., 501
Gummer, P. R., 228
Gunter-Smith, P. J., 317
Gunter-Smith, P., 391
Gunther, G. R., 275, 276(25)
Gunther, R. D., 404
Gupta, B. J., 692
Gupta, B. L., 608
Gurll, N. J., 150
Guttman, L., 583
Gylfe, E., 205

H

Haag, K., 16, 21, 22, 24(6), 25(6), 466, 468(20, 21)
Haase, W., 175, 266, 280, 288(2), 292(2), 293(2), 396, 517
Habener, J. F., 235
Haberich, F. J., 256, 258(4)
Haffen, K., 356
Haftek, M., 224
Hagen, D. L., 463
Hagen, S. J., 410, 411(19), 416(19)
Hagenbuch, B., 394
Hajek, S. V., 204
Hakanson, R., 188, 199(4)
Halban, P. A., 195, 197, 198(28), 199(28), 202(28), 208, 220, 221, 224, 227, 228, 233, 234, 235, 274
Halban, P., 221, 233
Halenda, S. P., 273
Hales, C. N., 224, 236
Hall, H. D., 7
Hall, R. D., 38
Hall, T. A., 692
Hallac, R., 756
Hallam, T. J., 69, 71, 72, 123
Halm, D. R., 735, 736, 739(8), 740, 744(21), 747, 748(5), 749(11), 751
Halsted, C. H., 429
Hamamoto, S. T., 356
Hamann, K. F., 7
Hamer, C. M., 419
Hamill, O. P., 92, 311, 496, 633, 679
Hamill, V. P., 323
Hamilton, J. A., 518, 519(15), 529(15)
Hamilton, R. L., 488
Hammill, O. P., 693
Hampson, S. E., 140
Hanahan, D., 226
Handler, J. S., 356, 359, 481
Hanke, W., 715
Hanna, S. D., 344
Hannafin, J., 754
Hanoune, J., 496
Hanozet, G. M., 600, 601
Hanrahan, J. W., 633
Hansen, D., 97
Hansen, U.-P., 423
Hanzel, D. K., 151, 163
Harig, J. M., 391

Subject Index

D

E

F

H

oxyntic cells, cell membrane potentials
and voltage divider ratios of, 88
stomach, cell-to-cell coupling measurements in, 90
Necturus maculosus, surface epithelial cells,
cell membrane potentials and voltage
divider ratios of, 87
Nephron. *See* Amphibian nephron
Nernst equation, 420–421
Neurotransmitter receptors, salivary gland,
28–31
radioligand binding assay, 29–31
Nicolsky–Eisenman equation, 667
NL/T4 cells, 566
Noise analysis, 84
Nuclear magnetic resonance spectroscopy,
691–692
Nuclepore filters
coating, procedure for, 387
collagen-coating, 359
Nycodenz one-step technique, for separation
of gastric cells, 102
Nyqvist plot, of measured impedance data
in complex impedance plane, 473–474

O

OC. *See* Oxyntic cells
Ohm's law, in measurement of epithelial
resistivity, 466
Onymacris plana, Malpighian tubules,
lumenal resistances, 630
Open circuit potential, and edge leaks, in
Ussing chamber, 697–698
Optical methods, for transport studies, 413
Ouabain, inhibition of Na^+/K^+ pump with,
78
Overshoot phenomenon, 431
Oxygen consumption
in isolated parietal cells, 103–104
of oxyntic cell, 151
Oxygen electrodes, 104
Oxyntic cells, 82
apical and basolateral cell membranes,
ionic conductances of, 88–90
apical membrane vesicles from
f-actin in, 162–163
H^+,K^+-ATPase activity, 160–161
determination of, 155–157
orientation, 162–163

preparation, 152–155
proteins in, 163–164
size, 162
stimulated, 152
cell membrane potential measurements
of, 88
H^+,K^+-ATPase-containing membranes
isolation of, 151–165
specific ion permeabilities of, 151–152,
165
impalement of, with microelectrodes, 87
impedance properties of, 85
microdissection procedures, 87–89
tubulovesicles, 151
H^+,K^+-ATPase in, 160–161
determination of, 155–157
preparation of, 152–155
proteins in, 163–164
vesicles, pH gradient formation by,
161–162
voltage divider ratio measurements of, 88

P

Pancreas
acinar cell atrophy, 259–260
cat
vascular perfused preparations,
263–266
vascular supply of, 263–264
collagenase digestion
application to larger animals or man,
195
in shaking water bath, 192–194
static, 194–195
dog, vascular perfused preparation, 263
duodenal, 188
function, variations in, 256
guinea pig, vascular perfused preparation,
263
perfused, studies using, 263–264
pig, vascular perfused preparation, 263
rabbit
isolated, 263
studies using, 262
isolation, 262
rat
vascular perfused preparation, 263
whole, collagenase digestion of,
192–195